Lecture Notes in Computer Science   14413

More information about this series at https://link.springer.com/bookseries/558

Jugal Garg · Max Klimm · Yuqing Kong
Editors

Web and Internet Economics

19th International Conference, WINE 2023
Shanghai, China, December 4–8, 2023
Proceedings

 Springer

Editors
Jugal Garg (iD)
Department of Industrial and Enterprise
Systems Engineering
University of Illinois System
Urbana, IL, USA

Max Klimm (iD)
Institute of Mathematics
Technische Universität Berlin
Berlin, Germany

Yuqing Kong (iD)
Center on Frontiers of Computing Studies
Peking University
Beijing, China

ISSN 0302-9743   ISSN 1611-3349 (electronic)
Lecture Notes in Computer Science
ISBN 978-3-031-48973-0   ISBN 978-3-031-48974-7 (eBook)
https://doi.org/10.1007/978-3-031-48974-7

This Springer imprint is published by the registered company Springer Nature Switzerland AG
The registered company address is: Gewerbestrasse 11, 6330 Cham, Switzerland

Paper in this product is recyclable.

Preface

This volume contains the papers and extended abstracts presented at the 19th Conference on Web and Internet Economics (WINE 2023), held during December 4–8, 2023, at ShanghaiTech University. The WINE conference series is a leading forum to exchange research ideas combining incentives and computational aspects as they appear in a diverse area of applications at the interface of theoretical computer science, artificial intelligence, operations research, and economics. The growing importance of these topics is shown by the fact that for WINE 2023, there was a record number of 221 submissions. We invited 124 researchers to the program committee, including 39 senior program committee members. Each submission was reviewed by three program committee members, and an additional meta-review was provided by one senior program committee member. We are very grateful for their thorough, insightful, and fair reviews and discussions that helped to make the difficult decisions about which papers to accept for the conference. As has become standard for many scientific conferences in the area, this year, WINE moved to a double-blind review format where the authors' identities were hidden from the senior program committee and program committee. In the end, we selected 66 papers for presentation at WINE 2023, leading to a very competitive acceptance rate of just below 30%.

The works accepted for publication in this volume cover a wide range of topics such as computational social choice, fair division, stopping problems, online resource allocation, information design, recommendation systems, blockchains, auctions, matching markets, games on networks, Nash equilibria, learning in games, and privacy. To accommodate the publishing traditions of different fields, authors of accepted papers could ask that only a one-page abstract of the paper appear in the proceedings.

Furthermore, due to the generous support by Springer, we were able to provide two Best Paper Awards that we decided to give to the papers "Stable Dinner Party Seating Arrangements" and "Buy-Many Mechanisms for Many Unit-Demand Buyers" appearing first in this volume.

Besides the scientific talks on the papers contained in this volume, the conference program included four invited talks by leading researchers in the field: Xiaotie Deng (Peking University, China), Jason Hartline (Northwestern University, USA), Hervé Moulin (University of Glasgow, UK), and Inbal Talgam-Cohen (Technion, Israel). Furthermore, the conference program featured six tutorial sessions to showcase and explore emerging topics related to incentives and computation.

We would like to thank all the authors for their interest in submitting their work to WINE 2023. Our special thanks go to the steering committee of WINE and to the local organizers.

October 2023

Jugal Garg
Max Klimm
Yuqing Kong

Organization

General Chairs

Shouyang Wang   ShanghaiTech University, China
Jingyi Yu   ShanghaiTech University, China

Conference Chairs

Andrew Chi-Chih Yao   Tsinghua University, China
Yinyu Yu   Stanford University, USA

Steering Committee

Xiaotie Deng   Peking University, China
Paul Goldberg   University of Oxford, UK
Nicole Immorlica   Microsoft Research, USA
Scott Kominers   Harvard University, USA
Katrina Ligett   Hebrew University, Israel
Christos Papadimitriou   Columbia University, USA
Paul Spirakis   University of Liverpool, UK
Rakesh Vohra   University of Pennsylvania, USA
Andrew Yao   Tsinghua University, China
Yinyu Ye   Stanford University, USA

Local Organization Committee

Yukun Cheng   Suzhou University of Science and Technology, China
Tracy Xiao Liu   Tsinghua University, China
Biaoshuai Tao   Shanghai Jiao Tong University, China
Dengji Zhao   ShanghaiTech University, China

Program Committee Chairs

Jugal Garg   University of Illinois at Urbana-Champaign, USA
Max Klimm   Technische Universität Berlin, Germany
Yuqing Kong   Peking University, China

Senior Program Committee

Elliot Anshelevich   Rensselaer Polytechnic Institute, USA
Siddhartha Banerjee   Cornell University, USA
Siddharth Barman   Indian Institute of Science, India

Xiaohui Bei	Nanyang Technological University, Singapore
Ozan Candogan	University of Chicago, USA
Zhigang Cao	Beijing Jiaotong University, China
Ioannis Caragiannis	Aarhus University, Denmark
Xujin Chen	Chinese Academy of Sciences, China
Shahar Dobzinski	Weizmann Institute of Science, Israel
Aris Filos-Ratsikas	University of Edinburgh, UK
Dimitris Fotakis	National Technical University of Athens, Greece
Vasilis Gkatzelis	Drexel University, USA
Nikolai Gravin	Shanghai University of Finance and Economics, China
Nima Haghpanah	Pennsylvania State University, USA
Kristoffer Arnsfelt Hansen	Aarhus University, Denmark
Tobias Harks	Augsburg University, Germany
Martin Hoefer	Goethe University Frankfurt, Germany
Zhiyi Huang	University of Hong Kong, China
Stefano Leonardi	Sapienza University of Rome, Italy
Minming Li	City University of Hong Kong, China
Tracy Liu	Tsinghua University, China
Brendan Lucier	Microsoft Research, USA
Will Ma	Columbia University, USA
Ali Makhdoumi	Duke University, USA
Azarakhsh Malekian	University of Toronto, Canada
Ruta Mehta	University of Illinois at Urbana-Champaign, USA
Thanh Nguyen	Purdue University, USA
Mallesh Pai	Rice University, USA
Qi Qi	Renmin University of China, China
Daniela Saban	Stanford University, USA
Grant Schoenebeck	University of Michigan, USA
Balasubramanian Sivan	Google, USA
Carmine Ventre	King's College London, UK
Adrian Vetta	McGill University, Canada
Lirong Xia	Rensselaer Polytechnic Institute, USA
Dengji Zhao	ShanghaiTech University, China

Program Committee

Johannes Gerhardus Benade	Boston University, USA
Martin Bullinger	Technical University of Munich, Germany
Mingliu Chen	University of Texas at Dallas, USA
Yukun Cheng	Suzhou University of Science and Technology, China
Yun Kuen Cheung	Australian National University, Australia
Bart de Keijzer	King's College London, UK
David Delacretaz	University of Manchester, UK
Argyrios Deligkas	Royal Holloway University of London, UK
Yuan Deng	Google Research New York, USA
Battal Dogan	University of Bristol, UK

Umut Dur	North Carolina State University, USA
Marcin Dziubiński	University of Warsaw, Poland
Hadi Elzayn	Meta, USA
Tomer Ezra	Sapienza University of Rome, Italy
Alireza Fallah	Massachusetts Institute of Technology, USA
Zhixuan Fang	Tsinghua University, China
Yiding Feng	Microsoft Research & Northwestern University, USA
Diodato Ferraioli	Università di Salerno, Italy
Matheus V. X. Ferreira	Princeton University, USA
Jessica Finocchiaro	University of Colorado Boulder, USA
Federico Fusco	Sapienza University of Rome, Italy
Ganesh Ghalme	Indian Institute of Technology Hyderabad, India
Paul Golz	Carnegie Mellon University, USA
Michael Hamilton	University of Pittsburgh, USA
Kevin He	University of Pennsylvania, USA
Alexandros Hollender	University of Oxford, UK
Xin Huang	Technion–Israel Institute of Technology, Israel
Edin Husic	IDSIA, Switzerland
Krishnamurthy Iyer	University of Minnesota, USA
Anson Kahng	University of Toronto, Canada
Gregory Kehne	Harvard University, USA
Rucha Kulkarni	University of Illinois at Urbana-Champaign, USA
Philip Lazos	University of Oxford, UK
Xiao Lei	University of Hong Kong, China
Pascal Lenzner	Hasso Plattner Institute Potsdam, Germany
Stefanos Leonardos	King's College London, UK
Bo Li	Hong Kong Polytechnic University, China
Bo Li	Peking University, China
Jiangtao Li	Singapore Management University, Singapore
Mengling Li	Xiamen University, China
Yingkai Li	Yale University, USA
Irene Lo	Stanford University, USA
Jiaqi Lu	Columbia University, USA
Suraj Malladi	Northwestern University, USA
Simon Mauras	Tel Aviv University, Israel
Evi Micha	University of Toronto, Canada
Divyarthi Mohan	Tel Aviv University, Israel
Vishnu Narayan	Tel Aviv University, Israel
Kim Thang Nguyen	Grenoble Alpes University, France
Dario Paccagnan	Imperial College London, UK
Ioannis Panageas	University of California Irvine, USA
Harry Pei	Northwestern University, USA
Chara Podimata	Massachusetts Institute of Technology, USA
Emmanouil Pountourakis	Drexel University, USA
Alexandros Psomas	Purdue University, USA
Nidhi Rathi	Max Planck Institute for Informatics, Germany

Contents

Best Paper Awards

Stable Dinner Party Seating Arrangements

Damien Berriaud[ID], Andrei Constantinescu[(✉)][ID], and Roger Wattenhofer[ID]

ETH Zurich, Rämistrasse 101, 8092 Zurich, Switzerland
{dberriaud,aconstantine,wattenhofer}@ethz.ch

Abstract. A group of n agents with numerical preferences for each other are to be assigned to the n seats of a dining table. We study two natural topologies: circular (cycle) tables and panel (path) tables. For a given seating arrangement, an agent's utility is the sum of their preference values towards their (at most two) direct neighbors. An arrangement is envy-free if no agent strictly prefers someone else's seat, and it is stable if no two agents strictly prefer each other's seats. Recently, it was shown that for both paths and cycles it is NP-hard to decide whether an envy-free arrangement exists, even for symmetric binary preferences. In contrast, we show that, if agents come from a bounded number of classes, the problem is solvable in polynomial time for arbitrarily-valued possibly asymmetric preferences, including outputting an arrangement if possible. We also give simpler proofs of the previous hardness results if preferences are allowed to be asymmetric. For stability, it is known that deciding the existence of stable arrangements is NP-hard for both topologies, but only if sufficiently-many numerical values are allowed. As it turns out, even constructing unstable instances can be challenging in certain cases, e.g., binary values. We propose a complete characterization of the existence of stable arrangements based on the number of distinct values in the preference matrix and the number of agent classes. We also ask the same question for non-negative values and give an almost-complete characterization, the most interesting outstanding case being that of paths with two-valued non-negative preferences, for which we experimentally find that stable arrangements always exist and prove it under the additional constraint that agents can only swap seats when sitting at most two positions away. Similarly to envy-freeness, we also give a polynomial-time algorithm for determining a stable arrangement assuming a bounded number of classes. We moreover consider the swap dynamics and exhibit instances where they do not converge, despite a stable arrangement existing.

Keywords: Hedonic Games · Stability · Computational Complexity

1 Introduction

Your festive dinner table is ready, and the guests are arriving. As soon as your guests take their assigned seats, two of them are unhappy about their neighbors and rather want to switch seats. Alas, right after the switch, two other guests

© The Author(s), under exclusive license to Springer Nature Switzerland AG 2024
J. Garg et al. (Eds.): WINE 2023, LNCS 14413, pp. 3–20, 2024.
https://doi.org/10.1007/978-3-031-48974-7_1

become upset, and then pandemonium ensues! Could you have prevented all the social awkwardness by seating your guests "correctly" from the get-go?

In this paper, we study the difficulty of finding a *stable* seating arrangement; i.e. one where no two guests would switch seats. We focus on two natural seating situations: a round table (cycle), and an expert panel (path). In either case, we assume guests only care about having the best possible set of direct left and right neighbors. In certain cases, not even a stable arrangement might make the cut, as a single guest envying the seat of another could potentially lead to trouble. Therefore, we are also interested in finding *envy-free* arrangements.

Formally, n guests have to be assigned bijectively to the n seats of a dining table: either a path or a cycle graph. Guests express their preferences for the other guests numerically, with higher numbers corresponding to a greater desire to sit next to the respective guest. The utility of a guest g for a given seating arrangement is the sum of g's preference values towards g's neighbors. A guest g envies another guest if g's utility would strictly increase if they swapped places. Two guests want to swap places whenever they envy each other. Our goal is to compute a stable (no two guests want to swap) respectively envy-free (no guest envies another) seating arrangement.

Besides the table topology, we conduct our study in terms of two natural parameters. The first parameter is the number of numerical values guests can choose from when expressing their preferences for other guests. For instance, an example of two-valued preferences would be when all preference values are either zero or one (i.e., *binary*, also known as *approval* preferences), in which case every guest has a list of "favorite" guests they want to sit next to, and is indifferent towards the others. In contrast, if the values used were ± 1; i.e., every guest either likes or dislikes every other guest; then the preferences are still two-valued, but no longer binary. Increasing the number of allowed values allows for finer-grained preferences. It is also interesting to distinguish the special case of non-negative preferences; i.e., no guest can lose utility by gaining a neighbor.

Our second parameter is the number of different guest classes. In particular, dinner party guests can often be put into certain categories, e.g., charmer, entertainer, diva, politico, introvert, outsider. Each class has its own preferences towards other classes, e.g., outsiders would prefer to sit next to a charmer, but not next to an introvert.

Our Contribution. We study the existence and computational complexity of finding stable/envy-free arrangements on paths and cycles. Some of our results can be surprising. For instance, six people with binary preferences can always be stably seated at a round table, while for five (or seven) guests some preferences are inherently unstable, so we better invite (or uninvite) another guest. However, even for six people with binary preferences, for which a stable arrangement always exists, the swap dynamics might still never converge to one.

Our computational results are exhibited in Table 1. In summary, if the number of guest classes is bounded by a constant (which can be arbitrary), then both stable and envy-free arrangements can be computed in polynomial time whenever they exist with no assumption on the preference values, while dropping this

Table 1. Summary of computational results for envy-free (EF) and stable (STA) arrangements. We distinguish between two cases, depending on whether the total number of guest classes is bounded by a constant or not. Hardness results are for the existential questions "do such arrangements exist?" Polynomial-time algorithms recover an envy-free/stable arrangement whenever one exists.

	No. of Classes				No. of Classes	
	Bounded	Unbounded			Bounded	Unbounded
EF	Poly*	NP-hard†		EF	Poly*	NP-hard†
STA	Poly*	NP-hard*		STA	Poly*	NP-hard§

(a) Complexity results for cycles.   (b) Complexity results for paths.

* are shown in Theorems 4 and 5 (working for arbitrary values).
† are shown in Theorems 1 and 2 (using binary values). They were also recently shown in [11] for symmetric binary preferences, although with more complex proofs.
* is shown in [11] using four non-negative values, the binary case is open.
§ is shown in [11] using six values including negatives, the non-negative case is open.

assumption makes the two problems difficult even for very constrained preference values. For envy-freeness, this already happens for binary preferences, arguably the most prominent case. For stability, on the other hand, this requires more contrived values (four non-negative values for cycles and six values including negatives for paths), so it would be interesting to have a finer-grained understanding of stability in terms of the values allowed when expressing one's preferences. For instance, is it still hard to find stable arrangements for binary preferences? As we show, it turns out to be surprisingly difficult to even construct unstable binary preferences, setting aside the computational considerations. To this end, we conduct a fine-grained study aiming to answer for which combinations of our two parameters, i.e., number of preference values and guest classes, do stable arrangements always exist and for which combinations this is not the case. Our results are exhibited in Table 2. Notably, for cycles we give a full characterization, either showing guaranteed stability with arbitrary values or giving unstable instances with (simple) non-negative values. Similarly, for paths, we close all cases, with the notable exception of two-valued preferences with three or more classes being permitted, where all the unstable instances we found require negative values. We conjecture that for non-negative values stability on paths is guaranteed in the two-valued case. We support this conjecture with experimental evidence, as well as a partial result under the additional assumption of guests only being willing to swap seats if they are separated by at most one seat (see Table 2 for more details).

Full Version. In the full version of the paper [3], we supply the proofs omitted from the main text, as well as supporting material. Moreover, we show that stability is a highly fragile notion, being non-monotonic with respect to adding/removing guests. We also give evidence that knowledge about stability

Table 2. Summary of results characterizing the existence of unstable instances for different combinations of constraints on the number of preference values and classes of guests. Our stability (S) results mean that all instances satisfying the constraints admit a stable arrangement, and hold for arbitrary preference values. Our constructions with no stable arrangements (U) only use small non-negative values (except •, discussed below), often $0, 1, 2, \ldots$, and work for any large enough number of guests.

Values \ Classes	$\leqslant 2$	3	4	$\geqslant 5$
$\leqslant 2$	S	**S**•	**U**§	U
3	S	**U**†	U	U
$\geqslant 4$	**S**•	U	U	U

(a) Characterization results for cycles.

Values \ Classes	$\leqslant 2$	3	$\geqslant 4$
$\leqslant 2$	S	**U**•	**U**•
3	S	**U**†	U
$\geqslant 4$	**S**•	U	U

(b) Characterization results for paths.

• are shown in Theorems 6 and 7 (working for arbitrary values).
† are shown in Theorems 8 and 9 (using non-negative values).
* is shown in Theorem 11 (working for arbitrary values).
§ is shown in Theorem 10 (using binary values).
• is shown in Theorem 12 (using also negative values). The case of non-negative values is open, and we believe that the answer is **S**. We have exhaustively established this for 4-class instances with at most ten guests per class and 5-class instances with at most four guests per class, as well as all 7-guest instances. Moreover, it holds irrespective of the number of classes assuming that guests are only willing to swap places with other guests that they are separated from by at most one seat (Theorem 13).

on paths is unlikely to transfer to computing cycle-stable arrangements. Finally, we use probabilistic tools to study the expected number of stable arrangements of Erdős-Rényi binary preferences.

2   Related Work

The algorithmic study of stability in collective decision-making has its roots in the seminal paper of Gale and Shapley [13], introducing the now well-known *Stable Marriage* and *Stable Roommates* problems. Classically, the former is presented as follows: an equal number of men and women want to form couples such that no man and woman from different couples strictly prefer each other over their current partners, in which case the matching is called *stable*. The authors give the celebrated Gale-Shapley deferred acceptance algorithm showing that a stable matching always exists and can be computed in linear time. Irving [15] later extended the algorithm to also handle preferences with ties; i.e., a man (woman) being indifferent between two women (men). The *Stable Roommates* problem is the non-bipartite analog of Stable Marriage: an even number of students want to allocate themselves into identical two-person rooms in a dormitory. A matching is stable if no two students allocated to different rooms prefer each other over their current roommates. In this setting, stable matchings might no longer exist, but a polynomial-time algorithm for computing one if any exist is known [14]. However, when ties are allowed, the problem becomes NP-hard [17].

The seating arrangement problem that we study is, in fact, well-connected with Stable Roommates. Instead of one table with n seats, the latter considers $n/2$ tables with two seats each. However, there is another more subtle difference: in Stable Roommates, two people unhappy with their current roommates can choose to move into any free room. This is not possible if there are exactly $n/2$ rooms. Instead, the stability notion that we study corresponds to the distinct notion of *exchange-stability* in the Stable Roommates model, where unhappy students can agree to exchange roommates. Surprisingly, under exchange-stability, finding a stable roommate allocation becomes NP-hard even without ties [10].

One can also see our problem through the lens of coalition formation. In particular, *hedonic games* [2] consider the formation of coalitions under the assumption that individuals only care about members in their own coalition. Then, fixing the sizes of the coalitions allows one to generalize from tables of size two and study stability more generally. Bilò et al. [6] successfully employ this approach to show a number of attractive computational results concerning exchange-stability. The main drawback of this approach, however, is that it assumes that any two people sitting at the same table can communicate, which is not the case for larger tables. Our approach takes the topology of the dining table into account.

Some previous works have also considered the topology of the dining table. Perhaps closest to our paper is the model of Bodlaender et al. [7], in which n individuals are to be assigned to the n vertices of an undirected seating graph. The authors prove a number of computational results regarding both envy-freeness and exchange-stability, among other notions. However, we found some of the table topologies considered to be rather unnatural, especially in hardness proofs (e.g., trees or unions of cliques and independent sets). Bullinger and Suksompong [9] also conduct an algorithmic study of a similar problem, but with a few key differences: (i) individuals are seated in the nodes of a graph, but there may be more seats than people; (ii) for the stability notion, they principally consider *jump-stability*, where unhappy people can choose to move to a free seat; (iii) individuals now contribute to everyone's utility, although their contribution decreases with distance.

Last but not least, studying stability in the context of *Schelling games* has recently been a popular area of research [1,4,5,12,16]. In Schelling games, individuals belong to a fixed number of classes. However, unlike in our model, agents from one class only care about sitting next to others of their own class. This additional assumption often allows for stronger results; e.g., in [4] the authors prove the existence of exchange-stable arrangements on regular and almost regular topological graphs such as cycles and paths, and show that the swap dynamics are guaranteed to converge in polynomial time on such topologies.

Overall, it seems that exchange stability has been studied in the frameworks of both hedonic and Schelling games. However, both approaches present some shortcomings: on the one hand, Schelling games inherently consider a topology on which agents evolve, but, being historically motivated by the study of segregation (e.g., ethnic, racial), they usually restrict themselves to very simple preferences. On the other hand, works on hedonic games are accustomed to considering diverse preferences. However, while multiple works have introduced topological considerations, their analysis is usually constrained to graphs that can be interpreted as non-overlapping coalitions, e.g., with multiple fully connected components. One notable exception is the very recent work of Ceylan, Chen and Roy [11], appearing in IJCAI'23, also building on the model of Bodlaender et al. [7]. In comparison to us, they also prove the NP-hardness of deciding the existence of envy-free arrangements for binary preferences for both topologies, in their case for symmetric preferences, but at the expense of more complex proofs. They moreover show that hardness holds for stability, although the presented proofs requires four non-negative values for cycles and six values including negatives for paths. In our work, we aim to understand stability under more natural preference values, such as approval preferences.

3   Preliminaries

We write $[n] = \{1, \ldots, n\}$. Given an undirected graph $G = (V(G), E(G))$, we write $N_G(v)$ for the set of neighbors of vertex $v \in V(G)$. When clear from context, oftentimes we will simply write V, E and $N(v)$ respectively.

The model we describe next is similar to the one in [7]. A group of n agents (guests) $\mathcal{A}$ has to be seated at a dining table represented by an undirected graph $G = (V, E)$, where vertices correspond to seats. We will be interested in the cases of G being a cycle or a path. We assume that $|V| = n$ and that no two agents can be seated in the same place, from which it also follows that all the seats have to be occupied. Agents have numerical preferences over each other, corresponding to how much utility they gain from being seated next to other agents. In particular, each agent $i \in \mathcal{A}$ has a *preference* over the other agents expressed as a function $p_i : \mathcal{A} \setminus \{i\} \to \mathbb{R}$, where $p_i(j)$ denotes the utility gained by agent i when sitting next to j. Note that we do not assume symmetry; i.e., it might be that $p_i(j) \neq p_j(i)$. We denote by $\mathcal{P} = (p_i)_{i \in \mathcal{A}}$ the collection of agent preferences, or *preference profile*, of the agents. A number of different interpretations can be associated to $\mathcal{P}$. In particular, we will usually see $\mathcal{P}$ as a matrix $\mathcal{P} = (p_{ij})_{i,j \in \mathcal{A}}$, where $p_{ij} = p_i(j)$. Note that the diagonal entries are not defined, but, for convenience, we will oftentimes use the convention that $p_{ii} = 0$. Using the matrix notation, we say that the preferences in $\mathcal{P}$ are *binary* when $\mathcal{P} \in \{0, 1\}^{n \times n}$ and *k-valued* if there exists $\Gamma \subseteq \mathbb{R}$, $|\Gamma| = k$, such that $\mathcal{P} \in \Gamma^{n \times n}$ (disregarding diagonal entries, since they are undefined). Note that binary preferences are two-valued, but two-valued preferences are not necessarily binary. Moreover, we will often represent binary preferences as a directed graph, where a directed edge between two agents signifies that the first agent approves of the second. Finally, when $p_{ij} \geqslant 0$ for any two agents $i, j \in \mathcal{A}$ we say that the preferences are *non-negative*.

We define a *class of agents* to be a subset of indistinguishable agents $\mathcal{C} \subseteq \mathcal{A}$. More formally, all agents in $\mathcal{C}$ share a common preference function $p_{\mathcal{C}} : \mathcal{A} \to \mathbb{R}$

and no agent in $\mathcal{A}$ discriminates between two agents in $\mathcal{C}$. Note that this implies that the lines and columns of the preference matrix corresponding to agents in $\mathcal{C}$ are identical, if we adopt the convention that diagonal entries inside a class are all equal but not necessarily zero. We say that preference profile $\mathcal{P}$ has k *classes*, or is a *k-class profile*, if $\mathcal{A}$ can be partitioned into k classes $\mathcal{C}_1 \cup \ldots \cup \mathcal{C}_k = \mathcal{A}$.

We define an *arrangement* of the agents on G to be a bijection $\pi : \mathcal{A} \to V(G)$, i.e., an assignment of each agent to a unique vertex of the seating graph (and vice-versa). For a given arrangement π, we define for each agent $i \in \mathcal{A}$ their *utility* $U_i(\pi) = \sum_{v \in N_G(\pi(i))} p_i(\pi^{-1}(v))$ to be the sum of agent i's preferences towards their graph neighbors in the arrangement. We say that agent i *envies* agent j whenever $U_i(\pi) < U_i(\pi')$, where π' is π with $\pi(i)$ and $\pi(j)$ swapped. We further say that (i,j) is a *blocking pair* if both i envies j and j envies i; i.e., they would both strictly increase their utility by exchanging seats. An arrangement is *envy-free* if no agent envies another, and it is *stable* if it induces no blocking pairs. Note that envy-freeness implies stability, but the converse is not necessarily true. By extension, we call preference profile $\mathcal{P}$ stable (respectively envy-free) on G if there exists a stable (respectively envy-free) arrangement π on G. Profile $\mathcal{P}$ is unstable if it is not stable.

A few preliminary observations follow. Note that, for *symmetric* preferences; i.e., $p_{ij} = p_{ji}$ for any two agents $i, j \in \mathcal{A}$; a stable arrangement always exists, in fact on any graph G, not just cycles and paths. This is because swaps in that case strictly increase the total sum of agent utilities, and hence the swap dynamics will converge to a stable arrangement. Hence, the interesting case is the asymmetric one. Observe that envy-free arrangements need not exist for symmetric preferences (deciding existence is NP-hard for both paths and cycles [11]). Moreover, note that, for cycles, and in fact any regular graph G, agents have the same number of neighbors, so transforming the preferences of an agent i by adding/subtracting/multiplying by a positive constant the values p_{ij} does not inherently change the preferences. This implies that for cycles the two-valued case coincides with the binary case and the non-negative case coincides with the unconstrained case. It also shows that for cycles the definition of k-valued preferences: $\mathcal{P} \in \Gamma^{n \times n}$ where $|\Gamma| = k$; can be restated equivalently to require that every row of $\mathcal{P}$ consists of at most k different values.

4   Envy-Freeness

It is relatively easy to construct preferences with no envy-free arrangements: for paths, even if all agents like each other, the agents sitting at the endpoints will envy the others; for cycles, add an agent despised by everyone, agents sitting next to them will envy their peers. We now show that, furthermore, it is NP-hard to decide whether envy-free arrangements exist, for both paths and cycles, even under binary preferences. This has also been recently shown in [11] in a stronger form, using only symmetric binary preferences, but we found the construction rather involved and the correctness argument based on careful counting delicate.

Theorem 1. *For binary preferences, deciding whether an envy-free arrangement on a cycle exists is NP-hard.*

Proof. We proceed by reduction from Hamiltonian Cycle on directed graphs. Let $G = (V, E)$ be a directed graph[1] such that, without loss of generality, $V = [n]$. If G has any vertices with no outgoing edges, then map the input instance to a canonical no-instance, unless $n = 1$, in which case we map it to a canonical yes-instance. Hence, from now on, assume that all vertices have outgoing edges. For each vertex $v \in V$ introduce three agents x_v, y_v, z_v such that agent x_v only likes y_v and dislikes[2] everyone else, agent y_v only likes z_v and dislikes everyone else, and agent z_v likes agent x_u for all $u \in V$ such that $(v, u) \in E$, and dislikes everyone else. We claim that the so-constructed preference profile $\mathcal{P}$ has an envy-free arrangement on a cycle precisely when G has a Hamiltonian cycle. To show this, first assume without loss of generality that $1 \to 2 \to \ldots \to n \to 1$ is a Hamiltonian cycle in G. Then, arranging agents around the cycle in the order $x_1, y_1, z_1, x_2, y_2, z_2, \ldots, x_n, y_n, z_n$ is an envy-free arrangement. To see why, notice that in this arrangement all agents get utility 1, so envy could only potentially stem from an agent being able to swap places with another agent to get utility 2. To prove this is not possible, first notice that agents $(x_i)_{i \in [n]}$ and $(y_i)_{i \in [n]}$ each only like one other agent, so they can never get a utility of more than 1 in any arrangement. Moreover, no agent z_i can get to a utility of 2 by a single swap because any two agents they like are seated at least three positions from each other on the cycle. Conversely, assume that an envy-free arrangement π exists. First, if x_i is not sitting next to y_i in π, then x_i could improve by swapping to a place next to y_i (also similarly for y_i and z_i). Therefore, in arrangement π agent x_i is seated next to y_i and y_i is seated next to z_i. Moreover, consider the other neighbor of z_i in π. Since z_i does not like y_i, it follows that if z_i also does not like their other neighbor, then z_i could strictly improve their utility by swapping next to some agent they like, which is always possible because all vertices in G have outgoing edges. Therefore, the other neighbor of z_i has to be some agent that they like, hence being of the form x_j, where $j \neq i$. Note that, by construction, $(i, j) \in E$. Putting together what we know, we get that under π the agents are arranged around in the cycle in some order $x_{\sigma_1}, y_{\sigma_1}, z_{\sigma_1}, x_{\sigma_2}, y_{\sigma_2}, z_{\sigma_2}, \ldots, x_{\sigma_n}, y_{\sigma_n}, z_{\sigma_n}$, where σ is a permutation of the n agents such that $(\sigma_i, \sigma_{i+1}) \in E$ holds for all $i \in [n]$.[3] Therefore, a Hamiltonian cycle $\sigma_1 \to \sigma_2 \to \ldots \to \sigma_n \to \sigma_1$ exists in G. $\square$

A similar proof can be used to show hardness for the case of paths. We outline the changes required in the full version.

Theorem 2. *For binary preferences, deciding whether an envy-free arrangement on a path exists is NP-hard.*

Proving matching hardness results for stability under binary preferences would be highly desirable but for the time being is left open. We in fact conjecture that all instances with binary preferences are stable on a path (see Sect. 6.3).

---

[1] In this proof G is not the seating graph but rather an arbitrary graph.

[2] Technically, is indifferent to everyone else, but found this formulation reads better.

[3] Assuming that addition is performed with wrap-around using $n + 1 \equiv 1 \pmod{n}$.

Moreover, one might ask how does the number of agent classes affect the computational complexity of our problems. In the next section, we address this question, showing that limiting the number of agent classes renders the problems that we consider polynomial-time solvable, even for arbitrary preference values.

5   Polynomial Solvability for k-Class Preferences

In this section, we show that deciding whether envy-free and stable arrangements exist for a given preference profile can be achieved in polynomial time, for both paths and cycles, assuming that the number of agent classes is bounded by a number k. Note that preferences in this case are not constrained to being binary, and can in fact be arbitrary. By extension, our algorithms can also be used to construct such arrangements whenever they exist.

We begin with the case of paths. For simplicity, we assume that $n \geqslant 3$, as for $n \leqslant 2$ any arrangement is both stable and envy-free. Assume that the agent classes are identified by the numbers $1, \ldots, k$ and that $n_1, n_2, \ldots, n_k$ are the number of agents of each class in our preference profile, where $n_1 + \ldots + n_k = n$. For ease of writing, we will see arrangements as sequences $s = (s_i)_{i \in [n]}$, where $s_i \in [k]$ and for any agent class $j \in [k]$ the number of values j in s is n_j. Moreover, for brevity, we lift agent preferences to class preferences, in order to give meaning to statements such as "class a likes class b." To simplify the treatment of agents sitting at the ends of the path, we introduce two agents of a dummy class 0 with preference values 0 from and towards the other agents. We require the dummy agents to sit at the two ends of the path; i.e., $s_0 = s_{n+1} = 0$. In order to use a common framework for stability and envy-freeness, we define the concept of *compatible triples* of agent classes, as follows. First, for envy-freeness, let a, b, c, d, e, f be agent classes, then we say that triples (a, b, c) and (d, e, f) are *long-range compatible* if $p_b(a) + p_b(c) \geqslant p_b(d) + p_b(f)$ and $p_e(d) + p_e(f) \geqslant p_e(a) + p_e(c)$; intuitively, neither b wants to swap with e, nor vice-versa. Furthermore, for a, b, c, d agent classes, we say that triples (a, b, c) and (b, c, d) are *short-range compatible* if $p_b(a) \geqslant p_b(d)$ and $p_c(d) \geqslant p_c(a)$; intuitively, if a, b, c, d are consecutive in the arrangement, then neither b wants to swap with c, nor vice-versa. For stability, we keep the same definitions but use "or" instead of "and." Note that long-range and short-range compatibility do not imply each other. We call an arrangement s *compatible* if for all $1 \leqslant i < j \leqslant n$ the triplets (s_{i-1}, s_i, s_{i+1}) and (s_{j-1}, s_j, s_{j+1}) are long-range compatible when $j - i > 1$ and short-range compatible when $j - i = 1$. Note that arrangement s is envy-free (resp. stable) if and only if it is compatible. In the following, we explain how to decide the existence of a compatible arrangement.

Lemma 3. *Deciding whether compatible arrangements exist can be achieved in polynomial time.*

Proof. We first present a nondeterministic algorithm (i.e., with guessing) that solves the problem in polynomial time. The algorithm builds a compatible arrangement s one element at a time. Initially, the algorithm sets $s_0 \leftarrow 0$ and guesses the values of s_1 and s_2. Then, at step i, for $3 \leqslant i \leqslant n+1$, the algorithm will guess s_i (except for $i = n + 1$, where we enforce that $s_i \leftarrow 0$) and check whether $(s_{i-3}, s_{i-2}, s_{i-1})$ is short-range conflicting with (s_{i-2}, s_{i-1}, s_i), rejecting if so. Moreover, the algorithm will check whether (s_{i-2}, s_{i-1}, s_i) is long-range conflicting with any (s_{j-2}, s_{j-1}, s_j) for $2 \leqslant j \leqslant i - 2$, again rejecting if so. At the end, the algorithm checks whether for each $i \in [k]$ value i occurs in s exactly n_i times, accepting if so, and rejecting otherwise.

Alone, this algorithm only shows containment in NP, which is not a very attractive result. Next, we show how the same algorithm can be implemented with only a constant number of variables, explaining afterward why this implies our result. First, to simulate the check at the end of the algorithm without requiring knowledge of the whole of s, it is enough to maintain throughout the execution counts $(x_j)_{j \in [k]}$ such that at step i in the algorithm x_j gives the number of positions $1 \leqslant \ell \leqslant i$ such that $s_\ell = j$. To simulate the short-range compatibility check, it is enough that at step i we have knowledge of $s_{i-3}, \ldots, s_i$. Finally, for the long-range compatibility check, a more insightful idea is required. In particular, we make the algorithm maintain throughout the execution counts $m_{a,b,c}$ for each triple (a, b, c) of agent classes, such that at step i value $m_{a,b,c}$ gives the number of positions $2 \leqslant \ell \leqslant i$ such that $(s_{\ell-2}, s_{\ell-1}, s_\ell) = (a, b, c)$. Using this information, to check at step i whether (s_{i-2}, s_{i-1}, s_i) long range conflicts with any (s_{j-2}, s_{j-1}, s_j) for $2 \leqslant j \leqslant i - 2$, it is enough to temporarily decrease by one the values $m_{s_{i-3}, s_{i-2}, s_{i-1}}$ and $m_{s_{i-2}, s_{i-1}, s_i}$ and then check whether there exists a triple (a, b, c) of agent classes such that $m_{a,b,c} > 0$ and (a, b, c) long-range conflicts with (s_{i-2}, s_{i-1}, s_i). In total, at step i, the algorithm only needs to know the values $s_{i-3}, \ldots, s_i$, as well as $(x_j)_{j \in [k]}$ and the counts $m_{a,b,c}$ for all triples (a, b, c) of agent classes. Since $k + 1$ bounds the total number of agent classes, this is only a constant number of variables. As each variable can be represented with $O(\log n)$ bits, it follows that our nondeterministic algorithm uses only logarithmic space, implying containment in the corresponding complexity class NL. It is well known that NL $\subseteq$ P, from which our conclusion follows. For readers less familiar with this result, we give a short overview of how our algorithm can be converted into a deterministic polynomial-time algorithm, as follows. Since our NL algorithm uses only logarithmic space, it follows that the space of algorithm states that can be reached depending on the nondeterministic choices is of at most polynomial size, since $2^{O(\log n)}$ is polynomial. Therefore, building a graph with vertices being states and oriented edges corresponding to transitions between states, the problem reduces to deciding whether an accepting state can be reached from the initial state, which can be done with any efficient graph search algorithm.

Theorem 4. *Fix $k \geqslant 1$. Then, for k-class preferences, there are polynomial-time algorithms computing an envy-free/stable arrangement on a path or reporting the nonexistence thereof.*

For the case of cycles, a similar approach can be used, although with rather tedious, yet minor tweaks, deferred to the full version.

Theorem 5. *Fix $k \geqslant 1$. Then, for k-class preferences, there are polynomial-time algorithms computing an envy-free/stable arrangement on a cycle or reporting the nonexistence thereof.*

6 A Fine-Grained Analysis of Stability

Previously, we showed that deciding whether envy-free arrangements exist is NP-hard for binary preferences on both topologies. For stability, in [11] it is shown that similar hardness results hold, but this time more than two values are needed in the proofs, namely four values for cycles and six for paths, the latter also requiring negative numbers. It is unclear whether hardness is retained without assuming this level of preference granularity, and a first step towards understanding the difficulty of the problem constrained to fewer/simpler allowed values is being able to construct instances where no stable arrangements exist; after all, a problem where the answer is always "yes" cannot be NP-hard.

In this section, we conduct a fine-grained analysis of the conditions allowing for unstable instances. In particular, for both topologies we consider how different constraints on the number of agent classes as well as the number of different values allowed in the preferences influence the existence of unstable preferences The non-negativity of the values needed is also taken into account. Table 2 summarizes our results.

6.1 Two-Class Preferences

As a warm-up, note that when all agents come from a single class, any arrangement on any given seating graph is stable. In the following, we extend this result to two classes of agents for cycles and paths. We begin with cycles:

Theorem 6. *Two-class preferences always induce a stable arrangement on a cycle.*

Proof. Suppose there are two classes of agents, say Blues and Reds. Without loss of generality, preferences can be assumed to be binary, since for cycles one can normalize the preference values as described towards the end of the preliminaries section. First, note that any blocking pair must consist of one Blue and one Red. Moreover, note that any arrangement is stable whenever one of the classes likes the two classes equally. Now, suppose this is not the case, meaning that each class has a preferred class to sit next to. There are only two cases to consider:

If one class, say Blue, prefers its own class, then sit all Blues together and give the remaining seats to Reds: all Blues but two, say B_1 and B_2, get maximum utility, and neither B_1 nor B_2 can improve since no Red has more than one Blue neighbor. Hence, no Blue is part of a blocking pair, so the arrangement is stable.

If both classes prefer the opposite class, we may assume there are at least as many Reds as Blues. Then we alternate between Reds and Blues for as long as there are Blues without a seat, then seat all the remaining Reds next to each other. Every Blue has maximum utility, and hence cannot be part of a blocking pair, so the arrangement is stable.

The following extends the result to the case of paths. The proof is largely similar, but the case analysis becomes more involved, because preferences can no longer be assumed to be binary, so we present it in the full version.

Theorem 7. *Two-class preferences always induce a stable arrangement on a path.*

Note that a path of n is equivalent to a cycle of $n + 1$ where an agent with null preferences is added. This explains why the case of paths is harder to study than that of cycles, as it corresponds to having one more class of agents and potentially one more value (zero).

6.2   Three-Class Three-Valued Preferences

We now consider the case of three-valued preferences with three agent classes, exhibiting unstable non-negative preferences both for paths and for cycles. We begin with the case of cycles.

Theorem 8. *For $n \geq 4$, there exist three-class three-valued non-negative preferences such that all arrangements on a cycle are unstable.*

Proof. Consider three classes of agents: Alice, Bob, and $n - 2$ of Bob's friends. The story goes as follows: Alice and Bob broke up. Alice does not want to hear about Bob and would hence prefer to sit next to any of his friends rather than Bob. On the other hand, Bob wants to win her back, so he would above all want to sit next to Alice. Finally, Bob's friends prefer first Bob, then the other friends, and finally Alice. Used preference values can be arbitrary, so to get the required conclusion, we make sure that they are non-negative. To show that these preferences are unstable on a cycle, there are two cases: Alice and Bob can either sit next to each other, or separately.[4]

In the first case, Alice and her second neighbor, who is one of Bob's friends, would exchange seats. After the switch, Alice is better as she no longer sits next to Bob, and the friend is better because he sits next to Bob.

In the second case, Bob and one of Alice's neighbors would exchange seats. Bob is better because he now sits next to Alice. To see that the neighbor, who is one of Bob's friends, is also better, distinguish two sub-cases: if the friend sits right between Alice and Bob, then he is better because he now no longer sits next to Alice, while if this is not the case, he is better because before he was sitting next to Alice and a friend, while now he is sitting next to two friends.

It is possible to use the same construction for paths by allowing negative preference values, but otherwise, the proof of Theorem 8 does not directly transfer to paths; e.g., for $n = 5$, path arrangement (F, F, B, A, F) is stable. Negative preferences turn out to not be necessary for n large enough. The main trick here is to use three copies of Bob to ensure that at least one of them does not sit at either end of the path. In particular, we show the following in the full version:

---

[4] The swap dynamics here will exhibit so-called run-and-chase behaviour [8], which is common to many classes of hedonic games.

Table 3. The number of non-isomorphic families of unstable non-negative two-valued preferences. For cycles, this coincides with the case of binary values, and also with that of general values.

n	3	4	5	6	7
Cycle	0	0	1	0	2
Path	0	0	0	0	0

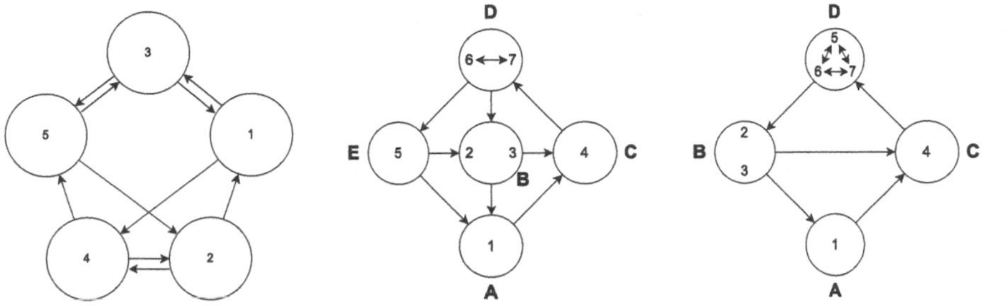

(a) $\mathcal{P}_5$ as a directed graph.   (b) $\mathcal{P}_7^{(1)}$ as a directed graph.   (c) $\mathcal{P}_7^{(2)}$ as a directed graph.

Fig. 1. The three families of cycle unstable binary preferences for $n \leqslant 7$.

Theorem 9. *For $n \geqslant 12$, there exist three-class three-valued non-negative preferences such that all arrangements on a path are unstable.*

6.3   Two-Valued Preferences

It remains to study what happens for two-valued preferences with three or more classes of agents. For cycles, we show that three classes always yield a stable arrangement, while four classes allow for a counterexample with binary values. For paths, we exhibit a counterexample with already three classes, but using some negative values. For non-negative values, we conjecture that a stable arrangement always exists, but have not been able to prove it. Instead, we gather both experimental and theoretical evidence to support it.

Two-Valued Preferences on Cycles. Cycles being regular graphs, it is sufficient to study binary preferences. We exhausted all binary preferences for $n \leqslant 7$ using a Z3 Python solver (see full version). Unstable preferences were found for $n = 5$ and $n = 7$, with one and, respectively, two non-isomorphic families of unstable preferences (see Table 3). Examples of such preferences from each family $\mathcal{P}_5, \mathcal{P}_7^{(1)}$ and $\mathcal{P}_7^{(2)}$ are illustrated in Fig. 1. Analyzing why $\mathcal{P}_5$ is unstable turns out to be quite complex (see full version), and, as of our current understanding, its instability seems to be more of a "small size artifact" than anything else. In contrast, with their highly regular structure, the two instances with $n = 7$ seem more promising. In particular, profile $\mathcal{P}_7^{(2)}$ consists of only four classes, denoted by A, B, C, D in Fig. 1c. In the following, we show that $\mathcal{P}_7^{(2)}$ can be extended to unstable preferences of any $n \geqslant 7$.

Theorem 10. *For $n \geqslant 7$, there exist four-class binary preferences such that all arrangements on a cycle are unstable.*

Proof. For $n \geqslant 7$, we consider four classes A, B, C and D, as well as their respective members a, b_1, b_2, c and $d_1, \ldots, d_{n-4}$. Similarly to Fig. 1c, we suppose that: (i) a only likes c; (ii) b_1 and b_2 both only like a and c; disliking each other; (iii) c only likes members of D; (iv) members of D all like each other, as well as b_1 and b_2, only disliking a and c. We show in the full version why such preferences induce no stable arrangements on a cycle.

This result shows that four classes of agents are sufficient to make all arrangements unstable on a cycle for two-valued preferences; we show in the following it is also necessary, as three classes of agents with two-valued preferences always induce a stable arrangement on a cycle.

Theorem 11. *Three-class two-valued preferences always induce a stable arrangement on a cycle.*

Proof. We consider three classes: Reds, Greens, and Blues, each containing r, g, and b agents respectively. Without loss of generality, assume that $r, g, b \geqslant 1$. Since the seating graph is regular, recall that we may assume the preferences to be binary. Note moreover that any blocking pair must have two agents of different colors. A principled case distinction now allows us to relatively quickly exhaust over all the possibilities for the preferences. We present the case distinction here and delegate the proofs themselves to specialized lemmas in the full version.

First, whenever at least one class likes its own kind, Lemmas 23, 24 and 25 together show the existence of a stable arrangement. Note that the proofs are constructive and all the stable arrangements presented intuitively seat the self-liking class consecutively.

Then, if no class likes itself, Lemmas 26 and 27 show the existence of a stable arrangement whenever one class likes or is disliked by every other class. Note this time that intuitively all constructions present an alternation of two classes.

This only leaves us to handle the case where each class likes and is liked by exactly another class: this case is treated separately in Lemma 29, where we show that an arrangement maximizing utilitarian social welfare is always stable. The proof also shows that a modified variant of the swap dynamics converges.

Two-Valued Preferences on Paths. We now consider the more complex case of paths, where the endpoints and the associated loss of regularity artificially introduce an implicit comparison with zero, hence giving rise to a change of behavior between positive and negative preferences. In general, if we allow for negative values in the preferences, then there exist three-class two-valued preferences such that no arrangement on a path is stable:

Theorem 12. *For $n \geq 3$, there exist three-class two-valued preferences such that all arrangements on a path are unstable.*

Proof. Consider a variant of "Alice, Bob and Friends" where the preferences of Bob towards Alice, Alice towards Friends, and Friends towards Bob and themselves are all one; the preferences of Alice towards Bob, Bob towards Friends, and Friends towards Alice are all minus one. We show in the full version why all arrangements on a path are unstable.

For non-negative preferences, on the other hand, exhaustion for $n \leq 7$ using a similar solver[5] yields no unstable instances (see Table 3). Surprised by the outcome, we also wrote C++ code to test all k-class instances with at most b agents per class for $(k, b) \in \{(4, 10), (5, 4)\}$, also leading to no unstable instances. *We conjecture that two-valued instances with non-negative preferences always induce a stable arrangement on a path.* In the following, we show that this is true under the additional assumption that two agents are only willing to swap seats when they are at most two positions away on the path, no matter how much their utilities would increase otherwise. This can be thought of as a practical constraint: once the agents are seated, each agent knows which other agents they envy, but finding out whether envy is reciprocal would be too cumbersome if the other agent is seated too far away. For this setup, we prove that the swap dynamics always converge, so a stable arrangement can be found by starting with an arbitrary arrangement and swapping blocking pairs until the arrangement becomes stable. This can be seen as a generalization of a result from [5], where agents only have preference for others of their own kind and swaps are only with adjacent agents. This is stated below and proven next.

Theorem 13. *Two-valued non-negative preferences always induce a stable arrangement on a path assuming that agents are only willing to exchange seats with other agents sitting at distance at most two on the path. Moreover, the swap dynamics converge in this case.*

To begin showing this, note that no local swap could occur with an agent seated at either endpoint, since the preferences are non-negative. Since the two endpoints are the only irregularities, removing them from consideration, we can restrict our analysis from non-negative two-valued to binary without loss of generality. For any arrangement π, define the utilitarian social welfare $W(\pi) = \sum_{i \in \mathcal{A}} U_i(\pi)$. Moreover, to each arrangement π we associate a sequence $S(\pi)$ of length $n - 1$ with elements in $\{0, 1, 2, 3\}$, constructed as follows. Let π_i and π_{i+1} be the agents sitting at positions i and $i + 1$ on the path: if they do not like each other, then $S(\pi)_i = 0$, if they both like each other, then $S(\pi)_i = 3$, if only π_i likes π_{i+1}, then $S(\pi)_i = 1$, otherwise $S(\pi)_i = 2$. To prove that the swap dynamics converge, we define the potential $\Phi(\pi) = (W(\pi), S(\pi))$, where sequences are compared lexicographically, and prove that swapping blocking pairs always strictly increases the potential. The following two lemmas show this for swaps at distances one and two, respectively.

---

[5] One can show that it suffices to try the cases $\Gamma \in \{\{0, 1\}, \{1, 2\}, \{1, 3\}, \{2, 3\}\}$.

Lemma 14. *Let π be an arrangement where a and b form a blocking pair and sit in adjacent seats. Let π' be π with a and b's seats swapped. Then, $\Phi(\pi') > \Phi(\pi)$.*

Proof. When $n \leq 2$, there are no blocking pairs, so assume $n \geq 3$. First, notice that swapping the places of a and b keeps a and b adjacent, from which the swap changes the utility of any agent by at most one. Since a and b's utilities have to increase, they have to each change by exactly one. Moreover, note that neither a nor b can be seated at the ends of the table, as otherwise swapping would make one of them lose a neighbor while keeping the other one, hence not increasing their utility. Hence, assume that x is the other neighbor of a and y is the other neighbor of b; i.e., x, a, b, y are seated consecutively in this order on the path, at positions say $i, \ldots, i + 3$. If either $U_x(\pi') \geq U_x(\pi)$ or $U_y(\pi') \geq U_y(\pi)$, it follows that $W(\pi') - W(\pi) = U_x(\pi') - U_x(\pi) + U_y(\pi') - U_y(\pi) + 2 \geq 1$, so $\Phi(\pi') > \Phi(\pi)$. Otherwise, we know that $U_x(\pi') - U_x(\pi) = U_y(\pi') - U_y(\pi) = -1$, from which $W(\pi') = W(\pi)$. Together with $U_a(\pi') - U_a(\pi) = U_b(\pi') - U_b(\pi) = 1$, this means that preferences satisfy $a \to y \to b \to x \to a$, where an arrow $u \to v$ indicates that agent u likes v but not the other way around. Therefore, $S(\pi)_i = 1$ and $S(\pi')_i = 2$. Since $S(\pi)$ and $S(\pi')$ only differ at positions $i, \ldots, i + 2$, this means that $S(\pi') > S(\pi)$, so $\Phi(\pi') > \Phi(\pi)$, as required.

Lemma 15. *Let π be an arrangement where a and b form a blocking pair and sit two seats away. Let π' be π with a and b's seats swapped. Then, $\Phi(\pi') > \Phi(\pi)$.*

Proof. The same argument works, except that now we consider five agents x, a, z, b, y seated at positions $i, \ldots, i + 4$. This is because agent z remains a common neighbor to a and b when swapping places, and can, essentially, be ignored.

Therefore, since the potential is upper-bounded, we get that the swap dynamics have to converge. One might now rightfully ask whether convergence is guaranteed to take polynomial time. While we could neither prove nor disprove this, in the full version we give evidence of why exponential time might be required. Moreover, note that for cycles convergence is not guaranteed even for swaps at distance at most two; e.g., $\mathcal{P}_5$ in Fig. 1a, where any two agents are seated at most two seats away anyway. For paths, on the other hand, one could still hope that the result generalizes beyond distance at most two when the preferences are non-negative. This is however not the case, even when stable arrangements exist, as we show next (details in the full version).

Lemma 16. *Consider profile $\mathcal{P}_4$ in Fig. 2. A path stable arrangement exists, yet the swap dynamics started from certain arrangements cannot converge.*

This generalizes to $n \geq 4$ agents by adding $n - 4$ dummy agents to $\mathcal{P}_4$ liking nobody and being liked by nobody and seating them at positions $5, \ldots, n$ on the path. Hence, non-convergence for distance ≤ 3 is not a small-n artifact.

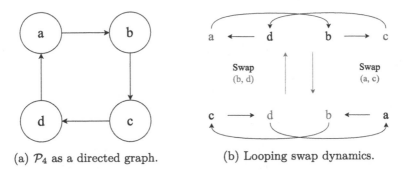

(a) $\mathcal{P}_4$ as a directed graph.   (b) Looping swap dynamics.

Fig. 2. Four-agent binary preferences $\mathcal{P}_4$ with path stable arrangement $\pi^* = (a, b, c, d)$ but where the swap dynamics necessarily alternate between $\pi_1 = (a, d, b, c)$ and $\pi_2 = (c, d, b, a)$, up to reversal, when started in either of them.

7   Conclusions and Future Work

We studied envy-freeness and exchange-stability on paths and cycles. For both topologies, we showed that finding envy-free/stable arrangements can be achieved in polynomial time when the number of agent classes is bounded. For envy-freeness the problem becomes NP-hard without this restriction, even for binary preferences. For stability, it is known that for sufficiently many values the problem is also NP-hard [11]. It would be interesting to see, e.g., if the same holds for binary preferences. We believe this to be the case at least for cycles, but were unable to prove it. In part, this is because of the difficulty of constructing unstable instances in the first place. Moreover, for both topologies, we gave a full characterization of the pairs (k, v) such that k-class v-valued unstable preferences exist. For paths, the characterization requires negative values in the two-valued case, and we are still unsure whether two-valued non-negative preferences that are unstable on a path exist. We, however, partially answer this in the negative by showing that the swap dynamics are guaranteed to converge if agents can only swap places with other agents seated at most two positions away from them. Without this assumption, convergence might not be guaranteed even when stable arrangements exist, so a different approach would be required to prove existence. It would also be interesting to know if unstable preferences are exceptions or the norm. We give a probabilistic treatment of this question for random preference digraphs of average degree $O(\sqrt{n})$ in the full version. As an avenue for future research, it would be attractive to consider other kinds of tables commonly used in practice, like the $2 \times n$ grid with guests on either side facing each other. It would additionally be interesting to consider non-additive utilities or to enforce additional constraints on the arrangement, such as certain people coming "in groups" and hence having to sit consecutively at the table. It would also be worth investigating replacing addition by taking minimum in the agents' utilities (and also the more general lexicographical variant).

References

1. Agarwal, A., Elkind, E., Gan, J., Igarashi, A., Suksompong, W., Voudouris, A.A.: Schelling games on graphs. Artif. Intell. **301**, 103576 (2021)
2. Aziz, H., Savani, R.: Hedonic games. In: Brandt, F., Conitzer, V., Endriss, U., Lang, J., Procaccia, A.D. (eds.) Handbook of Computational Social Choice, chap. 15, 1st edn. Cambridge University Press, Cambridge (2016)
3. Berriaud, D., Constantinescu, A., Wattenhofer, R.: Stable dinner party seating arrangements. arXiv:2305.09549 (2023)
4. Bilò, D., Bilò, V., Lenzner, P., Molitor, L.: Topological influence and locality in swap Schelling games. Auton. Agent. Multi-Agent Syst. **36**, 47 (2022)
5. Bilò, V., Fanelli, A., Flammini, M., Monaco, G., Moscardelli, L.: Nash stable outcomes in fractional hedonic games: existence, efficiency and computation. J. Artif. Intell. Res. **62**, 315–371 (2018)
6. Bilò, V., Monaco, G., Moscardelli, L.: Hedonic games with fixed-size coalitions. In: AAAI 2022, vol. 36, no. 9, pp. 9287–9295, June 2022
7. Bodlaender, H.L., Hanaka, T., Jaffke, L., Ono, H., Otachi, Y., van der Zanden, T.C.: Hedonic Seat Arrangement Problems. arXiv, February 2020
8. Boehmer, N., Bullinger, M., Kerkmann, A.M.: Causes of stability in dynamic coalition formation. In: AAAI 2023, vol. 37, no. 5, pp. 5499–5506, June 2023
9. Bullinger, M., Suksompong, W.: Topological distance games. In: AAAI 2023, vol. 37, no. 5, pp. 5549–5556, June 2023
10. Cechlárová, K.: On the complexity of exchange-stable roommates. Discret. Appl. Math. **116**(3), 279–287 (2002)
11. Ceylan, E., Chen, J., Roy, S.: Optimal seat arrangement: what are the hard and easy cases? In: Elkind, E. (ed.) IJCAI 2023, pp. 2563–2571, August 2023
12. Chauhan, A., Lenzner, P., Molitor, L.: Schelling segregation with strategic agents. In: Deng, X. (ed.) SAGT 2018. LNCS, vol. 11059, pp. 137–149. Springer, Cham (2018). https://doi.org/10.1007/978-3-319-99660-8_13
13. Gale, D., Shapley, L.S.: College admissions and the stability of marriage. Am. Math. Mon. **69**(1), 9–15 (1962)
14. Irving, R.W.: An efficient algorithm for the "stable roommates" problem. J. Algorithms **6**(4), 577–595 (1985)
15. Irving, R.W.: Stable marriage and indifference. Discret. Appl. Math. **48**(3), 261–272 (1994)
16. Kreisel, L., Boehmer, N., Froese, V., Niedermeier, R.: Equilibria in Schelling games: computational hardness and robustness. In: AAMAS 2022, pp. 761–769 (2022)
17. Ronn, E.: NP-complete stable matching problems. J. Algorithms **11**(2), 285–304 (1990)

Buy-Many Mechanisms for Many Unit-Demand Buyers

Shuchi Chawla[1] , Rojin Rezvan[1]([✉]) , Yifeng Teng[2] ,
and Christos Tzamos[3]

[1] University of Texas at Austin, Austin, TX 78712, USA
`rojinrezvan@gmail.com`
[2] Google Research, New York, NY, USA
[3] University of Athens, Athens, Greece

Abstract. A recent line of research has established a novel desideratum for designing approximately-revenue-optimal multi-item mechanisms, namely the buy-many constraint. Under this constraint, prices for different allocations made by the mechanism must be subadditive, implying that the price of a bundle cannot exceed the sum of prices of individual items it contains. This natural constraint has enabled several positive results in multi-item mechanism design bypassing well-established impossibility results. Our work addresses the main open question from this literature of extending the buy-many constraint to multiple buyer settings and developing an approximation.

We propose a new revenue benchmark for multi-buyer mechanisms via an ex-ante relaxation that captures several different ways of extending the buy-many constraint to the multi-buyer setting. Our main result is that a simple sequential item pricing mechanism with buyer-specific prices can achieve an $O(\log m)$ approximation to this revenue benchmark when all buyers have unit-demand or additive preferences over m items. This is the best possible as it directly matches the previous results for the single-buyer setting where no simple mechanism can obtain a better approximation.

From a technical viewpoint we make two novel contributions. First, we develop a supply-constrained version of buy-many approximation for a single buyer. Second, we develop a multi-dimensional online contention resolution scheme for unit-demand buyers that may be of independent interest in mechanism design.

Keywords: Buy-many Mechanisms · Sequential Item Pricing · Multi-item Mechanism Design

1 Introduction

Revenue maximization in multi-parameter settings is notoriously challenging. It is known, for example, that in the absence of strong assumptions on the buyer's value distribution, the optimal revenue cannot be approximated within any finite factor by any mechanism with finite description complexity even for the simplest possible

J. Garg et al. (Eds.): WINE 2023, LNCS 14413, pp. 21–38, 2024.
https://doi.org/10.1007/978-3-031-48974-7_2

setting of two items and a single unit-demand buyer [4,17]. These impossibility results motivate the search for a different benchmark that captures salient features of the problem space while also permitting non-trivial approximation.

For single-agent settings one such benchmark was proposed by Briest et al. [4] and Chawla et al. [10]. These works showed that the infinite gap between the revenue of the optimal mechanism and any simple mechanism arises precisely when the mechanism offers "super-additive" options that charge for some bundles of items more than the sum of prices of their individual components. The so-called "buy-many" constraint disallows such exploitative behavior: when interpreted as a pricing over all possible allocations, the mechanism should be subadditive (defined appropriately over randomized allocations, as described in Sect. 2). An alternate motivation for this constraint arises as a consequence of buyer behavior in scenarios where the buyer can interact with the mechanism multiple times, purchasing an option from the menu each time. The buyer can then construct an allocation by purchasing a multiset of options from the menu in the cheapest possible manner. The buy-many constraint restricts the kinds of mechanisms the seller can use to extract revenue from the buyer. The corresponding benchmark is the optimal revenue that can be obtained from any mechanism satisfying the constraint. [4] and [10] showed that imposing such a constraint enables positive results even without requiring any assumptions on the buyer's valuation function or the value distribution: for settings with a single buyer and m items, the optimal revenue obtained from any mechanism satisfying the buy-many constraint is no more than $O(\log m)$ times the revenue obtained from an item pricing.

Our work addresses the primary direction left open by these works and their followups, namely extending the buy-many constraint and revenue benchmark to settings with multiple buyers. Obtaining such an extension is challenging, however, as it depends on the particular implementation of the mechanism, differences in these details can lead to very different benchmarks. Indeed while the two approaches described above for formalizing the buy-many constraint – as a restriction on the pricing function and as a consequence of buyer behavior – lead to equivalent definitions in the single-buyer setting, they turn out to be very different in the multi-buyer setting. In fact, the latter approach of allowing buyers to interact with the mechanism multiple times provides no meaningful restriction on mechanisms at all and once again allows for unbounded revenue gaps between simple and optimal mechanisms.[1] We instead define the buy-many

---

[1] Consider, for example, multiple interactions of a buyer with the mechanism interleaved by purchases made by other buyers. As the item supply changes, the mechanism can update the prices on its menu, and no longer necessarily needs to satisfy a subadditivity constraint on the final pricing observed by the buyer. In fact, by exploiting this supply-based pricing approach, a multi-buyer buy-many mechanism can simulate any single-agent non-buy-many mechanism, inheriting the unbounded simple-versus-optimal revenue gaps of the latter setting. Sybil-proofness or false-name-proofness is even easier to achieve in principle, unless some symmetry-type restrictions are placed on the mechanism (as in [18,25], for example), as the mechanism can simply refuse to make any allocations unless the number of agents is exactly n.

constraint as a restriction on the effective pricing faced by any individual buyer participating in the mechanism. We view the buy-many constraint as a standalone desideratum for the design of mechanisms that disallows exploitative behavior on part of the seller. This approach leads to a non-trivial restriction, while continuing to permit a range of different designs. We show that it becomes possible to recover upper bounds on the revenue gap between simple and optimal mechanisms without assumptions about buyers' value distributions, just as in the single-buyer setting.

Even as a restriction on the mechanism's pricing, the buy-many constraint can take many different forms depending on the information available to buyers at different stages in the mechanism. To obtain a comprehensive theory of multi-buyer buy-many mechanisms without going into the intricacies of specific applications, we avoid choosing any particular such extension, and instead define an ex-ante relaxation of the optimal buy-many revenue that simultaneously captures a broad range of settings. In Sect. 2, we describe the different forms of the buy-many constraint encompassed by this relaxation.

An Ex-Ante Relaxation of Buy-Many Revenue

We view multi-buyer settings as a collection of single-buyer instances via an ex-ante relaxation that allows the mechanism to allocate each item multiple times as long as the *expected* number of buyers each item is allocated to is at most 1. To be specific, let x_i be an m-dimensional allocation vector, with $x_{ij} \in [0, 1]$ denoting the probability of allocation of item j to a particular buyer i. We consider single-buyer mechanisms satisfying two restrictions. First, over the randomness in the buyer's valuation function, we require that the probability of allocation of each item j to the buyer is at most x_{ij}. We say that the mechanism satisfies the ex-ante constraint x_i. Second, we require that the mechanism satisfies the single-buyer buy-many constraint. We then consider the maximum revenue that can be obtained from any mechanism for buyer i that satisfies both of the aforementioned constraints. Note that the buy-many constraint is not closed over convex combinations, so distributions over buy-many menus can potentially obtain higher revenue than buy-many menus themselves. We accordingly consider the maximum revenue obtainable from individual buy-many mechanisms or distributions over buy-many mechanisms that satisfy the ex-ante constraint x_i. Let $\text{BUYMANYREV}_i(x_i)$ denote this upper bound. The following program then gives an upper bound on the revenue of multi-buyer buy-many mechanisms:

$$\text{EXANTE-BUYMANYREV} := \max_{x_1,\ldots,x_n \geq 0} \sum_i \text{BUYMANYREV}_i(x_i) \text{ s.t. } \sum_i x_i \preceq 1.$$

Here "$\preceq$" means pointwise dominance: for any two vectors $x, y \in \mathbb{R}^m$, $y \preceq x$ means $y_j \leq x_j$ for every $j \in [m]$. Our goal in this work is to design simple multi-buyer mechanisms that are clearly buy-many but at the same time are competitive with this new benchmark.

Approximating the Buy-Many Revenue by Item Pricings

Although buy-many mechanisms can, in general, be quite complicated, we show that when all buyers are unit-demand or additive, the optimal buy-many revenue (as well as the ex-ante upper-bound) can be approximated via a simple class of mechanisms, namely sequential item pricings. We obtain an $O(\log m)$ approximation where m is the number of items, matching within constant factors the approximation achieved by Chawla *et al.* [10] and Briest *et al.* [4] for single buyer settings.

Theorem 1. *For n independent buyers that are unit-demand or additive over m items with value distribution $\mathfrak{D}$,*

$$\text{ExAnte-BuyManyRev}(\mathfrak{D}) \leq O(\log m)\text{SRev}(\mathfrak{D}).$$

Here SRev denotes the optimal revenue achievable through sequential item pricings. A sequential item pricing mechanism interacts with buyers sequentially in a particular order. It offers to each buyer the set of remaining items at predetermined item prices and allows the buyer to purchase any item of her choice. Notably, our approximation result holds for a worst-case order of arrival of the buyers. Furthermore, the item prices offered to each buyer are non-adaptive in that prices are computed once at the beginning of the mechanism before buyer values are instantiated, and do not change based on instantiated values and purchasing decisions of buyers that arrive earlier in the ordering. Interestingly, sequential item pricing was previously shown in [7] to obtain a constant-factor approximation to the overall (non-buy-many) optimal revenue for unit-demand buyers when buyers' values are independent across items.

Our Techniques

Our main approximation result consists of two parts, each of which is of independent interest. First, we define an ex-ante supply-constrained relaxation of (distributions over) item pricings, similar to the ex-ante relaxation of general buy-many mechanisms discussed above. We show that the ex-ante item pricing revenue provides a logarithmic approximation to the ex-ante buy-many revenue for buyers with any combinatorial valuation function.

Focusing on ex-ante relaxations allows us to revert back to the single-agent setting. We extend the logarithmic upper bound on the gap between buy-many and item pricing revenues proved by [10] to the single buyer setting with a supply constraint. However, extending [10]'s argument is not straightforward because it does not provide any control on how the expected allocation of the item pricing relates to that of the optimal buy-many mechanism. We instead consider a Lagrangian version of the limited supply setting: in this setting, we are given a production cost for each item and our goal is to maximize the mechanism's profit, namely its revenue minus the expected cost it has to pay to produce the items it sells. Unfortunately, it turns out that for some cost vectors, the expected profit of a buy-many mechanism can be an $\Omega(m)$ factor larger than the expected profit of any item pricing. We show instead that the optimal item pricing profit approximates the profit of an optimal buy-many mechanism that faces slightly higher (in multiplicative terms) production costs. This allows us to obtain a logarithmic bound on the gap between the two ex-ante relaxations.

Theorem 2. *For any joint value distribution* $\mathfrak{D}$ *over* n *buyers and* m *items,*

$$\textsc{ExAnte-BuyManyRev}(\mathfrak{D}) \leq O(\log m)\textsc{ExAnte-SRev}(\mathfrak{D}).$$

A Multi-dimensional Contention Resolution Scheme. The second part of our argument relates the ex-ante item pricing revenue to the revenue of a non-adaptive sequential item pricing for unit-demand or additive buyers. This part employs an argument reminiscent of prophet inequalities and contention resolutions schemes (see, e.g., Feldman *et al.* [12]). Specifically, let x_i denote the allocation vector for agent i in the ex-ante-optimal item pricing revenue. For any given fixed ordering over the buyers, we construct pricings $\{q_i\}$ such that: (1) every buyer i obtains an expected allocation of at least $x_i/2$, and (2) the revenue obtained by the item pricing q_i from agent i is at least half the agent's contribution to the ex-ante item pricing revenue. Both of these properties are easy to observe for additive buyers, but challenging for unit-demand buyers. The challenge in showing (1) is that the set of items available when agent i arrives depends upon the instantiations of previous agents' values: as this set changes, the buyer's choice of what to buy also changes. (2) is challenging because revenue is not linear in allocation probability. Our key observation is that for unit-demand buyers, item pricing revenue exhibits concavity as a function of allocation probabilities: given any item pricing p with allocation probabilities x and any $\alpha < 1$, we can find another item pricing that allocates items with probabilities at most αx and obtains at least an α fraction of $\mathbf{p}$'s revenue. We leave open the question of extending this type of multi-dimensional contention resolution scheme (and in particular, achieving the following theorem) to other valuation functions.

Theorem 3. *For any joint value distribution* $\mathfrak{D}$ *over* n *unit-demand or additive buyers and* m *items,*

$$\textsc{ExAnte-SRev}(\mathfrak{D}) \leq 2\textsc{SRev}(\mathfrak{D}).$$

Theorem 1 follows immediately from Theorems 2 and 3. We prove Theorem 2 in Sect. 3 and Theorem 3 in Sect. 4.

In Sect. 5, we mention some **motivating examples** regarding multi-buyer buy-many mechanisms, and address what are the main difficulties of extending our results to all valuation functions.

Further Related Work

The buy-many constraint was first proposed as an alternative to unconstrained revenue maximization by Briest *et al.* [4] in the context of a single unit-demand buyer. Chawla *et al.* [10] extended the notion to arbitrary single buyer settings. Chawla *et al.* chawla2021pricing present improved approximations for buy-many mechanisms in single buyer settings where buyers' values satisfy an additional ordering property that includes, for example, the so-called FedEx setting. In a different direction, Chawla *et al.* chawla2020menu study the menu size complexity and revenue continuity of single buyer buy-many mechanisms. All of these works focus on single buyer problems.

There is extensive literature on approximating the optimal non-buy-many revenue under various constraints on the buyers' value distributions in both single-buyer and multi-buyer settings. Interestingly, many of these approximation results are achieved through sequential item pricing (or, in some cases, grand bundle pricing) mechanisms. For example, Chawla *et al.* [7] show that for multiple settings of matroid feasibility constraints, sequential item pricings give a constant approximation to the optimal non-buy-many revenue. In particular, for multiple unit-demand buyers with independent values over all items, sequential item pricing gives a 6.75-approximation to the optimal revenue, and this competitive ratio was improved to 4 by Alaei *et al.* [1] and generalized to a setting where each item has multiple units. For buyers with more general distribution, either a sequential pricing mechanism with personalized item prices or a sequential item pricing with anonymous prices and entry fee gives a constant approximation to the optimal revenue when there are multiple fractional subadditive buyers [5] and an $O(\log \log m)$ approximation when buyers are subadditive [5,11]. Ma and Simchi-Levi [21] consider additive-valued buyers and a seller facing production costs, a setting that we revisit in our proof of Theorem 4, and achieve approximations using two-part tariffs. Also see [3,6,8,17,19,20,22,24] for some other previous work on using simple mechanisms to approximate the optimal non-buy-many revenue. For sequential item pricings to be able to approximate the revenue of the optimal non-buy-many mechanisms, an important assumption is that each buyer should have independent item values, which we do not assume in this work.

Finally, ex-ante relaxations, first introduced to multi-buyer mechanism design by Alaei *et al.* [1] and Yan and Qiqi [23], have emerged as a powerful tool for simplifying multi-buyer mechanism design problems by breaking them up into their single-agent counterparts. For example, Chawla *et al.* [9] and Cai *et al.* [5] employ this approach for designing sequential mechanisms that approximate the optimal revenue for multiple agents with subadditive values. Such a technique is also used for analyzing the setting where buyers have non-linear utilities [2,13,15], and determining the revelation gap of the optimal mechanisms [14,16].

2   Definitions

We study the multidimensional mechanism design problem where the seller has m heterogeneous items to sell to n buyers with independent value distributions, and aims to maximize the revenue. The buyer i's value vector over items is specified by a distribution $\mathcal{D}_i$ over all value functions $v_i : 2^{[m]} \to \mathbb{R}^{\geq 0}$. Each of these such functions v_i assigns a non-negative value to any subset of items $S \subseteq [m]$. Let $\mathfrak{D} = \mathcal{D}_1 \times \cdots \times \mathcal{D}_n$ denote the joint distribution over all values. Let $\Delta_m = [0,1]^m$ be the set of all possible randomized allocations.

Unit-Demand and Additive Buyers. In this paper, we focus on settings where every buyer is unit-demand or where every buyer is additive valued. We say that a buyer is unit-demand over all items, if the buyer is only interested in purchasing one item, and her value for any set of items is solely determined by the item that is most valuable to her. In other words, for any set of items $S \subseteq [m]$, $v_i(S) = \max_{j \in S} v_i(\{j\})$. We say that a buyer is additive if for all $S \subseteq [m]$, $v_i(S) = \sum_{j \in S} v_i(\{j\})$. For ease of notation, let $v_i(\{j\}) = v_{ij}$. We make no assumptions on each buyer's value distribution $\mathcal{D}_i$ – the buyer's values for different items can be arbitrarily correlated.

Single-Buyer Mechanisms. By the taxation principle, any single-buyer mechanism is equivalent to a menu of options, and the buyer can select an outcome that maximizes her utility. Without loss of generality, we can assume that the menu assigns a price to *every* randomized allocation, a.k.a. lottery, $\lambda \in \Delta_m$, and can therefore simply represent the mechanism by a pricing function p. $p(\lambda)$ denotes the price of the lottery λ. For a buyer of type v_i, her (expected) value for a lottery λ is defined by $v_i(\lambda) := \sum_j v_{ij} \lambda_j$, and her utility for the lottery is defined to be $u_{v_i,p}(\lambda) := v_i(\lambda) - p(\lambda)$. When p is clear from the context, we drop the subscript and write the buyer's utility as $u_{v_i}(\lambda)$. Henceforth, we will refer interchangeably to mechanisms as pricing functions.

Given a pricing function p, a buyer of type v chooses the utility-maximizing lottery, denoted $\lambda_{v,p} := \arg\max_\lambda u_{v,p}(\lambda)$. The buyer's utility in the mechanism p is given by $u_p(v) := v(\lambda_{v,p}) - p(\lambda_{v,p})$; the buyer's payment is $\mathrm{REV}_p(v) := p(\lambda_{v,p})$. We write the revenue of the mechanism under buyer distribution $\mathcal{D}$ as $\mathrm{REV}_p(\mathcal{D}) = \mathbb{E}_{v \sim \mathcal{D}} \mathrm{REV}_p(v)$. The mechanism is called a buy-one mechanism since the buyer only purchases one option from the menu.

Single-Buyer Buy-Many Mechanisms. In single buyer settings, if the buyer is allowed to purchase multiple options from a mechanism's menu, we call the mechanism buy-many. In particular, in a buy-many mechanism the buyer can purchase a (random) sequence of lotteries, where each lottery in the sequence can depend adaptively on the instantiations of previous lotteries. At the end of the process, the buyer gets the union of all allocated items in each step and pays the sum of the prices of all purchased lotteries.

Any buy-many mechanism can be described by a buy-one pricing function that satisfies a buy-many constraint. Intuitively, a buy-one pricing function satisfies the buy-many constraint if the buyer always prefers to purchase a single option from the menu, even if she has the option to adaptively interact with the mechanism multiple times.[2] For example, for a pricing function p, let $\hat{p}_i$ denote the minimum cost of acquiring item i by repeatedly purchasing some lottery until the item is instantiated. Then, the buy-many constraint implies that for any λ,

---

[2] For formal definitions, see [10].

$p(\lambda) \leq \hat{p} \cdot \lambda$. Pricing functions that satisfy the buy-many constraint are called buy-many pricings, and we use BuyMany to denote the set of such functions.

Let $\text{BuyManyRev}(\mathcal{D}) := \max_{p \in \text{BuyMany}} \text{Rev}_p(\mathcal{D})$ denote the revenue of the optimal buy-many mechanism for a buyer with value distribution $\mathcal{D}$.

Item Pricings and Sequential Item Pricings. In a single-buyer setting, a (deterministic) item pricing mechanism sets an item price $p_j \in \mathbb{R}_+$ for every item j; a buyer with value function v_i purchases the item j that maximizes $v_{ij} - p_j$. We use the price vector $p = (p_1, \cdots, p_m)$ to denote the item pricing. Denote by $\text{SRev}(\mathcal{D})$ the optimal revenue obtained by any item pricing for a buyer with distribution $\mathcal{D}$. We also define random item pricing mechanisms as distributions over deterministic item pricings; we continue to use p to denote such pricings, although p is now a random variable.

A sequential item pricing for a multi-buyer setting specifies a serving order σ being a permutation over the n buyers, and n item pricings $p_1, \cdots, p_n$, such that at step i the seller posts the pricing $p_{\sigma(i)}$ to buyer $\sigma(i)$, and the buyer purchases her favorite among the items that are still available. We use $\text{SRev}(\mathfrak{D})$ to denote the optimal revenue obtained by a sequential item pricing mechanism for buyers with values drawn from the joint distribution $\mathfrak{D}$.

Profit Maximization Under Production Costs. For a single-buyer setting, we study the profit-maximizing problem of the seller where each item has a production cost. Let $c = (c_1, c_2, \cdots, c_m) \in \mathbb{R}_+^m$ denote the vector of production costs. Then the profit of the single-buyer mechanism given by pricing p is defined to be the revenue minus the production costs of the items:

$$\text{Profit}_{p,c}(\mathcal{D}) = \mathbb{E}_{v \sim \mathcal{D}}[p(\lambda_{v,p}) - \lambda_{v,p} \cdot c].$$

We denote by $\text{SProfit}_c(\mathcal{D}) = \max_{\text{item pricing } p} \text{Profit}_{p,c}(\mathcal{D})$ the optimal profit achievable by item pricings for the buyer with distribution $\mathcal{D}$ when there are production costs c for all items. Similarly define

$$\text{BuyManyProfit}_c(\mathcal{D}) = \max_{\text{buy-many } p} \text{Profit}_{p,c}(\mathcal{D})$$

to be the optimal profit achievable by buy-many mechanisms for the buyer with distribution $\mathcal{D}$ for the setting with production costs c.

Ex-Ante Constrained Revenue

Ex-ante relaxations are a powerful technique for reducing multi-buyer mechanism design problems to their single-buyer counterparts. The key idea is to relax the ex-post supply constraint on items to an ex-ante feasibility constraint that requires each item to be sold at most once in expectation.

Recall that x_{ij} denotes the probability of allocating item j to buyer i; $x_i = (x_{i1}, \cdots, x_{im})$ denotes the vector of allocations of all items to buyer i; and $x = (x_1, \cdots, x_n)$ denote the vector of all allocations.

We say that a single-buyer pricing function p_i satisfies the ex-ante constraint x_i with respect to the value distribution $\mathcal{D}_i$ if for all $j \in [m]$, the expected allocation of item j to the buyer is no more than x_{ij}: $\mathbb{E}_{v \sim \mathcal{D}_i}[\lambda_{v,p_i}] \leq x_i$. Note that this definition extends to random pricing functions p in the straightforward manner: we want the expected allocation to be bounded by x_i where the expectation is taken over both the buyer's value and the randomness in the mechanism. That is, $\mathbb{E}_{v \sim \mathcal{D}_i, p}[\lambda_{v,p_i}] \leq x_i$.

For a single buyer i and an ex-ante constraint x_i, we can now define the optimal revenue that can be obtained from the buyer subject to an ex-ante constraint for various classes of mechanisms. In particular, let Δ_{IP} denote the space of all distributions over item pricings, and Δ_{BM} denote the space of all distributions over buy-many pricings. Then we define:

$$\mathrm{SREV}(\mathcal{D}_i, x_i) = \max_{p \in \Delta_{IP}: p \text{ satisfies } x_i \text{ w.r.t. } \mathcal{D}_i} \mathrm{REV}_p(\mathcal{D}_i), \qquad \text{and,}$$

$$\mathrm{BUYMANYREV}(\mathcal{D}_i, x_i) = \max_{p \in \Delta_{BM}: p \text{ satisfies } x_i \text{ w.r.t. } \mathcal{D}_i} \mathrm{REV}_p(\mathcal{D}_i).$$

Given a combined vector x of ex-ante constraints for every buyer $i \in [m]$, we can write the collective revenue of the optimal single-buyer mechanisms that satisfy these constraints as:

$$\mathrm{EA\text{-}SREV}(\mathfrak{D}, x) = \sum_i \mathrm{SREV}(\mathcal{D}_i, x_i), \qquad \text{and,}$$

$$\mathrm{EA\text{-}BUYMANYREV}(\mathfrak{D}, x) = \sum_i \mathrm{BUYMANYREV}(\mathcal{D}_i, x_i),$$

Finally, we can define the ex-ante relaxation for each class of mechanisms.

$$\mathrm{EA\text{-}SREV}(\mathfrak{D}) = \max_{x: \sum_i x_{ij} \leq 1 \forall j \in [m]} \mathrm{EA\text{-}SREV}(\mathfrak{D}, x), \qquad \text{and,}$$

$$\mathrm{EA\text{-}BUYMANYREV}(\mathfrak{D}) = \max_{x: \sum_i x_{ij} \leq 1 \forall j \in [m]} \mathrm{EA\text{-}BUYMANYREV}(\mathfrak{D}, x),$$

Some Settings Captured by the Ex-Ante Relaxation

Our ex-ante relaxation captures approaches that define the buy-many constraint as a restriction on the effective pricing faced by any individual buyer participating in the mechanism. We now describe some specific such settings. We start with one of the simplest settings, in which buyers arrive one after the other and each buyer faces a single-buyer mechanism.

Sequential Buy-Many Mechanisms. Sequential mechanisms make offers to each buyer sequentially in a pre-specified order. The i-th buyer in the order is asked to choose which items to purchase among the subset of items that remain after buyers 1 through $i - 1$ have made their choices. A natural definition for the buy-many constraint for this multi-buyer setting boils down to offering a buy-many constrained mechanism to each buyer for any subset of items remaining, i.e. the full-mechanism can be defined as a collection of single-buyer mechanisms $\mathcal{M}_{i,S} \in \textsc{BuyMany}$ for any buyer i and any subset S of remaining items.

Ex-Post Buy-Many Mechanisms. A more flexible design space for multi-buyer buy-many mechanisms is to consider direct mechanisms in which buyers truthfully submit their valuations and conditional on the valuations of other buyers each buyer is faced with a single buyer buy-many mechanism. Buy-many mechanisms for this setting are specified as a collection of single buyer buy-many mechanisms $\mathcal{M}_{i,\vec{v}_{-i}} \in \textsc{BuyMany}$ for any buyer i and any combination of valuation functions v_{-i} for the other buyers.

Bayesian Buy-Many Mechanisms. Another potential definition of multi-buyer buy-many mechanisms is to consider Bayesian settings in which we require the options any single buyer faces to be buy-many in expectation over the valuations of other buyers. If $q_i(v_i, \vec{v}_{-i})$ and $p_i(v_i, \vec{v}_{-i})$ is the allocation and price offered to buyer i when her value is v_i and the other buyers have valuations $\vec{v}_{-i}$, we would require that the single buyer mechanism with allocation probabilities $q_i(v_i) \triangleq \mathbb{E}_{\vec{v}_{-i}} q_i(v_i, \vec{v}_{-i})$ and price $p_i(v_i) \triangleq \mathbb{E}_{\vec{v}_{-i}} p_i(v_i, \vec{v}_{-i})$ to satisfy the buy-many constraint.

3 Relating the Ex-Ante Relaxations

In this section we bound the gap between the ex-ante optimal buy many revenue and the ex-ante optimal item pricing revenue when the seller faces many buyers. This is a generalization of single-buyer buy-many revenue approximations to supply constrained settings. We emphasize that for the results in this section we *do not* require any assumptions on the buyer's valuation function, such as that it is unit-demand or additive.

A note on notation: since we consider a single-buyer problem in this section, we drop the subscript i from most notation and simply denote buyer i's valuation function as v; her allocation vector as x; the probability that item j is allocated to the buyer as x_j; etc. We will write the ex-ante buy many and item pricing revenues of this buyer simply as $\textsc{BuyManyRev}(\mathcal{D}, x)$ and $\textsc{SRev}(\mathcal{D}, x)$ respectively.

Theorem 4. *For any single buyer with value distribution $\mathcal{D}$ over m items and any ex-ante supply constraint $x \in \Delta_m$,*

$$\textsc{EA-BuyManyRev}(\mathcal{D}, x) \leq O(\log m)\textsc{EA-SRev}(\mathcal{D}, x).$$

Applying this theorem to each of n buyers, we obtain the following corollary.

Theorem 2. *For any joint value distribution $\mathfrak{D}$ over n buyers and m items,*

$$\textsc{EA-BuyManyRev}(\mathfrak{D}) \leq O(\log m)\textsc{EA-SRev}(\mathfrak{D}).$$

Before we present a formal proof of Theorem 4, let us describe the main ideas. Theorem 4 is a generalization of Theorem 1.1 from [10], which shows that the ratio of BuyManyRev and SRev is bounded by $O(\log m)$ in the absence of an ex-ante supply constraint. The proof technique of [10] does not directly lend itself to the ex-ante setting because it does not provide much control over the allocation probability of the random item pricing it produces. Indeed, the (random) item pricing it returns is independent of the buyer's value distribution, whereas the allocation probabilities (that are expectations over values drawn from the distribution) necessarily depend on the value distribution. Instead of applying the approach of [10] directly, we first consider a Lagrangian version of the supply constrained problem.

To simplify the following discussion, we will hide the argument $\mathcal{D}$ from the respective revenue benchmarks. Viewing $\text{SRev}(x)$ and $\text{BuyManyRev}(x)$ as two multi-variate functions over $x \in \Delta_m$, we first observe that for any fixed ex-ante constraint x^o, there exists a cost vector c^o such that x^o is the solution to the optimization problem $\max_x(\text{SRev}(x) - c^o \cdot x)$. Indeed, because $\text{SRev}(x)$ is concave[3], $c^o = \nabla\text{SRev}(x^o)$ is such a function. Furthermore, c^o is a non-negative vector, and so the gap between $\text{SRev}(x^o)$ and $\text{BuyManyRev}(x^o)$ is bounded by the gap between $\text{SRev}(x^o) - c \cdot x^o$ and $\text{BuyManyRev}(x^o) - c \cdot x^o$. This motivates studying the Lagrangian problem of maximizing the profit of a mechanism subject to production costs c^o. In particular, for any value distribution $\mathcal{D}$, we have:

$$\max_x \frac{\text{EA-BuyManyRev}(\mathcal{D}, x)}{\text{EA-SRev}(\mathcal{D}, x)} < \max_c \frac{\text{BuyManyProfit}_c(\mathcal{D})}{\text{SProfit}_c(\mathcal{D})}$$

Unfortunately, the gap on the right hand side can be very large:

Theorem 5. *There exists a unit-demand value distribution $\mathcal{D}$ over m items and a cost vector $c \in \mathbb{R}^m$, such that $\text{BuyManyProfit}_c = \Omega(m)\text{SProfit}_c(\mathcal{D})$.*

We instead provide a bi-criteria approximation for the Lagrangian problem. In particular, we compare the profit of the optimal item pricing for cost vector c with the profit of the optimal buy many mechanism with production costs $2c$. This suffices to imply Theorem 4 with a slight worsening in the approximation factor.

Theorem 6. *For any single buyer with value distribution $\mathcal{D}$ over m-items and production costs vector c,*

$$\text{BuyManyProfit}_{2c}(\mathcal{D}) \leq 2\ln 4m\,\text{SProfit}_c(\mathcal{D}).$$

The rest of the section is organized as follows. We first show a complete proof of Theorem 4 based on Theorem 6. We then describe and verify the gap example in Theorem 5. Each of these components is self-contained.

---

[3] In fact, the function $\text{Rev}(x)$ defined as maximum revenue from any restricted set of mechanism with allocations at most x is concave. This is because for any two allocations x and y and coefficient $1 \geq \alpha \geq 0$, one can consider mechanisms $M(x)$ and $M(y)$ that define $\text{Rev}(x)$ and $\text{Rev}(y)$, and run the former with probability α and the latter with probability $(1-\alpha)$. Then $\text{Rev}(\alpha x + (1-\alpha)y) \geq \alpha\text{Rev}(x) + (1-\alpha)\text{Rev}(y)$.

3.1   Proof of Theorem 4

Proof (Proof of Theorem 4).

We first note that $\mathrm{SREV}(x)$ is a concave function over x because it optimizes for revenue over random pricings. Fix an ex-ante constraint x^o and consider the function $g(x) := \mathrm{SREV}(x) - x \cdot \nabla \mathrm{SREV}(x^o)$. This function is maximized at $x = x^o$. Furthermore, since $\mathrm{SREV}(x)$ is monotone non-decreasing, $\nabla \mathrm{SREV}(x^o) \geq \mathbf{0}$, and so $c := \nabla \mathrm{SREV}(x^o)$ can be thought of as a vector of production costs. Since $g(x)$ is exactly the profit of an item pricing for buyer distribution $\mathcal{D}$ with allocation x and production cost vector c for all items, we know that the random item pricing p that achieves $\mathrm{SREV}(x^o)$ is also the optimal item pricing for a buyer with value distribution $\mathcal{D}$ and item production costs c, without any allocation constraint. [4]

Now consider the buy-many profit optimization problem with production costs $2c$. The optimal profit is given by $\mathrm{BUYMANYPROFIT}_{2c}$. Restricting attention to buy many mechanisms that satisfy the ex-ante constraint x^o, we let $\mathrm{BUYMANYPROFIT}_{2c}(x^o)$ denote the optimal profit obtained over that set of mechanisms. Then, we can apply Theorem 6 to obtain:

$$
\begin{aligned}
\mathrm{SREV}(x^o) \quad &= \mathrm{SPROFIT}_c(x^o) + c \cdot x^o \\
&= \mathrm{SPROFIT}_c(\mathcal{D}) + c \cdot x^o \\
&\geq \tfrac{1}{2\ln 4m} \mathrm{BUYMANYPROFIT}_{2c}(\mathcal{D}) + c \cdot x^o \\
&\geq \tfrac{1}{2\ln 4m} \mathrm{BUYMANYPROFIT}_{2c}(x^o) + c \cdot x^o \\
&= \tfrac{1}{2\ln 4m}\left(\mathrm{BUYMANYPROFIT}_{2c}(x^o) + 2c \cdot x^o\right) + c \cdot x^o \left(1 - \tfrac{1}{\ln 4m}\right) \\
&\geq \tfrac{1}{2\ln 4m} \mathrm{BUYMANYREV}(x^o),
\end{aligned}
$$

Here the first line is true by extracting the terms of item costs; the second line is true since $x = x^o$ is optimal for profit under item costs c; the third line is true by Theorem 6; the fourth line is true since adding an allocation restriction cannot increase profit; the last line is true since $\mathrm{BUYMANYPROFIT}_{2c}(x^o) + 2c \cdot x^o \geq \mathrm{BUYMANYREV}(x^o)$. This finishes the proof of the theorem.

3.2   Proof of Theorem 5

Proof (Proof of Theorem 5). Let $c = (0, 2^m, 2^m, \cdots, 2^m)$. For every j such that $2 \leq j \leq m$, let $v^{(j)}$ be the following unit-demand value function: $v_1^{(j)} = 2^j$; $v_j^{(j)} = 2^m$; $v_k^{(j)} = 0$ for $k \notin \{1, j\}$. In other words, the buyer with value $v^{(j)}$ is only interested in two items 1 and j, with the value for the first item being 2^j, and that for item j being 2^m. Consider the following value distribution $\mathcal{D}$: for every j such that $2 \leq j \leq m$, with probability 2^{-j}, $v = v^{(j)}$; for the remaining probability, $v = \mathbf{0}$. Now we analyze $\mathrm{SPROFIT}_c$ and $\mathrm{BUYMANYPROFIT}_c$.

---

[4] For any buyer i, it is possible that p has allocation $y \leq x^o$, with $\mathrm{SREV}(y) = \mathrm{SREV}(x^o)$. However, for any item j such that $y_j < x_j^o$, since $\mathrm{SREV}(y) = \mathrm{SREV}(x^o)$, the gradient $\nabla \mathrm{SREV}(x^o)$ has value $c_j = 0$ on the jth component. Thus the profit of item pricing p for the buyer with production costs c is still $\mathrm{SREV}(x^o) - c \cdot x^o$, although the actual allocation is less than x^o.

For any item pricing p, consider its profit contribution from the first item and the rest of the items. Since the buyer's value for the first item forms a geometric distribution, the profit contribution from the first item is upper bounded by the revenue of selling only the first item, which is $O(1)$. For the rest of the items, since when the buyer purchases some item $j > 1$, the item must have a price at most $2^m = c_j$, this means that the profit contribution of item j is at most 0. Thus $\text{SPROFIT}_c = O(1)$.

Consider the following buy-many mechanism: for every $j \geq 2$, there is a menu entry with allocation $\lambda^{(j)}$ and price $p^{(j)} = 2^{j-1} + 2^{m-1}$, where $\lambda_1^{(j)} = \lambda_j^{(j)} = 0.5$. For any $\lambda \in \Delta_m$ that is not some $\lambda^{(j)}$, its price is determined by the cheapest way to adaptively purchase it with $\lambda^{(2)}, \cdots, \lambda^{(m)}$. For every buyer of type $v^{(j)}$, she buys lottery $(\lambda^{(j)}, p^{(j)})$ in the mechanism with utility 0.[5] Since item j is only of interest to buyer $v^{(j)}$, the buyer would not purchase any other set of lotteries in the mechanism. Since profit from buyer $v^{(j)}$ is 2^{j-1}, and she gets realized in $\mathcal{D}$ with probability 2^{-j}, the expected profit of the buy-many mechanism is $\Omega(m)$.

4   Approximation via Sequential Item Pricing

We will now focus on item pricings and prove Theorem 3. In particular, we show that non-adaptive sequential item pricings can obtain half of the ex-ante optimal item pricing revenue, regardless of the order in which buyers are served.

Theorem 3. (Restatement) *For any joint distribution $\mathfrak{D} = (\mathcal{D}_1, \mathcal{D}_2, \cdots, \mathcal{D}_n)$ over n unit-demand or additive buyers and any order σ on arrival of buyers, there exists a deterministic sequential item pricing q with buyers arriving in order σ, such that*

$$\text{REV}_\mathcal{D}(q) \geq \frac{1}{2}\text{EA-SREV}(\mathfrak{D}).$$

For additive buyers, the items impose no externalities on each other, and so the theorem follows immediately from the single-unit prophet inequality, Henceforth we focus on unit-demand buyers. Our argument is based loosely around online contention resolution schemes (OCRS) [12] and prophet inequality arguments. The idea is to start with the optimal solution to the ex ante item pricing revenue: $x^* := \arg\max_{x: \sum_i x_{ij} \leq 1 \forall j \in [m]} \sum_i \text{SREV}(\mathcal{D}_i, x_i)$. Then, given some ordering σ over the buyers, we try to mimic this allocation by choosing pricings for each buyer that ensure that the buyer receives allocation comparable to x_i^*. As in OCRS, we tradeoff assigning enough allocation to a buyer with maintaining a good probability that items remain available for future buyers.

A key difference in our setting relative to work on OCRS is that the latter mostly focuses on utilitarian objectives, e.g. social welfare, so that the tradeoff is easily quantified: choosing an alternative with half the probability of the ex ante optimum, for example, provides half its contribution to the objective. Chawla *et*

---

[5] We can reduce the price of $\lambda^{(j)}$ by some small $\epsilon > 0$ to make each buyer type's utility be strictly positive.

al. [6] show how to apply this approach to revenue for unit demand buyers with values independent across items by transforming values to Myersonian virtual values. Unfortunately this approach does not extend to values correlated across items because in correlated settings it is not possible to assign virtual values to each individual value independent of other values.

We develop an alternate argument. For any single unit-demand buyer, we consider how the item pricing revenue changes as the allocation of the buyer is decreased from some intended allocation x^* to a new allocation y that is component-wise smaller. We show that if x^* is realized by item pricing p, then we can realize allocation y while obtaining revenue at least $y \cdot p$. In particular, uniformly scaling down allocations by some factor scales down revenue by no more than the same factor. This allows us to carry out the OCRS-style argument. We formalize the above claim as a lemma before providing a proof of Theorem 3.

In the following discussion, for any unit-demand buyer with value distribution $\mathcal{D}$, let $\mathbf{x}_{p,S}(\mathcal{D})$ be the allocation vector of item pricing p over the set of available items $S \subseteq [m]$. We remove the distribution $\mathcal{D}$ whenever it is clear from the context. When $S = [m]$, we use $\mathbf{x}_p$ instead of $\mathbf{x}_{p,[m]}$.

Lemma 1. *For any unit-demand buyer, any deterministic item pricing p, and any distribution over set S of available items, let $x^* = \mathbb{E}_S[\mathbf{x}_{p,S}]$ be the expected allocations of p conditioned on the available set of items being S. Then for any allocation vector $y \in \Delta_m$ such that $y \preceq x^*$, there exists a random item pricing q such that*

$$\mathbb{E}_{q,S}[\mathbf{x}_q(S)] = y, \quad and \quad \mathrm{REV}_{q,S} = y \cdot p.$$

Proof. We first prove the theorem when both S and p are deterministic. Then, we extend the proof to the case where the available set S can be possibly randomized.

For $p = (p_1, \ldots, p_m) \in \mathbb{R}_+^m$ being a deterministic item pricing, assume that the set of available items is fixed to be some deterministic set S. For any set $T \subseteq S$, define item pricing p_T to be the pricing p restricted to items in T. In other words, $p_{T,j} = p_j$ for all $j \in T$, and $p_{T,j} = \infty$ otherwise. For ease of notation, let $x_T = \mathbf{x}_{p,T}$ be the allocation under available set T. Observe that for any $j \notin T$, $x_j^* = 0$ and for any $j \in T$, $x_T(j) \geq x_j^*$: the latter is true since under the same pricing, when fewer items are available, buyer types that purchase an item not in T may switch to purchase some item in T, while the other buyer types' incentives remain unchanged. Thus for $y \preceq x^*$, y is in the convex hull of the set of $2^{|S|}$ points $X = \{x_T | T \subseteq S\}$. Write $y = \sum_{T \subseteq S} \alpha_T x_T$ as a convex combination of vectors in X, here $\alpha_T \in [0,1]$ for every T, and $\sum_T \alpha_T = 1$. Consider the following randomized item pricing q: with probability α_T, $q = p_T$, $\forall T \subseteq S$. Then the expected allocation of q is exactly y. On the other hand, the expected revenue is $\mathrm{REV}_{q,S} = p \cdot y$ because whenever item j is sold in q, it is sold at a price of p_j.

Now let S be a random variable over sets of available items. By defining $x_T = \mathbb{E}_S \mathbf{x}_{p,T \cap S}$ to be the expected allocation of item pricing p under available item set $T \cap S$, the above proof still goes through. This finishes the proof of the lemma.

With the help of the lemma above, we are now ready to prove Theorem 3.

Proof (Proof of Theorem 3). For every buyer i, let p_i be the (randomized) item pricing defining EA-SREV$(\mathcal{D})$, and x_i be the corresponding allocation vector. We may assume $\sigma_i = i$ without loss of generality. We build the desired deterministic sequential item pricing q incrementally for every buyer. At arrival of buyer i, we use the allocation vector x_i, the distribution over currently available items (where the randomness comes from previous buyers and item pricings) and Lemma 1 to produce a (randomized) item pricing vector q_i with the property that for S_i being the random set of available items to buyer i,

$$\mathbb{E}_{q_i, S_i}[x_{q_i, S_i}(\mathcal{D}_i)] = \frac{x_i}{2}, \text{ and, } \mathbb{E}_{q_i, S_i}[\text{REV}_{q_i}(\mathcal{D}_i, S_i)] = \frac{x_i}{2} \cdot p_i = \frac{1}{2}\text{EA-SREV}(\mathcal{D}_i).$$
$$(1)$$

Here $\text{REV}_{q_i}(\mathcal{D}_i, S_i)$ denotes the revenue of item pricing q_i for buyer i conditioned on the available item set being S_i when i arrives. If such q_i exists for every i, then the random sequential item pricing q satisfies $\text{REV}_{\mathcal{D}}(q) \geq \frac{1}{2}\text{EA-SREV}(\mathcal{D})$. Thus there must exist a realization of q being a deterministic sequential item pricing satisfying the requirement of the theorem.

Now it suffices to show that item pricing q_i exists for (1), and the rest of the proof is dedicated to proving this. An important observation is that if by induction the item prices $q_{i'}$ satisfying (1) exist for every $i' < i$, then every item belongs to S_i with probability at least $\frac{1}{2}$. Such a observation is true by noticing that by union bound, the allocation of item j in the first $i - 1$ steps is at most $\sum_{i' < i} \frac{x_{i'j}}{2} \leq \sum_{i' \leq m} \frac{x_{i'j}}{2} \leq \frac{1}{2}$ since item j has a total ex-ante allocation at most 1.

When p_i is deterministic, since every element in $[m]$ exists with probability at least $\frac{1}{2}$ in S_i, we know that for any realized buyer type, her favorite item still remains with probability at least $\frac{1}{2}$ and she would not deviate to purchase something else. Thus $\mathbb{E}_{p_i, S_i}[\mathbf{x}_{p_i, S_i}] \succeq \frac{1}{2}x_i$. By Lemma 1, there exists a random item pricing q, such that $\mathbb{E}_{q, S_i}[\mathbf{x}_{q, S_i}] = \frac{1}{2}x_i$, and $\mathbb{E}_{q, S_i}[\text{REV}_{q, S_i}] = \frac{1}{2}x_i \cdot p_i = \frac{1}{2}\text{EA-SREV}(\mathcal{D}_i)$. Thus (1) is satisfied.

When p_i is random, consider any instantiation of p_i. The same as the reasoning in the previous paragraph, $\mathbb{E}_{p_i, S_i}[\mathbf{x}_{p_i, S_i}] \succeq \frac{1}{2}\mathbf{x}_{p_i}$ still holds, and there exists a random item pricing q_{p_i} such that $\mathbb{E}_{q_{p_i}, S_i}[\mathbf{x}_{q_{p_i}, S_i}] = \frac{1}{2}\mathbf{x}_{p_i}$, and $\mathbb{E}_{q_{p_i}, S_i}[\text{REV}_{q_{p_i}, S_i}] = \frac{1}{2}\mathbf{x}_{p_i} \cdot p_i$. Consider the following random item pricing q_i: firstly generate a realization of random item pricing p_i, then generate a realization of random item pricing q_{p_i} defined above. The expected allocation of q_i is

$$\mathbb{E}_{p_i}\mathbb{E}_{q_{p_i}, S}[\mathbf{x}_{q_{p_i}, S}] = \frac{1}{2}x_i,$$

while the expected revenue is

$$\mathbb{E}_{p_i}\mathbb{E}_{q_{p_i}, S}[\text{REV}_{q_{p_i}, S}] = \mathbb{E}_{p_i}\left[\frac{1}{2}\mathbf{x}_{p_i} \cdot p_i\right] = \frac{1}{2}\text{EA-SREV}(\mathcal{D}_i).$$

Thus q_i satisfies (1), which finishes the proof of the theorem.

5   Discussion

In this section, we mention some examples that motivate definition and further work in multi-buyer buy-many mechanisms. Moreover, we discuss why extension of our results to any valuation function is challenging.

Motivating Examples. Consider a seller who is selling multiple items in multiple markets, and faces a common supply constraint across these markets. The seller is free to choose a different selling mechanism in each market, but interacts with a buyer from each market in just one go. For example, imagine Amazon selling rare books in multiple markets (US, Europe, India). Within each market, Amazon will display a price schedule that does not change frequently based on purchase decisions in other markets (but updates availability). This price schedule could be different for different markets based on local preferences and demand. Within each market individually, given that prices will remain static over short periods of time, a buy-many constraint is a natural property to satisfy. This scenario fits directly within our model.

For another similar scenario consider a travel website like hotwire.com which offers deals on airline tickets, hotel rooms, etc., without revealing complete vendor information. In effect, it sells lotteries. This is another example with supply constraints where the mechanism may personalize prices for each potential buyer (e.g. based on which browser the buyer is using). If the seller uses a non-buy-many mechanism (e.g. if lotteries on multiple items are generally more expensive than individual prices on the items they contain) it would lose customers over time.

Difficulties of Extending the Results to All Valuations. One component of our argument, namely approximating the ex-ante buy many revenue by the ex-ante SRev holds for every possible value function. However, we don't know how to extend the second part of the argument - approximating the ex-ante SRev using sequential item pricing - for value functions that are not unit-demand or additive. This requires constructing a multi-dimensional prophet inequality. The key technical challenge for non-unit-demand valuations is in keeping track of and controlling how the probability that a particular subset of items is available to an agent depends on decisions of other buyers.

References

1. Alaei, S.: Bayesian combinatorial auctions: expanding single buyer mechanisms to many buyers. SIAM J. Comput. **43**(2), 930–972 (2014)
2. Alaei, S., Fu, H., Haghpanah, N., Hartline, J.: The simple economics of approximately optimal auctions. In: 2013 IEEE 54th Annual Symposium on Foundations of Computer Science, pp. 628–637. IEEE (2013)
3. Babaioff, M., Immorlica, N., Lucier, B., Weinberg, S.M.: A simple and approximately optimal mechanism for an additive buyer. In: 2014 IEEE 55th Annual Symposium on Foundations of Computer Science, pp. 21–30. IEEE (2014)

4. Briest, P., Chawla, S., Kleinberg, R., Weinberg, S.M.: Pricing lotteries. J. Econ. Theory **156**, 144–174 (2015)
5. Cai, Y., Zhao, M.: Simple mechanisms for subadditive buyers via duality. In: Proceedings of the 49th Annual ACM SIGACT Symposium on Theory of Computing, pp. 170–183 (2017)
6. Chawla, S., Hartline, J.D., Kleinberg, R.: Algorithmic pricing via virtual valuations. In: Proceedings of the 8th ACM Conference on Electronic Commerce, pp. 243–251 (2007)
7. Chawla, S., Hartline, J.D., Malec, D.L., Sivan, B.: Multi-parameter mechanism design and sequential posted pricing. In: Proceedings of the forty-second ACM symposium on Theory of computing, pp. 311–320 (2010)
8. Chawla, S., Malec, D., Sivan, B.: The power of randomness in Bayesian optimal mechanism design. Games Econom. Behav. **91**, 297–317 (2015)
9. Chawla, S., Miller, J.B.: Mechanism design for subadditive agents via an ex ante relaxation. In: Proceedings of the 2016 ACM Conference on Economics and Computation, pp. 579–596 (2016)
10. Chawla, S., Teng, Y., Tzamos, C.: Buy-many mechanisms are not much better than item pricing. In: Proceedings of the 2019 ACM Conference on Economics and Computation, pp. 237–238 (2019)
11. Dütting, P., Kesselheim, T., Lucier, B.: An o(log log m) prophet inequality for subadditive combinatorial auctions. In: 2020 IEEE 61st Annual Symposium on Foundations of Computer Science (FOCS), pp. 306–317 (2020). https://doi.org/10.1109/FOCS46700.2020.00037
12. Feldman, M., Svensson, O., Zenklusen, R.: Online contention resolution schemes. In: Proceedings of the Twenty-Seventh Annual ACM-SIAM Symposium on Discrete Algorithms, pp. 1014–1033. SIAM (2016)
13. Feng, Y., Hartline, J., Li, Y.: Simple mechanisms for non-linear agents. arXiv preprint arXiv:2003.00545 (2020)
14. Feng, Y., Hartline, J.D.: An end-to-end argument in mechanism design (prior-independent auctions for budgeted agents). In: 2018 IEEE 59th Annual Symposium on Foundations of Computer Science (FOCS), pp. 404–415. IEEE (2018)
15. Feng, Y., Hartline, J.D., Li, Y.: Optimal auctions vs. anonymous pricing: beyond linear utility. In: Proceedings of the 2019 ACM Conference on Economics and Computation, pp. 885–886 (2019)
16. Feng, Y., Hartline, J.D., Li, Y.: Revelation gap for pricing from samples. In: Proceedings of the 53rd Annual ACM SIGACT Symposium on Theory of Computing, pp. 1438–1451 (2021)
17. Hart, S., Nisan, N.: Selling multiple correlated goods: revenue maximization and menu-size complexity. J. Econ. Theory **183**, 991–1029 (2019)
18. Iwasaki, A., et al.: Worst-case efficiency ratio in false-name-proof combinatorial auction mechanisms. In: van der Hoek, W., Kaminka, G.A., Lespérance, Y., Luck, M., Sen, S. (eds.) 9th International Conference on Autonomous Agents and Multiagent Systems (AAMAS 2010), Toronto, Canada, 10–14 May 2010, vol. 1–3. pp. 633–640. IFAAMAS (2010)
19. Kleinberg, R., Weinberg, S.M.: Matroid prophet inequalities. In: Proceedings of the Forty-Fourth Annual ACM Symposium on Theory of Computing, pp. 123–136 (2012)
20. Li, X., Yao, A.C.C.: On revenue maximization for selling multiple independently distributed items. Proc. Nat. Acad. Sci. **110**(28), 11232–11237 (2013)

21. Ma, W., Simchi-Levi, D.: Reaping the benefits of bundling under high production costs. In: Banerjee, A., Fukumizu, K. (eds.) The 24th International Conference on Artificial Intelligence and Statistics, AISTATS 2021, 13–15 April 2021, Virtual Event. Proceedings of Machine Learning Research, vol. 130, pp. 1342–1350. PMLR (2021)
22. Rubinstein, A., Weinberg, S.M.: Simple mechanisms for a subadditive buyer and applications to revenue monotonicity. ACM Trans. Econ. Comput. (TEAC) **6**(3–4), 1–25 (2018)
23. Yan, Q.: Mechanism design via correlation gap. In: Proceedings of the Twenty-Second Annual ACM-SIAM Symposium on Discrete Algorithms, pp. 710–719. SIAM (2011)
24. Yao, A.C.C.: An n-to-1 bidder reduction for multi-item auctions and its applications. In: Proceedings of the Twenty-Sixth Annual ACM-SIAM Symposium on Discrete Algorithms, pp. 92–109. SIAM (2014)
25. Yokoo, M., Sakurai, Y., Matsubara, S.: The effect of false-name bids in combinatorial auctions: new fraud in internet auctions. Games Econ. Behav. **46**(1), 174–188 (2004)

Full Papers

Partial Allocations in Budget-Feasible Mechanism Design: Bridging Multiple Levels of Service and Divisible Agents

Georgios Amanatidis[1,4], Sophie Klumper[2,5], Evangelos Markakis[3,4,6], Guido Schäfer[2,7], and Artem Tsikiridis[2(✉)]

[1] University of Essex, Colchester, UK
[2] Centrum Wiskunde & Informatica (CWI), Amsterdam, The Netherlands
artem.tsikiridis@cwi.nl
[3] Athens University of Economics and Business, Athens, Greece
[4] Archimedes/Athena RC, Athens, Greece
[5] Vrije Universiteit, Amsterdam, The Netherlands
[6] Input Output Global (IOG), Athens, Greece
[7] University of Amsterdam, Amsterdam, The Netherlands

Abstract. Budget-feasible procurement has been a major paradigm in mechanism design since its introduction by Singer [24]. An auctioneer (buyer) with a strict budget constraint is interested in buying goods or services from a group of strategic agents (sellers). In many scenarios it makes sense to allow the auctioneer to only partially buy what an agent offers, e.g., an agent might have multiple copies of an item to sell, they might offer multiple levels of a service, or they may be available to perform a task for any fraction of a specified time interval. Nevertheless, the focus of the related literature has been on settings where each agent's services are either fully acquired or not at all. A reason for this is that in settings with partial allocations, without any assumptions on the costs, there are strong inapproximability results (see, e.g., Chan and Chen [10], Anari et al. [5]). Under the mild assumption of being able to afford each agent entirely, we are able to circumvent such results. We design a polynomial-time, deterministic, truthful, budget-feasible, $(2 + \sqrt{3})$-approximation mechanism for the setting where each agent offers multiple levels of service and the auctioneer has a discrete separable concave valuation function. We then use this result to design a deterministic, truthful and budget-feasible mechanism for the setting where any fraction of a service can be acquired and the auctioneer's valuation function is separable concave (i.e., the sum of concave functions). The approximation ratio of this mechanism depends on how "nice" the concave functions are, and is $O(1)$ for valuation functions that are sums of $O(1)$-regular functions (e.g., functions like $\log(1+x)$). For the special case of a linear valuation function, we improve the best known approximation ratio from $1 + \phi$ (by Klumper and Schäfer [17]) to 2. This establishes a separation result between this setting and its indivisible counterpart.

Keywords: Procurement Auctions · Budget-Feasible Mechanism Design · Multiple Levels of Service

J. Garg et al. (Eds.): WINE 2023, LNCS 14413, pp. 41–58, 2024.
https://doi.org/10.1007/978-3-031-48974-7_3

1   Introduction

Consider a procurement auction where the agents have *private* costs on the services that they offer, and the auctioneer associates a value for each possible set of selected agents. This forms a single parameter auction environment, where the agents may strategically misreport their cost to their advantage: obtaining higher payments. Imagine now that the auctioneer additionally has a strict budget constraint that they cannot violate. Under these considerations, a natural goal for the auctioneer is to come up with a truthful mechanism for hiring a subset of the agents, that maximizes her procured value, such that the total payments to the agents respect the budget limitation. This is precisely the model that was originally proposed by Singer [24] for *indivisible* agents, i.e., with a binary decision to be made for each agent (to be hired or not). Given also that even the non-strategic version of such budget-constrained problems tend to be NP-hard, the main focus is on providing budget-feasible mechanisms that achieve approximation guarantees on the auctioneer's optimal potential value.

Ever since the work of Singer [24], a large body of works has emerged, devoted to obtaining improved results on the original model, as well as proposing a number of extensions. These extensions include, among others, additional feasibility constraints, richer objectives, more general valuation functions and additional assumptions, such as Bayesian modeling. Undoubtedly, all these results have significantly enhanced our understanding for the indivisible scenario. In this paper, we move away from the case of indivisible agents and concentrate on two settings that have received much less attention in the literature. In both of the models that we study, instead of hiring agents entirely or not at all, the auctioneer has more flexibility and is allowed to partially procure the services offered by each agent. We assume that the auctioneer's valuation function is the sum of individual valuation functions, each associated with a particular agent.

Agents with Multiple Levels of Service: In this setting, each agent offers a service that consists of multiple levels. We can think of the levels as corresponding to different qualities of service. Hence, the auctioneer can choose not to hire an agent, or hire the first x number of levels of an agent, for some integer x, or hire the agent entirely, i.e., for all the levels that she is offering. Furthermore, the valuation function associated with each agent is concave, meaning that the marginal value of each level of service is non-increasing. This setting was first introduced by Chan and Chen [10] in the context of each agent offering multiple copies of the same good and each additional copy having a smaller marginal value. In their work it is assumed that the cost of a single level is arbitrary, meaning that it is plausible that the auctioneer can only afford to hire a single level of service of a single agent. Chan and Chen [10] proposed randomized, truthful, and budget-feasible mechanisms for this setting, with approximation guarantees that depend on the number of agents. The crucial difference with our setting is that we assume that the auctioneer's budget is big enough to afford any single individual agent entirely, which is in line with the indivisible setting in which the auctioneer can afford to hire any individual agent.

Divisible Agents: Another relevant setting is the setting in which agents are offering a divisible service, e.g., offering their time. In this case, it is reasonable to assume that the auctioneer can hire each agent for any fraction of the service that they are offering. Again, the valuation function associated with each agent is assumed to be concave, meaning that the marginal gain is non-increasing in the fraction of the acquired service. Note that this problem is the fractional relaxation of the problem introduced by Singer [24], when it is assumed that the auctioneer can afford to hire any individual agent entirely. Anari et al. [5] were the first to study the divisible setting. In their work they employed a *large market* assumption, which, in the context of budget-feasible mechanism design, roughly means that the cost of each agent for their entire service is insignificant, compared to the budget of the auctioneer. Additionally, they notice that in the divisible setting, no truthful mechanism with a finite approximation guarantee exists without any restriction on the costs. Very recently, Klumper and Schäfer [17] revisited this problem without the large market assumption but under the much milder assumption that the auctioneer can afford to hire any individual agent entirely (which is standard in the literature for the indivisible setting, but here it does restrict the bidding space). They present a deterministic, truthful and budget-feasible mechanism that achieves an approximation ratio of $1 + \phi \approx 2.62$ for linear valuation functions and extend it to the setting in which all agents are associated with the same concave valuation function.

The two discussed settings of procurement auctions have a number of practical applications in various domains. As previously mentioned, the divisible setting would for example be useful to model the time availability of a worker in the context of crowdsourcing. Moreover, these types of auctions can also be applied to other industries, such as transportation and logistics, where the delivery of goods and services can be broken down into multiple levels of service. For instance, in the transportation industry, the first level of service can represent the basic delivery service, while the higher levels can represent more premium and specialized services, such as express delivery or temperature-controlled shipping. The auctioneer can then choose to hire each agent up to an available level of service, not necessarily the best offered, based on the budget constraint and the value of the services provided.

Our Contributions. In this work, we propose deterministic, truthful and budget-feasible mechanisms for settings with partial allocations. Specifically:

- We present a mechanism, SORT-&-REJECT(k) (Mechanism 1), with an approximation ratio of $2 + \sqrt{3} \approx 3.73$ for the indivisible agent setting with multiple levels of service and concave valuation functions (Sect. 3, Theorem 2). The main idea behind our novel mechanism is to apply a backwards greedy approach, in which we start from an optimal fractional solution and we discard single levels of service one by one, until a carefully chosen stopping condition is met. For this setting, no constant-factor approximation mechanism was previously known.
- We use SORT-&-REJECT(k) as a subroutine in order to design the mechanism for the setting with divisible agents, CHUNK-&-SOLVE (Mechanism 2), that

achieves an approximation ratio of $L(1 + \phi + o(k^{-1}))$ for L-regular concave valuation functions (Sect. 4.1, Theorem 3), where k is a discretization parameter. This is the first result for the problem that is independent of the number of agents n. Note that L-regularity is a Lipschitz-like condition and for $L = 1$ the problem reduces to the setting with linear valuation functions. In this case, our ratio retrieves the best known guarantee of Klumper and Schäfer [17] as k grows. On a technical level, we exploit the correspondence between the discrete and the continuous settings; as the number of service levels grows large, the former converges to the latter.

- We improve on the aforementioned best known result for $L = 1$, by suggesting a 2-approximation mechanism, PRUNE-&-ASSIGN (Mechanism 4), for the divisible setting with linear valuation functions (Sect. 4.2, Theorem 4). This mechanism is inspired by the randomized 2-approximate mechanism proposed by Gravin et al. [14] for the indivisible setting.

As we mentioned above, all our results are under the mild assumption that we can afford each agent entirely. For the setting with divisible agents this is necessary in order to achieve any non-trivial factor [5], and it was also assumed by Klumper and Schäfer [17]. Even for the discrete setting with multiple levels of service this assumption circumvents a strong lower bound of Chan and Chen [10] (see also Remark 1). In both settings our assumptions are much weaker than the large market assumptions often made in the literature (see, e.g., [5,16]).

Further Related Work. The design of truthful budget-feasible mechanisms for indivisible agents was introduced by Singer [24], who gave a deterministic mechanism for additive valuation functions with an approximation guarantee of 5, along with a lower bound of 2 for deterministic mechanisms. This guarantee was subsequently improved to $2 + \sqrt{2} \approx 3.41$ by Chen et al. [11], who also provided a lower bound of 2 for randomized mechanisms and a lower bound of $1 + \sqrt{2} \approx 2.41$ for deterministic mechanisms. Gravin et al. [14] gave a 3-approximate deterministic mechanism, which is the best known guarantee for deterministic mechanisms to this day, along with a lower bound of 3 when the guarantee is with respect to the optimal non-strategic fractional solution. Regarding randomized mechanisms, Gravin et al. [14] settled the question by providing a 2-approximate randomized mechanism, matching the lower bound of Chen et al. [11]. Finally, the question has also been settled under the large market assumption by Anari et al. [5], who extended their $\frac{e}{e-1} \approx 1.58$ mechanism for the setting with divisible agents to the indivisible setting. As mentioned earlier, Klumper and Schäfer [17] study the divisible setting without the large market assumption, but under the assumption that the private cost of each agent is bounded by the budget and give, among other results, a deterministic $(1 + \phi)$-approximate mechanism for linear valuation functions, i.e., non-identical valuations.

For indivisible agents, the problem has also been extended to richer valuation functions. This line of inquiry also started by Singer [24], who gave a randomized algorithm with an approximation guarantee of 112 for a monotone submodular objective. Once again, this result was improved by Chen et al. [11]

to a 7.91 guarantee, and the same authors devised a deterministic mechanism with a 8.34 approximation. Subsequently, the bound for randomized mechanisms was improved by Jalaly and Tardos [16] to 5. Very recently, Balkanski et al. [8] proposed a new method of designing mechanisms that goes beyond the sealed-bid auction paradigm. Instead, Balkanski et al. [8] presented mechanisms in the form of deterministic clock auctions and, for the monotone submodular case, present a deterministic clock auction which achieves a 4.75 guarantee.

Beyond monotone submodular valuations, it becomes much harder to obtain truthful mechanisms with small constants as approximation guarantees. Namely, for non-monotone submodular objectives, the first randomized mechanism that runs in polynomial time is due to Amanatidis et al. [3] and its approximation guarantee is 505. This guarantee was improved to 64 by Balkanski et al. [8] who provided a deterministic mechanism for the problem and Huang et al. [15] who gave a further improvement of $(3 + \sqrt{5})^2$ for randomized mechanisms. In both [8] and [15] the mechanisms take the form of clock auctions, procedures in which bidders are offered prices in multiple rounds, see also [21].

Richer valuations that have been studied are XOS valuation functions (see Bei et al. [9], Amanatidis et al. [2]) and subadditive valuation functions (see Dobzinski et al. [12], Bei et al. [9], Balkanski et al. [8]). For subadditive valuation functions, no mechanism achieving a constant approximation is known. However, Bei et al. [9] have proved that such a mechanism should exist, using a non-constructive argument. Finding such a mechanism is an intriguing open question.

Other settings that have been studied include environments with underlying feasibility constraints, such as downward-closed environments (Amanatidis et al. [1], Huang et al. [15]) and matroid constraints (Leonardi et al. [18]). Other environments in which the auctioneer wants to get a set of heterogeneous tasks done and each task requires that the hired agent has a certain skill, have been studied as well, see Goel et al. [13], Jalaly and Tardos [16]. Recently, Li et al. [19] studied facility location problems under the lens of budget-feasibility, in which facilities have private facility-opening costs. Finally, the problem has been studied in a beyond worst-case analysis setting by Rubinstein and Zhao [23].

2   Model and Preliminaries

We first define the standard budget-feasible mechanism design model below which constitutes the basis of the more general models considered in this paper. The multiple service level model is introduced in Sect. 2.2 and the divisible agent model in Sect. 2.3.

2.1   Basic Model

We consider a procurement auction consisting of a set of agents $N = \{1, \ldots, n\}$ and an auctioneer who has an available budget $B \in \mathbb{R}_{>0}$. Each agent $i \in N$ offers a service and has a private cost parameter $c_i \in \mathbb{R}_{>0}$, representing their true cost for providing this service. The auctioneer derives some value $v_i \in \mathbb{R}_{\geq 0}$ from the service of agent i which is assumed to be public information.

A deterministic mechanism $\mathcal{M}$ in this setting consists of an allocation rule $\mathbf{x} : \mathbb{R}_{\geq 0}^n \to \mathbb{R}_{\geq 0}^n$ and a payment rule $\mathbf{p} : \mathbb{R}_{\geq 0}^n \to \mathbb{R}_{\geq 0}^n$. To begin with, the auctioneer collects a profile $\mathbf{b} = (b_i)_{i \in N} \in \mathbb{R}_{\geq 0}^n$ of declared costs from the agents. Here, b_i denotes the cost declared by agent $i \in N$, which may differ from their true cost c_i. Given the declarations, the auctioneer determines an allocation (hiring scheme) $\mathbf{x}(\mathbf{b}) = (x_1(\mathbf{b}), \ldots, x_n(\mathbf{b}))$, where $x_i(\mathbf{b}) \in \mathbb{R}_{\geq 0}$ is the allocation decision for agent i, i.e., to what extent agent i is hired. Generally, we distinguish between the *divisible* and *indivisible* agent setting by means of the corresponding allocation rule. In the divisible setting, each agent i can be allocated fractionally, i.e., $x_i(\mathbf{b}) \in \mathbb{R}_{\geq 0}$. In the indivisible setting, each agent i can only be allocated integrally, i.e., $x_i(\mathbf{b}) \in \mathbb{N}_{\geq 0}$. Given an allocation $\mathbf{x}$, we define $W(\mathbf{x}) = \{i \in N \mid x_i > 0\}$ as the set of agents who are positively allocated under $\mathbf{x}$. The auctioneer also determines a vector of payments $\mathbf{p}(\mathbf{b}) = (p_1(\mathbf{b}), \ldots, p_n(\mathbf{b}))$, where $p_i(\mathbf{b})$ is the payment agent i will receive for their service.

We assume that agents have quasi-linear utilities, i.e., for a deterministic mechanism $\mathcal{M} = (\mathbf{x}, \mathbf{p})$, the utility of agent $i \in N$ for a profile $\mathbf{b}$ is $u_i(\mathbf{b}) = p_i(\mathbf{b}) - c_i \cdot x_i(\mathbf{b})$. We are interested in mechanisms that satisfy three properties for any true profile $\mathbf{c}$ and any declared profile $\mathbf{b}$:

- *Individual rationality:* Each agent $i \in N$ receives non-negative utility, i.e., $u_i(\mathbf{b}) \geq 0$.
- *Budget-feasibility:* The sum of all payments made by the auctioneer does not exceed the budget, i.e., $\sum_{i \in N} p_i(\mathbf{b}) \leq B$.
- *Truthfulness:* Each agent $i \in N$ does not have and incentive to misreport their true cost, regardless of the declarations of the other agents, i.e., $u_i(c_i, \mathbf{b}_{-i}) \geq u_i(\mathbf{b})$ for any b_i and $\mathbf{b}_{-i}$.

Given an allocation $\mathbf{x}$, the total value that the auctioneer obtains is denoted by $v(\mathbf{x})$. The exact form of this function depends on the respective model we are studying and will be defined in the subsections below.

All the models that are studied in this paper are single-parameter settings and so the characterization of Myerson [22] applies.[1] It is therefore sufficient to focus on the class of mechanisms with *monotone non-increasing* (called *monotone* for short) allocation rules. An allocation rule is monotone non-increasing if for every agent $i \in N$, every profile $\mathbf{b}$, and all $b_i' \leq b_i$, it holds that $x_i(\mathbf{b}) \leq x_i(b_i', \mathbf{b}_{-i})$. We will use this together with Theorem 1 below to design truthful mechanisms.

Theorem 1 ([7,22]). *A monotone non-increasing allocation rule $\mathbf{x}(\mathbf{b})$ admits a payment rule that is truthful and individually rational if and only if for all agents $i \in N$ and all bid profiles $\mathbf{b}_{-i}$, we have $\int_0^\infty x_i(z, \mathbf{b}_{-i})dz < \infty$. In this case, we can take the payment rule $p(\mathbf{b})$ to be*

$$p_i(\mathbf{b}) = b_i x_i(\mathbf{b}) + \int_{b_i}^\infty x_i(z, \mathbf{b}_{-i})dz . \tag{1}$$

---

[1] We refer the reader to [6] for a rigorous treatment of the uniqueness property of Myerson's characterization result.

In this paper, we will exclusively derive monotone allocation rules that are implemented with the payment rule as defined in (1). Therefore, in the remainder of this paper, we adopt the convention of referring to the true cost profile $\mathbf{c}$ of the agents as input (rather than distinguishing it from the declared cost profile $\mathbf{b}$). Also, throughout the paper we will omit the explicit reference to the respective cost profile $\mathbf{c}$ whenever it is clear from the context.

2.2 k-Level Model

We consider the following multiple service level model as a natural extension of the standard model introduced above (see also [10]). Throughout the paper, we refer to this model as the k-*level model* for short: Suppose each agent $i \in N$ offers $k \geq 1$ levels of service and has an associated valuation function $v_i : \{0, \ldots, k\} \to \mathbb{R}_{\geq 0}$ which is public information.[2] Here, $v_i(j)$ denotes the auctioneer's valuation of the first j levels of service of agent i for the auctioneer. Observe that in this setting each agent $i \in N$ is indivisible and the range of the allocation rule is constrained to $\{0, \ldots, k\}$, i.e., $x_i : \mathbb{R}_{\geq 0}^n \to \{0, \ldots, k\}$. Note also that the total cost of agent i is linear (as defined above), i.e., the cost of using $x_i = j$ service levels of agent i is $j \cdot c_i$.

Valuation Functions: Without loss of generality, we assume that each v_i is normalized such that $v_i(0) = 0$. We study the general class of *concave* valuation functions, i.e., for each agent i, $v_i(j) - v_i(j-1) \geq v_i(j+1) - v_i(j)$ for all $j = 1, \ldots, k-1$. We also define the j-*th marginal valuation* of agent i as $m_i(j) := v_i(j) - v_i(j-1)$ for $j \in \{1, \ldots, k\}$. Given a profile $\mathbf{c}$, the total value that the auctioneer derives from an allocation $\mathbf{x}$ is defined by the separable concave function $v(\mathbf{x}(\mathbf{c})) = \sum_{i \in N} v_i(x_i(\mathbf{c}))$.

Cost Restrictions: We consider different assumptions with respect to the ability of the auctioneer to hire multiple service levels. In the *all-in setting*, we assume that the auctioneer can afford to hire all levels of each single agent, i.e., given a cost profile $\mathbf{c}$, for every agent $i \in N$ it holds that $k \cdot c_i \leq B$. In contrast, in the *best-in setting*, which is equivalent to the setting of Chan and Chen [10], the auctioneer is guaranteed only to be able to afford the first service level, i.e., given a cost profile $\mathbf{c}$, for every agent $i \in N$ it holds that $c_i \leq B$.[3] We focus on the all-in setting throughout this extended abstract; see the full version of our work [4] for an almost tight result on the best-in setting.

Remark 1. For the best-in setting, Chan and Chen [10] show a lower bound of k for the approximation guarantee of any deterministic, truthful, budget-feasible mechanism and a lower bound of $\ln(k)$ for the approximation guarantee of any

---

[2] Our results very easily extend to the setting where there is a different (public) k_i associated with each agent i. We use a common k for the sake of presentation.

[3] Whenever we use one of these assumptions, we implicitly constrain the space of the (declared) cost profiles. That is, we assume that any agent who violates the respective condition is discarded up front from further considerations, e.g., by running a pre-processing step that removes such agents.

randomized, universally truthful, budget-feasible mechanism. For these bounds, a single agent is used and then it is claimed that they generalize to nk and $\ln(nk)$, respectively, for n agents. The former is not correct, as we show in the full version of this paper [4], where we present a $(k + 2 + o(1))$-approximate mechanism, named GREEDY-BEST-IN(k), almost settling the deterministic case. Note that the mechanism suggested by Chan and Chen [10] is $4(1 + \ln(nk))$-approximate.

Benchmark: The performance of a mechanism is measured by comparing $v(\mathbf{x}(\mathbf{c}))$ with the underlying (non-strategic) combinatorial optimization problem, which is commonly referred to as the *k-Bounded Knapsack Problem*[4] (see, e.g., [20] for a classification of knapsack problems):

$$\mathrm{OPT}_I^k(\mathbf{c}, B) := \max \sum_{i=1}^{n} v_i(x_i), \quad \text{s.t.} \quad \sum_{i=1}^{n} c_i x_i \leq B, \quad x_i \in \{0, \ldots, k\}, \forall i \in N.$$

(2)

The k-Bounded Knapsack Problem is NP-hard in general. We say that a mechanism $\mathcal{M} = (\mathbf{x}, \mathbf{p})$ is *α-approximate* with $\alpha \geq 1$ if $v(\mathbf{x}(\mathbf{c})) \geq \frac{1}{\alpha} \mathrm{OPT}_I^k(\mathbf{c}, B)$. We also consider the relaxation of the above problem as a proxy for $\mathrm{OPT}_I^k(\mathbf{c}, B)$. The definition and further details about this are deferred to Sect. 2.4 below.

An instance I of the k-level model will be denoted by a tuple $I = (N, \mathbf{c}, B, k, (v_i)_{i \in N})$. Whenever part of the input is clear from the context, we omit its explicit reference for conciseness (e.g., often we refer to instance simply by its corresponding cost vector $\mathbf{c}$).

2.3   Divisible Agent Model

Next, we introduce the fractional model that we study in this paper. Throughout the paper, we refer to this model as the *divisible agent model*: Here, the auctioneer is allowed to hire each agent for an arbitrary fraction of the full service. More precisely, each agent $i \in N$ is divisible and the range of the allocation rule is constrained to $[0, 1]$, i.e., $x_i : \mathbb{R}_{\geq 0}^n \to [0, 1]$. Each agent $i \in N$ has an associated valuation function $\bar{v}_i : [0, 1] \to \mathbb{R}_{\geq 0}$ (which is public information), where $\bar{v}_i(x)$ represents how valuable a fraction $x \in [0, 1]$ of the service of agent i is to the auctioneer.

Valuation Functions: Also here, we assume without loss of generality that each v_i is normalized such that $\bar{v}_i(0) = 0$. We focus on the general class of nondecreasing and concave valuation functions. The total value that the auctioneer derives from an allocation $\mathbf{x}(\mathbf{c})$ is defined as $v(\mathbf{x}(\mathbf{c})) = \sum_{i \in N} \bar{v}_i(x_i(\mathbf{c}))$.

L-Regularity Condition: We introduce the following regularity condition for the valuation functions which will be crucial in our analysis of the divisible agent model below. Given a function $f : [0, 1] \to \mathbb{R}_{\geq 0}$, we say that f is *L-regular Lipschitz* (or just *L-regular* for short) for $L \geq 1$ if

$$f(x) \leq xLf(1) \quad \forall x \in [0, 1).$$

(3)

---

[4] Note that for $k = 1$, the problem reduces to the well-known *0-1 Knapsack Problem*.

Note that if we remove $f(1)$ from the above definition, then this definition coincides with the standard Lipschitz definition. We say that an instance of the divisible agent model is *L-regular* for some $L \geq 1$, if for each agent $i \in N$ the valuation function $\bar{v}_i$ is L-regular as defined in (3).

Cost Restrictions: We assume that the auctioneer can afford each agent to the full extent. More formally, given a cost profile $\mathbf{c}$ it must hold that for each agent $i \in N$, $c_i \leq B$. With respect to this assumption, the same remarks as given above apply (see Footnote 3).

Benchmark: As above, the performance of a mechanism is measured by comparing $v(\mathbf{x}(\mathbf{c}))$ with the underlying (non-strategic) combinatorial optimization problem, which we refer to as the *Fractional Concave Knapsack Problem*:

$$\text{OPT}_\text{F}(\mathbf{c}, B) := \max \sum_{i=1}^{n} \bar{v}_i(x_i) \quad \text{s.t.} \quad \sum_{i=1}^{n} c_i x_i \leq B, \quad x_i \in [0, 1] \, \forall i \in N. \quad (4)$$

In the divisible agent model, a mechanism $\mathcal{M} = (\mathbf{x}, \mathbf{p})$ is α-approximate with $\alpha \geq 1$ if $v(\mathbf{x}(\mathbf{c})) \geq \frac{1}{\alpha} \text{OPT}_\text{F}(\mathbf{c})$.

An instance I of the divisible agent model will be denoted by a tuple $I = (N, \mathbf{c}, B, (\bar{v}_i)_{i \in N})$. As mentioned before, we will omit the explicit reference of certain input parameters if they are clear from the context.

2.4 Fractional k-Bounded Knapsack Problem

We also consider the *Fractional k-Bounded Knapsack Problem* that follows from the k-Bounded Knapsack Problem in (2) by relaxing the integrality constraint:

$$\text{OPT}_\text{F}^\text{k}(\mathbf{c}, B) := \max \sum_{i=1}^{n} v_i(\lfloor x_i \rfloor) + m_i(\lceil x_i \rceil)(x_i - \lfloor x_i \rfloor),$$

$$\text{such that} \quad \sum_{i=1}^{n} c_i x_i \leq B, \quad x_i \in [0, k] \, \forall i \in N.$$

Naturally, it also holds that $\text{OPT}_\text{F}^\text{k}(\mathbf{c}, B) \geq \text{OPT}_\text{I}^\text{k}(\mathbf{c}, B)$. Note that $\text{OPT}_\text{F}^1(\mathbf{c}, B)$ is the fractional relaxation of the well-known *Knapsack Problem*. It is not hard to see that $\text{OPT}_\text{F}^\text{k}(\mathbf{c}, B)$ inherits the well-known properties of its one-dimensional analogue, including the fact that an optimal solution can be computed by a simple greedy algorithm in polynomial time.

We state the following as a fact and refer the reader to [4] for details. More generally, all the proofs that are omitted here can be found in [4].

Fact 1. *Given an instance $I = (N, \mathbf{c}, B, k, (v_i)_{i \in N})$ of the Fractional k-Bounded Knapsack Problem, a simple greedy algorithm computes in time $O(kn \log(kn))$ an optimal solution $\mathbf{x}^*$ that has at most one coordinate with a non-integral value.*

The next fact relates the values of instances which only differ with respect to their budget.

Fact 2. *Let* $I = (N, \mathbf{c}, B, k, (v_i)_{i \in N})$ *and* $I' = (N, \mathbf{c}, B', k, (v_i)_{i \in N})$ *with* $B < B'$ *be two instances of the Fractional k-Bounded Knapsack Problem. Then,* $\mathrm{OPT}_{\mathrm{F}}^{\mathrm{k}}(\mathbf{c}, B)/B \geq \mathrm{OPT}_{\mathrm{F}}^{\mathrm{k}}(\mathbf{c}, B')/B'$.

In most cases the budget is going to be clear from the context, so usually we are going to omit B from $\mathrm{OPT}_{\mathrm{F}}^{\mathrm{k}}(\mathbf{c}, B)$ and $\mathrm{OPT}_{\mathrm{I}}^{\mathrm{k}}(\mathbf{c}, B)$, and simply write $\mathrm{OPT}_{\mathrm{F}}^{\mathrm{k}}(\mathbf{c})$ and $\mathrm{OPT}_{\mathrm{I}}^{\mathrm{k}}(\mathbf{c})$, respectively.

3   Budget-Feasible Mechanism for Multiple Service Levels

We derive a natural truthful and budget-feasible greedy mechanism for the k-level model. This mechanism will also be used in our CHUNK-&-SOLVE mechanism for the divisible agent model (see Sect. 4.1).

3.1   A Truthful Greedy Mechanism

The main idea underlying our mechanism is as follows: If there is an agent i^* whose maximum valuation $v_{i^*}(k)$ is valuable enough (in a certain sense), then we simply pick all service levels of this agent. Otherwise, we compute an allocation using the following greedy procedure: We first compute an optimal allocation $\mathbf{x}^*(\mathbf{c})$ to the corresponding Fractional k-Bounded Knapsack Problem (which can be done in polynomial time) and use the integral part of this solution as an initial allocation. Note that this allocation is close to the optimal fractional solution because $\mathbf{x}^*(\mathbf{c})$ has at most one fractional component (Fact 1). We then repeatedly discard the worst service level (in terms of marginal value-per-cost) of an agent from this allocation, until the total value of our allocation would drop below an α-fraction of the optimal solution.

We need some more notation for the formal description of our mechanism: Given an allocation $\mathbf{x}$, we denote by $\ell(\mathbf{x})$ the agent whose $x_{\ell(\mathbf{x})}$-th level of service is the least valuable in $\mathbf{x}$, in terms of their marginal value-per-cost ratio. Notice that due to the fact that the valuation functions are concave, the worst case marginal value-per-cost ratio indeed corresponds to the $x_{\ell(\mathbf{x})}$-th ratio of agent $\ell(\mathbf{x})$. When $\mathbf{x}$ is clear from the context, we refer to this agent simply as ℓ. A detailed description of our greedy mechanism is given in Mechanism 1.

The main result of this section is the following theorem:

Theorem 2. SORT-&-REJECT(k) *with* $\alpha = \frac{1}{2+\sqrt{3}}$ *is truthful, individually rational, budget-feasible and* $(2+\sqrt{3})$-*approximate for instances of the k-level model, and runs in time polynomial in n and k.*

The polynomial running time for the allocation is straightforward. In the remainder of this section, we prove several lemmata to establish the properties stated in Theorem 2. Technically, the most challenging part is to prove that the mechanism is budget-feasible (see Sect. 3.2).

We first show that the allocation rule of SORT-&-REJECT(k) is monotone.

Lemma 1. *The allocation rule of* SORT-&-REJECT(k) *is monotone.*

Mechanism 1: SORT-&-REJECT(k)

 ▷ **Input:** A profile **c** and a parameter $\alpha \in (0, 1)$;

1 Set $i^* = \arg\max_{i \in N} v_i(k) / \mathrm{OPT}_{\mathrm{F}}^{\mathrm{k}}(\mathbf{c}_{-i})$

2 **if** $v_{i^*}(k) \geq \frac{\alpha}{1-\alpha} \cdot \mathrm{OPT}_{\mathrm{F}}^{\mathrm{k}}(\mathbf{c}_{-i^*})$ **then**

3 ⌊ set $x_{i^*} = k$ and $x_i = 0$ for all $i \in N \setminus \{i^*\}$

4 **else**

5 Compute an optimal allocation $\mathbf{x}^*(\mathbf{c})$ of $\mathrm{OPT}_{\mathrm{F}}^{\mathrm{k}}(\mathbf{c})$. ;

6 Initialize $\mathbf{x} = (\lfloor x_1^*(\mathbf{c}) \rfloor, \ldots, \lfloor x_n^*(\mathbf{c}) \rfloor)$. ;

7 **for** $i \in N$ *and* $j = 1, \ldots, x_i$ **do**

8   ⌊ add the marginal value-per-cost ratio $m_i(j)/c_i$ to a list $\mathcal{L}$.

9 Sort $\mathcal{L}$ in non-increasing order and let ℓ be the index of the last agent of $W(\mathbf{x})$ in $\mathcal{L}$.

10 **while** $v(\mathbf{x}) - m_\ell(x_\ell) \geq \alpha \mathrm{OPT}_{\mathrm{F}}^{\mathrm{k}}(\mathbf{c})$ **do**

11   Set $x_\ell = x_\ell - 1$. ;

12   Remove the last element from $\mathcal{L}$ and update ℓ. ;

13 Allocate $\mathbf{x}$ and determine the payments $\mathbf{p}$ according to (1).

Since the payments are computed according to (1), we conclude that the mechanism is truthful and individually rational. We continue by showing that SORT-&-REJECT(k) achieves the claimed approximation guarantee.

Lemma 2. *Let* $\mathbf{x}(\mathbf{c})$ *be the allocation computed by* SORT-&-REJECT(k) *for a cost profile* **c**. *It holds that* $v(\mathbf{x}(\mathbf{c})) \geq \alpha \mathrm{OPT}_{\mathrm{I}}^{\mathrm{k}}(\mathbf{c})$.

3.2 Making Sort-&-Reject(k) Budget-Feasible

It remains to prove that SORT-&-REJECT(k) is budget-feasible. We introduce some auxiliary notation: Consider a cost profile **c** and an agent $i \in W(\mathbf{x}(\mathbf{c}))$. Let $j \in \{1, \ldots, x_i(\mathbf{c})\}$ be an arbitrary service level allocated to i. Intuitively, we refer to the *critical payment* $p_{ij}(\mathbf{c}_{-i})$ for service level j of i as the largest cost q that i can declare and still obtain service level j (see Fig. 1 for an illustration). More formally, we define $Q_{ij}(\mathbf{c}_{-i})$ as the set of all points q satisfying $\lim_{z \to q^-} x_i(z, \mathbf{c}_{-i}) \geq j$ and $\lim_{z \to q^+} x_i(z, \mathbf{c}_{-i}) \leq j$ and let $p_{ij}(\mathbf{c}_{-i}) = \sup(Q_{ij}(\mathbf{c}_{-i}))$. Note that such a point q must always exist and $c_i \leq q \leq \frac{B}{k}$. To see this, note that $x_i(c_i, \mathbf{c}_{-i}) \geq j$ which implies that $c_i \leq q$ and that $x_i(z, \mathbf{c}_{-i}) = 0 < j$ for all $z > \frac{B}{k}$ (by our assumption that $c_i \leq \frac{B}{k}$) which implies that $q \leq \frac{B}{k}$.[5]

It is easy to see that the payment of an agent i can be written as the sum over these critical payments for the levels of service i was hired for.

Fact 3. *Let* **c** *be a cost profile and let* $i \in W(\mathbf{x}(\mathbf{c}))$. *It holds that* $p_i(\mathbf{c}) = $

$$\sum_{j=1}^{x_i(\mathbf{c})} p_{ij}(\mathbf{c}_{-i}).$$

---

[5] It is not hard to see that the set $Q_{ij}(\mathbf{c}_{-i})$ is also closed and thus the supremum always exists.

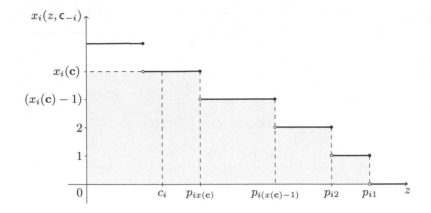

Fig. 1. Illustration of the critical payments of agent i.

Lemma 3 is the main technical tool needed to establish budget-feasibility for the **else** part of SORT-&-REJECT(k). It is also used in the proof of Theorem 3 in the divisible agent setting.

Lemma 3 (Budget Feasibility Lemma). *Let* **c** *be a cost profile such that* $v_{i^*}(k) < \frac{\alpha}{1-\alpha} \mathrm{OPT}_\mathrm{F}^\mathrm{k}(\mathbf{c}_{-i^*})$. *Then,*

$$\sum_{i=1}^{n} p_i(\mathbf{c}) \le \frac{B}{1-\alpha}\left(\frac{m_\ell(x_\ell(\mathbf{c}))}{\mathrm{OPT}_\mathrm{F}^\mathrm{k}(\mathbf{c}_{-\ell})} + \frac{\alpha}{1-\alpha} \right).$$

Observe that using Lemma 3, we can determine a range for values of α, for which SORT-&-REJECT(k) is budget-feasible:

Lemma 4. SORT-&-REJECT(k) *is budget-feasible for* $\alpha \le \frac{1}{2+\sqrt{3}}$.

It is clear that Lemma 4 along with Lemmata 1 and 2 conclude the proof of Theorem 2.

4 Two Budget-Feasible Mechanisms for Divisible Agents

We consider the divisible agent model and derive two truthful and budget-feasible mechanisms. The first one is obtained by discretizing the valuation functions and reducing the problem to the k-level model (Sect. 4.1). The second one is an improved 2-approximate mechanisms for the divisible agent model with linear valuation functions (Sect. 4.2).

4.1 Using Sort-&-Reject(k) for Divisible Agents

Recall that in the divisible agent model, we have $\mathbf{x}(\mathbf{c}) \in [0,1]^n$ and concave non-decreasing valuation functions $\bar{v}_i : [0,1] \to \mathbb{R}_{\ge 0}$ with $\bar{v}_i(0) = 0$ for all $i \in N$.

Mechanism 2: CHUNK-&-SOLVE

    ▷ **Input:** A profile **c** and a positive integer k

1 Initialize for each $i \in N$, $v_i : \{0, \ldots, k\} \rightarrow \mathbb{R}_+$ with $v_i(0) = 0$. ;

2 **for** $i \in N$ *and* $j \in \{1, \ldots, k\}$ **do** set $v_i(j) := \bar{v}_i(j/k)$.

3 Set $\alpha(L, k) = \big(3k + L - \sqrt{5k^2 + 6kL + L^2}\big)/2k$.

4 Let $\tilde{I} = (N, \mathbf{c}, kB, k, (v_i)_{i \in N})$ denote the resulting discretized instance of the k-level model. ;

5 Compute **x** by running SORT-&-REJECT$(k)(\mathbf{c}, \alpha(L, k))$ on $\tilde{I}$. ;

6 **for** $i \in N$ **do** set $\bar{x}_i = x_i/k$.

7 Allocate $\bar{\mathbf{x}}$ and determine the payments $\bar{\mathbf{p}}$ according to (1).

Throughout this section, we assume that all valuation functions are L-regular for some $L \geq 1$ as defined in (3).

There is a natural correspondence between the setting with $k \geq 1$ levels of service and the setting with divisible agents: If we subdivide the $[0, 1]$ interval into k chunks of length $\frac{1}{k}$ and evaluate the $\bar{v}_i(\cdot)$'s at $\frac{1}{k}, \frac{2}{k}, \ldots, \frac{k}{k}$, then this can be interpreted as the value of hiring $1, 2, \ldots k$ levels of service, respectively. We can then obtain results for the setting with divisible agents by applying this discretization, using SORT-&-REJECT(k) from Sect. 3 and letting k grow. Our CHUNK-&-SOLVE mechanism basically exploits this idea. A detailed description is given in Mechanism 2.

The following is the main result of this section:

Theorem 3. CHUNK-&-SOLVE *is truthful, individually rational, budget-feasible and $(L(1 + \frac{1}{k})/\alpha(L, k))$-approximate for L-regular instances of the divisible agent model.*

It is a matter of using simple calculus to show that $\lim_{k \to \infty} \alpha(L, k) = \frac{1}{\phi+1}$, and thus the approximation ratio of Theorem 3 goes to $(\phi + 1)L$. Given that the running time of SORT-&-REJECT(k) is polynomial in k, a reasonable question is whether we can have a good approximation guarantee for 'small' k when L is $O(1)$. Again, it is a matter of calculations to show that using $k = O(L)$ suffices. For instance, taking $k = 11L$ implies an approximation ratio of $3L$ for any $L \geq 1$. Qualitatively, this means that, for $L \in O(1)$, CHUNK-&-SOLVE achieves a constant approximation ratio in polynomial time.

4.2 A Mechanism for Divisible Agents and Linear Valuations

CHUNK-&-SOLVE retrieves the best known approximation guarantee of $1 + \phi$ for $L = 1$ and as $k \to \infty$ [17] (i.e., for divisible agents with linear valuations). Below, we improve upon this and give a simple 2-approximate budget-feasible mechanism for this setting. Our mechanism is inspired by the randomized 2-approximate budget-feasible mechanism by Gravin et al. [14] for indivisible agents. We also prove a lower bound of 2 for deterministic, truthful, individually rational and budget feasible mechanisms with independent allocation rules (as defined below).

Mechanism 3: PRUNING by Gravin et al. [14]

▷ **Input:** A profile $\mathbf{c}$ with the agents relabeled so that $\frac{v_1}{c_1} \geq \cdots \geq \frac{v_n}{c_n}$.

1 Let $r := \frac{1}{B} \max\{v_i \mid i \in N\}$.

2 **foreach** $i \in N$ **do** set $\bar{x}_i = 1$ if $\frac{v_i}{c_i} \geq r$ and $\bar{x}_i = 0$ otherwise.

3 Let $\ell := \arg\max\{i \mid \bar{x}_i = 1\}$.

4 **while** $rB < \sum_{i=1}^{\ell} v_i - \max\{v_i \mid i = 1, \ldots, \ell\}$ **do**

5 $\quad$ Continuously increase rate r.

6 $\quad$ **if** $\frac{v_\ell}{c_\ell} \leq r$ **then** set $\bar{x}_\ell = 0$ and $\ell = \ell - 1$.

7 **return** $(r, \bar{\mathbf{x}})$

Phase 1: Pruning Mechanism for Divisible Agents. We first extend the PRUNING mechanism of Gravin et al. [14] to the divisible setting. This mechanism constitutes a crucial building block for both their deterministic 3-approximate mechanism and their randomized 2-approximate mechanism for indivisible agents [14]. As we show below, it serves as a useful starting point for the divisible setting as well.

Given a profile $\mathbf{c}$, this mechanism computes an allocation $\bar{\mathbf{x}}(\mathbf{c})$, which we refer to as the *provisional allocation*, and a positive quantity $r(\mathbf{c})$, which we refer to as the *rate*. We assume that the agents are initially relabeled by their decreasing value-per-cost ratio, i.e., $\frac{v_1}{c_1} \geq \frac{v_2}{c_2} \geq \cdots \geq \frac{v_n}{c_n}$. The mechanism proceeds as described in Mechanism 3.

Gravin et al. [14] showed that PRUNING is monotone. In fact, an even stronger *robustness* property holds (and is implicit in the proof of Lemma 3.1 in [14]):

each agent i that is a winner in the provisional allocation cannot alter the outcome of PRUNING unilaterally while remaining a winner in the provisional allocation.

Lemma 5 (implied by Lemma 3.1 of [14]). *Let $\mathbf{c}$ be a profile. Consider an agent $i \in N$ with $\bar{x}_i(\mathbf{c}) = 1$. Then, for all c_i' such that $\bar{x}_i(c_i', \mathbf{c}_{-i}) = 1$, it holds that $\bar{\mathbf{x}}(c_i', \mathbf{c}_{-i}) = \bar{\mathbf{x}}(\mathbf{c})$ and $r(c_i', \mathbf{c}_{-i}) = r(\mathbf{c})$.*

Given this robustness property, PRUNING can be used as a first filtering step to discard inefficient agents, followed by a subsequent allocation scheme which takes $(r(\mathbf{c}), \bar{\mathbf{x}}(\mathbf{c}))$ as input. If the subsequent allocation scheme is monotone, then the sequential composition of PRUNING with this allocation scheme is monotone as well. This *composability property* is proven in Lemma 3.1 of [14].

Let $(r, \bar{\mathbf{x}})$ be the output of PRUNING for a cost profile $\mathbf{c}$. Given $\bar{\mathbf{x}}$, we define S as the set of agents that are provisionally allocated, i^* as the highest value agent in S, and T as the set of remaining agents. More formally, we define

$$S = \{i \in N \mid \bar{x}_i = 1\}, \quad i^* = \arg\max\{v_i \mid i \in S\} \quad \text{and} \quad T = S \setminus \{i^*\}. \quad (5)$$

Note that the definitions of S, i^* and T depend on $\bar{\mathbf{x}}$ (and thus the cost profile $\mathbf{c}$). For notational convenience, we do not state this reference explicitly if it is clear from the context.

The following properties were proved in [14] and are useful in our analysis.

Lemma 6 (Lemma 3.2 of [14]). *Given a profile* **c***, let* $(r, \bar{\mathbf{x}})$ *be the output of* PRUNING. *Let* $S = T \cup \{i^*\}$ *be defined as in* (5) *with respect to* $\bar{\mathbf{x}}$. *We have*

1. $c_i \leq \frac{v_i}{r} \leq B$ *for all* $i \in S$.
2. $v(T) \leq rB < v(S)$.
3. $\text{OPT}_F \leq v(S) + r \cdot (B - c(S))$.

Phase 2: Independent Allocation Schemes. Our mechanism combines the PRUNING mechanism above with the allocation schemes defined in (6) below. We refer to the resulting mechanism as PRUNE-&-ASSIGN (see Mechanism 4).

First, we need to define the following constants:

$$q_{i^*} = \begin{cases} \frac{1}{2} - q \text{ if } v_{i^*} \leq v(T) \\ \frac{1}{2} \quad \text{otherwise} \end{cases}, \quad q_i = 1 - q_{i^*} - q, \forall i \in T, \quad \text{where } q = \frac{1}{2} \frac{v(S) - rB}{\min\{v_{i^*}, v(T)\}}.$$

Note that the constant q_i for all agents $i \in T$ is the same. It is not hard to prove that $q \in [0, \frac{1}{2}]$ (see [14, Lemma 5.1]). The constants above are chosen so that $rB/2 = q_{i^*} v_{i^*} + (1 - q_{i^*} - q)v(T)$.

We can now define our (fractional) allocation function $x_i(\mathbf{c}) = x_i(c_i)$ for each agent $i \in T \cup \{i^*\}$:

$$x_i(z) = q_i + \frac{v_i - rz}{2v_i} \qquad \text{for } z \in \left[0, \frac{v_i}{r}\right]. \tag{6}$$

Note that $x_i(c_i, \mathbf{c}_{-i}) = x_i(c_i)$ only depends on agent i's cost c_i. We call such allocation rules *independent*. Further, note that by property (1) of Lemma 6, the cost c_i of each agent $i \in S$ after pruning is at most $\frac{v_i}{r}$, i.e., $x_i(z)$ will be determined by some value $z \in [0, \frac{v_i}{r}]$. It is not hard to verify that x_i is well-defined (given the chosen parameters q_{i^*}, q_T and q above).

Theorem 4. PRUNE-&-ASSIGN *is truthful, individually rational, budget-feasible and 2-approximate for instances of the divisible agent model with linear valuations.*

Given the lower bound of $1 + \sqrt{2} \approx 2.41$ by Chen et al. [11] in this setting with indivisible agents, Theorem 4 establishes a separation between the indivisible agent model and the divisible agent model with linear valuations.

Finally, we give a lower bound of 2 for deterministic, truthful, individually rational and budget feasible mechanisms with independent allocation rules. Note that PRUNE-&-ASSIGN does not belong to this class of mechanisms (due to PRUNING). However, our analysis of PRUNE-&-ASSIGN is tight (see [4]).

Theorem 5. *Let* $\mathcal{M}$ *be a deterministic, truthful, individually rational and budget feasible mechanism with an approximation guarantee of* α. *If* $\mathcal{M}$ *has independent allocation rules, then* $\alpha \geq 2$.

Mechanism 4: PRUNE-&-ASSIGN for Divisible Agents

  ▷ **Input:** A profile $\mathbf{c}$ with the agents relabeled so that $\frac{v_1}{c_1} \geq \frac{v_2}{c_2} \geq \cdots \geq \frac{v_n}{c_n}$

1 Obtain $(r, \bar{\mathbf{x}})$ by running PRUNING for profile $\mathbf{c}$.
2 Let $S = T \cup \{i^*\}$ be as defined in (5).
3 Determine the fractional allocation: $x_{i^*}(c_{i^*})$ and $x_i(c_i), \forall i \in T$ as in (6).
4 Allocate $\mathbf{x}$ and determine the payments $\mathbf{p}$ as in (1).

---

5 Conclusion and Future Work

In this work we revisited two settings where partial allocations are allowed and draw clear connections between them. Under mild assumptions like being able to afford each agent entirely and having "nice" concave valuation functions (i.e., $O(1)$-regular), we give deterministic, truthful and budget-feasible mechanisms with constant approximation guarantees. We believe these are settings that are both interesting and relevant to applications and there are several open questions we do not settle here. A natural direction, not considered at all in this work, is to deal with additional combinatorial constraints, like matching, matroid, or even polymatroid (for the k-level setting) constraints. For the k-level setting, it would be interesting to understand whether we can obtain mechanisms with approximation guarantees closer to those possible for single-level settings, or alternatively, determine whether allowing multiple levels of service is an inherently harder problem. As far as simple settings are concerned, the most important open problem is still the indivisible agents case with additive valuations, for which the best-possible approximation ratio is in $[1 + \sqrt{2}, \ 3]$ (due to [11,14]). The corresponding range for the divisible agent setting is $[e/(e - 1), \ 2]$ (due to [5] and our Theorem 4). Any progress on these fronts may give rise to novel techniques, which may be also used for problems in richer environments.

Acknowledgment. This work was supported by the Dutch Research Council (NWO) through its Open Technology Program, proj. no. 18938, and the Gravitation Project NETWORKS, grant no. 024.002.003. It has also been partially supported by project MIS 5154714 of the National Recovery and Resilience Plan Greece 2.0, funded by the European Union under the NextGenerationEU Program and the EU Horizon 2020 Research and Innovation Program under the Marie Skodowska-Curie Grant Agreement, grant no. 101034253.

    Moreover, it was supported by the "1st Call for HFRI Research Projects to support faculty members and researchers and the procurement of high-cost research equipment" (proj. no. HFRI-FM17-3512). Finally, part of this work was done during AT's visit to the University of Essex that was supported by COST Action CA16228 (European Network for Game Theory).

References

1. Amanatidis, G., Birmpas, G., Markakis, E.: Coverage, matching, and beyond: new results on budgeted mechanism design. In: Cai, Y., Vetta, A. (eds.) WINE 2016.

LNCS, vol. 10123, pp. 414–428. Springer, Heidelberg (2016). https://doi.org/10. 1007/978-3-662-54110-4_29

2. Amanatidis, G., Birmpas, G., Markakis, E.: On budget-feasible mechanism design for symmetric submodular objectives. In: Devanur, N.R., Lu, P. (eds.) WINE 2017. LNCS, vol. 10660, pp. 1–15. Springer, Cham (2017). https://doi.org/10.1007/978-3-319-71924-5_1

3. Amanatidis, G., Kleer, P., Schäfer, G.: Budget-feasible mechanism design for non-monotone submodular objectives: offline and online. In: Proceedings of the 2019 ACM Conference on Economics and Computation, EC 2019, pp. 901–919 (2019)

4. Amanatidis, G., Klumper, S., Markakis, E., Schäfer, G., Tsikiridis, A.: Partial allocations in budget-feasible mechanism design: bridging multiple levels of service and divisible agents. CoRR/arXiv abs/2307.07385 (2023)

5. Anari, N., Goel, G., Nikzad, A.: Budget feasible procurement auctions. Oper. Res. **66**(3), 637–652 (2018)

6. Apt, K.R., Heering, J.: Characterization of incentive compatible single-parameter mechanisms revisited. J. Mech. Inst. Design **7**(1), 113–129 (2022)

7. Archer, A., Tardos, É.: Truthful mechanisms for one-parameter agents. In: Proceedings of the 42nd IEEE Symposium on Foundations of Computer Science, pp. 482–491 (2001)

8. Balkanski, E., Garimidi, P., Gkatzelis, V., Schoepflin, D., Tan, X.: Deterministic budget-feasible clock auctions. In: Proceedings of the 2022 Annual ACM-SIAM Symposium on Discrete Algorithms, SODA 2022, pp. 2940–2963 (2022)

9. Bei, X., Chen, N., Gravin, N., Lu, P.: Worst-case mechanism design via Bayesian analysis. SIAM J. Comput. **46**(4), 1428–1448 (2017)

10. Chan, H., Chen, J.: Truthful multi-unit procurements with budgets. In: Liu, T.-Y., Qi, Q., Ye, Y. (eds.) WINE 2014. LNCS, vol. 8877, pp. 89–105. Springer, Cham (2014). https://doi.org/10.1007/978-3-319-13129-0_7

11. Chen, N., Gravin, N., Lu, P.: On the approximability of budget feasible mechanisms. In: Proceedings of the 22nd ACM-SIAM Symposium on Discrete Algorithms, SODA 2011, pp. 685–699 (2011)

12. Dobzinski, S., Papadimitriou, C.H., Singer, Y.: Mechanisms for complement-free procurement. In: Proceedings of the 12th ACM Conference on Electronic Commerce, EC 2011, pp. 273–282 (2011)

13. Goel, G., Nikzad, A., Singla, A.: Allocating tasks to workers with matching constraints: truthful mechanisms for crowdsourcing markets. In: Proceedings of the 23rd International Conference on World Wide Web, WWW 2014, pp. 279–280 (2014)

14. Gravin, N., Jin, Y., Lu, P., Zhang, C.: Optimal budget-feasible mechanisms for additive valuations. ACM Trans. Econ. Comput. (TEAC) **8**(4), 1–15 (2020)

15. Huang, H., Han, K., Cui, S., Tang, J.: Randomized pricing with deferred acceptance for revenue maximization with submodular objectives. In: Proceedings of the ACM Web Conference 2023, pp. 3530–3540 (2023)

16. Jalaly Khalilabadi, P., Tardos, É.: Simple and efficient budget feasible mechanisms for monotone submodular valuations. In: Christodoulou, G., Harks, T. (eds.) WINE 2018. LNCS, vol. 11316, pp. 246–263. Springer, Cham (2018). https://doi.org/10. 1007/978-3-030-04612-5_17

17. Klumper, S., Schäfer, G.: Budget feasible mechanisms for procurement auctions with divisible agents. In: Kanellopoulos, P., Kyropoulou, M., Voudouris, A. (eds.) SAGT 2022. LNCS, vol. 13584, pp. 78–93. Springer, Cham (2022). https://doi. org/10.1007/978-3-031-15714-1_5

18. Leonardi, S., Monaco, G., Sankowski, P., Zhang, Q.: Budget feasible mechanisms on matroids. In: Eisenbrand, F., Koenemann, J. (eds.) IPCO 2017. LNCS, vol. 10328, pp. 368–379. Springer, Cham (2017). https://doi.org/10.1007/978-3-319-59250-3_30
19. Li, M., Wang, C., Zhang, M.: Budget feasible mechanisms for facility location games with strategic facilities. Auton. Agent. Multi-Agent Syst. **36**(2), 35 (2022)
20. Martello, S., Toth, P.: Knapsack Problems: Algorithms and Computer Implementations. Wiley, Hoboken (1990)
21. Milgrom, P., Segal, I.: Clock auctions and radio spectrum reallocation. J. Polit. Econ. **128**(1), 1–31 (2020)
22. Myerson, R.: Optimal auction design. Math. Oper. Res. **6**(1), 58–73 (1981)
23. Rubinstein, A., Zhao, J.: Beyond worst-case budget-feasible mechanism design. In: 14th Innovations in Theoretical Computer Science Conference (ITCS 2023). Schloss Dagstuhl-Leibniz-Zentrum für Informatik (2023)
24. Singer, Y.: Budget feasible mechanisms. In: In Proceedings of the 51st Annual Symposium on Foundations of Computer Science, FOCS 2010, pp. 765–774 (2010)

High-Welfare Matching Markets
via Descending Price

Robin Bowers(✉) and Bo Waggoner

University of Colorado at Boulder, 1111 Engineering Drive, Boulder, CO 80309, USA
robin.bowers@colorado.edu

Abstract. We consider the design of monetary mechanisms for two-sided matching. Mechanisms in the tradition of the deferred acceptance algorithm, even in variants incorporating money, tend to focus on the criterion of stability. In this work, instead, we seek a simple auction-inspired mechanism with social welfare guarantees. We consider a descending-price mechanism called the Marshallian Match, proposed (but not analyzed) by [22]. When all values for potential matches are positive, we show the Marshallian Match with a "rebate" payment rule achieves constant price of anarchy. This result extends to models with costs for acquiring information about one's values, and also to group formation, i.e. matching on hypergraphs. With possibly-negative valuations, which capture e.g. job markets, the problem becomes harder. We introduce notions of approximate stability and show that they have beneficial welfare implications. However, the main problem of proving constant factor welfare guarantees in "ex ante stable equilibrium" remains open.

Keywords: Price of Anarchy · matching markets · auction design

1 Introduction

A primary goal of designing mechanisms is to coordinate groups to arrive at collectively good allocations or outcomes. For example, in auctioning a set of items to unit-demand buyers, the problem is to coordinate among the varied preferences of the buyers to achieve an overall good matching of buyers to items. In such auction settings, "good" is usually formalized via *price of anarchy* [21]: in any equilibrium of the auction game, the expected social welfare (total utility) should be approximately optimal.

Matching people to people, with preferences on both sides, appears to require even more coordination. [8] introduced the foundational deferred-acceptance algorithm – a matching mechanism without money – for participants with ordinal preferences. The key "good" property it achieves (when participants are truthful) is *stability*: no pair prefers to switch away from their given matchings and match to each other instead. Variants of deferred acceptance have had significant impact in applications from kidney exchange to the National Residency Matching Program (NRMP) for doctors and hospitals (e.g. [14]).

© The Author(s), under exclusive license to Springer Nature Switzerland AG 2024
J. Garg et al. (Eds.): WINE 2023, LNCS 14413, pp. 59–76, 2024.
https://doi.org/10.1007/978-3-031-48974-7_4

We are motivated by two drawbacks of deferred-acceptance-style approaches. First, the *social welfare* of such mechanisms is unclear, even in settings such as matching with contracts [10] which explicitly models money. While their criterion of stability is a nice property, its relationship to welfare is not obvious and we would like a mechanism with explicit welfare guarantees.

Second, it is unclear the extent to which such mechanisms are compatible with *inspections* in which participants must invest effort to discover their preferences. For example, the design of the NRMP requires relatively expensive and time-constrained interviews, which must be completed before matching begins. While such concerns have motivated significant work on information acquisition in matching markets, particularly variants of deferred acceptance (e.g. [13]; see Sect. 1.2), to our knowledge none of it incorporates monetary mechanisms with quantitative welfare guarantees. On the other hand, prior work of [16] has shown that even in the special case of matching people to items (which have no preferences), approximately optimal welfare *requires* a market design with dynamically interspersed matching and information acquisition. [16] showed that descending-price mechanisms tend to be compatible with costly inspection stages and still yield high social welfare, due to a connection with the Pandora's box problem [23].

The 1/4-Rebate Marshallian Match. Inspired by [16,22] propose the "Marshallian Match" (MM) for two-sided matching with money. In [19], its namesake describes a theory of market clearing in which the buyer-seller matches that generate the largest surplus – i.e. the most net utility between the pair – occur first, and so on down.[1] A similar dynamic is observed in decentralized matching markets [3], yet it does not directly underlie standard centralized designs such as deferred acceptance.

Similarly, the MM begins with a high price that descends over time. Participants maintain a bid on each of their potential matches, with the sum of the two bids ideally representing the total surplus generated by the match. When the price reaches the sum of any pair's bids on each other, that pair is matched. They pay their bids and drop out of the mechanism, which continues. In the "1/4-rebate" variant studied in this paper, the mechanism only keeps half of the sum of the bids and the participants split the other half, each receiving a rebate of 1/4 of the sum of the bids. [22] speculate on the dynamics, strategy, and benefits of this mechanism and variants, but do not obtain theoretical results.

1.1 Our Results

Nonnegative values. We first consider a setting where all participants' values are nonnegative. Under this restriction, we show a general *price of anarchy (PoA)*

---

[1] This dynamic eventually leads to the market clearing price (e.g. [20]), after which no more positive-surplus matches are possible. Indeed, in a simple commodity market it is possible to ignore Marshall's dynamics and focus on the calculation of the clearing price. But in a more complex two-sided matching problem, we appear to require dynamics in order to properly coordinate matches.

guarantee for the Marshallian Match, i.e. in any Bayes-Nash equilibrium the expected social welfare is within a constant factor of the optimal. The positive-bids setting can model, for example, matching of industrial plants to geographic areas (an application of [17]) or matching local businesses to municipal-owned locations.

Next, we extend the result to a *group formation* setting: agents must be partitioned into subsets of size at most k, with private valuations for joining each possible subset. We modify the MM by clearing a subset when the price reaches the sum of its bids. We obtain a $\Omega(1/k^2)$ price of anarchy for this problem.

We next show that the welfare guarantee also extends to a model with *information acquisition costs*. Participants may not initially know their values for a potential match and must expend time, effort, and/or money to investigate and discover one's value. A good mechanism should carefully coordinate these investigations to happen at appropriate times, or risk losing significant welfare: participants will waste too much utility on unnecessary inspections, or else forego valuable matches due to the inspection cost and uncertainty about the match. Here, although the optimal first-best is unknown and likely NP-hard, we obtain the same PoA guarantee.

Theorem 1. *With nonnegative values, the 1/4-rebate Marshallian Match has the following guarantees:*

1. *For matching on general graphs, a Bayes-Nash price of anarchy of at least 1/8.*
2. *For matching on hypergraphs with group size at most k, a Bayes-Nash price of anarchy of at least $\frac{1}{2k^2}$.*
3. *For matching on general graphs with inspection costs, a Bayes-Nash price of anarchy of at least 1/8.*

Proof Ingredients. The proofs rely on a smoothness approach (e.g. [21]) along with several key properties of our variant of the MM. First, the MM limits information leakage: a participant cannot observe others' strategies until they themselves are matched, at which point it is too late to react. This controls the otherwise complex strategic interactions of dynamic mechanisms.

Next, the "rebate" payment rule of the MM crucially allows participants to align their personal utility with the order of market clearing. A participant who deviates to truthful bidding will always receive utility equal to their rebate, 1/4 of the current descending price. So the earlier the participant is cleared, the higher their utility.

Finally, the descending-price structure of the MM is compatible with information acquisition and the "Pandora's Box" problem [23]. Participants are able to manage the risk-reward tradeoff for investing in information acquisition, because once prices have descended to a low point, they know that they can lock in available matches for a bounded cost. To prove PoA in this setting, we adapt techniques of [16] for analyzing welfare in models of inspection.

Possibly-Negative Values. We then consider a general two-sided matching setting where values may be negative. This setting more accurately captures job markets, where workers incur a cost (i.e. negative value) for being matched to a job and must be compensated more than that cost. For natural reasons, a PoA result for the MM in this setting is impossible: if all participants set their bids so as to refuse all matches, no single participant can deviate to cause any change and the equilibrium obtains zero welfare. This suggests that a kind of stability condition may be natural and necessary for high-welfare matching mechanisms of any kind. We observe that *approximate ex post* stability implies approximately optimal welfare, and show that the MM achieves approximate ex post stability if participants bid truthfully.

Proposition 1. *In any k-approximate ex post stable strategy profile, the 1/4-rebate Marshallian Match (in fact, any mechanism) achieves at least 1/k of the optimal expected welfare. Truthful bidding in the 1/4-rebate Marshallian Match is 4-approximately ex post stable.*

However, truthfulness is not generally an equilibrium.[2] This raises the question of whether it is reasonable to assume participants will adopt approximately stable strategy profiles. We argue that ex post stability is too strong an assumption, and instead propose *ex ante* stability. We show that in a Nash setting (not Bayes-Nash) with fixed valuations, if strategies are approximately *ex ante stable*, then the welfare of MM is approximately optimal.

Theorem 2. *In the general Nash setting with fixed valuations, in any strategy profile that is k-approximate ex ante stable, the 1/4-rebate Marshallian Match achieves at least a $\frac{1}{4k}$ fraction of the optimal welfare.*

In fact, the stability property also ensures that participants keep a large fraction of the welfare; the proof uses that participant surplus alone (i.e. welfare minus payments) is at least $\frac{1}{4k}$ of the optimal welfare.

Unfortunately, this stable-price-of-anarchy result is fragile and the proof does not extend to the *Bayes-Nash* setting where private valuations are drawn from a common-knowledge prior. This is roughly due to the difficulty of coordinating and communicating deviations between "blocking pairs". Therefore, the main problem of proving a welfare guarantee in a general negative-bids setting and under a reasonable stability assumption remains open.

Open Problem 1. *Give a well-justified stability assumption and a mechanism such that, in the* Bayes-Nash setting *with general values, stable strategy profiles guarantee a constant factor of the optimal expected welfare.*

---

[2] We note that even for deferred acceptance, which is also stable *if all participants are truthful*, in general only one side of the market optimizes their outcomes by being truthful.

1.2 Related Work

We have not found any mechanisms in the literature involving a two-sided matching market with money and quantifiable welfare results.[3] However, the literature on matching with strategic agents is very broad, including with inspection stages, and we highlight a number of related papers.

[15] explores a mechanism for two-sided monetary matching under full information, inspired by job markets, through a process reminiscent of the deferred-acceptance mechanism. They provide stability results for this mechanism given truthful behavior from all agents, but do not analyze the welfare of these outcomes. Our approach, in contrast, utilizes descending-price dynamics and emphasizes welfare over stability.

Probably closest to our work is [13], which considers design of a platform to coordinate two-sided matchings with inspection costs and quantifiable welfare guarantees, as in e.g. a dating app. The paper studies agents coming from specific populations with a known distribution of types, and uses knowledge of the type distributions to compute strategies for directing inspection stages. [13] is able to show the structure of equilibria and use this to give welfare guarantees, all in a setting without money. In contrast, we are interested in monetary mechanisms, motivated (eventually) by e.g. labor markets. We consider a very simple descending-price mechanism that has no access to knowledge about the agent types or distributions.

Beyond [13] there is an extensive literature on matching marketplaces and dynamics. Work in that literature involving money (transferable utility) historically often takes a Walrasian equilibrium perspective, while ours is in the tradition of auction design. We refer to the survey of [3] for more on this literature.

As mentioned above, a number of recent works study matching markets with information acquisition, but generally consider variants of deferred acceptance without money. Works of this kind include [1,4,5,7,9,11,12].

Among these, [12] uses a lens of optimal search theory similar to ours. It studies matching of students to schools. Its matching problem is almost one-sided, in the sense that schools have known and fixed preferences. The focus is on coordinating efficient acquisition of information by students. Unlike our social-welfare perspective, that work focuses on the more standard criterion of stability and introduces regret-free stable outcomes. In particular, it does not involve money. Similarly, [9] study a serial-dictator mechanism for coordinating student inspection without money. We refer to [12] for an extensive discussion of further work related to information acquisition in matching markets.

Our notions of stability are naturally closely related to others in the literature. Ex post stability is only a quantitative version of stability in matching; a more sophisticated version of it is used by [12], for example. [7] also utilizes a similar notion of stability, in a setting of incomplete information.

---

[3] One can always apply a general Vickrey-Clarke-Groves (VCG) mechanism, which has an equilibrium with optimal social welfare. But VCG is undesirable because it appears incompatible with both price of anarchy results and models with inspection costs, see [16].

2   Preliminaries

We now define the model and the variant of the Marshallian Match mechanism studied in this paper, originally described by [22]. We define a general model, in which the graph is possibly non-bipartite and bids and values are possibly negative. Later sections will consider specific restrictions.

There is a finite set of n agents, forming vertices of an undirected graph $G = (\{1, \ldots, n\}, E)$. For now, we do not assume that G is bipartite. The presence of an edge $\{i, j\} \in E$ represents that it is feasible to match agents i and j. In this case, agent i has a possibly negative *value* $v_{ij} \in \mathbb{R}$ for being matched to j, and symmetrically, j has a value v_{ji} for being matched to i. If i and j are neighbors, we let $s_{ij} = v_{ij} + v_{ji}$ denote the *surplus* of the edge $\{i, j\}$.

An agent i's *type* consists of their values v_{ij} for each feasible partner j. In the *Nash* setting, each agent has a fixed type, and types are common knowledge. In the better-motivated *Bayes-Nash* setting, types are drawn from a common-knowledge joint distribution $\mathcal{D}$ and each agent observes their own type.

2.1   The Marshallian Match

In the MM, a global price $p(t)$ begins at $+\infty$, i.e. $p(0) = \infty$, and descends continuously in time until it reaches zero at time 1, i.e. $p(1) = 0$.[4] Each agent i maintains, for all neighbors j, a bid $b_{ij}(t)$ at each time t. The mechanism can observe all bids at all times, but agents cannot observe any bids except their own. For convenience, we may drop the dependence on t from the bid notation. When the sum of bids on any edge exceeds the global price, i.e. $b_{ij} + b_{ji} \geq p(t)$, then i and j are immediately matched to each other. Each agent pays their respective bid to the mechanism.

The mechanism keeps half of this total payment and returns one-fourth to each player. Therefore, we call this variant the 1/4-rebate Marshallian Match. We discuss other variants in Sect. 5.

Intuition for the Mechanism. Why might this mechanism be good, and how might participants strategize? We briefly describe some intuition, referring the reader to [22] for more detailed discussion. Social welfare and price of anarchy will be formally defined below.

First, the edges of the graph are in competition with each other to match first. When an edge $\{i, j\}$ is matched, it produces a surplus of $s_{ij} = v_{ij} + v_{ji}$. The first-best solution, i.e. the optimal solution for a central planner who holds all information, is to select a maximum matching where s_{ij} are the edge weights. However, as discussed in [16], algorithms for maximum matching are complex and appear to interact poorly with inspection stages. More robust is an approximate first-best solution: the *greedy* matching, where the highest-weight edge is matched first, and so on down. This procedure obtains at least half of

---

[4] This can be accomplished in theory by letting the price be e.g. $p(t) = \frac{1}{2t}$ for $t \in [0, 1/2]$ and $p(t) = 2 - 2t$ for $t \in [1/2, 1]$.

the optimal welfare, and can be simulated by a Marshallian Match in which all participants bid truthfully. In [16], this fact was used to obtain a constant price of anarchy for matching people to items, including in the presence of inspection costs.

However, two-sided matching introduces new strategic considerations. "Within" an edge $\{i, j\}$, there is a competition or bargaining for how to split the surplus generated. An agent i with many edges (many outside options) may be able to underbid significantly, while her counterpart j with very few outside options must offer a very high bid along that edge. The question is whether this strategizing and competition between i and j destroys the cooperation incentives for the overall bid on the edge $\{i, j\}$.

2.2 Notation, Strategies, and Welfare

A strategy b_i for player i consists of a plan[5] $b_{ij}(t)$ for how to set bids over time, for each neighbor j.

A full strategy profile is denoted $b = (b_1, \ldots, b_n)$. We let $v_i(b)$ be i's value for their match when b is played (zero if none), a random variable depending on the randomization in the strategies and, in the Bayes-Nash setting, on the random draw of the types. Next, $\pi_i(b)$ denotes i's net payment to the mechanism, i.e. bid minus rebate. Finally, $p_i(b)$ denotes the *total* net payment on the edge that i is matched along, i.e. $p_i(b) = \pi_i(b) + \pi_j(b)$ when i is matched to j.

We assume all agents are Bayesian, rational, and have preferences quasilinear in payment. That is, given a mechanism and a particular strategy profile b, the *utility* of a participant i is the random variable

$$u_i(b) = v_i(b) - \pi_i(b).$$

For example, if under profile b, i matches to j at time t, then $\pi_i(b) = b_{ij}(t) - \frac{p(t)}{4}$ and $u_i(b) = v_{ij} - b_{ij}(t) + \frac{p(t)}{4}$.

A *Nash equilibrium* of a mechanism with given types $\{v_{ij}\}$ is a strategy profile b, consisting of a plan for how to set bids at each moment in time, where each participant maximizes their expected utility, i.e.

$$\mathbb{E}u_i(b) \geq \mathbb{E}u_i(b_{-i}, b_i')$$

for all i and for any other strategy b_i' of i, where the randomness is taken over the strategies. A *Bayes-Nash equilibrium* is defined in exactly the same way, but in a setting consisting of a joint distribution over types. In that case, the randomness is taken over both types and strategies (which are maps from an agent's type to a plan for bidding over time).

---

[5] We observe that usual nuances around equilibrium in dynamic games, such as non-credible threats, refinements such as subgame perfect equilibrium, etc., do not arise here. In our variant of the MM, each agent i observes nothing until they are matched, after which they can no longer affect the game. So without loss of generality, i commits in advance to a plan $\{b_{ij}(t)\}$ and follows it until matched.

We use Welfare(b) to denote the expected social welfare, i.e. sum of utilities and payments:

$$\text{Welfare}(b) = \mathbb{E}\left[\sum_i v_i(b) + \sum_j v_j(b)\right],$$

where the expectation is over all randomness. In the setting with inspection costs, utility and social welfare also accounts for the loss of utility from the inspection processes; this will be formalized at the relevant point, Sect. 3.3.

Welfare(OPT) refers to the optimal social welfare. In the Nash and Bayes-Nash settings that is, respectively,

$$\text{Welfare}_{NE}(\text{OPT}) = \max_M \sum_{\{i,j\}\in M} v_{ij} + v_{ji}, \text{ and}$$

$$\text{Welfare}_{BNE}(\text{OPT}) = \mathbb{E}\left[\max_M \sum_{\{i,j\}\in M} v_{ij} + v_{ji}\right],$$

where the maximum is over all matchings M and the expectation in the Bayes-Nash setting is over the realizations of types.

The *price of anarchy* measures the worst-case ratio of Welfare(b) to Welfare(OPT) in any equilibrium b. For Nash equilibrium, we have

$$\text{PoA} = \min \frac{\text{Welfare}(b)}{\text{Welfare}(\text{OPT})},$$

where the minimum is taken over all settings (i.e. all types of the participants) and all Nash equilibria b. The Bayes-Nash price of anarchy is defined in exactly the same way, but the minimum is now over all Bayes-Nash settings (i.e. joint distributions on types) and all Bayes-Nash equilibrium strategy profiles b. Note that a Nash equilibrium is a special case of Bayes-Nash where the type distributions are degenerate. Therefore, a Bayes-Nash price of anarchy result immediately implies a Nash price of anarchy.

3   Results for Positive Valuations

In this section, we consider a restriction of the general setting where all values v_{ij} are nonnegative. First, we show that the 1/4-rebate Marshallian Match achieves a constant approximation of optimal welfare for matching. Second, we extend the result to the group formation (i.e. matchings on hypergraphs) setting. Finally, we extend the result in a different direction to the case where participants do not initially know their valuations and can choose to expend effort to discover them.

3.1   The Vanilla Positive Valuations Model

Here, we take the general model of Sect. 2 and assume that each valuation satisfies $v_{ij} \geq 0$. We modify the Marshallian Match to require all bids b_{ij} to be nonnegative at all times.

Intuition. The nonnegative MM is similar to running multiple interlocking descending-price unit-demand auctions simultaneously.[6] This parallel is most obvious in the bipartite case: any bidder i competes against other bidders in the same set for her favorite matches. Extending this perspective, each bidder i could hypothetically bid as though her neighbors j were simply items with no preferences. By analyzing the failure of this hypothetical strategy, we find in the MM that it is connected with a different high-welfare event, namely j matching early. This intuition underlies our smoothness lemma, discussed next.

Smoothness for Two-Sided Matching. We give a smoothness lemma that powers our price of anarchy result. Recall (e.g. [21]) that smoothness proofs of PoA proceed by guaranteeing high-welfare events in a counterfactual world where i deviates to a less-preferred strategy. One challenge is that in a dynamic mechanism that proceeds over time, a deviation could cause chain reactions that make it impossible to reason about the outcomes. Here, we rely on our variant of MM that does not leak any information about bids or strategies. The only piece of information the agent receives from the mechanism comes at the moment they are matched, after which they cannot react.

A second key challenge is that in a matching market, i may not match j *and* the prices that i and j each pay may still be low, obstructing an adaptation of a standard smoothness proof. In the MM there is, however, a high *total* price paid for an edge that obstructs the match. Recall that while $\pi_i(b)$ is i's payment, $p_i(b)$ is the *total* payment on the edge containing i that is matched (zero if i is unmatched).

The Deviation and Smoothness Lemma. Let b be any strategy profile. For any bidder i, define the deviation strategy b_i' as $b_{ij}'(t) = v_{ij}$ for all feasible neighbors j and all times t. That is, the deviation strategy is simply truthful bidding.

Recall that for the 1/4-rebate MM, i's utility in deviation when matching to j at bid $b_{ij}'(t)$ and price $p(t)$ is $u_i(b_{-i}, b_i') = v_{ij} - b_{ij}'(t) + \frac{p(t)}{4} = \frac{p(t)}{4}$.

Lemma 1. *In the nonnegative values setting, for any feasible pair $\{i, j\}$, any strategy profile b, and any realization of types, $u_i(b_{-i}, b_i') + \frac{p_i(b)}{4} + \frac{p_j(b)}{4} \geq \frac{v_{ij}}{8}$.*

Proof. All three terms on the left-hand side are nonnegative. Therefore, if $p_j(b) \geq v_{ij}/2$ or $p_i(b) \geq v_{ij}/2$, the result is immediate.

Otherwise, under b, neither i nor j is matched on an edge with net payment at least $v_{ij}/2$. So they are both unmatched by the time the price has dropped to $p(t) = v_{ij}$. The behavior of the mechanism and all participants is identical under b and (b_{-i}, b_i') until i matches, because the two profiles cannot be distinguished. So if i is still unmatched under (b_{-i}, b_i') when $p(t) = v_{ij} = b_{ij}'(t)$, then i matches to j at this price. We conclude that i matches at some price $p(t) \geq v_{ij}$. Therefore, $u_i(b_{-i}, b_i') \geq p(t)/4 \geq v_{ij}/4$.

---

[6] Although general price of anarchy results are available for these kinds of auctions for goods, e.g. [6,18], we do not know of any that apply to two-sided matching.

Theorem 3. *In the nonnegative values setting, the 1/4-rebate Marshallian Match has a Bayes-Nash price of anarchy of at least 1/8.*

Proof. Let b be any Bayes-Nash equilibrium and M^* be the optimal matching (a random variable). The definition of equilibrium give us that i prefers b_i to any b_i', so

$$\text{Welfare}(b) \geq \mathbb{E} \sum_i \left(u_i(b_{-i}, b_i') + \frac{p_i(b)}{2} \right)$$

$$\geq \mathbb{E} \left[\sum_{\{i,j\} \in M^*} \left(u_i(b_{-i}, b_i') + u_j(b_{-j}, b_j') + \frac{p_i(b)}{2} + \frac{p_j(b)}{2} \right) \right]$$

$$\geq \mathbb{E} \left[\sum_{\{i,j\} \in M^*} \left(\frac{v_{ij}}{8} + \frac{v_{ji}}{8} \right) \right] = \frac{1}{8} \text{Welfare}(\text{OPT}).$$

The second line follows as M^* contains at most every participant, and all utilities and payments are nonnegative. Lemma 1 gives us the last inequality after rearranging utilities and payments.

3.2 Group Matchings

We also provide a modified MM which obtains a constant price of anarchy in the setting of group formation, with groups up to size k. A formalization of the MM in the group matching setting is given in Appendix A, along with all associated proofs. The approach is similar to the matching case above, i.e. the special case where all allowable groups have size exactly 2.

We wish to form groups of agents of size at most k, with some collection of allowable groups S. A "matching" or assignment M consists of a set of disjoint groups from S. Each agent has some (nonnegative) value for each group to which they could potentiallly, and we wish to maximize the sum of agents' values for their assigned matches in M.

3.3 Inspection

We now extend our welfare result for graphs to a model with information acquisition costs, again without the requirement that the graph be bipartite. In the interest of space, we will defer all proofs regarding the inspection model to Appendix A. The model is augmented as follows, following e.g. [16]. The type of i consists of, for each feasible partner j, a cost of inspection $r_{ij} \geq 0$ and a distribution D_{ij} over the nonnegative reals. Our setting is Bayes-Nash, i.e. all types are drawn jointly from a common-knowledge prior. When i inspects an edge $\{i, j\}$, they incur a cost of r_{ij} and observe a value $v_{ij} \sim D_{ij}$ independently of all other randomness in the game. The cost r_{ij} can model a financial investment, or the cost of time or effort required for i to learn their value v_{ij}.

Let $I_{ij} \in \{0,1\}$ be the random variable indicator that i inspects j and let $A_{ij} \in \{0,1\}$ be the indicator that i is matched to j. We adopt the standard assumption (although recent algorithmic work of [2] has weakened it) that i must inspect j prior to being matched to j; i.e. if the match occurs, i must inspect and incur cost r_{ij} if they haven't yet. In other words, $A_{ij} \leq I_{ij}$ pointwise. We also assume that, in the game, inspection is instantaneous with respect to the movement of the price $p(t)$.

An agent i's utility is their value for their match (if any) minus the sum of all inspection costs and their net payment. Formally, we have $u_i(b) = \sum_j (A_{ij}v_{ij} - I_{ij}r_{ij}) - \pi_i(b)$, where the sum is over feasible neighbors. Social welfare is the sum of all agent utilities and revenue, i.e. Welfare$(b) = \sum_i (u_i(b) + \pi_i(b))$.

Covered Call Values and Exercising in the Money. [16] give technical tools, based on a solution of the Pandora's box problem [23], utilizing finance-inspired definitions.[7]

Definition 1. *Given a cost r_{ij} and distribution D_{ij}, the* strike price *is the unique value σ_{ij} satisfying*

$$\mathbb{E}_{v_{ij} \sim D_{ij}} \left(v_{ij} - \sigma_{ij}\right)^+ = r_{ij},$$

where $(\cdot)^+ = \max\{\cdot\,,\, 0\}$. The covered call value *is the random variable $\kappa_{ij} = \min\{\sigma_{ij}, v_{ij}\}$.*

We assume $\mathbb{E}_{v_{ij} \sim D_{ij}} v_{ij} \geq r_{ij}$ if $\{i,j\}$ is a feasible match. As we have nonnegative values, this is equivalent to the condition $\sigma_{ij} \geq 0$.

A *matching process* is any procedure that involves sequentially inspecting some of the potential matches and making matches, according to any algorithm or mechanism. The following property, along with Lemma 2, allows us to relate surplus in a matching process to that of a world with zero inspection costs and values κ_{ij}.

Definition 2. *In any matching process, we say that the ordered pair (i,j) exercises in the money if, for all realizations of the process, if $v_{ij} > \sigma_{ij}$ and $I_{ij} = 1$ then $A_{ij} = 1$. We say that agent i exercises in the money if (i,j) exercises in the money for all feasible partners j.*

Lemma 2 (Immediate extension of [16]). *For any feasible partners $\{i,j\}$, any fixed types of all agents, and any matching process, $\mathbb{E}[A_{ij}v_{ij} - I_{ij}r_{ij}] \leq \mathbb{E}[A_{ij}\kappa_{ij}]$, with equality if and only if (i,j) exercises in the money.*

---

[7] We refer the reader to [16] for explanation of the terminology, but in brief, the idea is to imagine that when i inspects j, the cost r_{ij} is subsidized by an investor in return for a "call option", i.e. the right to the excess surplus $v_{ij} - \sigma_{ij}$ beyond a threshold σ_{ij}, if any. The agent's surplus for that match becomes κ_{ij}, and the investor breaks even if (i,j) exercises in the money, otherwise loses money.

Lemma 2 follows from [16], with the proof replicated in Appendix B.

Informally, Lemma 2 states that the utility of any strategy is upper-bounded by the covered call utility of the same strategy, with equality if and only if the strategy exercises in the money. To analyze the mechanism using the smoothness framework, we propose and analyze the utility of a deviation strategy, which exercises in the money.

The Deviation Strategy. Define the deviation strategy b'_i for agent i as follows: Initially bid 0 on each neighbor j; when the clock reaches σ_{ij}, inspect neighbor j and update bid to $\kappa_{ij} = \min\{\sigma_{ij}, v_{ij}\}$. For a strategy profile b, define the random variable $\kappa_i(b)$ to be the covered call value of i for its match in profile b, i.e. $\min\{\sigma_{ij}, v_{ij}\}$ when i is matched to j. Let $\bar{u}_i(b)$ denote i's "covered call utility", i.e.

$$\bar{u}_i(b) = \kappa_i(b) - \pi_i(b).$$

Lemma 3. *The deviation strategy b'_i exercises in the money and ensures $\bar{u}_i(b_{-i}, b'_i) \geq 0$, for any b_{-i}.*

This allows us to evaluate the strategy in the covered-call world of Lemma 2.

Lemma 4 (Covered call smoothness). *For any feasible neighbors $\{i, j\}$ and any strategy profile b, for all realizations of types and values, $\bar{u}_i(b_{-i}, b'_i) + \frac{p_i(b)}{4} + \frac{p_j(b)}{4} \geq \frac{\kappa_{ij}}{8}$.*

Theorem 4. *In the inspection setting with nonnegative values, the 1/4-rebate Marshallian Match guarantees a Bayes-Nash price of anarchy of at least 1/8.*

This proof follows the same general structure as the proof of Theorem 3 by converting to and analyzing the covered call utility of deviation strategies, and comparing them to an upper-bound on the optimal welfare. Proofs of Lemmas 2 and 3, alongside Theorem 4, are given in Appendix B.

4   General Bids

It is natural to consider negative values and bids in a matching market. Negative values model *costs* incurred for a match, such as in a job market. When a worker is matched as an employee to a company, she experiences some cost for which she must be compensated. The goal of the matching market is to find an efficient price for her labor. However, negative costs complicate matters because they also introduce negative bids. A participant can make it difficult for a match to occur by bidding an arbitrary negative amount. The result is that equilibrium is no longer sufficient for good welfare, even in bipartite graphs.

In this section, we first formalize the failure of equilibrium. We then turn to *stability* as a possible saviour. We observe that approximate *ex post* stability indeed implies good welfare in *any mechanism*, and that the MM satisfies approximate ex post stability when participants are truthful.

However, ex post stability is a very strong requirement. We formulate an alternative, *ex ante* stability, and show that the MM has approximately optimal welfare in any ex-ante stable *Nash* equilibrium. Finally, however, we observe that this result does not naturally extend to Bayes-Nash equilibrium and illustrate the apparent difficulty involving coordinated communication.

4.1   Model and Failure of Equilibrium

We make one restriction on the general model of Sect. 2: we assume the graph is bipartite. This is primarily for notational and narrative convenience. To make our presentation more intuitive, we adopt terminology in which the two sides of the bipartite market are asymmetric: One side (e.g. employers) are *bidders*, while the other side (e.g. workers) are *askers*. The bidders are indexed by i. They have values v_{ij} and make bids b_{ij}.

The askers are indexed by j. We assume they have *costs* c_{ij} and make *asks* a_{ij}. The costs and asks are simply the negative of their values and bids under the previous section's model. Now, for example the surplus of a match between i and j is $s_{ij} = v_{ij} - c_{ij}$. We still use π_j to denote the net payment made by an asker. So the utility of an asker j for being matched to i is $u_j = -c_{ij} - \pi_i$, and so on.

We generally picture values, bids, costs, and asks all as positive numbers, so a bidder has a positive value for matching to an asker, who incurs a positive cost from the match. However, our results are all fully general and would allow for any value, bid, cost, or ask to be either positive or negative.

Failure of Equilibrium. When asks are allowed, equilibrium becomes insufficient to provide welfare guarantees. Participants in the mechanism can place bids and asks such as to effectively refuse matches with one another, by asking above value or bidding below cost. If two players both "refuse" matches with one another, neither can unilaterally fix the situation. We prove this result for the MM with 1/4 rebate, but the same profile is also a zero welfare equilibrium with no rebate.

Proposition 2. *In the bipartite setting with general values and costs, there always exists a Nash equilibrium of the Marshallian Match with zero welfare.*

Proof. Let $x = \max_{i,j} \max\{|v_{ij}|, |c_{ij}|\}$. Consider the strategy profile where every bidder i bids $b_{ij} = -2x$ on all neighbors j, and every asker j asks $a_{ij} = 2x$ on all neighbors i. This is an equilibrium in which no matches occur. For a unilateral deviation to cause a match to occur, e.g. a bidder would have to change a bid to at least $b_{ij} = 2x$, resulting in a net payment of at least $2x$ after the rebate, giving negative utility. The asker's case is analogous.

In this example, any single player cannot deviate alone to improve her welfare. However, any pair of bidders sharing an edge can coordinate a deviation together and guarantee themselves higher welfare. This suggests that the bad equilibrium profile lacks *stability*, a key concept in matching algorithm design.

4.2   Ex Post Stability

In classical "stable matching" problems [8], the goal is that, once the mechanism produces a final matching, no two participants i, j both prefer to leave their assigned partners and switch to matching each other instead. We use *ex post* to refer to the fact that this evaluation occurs after all randomness and the matching's outcome have been realized. In our setting, if i and j chose to match each other, they would obtain a net utility of their surplus $s_{ij} = v_{ij} - c_{ij}$. If this amount is larger than their total utility in the mechanism, then they could switch to each other and split the surplus so as to make them both better off. On the other hand, making this switch presumably involves some amount of friction. Therefore, we introduce an approximate version of stability, as a more lenient requirement of a mechanism.

Definition 3 (Approximate ex post stability). *A strategy profile (b, a) in a mechanism is k-ex post stable if, for all realizations of costs and values and for all feasible pairs of bidder i and asker j,*

$$u_i(b, a) + u_j(b, a) \geq \frac{1}{k} s_{ij}.$$

Ex post stability is an extremely strong notion. For any matching mechanism, an ex post stable strategy profile produces an approximately optimal matching.

Observation 1. *For any matching mechanism, any k-ex post stable strategy (if one exists) is a $\frac{1}{k}$-approximation of the first-best welfare.*

A proof of this observation is presented in Appendix C. We note that while there exist deferred-acceptance style "stable" matching mechanisms that technically involve money, such as matching with contracts [10], we have not ascertained if they can be made to satisfy approximate ex post stability in our sense.

Ex Post Stability of the Marshallian Match Under Truthfulness. Ideally, a matching mechanism would admit equilibria in ex post stable strategies, so that welfare would be high and participants would adhere to the outcomes of the mechanism. But as a weaker requirement, we would like an indication of whether approximately ex post stable strategies might be reasonable in a mechanism. We present a simple 4-ex post stable strategy profile for the 1/4-rebate MM, for all realizations of costs and values.

Proposition 3. *The truthful strategy in which all participants bid or ask their true value or cost is 4-ex post stable in the 1/4-rebate Marshallian Match.*

The proof of this lemma is deferred to Appendix C. Combined, Observation 1 and Proposition 3 show that any truthful strategy is approximately optimal, but it is not in general even an approximate equilibrium. We provide an example in the 1/4-rebate MM setting in which a player can increase her welfare arbitrarily by overstating her true value for a match. In such situations, it is likely that they will not adhere to an ex post stable profile.

Example 1 (Figure 1). Consider the bipartite graph with three participants: A and B place bids on matching to C. C has cost $c_{CA} = c_{CB} = 0$ for both matches, A has value $v_{AC} = k + 1$ and B has value $v_{BC} = k$ for matching with C. In the truthful profile, A is matched to C, and B goes unmatched with welfare 0. If B deviates to the non-truthful strategy of bidding $b'_{BC} = k+2$, B would be matched with C and would receive utility $u_B(b^*_{-B}, a^*, b') = k - (k + 2) + (k + 2)/4 = k/4 - 3/2$. Thus, picking k appropriately, B can benefit arbitrarily by deviating from the truthful strategy.

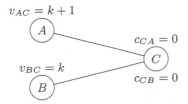

Fig. 1. In the 1/4-rebate MM, B is incentivized to deviate from truthfulness to the non-truthful bid $k + 2$.

Drawbacks of Ex Post Stability. If one can prove that a mechanism is ex post stable in equilibrium, that is an ideal result as a welfare guarantee immediately follows. However, without such a result, the value of ex post stability is questionable.

Consider the following somewhat subtle point. Intuitively, it may seem reasonable to view stability as a sort of equilibrium refinement. In particular, if strategy profile b is not approximately stable, then (one would think) there are two participants i and j who would prefer to jointly switch their strategies so as to match to each other. However, *ex post* stability does not give this kind of guarantee. It can only tell i and j whether they are satisfied after the mechanism happens. To capture the above intuition, we turn to *ex ante* stability.

4.3   Ex Ante Stability

We now consider a model of stability that generalizes equilibrium by supposing no *pair* of participants has an incentive to *bilaterally* deviate. Importantly, the incentive is relative to expected utility, so it involves an *ex ante* calculation by the participants rather than ex post.

Definition 4 (ex ante stability). *A strategy profile (b, a) is k-ex ante stable if, for all feasible pairs of bidder i and asker j, for all strategies b'_i of i and a'_j of j,*

$$\mathbb{E}[u_i(b, a) + u_j(b, a)] \geq \frac{1}{k}\mathbb{E}[u_i(b_{-i}, b'_i, a_{-j}, a'_j) + u_j(b_{-i}, b'_i, a_{-j}, a'_j)],$$

with randomness taken over realizations of types and strategies.

That is, a profile is k-ex ante stable if there exists no deviation for any pair of players by which they could expect to increase their collective welfare by a factor of more than k. There are two primary differences between ex ante and ex post stability. First, ex ante applies to preferences "before the fact", i.e. in expectation, while ex post applies to preferences "after the fact". Second, ex post stability postulates the ability of two participants to completely bypass the rules of the mechanism and match to each other. In ex ante stability, the participants are limited to deviating to strategies actually allowed by the mechanism.

We observe that an ex ante stable profile is by definition a Bayes-Nash equilibrium, as one can in particular consider profiles where only one of the two participants deviates. We show that for deterministic values and costs, ex ante stability is in fact sufficient to guarantee an approximation of optimal welfare.

Smoothness and Deviation Strategies. We define a pairwise deviation of $\{i, j\}$ to truthfulness for bids and asks as follows: $b'_{i\ell} = v_{i\ell}$ for all feasible neighbors ℓ, and similarly $a'_{j\ell} = c_{j\ell}$ for all neighbors ℓ. For any pairwise deviation to truthfulness, we show that that pair collectively achieves a constant fraction of their welfare in the 1/4-rebate MM.

Lemma 5. *For any pair of feasible neighbors i and j, for any strategy profile (b, a), $u_i(b_{-i}, b'_i, a_{-j}, a'_j) + u_j(b_{-i}, b'_i, a_{-j}, a'_j) \geq \frac{s_{ij}}{4}$.*

This proof is given in Appendix C.

Theorem 5. *For deterministic values and costs, and strategy profile (b, a) that is k-ex ante stable, the 1/4-rebate Marshallian Match achieves a $\frac{1}{4k}$-approximation of the optimal expected welfare.*

The proof given in Appendix C surprisingly shows that the contribution of payments to overall welfare can be ignored: we actually obtain that total participant surplus is a constant fraction of the optimal welfare.

This result relies heavily on participants' ability to calculate the fixed matching with optimal social welfare, and coordinate deviations with their partner. We exploit the deterministic costs and values considered to fix the optimal matching M^* across realizations of potentially mixed strategies.

In the Bayes-Nash setting, agents cannot coordinate deviations in the same way. The question of welfare guarantees in this setting remains open, and is discussed in Appendix D.

5   Conclusion and Future Work

Two-sided matching is difficult, even in such simplified abstract models as in this paper, for at least three reasons: 1) the process should accommodate information acquisition in a way compatible with optimal search theory, 2) the problem is generally dynamic and sequential (for the previous reason), complicating strategic behavior, and 3) the constraints are complex and interlocking, i.e. i can match to j if and only j matches to i, yet their preferences over this event can be conflicting and contextualized by their other options.

The variants of the Marshallian Match studied in this paper address each challenge to some extent. In the nonnegative values setting, the MM is compatible with optimal search because matches seem to occur approximately in order from highest surplus to lowest, allowing inspections to occur approximately in order of their "index" ("strike price") from the Pandora's box problem. In particular, it enables the useful technical property of "exercising in the money", i.e. a bidder who inspects at a late stage and discovers a very valuable match is able to unilaterally lock in that match.

Dynamic strategizing is addressed by strictly limiting information leakage, i.e. each participant can only see their own bid and learn when they match. It is unclear whether this feature actually makes the mechanism better, but it does make it easier to analyze. We believe that all of our results extend if participants are able to observe when any match occurs, but this would require careful formalization as the game becomes truly dynamic in that case.

The complex constraints are addressed to an extent by the coordination of matches in order of value. A bidder i in the 1/4-rebate MM has deviation strategies available in which the timing of their match corresponds to their utility. If another participant bids high enough to lock in a match with i early, this is out of i's direct control, but i can bid so that they are assured of enough utility to make the match worthwhile.

Future Work. There are a number of appealing variants on the model and directions for future investigation. In the job market application, an interview is a *simultaneous* inspection event between a worker and employer. Can the MM's welfare guarantees extend to a model where inspection is simultaneous, even in the positive-values setting? Other variations can include multiple stages of inspection (an extension in [16]) or matching where inspection is optional (studied algorithmically by [2]).

References

1. Ashlagi, I., Braverman, M., Kanoria, Y., Shi, P.: Clearing matching markets efficiently: informative signals and match recommendations. Manage. Sci. **66**(5), 2163–2193 (2020)
2. Beyhaghi, H., Kleinberg, R.: Pandora's problem with nonobligatory inspection. In: Proceedings of the 2019 ACM Conference on Economics and Computation, EC 2019, pp. 131–132. Association for Computing Machinery (2019). https://doi.org/10.1145/3328526.3329626
3. Chade, H., Eeckhout, J., Smith, L.: Sorting through search and matching models in economics. J. Econ. Lit. **55**(2), 493–544 (2017)
4. Che, Y.K., Tercieux, O.: Efficiency and stability in large matching markets. J. Polit. Econ. **127**(5), 2301–2342 (2019)
5. Chen, Y., He, Y.: Information acquisition and provision in school choice: a theoretical investigation. Econ. Theory **74**, 1–35 (2021)

6. Feldman, M., Immorlica, N., Lucier, B., Roughgarden, T., Syrgkanis, V.: The price of anarchy in large games. In: Proceedings of the Forty-Eighth Annual ACM Symposium on Theory of Computing, STOC 2016, pp. 963–976. Association for Computing Machinery, New York (2016). https://doi.org/10.1145/2897518.2897580
7. Fernandez, M.A., Rudov, K., Yariv, L.: Centralized matching with incomplete information. Working Paper 29043, National Bureau of Economic Research (2021). https://doi.org/10.3386/w29043, http://www.nber.org/papers/w29043
8. Gale, D., Shapley, L.S.: College admissions and the stability of marriage. Am. Math. Mon. **69**(1), 9–15 (1962)
9. Hakimov, R., Kübler, D., Pan, S.: Costly information acquisition in centralized matching markets. Rationality and Competition Discussion Paper Series 280, CRC TRR 190 Rationality and Competition (2021). https://ideas.repec.org/p/rco/dpaper/280.html
10. Hatfield, J.W., Milgrom, P.R.: Matching with contracts. Am. Econ. Rev. **95**(4), 913–935 (2005)
11. He, Y., Magnac, T.: Application costs and congestion in matching markets (2020)
12. Immorlica, N., Leshno, J., Lo, I., Lucier, B.: Information acquisition in matching markets: the role of price discovery. Available at SSRN 3705049 (2020)
13. Immorlica, N., Lucier, B., Manshadi, V., Wei, A.: Designing approximately optimal search on matching platforms, pp. 632–633. Association for Computing Machinery, New York (2021). https://doi.org/10.1145/3465456.3467530
14. Iwama, K., Miyazaki, S.: A survey of the stable marriage problem and its variants. In: International Conference on Informatics Education, ICKS 2008, pp. 131–136. IEEE (2008)
15. Kelso, A.S., Crawford, V.P.: Job matching, coalition formation, and gross substitutes. Econometrica **50**(6), 1483–1504 (1982). http://www.jstor.org/stable/1913392
16. Kleinberg, R., Waggoner, B., Weyl, E.G.: Descending price optimally coordinates search. In: Proceedings of the 2016 ACM Conference on Economics and Computation, EC 2016, pp. 23–24. Association for Computing Machinery, New York (2016). https://doi.org/10.1145/2940716.2940760
17. Koopmans, T.C., Beckmann, M.: Assignment problems and the location of economic activities. Econom.: J. Econom. Soc. 53–76 (1957)
18. Lucier, B., Syrgkanis, V.: Greedy algorithms make efficient mechanisms. In: Proceedings of the Sixteenth ACM Conference on Economics and Computation, pp. 221–238 (2015)
19. Marshall, A.: Principles of Economics, 8th edn. (1920)
20. Plott, C., Roy, N., Tong, B.: Marshall and Walras, disequilibrium trades and the dynamics of equilibration in the continuous double auction market. J. Econ. Behav. Organ. **94**, 190–205 (2013). https://doi.org/10.1016/j.jebo.2012.12.002
21. Roughgarden, T., Syrgkanis, V., Tardos, E.: The price of anarchy in auctions. J. Artif. Intell. Res. **59**, 59–101 (2017)
22. Waggoner, B., Weyl, E.G.: Matching markets via descending price. Available at SSRN 3373934 (2019)
23. Weitzman, M.L.: Optimal search for the best alternative. Econometrica **47**(3), 641–654 (1979)

Fair Division with Allocator's Preference

Xiaolin Bu[1], Zihao Li[2], Shengxin Liu[3(✉)], Jiaxin Song[1], and Biaoshuai Tao[1(✉)]

[1] Shanghai Jiao Tong University, Shanghai, China
{lin_bu,sjtu_xiaosong,bstao}@sjtu.edu.cn
[2] Nanyang Technological University, Singapore, Singapore
zihao004@e.ntu.edu.sg
[3] Harbin Institute of Technology, Shenzhen, Guangdong, China
sxliu@hit.edu.cn

Abstract. We consider the problem of fairly allocating indivisible resources to agents, which has been studied for years. Most previous work focuses on fairness and/or efficiency *among agents* given agents' preferences. However, besides the agents, the allocator as the resource owner may also be involved in many real-world scenarios (e.g., government resource allocation, heritage division, company personnel assignment, etc.). The allocator has the inclination to obtain a fair or efficient allocation based on her own preference over the items and to whom each item is allocated. In this paper, we propose a new model and focus on the following two problems concerning the allocator's fairness and efficiency:

1. Is it possible to find an allocation that is fair for both the agents and the allocator?
2. What is the complexity of maximizing the allocator's social welfare while satisfying the agents' fairness?

We consider the two fundamental fairness criteria: *envy-freeness* and *proportionality*. For the first problem, we study the existence of an allocation that is envy-free up to c goods (EF-c) or proportional up to c goods (PROP-c) from both the agents' and the allocator's perspectives, in which such an allocation is called *doubly EF-c* or *doubly PROP-c* respectively. When the allocator's utility depends exclusively on the items (but not to whom an item is allocated), we prove that a doubly EF-1 allocation always exists. For the general setting where the allocator has a preference over the items *and* to whom each item is allocated, we prove that a doubly EF-1 allocation always exists for two agents, a doubly PROP-2 allocation always exists for binary valuations, and a doubly PROP-$O(\log n)$ allocation always exists in general.

For the second problem, we provide various (in)approximability results in which the gaps between approximation and inapproximation ratios are asymptotically closed under most settings.

Most of our results are based on some novel technical tools including the chromatic numbers of the Kneser graphs and linear programming-based analysis.

Keywords: Fair Division · Allocator's Preference · EF-c/PORP-c

The full version of this paper is available on arXiv [11]. Due to the space limit, many technical details are omitted in this version.

© The Author(s), under exclusive license to Springer Nature Switzerland AG 2024
J. Garg et al. (Eds.): WINE 2023, LNCS 14413, pp. 77–94, 2024.
https://doi.org/10.1007/978-3-031-48974-7_5

1 Introduction

Fair division studies how to fairly allocate a set of resources to a set of agents with heterogeneous preferences. It is becoming a valuable instrument in solving real-world problems, e.g., Course Match for course allocation at the Wharton School [13], and the website Spliddit (spliddit.org) for fair division of rent, goods, credit, and so on [20]. The construct of fair division was first articulated by Steinhaus [41,42] in the 1940s, and has become an attractive topic of interest in wide range of fields, such as mathematics, economics, computer science, and so on (see, e.g., [1,2,9,29,35,37,38,44] for a survey).

The classic fair division problem mainly focuses on finding fair and/or efficient allocations *among agents given agents' preferences*. However, in many real-world scenarios, the allocator as the resource owner may also be involved, and, particularly, may have the inclination to obtain a fair or efficient allocation based on her own preference. For example, consider the division of inheritances, e.g., multiple companies and multiple houses, from the parent to two children. Both children would prefer the companies as they believe the market value of the companies will be increased more than the houses in the future. At the same time, the parent may want to allocate the companies to the elder child since the parent thinks the elder child has a better ability to run the companies. The final allocation should be fair for children and may also need to incorporate the parent's ideas about the allocation. Another example is the government distributing educational resources (e.g., land, funding, experienced teachers or principals) among different schools. Some well-established schools may prefer land to build a new campus, while some new schools may need experienced teachers. On the other hand, the government may also have a preference (over the resources and to whom each resource is allocated) based on macroeconomic policy and may want the resulting distribution to be efficient on top of each school feels that it gets a fair share. Other examples abound: a company allocates resources to multiple departments, an advisor allocates tasks/projects to students, a conference reviewer assignment system allocates papers to reviewers, etc.

We focus on the allocation of indivisible goods in this work. To measure fairness, the two most fundamental criteria in the literature are *envy-freeness* and *proportionality*, respectively [18,41,42,45]. In particular, an allocation is said to be envy-free if each agent weakly prefers her bundle over any other agent's based on her own preference, and proportional if each agent values her bundle at least $1/n$ of her value for the whole resources, where n is the number of agents. Both fairness criteria can always be achieved in divisible resource allocation but it is not the case for indivisible resources (say, a simple example with two agents and one good). This triggers an increasing number of research work to consider relaxing exact fairness notions of envy-freeness and proportionality to *envy-freeness up to c goods (EF-c)* and *proportionality up to c goods (PROP-c)* (see, e.g., [12,16,28]). Specifically, an allocation is said to be EF-c if any agent's envy towards another agent could be eliminated by (hypothetically) removing at most c goods in the latter's bundle, and PROP-c if any agent's fair share of $1/n$ could be guaranteed by (hypothetically) adding at most c goods that are allocated to other agents, where c is a positive integer. Besides fairness, another important issue of fair division is *(economic) efficiency* (e.g., social welfare), which is used to measure the total happiness of the agents [3,6,8,15].

The fair division problem with allocator's preference presents new challenges compared to the classic fair division problems. With indivisible goods, it is well known that the *round-robin algorithm*[1] can return a fair, i.e., EF-1, allocation from the agents' perspective. However, this algorithm cannot be easily adapted to the problem where both agents and the allocator have preferences over items. Specifically, an agent's preference describes how much this agent values each item, while the allocator's preference describes how much the allocator regards each item values for each agent. Consider the instance with both agents' and the allocator's preferences shown in Tables 1 and 2.

Table 1. Agents' Preferences

	Item 1	Item 2	Item 3
Agent 1	2	1	0
Agent 2	0	1	2

Table 2. Allocator's Preferences

	Item 1	Item 2	Item 3
Allocator for Agent 1	0	2	1
Allocator for Agent 2	1	2	0

Suppose, w.l.o.g, agent 1 is before agent 2 in the ordering of the round-robin algorithm. When performing the algorithm without considering the allocator's preference, agent 1 gets a bundle of items 1 and 2 while agent 2 gets item 3. From the allocator's perspective, this allocation is not EF-1 since the allocator thinks agent 2 will envy agent 1 even when an arbitrary item is removed from agent 1's bundle. One can also verify that the above allocation is not social welfare maximizing based on the allocator's viewpoint, i.e., the allocator thinks there is another allocation such that the total happiness of the agents is larger. On the other hand, performing the round-robin algorithm based solely on the allocator's preference will return an allocation where agent 1 gets items 2 and 3 while agent 2 gets item 1 (assuming agent 1 has a higher priority in the ordering). Specifically, we want to answer the following two questions in this paper.

Question 1: Is it possible to find an allocation that guarantees both the allocator's and agents' fairness?
Question 2: What is the complexity of maximizing the allocator's efficiency while ensuring agents' fairness?

1.1 Our Results

We initiate the study of fair division with allocator's preference and address the two research questions above in this paper. We focus on the allocation of *indivisible* resources and discuss the *divisible* resources in the full version of our paper [11].

---

[1] The round-robin algorithm works as follows: Given an ordering of agents, each agent picks her favorite item among the remaining items to her bundle following the ordering in rounds until there is no remaining item.

Table 3. Positive and Negative Results of Maximizing Allocator's Efficiency. The numbers of agents and items are denoted by n and m, respectively. For each agent i, v_i represents her utility function while u_i represents how much the allocator regards each item values for agent i. Numbers α for negative results indicate that the problem is NP-hard to approximate to within the ratio α; numbers α for positive results indicate that the problem admits a polynomial time α-approximation algorithm. All our negative results hold for the special case $c = 1$. The theorem statements and the proofs for these negative results are only available in the full version of this paper [11].

n	Fairness	v_i	u_i	Negative Results	Positive Results
2	EF-c	arbitrary	arbitrary	2	2 (Theorem 7)
	EF-c	arbitrary	binary	2	2 (Theorem 7)
	EF-c	binary	arbitrary	–	1 (Theorem 8)
constant	EF-c	arbitrary	binary	$\left\lfloor \frac{1+\sqrt{4n-3}}{2} \right\rfloor$ [10]	?
	EF-c	binary	arbitrary	–	1 (Theorem 8)
general	EF-c	binary	binary	$m^{1-\epsilon}, n^{1/2-\epsilon}$	m (Theorem 9)
	EF-c	arbitrary	arbitrary	$m^{1-\epsilon}, n^{1/2-\epsilon}$	m (Theorem 9)
	PROP-c	arbitrary	binary	2	?
	PROP-c	binary	arbitrary	–	1 (Theorem 10)

For the first problem, we propose new fairness notions *doubly EF-c* and *doubly PROP-c* that extend EF-c and PROP-c to our setting with regard to the allocator's preference. We first consider the setting where the allocator's utility only depends on the items (but not to whom an item is allocated), and we show that a doubly EF-1 allocation always exists. We then consider the general setting where the allocator's utility depends on both the items and the allocation. For two agents, we show that 1) a doubly EF-1 allocation always exists, and 2) a doubly EF-2 allocation and a doubly PROP-1 allocation can be computed in polynomial time. For a general number of agents, we show that a doubly PROP-$\log_2 n$ allocation always exists for n being an integer power of 2, and we show that a doubly PROP-$(2\lceil \log n \rceil)$ allocation always exists and can be computed in polynomial time. If we restrict to binary valuations, we show that a doubly PROP-2 allocation always exists and can be computed in polynomial time.

For the second problem, we study its complexity and approximability for both binary and general (additive) valuations. Our results are presented in Table 3. The gap between the approximation ratio and the inapproximability ratio is closed, or asymptotically closed, under most settings. If agents' valuations are binary, this problem is tractable for EF-c with a constant number of agents and for PROP-c with a general number of agents. Under most other settings, this problem admits strong inapproximability ratios even for $c = 1$.

Our results use many technical tools that are uncommon in the fair division literature, including i) the chromatic numbers of generalized Kneser graphs and ii) some linear programming-based analyses.

For i), we use a generalized Kneser graph to model a set of allocations and the relations between the allocations. Specifically, the set of allocations that are not fair based on an agent's valuation form an independent set in the graph. The existence of a doubly fair allocation is built upon the argument that there are still remaining vertices after removing all vertices that correspond to unfair allocations. Since the set of unfair allocations for each agent forms an independent set, the chromatic number of the graph plays an important role in our analysis.

For ii), we use linear programs to formulate our problems. The solution to the linear program naturally corresponds to a *fractional* allocation. Our technique is mainly based on the analysis of the vertices of the polytope defined by the linear program. In some applications, we bound the number of the fractional items in an allocation given by a vertex solution of the linear program, and then handle those few fractional items separately. In other applications, we prove that all the vertex solutions of the linear program are integral.

1.2 Further Related Work

Conceptually, our model with allocator's preference shares similarities with recent research work on fair division with two-sided fairness, e.g., [19,21,25,36]. The existing two-sided fairness literature studies the fair division problem where there are two disjoint groups of agents and each agent in one group has a preference over the agents of the other group. The objective is then to find a (many-to-many) matching that is fair to each agent with respect to her belonging group. We remark these two models are different due to the following major reasons:

- In their model, there are two disjoint sets of agents, and each group of preferences is defined from one set of agents to the other set of agents (viewed as a set of "goods"). On the other hand, the two groups of preferences (one is from the agents and the other one is from the allocator) in our setting are both defined on a single set of agents and a single set of goods.
- In their model, each agent will be allocated (or matched) a set of agents from the other group which is different from ours, whereas the allocator in our model will not receive any resource in the allocation.

As we can see, our model with allocator's preference reduces to the standard setting of indivisible goods when the allocator's preference coincides with agents' preferences. Our first research question reduces to find EF-c or PROP-c allocations in indivisible fair allocation, where the fairness notions of EF-1 and PROP-1 are extensively studied. In particular, an EF-1 allocation always exits and can be computed in polynomial time [14,28]. For PROP-1, an allocation that is PROP1 and Pareto optimal always exits and can be computed in polynomial time [4,7,16,34]. When considering the issue of economic efficiency, the problem in our second research question could be mapped to the problem of maximizing social welfare within either EF-1 or PROP-1 allocations in the indivisible goods setting. Aziz et al. [3] showed that the problem with either the EF-1 or the PROP-1 condition is NP-hard for $n \geq 2$ and Barman et al. [6] showed that the problem with the EF-1 requirement is NP-hard to approximate to within a factor of $1/m^{1-\varepsilon}$ for any $\varepsilon > 0$ for general numbers of agents n and items m. Later, Bu et al. [10] gave a complete landscape for the approximability of the problem with the EF-1 criterion.

Moreover, several works studied the fair division problem where the resources need to be allocated among *groups* of agents and the resources are shared among the agents within each predefined group [31,39,40,43]. In their model, $n = n_1 + \cdots + n_k$ agents will be divided into $k \geq 2$ groups, where group i contains $n_i \geq 1$ agents. An allocation is a partition of goods into k groups. Each agent in the i-th group extracts utilities according to the i-th bundle. Kyropoulou et al. [27] also generalized the classic EF-c to the group setting: An agent's envy towards another group could be eliminated by removing at most c goods from that group's bundle. PROP-c could be defined similarly [32]. With binary valuations, Kyropoulou et al. [27] gave the characterization of the cardinalities of the groups for which a group EF-1 allocation always exits. In particular, they showed that a group EF-1 allocation always exists when there are two groups and each group contains two agents with binary valuations. Later, Manurangsi and Suksompong [32] showed via the discrepancy theory that EF-$O(\sqrt{n})$ and PROP-$O(\sqrt{n})$ allocations always exist in the group setting. Note that, when each group contains exactly two agents, i.e., $n_1 = \ldots = n_k = 2$, the fair division problem in the predefined group setting coincides with our model (where each group could be considered to have an agent and the allocator). However, we obtain improved results in this particular setting through different technical tools.

2   Preliminaries

Let $[k] = \{1, \ldots, k\}$. Our model consists of a set of agents $N = [n]$, a set of indivisible items $M = \{g_1, \ldots, g_m\}$, and *the allocator*. Each agent i has a nonnegative *utility function* $v_i : \{0,1\}^m \to \mathbb{R}_{\geq 0}$. In addition, the allocator has her own preference in our model. The allocator's preference is composed by n utility functions $u_i : \{0,1\}^m \to \mathbb{R}_{\geq 0}$ where each u_i is used to describe how much the allocator regards each item values for agent i. We assume both utility functions u_i and v_i are *additive*, which means $v_i(X) = \sum_{g \in X} v_i(g)$ and $u_i(X) = \sum_{g \in X} u_i(g)$ for any bundle $X \subseteq M$. A utility function v_i (or u_i) is said to be *binary* if $v_i(g) \in \{0,1\}$ for any item $g \in M$. An *allocation* of the items $\mathcal{A} = (A_1, A_2, \ldots, A_n)$ is an ordered partition of M, where A_i is the bundle of items allocated to agent i.

An allocation $\mathcal{A}$ is said to be *envy-free up to c goods (EF-c)* if for all pairs of agents $i \neq j$, there exists a set $B \subseteq A_j$ such that $|B| \leq c$ and $v_i(A_i) \geq v_i(A_j \setminus B)$ (or $v_i(A_i) \geq v_i(A_j) - v_i(B)$ for additive utility functions). An allocation $\mathcal{A}$ is said to be *proportional up to c goods (PROP-c)* if for any agent i, there exists a set $B \subseteq M \setminus A_i$ such that $|B| \leq c$ and $v_i(A_i \cup B) \geq \frac{1}{n} v_i(M)$ (or $v_i(A_i) \geq \frac{1}{n} v_i(M) - v_i(B)$ for additive utility functions).

EF-c implies PROP-c for additive utility functions. An EF-1 (hence, PROP-1) allocation always exists and can be computed in polynomial time [14,28]. In our model, besides ensuring fairness among agents, we also consider allocator's fairness. Thus, we generalize the above fairness criteria in the following.

Definition 1 (Doubly Envy-free up to c goods). *An allocation $\mathcal{A}$ is said to be doubly envy-free up to c goods (Doubly EF-c) if for all pairs of agents $i \neq j$, there exist sets $B_1, B_2 \subseteq A_j$ such that $|B_1|, |B_2| \leq c$, $v_i(A_i) \geq v_i(A_j \setminus B_1)$, and $u_i(A_i) \geq u_i(A_j \setminus B_2)$.*

Definition 2 (Doubly Proportional up to c goods). *An allocation $\mathcal{A}$ is said to be doubly proportional up to c goods (Doubly PROP-c) if for any $i \in N$, there exist sets $B_1, B_2 \subseteq M \setminus A_i$ such that $|B_1|, |B_2| \leq c$, $v_i(A_i \cup B_1) \geq \frac{1}{n} v_i(M)$, and $u_i(A_i \cup B_2) \geq \frac{1}{n} u_i(M)$.*

When the allocator's utility functions are identical to agents' utility functions, it is easy to see that doubly EF-c and doubly PROP-c degenerate to EF-c and PROP-c respectively. The above defined double fairness notions with the allocator's preference can also be interpreted as: There are two groups of valuation functions u and v where one is from the agents and the other one is from the allocator. A single allocation is said to satisfy double fairness if such an allocation is fair, e.g., doubly EF-c/PROP-c, with respect to both valuation functions u and v.

To measure the economic efficiency for the allocator, we consider *allocator's efficiency*:

Definition 3. *The* allocator's efficiency *of an allocation $\mathcal{A} = (A_1, \ldots, A_n)$, denoted by* EFFICIENCY($\mathcal{A}$), *is the summation of the allocator's utilities of all the agents* EFFICIENCY($\mathcal{A}$) $= \sum_{i=1}^{n} u_i(A_i)$.

In this paper, we are interested in the following two problems.

Problem 1. Given a set of indivisible items M, a set of agents $N = [n]$ with their utility functions $(v_1, \ldots, v_n)$, and the allocator with her preference $(u_1, \ldots, u_n)$, determine whether there exists an allocation $\mathcal{A} = (A_1, \ldots, A_n)$ that is doubly EF-c/PROP-c.

Problem 2. Given a set of indivisible items M, a set of agents $N = [n]$ with their utility functions $(v_1, \ldots, v_n)$, and the allocator with her preference $(u_1, \ldots, u_n)$, the problem of *maximizing allocator's efficiency subject to EF-c/PROP-c* aims to find an allocation $\mathcal{A} = (A_1, \ldots, A_n)$ that maximizes allocator's efficiency EFFICIENCY subject to that $\mathcal{A}$ is EF-c/PROP-c.

2.1 Kneser Graph and Chromatic Number

Let n, k be two integers. The Kneser graph $\mathcal{K}(n, k)$ is the graph with the set of all the k-element subsets of $[n]$ as its vertex set and two vertices are adjacent if their intersection is empty. It was further extended to the following generalized version. Given three integers $n, k, s \in \mathbb{Z}^+$, in the generalized Kneser graph $\mathcal{K}(n, k, s)$, two vertices are adjacent if and only if their corresponding subsets intersect in s or fewer elements.

The *chromatic number* of a graph is the minimum number of colors needed to color the vertices such that no two adjacent vertices have the same color. In other words, the vertices with the same color form an independent set. We denote the chromatic number of a kneser graph $\mathcal{K}(n, k, s)$ by $\chi(n, k, s)$. For instance, when $n = 4, k = 3, s = 2$, the kneser graph has $\binom{4}{3} = 4$ vertices and every two vertices are adjacent. Thus, $\mathcal{K}(4, 3, 2)$ is a clique and $\chi(4, 3, 2) = 4$.

For the chromatic number of the Kenser graph, when $n \geq 2k$, it is exactly equal to $n - 2k + 2$ [5,22,30,33]. For the generalized Kneser graph, Jafari and Moghaddamzadeh [26] gave the following lower bounds.

Lemma 1 ([26]). *For any positive integers $s < k < n$,*

$$\chi(n, k, s) \geq n - 2k + 2s + 2.$$

Lemma 2 ([26]). *For any $k \in \mathbb{Z}^+ \geq 2$, $\chi(2k, k, 1) = 6$.*

2.2 Totally Unimodular Matrix and Linear Programming

Totally unimodular matrix is a special family of matrices that can be used to check whether a linear programming is *integral*, i.e., there exists one optimal solution such that all decision variables are integers.

Definition 4 (Totally Unimodular Matrix). *A matrix $\mathbf{A}_{m \times n}$ is a totally unimodular matrix (TUM) if every square submatrix of $\mathbf{A}$ has determinant 0, $+1$ or -1.*

To determine whether a matrix is TUM, we have the following lemma.

Lemma 3. *Given a matrix $\mathbf{A} \in \{0, \pm 1\}^{m \times n}$, $\mathbf{A}$ is TUM if it can be written as the form of $\begin{bmatrix} \mathbf{A}_1 \\ \mathbf{A}_2 \end{bmatrix}$, where $\mathbf{A}_1 \in \{0, 1\}^{r \times n}$ (or $\{0, -1\}^{r \times n}$), $\mathbf{A}_2 \in \{0, 1\}^{(m-r) \times n}$ (or $\{0, -1\}^{(m-r) \times n}$), $1 \leq r \leq m$ and there is at most one nonzero number in every column of $\mathbf{A}_1$ or $\mathbf{A}_2$.*

Lemma 4 ([24]). *If $\mathbf{A}$ is totally unimodular and $\mathbf{b}$ is an integer vector, then each vertex of the polytope $\{\mathbf{Ax} \leq \mathbf{b}, \mathbf{x} \geq \mathbf{0}\}$ has integer coordinates.*

We can further show there exist polynomial-time algorithms to find the optimal vertex solution for such a linear program by the following lemma.

Lemma 5 ([23]). *For a linear program $\max\{\mathbf{c}^\top \mathbf{x} : \mathbf{Ax} \leq \mathbf{b}, \mathbf{x} \geq \mathbf{0}\}$, if optimal solutions exist, an optimal vertex solution can be found in polynomial time. In particular, we can find an (integral) vertex of the polytope $\{\mathbf{Ax} \leq \mathbf{b}, \mathbf{x} \geq \mathbf{0}\}$ in polynomial time.*

3   Double Fairness

In this part, we present the results for the existence of double fairness. In the first part, we assume the allocator's utility depends exclusively on the item (rather than to whom an item is allocated). That is, we assume the allocator's utility functions are identical $u_1 = \cdots = u_n$. We show that a doubly EF-1 allocation always exists in this case by adapting the envy-cycle procedure. After that, we consider the general setting without $u_1 = \cdots = u_n$. In this case, we first show that a doubly EF-1 allocation always exists for $n = 2$ based on the chromatic number of the generalized Kneser graph $\mathcal{K}(m, m/2, 1)$. However, the existence of doubly EF-1 allocations for $n > 3$ is highly non-trivial. For this reason, we consider relaxing the fairness constraint to doubly EF-c or PROP-c and try to minimize the value of c. We show that a doubly PROP-$O(\log n)$ allocation always exists for any number of agents via the techniques based on the generalized Kneser graph and linear programming. Finally, we also consider another common setting, where both the allocator and agents' utility functions are binary (the utility value can only be 0 or 1). This relaxation makes the problem tractable and we demonstrate a doubly PROP-2 allocation always exists in this setting.

3.1 Identical Allocator's Utility Function

This section considers the case when the allocator's utility functions $u_1, \ldots, u_n$ are identical. Let $u = u_1 = \cdots = u_n$.

We first give a brief introduction of the techniques used in this section. The envy-cycle procedure was first proposed by [28] to compute an EF-1 allocation for general valuations. In the envy-cycle procedure, an *envy-graph* is constructed for a partial allocation. Each vertex in the envy-graph represents an agent and each directed edge (u, v) means that agent u envies agent v in the current allocation. When there is a cycle in an envy-graph, we use the cycle-elimination algorithm to eliminate this cycle.

Definition 5 (Cycle-elimination Algorithm). *Given an envy-graph with a cycle $u_1 \rightarrow \ldots \rightarrow u_n \rightarrow u_1$, shift the agents' bundles along the cycle ($A_{u_i} \leftarrow A_{u_{i+1}}$ for $i = 1, \ldots, n-1$ and $A_n \leftarrow A_1$).*

Theorem 1. *When the allocator's utility functions are identical, a doubly EF-1 allocation always exists for any number of agents n, and can be found by Algorithm 1 in polynomial time.*

Our algorithm is presented in Algorithm 1. At the beginning of the algorithm, we construct an envy-graph G with n vertices and no edges and sort the items according to the allocator's utility function in descending order. Then, we divide the sorted items into $\left\lceil \frac{m}{n} \right\rceil$ groups where each group contains n items. In each round, we allocate a group of items to the agents such that each agent receives exactly one item. This can guarantee the EF-1 property from the allocator's perspective. To allocate the group of n items to the n agents in each round, each agent takes away her favorite item from the group, where the agents are sorted in the topological order of G before the iteration begins. After all these n items are allocated, we update the envy-graph and run the cycle-elimination algorithm, so that the envy-graph contains no cycle and a topological order of the agents can be successfully found in the next round. By an induction-based argument showing the EF-1 property of the well-known round-robin algorithm, the EF-1 property of our algorithm in the agents' perspectives can be proved.

Algorithm 1: Finding Doubly EF-1 Allocation for Identical Utility Function

---

Input: the set of agents N, the set of items M, agents' utility functions v_i, allocator's utility function u

Output: a doubly EF-1 allocation

1   Let $G = (V, E)$ be the envy-graph where each vertex represents an agent and $E \leftarrow \emptyset$;

2   Initialize $\mathcal{A} = (\emptyset, \ldots, \emptyset)$;

3   **if** $n \nmid m$ **then**

4   Add dummy items to M such that $n \mid m$ and set the utility of each dummy item as 0;

5   Let M_s be the sorted array of the items according to allocator's utility function u in descending order;

6   **for** *every n items $M_n \subseteq M_s$* **do**

7   Let $\{i_1, \ldots, i_n\}$ be the agents in topological order of graph G;

8   **for** *each $j \in \{1, \ldots, n\}$* **do**

9   Allocate agent i_j's favorite item $g \in M_n$ to i_j: $A_{i_j} \leftarrow A_{i_j} \cup \{\text{argmax}_{g \in M_n} v_{i_j}(g)\}$;

10        $M_n \leftarrow M_n \setminus \{g\}$;

11   Update the envy-graph G;

12   Iteratively run the cycle-elimination algorithm and update G until G contains no cycle;

13   Remove the dummy items from the allocation $\mathcal{A}$;

14   **return** *the allocation $\mathcal{A}$*

3.2   General Additive Valuations with Two Agents

For general (monotone) valuations with two agents, the existence of a doubly EF-1 allocation can be proved with the help of the generalized Kneser graph.

Theorem 2. *When $n = 2$, there always exists a doubly EF-1 allocation.*

Proof. Our high-level idea is to consider the allocations that some agents or the allocator do not regard as EF-1. We then use the Kneser graph to demonstrate that the union of these allocations cannot cover the entire space of all possible allocations. We assume the number of items, m, is even. Otherwise, we can add a dummy item g such that $v_i(g) = u_i(g) = 0$. We denote the set of allocations where each bundle's size is exactly equal to $\frac{m}{2}$ by Π. For $i = 1, 2$, Let $\mathcal{V}_i$ be the set of allocations that agent i does not regard as EF-1. Besides, $\mathcal{U}_i$ represents the set of allocations that the allocator does not regard as EF-1 for agent i. Formally, they are given by the following formulas:

$$\mathcal{V}_i \triangleq \{\mathcal{A} \in \Pi : v_i(A_i) < v_i\left((M \backslash A_i) \backslash \{g\}\right), \forall g \in (M \backslash A_i)\},$$
$$\mathcal{U}_i \triangleq \{\mathcal{A} \in \Pi : u_i(A_i) < u_i\left((M \backslash A_i) \backslash \{g\}\right), \forall g \in (M \backslash A_i)\}.$$

For the existence of a doubly EF-1 allocation, it suffices to show that $\mathcal{V}_1 \cup \mathcal{V}_2 \cup \mathcal{U}_1 \cup \mathcal{U}_2 \subsetneq \Pi$. Let $\mathcal{V}_1^{(1)} = \{A_1 : (A_1, A_2) \in \mathcal{V}_1\}$, and define $\mathcal{V}_2^{(1)}, \mathcal{U}_1^{(1)}, \mathcal{U}_2^{(1)}$ analogously. We give the proposition for $\mathcal{V}_1^{(1)}$ below, which also works for the other three sets.

Proposition 1. *For each $A_1, A_1' \in \mathcal{V}_1^{(1)}$, $|A_1 \cap A_1'| \geq 2$.*

Proof. For the sake of contradiction, we assume $|A_1 \cap A_1'| \leq 1$. If $A_1 \cap A_1' = \emptyset$, (A_1, A_1') is a valid allocation. If (A_1, A_1') is not EF1 according to v_1, then (A_1', A_1) is envy-free, which means $A_1' \notin \mathcal{V}_1^{(1)}$.

If $|A_1 \cap A_1'| = 1$, let g_1 be the item in $A_1 \cap A_1'$ and g_2 be the only item in $M \setminus (A_1 \cup A_1')$. According to the definition of $\mathcal{V}_1$, we have

$$v_1(A_1) < v_1(M \setminus A_1) - v_1(g_2) = v_1(A_1') - v_1(g_1),$$
$$v_1(A_1') < v_1(M \setminus A_1') - v_1(g_2) = v_1(A_1) - v_1(g_1).$$

Combining the above two inequalities yields a contradiction. $\qquad\square$

Return to the proof of Theorem 2. We consider the generalized Kneser graph $\mathcal{H} = \mathcal{K}\left(m, \frac{m}{2}, 1\right)$. Each vertex of the graph defines a bundle B of $m/2$ items, and it defines an allocation (A_1, A_2) where $A_1 = B$ and $A_2 = M \setminus B$. Due to Proposition 1, each of $\mathcal{V}_1^{(1)}, \mathcal{V}_2^{(1)}, \mathcal{U}_1^{(1)}, \mathcal{U}_2^{(1)}$ cannot contain two adjacent vertices of $\mathcal{H}$ and is thus an independent set.

Finally, we have $\mathcal{V}_1 \cup \mathcal{V}_2 \cup \mathcal{U}_1 \cup \mathcal{U}_2 \subsetneq \Pi$. Otherwise, $\mathcal{H}$ can be decomposed into four independent sets, which contradicts to $\chi(\mathcal{H}) = 6$ (Lemma 2). $\qquad\square$

Remark 1. Theorem 2 also holds for general monotone utility functions that are not necessarily additive, with the same proof above.

Theorem 2 is non-constructive. For constructive results, we use linear programming to construct a doubly EF-2 allocation in Theorem 3.

Theorem 3. *When $n = 2$, there always exists a doubly EF-2 allocation that can be computed in polynomial time.*

Proof. For each item $g_j \in M$, we use one decision variable x_j to represent the fraction of item g_j allocated to group N_1. Consider the following linear program:

$$\max_{\mathbf{x}} \left(\sum_{j=1}^{m} v_1(g_j) x_j - \sum_{j=1}^{m} v_1(g_j)(1 - x_j) \right)$$

$$\text{subject to} \quad 0 \leq x_j \leq 1, \forall 1 \leq j \leq m \tag{1}$$

$$\sum_{j=1}^{m} u_1(g_j) x_j \geq \sum_{j=1}^{m} u_1(g_j)(1 - x_j) \tag{2}$$

$$\sum_{j=1}^{m} v_2(g_j) x_j \leq \sum_{j=1}^{m} v_2(g_j)(1 - x_j) \tag{3}$$

$$\sum_{j=1}^{m} u_2(g_j) x_j \leq \sum_{j=1}^{m} u_2(g_j)(1 - x_j) \tag{4}$$

Notice that a solution **x** with a non-negative objective provides a fractional doubly envy-free allocation. In addition, $\{x_j = 0.5\}_{j=1,\ldots,m}$ is such a solution.

Consider an optimal *vertex solution* $\mathbf{x}^*$. Lemma 5 implies such a solution can be found in polynomial time. Among all the constraints given in (1), (2), (3), and (4), a vertex solution is obtained by setting m of them being tight. Even if all of the three constraints (2), (3), and (4) are tight, at least $m - 3$ constraints are tight in (1). This implies at most three variables have value in the open interval $(0, 1)$. Let x_1^*, x_2^*, x_3^* be them without loss of generality. This implies we have a doubly envy-free allocation where only the first three items may be fractionally allocated. In the remaining part of the proof, we show that how to carefully decide the (integral) allocation of the first three items to make the allocation doubly EF-2.

Let (O_1, O_2) be the allocation of the remaining $m - 3$ items indicated by the LP solution $\mathbf{x}^*$. We consider the two cases of x_1^*, x_2^*, x_3^*.

Suppose at least two of them are no less than $\frac{1}{2}$, and assume $x_1^*, x_2^* \geq \frac{1}{2}$. Consider the allocation $(O_1 \cup \{g_1, g_2\}, O_2 \cup \{g_3\})$. For agent 1, for each $v \in \{u_1, v_1\}$, we have $v(A_1) + v(g_3) \geq \sum_{g_j \in M} v(g_j) x_j^* \geq v(A_2)$. For agent 2, for each $v \in \{u_2, v_2\}$, assume $v(g_1) \geq v(g_2)$, then we have

$$\begin{aligned}
v(A_2) + v(g_1) &= v(A_2) + (1 - x_1^*)\, v(g_1) + x_1^* v(g_1) \\
&\geq v(A_2) + (1 - x_1)\, v(g_1) + (1 - x_2^*) v(g_1) \\
&\geq v(A_2) + (1 - x_1^*)\, v(g_1) + (1 - x_2^*)\, v(g_2) \\
&\geq \sum_{g_j \in M} v(g_j)\left(1 - x_j^*\right) \geq \frac{1}{2} v(M) \geq v(A_1).
\end{aligned}$$

Suppose at least two of them are no more than $\frac{1}{2}$, and assume $x_1^*, x_2^* \leq \frac{1}{2}$. Similarly, we can also verify that $(O_1 \cup \{g_3\}, O_2 \cup \{g_1, g_2\})$ is doubly EF-2. □

It is easy to see that an EF-2 allocation is always PROP-1 for two agents.

Corollary 1. *When $n = 2$, there always exists a doubly PROP-1 allocation that can be computed in polynomial time.*

3.3   General Additive Valuations with General Number of Agents

Next, we consider the lower bound of c when $n \geq 2$. Our results are shown in the two theorems below.

Theorem 4. *For any $n = 2^k$, there always exists a doubly PROP-k allocation.*

Theorem 5. *For any $n \geq 2$, there always exists a doubly PROP-$\left(2\lceil \log n \rceil\right)$ allocation and it can be computed in polynomial time.*

The proofs are available in the full version of our paper [11]. Here we briefly describe the high-level ideas. Both theorems are based on the idea of Even-Paz algorithm [17]. Given n agents, we first partition the agent set into two groups and try to allocate each group one bundle. After that, we fix the two bundles to the two groups and then do further allocating within groups recursively. To guarantee the property of PROP-c, in each recursive iteration, we ensure that each agent and the allocator believe the agent's group receives approximately a $\frac{1}{2}$ fraction of the total value. To be specific, the value should be at least $\frac{1}{2}$ up to d items, where d depends on the level of the recursion tree.

The ideas in the proofs of both Theorem 2 and Theorem 3 can be generalized to achieve this. The idea of Kneser graphs can give a better fairness guarantee with a smaller value of c. However, it only works when agents can be partitioned into two equal-sized groups, and it is non-constructive. This yields Theorem 4. The idea of linear programming analysis gives a slightly worse fairness guarantee, but it is constructive and it works for any number of agents.

3.4   Binary Valuations

As shown in Theorem 5, for general additive valuation, when $n \geq 2$, doubly PROP-$O(\log n)$ allocations always exist. In this section, we further consider another common setting where the utility functions are binary. We show that a doubly PROP-2 allocation always exists and can be found in polynomial time for any $n \geq 2$ in Theorem 6. The advantage of this setting is that an agent i only needs to focus on the items whose values are regarded as 1 by $v_i(\cdot)$ or $u_i(\cdot)$.

Theorem 6. *When u_i, v_i are both binary for any $i \in N$, a doubly PROP-2 allocation always exists for any $n \geq 2$ and can be computed in polynomial time.*

Proof. For each agent $i \in N$, we define the following three item sets: $\mathcal{I}_i^{(1)} \triangleq \{g \in M : v_i(g) = 1 \wedge u_i(g) = 0\}$, $\mathcal{I}_i^{(2)} \triangleq \{g \in M : v_i(g) = u_i(g) = 1\}$, $\mathcal{I}_i^{(3)} \triangleq \{g \in M : u_i(g) = 1 \wedge v_i(g) = 0\}$. Then, we formulate this problem by a linear program. For each agent $i \in N$, we define a vector $\mathbf{x}_i = (x_{i,j})_{j \in [m]}$, where $x_{i,j}$ represents the fraction of item g_j allocated to agent i. Denote $(\mathbf{x}_1, \ldots, \mathbf{x}_n)$ by $\mathbf{x}$. Hence $\mathbf{x}$ is a vector with $n \times m$ variables. Consider the polytope $P = \{\mathbf{x} : \mathbf{A}\mathbf{x}^\top \leq \mathbf{b}, \mathbf{x} \geq \mathbf{0}\}$, where $\mathbf{A} \in \mathbb{R}^{(3n+m) \times (n \times m)}$ and $\mathbf{A}\mathbf{x}^\top \leq \mathbf{b}$ is decomposed into two parts:

- For each agent $i \in N$ and $k \in \{1, 2, 3\}$, $\sum_{j \in \mathcal{I}_i^{(k)}} x_{i,j} \geq \lfloor \frac{1}{n} \cdot |\mathcal{I}_i^{(k)}| \rfloor$.
- For each item $g_j \in M$, $\sum_{i \in N} x_{i,j} \leq 1$.

The second part says that a total amount of at most one unit can be allocated for each item j. The first part gives a sufficient condition for the allocation being PROP-2. Specifically, for each agent i, it implies $1 + \sum_{j \in \mathcal{I}_i^{(k)}} x_{i,j} \geq 1/n \cdot |\mathcal{I}_i^{(k)}|$ for $k = 1, 2, 3$. For $k = 1, 2$, this implies the allocation is PROP-2 with respect to v_i, $2 + \sum_{j \in \mathcal{I}_i^{(1)}} x_{i,j} + \sum_{j \in \mathcal{I}_i^{(2)}} x_{i,j} \geq 1/n \cdot (|\mathcal{I}_i^{(1)}| + |\mathcal{I}_i^{(2)}|)$; for $k = 2, 3$, this implies the allocation is PROP-2 with respect to u_i, $2 + \sum_{j \in \mathcal{I}_i^{(2)}} x_{i,j} + \sum_{j \in \mathcal{I}_i^{(3)}} x_{i,j} \geq 1/n \cdot (|\mathcal{I}_i^{(2)}| + |\mathcal{I}_i^{(3)}|)$.

Notice that $\mathbf{A}$ can also be written as the form $\begin{bmatrix} \mathbf{A}_1 \\ \mathbf{A}_2 \end{bmatrix}$, where $\mathbf{A}_1$ and $\mathbf{A}_2$ correspond to the two parts of the constraints. It is easy to verify that $\mathbf{A}_1$ is a matrix containing only 0 and -1 and $\mathbf{A}_2$ is a matrix containing only 0 and 1. Moreover, each column of $\mathbf{A}_1$ and $\mathbf{A}_2$ contains at most one non-zero entry. According to Lemma 3, $\begin{bmatrix} \mathbf{A}_1 \\ \mathbf{A}_2 \end{bmatrix}$ is TUM.

Since $\mathbf{x} = (x_{i,j}) = \left(\frac{1}{n}\right)$ is in the polytope, P is nonempty. In addition, since A is TUM and $\mathbf{b}$ is an integer vector, by Lemma 4, all vertices of P are integral. By Lemma 5, we can find a vertex $\mathbf{x}^* \in \{0,1\}^{n \times m}$ in polynomial time. Then for each agent $i \in N$, allocate the bundle $A_i = \{g_j \in M : x_{i,j}^* = 1\}$ to her.

Thus, by the definition of the above linear program,

$$v_i(A_i) \geq v_i\left(\mathcal{I}_i^{(1)}\right) + v_i\left(\mathcal{I}_i^{(2)}\right) \geq \frac{1}{n}\left|\mathcal{I}_i^{(k)}\right| - 1 + \frac{1}{n}\left|\mathcal{I}_i^{(k)}\right| - 1 = \frac{v_i(M)}{n} - 2.$$

Similarly, we can verify the above inequality for u_i. If there are no less than two items with value 1 outside A_i, this allocation is already PROP-2. Otherwise, if there is at most one item with value 1 outside A_i, then $v_i(M) \leq v_i(A_i) + 1$. It is easy to verify any bundle A_i satisfying this condition is PROP-2. □

4   Allocator's Efficiency

In this section, we consider the problem of maximizing allocator's efficiency subject to EF-c or PORP-c constraint for the agents. Other than general additive utility functions, we also consider the special case of binary utility functions. Note that we no longer consider the special case with identical allocator's utility $u_1 = \cdots = u_n$ since the problem becomes trivial otherwise (all allocations have the same allocator's efficiency).

All the negative results (corresponding to the fifth column in Table 3), including theorem statements and proofs, are available in the full version of our paper and are omitted here [11].

When the number of agents is 2, the problem is NP-hard to approximate to within a factor of 2 (see the full version), and this is matched with the following positive result.

Theorem 7. *The problem of maximizing allocator's efficiency subject to EF-c for two agents has a polynomial time 2-approximation algorithm when the agents' utility functions are arbitrary.*

Proof ((sketch)). Our algorithm is described as follows. We initialize two empty bundles S_1 and S_2, and sort the items according to agent 1's utility in descending order. Assume the sorted items are $\{g_1, \ldots, g_m\}$, and use $G_i (i \geq 1)$ to denote a group of two items $\{g_{2i-1}, g_{2i}\}$. For each group $G_i (i \geq 1)$, we allocate one item to each bundle. In particular, without loss of generality, we assume $v_2(S_1) \geq v_2(S_2)$ before allocating group G_i. Then, if $v_2(g_{2i-1}) \geq v_2(g_{2i})$, we allocate g_{2i-1} to S_2 and g_{2i} to S_1. Otherwise, we allocate g_{2i-1} to S_1 and g_{2i} to S_2. Notice that, in this algorithm, agent 1's utility function is used exclusively for the ordering of the item, and agent 2's utility function is used exclusively for deciding the allocation of the two items in each group.

After all the items are allocated, we consider the two allocations (S_1, S_2) and (S_2, S_1), and output the allocation with a higher allocator's efficiency.

It is not hard to check that the allocation is always EF-1 and is thus EF-c for any $c \geq 1$. To see the approximation guarantee of 2, notice that $u_1(M) + u_2(M)$ is a trivial upper bound to the allocator's efficiency, and the social welfare of the output allocation is at least half of it.   □

For a constant number of agents, we have an inapproximability factor of about $\sqrt{n}$ (see the full version) even when the allocator has binary utility functions. However, when the agents have binary valuations, the problem becomes tractable.

Theorem 8. *The problem of maximizing allocator's efficiency subject to EF-c for any fixed $n \geq 3$ can be found in polynomial time when the agents' utility functions are binary.*

Proof ((sketch)). We can adopt the dynamic programming used in the proof of Theorem 7.5 in [3] to prove this theorem. The details are available in the full version of our paper [11].   □

For general numbers of agents, even when both the agents and the allocator have binary valuations, the optimization problem admits an inapproximability factor of $m^{1-\epsilon}$ or $n^{1/2-\epsilon}$. This is complemented by the following positive result.

Theorem 9. *The problem of maximizing allocator's efficiency subject to EF-c has a m-approximation algorithm when both the agents' and the allocator's utility functions are arbitrary.*

Proof. Let the allocator allocates a single item to a single agent with the highest value $u_i(g_j)$ for $1 \leq i \leq n, 1 \leq j \leq m$ to agent i. Then the agents use the round-robin algorithm to allocate the remaining items. The allocation is EF-1 (and is thus EF-c) guaranteed by the round-robin algorithm and is a trivial m-approximation to the optimal allocator's efficiency.   □

Finally, we turn our attention to PROP-c. If the agents' utility functions are binary, we can use linear programming to prove the following result.

Theorem 10. *When agents' utility functions are binary, the problem of maximizing allocator's efficiency subject to PROP-c can be solved exactly in polynomial time by linear programming.*

Proof. The problem can be formulated as a linear program. Moreover, a careful analysis with the help of Lemma 3 reveals that the coefficient matrix is totally unimodular and all values in the constraints are integers. The theorem follows from Lemma 4 and Lemma 5.   □

5   Conclusion and Future Work

In this paper, we initialize the study of a new fair division model that incorporates the allocator's preference. We focused on the indivisible goods setting and mainly studied two research questions based on the allocator's preference: 1) How to find a doubly fair allocation? 2) What is the complexity of the problem of maximizing the allocator's efficiency subject to agents' fairness constraint?

We believe this new model is worth more future studies. For example, could we extend our results to the setting with more general valuation functions, e.g., submodular valuations? It is also an interesting (and challenging) problem to study what is the minimum number of c where a doubly EF-c/PROP-c allocation is guaranteed to exist. Indeed, we do not know any lower bound for c. Specifically, we do not know if a doubly EF-1/PROP-1 allocation exists even for binary valuations. We have searched for a non-existence counterexample with the aid of computer programs, and a non-existence counterexample seems hard to find.

On the other hand, our current techniques about Kneser graph and linear programming seem to have their limitations for further reducing the upper bound of c. Our current technique with Kneser graph can only analyze a bi-partition of the items with an equal size $m/2$ (this is crucial for Propositions 1 in the proof of Theorem 2 and the proof of Theorem 4). In addition, the value of the bundle must be exactly *half* of the total value up to the addition of c items. This is why we need n to be an integer power of 2 in Theorem 4. Moreover, the nature of Kneser graph-based analysis makes the existence proof non-constructive. Our linear programming technique, on the other hand, provides a weaker bound on c. It seems to us that a Kneser graph captures more structural insights about our problem than a linear program. Nevertheless, linear programming-based techniques provide a constructive existence proof.

It is fascinating to see these techniques to be further exploited and the above-mentioned limitations to be bypassed. Unearthing new techniques for closing the gap between the upper bound and the lower bound of c may also be necessary.

In our double fairness setting, we aim to find an allocation that is fair with respect to *two* valuation profiles $(u_1, \ldots, u_n)$ and $(v_1, \ldots, v_n)$, one for the agents and one for the allocator. A natural generalization of this is to consider allocations that are fair with respect to t valuation profiles for general t. The problem of fair division with more than two sets of valuations is also well-motivated in many applications (e.g., there may be more than one "allocator" in many scenarios, and an agent's valuation of the items may be multi-dimensional). We discuss the setting with multiple sets of valuations in the full version of our paper [11].

Acknowledgment. The research of Biaoshuai Tao was supported by the National Natural Science Foundation of China (No. 62102252). The research of Shengxin Liu was partially supported by the National Natural Science Foundation of China (No. 62102117), by the Shenzhen Science and Technology Program (No. RCBS20210609103900003), and by the Guangdong Basic and Applied Basic Research Foundation (No. 2023A1515011188), and by CCF-Huawei Populus Grove Fund (No. CCF-HuaweiLK2022005).

References

1. Amanatidis, G., et al.: Fair division of indivisible goods: recent progress and open questions. Artif. Intell. 103965 (2023)
2. Aziz, H.: Developments in multi-agent fair allocation. In: AAAI, pp. 13563–13568 (2020)
3. Aziz, H., Huang, X., Mattei, N., Segal-Halevi, E.: Computing welfare-maximizing fair allocations of indivisible goods. Eur. J. Oper. Res. **307**(2), 773–784 (2023)
4. Aziz, H., Moulin, H., Sandomirskiy, F.: A polynomial-time algorithm for computing a Pareto optimal and almost proportional allocation. Oper. Res. Lett. **48**(5), 573–578 (2020)
5. Bárány, I.: A short proof of Kneser's conjecture. J. Combin. Theory Ser. A **25**(3), 325–326 (1978)
6. Barman, S., Ghalme, G., Jain, S., Kulkarni, P., Narang, S.: Fair division of indivisible goods among strategic agents. In: AAMAS, pp. 1811–1813 (2019)
7. Barman, S., Krishnamurthy, S.K.: On the proximity of markets with integral equilibria. In: AAAI, pp. 1748–1755 (2019)
8. Brams, S.J., Feldman, M., Lai, J.K., Morgenstern, J., Procaccia, A.D.: On maxsum fair cake divisions. In: AAAI, pp. 1285–1291 (2012)
9. Brams, S.J., Taylor, A.D.: An envy-free cake division protocol. Am. Math. Mon. **102**(1), 9–18 (1995)
10. Bu, X., Li, Z., Liu, S., Song, J., Tao, B.: On the complexity of maximizing social welfare within fair allocations of indivisible goods. arXiv preprint arXiv:2205.14296 (2022)
11. Bu, X., Li, Z., Liu, S., Song, J., Tao, B.: Fair division with allocator's preference. arXiv preprint arXiv:2310.03475 (2023)
12. Budish, E.: The combinatorial assignment problem: approximate competitive equilibrium from equal incomes. J. Polit. Econ. **119**(6), 1061–1103 (2011)
13. Budish, E., Cachon, G.P., Kessler, J.B., Othman, A.: Course match: a large-scale implementation of approximate competitive equilibrium from equal incomes for combinatorial allocation. Oper. Res. **65**(2), 314–336 (2017)
14. Caragiannis, I., Kurokawa, D., Moulin, H., Procaccia, A.D., Shah, N., Wang, J.: The unreasonable fairness of maximum Nash welfare. ACM Trans. Econ. Comput. **7**(3), 1–32 (2019)
15. Cohler, Y.J., Lai, J.K., Parkes, D.C., Procaccia, A.D.: Optimal envy-free cake cutting. In: AAAI, pp. 626–631 (2011)
16. Conitzer, V., Freeman, R., Shah, N.: Fair public decision making. In: EC, pp. 629–646 (2017)
17. Even, S., Paz, A.: A note on cake cutting. Discret. Appl. Math. **7**(3), 285–296 (1984)
18. Foley, D.K.: Resource allocation and the public sector. Yale Econ. Essays **7**(1), 45–98 (1967)
19. Freeman, R., Micha, E., Shah, N.: Two-sided matching meets fair division. In: IJCAI, pp. 203–209 (2021)
20. Goldman, J., Procaccia, A.D.: Spliddit: unleashing fair division algorithms. ACM SIGecom Exchanges **13**(2), 41–46 (2015)
21. Gollapudi, S., Kollias, K., Plaut, B.: Almost envy-free repeated matching in two-sided markets. In: Chen, X., Gravin, N., Hoefer, M., Mehta, R. (eds.) WINE 2020. LNCS, vol. 12495, pp. 3–16. Springer, Cham (2020). https://doi.org/10.1007/978-3-030-64946-3_1

22. Greene, J.E.: A new short proof of Kneser's conjecture. Am. Math. Mon. **109**(10), 918–920 (2002)
23. Güler, O., den Hertog, D., Roos, C., Terlaky, T., Tsuchiya, T.: Degeneracy in interior point methods for linear programming: a survey. Ann. Oper. Res. **46**(1), 107–138 (1993)
24. Hoffman, A.J., Kruskal, J.B.: Integral boundary points of convex polyhedra. In: Jünger, M., et al. (eds.) 50 Years of Integer Programming 1958-2008, pp. 49–76. Springer, Heidelberg (2010). https://doi.org/10.1007/978-3-540-68279-0_3
25. Igarashi, A., Kawase, Y., Suksompong, W., Sumita, H.: Fair division with two-sided preferences. In: IJCAI, pp. 2756–2764 (2023)
26. Jafari, A., Moghaddamzadeh, M.J.: On the chromatic number of generalized Kneser graphs and Hadamard matrices. Discret. Math. **343**(2), 111682 (2020)
27. Kyropoulou, M., Suksompong, W., Voudouris, A.A.: Almost envy-freeness in group resource allocation. Theoret. Comput. Sci. **841**, 110–123 (2020)
28. Lipton, R., Markakis, E., Mossel, E., Saberi, A.: On approximately fair allocations of indivisible goods. In: EC, pp. 125–131 (2004)
29. Liu, S., Lu, X., Suzuki, M., Walsh, T.: Mixed fair division: A survey. arXiv preprint arXiv:2306.09564 (2023)
30. Lovász, L.: Kneser's conjecture, chromatic number, and homotopy. J. Combin. Theory Ser. A **25**(3), 319–324 (1978)
31. Manurangsi, P., Suksompong, W.: Asymptotic existence of fair divisions for groups. Math. Soc. Sci. **89**, 100–108 (2017)
32. Manurangsi, P., Suksompong, W.: Almost envy-freeness for groups: Improved bounds via discrepancy theory. Theoret. Comput. Sci. **930**, 179–195 (2022)
33. Matoušek, J.: A combinatorial proof of Kneser's conjecture. Combinatorica **24**(1), 163–170 (2004)
34. McGlaughlin, P., Garg, J.: Improving Nash social welfare approximations. J. Artif. Intell. Res. **68**, 225–245 (2020)
35. Moulin, H.: Fair division in the Internet age. Annu. Rev. Econ. **11**(1), 407–441 (2019)
36. Patro, G.K., Biswas, A., Ganguly, N., Gummadi, K.P., Chakraborty, A.: FairRec: two-sided fairness for personalized recommendations in two-sided platforms. In: WWW, pp. 1194–1204 (2020)
37. Procaccia, A.D.: Cake cutting: not just child's play. Commun. ACM **56**(7), 78–87 (2013)
38. Robertson, J., Webb, W.: Cake-Cutting Algorithm: Be Fair If You Can. A K Peters/CRC Press (1998)
39. Segal-Halevi, E., Nitzan, S.: Fair cake-cutting among families. Soc. Choice Welfare **53**(4), 709–740 (2019)
40. Segal-Halevi, E., Suksompong, W.: Democratic fair allocation of indivisible goods. Artif. Intell. **277**, 103167 (2019)
41. Steinhaus, H.: The problem of fair division. Econometrica **16**(1), 101–104 (1948)
42. Steinhaus, H.: Sur la division pragmatique. Econometrica **17**, 315–319 (1949)
43. Suksompong, W.: Approximate maximin shares for groups of agents. Math. Soc. Sci. **92**, 40–47 (2018)
44. Suksompong, W.: Constraints in fair division. ACM SIGecom Exchanges **19**(2), 46–61 (2021)
45. Varian, H.R.: Equity, envy, and efficiency. J. Econ. Theory **9**(1), 63–91 (1974)

Optimal Stopping with Multi-dimensional Comparative Loss Aversion

Linda Cai[✉], Joshua Gardner, and S. Matthew Weinberg

Princeton University, Princeton, NJ 08544, USA
tcai@princeton.edu

Abstract. Motivated by behavioral biases in human decision makers, recent work by [11] explores the effects of loss aversion and reference dependence on the prophet inequality problem, where an online decision maker sees candidates one by one in sequence and must decide immediately whether to select the current candidate or forego it and lose it forever. In their model, the online decision-maker forms a reference point equal to the best candidate previously rejected, and the decision-maker suffers from loss aversion based on the quality of their reference point, and a parameter λ that quantifies their loss aversion. We consider the same prophet inequality setup, but with candidates that have *multiple features*. The decision maker still forms a reference point, and still suffers loss aversion in comparison to their reference point as a function of λ, but now their reference point is a (hypothetical) combination of the best candidate seen so far *in each feature*.

Despite having the same basic prophet inequality setup and model of loss aversion, conclusions in our multi-dimensional model differs considerably from the one-dimensional model of [11]. For example, [11] gives a tight closed-form on the competitive ratio that an online decision-maker can achieve as a function of λ, for any $\lambda \geq 0$. In our multi-dimensional model, there is a sharp phase transition: if k denotes the number of dimensions, then when $\lambda \cdot (k-1) \geq 1$, *no non-trivial competitive ratio is possible*. On the other hand, when $\lambda \cdot (k-1) < 1$, we give a tight bound on the achievable competitive ratio (similar to [11]). As another example, [11] uncovers an exponential improvement in their competitive ratio for the random-order vs. worst-case prophet inequality problem. In our model with $k \geq 2$ dimensions, the gap is at most a constant-factor. We uncover several additional key differences in the multi- and single-dimensional models.

Keywords: Optimal Stopping · Reference Dependence · Loss Aversion

1 Introduction

Reference dependence describes the tendency for human decision-makers to compare options against a previously seen or anticipated reference point, such that an option inferior to the reference point is perceived as being of a lower quality

© The Author(s), under exclusive license to Springer Nature Switzerland AG 2024
J. Garg et al. (Eds.): WINE 2023, LNCS 14413, pp. 95–112, 2024.
https://doi.org/10.1007/978-3-031-48974-7_6

than it would be had the person simply not been aware of the reference point [22]. Similarly, loss aversion describes the tendency of human decision-makers to be more sensitive to losses than to gains of equivalent magnitude. Both phenomena have robust empirical support [10], and have been observed in the context of online decision-making [19].

Recent work of [11], has considered these behavioral biases in the context of optimal stopping, constructing biased agents who suffer from loss aversion in comparing their chosen candidate against previously seen and foregone candidates. Prophet inequalities are a canonical optimal stopping problem, where an agent sees n candidates, each with a scalar value drawn independently from known distributions. The candidates are presented online, such that whenever the agent sees a candidate they must either accept it, halting their search, or forego it and thus lose the candidate forever. In the event they select a candidate, the utility they receive is exactly the value of the candidate. Motivated by the role of reference points and loss aversion, [11] pose a modified prophet inequality where the utility received by the agent is discounted based on their reference point. Specifically, the biased agent also remembers the best candidate they have previously seen, and perceives new candidates as having decreased utility if the best candidate previously seen had a higher value.

The elegant model of [11] captures scenarios in which candidates are well-represented by a single scalar (such as monetary value). Indeed, their model is partially motivated by the empirical findings of [19] concerning the unwillingness of investors to sell stocks below the price at which they were purchased. Similarly, their model captures the phenomenon of biased decision-makers stopping much sooner than optimal, once again mirroring the empirical results of [19]. Intuitively, due to the anticipation of experiencing regret over foregone candidates, biased decision-makers are likely to settle earlier than they should. In this respect, the theoretical model of [11] reflects human behavior relatively well, at least in the case of decisions involving scalar-valued candidates.

However, their model does not capture decisions in which candidates have multiple features of interest. Notably, many key decisions in life including career, romantic partner, or house typically have many dimensions. Moreover, even simple consumer decisions such as selecting which movie to watch or magazine subscription to purchase often have multiple dimensions. Experimental research by [2] suggests that human decision makers suffer from "comparative loss aversion" in which they compare options along multiple dimensions, and perceive disadvantages as being more significant than advantages due to loss aversion, thus decreasing received utility from the option chosen. Consider the case of selecting a house: suppose that house A is the most aesthetically pleasing, house B is in the best location, and house C has the most space. Regardless of which house a person selects, due to comparative loss aversion, the mere awareness of the other options decreases the satisfaction they experience. Empirical research [3, 8, 14] on the phenomenon of "choice overload" (where people become less satisfied with their choice when they are presented with a larger set of options) further supports the existence of comparative loss aversion.

One interpretation of this phenomenon that has been popularized in psychological literature like *The Paradox of Choice* [20] is that people combine the best features of options they have previously seen into an imagined alternative of superior quality. Later empirical work by [17] found that "regret is related to the comparison between the alternative chosen and the union of the positive attributes of the alternatives rejected". Intuitively, compared to an imagined alternative combining the best attributes seen, the option chosen is much less satisfying.

In our work, we model this phenomenon by associating each candidate i with a k-*dimensional vector* $\boldsymbol{\sigma}^{(i)} = \left(\sigma_1^{(i)}, \cdots, \sigma_k^{(i)}\right)$ rather than a single scalar. The agent's value for a single candidate is just the sum over the agent's value for the candidate on each dimension, namely, $\|\boldsymbol{\sigma}^{(i)}\|_1$. So for an unbiased decision-maker, this is still just an instance of the classic prophet inequality. As in [11], our agent is biased, and maintains a reference point $\boldsymbol{s}^{(i)}$ at time i based on options seen so far. The agent suffers from loss aversion parameterized by λ, and experiences (and anticipates) a loss equal to $\lambda \cdot \left(\|\boldsymbol{\sigma}^{(i)}\|_1 - \|\boldsymbol{s}^{(i)}\|_1\right)$ upon selecting candidate i. The key distinction between our model and that of [11] is in how the reference point is built. In [11], the reference point is simply the vector of a candidate $j \leq i$ maximizing the overall value $\|\boldsymbol{\sigma}^{(j)}\|_1$. In our model, the reference point is instead a hypothetical super-candidate that combines the best candidate seen so far for each feature. Formally, for each dimension ℓ, the value of the reference point is $s_\ell^{(i)} = \max_{j \leq i}\{\sigma_\ell^j\}$. As previously discussed, this enables our model to better capture the multi-dimensional problem of purchasing a house, versus the single-dimensional problem of buying/selling stocks that are well-captured by the model of [11]. To reiterate: our model is identical to that of [11], *except* for the choice of reference point. Surprisingly, this single change from a single-dimensional reference point to a multi-dimensional reference point drastically changes conclusions in the model.

1.1   Summary of Results

Our main findings are that an agent who suffers from comparative loss aversion (i.e. our model, the multi-dimensional case) suffers in a fundamentally different manner than an agent who only suffers from reference-dependent loss aversion (i.e. [11], the single-dimensional case). We now overview several results supporting this. In describing our results, "gambler" refers to the online decision-maker in the prophet inequality setting, and "prophet" refers to the offline decision-maker (who knows all options and simply picks the best).

Unbounded Loss. Perhaps our most surprising finding is a phase transition in λ as a function of k when comparing the utility of a biased gambler to that of an unbiased gambler or unbiased prophet. Specifically, [11] show that a biased gambler can always guarantee at least a $\frac{1}{2+\lambda}$-approximation to the expected utility achieved by an unbiased prophet, and a $\frac{1}{1+\lambda}$-approximation to that of an unbiased gambler (and both of these bounds are tight). When $\lambda \cdot (k - 1) < 1$, we find a natural generalization: the biased gambler can guarantee a $\frac{1-\lambda(k-1)}{2+\lambda}$-approximation to the unbiased prophet, and a $\frac{1-\lambda(k-1)}{1+\lambda}$-approximation to the unbiased gambler, and these are tight for all k (including $k = 1$, where it matches [11]). However, there is a phase transition at $\lambda \cdot (k - 1) = 1$: *for any $\lambda \cdot (k - 1) \geq 1$, and any C, there exist instances where the biased gambler achieves expected utility 1, but the unbiased gambler (and unbiased prophet) achieve expected utility at least C.* That is, there is no non-trivial guarantee on the competitive ratio as a function only of λ and k. Interestingly, this phase transition occurs for all $k > 1$, but not for $k = 1$. This marks a fundamental difference between online decision-making with comparative loss aversion versus only reference-dependent loss aversion.

Given the importance of the phase transition, we will call $\lambda \cdot (k - 1)$ the *feature-amplified bias*. Intuitively, feature-amplified bias measures how fast the gap between the utilities of unbiased and biased agents grows with the number of candidates. As we will see in an illustrative example in Sect. 3, when the feature-amplified bias is greater than one, the ratio between utility of an unbiased prophet and that of a biased gambler grows exponentially with the number of candidates. The formal statements of our main results when feature-amplified bias is less than one can be found in Theorems 2, 3 and 4. For the setting where feature-amplified bias is at least one, the main results can be found in Theorem 1.

No Improvement for Random Order, Optimal Order, and I.I.D. Distributions. One key finding of [11] is that the competitive ratio of a biased gambler compared to an unbiased gambler/prophet drastically improves (the loss is logarithmic in λ instead of linear in λ) when the candidates are revealed in random order. This of course holds in our model when $k = 1$ (because it is identical to [11]). When $k > 1$, however, this phenomenon no longer occurs. Specifically, for all $k \geq 2$, *even if the instance is i.i.d.*: (a) if feature-amplified bias $\lambda \cdot (k - 1) \geq 1$, then the expected reward of the biased gambler and unbiased prophet/gambler can be unbounded, and (b) when feature-amplified bias $\lambda \cdot (k - 1) < 1$, the worst-case ratio between the biased gambler and unbiased prophet/gambler is within a constant-factor of the worst-case ratio without the i.i.d. assumption.[1] The formal statements of our results for the i.i.d setting can be found in Theorems 5 and 6.

1.2 Related Work

Since its introduction, optimal stopping has been pertinent to the study of economic behaviors with immediate applications in a wide range of economic domains such as search theory, financial trading, auction design, etc. Various models have been proposed based on the optimal stopping paradigm including

---

[1] Of course, we do not nail down the tight competitive ratio for the gambler in the i.i.d. setting, because even the tight ratio between the gambler and prophet without loss aversion is quite involved. The tight competitive ratio between the unbiased gambler and unbiased prophet for the random order (but not i.i.d.) setting is still unknown.

prophet inequalities (value distribution known), secretary problems (value distribution unknown) and Pandora's box problem (where a cost is associated with inspection of options).

The model that is most relevant to our present work is that of the prophet inequality, where the agents are given the distribution of their value for the candidates a priori. In the seminal work of [13], it was shown that for prophet inequalities where the agent is limited to picking a single candidate, the ratio between the expected largest value (namely, the prophet's pick) and the expected value of the candidate an agent playing optimally selects online is at most 2. Furthermore, this ratio is tight. [18] further show that by using a threshold based algorithm, the online agent can also guarantee that the expected value of the candidate they select is a 1/2-approximation of the expected optimal value over all candidates. Prophet inequalities have since been further studied under a variety of feasibility constraints on the set of selected candidates [5,12,15,16], and a variety of arrival order constraints [1,4,6,7,9,23]).

Although optimal stopping may have direct applications in human decision making, very little research has devoted to behavioral models of optimal stopping. Recently, [11] constructed a model which factors in reference dependence and loss aversion, however to our knowledge there is no other prior work exploring behavioral phenomena in the context of optimal stopping.

1.3   Organization

The rest of the paper will be organized as follows. In Sect. 2, we discuss our model that incorporates comparative loss aversion and introduce notations that will be used throughout the paper. In Sect. 3, we present an illustrative example that highlights the difference between our model and the single dimensional model presented in [11]. In Sect. 4, we analyze our model under various ordering constraints, with a focus on the gap between the utility of the biased *gambler* and that of rational agents. In Sect. 5 we discuss the limitations of our model and possible directions for future work.

2   Model and Preliminary

2.1   Modeling Comparative Loss Aversion

Multidimensional Optimal Stopping. We now construct a multidimensional model of the optimal stopping problem such that candidates have multiple dimensions of value. First, we must formally define the multidimensional optimal stopping problem, using slight modifications of the classic definition of the problem in one dimension. Specifically, let there be n candidates each of which has k dimensions of value that are of interest to a decision making agent, i.e. there are n vector valued candidates, such that their value vectors are k-dimensional. Note that the use of vector values differs from traditional constructions of the problem in which scalar values are used, i.e. in the traditional problem k is 1.

We will use $\mathcal{F}_t$ with support on $\mathbb{R}_+^k$ to denote the prior distribution of the gambler's value vector for the tth candidate. Let σ be an ordered sequence of n candidates, where the tth candidate has a k-dimensional vector value $\sigma^{(t)} \in \mathbb{R}^k$ drawn from $\mathcal{F}_t$. Formally, we will use $\mathcal{F} = \times_{t \in [n]} \mathcal{F}_t$ to denote the joint distribution of the value for all candidates, such that σ is drawn from $\mathcal{F}$. Moreover, let it be the case that the agent sees the realized value of $\sigma^{(t)}$ only when they reach step t in the sequence. Upon reaching the tth candidate, $\sigma^{(t)}$, the agent must either accept them - thus halting their search - or decline them, continuing their search and losing the candidate forever. The agent attempts to select $\sigma^{(t)}$ to maximize their utility. We now need to reason about what an agent's utility function should look like.

The Utility of Agents with Comparative Loss Aversion. In keeping with the notation of [11], let the parameter $\lambda \geq 0$ serve as a multiplier representing our agent's loss aversion. In particular, let λ be used to amplify the difference between the chosen candidate and the "reference candidate" along any dimensions for which the reference candidate is of higher value. Observe that in the case of our model, reasoning about the reference candidate is somewhat complex since an agent with comparative loss aversion compares each option against all previously seen along different dimensions independently. Regardless, it is without loss of generality to collapse the highest values previously seen along each dimension into single vector and consider this to be the reference candidate.

Accordingly, our agent remembers the best value previously seen along each dimension and combines these to form their reference candidate, which we refer to as the "super candidate", since is an upperbound on the quality of all previously seen candidates and combines all of their best features. Observe that this super candidate changes over time as we see new candidates that are better than all previously seen along some dimension. Due to the agent's reference dependence, when the super candidate is better than the agent's chosen candidate on any dimension, this detracts from the utility they receive. Let us denote the super candidate at step t in the sequence as $s^{(t)}$ and the jth entry of the super candidate at step t as $s_j^{(t)}$. We define the super candidate in step t element-wise as follows:

$$s_j^{(t)} = \max_{w : w \in [t]} (\sigma_j^{(w)})$$

We define the utility received by a biased gambler, who suffers from loss aversion λ and who selects the tth candidate in sequence σ, as follows:

$$U_{g_b}(\sigma, t) = \sum_j \left(\sigma_j^{(t)} - \lambda(s_j^{(t)} - \sigma_j^{(t)}) \right) = \|\sigma^{(t)}\|_1 - \lambda \left(\|s^{(t)}\|_1 - \|\sigma^{(t)}\|_1 \right),$$

and we define the difference between the value selected by the biased gambler and their utility as the biased gambler's *regret*.

Recall that, by construction, we have already restricted all values to be nonnegative, allowing us to make use of L1 norms. Moreover, let's compare this to the utility received by biased prophet who selects option t:

$$U_{p_b}(\sigma, t) = \sum_j \left(\sigma_j^{(t)} - \lambda(s_j^{(n)} - \sigma_j^{(t)}) \right) = \|\sigma^{(t)}\|_1 - \lambda\left(\|s^{(n)}\|_1 - \|\sigma^{(t)}\|_1 \right)$$

Note that the only difference between the utility of the biased gambler and biased prophet is that the super option of the biased prophet is constructed from all n candidates, whereas that of the biased gambler is constructed only from the t candidates it has seen so far.

Next let us reason about the utilities received by a rational gambler and rational prophet, respectively. In particular, observe that since they do not suffer from any reference dependence or loss aversion, their payoff is just equal to the value of the option they select:

$$U_{p_r}(\sigma, t) = U_{g_r}(\sigma, t) = \sum_j \sigma_j^{(t)} = \|\sigma^{(t)}\|_1$$

Furthermore, assuming that our agents seek to maximize their utilities, the payoff of a rational prophet does not need parameter t since it selects t deterministically to maximize realized utility. Namely, it suffices to denote the utility of a rational prophet as follows:

$$U_{p_r}(\sigma) = \max_{t:t\in[n]} (U_{p_r}(\sigma, t))$$

Similarly, since the biased prophet sees all options and selects t deterministically to maximize realized utility, we can remove the parameter t here as well:

$$U_{p_b}(\sigma) = \max_{t:t\in[n]} (U_{p_b}(\sigma, t))$$

Now that the model has been developed, we can proceed to begin analyzing it. Next we will develop notations to describe the expected utility of biased and rational agents, and their relative differences. These notations will be used throughout the paper.

2.2 Preliminaries

In keeping with standard conventions, we use $\mathbb{E}_{\mathcal{F}}$ to denote the expectation with respect to the distribution $\mathcal{F}$. To abbreviate notation, we will use $\mathbb{E}_{\mathcal{F}}[U_{g_b}^*]$ and $\mathbb{E}_{\mathcal{F}}[U_{g_r}^*]$ to denote the expected utility of the biased gambler and a rational gambler, respectively, when they use a utility optimal online algorithm. Similarly, we will use $\mathbb{E}_{\mathcal{F}}[U_{p_b}]$ and $\mathbb{E}_{\mathcal{F}}[U_{p_r}]$ to denote the expected utility of the biased prophet and rational prophet, respectively. In addition, we will use V^* to denote the maximum value of all candidates, namely, $V^* = \max_{t\in[n]} \|\sigma^{(t)}\|_1$. We will use S_j^* to the denote the maximum value on the jth dimension, namely, $S_j^* = \max_{t\in[n]} \sigma_j^{(t)}$.

Next, we will create several definitions to describe the relationship between utility of the biased gambler and the utility of rational agents, which will be frequently referenced in subsequent sections. Informally, the prophet/online utility ratio describes how well the biased gambler can do compared to the rational prophet/gambler *given a specific prior distribution $\mathcal{F}$*. On the other hand, the prophet/online competitive ratio describes how well the biased gambler can do compared to the rational prophet/gambler *given the most adversarial prior distribution $\mathcal{F}$*.

Definition 1 (Prophet Utility Ratio). *We will use the term prophet utility ratio to denote the ratio between the expected utility of a rational prophet and the expected optimal utility from a biased gambler under the prior distribution $\mathcal{F}$. More formally, we define the prophet utility ratio as $\frac{\mathbb{E}_{\mathcal{F}}[U_{pr}]}{\mathbb{E}_{\mathcal{F}}[U_{g_b}^*]}$.*

Note that the Prophet Utility Ratio captures the loss that the biased gambler experiences *both due to bias and due to making decisions online*.

Definition 2 (Online Utility Ratio). *We will use the term online utility ratio to denote the ratio between the expected optimal utility of a rational gambler and the expected optimal utility from a biased gambler under the prior distribution is $\mathcal{F}$. Formally, we define the online utility ratio as $\frac{\mathbb{E}_{\mathcal{F}}[U_{g_r}^*]}{\mathbb{E}_{\mathcal{F}}[U_{g_b}^*]}$.*

Note that the Online Utility Ratio captures the loss that the biased gambler experiences *just due to bias*.

Definition 3 (Prophet/Online Competitive Ratio). *Given a setting (such as adversarial arrival order or i.i.d prior distribution), we define the terms prophet competitive ratio and online competitive ratio as the maximum prophet and online utility ratio possible in the setting. Specifically, in the adversarial arrival order setting, the prophet and online competitive ratio are formally defined as $\max_{\mathcal{F}} \frac{\mathbb{E}_{\mathcal{F}}[U_{pr}]}{\mathbb{E}_{\mathcal{F}}[U_{g_b}^*]}$ and $\max_{\mathcal{F}} \frac{\mathbb{E}_{\mathcal{F}}[U_{g_*}]}{\mathbb{E}_{\mathcal{F}}[U_{g_b}^*]}$, respectively. In the i.i.d prior distribution setting, we will denote $\mathcal{U}$ as the set of all i.i.d distributions. The prophet and online competitive ratio in the i.i.d prior distribution setting are then formally defined as $\max_{\mathcal{F} \in \mathcal{U}} \frac{\mathbb{E}_{\mathcal{F}}[U_{pr}]}{\mathbb{E}_{\mathcal{F}}[U_{g_b}^*]}$ and $\max_{\mathcal{F} \in \mathcal{U}} \frac{\mathbb{E}_{\mathcal{F}}[U_{g_r}^*]}{\mathbb{E}_{\mathcal{F}}[U_{g_b}^*]}$.*

As stated in the introduction, we define $\lambda \cdot (k-1)$ to be the *feature-amplified bias*. We will soon see that whether the feature-amplified bias $\lambda \cdot (k-1)$ is greater or less than one is critical in determining whether the biased gambler has a non trivial prophet/online competitive ratio.

Finally, the following claim that directly follows from classical prophet inequality results will be referenced and used throughout proofs in the rest of the paper.

Proposition 1. *For any prior distribution $\mathcal{F}$ and any assumption about candidate arrival order, $\mathbb{E}_{\mathcal{F}}[U_{g_r}^*] \geq \frac{1}{2} \cdot \mathbb{E}_{\mathcal{F}}[U_{pr}]$.*

Proof. In our model, rational agents' utilities are simply their value for the option they select. Therefore our claim follows directly from the classical prophet inequality result [12,13] that the expected value from a value-optimal online algorithm is a $\frac{1}{2}$ approximation of the expected value from an offline prophet.

3   Unbounded Loss: A Motivating Example

Prior to proving more general claims, we will first show an adversarial example that illustrates a fundamental difference between our multi-dimensional optimal stopping problem with biased agents and the single dimensional model. Specifically, in the single dimensional model, the biased gambler attain worse utility than rational agents only when there is *uncertainty*. (Imagine if the biased agent already knows all values of the candidates, they can simply pick the one with the largest value and suffers no regret). However, in our multi-dimensional model, we will show an example where the biased gambler attains arbitrarily worse utility than rational agents, even when the prior distribution is *deterministic*.

Example 1. Consider n candidates arriving in adversarial order, where the biased gambler considers $k = 2$ features of candidates and has loss aversion parameter $\lambda = 2$. Furthermore, let σ be the following sequence of candidate values:

$$(1, 0), (0, 1), (2, 0), (0, 2), (2^2, 0), (0, 2^2), \cdots, (2^{n/2-1}, 0), (0, 2^{n/2-1}),$$

and let the prior distribution $\mathcal{F}$ be deterministic – where the realized sequence from $\mathcal{F}$ is exactly σ with probability one.

In our example, the overall value of the candidates increases exponentially over time. Consequently, the rational prophet will receive high utility by simply selecting one of the last two candidates. However, the fact that the candidates are excellent on alternating dimensions causes the biased gambler to suffer from increasing regret as they see more candidates. In fact, should the biased gambler choose a later candidate rather than an earlier candidate, the increment in their regret roughly cancels out the increment in their value for the candidate.

   With this intuition in mind, let's formally reason about the behaviors of our agents. Recall that the gambler is aware of the prior distribution $\mathcal{F}$ and thus is aware of σ a priori for this example. The rational gambler is essentially the same as the rational prophet in this case since they know σ a priori and experience no regret. Thus, the expected utility of both the rational prophet and the rational gambler is simply $2^{n/2-1}$.

   Under our stylized model, the biased gambler only experiences regret for realized options, completely disregarding any future options they may be aware of a priori. Consequently, the biased gambler has utility 1 if they just take the first candidate and are not presented other candidates, and they have utility ≤ 0 if they take any other candidate. (e.g. If the gambler decides to select let's say the candidate $(4, 0)$, then the super candidate has value $(4, 2)$, which means that the biased gambler's utility is $\|(4, 0)\|_1 - 2 \cdot (\|(4, 2)\|_1 - \|(4, 0)\|_1) = 4 - 2 \cdot (6 - 4) = 0$). Therefore we know $\mathbb{E}_{\mathcal{F}}[U_{g_b}^*] \leq 1$. As a result, the biased gambler's prophet utility ratio, $\frac{\mathbb{E}_{\mathcal{F}}[U_{pr}]}{\mathbb{E}_{\mathcal{F}}[U_{g_b}^*]} = \frac{U_{pr}(\sigma)}{U_{g_b}(\sigma)}$, and online utility ratio, $\frac{\mathbb{E}_{\mathcal{F}}[U_{gr}^*]}{\mathbb{E}_{\mathcal{F}}[U_{g_b}^*]} = \frac{U_{gr}(\sigma)}{U_{g_b}(\sigma)}$, are both equal to $2^{n/2-1}$, and thus tend toward ∞ as n increases. Recall that prophet and online utility ratios lower bound the prophet and online competitive ratio, respectively, implying that both competitive ratios must also be ∞.

As we will soon show in Proposition 2 and Proposition 3, our present adversarial example with candidates that are excellent on alternating dimensions can be generalized to any k and λ such that the feature-amplified bias $\lambda \cdot (k-1)$ is at least one in the adversarial setting. Furthermore, as we will show in Proposition 5 and Proposition 6, we can extend our example to i.i.d setting (and thus to random and optimal candidate ordering setting) by carefully constructing the prior distribution so that with high probability, the sequence of realized candidate values presented to the agent is adversarial and (after deleting subsequent duplicates) similar to the sequence σ in this example. Our capability to generalize the adversarial example under *any* arrival order constraint drives our model's departure from the single dimensional model. Further, our adversarial example for the i.i.d setting is highly randomized, which shows that the adversarial examples are not an artifact of deterministic distributions.

Lastly, this example illustrates an additional property unique to our model: the biased gambler may outperform the biased prophet. Notice that in our example, the biased prophet is forced to see both $(2^{n/2-1}, 0)$ and $(0, 2^{n/2-1})$ as realized values, resulting in a reference point $(2^{n/2-1}, 2^{n/2-1})$ – much larger than any value they could have chosen. Consequently, the biased prophet has negative utility and is worse off compared to the biased gambler. In contrast, in [11]'s model (where $k = 1$), the biased prophet is guaranteed to attain at least as much utility as that of online agents.

4   Effects of Comparative Loss Aversion

In this section we formally analyze the effect of comparative loss aversion on the biased gambler's utility. Specifically, we compare the biased gambler's utility with those of both the rational prophet and rational gambler. We will focus on the two extremes of arrival order constraints: adversarial order and optimal order (even more specifically, we consider sequences drawn from an i.i.d. prior distribution, where any arrival order is optimal). Under both ordering constraints, we show a phase transition when the feature-amplified bias $\lambda \cdot (k-1)$ is equal to one. When the feature-amplified bias $\lambda(k-1) \geq 1$, the biased gambler's prophet and online competitive ratio are both equal to infinity. On the other hand, when the feature-amplified bias $\lambda(k-1) < 1$, the biased gambler can achieve a competitive ratio that is only dependent on λ and k, and of similar magnitude under both ordering constraints.

It is surprising that our model exhibits similar behavior in terms of competitive ratios regardless of constraints on arrival order. In contrast, [11] shows that for the single dimensional model, assuming the arrival order to be random rather than adversarial results in an exponential improvement in utility guarantees in terms of λ.

Missing proofs in Sect. 4.1 and Sect. 4.2 can be found in the full version of the paper[2].

---

[2] https://arxiv.org/abs/2309.14555.

4.1   Utility Loss Under Adversarial Arrival Order

Firstly, we generalize Example 1 to show that when the feature-amplified bias $\lambda \cdot (k - 1) \geq 1$, as n grows arbitrarily large, the prophet and online utility ratios (namely, the ratio between expected utility from the biased gambler and the expected utility from the rational prophet or gambler) can be arbitrarily large. In particular, we construct a prior distribution $\mathcal{F}_0(n)$, parameterized by the number of candidates n such that, when the feature-amplified bias $\lambda \cdot (k - 1) > 1$, both the prophet and online utility ratio are exponential in terms of n. When the feature-amplified bias $\lambda(k - 1) = 1$, we obtain a more conservative result that prophet and online utility ratio can be linear in terms of n.

Proposition 2. *When the feature-amplified bias $\lambda \cdot (k-1) > 1$, for any $n \in N^+$, there exists a sequence σ with n candidates such that when the distribution $\mathcal{F}_0(n)$ takes the values of σ deterministically, i.e. with probability 1, and when our prior distribution $\mathcal{F}$ equals $\mathcal{F}_0(n)$, the prophet utility ratio $\frac{\mathbb{E}_{\mathcal{F}}[U_{pr}]}{\mathbb{E}_{\mathcal{F}}[U_{g_b}^*]}$ and the online utility ratio $\frac{\mathbb{E}_{\mathcal{F}}[U_{g_r}^*]}{\mathbb{E}_{\mathcal{F}}[U_{g_b}^*]}$ are both equal to $\Omega\left((\lambda(k - 1))^{\frac{n}{k} - 1}\right)$.*

Proposition 3. *When the feature-amplified bias $\lambda \cdot (k-1) = 1$, for any $n \in \mathbb{N}^+$, there exists a sequence σ with n candidates such that when the distribution $\mathcal{F}_0(n)$ takes the values of σ deterministically, i.e. with probability 1, and when our prior distribution $\mathcal{F}$ equals $\mathcal{F}_0(n)$, the prophet utility ratio $\frac{\mathbb{E}_{\mathcal{F}}[U_{pr}]}{\mathbb{E}_{\mathcal{F}}[U_{g_b}^*]}$ and the online utility ratio $\frac{\mathbb{E}_{\mathcal{F}}[U_{g_r}^*]}{\mathbb{E}_{\mathcal{F}}[U_{g_b}^*]}$ are both equal to $\Omega\left(\frac{n}{k}\right)$.*

Theorem 1. *When the feature-amplified bias $\lambda \cdot (k - 1) \geq 1$, the prophet and online competitive ratio are both equal to ∞.*

On the other hand, when the feature-amplified bias $\lambda(k - 1) < 1$, we prove that the biased gambler, by using a simple threshold based algorithm, can get at least a $\frac{1 - \lambda(k-1)}{2 + \lambda}$ fraction of the prophet's utility and a $\frac{1 - \lambda(k-1)}{1 + \lambda}$ fraction of the optimal rational gambler's utility regardless of the value of n. We also show that both of these ratios are tight. More formally, we prove the following theorems.

Theorem 2. *When the feature-amplified bias $\lambda \cdot (k - 1) < 1$, for any prior distribution $\mathcal{F}$, the inverse of the prophet utility ratio $\frac{\mathbb{E}_{\mathcal{F}}[U_{g_b}^*]}{\mathbb{E}_{\mathcal{F}}[U_{pr}]} \geq (1 - \lambda(k - 1)) \cdot \max\left\{\frac{\gamma}{1 + \lambda + k}, \frac{1}{2 + \lambda}\right\}$, where $\gamma = \frac{\mathbb{E}_{\mathcal{F}}[\sum_{j=1}^{k} S_j^*]}{\mathbb{E}_{\mathcal{F}}[V^*]}$. Consequently, the prophet competitive ratio is at most $\frac{2 + \lambda}{1 - \lambda(k-1)}$.*

Theorem 3. *When the feature-amplified bias $\lambda \cdot (k - 1) < 1$, for any prior distribution $\mathcal{F}$, the online utility ratio $\frac{\mathbb{E}_{\mathcal{F}}[U_{g_r}^*]}{\mathbb{E}_{\mathcal{F}}[U_{g_b}^*]} \leq \frac{1 + \lambda}{(1 - \lambda(k-1))}$. Consequently, the online competitive ratio is at most $\frac{1 + \lambda}{1 - \lambda(k-1)}$.*

Theorem 4. *When the feature-amplified bias $\lambda \cdot (k - 1) < 1$, for any $\delta \in (0, 1)$, there exists a prior distribution $\mathcal{F}$ such that the prophet utility ratio $\frac{\mathbb{E}_{\mathcal{F}}[U_{pr}^*]}{\mathbb{E}_{\mathcal{F}}[U_{g_b}]} \geq$*

$\frac{2+\lambda}{1-\lambda(k-1)} \cdot (1-\delta)$ and that the online utility ratio $\frac{\mathbb{E}_{\mathcal{F}}[U_{g_r}^*]}{\mathbb{E}_{\mathcal{F}}[U_{g_b}^*]} \geq \frac{1+\lambda}{1-\lambda(k-1)} \cdot (1-\delta)$. Consequently, the prophet competitive ratio is at least $\frac{2+\lambda}{1-\lambda(k-1)}$ and the online competitive ratio is at least $\frac{1+\lambda}{1-\lambda(k-1)}$.

Our proof for Theorem 2 and Theorem 3 follows a similar outline to the single dimensional case presented in [11]. However, our analysis deals with the additional complexity from having multiple dimensions, and often requires a tighter analysis for intermediate terms in order to find the exact competitive ratios. For instance, in Theorem 2, we require an additional term when bounding prophet utility ratio (which will later be used in Theorem 3) compared to a similar claim in [11].

We will start by defining threshold based algorithms.

Definition 4 (Thresholding algorithms.). *We define a thresholding algorithm parameterized by $\alpha \in (0,1)$ and denoted A^α as follows. First, A^α sets threshold T such that $\alpha = \Pr[V^* \geq T]$ (recall that V^* is defined as the maximum value of all candidates). Next, during the selection process, the algorithm will select the first candidate whose value $\|\sigma^{(t)}\|_1$ is at least the threshold T. We will denote the expected utility from A^α as $\mathbb{E}_{\mathcal{F}}[U_{g_b}^\alpha]$.*

We will now reason about the biased gambler's utility from using a thresholding algorithm A^α. In the following two claims, we first bound the utility (and regret) of a biased gambler that uses the thresholding algorithm when they encounter a particular sequence of values σ. This utility bound can then be used to bound the expected utility of the biased gambler given the prior distribution of values.

Claim 1. *When the feature-amplified bias $\lambda \cdot (k-1) < 1$, for any sequence σ, for any threshold T, and for timestep t such that t is the smallest timestep where $\|\sigma^{(t)}\|_1 \geq T$, it must be the case that $U_{g_b}(\sigma, t) \geq \|\sigma^{(t)}\|_1 - \lambda(k-1) \cdot T$.*

Claim 2. *When the feature-amplified bias $\lambda \cdot (k-1) < 1$, for any prior distribution $\mathcal{F}$ and any $\alpha \in (0,1)$, $\mathbb{E}_{\mathcal{F}}[U_{g_b}^\alpha] \geq ((1+\lambda)\alpha - k\lambda) \cdot T + (1-\alpha) \cdot \sum_{t \in [n]} \mathbb{E}_{\sigma^{(t)} \sim \mathcal{F}_t}[(\|\sigma^{(t)}\|_1 - T)^+].$*

We can now bound the utility ratio between a biased gambler using algorithm A^α and the rational prophet, which can then be used to prove Theorem 2 and Theorem 3. It turns out that Theorem 2 and Theorem 3 requires two different ways of bounding the the utility ratio.

Lemma 1. *When the feature-amplified bias $\lambda \cdot (k-1) < 1$, for any prior distribution $\mathcal{F}$, there exists an α such that $\frac{\mathbb{E}_{\mathcal{F}}[U_{g_b}^\alpha]}{\mathbb{E}_{\mathcal{F}}[U_{p_r}]} \geq (1-\lambda(k-1)) \cdot \max\left\{\frac{\gamma}{1+\lambda+k}, \frac{1}{2+\lambda}\right\}$, where $\gamma = \frac{\mathbb{E}_{\mathcal{F}}[\sum_{j=1}^k S_j^*]}{\mathbb{E}_{\mathcal{F}}[V^*]}$ (recall that S_j^* denotes the maximum value in the sequence along dimension j and $V^* = \max_t \|\sigma^{(t)}\|_1$).*

We are now ready to prove Theorem 2 and Theorem 3.

Proof (of Theorem 2). The first line of the theorem follows directly by Lemma 1 and the fact that $\mathbb{E}_{\mathcal{F}}[U_{g_b}^*] \geq \mathbb{E}_{\mathcal{F}}[U_{g_b}^\alpha]$. (Because $\mathbb{E}_{\mathcal{F}}[U_{g_b}^*]$ is defined as the expected utility from a biased gambler that uses the utility optimal online algorithm, $\mathbb{E}_{\mathcal{F}}[U_{g_b}^*]$ is at least the expected utility from a biased gambler using some specific online algorithm).

Now we will derive the second line of the theorem. From the first line of the theorem we know that for any prior distribution $\mathcal{F}$,

$$\frac{\mathbb{E}_{\mathcal{F}}[U_{g_b}^*]}{\mathbb{E}_{\mathcal{F}}[U_{p_r}]} \geq (1 - \lambda(k-1)) \cdot \max\left\{\frac{\gamma}{1+\lambda+k}, \frac{1}{2+\lambda}\right\} \geq \frac{1 - \lambda(k-1))}{2+\lambda}.$$

Hence, for any prior distribution $\mathcal{F}$, the prophet utilit ratio $\frac{\mathbb{E}_{\mathcal{F}}[U_{p_r}]}{\mathbb{E}_{\mathcal{F}}[U_{g_b}^*]}$ is at most $\frac{2+\lambda}{1-\lambda(k-1))}$. We conclude that the prophet competitive ratio $\max\limits_{\mathcal{F}} \frac{\mathbb{E}_{\mathcal{F}}[U_{p_r}]}{\mathbb{E}_{\mathcal{F}}[U_{g_b}^*]}$ is at most $\frac{2+\lambda}{1-\lambda(k-1))}$.

Proof (of Theorem 3). Let $\beta = \frac{\mathbb{E}_{\mathcal{F}}[U_{p_r}]}{\mathbb{E}_{\mathcal{F}}[U_{g_b}^*]}$ and let $\Delta = \frac{\mathbb{E}_{\mathcal{F}}[U_{p_r}]}{\mathbb{E}_{\mathcal{F}}[U_{g_r}^*]}$. Since $\frac{\mathbb{E}_{\mathcal{F}}[U_{g_r}^*]}{\mathbb{E}_{\mathcal{F}}[U_{g_b}^*]} = \frac{\beta}{\Delta}$, proving our theorem is equivalent to proving that $\frac{\Delta}{\beta} \geq \frac{(1-\lambda(k-1))}{1+\lambda}$, and we will do exactly this for the rest of the proof.

Firstly, let $\gamma = \frac{\mathbb{E}_{\mathcal{F}}[\sum_{j=1}^{k} S_j^*]}{\mathbb{E}_{\mathcal{F}}[V^*]}$, then by Theorem 2 we know that $\frac{1}{\beta} = \frac{\mathbb{E}_{\mathcal{F}}[U_{g_b}^*]}{\mathbb{E}_{\mathcal{F}}[U_{p_r}]} \geq (1 - \lambda(k-1)) \cdot \frac{\gamma}{1+\lambda+k}$.

Next, because $\mathbb{E}_{\mathcal{F}}[U_{g_b}^*]$ represents the highest expected utility a biased gambler can gain from using an online algorithm, $\mathbb{E}_{\mathcal{F}}[U_{g_b}^*]$ must be at least the expected utility a biased gambler can gain from using the value-optimal online algorithm, which is equal to the optimal utility of a rational gambler minus the loss aversion the biased gambler experience from using the value-optimal online algorithm. Recall that S_j^* is defined as the maximum value on the jth dimension among all candidates, thus the value of super candidate the biased gambler experiences loss aversion against must be at most $\sum_{j=1}^{k} S_j^*$. Thus

$$\mathbb{E}_{\mathcal{F}}[U_{g_b}^*] \geq \mathbb{E}_{\mathcal{F}}\left[U_{g_r}^* - \lambda \cdot \left(\sum_{j=1}^{k} S_j^* - U_{g_r}^*\right)\right]$$

$$= (1+\lambda) \cdot \mathbb{E}_{\mathcal{F}}[U_{g_r}^*] - \lambda \cdot \mathbb{E}_{\mathcal{F}}\left[\sum_{j=1}^{k} S_j^*\right].$$

By the definition of Δ, we know that $\mathbb{E}_{\mathcal{F}}[U_{g_r}^*] = \frac{1}{\Delta} \cdot \mathbb{E}_{\mathcal{F}}[U_{p_r}]$. Similarly, by the definition of γ, $\mathbb{E}_{\mathcal{F}}[\sum_{j=1}^{k} S_j^*] = \gamma \cdot \mathbb{E}_{\mathcal{F}}[V^*] = \gamma \cdot \mathbb{E}_{\mathcal{F}}[U_{p_r}]$. Therefore combining these two statements with the previous inequality we know the following:

$$\mathbb{E}_{\mathcal{F}}[U_{g_b}^*] \geq \left(\frac{1+\lambda}{\Delta} - \lambda \cdot \gamma\right) \cdot \mathbb{E}_{\mathcal{F}}[U_{p_r}].$$

Re-expressing algebraically, we have $\frac{1}{\beta} = \frac{\mathbb{E}_{\mathcal{F}}[U_{g_b}^*]}{\mathbb{E}_{\mathcal{F}}[U_{p_r}]} \geq \left(\frac{1+\lambda}{\Delta} - \lambda \cdot \gamma\right) = \frac{1+\lambda-\lambda\cdot\gamma\cdot\Delta}{\Delta}$.

Now we have two bounds on $\frac{1}{\beta}$: we know that $\frac{1}{\beta} \geq (1 - \lambda(k-1)) \cdot \frac{\gamma}{1+\lambda+k}$ and that $\frac{1}{\beta} \geq \frac{1+\lambda-\lambda\cdot\gamma\cdot\Delta}{\Delta}$. By these two bounds, we conclude that

$$\frac{\Delta}{\beta} = \Delta \cdot \frac{1}{\beta} \geq \max\left\{(1 - \lambda(k-1)) \cdot \frac{\Delta \cdot \gamma}{1+\lambda+k}, \ 1+\lambda-\lambda\cdot\gamma\cdot\Delta\right\}.$$

Notice that inside the max bracket, the first term monotonically increases with Δ, while the second term monotonically decreases with Δ, therefore the minimum of the max of two terms occurs when the two terms are equal, namely, when $(1 - \lambda(k-1)) \cdot \frac{\Delta \cdot \gamma}{1+\lambda+k} = 1+\lambda-\lambda\cdot\gamma\cdot\Delta$. This is achieved when $\Delta = \frac{1+\lambda+k}{\gamma\cdot(1+\lambda)}$, at which point

$$(1 - \lambda(k-1)) \cdot \frac{\Delta \cdot \gamma}{1+\lambda+k} = 1+\lambda-\lambda\cdot\gamma\cdot\Delta = \frac{1-\lambda(k-1)}{1+\lambda}.$$

We thus conclude that it is always the case that $\frac{\mathbb{E}_{\mathcal{F}}[U_{g_b}^*]}{\mathbb{E}_{\mathcal{F}}[U_{g_r}^*]} = \frac{\Delta}{\beta} \geq \frac{1-\lambda(k-1)}{1+\lambda}$ regardless of the value of Δ.

We remain to discuss Theorem 4, which we prove by constructing a worst case prior distribution. Our construction combines a generalization of Example 1 with the canonical two item example showing a 1/2-competitive ratio for classical prophet inequality. In our construction, all but the last candidate has deterministic values constructed similarly to those in Example 1 (but the value of the candidates grow much slower due to the fact that the feature-amplified bias is less than one). The last candidate has a high value with a very small probability, which is similar to constructions for classical prophet inequality.

Proof (Proof Sketch) (of Theorem 4). Let the first $n-1$ candidates have the following deterministic value:

$$(1,0,\cdots,0), \qquad (0,1,0,\cdots,0), \qquad \cdots, \ (0,\cdots,0,1),$$
$$(1+\beta,0,\cdots,0), \qquad (0,1+\beta,0,\cdots,0), \qquad \cdots, \ (0,\cdots,0,1+\beta),$$
$$(1+\beta+\beta^2,0,\cdots,0), \quad (0,1+\beta+\beta^2,0,\cdots,0), \quad \cdots, \ (0,\cdots,0,1+\beta+\beta^2),$$
$$\cdots$$
$$(\textstyle\sum_{j=0}^{\lceil n/k\rceil-1}\beta^j,0,\cdots,0), \ (0,\sum_{j=0}^{\lceil n/k\rceil-1}\beta^j,0,\cdots,0), \ \cdots, \ (0,\cdots,0,\sum_{j=0}^{\lceil n/k\rceil-1}\beta^j).$$

Then, the last randomized candidate is equal to

$$\left((1-\epsilon)(1+\lambda) \cdot \sum_{j=0}^{\lceil n/k\rceil-1} \beta^j/\epsilon, 0, \cdots, 0\right)$$

with probability ϵ and is equal to $(0,0,\cdots,0)$ with probability $1-\epsilon$. One can then verify that the biased gambler can only get utility at most 1, since by the time the biased gambler reach the last randomized candidate, their imaginary super candidate is too strong for the biased gambler to gain positive expected utility from the randomized candidate. However, the rational prophet can simply select the maximum between the best deterministic option and the randomized option, getting utility that approaches $\frac{2+\lambda}{1-\lambda(k-1)}$ as ϵ approaches 0 and n approaches ∞. Similarly, the rational gambler can simply always select the randomized option and get utility that approaches and $\frac{1+\lambda}{1-\lambda(k-1)}$.

4.2 No Improvement Under Relaxed Arrival Order

We now show that most of our results for adversarial arrival order can be used to derive analogous results for the case where the prior distribution is i.i.d.

Specifically, in the next claim (Proposition 4), we will show that removing repeated candidate values from a sequence does not affect the biased gambler's optimal utility. We will then use the claim to establish in a subsequent lemma (Lemma 2) that for each deterministic distribution (i.e. each distribution such that values drawn from it are equal to some sequence σ with probability 1), we can find an i.i.d prior distribution $\mathcal{F}$ with a similar realized sequence (after deleting duplicates), and hence similar prophet and online utility ratio. As we have discussed in Sect. 3 and Sect. 4.1, the construction of the worst case instances for the adversarial arrival order setting is either a fully deterministic sequence, or with few randomized options. Consequently, Lemma 2 can be leveraged to translate all lower bounds on the utility ratios in adversarial setting into lower bounds in the i.i.d setting.

Definition 5. *A sequence σ is called succinct if there are no two identical candidates in the sequence and no candidate's value is a zero vector.*

Definition 6. *The representation $r(\sigma)$ of a sequence σ is the sequence of unique elements in σ ordered in the ascending order of their first occurance in σ. (e.g. $r(1, 2, 1, 2, 3) = 1, 2, 3$).*

Proposition 4. $U_{p_r}(\sigma) = U_{p_r}(r(\sigma))$ and $U_{g_b}(\sigma) = U_{g_b}(r(\sigma))$.

Lemma 2. *For any $\epsilon \in (0, 1)$, any succinct sequence σ such that its length $m = |\sigma|$ is greater than 1, there exists a distribution $\mathcal{F}$ where all candidates are i.i.d with support on the set $\{\sigma^{(1)}, \cdots, \sigma^{(m)}\}$ such that the prophet utility ratio $\frac{\mathbb{E}_{\mathcal{F}}[U_{p_r}]}{\mathbb{E}_{\mathcal{F}}[U_{g_b}^*]} \geq (1 - \epsilon) \cdot \frac{U_{p_r}(\sigma)}{U_{g_b}(\sigma)}$.*

We can now use Lemma 2 to prove that when the candidate values are i.i.d, it is also the case that when the feature-amplified bias $\lambda(k - 1) \geq 1$, prophet and online utility ratios can get progressively worse as the number of candidates n increases, and as a result, both the prophet and online competitive ratio are equal to ∞.

Proposition 5. *When the feature-amplified bias $\lambda \cdot (k - 1) > 1$, for all $n \in \mathbb{N}^+$, there exists a prior distribution $\mathcal{F}_0(n)$ where all candidates are i.i.d such that when $\mathcal{F} = \mathcal{F}_0(n)$, the prophet utility ratio $\frac{\mathbb{E}_{\mathcal{F}}[U_{p_r}]}{\mathbb{E}_{\mathcal{F}}[U_{g_b}^*]}$ and the online utility ratio $\frac{\mathbb{E}_{\mathcal{F}}[U_{g_r}^*]}{\mathbb{E}_{\mathcal{F}}[U_{g_b}^*]}$ are both equal to $\Omega\left((\lambda(k - 1))^{f(\lambda,k) \cdot \log^{\frac{1}{2}} n}\right)$, where $f(\lambda, k)$ is a function that only depends on λ and k.*

Proposition 6. *When the feature-amplified bias $\lambda \cdot (k-1) = 1$, for all $n \in \mathbb{N}^+$, there exists a prior distribution $\mathcal{F}_0(n)$ where all candidates are i.i.d such that when $\mathcal{F} = \mathcal{F}_0(n)$, the prophet utility ratio $\frac{\mathbb{E}_{\mathcal{F}}[U_{pr}]}{\mathbb{E}_{\mathcal{F}}[U_{g_b}^*]}$ and the online utility ratio $\frac{\mathbb{E}_{\mathcal{F}}[U_{g_r}^*]}{\mathbb{E}_{\mathcal{F}}[U_{g_b}^*]}$ are both equal to $\Omega\left(\frac{\log n}{k \cdot \log \log n}\right)$.*

Theorem 5. *In the i.i.d prior distribution setting, when the feature-amplified bias $\lambda \cdot (k-1) \geq 1$, the prophet and online competitive ratio are both equal to ∞.*

On the other hand, when the feature-amplified bias $\lambda(k-1) < 1$, we can utilize Lemma 2 to show a lower bound for the prophet and online competitive ratios that matches those in the adversarial setting within constant factors.

Proposition 7. *When the feature-amplified bias $\lambda(k-1) < 1$, for any $\delta \in (0, 1)$, there exists an i.i.d prior distribution $\mathcal{F}$ where the prophet utility ratio $\frac{\mathbb{E}_{\mathcal{F}}[U_{pr}]}{\mathbb{E}_{\mathcal{F}}[U_{g_b}^*]} \geq (1-\delta) \cdot \frac{1}{(1-\lambda(k-1))}$.*

Theorem 6. *When the feature-amplified bias $\lambda(k-1) < 1$ and $k > 1$, for any $\delta \in (0, 1)$, there exists an i.i.d prior distribution $\mathcal{F}$ where the prophet utility ratio $\frac{\mathbb{E}_{\mathcal{F}}[U_{pr}]}{\mathbb{E}_{\mathcal{F}}[U_{g_b}^*]} \geq \frac{(1-\delta)}{3} \cdot \frac{2+\lambda}{1-\lambda(k-1)}$ and the online utility ratio $\frac{\mathbb{E}_{\mathcal{F}}[U_{g_r}^*]}{\mathbb{E}_{\mathcal{F}}[U_{g_b}^*]} \geq \frac{(1-\delta)}{4} \cdot \frac{1+\lambda}{1-\lambda(k-1)}$. Consequently, the prophet competitive ratio is at least $\frac{1}{3} \cdot \frac{2+\lambda}{(1-\lambda(k-1))}$ and the online competitive ratio is at least $\frac{1}{4} \cdot \frac{1+\lambda}{(1-\lambda(k-1))}$.*

5   Conclusion and Directions for Future Work

We consider an online decision-maker who suffers from comparative loss aversion. Our model is identical to that of [11], *except* for how the reference point is formed. This change alone accounts for fundamental differences in conclusions. For example, even the prophet suffers loss in our model, and this enables novel phenomena to occur. Additionally, the loss due to bias in our model can be unboundedly bad, and is not significantly mitigated by assumptions on the prophet inequality arrival order.

Empirically observed biases have only recently been incorporated into research on optimal stopping. Our work takes a step beyond [11] and uncovers new phenomena by changing just one aspect of their model. An important direction for future work is to continue developing models that uncover additional phenomena of behavioral game theory (one such possibility is those studied in [21], where adding inferior options can influence decision-making).

References

1. Beyhaghi, H., Golrezaei, N., Leme, R.P., Pal, M., Sivan, B.: Improved approximations for free-order prophets and second-price auctions. CoRR abs/1807.03435 (2018). http://arxiv.org/abs/1807.03435

2. Brenner, L., Rottenstreich, Y., Sood, S.: Comparison, grouping, and preference. Psychol. Sci. **10**(3), 225–229 (1999)
3. Chernev, A., Böckenholt, U., Goodman, J.: Choice overload: a conceptual review and meta-analysis. J. Consum. Psychol. **25**(2), 333–358 (2015)
4. Correa, J.R., Foncea, P., Hoeksma, R., Oosterwijk, T., Vredeveld, T.: Posted price mechanisms for a random stream of customers. In: Proceedings of the 2017 ACM Conference on Economics and Computation, EC 2017, Cambridge, MA, USA, 26–30 June 2017, pp. 169–186 (2017). https://doi.org/10.1145/3033274.3085137
5. Duetting, P., Feldman, M., Kesselheim, T., Lucier, B.: Prophet inequalities made easy: stochastic optimization by pricing non-stochastic inputs. In: 58th IEEE Annual Symposium on Foundations of Computer Science, FOCS 2017, Berkeley, CA, USA, 15–17 October 2017, pp. 540–551 (2017). https://doi.org/10.1109/FOCS.2017.56
6. Ehsani, S., Hajiaghayi, M., Kesselheim, T., Singla, S.: Prophet secretary for combinatorial auctions and matroids. In: Proceedings of the Twenty-Ninth Annual ACM-SIAM Symposium on Discrete Algorithms, SODA 2018, New Orleans, LA, USA, 7–10 January 2018, pp. 700–714 (2018). https://doi.org/10.1137/1.9781611975031.46
7. Esfandiari, H., Hajiaghayi, M., Liaghat, V., Monemizadeh, M.: Prophet secretary. In: Proceedings of the Algorithms - ESA 2015–23rd Annual European Symposium, Patras, Greece, 14–16 September 2015, pp. 496–508 (2015). https://doi.org/10.1007/978-3-662-48350-3_42
8. Haynes, G.A.: Testing the boundaries of the choice overload phenomenon: the effect of number of options and time pressure on decision difficulty and satisfaction. Psychol. Mark. **26**(3), 204–212 (2009)
9. Hill, T.P., Kertz, R.P.: Comparisons of stop rule and supremum expectations of I.I.D. random variables. Ann. Probab. **10**, 336–345 (1982)
10. Kahneman, D., Tversky, A.: Prospect theory: an analysis of decision under risk. In: Handbook of the Fundamentals of Financial Decision Making: Part I, pp. 99–127. World Scientific (2013)
11. Kleinberg, J., Kleinberg, R., Oren, S.: Optimal stopping with behaviorally biased agents: the role of loss aversion and changing reference points. arXiv preprint arXiv:2106.00604 (2021)
12. Kleinberg, R., Weinberg, S.M.: Matroid prophet inequalities. In: Proceedings of the 44th Symposium on Theory of Computing Conference, STOC 2012, New York, NY, USA, 19–22 May 2012, pp. 123–136 (2012). https://doi.org/10.1145/2213977.2213991
13. Krengel, U., Sucheston, L.: On semiamarts, amarts, and processes with finite value. Adv. Probab. Relat. Top. **4**, 197–266 (1978)
14. Park, J.Y., Jang, S.S.: Confused by too many choices? Choice overload in tourism. Tour. Manag. **35**, 1–12 (2013)
15. Rubinstein, A.: Beyond matroids: secretary problem and prophet inequality with general constraints. In: Proceedings of the 48th Annual ACM SIGACT Symposium on Theory of Computing, STOC 2016, Cambridge, MA, USA, 18–21 June 2016, pp. 324–332 (2016). https://doi.org/10.1145/2897518.2897540
16. Rubinstein, A., Singla, S.: Combinatorial prophet inequalities. In: Proceedings of the Twenty-Eighth Annual ACM-SIAM Symposium on Discrete Algorithms, SODA 2017, Barcelona, Spain, Hotel Porta Fira, 16–19 January 2017, pp. 1671–1687 (2017). https://doi.org/10.1137/1.9781611974782.110
17. Sagi, A., Friedland, N.: The cost of richness: the effect of the size and diversity of decision sets on post-decision regret. J. Pers. Soc. Psychol. **93**(4), 515 (2007)

18. Samuel-Cahn, E.: Comparison of threshold stop rules and maximum for independent nonnegative random variables. Ann. Probab. **12**(4), 1213–1216 (1984)
19. Schunk, D., Winter, J.: The relationship between risk attitudes and heuristics in search tasks: a laboratory experiment. J. Econ. Behav. Organ. **71**(2), 347–360 (2009)
20. Schwartz, B.: The Paradox of Choice: Why More is Less. Ecco, New York (2004)
21. Simonson, I., Tversky, A.: Choice in context: tradeoff contrast and extremeness aversion. J. Mark. Res. **29**(3), 281–295 (1992)
22. Tversky, A., Kahneman, D.: Loss aversion in riskless choice: a reference-dependent model. Q. J. Econ. **106**(4), 1039–1061 (1991)
23. Yan, Q.: Mechanism design via correlation gap. In: Randall, D. (ed.) Proceedings of the Twenty-Second Annual ACM-SIAM Symposium on Discrete Algorithms, SODA 2011, San Francisco, California, USA, 23–25 January 2011, pp. 710–719. SIAM (2011). https://doi.org/10.1137/1.9781611973082.56

Selling to Multiple No-Regret Buyers

Linda Cai[1]([✉]), S. Matthew Weinberg[1], Evan Wildenhain[2], and Shirley Zhang[3]

[1] Princeton University, Princeton, NJ 08544, USA
tcai@princeton.edu
[2] Google, Mountain View, CA 94043, USA
[3] Harvard University, Cambridge, MA 02138, USA

Abstract. We consider the problem of repeatedly auctioning a single item to multiple i.i.d buyers who each use a no-regret learning algorithm to bid over time. In particular, we study the seller's optimal revenue, if they know that the buyers are no-regret learners (but only that their behavior satisfies some no-regret property—they do not know the precise algorithm/heuristic used).

Our main result designs an auction that extracts revenue equal to the *full expected welfare* whenever the buyers are "mean-based" (a property satisfied by standard no-regret learning algorithms such as Multiplicative Weights, Follow-the-Perturbed-Leader, etc.). This extends a main result of [4] which held only for a single buyer.

Our other results consider the case when buyers are mean-based but never overbid. On this front, [4] provides a simple LP formulation for the revenue-maximizing auction for a single-buyer. We identify several formal barriers to extending this approach to multiple buyers.

1 Introduction

Classical Bayesian auction design considers a static auction where buyers participate once. Here, the study of truthful auctions is ubiquitous following Myerson's seminal work [31]. But many modern auction applications (such as ad auctions) are *repeated*: the same buyers participate in many auctions over time. Moreover, the vast majority of auction formats used in such settings are *not* truthful (e.g. first-price auctions, generalized first-price auctions, generalized second-price auctions). Even those that are based on a truthful format (such as the Vickrey-Clarke-Groves mechanism [10,21,34]) are no longer truthful when the repeated aspect is taken into account (because the seller may increase or decrease reserves in later rounds based on buyers' behavior in earlier rounds). As such, it is imperative to have a study of non-truthful repeated auctions.

Over the past several years, this direction has seen significant progress on numerous fronts (we overview related work in Sect. 1.1). Our paper follows a recent direction initiated by [4] and motivated by empirical work of [32]. Specifically, [32] find that bidding behavior on Bing largely satisfies the no-regret guarantee (that is, there exist values for the buyers such that their bidding behavior

© The Author(s), under exclusive license to Springer Nature Switzerland AG 2024
J. Garg et al. (Eds.): WINE 2023, LNCS 14413, pp. 113–129, 2024.
https://doi.org/10.1007/978-3-031-48974-7_7

guarantees low regret—the paper makes no claims about any particular algorithm the buyers might be using). This motivates the following question: *if buyer behavior guarantees no-regret, what auction format for the designer maximizes her expected revenue?*

[4] initiated this study for a single buyer. The main focus of our paper is to initiate the study for multiple buyers. We formally pose the model in Sect. 2, and overview our main results here.

The concept of a "mean-based" no-regret learning algorithm appears in [4], and captures algorithms which pull an arm with very high probability when it is historically better than all other arms (formal definition in Sect. 2). While it is common to design non-mean-based algorithms for dynamic environments, standard no-regret algorithms such as Multiplicative Weights, EXP3, etc. are all mean-based.[1]

Main Result (Informal—See Theorem 3:) When any number of i.i.d. buyers use bidding strategies which satisfy the mean-based no-regret guarantee, there exists a repeated single-item auction for the seller which guarantees them expected revenue arbitrarily close to the optimal expected *welfare*.

One main result of [4] proves the special case with just a single buyer. While we defer technical details of our construction to Sect. 3, we briefly overview the main challenges here. The one-buyer [4] auction is already surprising, as it requires the seller to both (a) give the buyer the item every round, yet (b) charge them their full value (without knowing their value). The key additional challenge for the multi-buyer setting is that the seller must now give the item not just to *a* buyer in every round, but to *the* buyer with the highest value. This means, in particular, that we must set up the auction so that buyers will pull a distinct arm for each possible value, and yet we must still charge each buyer their full expected value by the end of the auction.

Our auction, like that of [4], is fairly impractical (for example, it alternates between running a second-price auction every round, and charging huge surcharges to the winner) and is not meant to guide practice. Still, Theorem 3 establishes that full surplus is possible for multiple mean-based buyers, and therefore sets a high benchmark for this setting without further modeling assumptions.

No Overbidding. Indeed, the main impracticality in our Full-Surplus-Extraction auction is that lures buyers into *overbidding* significantly, and eventually paying more than their value. In practice, it may be reasonable for buyers to be *clever*, and just remove from consideration all bids exceeding their value (but guarantee no-regret on the remaining ones). To motivate this, observe that overbidding is a dominated strategy in all the aforementioned non-truthful auctions. So we next turn to analyze auctions for clever, mean-based buyers.

---

[1] Note also that these canonical algorithms are mean-based even if the learning rate changes over time, as long as the learning rate is $\omega(1/T)$.

Here, the second main result of [4] characterizes the revenue-optimal repeated auction via a linear program, and shows that it takes a particularly simple form. On this front, we identify *several* formal barriers to extending this result to multiple buyers. Specifically:

- For a single buyer, [4] write a concise polytope (we call it the 'BMSW polytope') characterizing auctions which can be implemented for a single clever mean-based buyer (i.e. being in this polytope is necessary and sufficient to be implementable). We show that two natural extensions of this polytope to multiple buyers contain auctions which *cannot* be implemented for multiple clever mean-based buyers (we show that being in either natural polytope is necessary, but not sufficient). This is in Sect. 4.1.
- For a single buyer, [4] shows that the optimal auction is "pay-your-bid with declining reserve.²" We show that a natural generalization of this "pay-your-bid uniform auctions with declining reserve" to multiple buyers captures many extreme points of the multi-buyer BMSW polytope. But, we also show that such auctions are not necessarily optimal (meaning that this aspect of [4] does not generalize to multiple buyers either). This is in Sect. 4.2.
- Finally, we establish that not only does the particular multi-buyer BMSW polytope not capture all implementable auctions for clever mean-based buyers, but the space of implementable auctions is *not even convex*! This is in Sect. 4.3.

While our results are not a death sentence for the future work in the [4] model for clever mean-based buyers, the barriers do shut down most natural multi-buyer extensions of their approach. Still, these barriers also help focus future work towards other potentially fruitful approaches.

1.1 Related Work

There is a vast body of work at the intersection of learning and auction design. Much of this considers learning from the perspective of the *seller* (e.g. sample complexity of revenue-optimal auctions), and is not particularly related [5, 6, 18–20, 22–25]. More related is the recent and growing literature on dynamic auctions [1, 14, 17, 26, 28–30, 33] Like our model, the auction is repeated. The distinction between these works and ours is that they assume the buyer is *fully strategic* and processes fully how their actions today affect the seller's decisions tomorrow, perhaps the buyer needs to learn their (whereas we instead model buyers as no-regret learners).

---

² That is, each round there is a reserve. Any bid above the reserve wins the item, but pays their bid. The reserve declines over time.

The most related work to ours is in the [4] model itself, which studies the one seller one buyer scenario, where the buyer employs a mean-based no-regret algorithm. Follow-ups to [4] have extended the setting in [4] in a few different directions. First, [11,15,16] studies convergence of no-regret learning agents in a fixed mechanism such as first price auction, which diverges from the mechanism design perspective of [4]. More relevantly, [13,27] considers the problem of playing a two-player game against a no-regret learner. While technically not an auctions problem, there is thematic overlap with our main result. [12] extends the single-buyer results in [4] to be *prior-free*. Specifically, they show how to design auctions achieving the same guarantees as those in [4] but where the buyer's values are chosen adversarially. In comparison to these works, ours is the first to extend the model to consider multiple buyers.

Finally, [8] considers interaction between a learning buyer and a *learning* seller. Their seller does not have a prior against which to optimize, and instead itself targets a no-regret guarantee. In comparison, our seller (like the seller in all previously cited works) optimizes expected revenue with respect to a prior.

1.2   Organization

The rest of the paper will be organized as follows. In Sect. 2, we discuss our setting where buyer behavior falls under a broad class of no regret learning algorithms and introduce notations that will be used throughout the paper. In Sect. 3, we show that full surplus extraction is possible when buyers are using naive no-regret policies. In Sect. 4, we establish formal barriers in understanding optimal auction design when no regret buyers do not overbid their value. All missing proofs in Sect. 3 and Sect. 4 can be found in the full version of our paper[3].

2   Preliminaries

We consider the same setting as [4], extended to multiple buyers. Specifically, there are n buyers and T rounds. In each round, there is a single item for sale. Each buyer i has value $v_{i,t}$ for the item during round t, and each $v_{i,t}$ is drawn from $\mathcal{D}$ independently (that is, the buyers are i.i.d., and the rounds are i.i.d. as well). For simplicity of exposition (and to match prior work), we assume $\mathcal{D}$ has finite support $0 \leq w_1 < w_2 < \ldots < w_m \leq 1$ and we define q_j to be the probability w_j is drawn from $\mathcal{D}$.

Each round, the seller presents K arms for the buyers. Each arm is labeled with a bid, and we assume that one of the arms is labeled with 0 (to represent a bid of "don't participate"). Note that the same set of arms is presented to all buyers, and the same set of arms is presented in each round.

---

[3] https://arxiv.org/abs/2307.04175.

In each round t, the seller defines an anonymous auction. Specifically, for all i, t, the seller defines $a_{i,t}(\boldsymbol{b})$ to be the probability that buyer i gets the item in round t, and $p_{i,t}(\boldsymbol{b}) \in [0, b_i \cdot a_{i,t}(\boldsymbol{b})]$ to be the price buyer i pays, when each buyer j pulls the arm labeled b_j. To be anonymous, it must further be that for all permutations σ of the buyers that $(a_{\sigma(i),t}(\sigma(\boldsymbol{b})), p_{\sigma(i),t}(\sigma(\boldsymbol{b}))) = (a_{i,t}(\boldsymbol{b}), p_{i,t}(\boldsymbol{b}))$ (the auction is invariant under relabeling buyers). The only additional constraints on a are that $\sum_i a_{i,t}(\boldsymbol{b}) \leq 1$, for all $t, \boldsymbol{b}$ (item can be awarded at most once), and that $b_i' > b_i \Rightarrow a_{i,t}(b_i; \boldsymbol{b}_{-i}) \leq a_{i,t}(b_i'; \boldsymbol{b}_{-i})$ for all $i, \boldsymbol{b}_{-i}, b_i, b_i'$ (allocation is monotone). p must also be monotone ($b_i' > b_i \Rightarrow p_{i,t}(b_i; \boldsymbol{b}_{-i}) \leq p_{i,t}(b_i'; \boldsymbol{b}_{-i})$). When we state prior work in the single-buyer setting, we may drop the buyer subscript of i (for instance, we will write $a_{1,t}(b_1)$ as $a_t(b)$).

2.1 Contextual Bandits

Like [4], we model the buyers as online learners. Also like [4], our results apply equally well to the experts and bandits model, where $v_{i,t}$ serves as buyer i's context for round t. Specifically:

- For all subsequent definitions below, fix a buyer i, fix a bid vector $\boldsymbol{b}_{-i,t}$ for all rounds t, and fix $a_{i,t}(\cdot)$.
- For any bid b, buyer i, and round t, define $r_{ibt}(v) := v \cdot a_{i,t}(b; \boldsymbol{b}_{-i}) - p_{i,t}(b; \boldsymbol{b}_{-i})$. That is, define $r_{ibt}(v)$ to be the utility during round t that buyer i would enjoy by bidding b with value v.
- For an algorithm S (decides a bid for round t based only on what it observes through rounds $t - 1$, and its value $v_{i,t}$)[4] that submits bids b_{it} in round t, its total payoff is $P(S) := \mathbb{E}[\sum_t r_{ib_{it}t}(v_{i,t})]$. The expectation is over any randomness in the bids b_{it}, as S may be a randomized algorithm, and the randomness in $v_{i,t}$.
- An algorithm is fixed-bid if $v_{it} = v_{it'} \Rightarrow b_{it} = b_{it'}$. That is, the algorithm may make different bids in different rounds, but only due to changes in the buyer's value. Let $\mathcal{F}$ denote the set of all fixed-bid strategies.
- The *regret* of an online learning algorithm S is $\max_{F \in \mathcal{F}}\{P(F) - P(S)\}$.
- An algorithm is δ-low regret if it guarantees regret at most δ *on every fixed sequence of auctions, and fixed bids of the other players.* We say that an algorithm is *no-regret* if it is δ-low regret for some $\delta = o(T)$.

Like [4], we are particularly interested in algorithms "like Multiplicative Weights Update:"

Definition 1 (Mean-Based Online Learning Algorithm, [4]). *Let* $\sigma_{i,b,s}(v) := \sum_{t<s} r_{ibt}(v)$. *An algorithm is* γ-mean-based *if whenever* $\sigma_{i,b,s}(v_{i,s}) < \sigma_{i,b',s}(v_{i,s}) - \gamma T$ *(for any b, b'), then the probability that the algorithm bids b during round s is at most γ. An algorithm is* mean-based *if it is γ-mean-based for some $\gamma = o(1)$.*

---

[4] In the bandits model, buyer i learns only $r_{ibt}(v)$ for the bid $b := b_{it}$ after each round t (and all v). In the experts model, buyer i learns $r_{ibt}(v)$ for all b (and all v).

As noted in [4], natural extensions of Multiplicative Weights, EXP3, Follow the Perturbed Leader, etc. to the contextual setting are all mean-based online learning algorithms.

2.2   Learners and Benchmarks

Before formally stating our main results, we first provide relevant benchmarks. We use $\mathsf{Val}_n(\mathcal{D}) := \mathbb{E}_{v \leftarrow \mathcal{D}^n}[\max_i v_i]$ to denote the expected maximum value among the n buyers. We use $\mathsf{Mye}_n(\mathcal{D})$ to denote the expected revenue of the optimal truthful auction when n buyers have values drawn from $\mathcal{D}$. We make the following quick observation, which holds for *any* low regret learning algorithm (and extends an observation made in [4] for a single buyer).

Observation 1. *The seller cannot achieve expected revenue beyond $T \cdot \mathsf{Val}_n(\mathcal{D}) + o(T)$ when buyers guarantee no-regret, even if the seller knows precisely what algorithms the buyers will use.*

Finally, we will consider two types of no-regret learners. One type we will consider is simply no-regret learners who use a mean-based learning algorithm. Second, we will consider no-regret learners who use a no-regret learning algorithm but *never overbid*. Specifically, such learners immediately remove from consideration all bids $b_{it} > v_{i,t}$, but otherwise satisfy the no-regret guarantee. We refer to such learners are *clever*. [4] motivate such learners by observing that in most (perhaps all) standard non-truthful auction formats, overbidding is a *dominated strategy*. For example, it is always better to bid truthfully than to overbid in a first-price auction, generalized first-price auction, generalized second-price auction, and all-pay auction.

2.3   Border's Theorem

Some of our work will use Border's theorem [2], which considers the following. Consider a monotone, anonymous (not necessarily truthful) single-item auction, and a fixed strategy $s(\cdot)$ which maps values to actions. Let $x(w_j)$ denote the probability that a buyer using action $s(w_j)$ wins the item, assuming that all other buyers' values are drawn i.i.d. from $\mathcal{D}$ and use strategy s as well. Border's theorem asks the following: when given some vector $\langle x_1, \ldots, x_m \rangle$, does there exists a monotone anonymous (not necessarily truthful) single-item auction such that $x(w_j) = x_j$ for all j? If so, we say that $\boldsymbol{x}$ is *Border-feasible*. Below is Border's theorem. We will not actually use the precise Border conditions in any of our proofs, just the fact that they exist and are linear in $\boldsymbol{x}$.

Theorem 2 (Border's Theorem [3,7,9]). *When the buyers are drawn i.i.d from $\mathcal{D}$ (meaning each buyer's probability of valuing the item at w_j is q_j), $\boldsymbol{x}$ is Border-feasible if and only if it satisfies the Border conditions:*

$$n \sum_{\ell \geq j} q_j \cdot x_j \leq 1 - (1 - \sum_{\ell \geq j} q_j)^n \quad \forall j \in [m].$$

3 Full Surplus Extraction from Mean-Based Buyers

Here, we show a repeated auction which achieves expected revenue arbitrarily close to $T \cdot \mathsf{Val}_n(\mathcal{D})$ when buyers are mean-based (but consider overbidding). We also note that our auction does not depend on the particular mean-based algorithms used. The auction does *barely* depend on $\mathcal{D}$, but only in initial "setup rounds" (the auction during almost all rounds does not depend on $\mathcal{D}$). Recall this guarantee is the best possible, due to Observation 1.

Theorem 3. *Whenever n buyers use strategies satisfying the mean-based guarantee, there exists a repeated auction which obtains revenue $T \cdot (1 - \delta)\mathsf{Val}_n(\mathcal{D}) - o(T)$ for any constant $\delta < 1$.*

In this language, one main result of [4] proves Theorem 3 when $n = 1$. Before diving into our proof, we remind the reader of the main challenge. In order to possibly extract this much revenue, the auction must somehow both (a) charge each winning buyer their full value, leaving them with zero utility, yet also (b) figure out which buyer has the highest value in each round, and give them the item. The distinction between the $n = 1$ and $n > 1$ case is in (b). When $n = 1$, it is still challenging to give the buyer the item every round while charging their full value, but at least the buyer does not need to convey any information to the seller (so, for example, it is not necessary to incentivize the buyer to pull distinct arms for each possible value—the buyers just need to pay their full value on average by the end). When $n > 1$, we need the buyer to pull a distinct arm for each of their possible values, because we need to make sure that the highest buyer wins the item (and the only information we learn about each buyer's value is the arm they pull).

Additional Notation. We now provide our auction and analysis, beginning with some additional notation for this section. We will divide the T rounds of the auction into *phases* of $2R$ consecutive rounds, where $R = \Omega(T)$. There are $P := T/(2R)$ total phases (so P is a constant, but it will be a large constant depending on δ). In our contruction, the first $m - 1$ phases will be the setup phases and the last $P - m + 1$ phases will be the main phases. The goal of the setup phases is to align buyer's incentives so that they will behave in a particular manner in later phases. The main phases are where we will extract most of our revenue.

Recall that there are K non-zero arms labeled $b_1 < \ldots < b_K$. Our construction will use $K := P$ arms. Because the buyers consider overbidding, the precise bid labels are not important, so long as they are sufficiently large (concretely, we set $b_i := 2w_m + i$). We will sometimes index arms using $b_j^\tau := b_{P+j-\tau}$. This notation will be helpful to remind the reader that b_j^τ is the arm that we intend to be pulled by a buyer with value w_j during main phase τ.[5]

3.1   Defining the Auction

Intuitively, our auction tries to do the following. In each phase τ, there is a targeted arm b_j^τ for each possible value w_j, so there are m arms that are (intended to be) pulled during each phase. Ideally, since w_j needs to transition from pulling $b_j^{\tau-1}$ in phase $\tau - 1$ to pulling b_j^τ in phase τ, at the beginning of phase τ, w_j should be indifferent between pulling $b_j^{\tau-1}$, w_j's favourite arm in phase $\tau - 1$; and b_j^τ, w_j's intended arm for phase τ. Let us for now assume this is true and see how we design the auction during phase τ (which contains $2R$ rounds).

The base auction each round is just a second-price auction, where pulling arm b_j^τ submits a bid of w_j. For the first R rounds of each phase, this is exactly the auction executed. Because the second-price auction is dominant strategy truthful, this lures a mean-based buyer with value w_j into having high cumulative reward for arm b_j^τ (and in particular, strictly higher than any other arm). For the second R rounds of each phase, the base auction will still be the same second-price auction, except we will now overcharge each buyer so that their average utility during all $2R$ rounds of auction in phase τ is close to zero. In principle, this is possible because the buyers have high cumulative utility for this arm from the first R rounds, and are purely mean-based (and so they will pay more than their value to pull an arm which is historically much better than all others).

Now, by design our auction in phase τ gives the item to the highest buyer most of the time, therefore the expected welfare is almost optimal. Meanwhile, the expected utility is close to 0, which means we have managed to extract revenue that is almost the full welfare in phase τ. Lastly, notice that cumulative utility for arm $b_j^{\tau+1}$ increases during phase τ, so our phase cannot last forever. If we set the phase length to be too long, then $b_j^{\tau+1}$ will become w_j's favourite arm before phase τ ends. This is exactly why we need multiple phases instead of one phase. Let us set our phase length in such a way that at the end of τ, the increase in cumulative utility for arm $b_j^{\tau+1}$ is just enough for w_j to be indifferent between b_j^τ and $b_j^{\tau+1}$, then the exact condition we assume at the start of phase τ is satisfied, but for phase $\tau + 1$. Thus we can safely start a new phase $\tau + 1$ and extract almost full welfare by the same auction design.

Of course, this is just intuition for why an auction like this could possibly work—significant details remain to prove that it does in fact work (including precisely the choice of overcharges, analyzing incentives between phases, etc.). The formal description of our auction can be found in the full version of our paper.

Definition 2 (Full Surplus Extraction Auction). *The FSE Auction uses the following allocation and payment rule in each round. There are two steps in*

---

[5] So for example, arm b_{P-m} will first be (intended to be) pulled by buyers with value w_1 in phase $m + 1$, then by buyers with value w_2 in phase $m + 2$, etc.

each round. First, based on the arm pulled, a bid is submitted on behalf of the buyer into a secondary auction. Then, the secondary auction is resolved. There are three types of arms:

- *Some arms are* dormant. *These arms don't enter the secondary auction (i.e. no item and 0 payment).*
- *Some arms are* active. *Pulling arm $b_{P-\tau+j} = b_j^\tau$ enters a bid of w_j into a secondary auction.*
- *Some arms are* retired. *Pulling a retired arm enters a bid of $w_m + 1$ into a secondary auction.*

Which arms are dormant/active/retired change each phase. In addition, the secondary auction resolves differently for the first $m - 1$ phases (we call these the setup phases) versus the last $P - m + 1$ phases (we call these the main phases). Think of $P \gg m$, so the main phases are what matter most, the setup phases are just a technical setup to get incentives to work out. In any main phase ($\tau \geq m$):

- *Active arms: $b_{P-\tau+1} = b_1^\tau$ through $b_{P-\tau+m} = b_m^\tau$. Dormant: below $b_{P-\tau+1}$. Retired: above $b_{P-\tau+m}$. Note that by our definition, the index of active arms decreases as τ increases. For instance, if in n^{th} phase the active arms are $b_h, b_{h-1}, \cdots, b_l$, then in the $n + 1^{th}$ phase the active arms are $b_{h-1}, b_{h-2}, \cdots, b_{l-1}$.*
- *The secondary auction awards the item to a uniformly random buyer who submits the highest bid.*
- *If the winning arm was retired (i.e., submitted a bid of $w_m + 1$), they pay $2w_m$.*
- *If the winning arm was active, the winner pays the second-highest bid.*
- *Additionally, in the second R rounds of a phase, if the highest bid is w_j and the second-highest bid is w_ℓ, the winner pays an additional surcharge of $2(w_j - w_\ell)$.*

Due to space constraint, the description of the setup phase can be found in the full version of the paper.

We first quickly confirm that the FSE Auction is monotone.

Observation 4. *The allocation and payment rule for the FSE auction are both monotone.*

3.2   Mean-Based Behavior

Before analyzing the expected revenue of the seller, we first analyze the behavior of mean-based buyers. The challenge, of course, is that the payoff from each arm depends on the behavior of the other buyers, who are themselves mean-based. So our goal is to establish that mean-based learning in the FSE auction forms some sort of "equilibrium", in the sense that one mean-based buyer pulls the desired arm almost-always provided that all other buyers pull the desired arm almost-always. Our first step is characterizing a buyer's payoff for each arm at each round, assuming that all other buyers pull the intended arm almost always.

The main steps in our proof are as follows. First, we analyze the cumulative payoff for a buyer with each possible value for each possible arm, assuming that each other buyer pulls their intended arm. We then conclude that a buyer with value w_j has highest cumulative utility for their intended arm for the entirety of each phase. However, we also establish that the utility they enjoy during each phase for their intended arm is 0. This means that every buyer has 0 utility at the end (up to $o(T)$), meaning that the seller's revenue is equal to the expected welfare. Because we give the item to the highest value buyer whenever they pull the intended arm, the welfare is $T \cdot \mathsf{Val}(\mathcal{D})$. We now proceed with each step.

In each of the technical lemmas below, we let $H_s(v, b)$ denote the cumulative payoff during rounds 0 to s that a buyer with value v would have enjoyed in hindsight by pulling arm b in the FSE Auction, assuming that all other buyers pull their intended arm for at least a $1 - o(1)$ fraction of the rounds during every main phase τ, and that they pull either their intended arm (if it exists) or b_P (otherwise) during every setup phase τ. We let $X_{VCG}(v)$ denote the probability that a bidder with value v wins a second-price auction when bidding truthfully against $n - 1$ values drawn independently from $\mathcal{D}$ (ties broken randomly). And we let $P_{VCG}(v)$ denote the interim payment made by a bidder with value v to a second-price auction, in expectation over $n-1$ other values drawn independently from $\mathcal{D}$.[6]

Lemma 1. *At the end of phase τ, the change in cumulative payoff of a buyer with value v for each arm satisfies:*

- *For dormant arms b, $H_{2R\tau}(v, b) - H_{2R(\tau-1)}(v, b) = 0$.*
- *For active arms: $H_{2R\tau}(v, b_j^\tau) - H_{2R(\tau-1)}(v, b_j^\tau) = 2R \cdot (v - w_j) \cdot X_{VCG}(w_j) \pm o(T)$.*
- *For retired arms: $H_{2R\tau}(v, b_j) - H_{2R(\tau-1)}(v, b_j) = 2R \cdot (v - 2w_m) \pm o(T)$.*

Corollary 1. *At the end of phase τ, the cumulative payoffs for a buyer with value v satisfy:*

- *If b_j is dormant during phase τ ($j \leq P - \tau$): $H_{2R\tau}(v, b) = 0$.*
- *If b_j is active during phase τ ($j \in [P - \tau + 1, P - \tau + m]$):*

$$H_{2R\tau}(v, b_j) = 2R \cdot \left(\sum_{k=1}^{j+\tau-P} (v - w_k) \cdot X_{VCG}(w_k) \right) \pm o(T).$$

- *If b_j is retired during τ ($j \geq P - \tau + m + 1$):*

$$H_{2R\tau}(v, b) = 2R \cdot \left((\tau - m) \cdot (v - 2w_m) + \sum_{k=1}^{m} (v - w_k) \cdot X_{VCG}(w_k) \right) \pm o(T).$$

---

[6] Formally, let X_i be independent draws from $\mathcal{D}$ for $i = 1$ to $n - 1$. Define $X_0 := v$. Let $X := \max_{i \geq 1}\{X_i\}$, and Y be an indicator random variable for the event that a uniformly random element in $\arg\max_{i \geq 0}\{X_i\}$ is 0. Then $P_{VCG}(v) := \mathbb{E}[X \cdot Y]$.

Lemma 2. *For all τ, at the start of each phase τ, when $j \leq \tau$, a buyer with value w_j has highest cumulative utility for arm b_j^τ, and also $b_j^{\tau-1}$. Specifically, for all other arms b_ℓ:*

$$H_{2R(\tau-1)}(w_j, b_j^{\tau-1}) \pm o(T) = H_{2R(\tau-1)}(w_j, b_j^\tau) > H_{2R(\tau-1)}(w_j, b_\ell) + \Omega(T).$$

When $j > \tau$, for all $b_\ell \neq \tau$, $H_{2R(\tau-1)}(w_j, b_\tau^\tau) > H_{2R(\tau-1)}(w_j, b_\ell) + \Omega(T)$.

Lemma 3. *For all τ, assuming that all other buyers pull their intended arm except for $o(T)$ rounds, a mean-based buyer with value w_j pulls arm b_j^τ (if it exists) for the first R rounds, except for at most $o(T)$ rounds. Otherwise, they pull arm $b_P = b_\tau^\tau$ for the first R rounds, except for at most $o(T)$ rounds.*

Lemma 4. *For all τ, assuming that all other buyers pull their intended arm except for $o(T)$ rounds, a mean-based buyer with value w_j pulls arm b_j^τ (if it exists) for the last R rounds, except for at most $o(T)$ rounds. Otherwise, they pull arm $b_P = b_\tau^\tau$ for the last R rounds, except for at most $o(T)$ rounds.*

Finally, we combine everything together to conclude the following:

Proposition 1. *When all buyers are mean-based, they all pull their intended arm in the FSE Auction, except for at most $o(T)$ rounds.*

3.3 Analyzing the Revenue

Finally, we show that when all buyers pull their intended arm, the FSE auction extracts full surplus.

Proof (Proof of Theorem 3). Except for the setup phases, and for rounds where buyers do not pull their intended arm, the auction gives the item to the highest buyer. Therefore, the expected welfare of the auction is at least $(1 - m/P)T \cdot \mathsf{Val}(\mathcal{D}) - o(T)$. Moreover, Lemma 1 establishes that through an entire phase, the cumulative utility of a buyer for pulling their intended arm is $0 \pm o(T)$. Therefore, the total utility of the mean-based buyer is at most $o(T)$. Therefore, the revenue is at least $(1 - m/P)T \cdot \mathsf{Val}(\mathcal{D}) - o(T)$. Setting $P \geq m/\delta$ completes the proof.

4 Clever Mean-Based Buyers

In this section we consider clever mean-based buyers, and identify three formal barriers to developing optimal auctions for multiple clever mean-based buyers. We develop each barrier in the subsections below. Section 4.1 reminds the reader of the [4] Linear Program, which exactly captures the optimal auction for a single clever mean-based buyer, and provides a natural extension to multiple buyers.

4.1   A Linear Programming Upper Bound

We first remind the reader of the [4] Linear Program, and give a natural extension to multiple buyers. We first explicitly define variables for the results of a repeated auction.

Let A be a repeated auction with n i.i.d buyers of value distribution $\mathcal{D}$. For each buyer i, let S_i denote a strategy which takes as input a value v_{it} for round t (and all other information available from previous rounds) and outputs an arm $b_{it}^{S_i}(v_{it})$ to pull in round t. Let $\boldsymbol{v} := \langle v_{it} \rangle_{i \in [n], t \in T}$, which is drawn from the product distribution $\times_{nT} \mathcal{D}$. We use the following notation:

$$Rev_A(\mathcal{D}, S_1, \ldots, S_n) := \mathop{\mathbb{E}}_{\boldsymbol{v}} \left[\sum_{i=1}^{n} \sum_t p_{i,t} \left(b_{it}^{S_i}(v_{it}); b_{-it}^{S_{-i}}(v_{-it}) \right) \right]$$

$$Rev_n(\mathcal{D}, S_1, \ldots, S_n) := \max_A \{ Rev_A(\mathcal{D}, S_1, \ldots, S_n) \}$$

$$X_{ij}^A(\mathcal{D}, S_1, \ldots, S_n) = \frac{1}{T} \mathop{\mathbb{E}}_{\boldsymbol{v}} \left[\sum_t a_{it} \left(b_{it}^{S_i}(w_j); b_{-it}^{S_{-i}}(v_{-it}) \right) \right]$$

$$Y_{ij}^A(\mathcal{D}, S_1, \ldots, S_n) = \frac{1}{T} \mathop{\mathbb{E}}_{\boldsymbol{v}} \left[\sum_t a_{it} \left(w_j; b_{-it}^{S_{-i}}(v_{-it}) \right) \right]$$

$$U_{ij}^A(\mathcal{D}, S_1, \ldots, S_n) = \frac{1}{T} \mathop{\mathbb{E}}_{\boldsymbol{v}} \left[\sum_t w_j \cdot a_{it} \left(b_{it}^{S_i}(w_j); b_{-it}^{S_{-i}}(v_{-it}) \right) - p_{it} \left(b_{it}^{S_i}(w_j); b_{-it}^{S_{-i}}(v_{-it}) \right) \right]$$

Definition 3 (Auction Feasible). *A tuple of m-vectors (x^*, y^*, u^*) is n-buyer auction feasible if there exists a repeated auction A, such that for all $\gamma = o(T)$, whenever n buyers with values drawn i.i.d. from $\mathcal{D}$ run clever γ-mean-based strategies $S_1, \ldots, S_n$, then $\forall i, X_{ij}^A(\mathcal{D}, S_1, \ldots, S_n) = x_j^* \pm O(\gamma); Y_{ij}^A(\mathcal{D}, S_1, \ldots, S_n) = y_j^* \pm O(\gamma); U_{ij}^A(\mathcal{D}, S_1, \ldots, S_n) = u_j^* \pm O(\gamma)$. We call (x^*, u^*) n-buyer auction feasible if there exists y^* such that (x^*, y^*, u^*) is n-buyer auction feasible.*

One key insight in [4] is that the space of auction feasible tuples is convex and can be characterized by simple linear equations. Below, note that the "only if" direction is slightly non-trivial, and we rederive it later for arbitrary n. The "if" direction requires designing an auction (for which we refer the interested reader to [4, Theorem 3.4]). We will not rederive the "if" direction, although we define the relevant auction later as well.

Theorem 5 ([4]). *(x, u) is 1-buyer auction feasible if and only if it satisfies the BMSW constraints:*[7]

$$u_i \geq (w_i - w_j) \cdot x_j, \quad \forall i, j \in [m] : i > j,$$
$$x_i \geq x_j, \quad \forall i \in [m], i > j,$$
$$u_i \geq 0, x_i \in [0, 1], \quad \forall i \in [m].$$

---

[7] In fact, the 'only if' portion of this theorem holds when replacing the clever mean-based buyer with just a clever buyer. But the 'if' portion requires the stronger assumption of mean-based learning.

Intuitively, the first BMSW constraint is the interesting one, which is necessary for the buyer to not regret pulling arm b_j when their value is w_i (again recall this is non-trivial, but we argue this shortly as a special case for general n). The second constraint is necessary because the auction must be monotone. The final constraint is necessary because the auction must have a null arm, and because all allocation probabilities must be in $[0, 1]$ every round.

[4] also observe that the expected revenue of an auction A can be computed as a linear function of $X_{ij}^A(\mathcal{D}, S_1, \ldots, S_n)$ and $U_{ij}^A(\mathcal{D}, S_1, \ldots, S_n)$ (because revenue = welfare − utility). Therefore, Theorem 5 enables a simple LP formulation to find the optimal auction for clever buyers.

We consider two natural attempts to generalize Theorem 5, and show that both hold only in the 'only if' direction. The reason the BMSW constraints don't work verbatim for multiple buyers is that the feasibility constraints are wrong: it is not feasible to (for instance) have each buyer win the item with probability 1 every round. Indeed, there is only one copy of the item, implying (for instance) that $n \sum_i q_i \cdot x(w_i) \leq 1$, but also stronger conditions. These conditions are known as *Border's constraints* from Theorem 2 [2].

Proposition 2. *A tuple (x, y, u) is n-buyer auction feasible only if it satisfies the n-buyer BMSW constraints below. A tuple (x, u) is n-buyer auction feasible only if it satisfies the reduced n-buyer BMSW constraints.*[8]

n-buyer BMSW Constraints	**Reduced n-buyer BMSW Constraints**
$u_i \geq (w_i - w_j) \cdot y_j, \ \forall i > j \in [m],$	$u_i \geq (w_i - w_j) \cdot x_j, \ \forall i > j \in [m],$
$y_i \geq x_i, \ \forall i \in [m],$	
$u_i \geq 0, \ \forall i \in [m],$	$u_i \geq 0, \ \forall i \in [m],$
x satisfies Border's constraints,	*x satisfies Border's constraints,*
x, y monotone.	*x monotone.*

We next turn to see whether the other direction holds, as in Theorem 5 for the single-buyer case. If it did, then we could again write a linear program to find the optimal n-buyer feasible auction, because the expected revenue can be written as a function of (x, u). However, we provide an example showing that this extension is *false*.

Theorem 6. *There exist (x, y, u) that satisfy the n-buyer BMSW Constraints but are not n-buyer auction feasible, and (x, u) that satisfy the Reduced n-buyer BMSW Constraints but are not n-buyer auction feasible.*

4.2 Uniform Auctions with Declining Reserves

In this section, we consider the following possibility: although n-buyer BMSW constraints don't characterize the n-buyer feasible auctions, it is conceivable

---

[8] In fact, this claim holds when replacing mean-based clever buyers with just clever buyers, just like the 'only if' part of Theorem 5.

(although perhaps unlikely) that the Linear Programming solution (optimizing expected revenue subject to n-buyer BMSW constraints) happens to always yield an n-buyer feasible auction. The reason this is a priori possible is because the objective function and BMSW constraints are related: the objective function depends on $\mathcal{D}$, and so do the n-buyer Border constraints (this is another way in which n-buyer and 1-buyer auctions differ: 1-buyer Border constraints don't depend on $\mathcal{D}$).

For the single-buyer case, [4] shows that not only is every tuple satisfying the BMSW constraints 1-buyer auction feasible, but the auction witnessing this is particularly simple. First, whenever the buyer gets the item, they pay their bid (and in each round, each arm gives the item with probability 0 or 1). Second, the minimum winning bid is declining over time. We generalize both definitions below to multiple buyers, and show a connection between these auctions and certain types of tuples which satisfy the n-buyer BMSW conditions.

Definition 4 (Pay-your-bid). *A repeated auction is pay-your-bid if* $p_{i,t}(\boldsymbol{b}) = b_i \cdot a_{i,t}(\boldsymbol{b})$ *for all* i, t.

Definition 5 (Uniform Auction with Declining Reserve). *A repeated auction is a* uniform auction with declining reserve *when: (a) there exists a reserve* r_t *for every round* t *which is monotonically decreasing in* t, *and (b) in each round the item is awarded to a uniformly random buyer among those with* $b_{it} \geq r_t$.

Definition 6 (Correspondence). *We call* (x, y, u) *the* corresponding tuple *of repeated auction* A *if for 0-mean-based strategies* $S_1, \ldots, S_n$ *and all* i, j:
$$X_{ij}^A(\mathcal{D}, S_1, \ldots, S_n) = x_j; \quad Y_{ij}^A(\mathcal{D}, S_1, \ldots, S_n) = y_j; \quad U_{ij}^A(\mathcal{D}, S_1, \ldots, S_n) = u_j.[9]$$

In this language, [4] shows that when $n = 1$, every tuple which satisfies the BMSW conditions can be implemented as a pay-your-bid uniform auction with declining reserve (and this establishes the 'if' direction of Theorem 5). Due to Theorem 6, this claim clearly cannot extend to $n > 1$. However, we show that certain kinds of natural tuples can all be implemented as pay-your-bid uniform auctions with declining reserves.

Theorem 7. *Consider any repeated auction* A *and its corresponding tuple* (x, y, u). *If* (x, y, u) *satisfies the* n-buyer BMSW constraints, and $x = y$, and A is pay-your-bid, then A is a uniform auction with declining reserve.

With Theorem 7 in mind, another possible avenue towards characterizing optimal n-buyer feasible auctions would be through pay-your-bid uniform auctions with declining reserves. To this end, we first show that the optimal pay-your-bid uniform auction with declining reserve can be found by a linear program. However, we also show that examples exist where the optimal n-buyer feasible auction strictly outperforms the best pay-your-bid uniform auction with declining reserve.

---

[9] Observe that this is always well-defined, as the unique 0-mean-based strategy is Follow-the-leader.

Theorem 8. *The optimal[10] pay-your-bid uniform auction with declining reserve can be found by a linear program of size* Poly(m). *However, there exist 2-buyer instances where the optimal 2-buyer feasible auction strictly outperforms the best pay-your-bid uniform auction with declining reserve.*

4.3   Non-convexity of N-Buyer Feasible Auctions

Finally, we consider the possibility that while the n-buyer BMSW constraints don't capture the space of n-buyer feasible auctions, perhaps some other compact, convex space does. This too is not the case, as we show that the space of n-buyer feasible triples is non-convex (subject to one technical restriction).

Theorem 9. *Let P denote the set of all (x, y, u) that are n-buyer feasible auctions where the bid space is equal to the support of $\mathcal{D}$. Then P is not necessarily convex, even when $n = 2$.*

References

1. Ashlagi, I., Daskalakis, C., Haghpanah, N.: Sequential mechanisms with ex-post participation guarantees. In: Proceedings of the 2016 ACM Conference on Economics and Computation, EC 2016, Maastricht, The Netherlands, 24–28 July 2016, pp. 213–214 (2016). https://doi.org/10.1145/2940716.2940775
2. Border, K.C.: Implementation of reduced form auctions: a geometric approach. Econometrica **59**(4), 1175–1187 (1991)
3. Border, K.C.: Reduced form auctions revisited. Econ. Theor. **31**, 167–181 (2007)
4. Braverman, M., Mao, J., Schneider, J., Weinberg, M.: Selling to a no-regret buyer. In: Proceedings of the 2018 ACM Conference on Economics and Computation, Ithaca, NY, USA, 18–22 June 2018, pp. 523–538 (2018). https://doi.org/10.1145/3219166.3219233
5. Brustle, J., Cai, Y., Daskalakis, C.: Multi-item mechanisms without item-independence: learnability via robustness. In: Biró, P., Hartline, J., Ostrovsky, M., Procaccia, A.D. (eds.) EC 2020: The 21st ACM Conference on Economics and Computation, Virtual Event, Hungary, 13–17 July 2020, pp. 715–761. ACM (2020). https://doi.org/10.1145/3391403.3399541
6. Cai, Y., Daskalakis, C.: Learning multi-item auctions with (or without) samples. In: 58th IEEE Annual Symposium on Foundations of Computer Science, FOCS 2017, Berkeley, CA, USA, 15–17 October 2017, pp. 516–527 (2017). https://doi.org/10.1109/FOCS.2017.54
7. Cai, Y., Daskalakis, C., Weinberg, S.M.: A constructive approach to reduced-form auctions with applications to multi-item mechanism design. arXiv preprint arXiv:1112.4572 (2011)
8. Camara, M.K., Hartline, J.D., Johnsen, A.C.: Mechanisms for a no-regret agent: beyond the common prior. In: 61st IEEE Annual Symposium on Foundations of Computer Science, FOCS 2020, Durham, NC, USA, 16–19 November 2020, pp. 259–270. IEEE (2020). https://doi.org/10.1109/FOCS46700.2020.00033

---

[10] By optimal we mean the auction achieves the best revenue if all buyers run 0-regret algorithms. It is easy to see that when buyers have γ-regret, the revenue is within $O(n\gamma)$ of the revenue when buyers have 0-regret.

9. Che, Y.K., Kim, J., Mierendorff, K.: Generalized reduced-form auctions: a network-flow approach. Econometrica **81**, 2487–2520 (2013)
10. Clarke, E.H.: Multipart pricing of public goods. Public Choice **11**(1), 17–33 (1971)
11. Deng, X., Hu, X., Lin, T., Zheng, W.: Nash convergence of mean-based learning algorithms in first price auctions. In: Laforest, F., et al. (eds.) WWW 2022: The ACM Web Conference 2022, Virtual Event, Lyon, France, 25–29 April 2022, pp. 141–150. ACM (2022). https://doi.org/10.1145/3485447.3512059
12. Deng, Y., Schneider, J., Sivan, B.: Prior-free dynamic auctions with low regret buyers. In: Wallach, H.M., Larochelle, H., Beygelzimer, A., d'Alché-Buc, F., Fox, E.B., Garnett, R. (eds.) Advances in Neural Information Processing Systems 32: Annual Conference on Neural Information Processing Systems 2019, NeurIPS 2019, 8–14 December 2019, Vancouver, BC, Canada, pp. 4804–4814 (2019). http://papers.nips.cc/paper/8727-prior-free-dynamic-auctions-with-low-regret-buyers
13. Deng, Y., Schneider, J., Sivan, B.: Strategizing against no-regret learners. In: Wallach, H.M., Larochelle, H., Beygelzimer, A., d'Alché-Buc, F., Fox, E.B., Garnett, R. (eds.) Advances in Neural Information Processing Systems 32: Annual Conference on Neural Information Processing Systems 2019, NeurIPS 2019, 8–14 December 2019, Vancouver, BC, Canada, pp. 1577–1585 (2019). http://papers.nips.cc/paper/8436-strategizing-against-no-regret-learners
14. Deng, Y., Zhang, H.: Prior-independent dynamic auctions for a value-maximizing buyer. In: Ranzato, M., Beygelzimer, A., Dauphin, Y.N., Liang, P., Vaughan, J.W. (eds.) Advances in Neural Information Processing Systems 34: Annual Conference on Neural Information Processing Systems 2021, NeurIPS 2021, 6–14 December 2021, Virtual, pp. 13847–13858 (2021)
15. Feng, S., Yu, F., Chen, Y.: Peer prediction for learning agents. In: NeurIPS (2022)
16. Feng, Z., Guruganesh, G., Liaw, C., Mehta, A., Sethi, A.: Convergence analysis of no-regret bidding algorithms in repeated auctions. In: Thirty-Fifth AAAI Conference on Artificial Intelligence, AAAI 2021, Thirty-Third Conference on Innovative Applications of Artificial Intelligence, IAAI 2021, The Eleventh Symposium on Educational Advances in Artificial Intelligence, EAAI 2021, Virtual Event, 2–9 February 2021, pp. 5399–5406. AAAI Press (2021). https://doi.org/10.1609/aaai.v35i6.16680
17. Golrezaei, N., Jaillet, P., Liang, J.C.N., Mirrokni, V.: Bidding and pricing in budget and ROI constrained markets. arXiv preprint arXiv:2107.07725 (2021)
18. Gonczarowski, Y.A., Nisan, N.: Efficient empirical revenue maximization in single-parameter auction environments. In: Proceedings of the 49th Annual ACM SIGACT Symposium on Theory of Computing, STOC 2017, Montreal, QC, Canada, 19–23 June 2017, pp. 856–868 (2017). https://doi.org/10.1145/3055399.3055427
19. Gonczarowski, Y.A., Weinberg, S.M.: The sample complexity of up-to-ε multidimensional revenue maximization. In: 59th IEEE Annual Symposium on Foundations of Computer Science, FOCS (2018)
20. Gonczarowski, Y.A., Weinberg, S.M.: The sample complexity of up-to-ϵ multidimensional revenue maximization. J. ACM **68**(3), 15:1–15:28 (2021). https://doi.org/10.1145/3439722
21. Groves, T.: Incentives in teams. Econometrica **41**(4), 617–631 (1973)
22. Guo, C., Huang, Z., Tang, Z.G., Zhang, X.: Generalizing complex hypotheses on product distributions: auctions, prophet inequalities, and Pandora's problem. arXiv preprint arXiv:1911.11936 (2019)

23. Guo, C., Huang, Z., Zhang, X.: Settling the sample complexity of single-parameter revenue maximization. In: Proceedings of the 51st Annual ACM SIGACT Symposium on Theory of Computing, STOC 2019, Phoenix, AZ, USA, 23–26 June 2019, pp. 662–673 (2019). https://doi.org/10.1145/3313276.3316325
24. Guo, W., Jordan, M.I., Zampetakis, E.: Robust learning of optimal auctions. In: Ranzato, M., Beygelzimer, A., Dauphin, Y.N., Liang, P., Vaughan, J.W. (eds.) Advances in Neural Information Processing Systems 34: Annual Conference on Neural Information Processing Systems 2021, NeurIPS 2021, 6–14 December 2021, Virtual, pp. 21273–21284 (2021)
25. Hartline, J.D., Taggart, S.: Sample complexity for non-truthful mechanisms. In: Proceedings of the 2019 ACM Conference on Economics and Computation, EC 2019, Phoenix, AZ, USA, 24–28 June 2019, pp. 399–416 (2019). https://doi.org/10.1145/3328526.3329632
26. Liu, S., Psomas, C.: On the competition complexity of dynamic mechanism design. In: Proceedings of the Twenty-Ninth Annual ACM-SIAM Symposium on Discrete Algorithms, SODA 2018, New Orleans, LA, USA, 7–10 January 2018, pp. 2008–2025 (2018). https://doi.org/10.1137/1.9781611975031.131
27. Mansour, Y., Mohri, M., Schneider, J., Sivan, B.: Strategizing against learners in bayesian games. In: Loh, P., Raginsky, M. (eds.) Conference on Learning Theory, 2–5 July 2022, London, UK. Proceedings of Machine Learning Research, vol. 178, pp. 5221–5252. PMLR (2022). https://proceedings.mlr.press/v178/mansour22a.html
28. Mirrokni, V.S., Leme, R.P., Tang, P., Zuo, S.: Dynamic auctions with bank accounts. In: Proceedings of the Twenty-Fifth International Joint Conference on Artificial Intelligence, IJCAI 2016, New York, NY, USA, 9–15 July 2016, pp. 387–393 (2016). http://www.ijcai.org/Abstract/16/062
29. Mirrokni, V.S., Leme, R.P., Tang, P., Zuo, S.: Non-clairvoyant dynamic mechanism design. In: Proceedings of the 2018 ACM Conference on Economics and Computation, Ithaca, NY, USA, 18–22 June 2018, p. 169 (2018). https://doi.org/10.1145/3219166.3219224
30. Mirrokni, V.S., Leme, R.P., Tang, P., Zuo, S.: Optimal dynamic auctions are virtual welfare maximizers. In: The Thirty-Third AAAI Conference on Artificial Intelligence, AAAI 2019, The Thirty-First Innovative Applications of Artificial Intelligence Conference, IAAI 2019, The Ninth AAAI Symposium on Educational Advances in Artificial Intelligence, EAAI 2019, Honolulu, Hawaii, USA, 27 January–1 February 2019, pp. 2125–2132. AAAI Press (2019). https://doi.org/10.1609/aaai.v33i01.33012125
31. Myerson, R.B.: Optimal auction design. Math. Oper. Res. 6(1), 58–73 (1981)
32. Nekipelov, D., Syrgkanis, V., Tardos, E.: Econometrics for learning agents. In: Proceedings of the Sixteenth ACM Conference on Economics and Computation, EC 2015, pp. 1–18. ACM, New York (2015). https://doi.org/10.1145/2764468.2764522
33. Papadimitriou, C.H., Pierrakos, G., Psomas, C., Rubinstein, A.: On the complexity of dynamic mechanism design. In: Proceedings of the Twenty-Seventh Annual ACM-SIAM Symposium on Discrete Algorithms, SODA 2016, Arlington, VA, USA, 10–12 January 2016, pp. 1458–1475 (2016). https://doi.org/10.1137/1.9781611974331.ch100
34. Vickrey, W.: Counterspeculations, auctions, and competitive sealed tenders. J. Financ. 16(1), 8–37 (1961)

Penalties and Rewards for Fair Learning in Paired Kidney Exchange Programs

Margarida Carvalho[1], Alison Caulfield[2], Yi Lin[2], and Adrian Vetta[2,3(✉)]

[1] Department of Computer Science and Operations Research,
Université de Montréal, Montreal, Canada
`carvalho@iro.umontreal.ca`
[2] Mathematics and Statistics, McGill University, Montreal, Canada
`{alison.caulfield,yi.lin2}@mail.mcgill.ca`
[3] School of Computer Science, McGill University, Montreal, Canada
`adrian.vetta@mcgill.ca`

Abstract. A kidney exchange program, also called a kidney paired donation program, can be viewed as a repeated, dynamic trading and allocation mechanism. This suggests that a dynamic algorithm for transplant exchange selection may have superior performance in comparison to the repeated use of a static algorithm. We confirm this hypothesis using a full scale simulation of the Canadian Kidney Paired Donation Program: learning algorithms, that attempt to learn optimal patient-donor weights in advance via dynamic simulations, do lead to improved outcomes. Specifically, our learning algorithms, designed with the objective of fairness (that is, equity in terms of transplant accessibility across cPRA groups), also lead to an increased number of transplants and shorter average waiting times. Indeed, our highest performing learning algorithm improves egalitarian fairness by 10% whilst also increasing the number of transplants by 6% and decreasing waiting times by 24%. However, our main result is much more surprising. We find that the most critical factor in determining the performance of a kidney exchange program is **not** the judicious assignment of positive weights (rewards) to patient-donor pairs. Rather, the key factor in increasing the number of transplants, decreasing waiting times and improving group fairness is the judicious assignment of a negative weight (penalty) to the small number of non-directed donors in the kidney exchange program.

Keywords: Kidney exchange programs · Learning weights · Fairness · Canadian Kidney Paired Donation Program · Integer Programming

1 Introduction

The Canadian Kidney Paired Donation (KPD) program is a kidney exchange program (KEP) consisting of incompatible patient-donor pairs and non-directed anonymous donors, henceforth called *nodes*. The program uses a weighting (point) scheme to select compatible kidney exchanges between the donors and patients [8]. Our work was motivated by the accumulation over time of *hard-to-match* patients in the KPD program. We ask whether or not this phenomenon,

© The Author(s), under exclusive license to Springer Nature Switzerland AG 2024
J. Garg et al. (Eds.): WINE 2023, LNCS 14413, pp. 130–150, 2024.
https://doi.org/10.1007/978-3-031-48974-7_8

a recurring problem in KEPs [20], can be mitigated via the use of dynamic learning algorithms to select arc/node weights in the kidney exchange graphs used for transplant selection. We answer this question in the affirmative. Our key finding is that a judicious choice of weights can simultaneously improve performance with respect to multiple criteria including group fairness, waiting times, and the number of transplants. However, this conclusion requires that negative weights (penalties) are permitted in transplant selection, in addition to positive weights (rewards).

Background and Related Literature. Kidney exchanges were first proposed by Rapaport [33] and the first programs were instigated in South Korea in the 1990s [27,31]. Recently, these programs have spread across the globe: see [29] and [26] for details on the national kidney exchange programs in the UK and the Netherlands, respectively; see also [4] for an overview of KEPs in Europe. Canada started its own KEP, entitled the *Kidney Paired Donation (KPD) Program*, in 2008 [14].

The theoretical foundations underlying kidney exchange programs were provided by Roth et al. [34]; see also Sönmez and Ünver [37]. Succinctly, a KEP can be modelled by a directed graph, called a *kidney exchange graph* whose nodes consist of incompatible patient-donor pairs, denoted (p_i, d_i), and non-directed anonymous donors (NDADs), denoted $(\emptyset, d_j)$, also called altruistic donors. An arc in the graph indicates that a transplant is feasible between the corresponding donor and patient. We refer the reader to Sect. 2 for a detailed description of kidney exchange graphs. For now, we remark that kidney exchanges are selected using either a directed cycle or a directed simple path in the graph. For an illustration consider Fig. 1. Notice that when the cycle is selected, the altruistic donor d_4 is unused and remains available for future transplants. On the other hand, with the path, the donor d_1 is unused. The importance of directed paths and, thus, of altruistic donors in transplant selection was originally highlighted by Montgomery et al. [30] and Roth et al. [36]. Moreover, a detailed investigation into the most efficacious use of altruistic donors will be a key focus of this work.

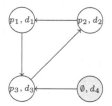 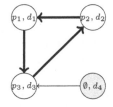 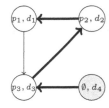

(a) A kidney exchange graph.   (b) A transplant cycle.   (c) A transplant path (chain).

Fig. 1. Illustration of feasible exchanges for a KEP.

The big question is what transplants (arcs) should be selected from the graph to optimize health care outcomes. At first thought, this appears obvious: maximize the number of transplants. This is the approach taken in KEPs that use the myopic algorithm for transplant selection. There are three major reasons why it is questionable. First, transplant selection is a multi-criteria decision problem, of which the quantity of transplants is just one aspect. For example, medical practitioners must also take into account the health of the patient and the urgency of the case, waiting times (both individual and collective), the quality of potential transplants, etc. Thus, choosing to maximize the number of transplants above other objectives is a subjective decision. Second, the myopic algorithm is implicitly unfair. It will prioritize patients who are easy-to-match (e.g., in terms of blood-group compatibility and antibody compatibility) at the expense of hard-to-match patients. Third, KEPs are repeated, dynamic mechanisms. This, in turn, has several consequences. Perhaps counter-intuitively, algorithms to optimize an objective in the short-term may not optimize the objective over the long-term. A striking example of this is the use of static pricing mechanisms in electricity markets. There, pricing for short-term efficiency can lead to huge inefficiencies in the long term [13]. For KEPs, this means the myopic approach of maximizing the number of transplants in each round may not maximize the number of transplants over many years. Our experiments will show this fact clearly. The dynamic nature of KEPs can also have deleterious effects. For example, a serious problem currently facing KEPs is the accumulation over time of hard-to-match patients.

A standard approach, and the one taken in the KPD program [14], to attempt to address these issues is to add weights (also called points or rewards) to prioritize certain types of exchanges. For instance, higher weight may be assigned to hard-to-match patients such as those who are highly-sensitized[1] or have O-blood type. Dickerson et al. [18] proposed the use of weights and in [20], they explain the usefulness of weights in dynamic contexts and in dealing with different objectives; see also [21,25]. Given a weighting, the resultant optimization problem is to find a maximum weight collection of node-disjoint cycles and paths in the kidney exchange graph. The optimization problem is NP-hard, but state-of-the-art software works well in practice. In addition to arc/node weightings, hierarchical optimization techniques have been used to incorporate prioritization criteria [15]; see Biro et al. [4,5] for a overview of weighted and hierarchical approaches in European KEPs.

So weights are important in KEPs, but what is the best choice of weights? The myopic approach corresponds to assigning equal weight to each patient-donor node. An alternative approach is expert-selected weights, where medical practitioners formulate a set of rules by which patient-donor nodes and/or arcs are prioritized (weighted). We apply a third approach: given the dynamic nature of a KEP, we use dynamic learning algorithms to select the weights. But, as alluded to, for multi-criteria decision problems such as in KEPs, the "best" choice of weights will depend upon the somewhat subjective choice of what criteria to optimize. Our focus in this paper is to learn weights to optimize *group fairness*. Informally, we will group the patients according to how hard they are to match and attempt to design transplant selection algorithms that are fair for each group

---

[1] Patients with high calculated Panel Reactive Antibody (cPRA) rates.

(see Sect. 4 for details). In particular, we wish to reduce the accumulation over time of hard-to-match patients. Fortunately, as will be shown, group fairness can be achieved whilst also balancing the objectives of maximizing transplant numbers and minimizing waiting times.

Overview and Results. In Sect. 2 we describe kidney exchange programs and explain how to model them using a kidney exchange graph. In Sect. 3 we present a collection of learning algorithms designed to learn weights with the aim of optimizing group fairness. The weights are learnt using a full-scale simulation of end-stage kidney disease patients in Canada. We also use the simulator to assess the performance of these algorithms in comparison to both the myopic algorithm and the Canadian KPD algorithm. For both tasks, learning the weights and performance testing, we use the implementation of the position-index formulations provided in [17].

In Sect. 4 we present a specific class of group fairness measures which will be used to analyze the performance of our transplant selection algorithms. The three most important measures in the class are *Nash group fairness*, *utilitarian group fairness*, and *egalitarian group fairness*; indeed, the latter measure is particularly suited for the task of evaluating equity in transplant accessibility.

The results from our experiments are described in Sect. 5. All of our learning algorithms significantly outperform the myopic algorithm and the Canadian KPD algorithm; see Table 1. In particular, our overall highest performing algorithm improves the egalitarian fairness measure by 10% over the Canadian KPD algorithm and by 21% over the myopic algorithm. That is perhaps unsurprising given that the algorithms were designed to optimize fairness. However, the learning algorithm also leads to an increase of 6% in the number of transplants, a decrease of 24% in waiting times, and a 2% reduction in the use of altruistic donors. This is much more surprising. In particular, the myopic algorithm was designed to maximize the number of transplants but the learning algorithms do significant better in the long run for that criterion. This illustrates the need to incorporate the repeated and dynamic nature of KEPs into transplant selection algorithms.

As stated, our experiments were tested using Canadian population statistics, but we believe the take-home lessons from these experiments apply to the design of KEPs more generally. These include the following six lessons:

Lesson I. *There is negligible benefit in allowing for paths of length greater than 5.*

Lesson II. *It is very important to allow for cycles of length 3. The improvements with cycles of lengths 4 or 5 are marginal.*

Lesson III. *Assigning an appropriate negative weight (penalty) to each altruistic donor is fundamental in determining the quality of outcomes.* [Table 2 and Table 3]

Lesson IV. *The marginal contribution in terms of lives saved per altruistic donor is non-increasing in the number of altruists. (At the current altruist arrival rate in Canada, each altruistic donor saves approximately two lives.)* [Table 4]

Lesson V. *Fair algorithms can be used to reduce the accumulation of hard-to-match patients in waiting pools.*   [Table 5]

Lesson VI. *Fair algorithms can be implemented without detrimental affects on utilitarian criteria such as waiting times and the number of transplants.* [Table 1 and Table 3]

We remark the first two lessons will not surprise practitioners. Whilst the potential importance of long paths was originally highlighted by Ashlagi et al. [2], early works gave theoretical and experimental evidence that consideration of only short cycles [35] or short paths [19] suffice.

The last two lessons validate the approach taken in this paper. Lesson V confirms that learning algorithms can reduce the accumulation of hard-to-match patients. Moreover improved fairness need not come at a cost; our fair learning algorithms can be used to simultaneously improve the performance of a KEP over a range a measures.

What is surprising, at least to us, is that Lesson III turns out to be by far the most important of the lessons. Put simply, weighting (negatively) altruistic donors has much greater beneficial effects than weighting (positively) all the other nodes combined! To our knowledge this fact has not previously been discovered. Indeed, as we show in Table 3, simply incorporating a negative weight or penalty for each altruistic donor in the myopic and KPD algorithms produces outcomes that are comparable to our learning algorithms. We remark that this lesson contrasts with recent work [22] giving theoretical arguments for the prioritization of paths over cycles.

Lesson IV is useful because it provides for an understanding of how to exploit Lesson III for maximum benefit. Furthermore, there is evidence [19] that some altruistic donors prefer a shorter active time on a deceased donor program list rather than a longer active time on a living donor program list. Knowledge that remaining in the KEP will save approximately two lives rather than just one in the deceased donor program may help alleviate this issue.[2]

2   The Graphical Kidney Exchange Model

A Static Model. A KEP can be modelled by a directed graph $G = (V, A)$, called an *exchange graph*. A node $i \in V$ is either a *patient-donor pair* or is a *non-directed anonymous donor* (NDAD), also called an *altruistic donor*. In the former case, $i = \{p_i, d_i\}$ consists of a patient p_i and an accompanying (incompatible) donor d_i. In the later case, $i = \{\emptyset, d_i\}$ consists simply of a donor with no accompanying patient. There is an arc $(i, j) \in A$ if the kidney of donor d_i is compatible with patient p_j.

It follows that a directed cycle in G of patient-donor pairs forms a feasible exchange of donors and thus, a feasible set of transplants. Furthermore, a

---

[2] Of course, if an altruistic donor requests only a short wait before donation then that can easily be accommodated in our model with the use of individual weight adjustments. However, such requests induce serious ethical questions [32,40].

directed path, also called a *chain*, whose source is an altruistic donor also forms a feasible set of transplants. Therefore, a feasible set of transplants is a node-disjoint collection of paths and cycles.

It will be important to understand the factors that determine transplant compatibility and, hence, the arcs in the exchange graph. The first major factor is *blood-group compatibility*. A patient of blood group O can only receive a transplant from a donor also of group O; a patient of group A or B can receive a transplant from a donor of group O or of their own blood group. a patient of group AB can receive a transplant from a donor of any blood group. The second major factor concerns *antibodies* and *antigens*. Antibodies protect against foreign pathogens. However, if an antibody in the patient recognises and binds to an antigen from the donor this can lead to graft failure. This is measured by the cPRA rate which estimates the proportion of donors against which the patient has antibodies rendering the match incompatible: easy-to-match patients have a cPRA close to zero and hard-to-match patients close to one.

Transplant Selection. The general approach for transplant selection is to add a weight w_i to each node i or a weight w_{ij} to each arc (i, j) in the graph. This induces an optimization problem whose optimal solution is a set of node-disjoint cycles and paths of maximum total weight. In addition, it is natural to impose a maximum cycle length, C, and a maximum path length, P, where the length of a cycle/path is the corresponding number of patients. Three basic methods are used in selecting the weights. First, the simplest approach uses $w_i = 1$ for every patient-donor pair i (and $w_i = 0$ for every altruistic donor i). The resultant objective is then to maximize the total number of transplants subject to the cycle and path length constraints; the corresponding algorithms are called *myopic*. The second approach uses *expert-defined* weights. Here the arc/node weights are prescribed in advance by medical experts to signify the relative priority and importance of a given transplant. This is the method implemented by the Canadian KPD program (Sect. 3). The third approach uses learned-weights. The weights are learnt internally by the system to optimize some performance objective [18]. This will be the approach taken in this paper where our objective will be group fairness.

A Dynamic Model. We can formulate and solve this kidney-exchange model. But this model is static, that is, it only encompasses one-time period. This is problematic, since non-selected patients are still waiting for transplants. Moreover, over time there are new node arrivals to the program and also unforeseen departures from the program (e.g. patients who die or receive a kidney from an alternate source, paired-donors who drop out, etc). Consequently, in reality, a KEP is dynamic with exchanges calculated at regular time intervals. For example, in Canada, matches are calculated every 4 months. In particular, in time period t there is an exchange graph $G_t = (V_t, A_t)$. In time period $t + 1$ there is a new exchange graph $G_{t+1} = (V_{t+1}, A_{t+1})$ including new arrivals but excluding nodes matched in the previous round and other departures.

To build a dynamic kidney-exchange model we need, in particular, arrival and departure rates for each type of patient-donor pairs and altruistic donors. To calculate these we used data from [8]. In particular, the expected arrival rate of patient-donor pairs per time-period (4 months) is 37 which we model with a random Poisson distribution with mean 37. We also model the arrival rate of altruistic donors through an independent Poisson process with mean $\lambda_A = 4.5625$. The blood type probabilities of donors and patients are $\mathbb{P}(A, B, AB, O) = (0.46, 0.42, 0.09, 0.03)$. Furthermore, the probabilities that the cPRA rate of a patient lie in specific intervals are $\mathbb{P}((0,0), (0.01, 0.50), (0.51, 0.94), (0.95, 0.96), (0.97, 1)) = (0.24, 0.29, 0.24, 0.10, 0.13)$.   Denote   these   interval probabilities by $(\alpha_1, \alpha_2, \alpha_3, \alpha_4, \alpha_5) = (0.24, 0.29, 0.24, 0.10, 0.13)$; these are used in [12] and correspond to the distribution among patients registered in the KPD program when it was launched [6].

In this way, we generate the new nodes in V_t with associated blood types and cPRA rates. In particular, for the latter, we first generate the cPRA interval of a patient and then, their actual cPRA number is drawn from the uniform distribution on that interval. Next, we generate the arcs of A_t. If donor d_i is blood type compatible with patient p_j, then the arc (i, j) is excluded from the exchange graph G_t with probability equal to the cPRA rate of the patient p_j.[3] It follows that, including altruistic donors, there are 84 types of nodes, distinguished by blood type and cPRA interval.

We remark it is natural to also incorporate a departure rate into the model. Departures may, for example, be due to a donor withdrawing from the program, a patient receiving a kidney from an alternative source, the death of a patient, etc. This is simple to include in a dynamic model, but we have chosen not to do so as we do not have accurate statistics on which to base the corresponding event probabilities. Such departures are excluded for every transplant selection algorithm we test and, a priori, there is no reason to suppose that such departures affect one algorithm more than another.

In each matching period, a transplant selection algorithm is used to select transplant cycles and paths. The corresponding patient-donor pairs and altruistic donors are then cleared from the market. Patients that did not receive a transplant remain in the pool. In addition, new patient-donor pairs and altruistic donors enter the system, according to the prescribed arrival rates, and a new set of transplant cycles and chains are selected in the next period. The process then continues and our interest is in understanding its long-run behaviour.

3   Algorithms and Experiments

Fair Learning Algorithms. Our learning algorithms will assign a weight to each node in the exchange graph. In particular, each patient-donor node is assigned a weight w_τ depending upon its *type* τ. The type is determined by

---

[3] We use the standard exclusion probability formula to map cPRA rate to compatibility. We remark, however, that cPRA is not the only factor determining compatibility and there is no perfect mapping formula; see Delorme et al. [16].

the blood group of the donor, the blood group of the patient, and the cPRA group of the patient. Because we consider 5 cPRA groups, there are 80 possible types of patient-donor node.[4]

Recall that our aim is to obtain a fair algorithm, specifically equity in terms of transplant accessibility. To do this, we desire a set of weights $\mathbf{w}$ that, in the long-run, contains each type in (roughly) the same proportion in the queue (program) as their proportion in the general population. Ergo, we define pop_τ to be the proportion of type τ in the population and $\text{que}_\tau(\mathbf{w})$ to be the proportion of type τ in the waiting pool in the long-run using the weighting system $\mathbf{w}$. For our purposes, we define the long-run to be 50 time periods, that is, $16\frac{2}{3}$ years with matching rounds every four months.

We pre-learn the weights using a sequence of 50 simulations each consisting of the aforementioned 50 matching periods. In each matching period, given the weights, we used the *position-indexed chain-edge formulation* by Dickerson et al. [17] to calculate the optimal choice of transplant paths and cycles.[5] We let $\mathbf{w}^t$ be the weights used throughout the 50 periods of the tth simulation. To begin we set $\mathbf{w} = \mathbf{w}^1$ where $w_\tau^1 = 1$ for each patient-donor type τ, that is, the myopic weighting system. At the end of the tth simulation we update $\mathbf{w}$ via $w_\tau^{t+1} = f\left(\frac{\text{que}_\tau(\mathbf{w}^t)}{\text{pop}_\tau}\right)$, where $f(x)$ is an update function. Evidently, the update function must have the property that it is monotonically increasing in x, that is, in $\frac{\text{que}_\tau(\mathbf{w}^t)}{\text{pop}_\tau}$.

To accomplish this we define two classes of update function, namely linear updates and exponential updates. The linear rule, $\text{Lin}(a)$, is defined as $f(x) = 1 + \frac{x}{a}$. The exponential rule, $\text{Exp}(a)$, is defined as $f(x) = (a+1) - ae^x$. Furthermore, we then scale the functions so that the minimum weight is at exactly 1.

For the linear rule we select $a \in \{1, 2\}$ and for the exponential rule we select $a \in \{1, 3, 5, 7, 9\}$. This produces seven learning algorithms. The performance of each algorithm will be evaluated (via experiments described below) using the weights $\mathbf{w} = \mathbf{w}^{51}$ obtained at the end of the sequence of simulations for that algorithm.

In addition to the 80 types of patient-donor node, there are 4 types of altruistic donor node. Until now the weight of such a node type has been set to $w_\tau = 0$. However, we rerun each learning algorithm over a range of different (predominantly, negative) weight values W for the altruistic donors. Similarly, weights were learnt for a range of maximum path and chain lengths and for various altruistic arrival rates.

The Canadian KPD Algorithm. Here we give a brief overview of the program; see [14] and [28] for detailed descriptions of the program. The KPD program uses an expert-defined weighting system. The nodes/arcs in the exchange graph are weighted according a point system; see the full paper [11] for details

---

[4] As stated, practitioners may also add individual weight adjustments to patient-donor nodes but this is irrelevant to the conclusions of this work.

[5] Code, by James Trimble, for the position-index formulations is available at https://github.com/jamestrimble/kidney_solver.

on the point system and on the accumulation of hard-to-match patients in the program.

Transplant selection occurs every 4 months in the KPD program. It is important to note that the matches proposed by the KPD algorithm (or, indeed, by any transplant selection algorithm) may not all be undertaken, e.g., due to surgical concerns or the withdrawal of a patient. To date, such factors cannot be accurately modelled. Furthermore, there are some aspects of the KPD scoring scheme that, whilst very simple to implement, we have excluded because we do not have the necessary statistical data to accurately incorporate them into our learning models. These include donor/recipient ages, geographical location, and time on dialysis. Regardless, a priori, there is no reason to suppose that these excluded elements will affect one transplant selection algorithm more than another; thus, for the purposes of algorithmic comparison we assume it is safe to omit these factors from our model.

Finally, we remark that for the KPD program the surplus donor at end of a path is transferred to the standard kidney donation program for unpaired patients as a donor for the medical centre that registered the altruistic donor [14], that is, as a living donor on the deceased donor list.

The Experiments. So we have described how weights are calculated for the myopic algorithm, the KPD algorithms, and our learning algorithms. Given these weighting systems how do the respective algorithms perform? We tested this using a large number of experiments run via the dynamic kidney exchange model. That model, described in Sect. 2, was implemented using Python 3. As stated, for these experiments, simulated patients and donors were generated using Canadian population data statistics based on publicly available data from Canadian Blood Services. For each algorithm, we conducted 50 random simulations, each lasting for 50 matching rounds, that is, over a time-span of 200 months. In each matching period, given the weights prescribed by the relevant algorithm, we again use the *position-indexed chain-edge formulation* by Dickerson et al. [17] to calculate the optimal choice of transplant paths and cycles. These experiments were carried out over a range of maximum cycle lengths ($C \leq 5$) and maximum path lengths ($P \leq 20$). Using these experiments we evaluate the long-term impact of each of these transplant selection algorithms. Since our algorithms are designed for group fairness, before describing the results of our learning experiments in Sect. 5 let us first discuss how to measure fairness in kidney exchange programs.

4   Fairness in Kidney Exchanges

We will present the results of our simulations in Sect. 5. These compare the performances of a variety of kidney exchange algorithms (i.e., transplant selection objectives) with respect to the number of transplants, waiting times, the utilization rate of altruistic donors, and fairness. The meanings of the first three of these criteria are self-evident. But what do we want from a "fair" kidney-exchange algorithm? We address this question in this section and, as a result, provide a class of measures of fairness by which we will compare our algorithms.

Measures of Fairness. An extremely important desideratum for kidney exchange mechanisms is fairness across groups and individuals. The first measures of fairness in kidney exchanges were provided by Dickerson et al. [20]. Motivated by the *price of fairness* for allocation problems [3,9] their fairness measures (lexicographic and weighted) concerned fairness across groups defined by cPRA rates. Fairness in kidney exchanges is now well-studied in the literature, typically based on measures of group fairness.[6]

There is a very natural way to evaluate group fairness using the weighted power mean. Specifically, given k groups with weights $\boldsymbol{\alpha} = (\alpha_1, \alpha_2, \ldots, \alpha_k)$ and utilities $\mathbf{u} = (u_1, u_2, \ldots, u_k)$, we may define group utility by the *weighted power mean*

$$M_{\boldsymbol{\alpha}}^{\rho}(\mathbf{u}) = \left(\sum_{i=1}^{k} \alpha_i \cdot u_i^{\rho} \right)^{\frac{1}{\rho}}. \tag{1}$$

We refer the reader to the book [7] by Bullen for technical details on weighted power means and their mathematical properties. Important here is the fact that (1) corresponds to the well-studied *constant elasticity of substitution* (CES) utility functions in economics [1,23,39], where the elasticity of substitution is $\sigma = \frac{1}{1-\rho}$. What is interesting about this class of functions is that ρ can be adjusted to give a variety of fairness measures over the groups. Most important are the special cases $\rho \in \{1, 0, -\infty\}$. The case $\rho = 1$ corresponds to *weighted utilitarian social welfare*, where we wish to maximize $\sum_{i=1}^{k} \alpha_i \cdot u_i$. The case $\rho = 0$ corresponds to *weighted Nash social welfare*, where we wish to maximize $\prod_{i=1}^{k} u_i^{\alpha_i}$, also known as the Cobb-Douglas utility. Finally, the case $\rho = -\infty$ corresponds to *egalitarian (Rawlsian) social welfare*, where we wish to maximize $\min_{i=1}^{k} u_i$, also known as the Leontif utility.

Thus (1) induces a class of fairness measures. However, it still remains to define the groups, the weights and utilities. Let us begin with the groups. Recall a major motivation for this work is the accumulation of hard-to-match patients in the waiting pool. Now the cPRA score estimates the proportion of donors against which the patient has antibodies rendering the match incompatible; thus, the higher the cPRA rate the harder it is to match a patient. Consequently, as in [20], it is natural to group the patients according to cPRA rate. We will use five groups for cPRA rates in the intervals $(0,0), (0.01, 0.50), (0.51, 0.94), (0.95, 0.96)$ and $(0.97, 1)$. We may then set the weight of a group to be the probability a patient lies in that group, that is $(\alpha_1, \alpha_2, \alpha_3, \alpha_4, \alpha_5) = (0.24, 0.29, 0.24, 0.10, 0.13)$. Finally, we need to define the utility u_j of a group j. This utility should be inversely related to the accumulation of the group in the patient pool, where we denote by q_j the number of patients of group j in the pool at the end of the experiments. In addition, we desire that the utility scales linearly; specifically if the quantity q_i of every group increases by a factor c then the utility of each

---

[6] However, measures based on individual fairness have been introduced; see Farnadi et al. [24]. Furthermore, the provision of fairer allocations via the incorporation of human values into matching algorithms has also been proposed; see, for example, Freedman et al. [25].

group should decrease by a factor c. These properties can be obtained by defining the utility of a group to be its weight divided by the number of its members in the queue. That is, $u_j = \frac{\alpha_j}{q_j}$. We can view this utility in the following way. Let $Q = \sum_{i=1}^{k} q_i$ be the total queue size, and let $\beta_j = \frac{q_j}{Q}$ be the proportion of the queue made up of members of group j. Then $u_j = \frac{1}{\beta_j/\alpha_j} \cdot \frac{1}{Q}$. Thus the utility of the group decreases as the group becomes over represented in the queue (when β_j grows relative to α_j) and as the queue size Q increases.

So, with these weights and utilities, the weighted power mean (1) induces three fairness scores for $\rho \in \{1, 0, -\infty\}$ by which we will evaluate our kidney exchange algorithms in Sect. 5. These scores illustrate varying trade-offs between the utilities of each group. At one extreme, utilitarian social welfare ($\rho = 1$), the objective is maximize collective welfare; all groups are important but specific groups may be penalized if this leads to greater benefits for the other groups. At the other extreme, egalitarian social welfare ($\rho = -\infty$), the objective is to maximize the utility of the worst-off group. For the purpose of KEPs, egalitarian social welfare is the most appropriate fairness objective with respect to the accumulation of hard-to-match patients in the waiting pool. So this measure will be our primary focus. However, Nash social welfare ($\rho = 0$) is also an extremely useful fairness measure for allocation problems (such as KEPs). This is highlighted by Caragiannis et al. [10] in their influential work on fairness. They state "the Nash social welfare solution exhibits an elusive combination of fairness and efficiency properties, and can easily be computed in practice. It provides the most practicable approach to date – arguably, the ultimate solution, for the division of indivisible goods under additive valuations". Informally, located between utilitarian and egalitarian social welfare, Nash social welfare encourages improvements in collective welfare whilst, at the same time, penalizing imbalances between groups.

5   Results

We now present the results of our experiments. First we provide a summary of the performance of the transplant selection algorithms for a variety of criteria including the number of transplants, waiting times and group fairness. These results highlight the value of learned-weights. Then we provide more in-depth analyses. In particular, we show the vital importance of assigning negative weights to altruistic donors. Finally we detail results on performance with respect to our three group fairness metrics.

Summary of Results. Table 1 shows the performance of the KPD and myopic algorithms, and the seven learning algorithms. There are four types of criteria for comparison. The three main criteria are:

1. *Number of Transplants.* This is shown in two equivalent ways. First the total number of transplants[7] achieved over the 50 time periods by each algorithm,

---

[7] The surplus donor at the end of a path may subsequently donate on the deceased donor program. Such matches are not counted in the total number of transplants.

Table 1. Performances of the transplant algorithms.

Periods: 50	# Participants: 1994.78		λ_A: 4.5625	C: 5	P: 5	
Alg.	W	#Matches	%Match	Wait Time (Recipients)	WaitTime	Egal.Fairness
Myop	0.0	1613.76	80.90	13.55	29.52	0.60
KPD	0.0	1612.64	80.84	13.58	29.56	1.00
Lin(1)	−2.0	1698.22	85.13	11.51	23.45	1.06
Lin(2)	−3.0	1700.38	85.24	11.47	23.45	0.85
Exp(1)	−1.5	1707.46	85.60	11.88	22.62	1.10
Exp(3)	−2.0	1694.72	84.96	11.69	23.56	1.17
Exp(5)	−3.0	1697.62	85.10	11.51	23.55	0.91
Exp(7)	−4.0	1705.88	85.52	10.89	22.86	0.61
Exp(9)	−5.0	1697.52	85.10	11.45	23.61	1.04

on average over all experiments. Second, the corresponding percentage of patients who received a transplant.

2. *Waiting Times.* This is measured in two different ways. The first is the average wait time of each transplant recipient in months. This is an important measure but it does not give the entire picture as it excludes those patients that remain on the waiting list, in particular, harder to match patients. Thus, the second way is the average wait time, which includes the waiting times of every patient.

3. *Group Fairness.* We measure fairness in three ways: utilitarian group fairness, Nash group fairness and egalitarian group fairness. Only our primary fairness measure, egalitarian group fairness, is shown in Table 1. The measure is scaled so that the KPD algorithm has measure 1 and the higher the measure the fairer the algorithm. (Detail discussion and our results concerning all three fairness measures will follow.)

The significance of the fourth criteria, *"Altruist Usage"*, (and that of the negative weight, W, assigned to each altruistic donor as in Table 1) will be shown in Tables 2 and 3.

The main observation from Table 1 is that all seven of the learning algorithms outperform the myopic algorithm for every single criteria (including utilitarian and Nash group fairness, not shown in the table). Four of the learning algorithms outperform the KPD algorithm for every single criteria; the remaining three learning algorithms outperform the KPD algorithm on every criteria except egalitarian fairness. The performance of each learning algorithm is comparable but, overall, Exp(1) and Exp(3) perform the best. The headline statistics are the following. In comparison to the KPD and myopic algorithms, the learning algorithm Exp(1) leads to a 6% increase in the number of transplants (from about 1613 to 1707), a decrease of 24% in the average waiting time (from about 29.5 to 22.6 months) and an improvement of 10% and 21%, respectively, in the egalitarian fairness measure. Furthermore, both Exp(1) and Exp(3) reduce by 2% the use of altruistic donors (see subsequent discussion).

Before presenting more detailed results and analyses, it is worth discussing whether even better algorithms may be obtainable. In fact, it will be hard to obtain better algorithms. To see this, consider the number of transplants. The %Match column shows the percentage of patients that receive transplants. The learning algorithm Exp(1) achieves 85.6%. This does not sound spectacular, but it is near optimal because, when we take these measurements after 50 periods, it is not realistic to provide transplants for all the recent arrivals. Indeed, it appears that about a 90% match rate is the maximum achievable. (An indication of this fact will be seen in Table 4 where a 90% match rate is achieved but only after tripling the number of altruistic donors in the KEP.)

Cycle and Path Lengths. The results shown in Table 1 are based on allowing cycles of length at most $C = 5$ and paths of length at most $P = 5$. These restrictions are justified by the following two lessons.

Lesson I. *There is negligible hypothetical benefit in allowing for paths of length greater than 5.*

Lesson II. *It is very important to allow for cycles of length 3. The hypothetical improvements with cycles length of 4 or 5 are marginal.*

We remark that these two lessons will not surprise practitioners, so we defer a detailed discussion and the presentation of our results on this topic to the full paper [11].

Altruist Weights. Our learning models learn positive weights for patient-donor nodes in the exchange graph. However, it turns out to be vital to also assign a negative node weight to an altruistic donor. Before investigating the effect of this modification, let us understand why it may be useful. The incorporation of positive weights on patient-donor nodes signifies the "value" of performing those transplants. In contrast, a negative weight can be interpreted as signifying the "price" of using an altruistic donor. An altruistic donor can be viewed as a scarce and valuable resource in an kidney exchange mechanism. They must be used prudently and the standard economic approach to enforce this is to assign them a price – the consequence of this price means an altruistic donor will only be used when the resultant benefit is significant. The reader may ask whether this is really helpful. The answer is yes: there is a dramatic difference in the performance of the kidney-exchange algorithm (in terms of the number of transplants, waiting times, and fairness) with different altruistic donor weights. Table 2 shows this emphatically for the learning model Exp(3). When the altruist weight is zero (or even positive) the number of matches is low, around 80.7%. As the weight falls, the number of matches increases dramatically to around 85% for the weight range $[-2, -3]$. Further decreases in weight beyond -3 then lead to a large fall in the number of transplants.

This is very natural. When the price is under 2 it is too low and the demand for altruists is too high. The result is that the altruists are used in a suboptimal manner, for example, to initiate short paths with easy-to-match patients. When

Table 2. The effect of varying the weight of altruistic donors.

Exp(3) Periods: 50   # Participants: 1994.78   C: 5 P: 5

W	Altruist % Usage	%Match
0.00	99.64	80.77
−0.75	98.21	83.69
−1.50	97.71	84.57
−2.00	97.32	84.96
−2.50	97.34	85.42
−3.00	96.12	85.18
−5.00	51.90	81.01
−10.00	40.23	75.45
−15.00	0.00	73.83

the price is above 3 it is too high and the demand for altruists is too low. The result is that the potential benefits of using the altruists are not all achieved. This then explains why a negative weight W for the altruists is given in Table 1 for each model. We remark that the optimal price for an altruistic varies relative to the other weights given by the model for the arcs or patient-donor pairs. Hence, it also explains why in Table 1 the best choice of altruist weight differs for each model. In this way, we obtain our most important lesson:

Lesson III. *Assigning an appropriate negative weight (price) to each altruistic donor is fundamental in the determining the quality of outcomes.*

To see why this lesson is critical, let us see what happens when we apply it to the myopic and KPD algorithms. Concretely, assigning a price to the altruists gives the results shown in Table 3, where we call the modified algorithms Myopic$^+$ and KPD$^+$. This trivial modification has a massive impact. Both Myopic$^+$ and KPD$^+$ now perform a comparable number of transplants to our learning algorithms roughly 85% each (up from 81%). There is also a very large improvement in fairness. The egalitarian group fairness measure increasing by 26% for the KPD algorithm and by 77% for the myopic algorithm. Finally, waiting times also improve significantly with one exception: for the KPD algorithm, the average waiting time decreases from 29.6 to 23.6 months but waiting times per recipient increase from 13.6 to 14.8 months. The explanation for this counter-intuitive behaviour is that KPD$^+$ performs many more transplants than KPD but many of these additional operations involve hard-to-match patients who have naturally spent longer than average time on the waiting lists.

Consequently, excluding waiting time per recipient, the KPD$^+$ algorithm has performance comparable to our learning algorithms. An important conclusion can be drawn from this: the expert-designed weights in the KPD system work very well; the inferior performance of the KPD system is almost entirely due to the lack of an altruistic donor price! A second important conclusion can be

Table 3. The impact of pricing altruistic donors.

Periods: 50   # Participants: 1994.78 λ_A: 4.5625   C: 5 P: 5

Alg.	W	%Match	Wait Time (Recipients)	Wait Time	Egal. Fairness
Myop	0.0	80.90	13.55	29.52	0.60
Myop+	−2.0	84.80	12.86	23.87	1.06
KPD	0.0	80.84	13.58	29.56	1.00
KPD+	−150.0	85.07	14.82	23.64	1.26
Exp(1)	−1.5	85.60	11.88	22.62	1.10
Exp(3)	−2.0	84.96	11.69	23.56	1.17

drawn by comparing Myopic$^+$ with KPD. We have seen that the choice of KPD weights are appropriate. However, Myopic$^+$ significantly outperforms KPD in all three measures. But Myopic$^+$ is an unweighted system except for the altruistic donor price. Consequently, choosing a (negative) weight for the altruists is more effective than assigning an accurate (positive) weight for *every* other patient-donor pair.

Given the huge impact of altruistic donors, a natural question is what would be the performance of a kidney exchange if the number of altruistic donors changed. We examine this in Table 4 where the altruist arrival rate λ_A varies from 0 to 10.5. Recall, in the Canadian KPD program the current altruist arrival rate is roughly $\lambda_A = 4.5$.

Table 4. The impact of the altruistic donors arrival rate.

Exp(3) Periods: 50   # Participants: 1994.78   C: 5 P: 5 W:-2

λ_A	%Match	Matches/ Altruist	#Altruists	#Paths	Donors/Altruist
0.0	73.74	–	0.00	0.00	–
1.5	78.02	1.14	76.88	75.24	0.98
3.0	81.47	0.92	150.12	146.38	0.98
4.5	85.03	0.95	224.92	219.94	0.98
6.0	87.85	0.75	299.16	281.62	0.94
7.5	89.14	0.34	371.48	311.86	0.84
9.0	89.71	0.15	448.96	324.96	0.72
10.5	90.04	0.09	521.76	335.50	0.64

Table 4 is extremely informative. For example it shows that, for the current altruist arrival rate $\lambda_A = 4.5$, each altruistic donor saves almost two lives. To see this note that each new altruist increases the number of transplants by 0.95 on average. In addition, the altruist usage rate is 98%. But the use of an altruist at the start of a path means the donor at the end of the path is unused. In the Canadian KPD, such donors are currently requested to then donate in the deceased donor program. This means each altruistic donor then creates on average an additional 0.98 transplants in deceased donor program. Thus $0.95 + 0.98 = 1.93$ is a good measure of the number of lives saved by each altruistic donor.

Instead, such a donor could be requested to remain in the paired donation program as a *defacto* altruistic donor, called a "bridge donor". Observe that the use of a bridge donor recursively then generates another bridge donor. Consequently, an important open question is to determine whether this *multiplier effect* implies the benefits of the donor remaining in the paired donation program are greater than the benefits of the donor moving to the deceased donor program.

Table 4 also shows that the marginal increase in transplants per altruist is *decreasing*. For example, between $\lambda_A = 0$ and $\lambda_A = 1.5$ each new altruist increases the number of transplants by 1.14 on average. The marginal increase remains around 1 until $\lambda_A = 6$ and then drops precipitously from $\lambda_A = 7.5$ upwards. Furthermore, when the altruist arrival rate becomes very high, the altruist usage rate drops noticeably. Indeed for $\lambda_A \geq 9$ the number of lives saved per altruist falls below 1. For example, for $\lambda_A = 10.5$ marginal increase in transplants is just 0.09 and the altruist usage rate is just 64%. This lead us to our next lesson:

Lesson IV. *The marginal contribution in terms of lives saved per altruistic donor is non-increasing in the number of altruists. (At the current altruist arrival rate in Canada, each altruistic donor saves approximately two lives.)* [Table 4]

Of course this lesson extends beyond the Canadian KPD program. For any kidney exchange program, it is important to determine how much more effective the program would be with additional altruistic donors. In particular, the key factor is where on the marginal contribution curve the current altruist arrival rate lies. We set aside the question of the ethicality of encouraging altruistic donors; see Patel et al. [32] and Woodle et al. [40] for discussions.

Fairness. In Sect. 4 we defined a class of welfare functions to evaluate group fairness using the weighted power mean. This class includes three important cases: utilitarian social welfare, Nash social welfare and egalitarian social welfare. To compute these welfare functions it is first necessary, for each of our transplant selection algorithms, to evaluate the waiting list distribution at the end of our experiments with respect to each group. These queue length results are provided in Appendix D. This produces the social welfare scores shown in Table 5. In addition, for the purposes of easy comparison, we create three group fairness measures by scaling the welfare scores by a fixed constant such that the score of the KPD algorithm is exactly one. This induces three group fairness measures: utilitarian, Nash and egalitarian group fairness.

To understand this table, recall that all three measures have the property that if the number of patients in *every* group change by an identical factor then the group fairness measure will change by the same factor. For example, if the number of patients in the waiting list falls by 10% for *every* group then

Table 5. Measures of fairness.

Periods: 50	# Participants: 1994.78		C: 5 P: 5
Alg.	Group Fairness Measure		
	Utilitarian	Nash	Egalitarian
Myop	1.14	1.10	0.60
Myop+	1.66	1.52	0.64
KPD	1.00	1.00	1.00
KPD+	1.28	1.27	1.22
Lin(1)	1.64	1.45	1.06
Lin(2)	1.76	1.54	0.85
Exp(1)	1.99	1.62	1.10
Exp(3)	1.51	1.39	1.17
Exp(5)	1.69	1.49	0.91
Exp(7)	4.34	2.41	0.61
Exp(9)	1.87	1.55	1.04

each measure will increase by 10%. However, the three measures differ in how they penalize imbalances, with extreme penalties in the case of egalitarian group fairness.

For the both utilitarian and Nash group fairness, we immediately see that algorithms give dramatic improvements in fairness over the KPD and myopic algorithms. All the learning algorithms also improve egalitarian group fairness over the myopic algorithm. But that is not the case for the KPD algorithm. The reason for this is the KPD point system heavily prioritizes hard-to-match patients (see Appendix A), that is, it is implicitly designed to provide high egalitarian group fairness. Despite this, four of the learning algorithms provide better egalitarian group fairness than the KPD algorithm, with Exp(3) proffering the best improvement with a fairness measure of 1.17.

Another valuable observation is that KPD$^+$ Pareto dominates KPD in terms of shorter queue sizes for every group (details are given in the full paper [11]). In fact, not only does each group improve but each improves by a large amount, with the worst case improvement being 18%. This algorithm then produces the best fairness of 1.22 with respect to our primary measure, egalitarian group fairness. Moreover, because KPD$^+$ uses the same weights for patient-donor nodes as KPD, the distribution of groups within the waiting lists are similar. This explains its similar fairness scores of 1.28 and 1.27 for utilitarian and Nash group fairness, respectively. In contrast, our learning algorithms lead to quite different distributions of groups within the waiting lists than the KPD algorithm; this leads to a greater fluctuations in their three group fairness measures. Four of the learning algorithms provide improved fairness over the KPD algorithm for all three measures. Thus, we obtain our next lessons:

Lesson V. *Fair algorithms can be used to reduce the accumulation of hard-to-match patients in waiting pools.*

Lesson VI. *Fair algorithms can be implemented without detrimental affects on utilitarian measures such as waiting times and the number of transplants.*

6   Conclusion

In this work, we proposed a kidney exchange algorithm based on learned weights. These weights were determined via a learning approach driven by group fairness defined in terms of blood type and cPRA rates. Then we presented results obtained from simulations of dynamic KEPs, comparing the KPD algorithm used in Canada, the myopic algorithm, and our learning algorithms. Our results provide lessons for increasing the number of transplants, decreasing waiting times, and improving group fairness. We hope this work helps contribute to enhance equity in transplantation access.

Many avenues to explore remain. First, we believe our learning algorithms can be improved. For example, the update rules for weights are rather ad-hoc. Do more structured approaches lead to better outcomes? For example, we used the weighted power mean to evaluate the fairness of our algorithms. So can optimal weights be learnt by optimizing the weighted power mean directly during updates?

Our algorithms were tested using realistic but simulated data. The next step is to confirm our results with real patient data. This would also allow for verification that the incorporation of omitted characteristics such as departure rates, geographic location, age, etc., do not affect our conclusions. It is also important to confirm our lessons apply to other kidney exchange programs based on weighting systems and whether or not they extend to programs based upon hierarchical optimization.

Improvements may also be possible by changing the set of objects that are weighted. Our algorithms learn node weights. Can better results be obtained by learning arc weights? That is, if a specific value (positive or negative) is given for the use of each specific donor-patient transplant. In this paper, we assigned a fixed (negative) weight for each altruist. This weight was obtained via experiments over a range of weights. This approach was vital in understanding the importance of altruistic donors and how best to use them. But can the most effective negative altruist weight be found applying the learning method used for the positive patient-donor node weights? Furthermore, would it be helpful to incorporate a more-refined weighting system for the altruists? For example, it would be natural for an altruistic donor of blood group O to have a higher price.

Of course, learning algorithms can be used for objectives other than fairness, or for a mix of objectives. Can dynamic learning algorithms be used to improve performance for other objective functions? Finally, the model can be used to test the effectiveness of other proposed modifications to a KEP. For example, currently if a donor-patient pair is compatible then that transplant is scheduled automatically. Thus, such pairs are not included in the KPD program. Theoretically, the inclusion of compatible donor-patient pairs may increase the chances of matching other patients in the program [38]. In particular, if the patient is easy-to-match then the corresponding donor may induce analogous benefits to that of an altruistic donor. In addition, as discussed, the donor at the end of a path is currently redesignated as a living donor in the deceased donor program. Quantifying the potential benefits of this donor remaining as a bridge donor in the KPD program is important.

Acknowledgment. The authors thank William Klement and Mike Gillissie of Canadian Blood Services for numerous discussions and expert advice. We are also extremely grateful to David Manlove and John Dickerson and detailed comments and advice. This project was partially supported by the Natural Sciences and Engineering Research Council of Canada and the Institut de valorisation des données and Fonds de recherche du Québec via an FRQ-IVADO Research Chair.

References

1. Arrow, K., Chenery, H., Minhas, B., Solow, R.: Capital-labor substitution and economic efficiency. Rev. Econ. Stat. **43**(3), 225–250 (1961)
2. Ashlagi, I., Gamarnik, D., Rees, M., Roth, A.: The need for (long) chains in kidney exchange. Technical report, National Bureau of Economic Research (2012)
3. Bertsimas, D., Farias, V., Trichakis, N.: The price of fairness. Oper. Res. **59**(1), 17–31 (2011)
4. Biro, P., et al.: Building kidney exchange programmes in Europe: an overview of exchange practice and activities. Transplantation **103**(7), 1514–1522 (2019)
5. Biro, P., et al.: Modelling and optimisation in European kidney exchange programmes. Eur. J. Oper. Res. **291**(2), 447–456 (2021)
6. Canadian Blood Services. Donation and transplantation kidney paired donation program data report 2009–2013 (2014). https://professionaleducation.blood.ca/sites/msi/files/Canadian-Blood-Services-KPD-Program-Data-Report-2009-2013.pdf
7. Bullen, P.S.: Handbook of Means and Their Inequalities. Springer, Dordrecht (2003). https://doi.org/10.1007/978-94-017-0399-4
8. Canadian Blood Services. Interprovincial organ sharing national data report: Kidney paired donation program 2009–2018 (2018). https://professionaleducation.blood.ca/sites/default/files/kpd-eng_2018.pdf
9. Caragiannis, I., Kaklamanis, C., Kanellopoulos, P., Kyropoulou, M.: The efficiency of fair division. Theory Comput. Syst. **50**, 589–610 (2012). https://doi.org/10.1007/s00224-011-9359-y
10. Caragiannis, I., Kurokawa, D., Moulin, H., Procaccia, A., Shah, N., Wang, J.: The unreasonable fairness of maximum Nash welfare. ACM Trans. Econ. Comput. **7**(3), 1–32 (2019)
11. Carvalho, M., Caulfield, A., Lin, Y., Vetta, A.: Penalties and rewards for fair learning in paired kidney exchange programs (2023). arXiv:2309.13421
12. Carvalho, M., Lodi, A.: A theoretical and computational equilibria analysis of a multi-player kidney exchange program. Eur. J. Oper. Res. **305**(1), 373–385 (2022)
13. Cho, I.-K., Meyn, S.: Efficiency and marginal cost pricing in dynamic competitive markets with friction. Theor. Econ. **5**, 215–239 (2010)

14. Cole, E., et al.: The Canadian kidney paired donation program: a national program to increase living donor transplantation. Transplantation **99**(5), 985–990 (2015)

15. Delorme, M., García, S., Gondzio, J., Kalcsics, J., Manlove, D., Pettersson, W.: New algorithms for hierarchical optimisation in kidney exchange programmes. Technical report ERGO 20-005, Edinburgh Research Group in Optimization (2020)

16. Delorme, M., et al.: Improved instance generation for kidney exchange programmes. Comput. Oper. Res. **141**, 105707 (2022)

17. Dickerson, J., Manlove, D., Plaut, P., Sandholm, T., Trimble, J.: Position-indexed formulations for kidney exchange. In: Proceedings of the 17th ACM Conference on Economics and Computation (EC), pp. 25–42 (2016)

18. Dickerson, J., Procaccia, A., Sandholm, T.: Dynamic matching via weighted myopia with application to kidney exchange. In: Proceedings of the 26th Conference on Artificial Intelligence (AAAI), pp. 1340–1346 (2012)

19. Dickerson, J., Procaccia, A., Sandholm, T.: Optimizing kidney exchange with transplant chains: theory and reality. In: Proceedings of the 11th International Conference on Autonomous Agents and Multiagent Systems (AAMAS), pp. 711–718 (2012)

20. Dickerson, J., Procaccia, A., Sandholm, T.: Price of fairness in kidney exchange. In: Proceedings of the 13th International Conference on Autonomous Agents and Multiagent Systems (AAMAS), pp. 1013–1020 (2014)

21. Dickerson, J., Sandholm, T.: FutureMatch: Combining human value judgments and machine learning to match in dynamic environments. In: Proceedings of the 29th Conference on Artificial Intelligence (AAAI), pp. 622–628 (2016)

22. Ding, Y., Ge, D., He, S., Ryan, C.: A nonasymptotic approach to analyzing kidney exchange graphs. Oper. Res. **66**(4), 918–935 (2018)

23. Dixit, A., Stiglitz, J.: Monopolistic competition and optimum product diversity. Am. Econ. Rev. **67**(3), 297–308 (1977)

24. Farnadi, G., St-Arnaud, W., Babaki, B., Carvalho, M.: Individual fairness in kidney exchange programs. In: Proceedings of the 35th Conference on Artificial Intelligence (AAAI), pp. 11496–11505 (2021)

25. Freedman, R., Borg, J., Sinnott-Armstrong, W., Dickerson, J., Conitzer, V.: Adapting a kidney exchange algorithm to align with human values. Artif. Intell. **283**, 103261 (2020)

26. Glorie, K., Van de Klundert, J., Wagelmans, A.: Kidney exchange with long chains: an efficient pricing algorithm for clearing barter exchanges with branch-and-price. Manuf. Serv. Oper. Manag. **16**(4), 498–512 (2014)

27. Huh, K., et al.: Exchange living-donor kidney transplantation: merits and limitations. Transpl. Proc. **86**, 430–435 (2008)

28. Malik, S., Cole, E.: Foundations and principles of the Canadian living donor paired exchange program. Can. J. Kidney Health Dis. **1**, 6 (2014)

29. Manlove, D., O'Malley, G.: Paired and altruistic kidney donation in the UK: algorithms and experimentation. ACM J. Exp. Algorithmics **19**(2), 663–668 (2014). Article No. 2.6

30. Montgomery, R., et al.: Domino paired kidney donation: a strategy to make best use of live non-directed donation. The Lancet **368**(9533), 419–421 (2006)

31. Park, K., Moon, J., Kim, S., Kim, Y.: Exchange-donor program in kidney transplantation. Transpl. Proc. **31**(1–2), 356–357 (1999)

32. Patel, S., Chadha, P., Papalois, V.: Expanding the live kidney donor pool: ethical considerations regarding altruistic donors, paired and pooled programs. Exp. Clin. Transplant. **1**, 181–186 (2011)

33. Rapaport, F.: The case for a living emotionally related international kidney donor exchange registry. Transpl. Proc. **18**, 5–9 (1986)
34. Roth, A., Sönmez, T., Ünver, U.: Pairwise kidney exchange. Quart. J. Econ. **119**(2), 457–488 (2004)
35. Roth, A., Sönmez, T., Ünver, U.: Efficient kidney exchange: coincidence of wants in markets with compatibility-based preferences. Am. Econ. Rev. **97**(3), 828–851 (2007)
36. Roth, A., Sönmez, T., Ünver, U., Delmonico, F., Saidman, S.: Utilizing list exchange and nondirected donation through 'chain' paired kidney donations. Am. J. Transplant. **6**, 2694–2705 (2006)
37. Sönmez, T., Ünver, U.: Market design for kidney exchange. In: Vulkan, N., Roth, A., Neeman, Z. (eds.) The Handbook of Market Design. Oxford University Press (2013). Chapter 4
38. Sönmez, T., Ünver, U., Yenmez, B.: Incentivized kidney exchange. Am. Econ. Rev. **110**(7), 2198–2224 (2020)
39. Solow, R.: A contribution to the theory of economic growth. Q. J. Econ. **70**(1), 65–94 (1956)
40. Woodle, E., et al.: Ethical considerations for participation of nondirected living donors in kidney exchange programs. Am. J. Transplant. **10**, 1460–1467 (2010)

Deterministic Impartial Selection with Weights

Javier Cembrano$^{(\boxtimes)}$, Svenja M. Griesbach, and Maximilian J. Stahlberg

Institute of Mathematics, Technische Universität Berlin, Straße des 17. Juni 136, Berlin, Germany
{cembrano,griesbach,stahlberg}@math.tu-berlin.de

Abstract. In the impartial selection problem, a subset of agents up to a fixed size k among a group of n is to be chosen based on votes cast by the agents themselves. A selection mechanism is *impartial* if no agent can influence its own chance of being selected by changing its vote. It is α-*optimal* if, for every instance, the ratio between the votes received by the selected subset is at least a fraction of α of the votes received by the subset of size k with the highest number of votes. We study deterministic impartial mechanisms in a more general setting with arbitrarily weighted votes and provide the first approximation guarantee, roughly $1/\lceil 2n/k \rceil$. When the number of agents to select is large enough compared to the total number of agents, this yields an improvement on the previously best known approximation ratio of $1/k$ for the unweighted setting. We further show that our mechanism can be adapted to the impartial assignment problem, in which multiple sets of up to k agents are to be selected, with a loss in the approximation ratio of $1/2$.

Keywords: Impartial selection · Mechanism design · Social choice

1 Introduction

Votes and referrals are a key mechanism in the self-organization of communities: political parties elect their representatives, researchers review and rate each other's manuscripts, and hyperlinks on the web attribute topical relevance to an external resource. Oftentimes, the agents who give the recommendations are themselves interested in being within a top-rated fraction of their group: to occupy a prestigious position, be invited to a conference, or to have a website appear more prominently in search results. Objectives like these provide an incentive to deviate from a fair evaluation of one's peers. In particular, agents might omit a recommendation for an immediate contender in order to be ranked above them when the votes are counted.

In a seminal work, Alon et al. [1] initiated the search for impartial mechanisms to aggregate the votes cast by n agents who want to elect k individuals among them, which we refer to as the exact (n, k)-selection problem. The authors require that no agent is able to influence their own chance of being selected by

© The Author(s), under exclusive license to Springer Nature Switzerland AG 2024
J. Garg et al. (Eds.): WINE 2023, LNCS 14413, pp. 151–168, 2024.
https://doi.org/10.1007/978-3-031-48974-7_9

adjusting the subset of peers that they vote for, while, at the same time, the agents selected by the mechanism should receive an expected sum of votes that is close to that of the highest voted subset of size k. We refer to the first condition as *impartiality* and to the second as α-*optimality*, where $\alpha \in [0,1]$ denotes the performance guarantee. If the mechanism is allowed to make use of random choice and agents may vote for any subset of their peers, then the best known performance guarantee is $\frac{k}{k+1}\left(1 - \left(\frac{k-1}{k}\right)^{k+1}\right)$, which gives $1/2$ for the selection of a single agent and approaches $1 - 1/e$ as $k \to \infty$ [4]. It is further known that no impartial mechanism can be better than $k/(k+1)$-optimal, which is tight only for $k = 1$. We discuss variants with a limited number of votes per participant as related work.

The problem only becomes more difficult in the deterministic setting, where the mechanism is forced to choose one agent over another even for highly symmetric input. The instance in which two agents vote for each other and one of them shall be selected requires the mechanism to break the tie, based on an external preference list, in favor of one of the agents. Impartiality demands that the same agent must be selected also when the other agent withdraws its vote. But then, an agent with no votes is selected, even though the other agent still receives one. This yields a performance guarantee of zero for the selection of a single agent in the worst case. Even for $k > 1$, no positive performance guarantee is possible [1], unless, surprisingly, when the mechanism is allowed to select less than k agents in some instances. In this case an algorithm achieving $\alpha = 1/k$ is known [4]. We refer to this relaxation as the *inexact* (n, k)-*selection problem*. Since this insight, the gap towards the best known upper bound, which is $(k - 1)/k$ in the inexact selection setting, remained remarkably wide.

More generally, the selection problem allows for votes to be weighted: one then compares the total weight of the selected agents to that of the maximum-weight subset of size k. In a peer review setting, reviewers are often asked to rate the manuscript under consideration on a point scale that ranges from a recommendation to reject to a claim of excellence. An editor or program chair would then aggregate these scores and accept a limited number of highly rated submissions. While the established rule to disclose any conflicts of interest protects, if obeyed, against abuse based on personal ties, authors whose papers are on the verge of selection might still profit from giving ratings below their honest estimate, unless the selection mechanism is impartial. In this setting, although computational studies have been made [2], no deterministic mechanism providing a worst-case guarantee was known to date.

1.1 Our Contribution

We propose a deterministic impartial mechanism that can be applied in the weighted setting and which achieves a performance guarantee of $1/\lceil 2n/k \rceil$, for $k \geq 2\sqrt{n}$ even, and $(k-1)/(k\lceil 2n/(k-1) \rceil)$, for $k \geq 2\sqrt{n} + 1$ odd. In particular, it achieves asymptotically a guarantee of $\alpha = 1/4$ for selecting at most half and $\alpha = 1/3$ for selecting at most two thirds of the agents. These are the first lower bounds for deterministic selection with weights. In its applicable range, the mechanism further improves upon the previous best bound of $1/k$ in the unweighted setting. The improvement is most noticeable when k is large, where the gap between the previously best known lower and upper bounds of $1/k$ and $(k-1)/k$, respectively, has been widest. The construction is best behaved whenever $b := 2n/k \in \mathbb{N}$ and $b \le k/2 \in \mathbb{N}$: here a guarantee of $\alpha = 1/b$ is provided and the analysis of the mechanism is tight. The mechanism uses a well-structured set of partitions of the agents, whose existence we study in Sect. 3 using a connection to hypergraph theory and graph coloring. The mechanism itself and the proof of the approximation guarantee are presented in Sect. 4.

In Sect. 5, we show how the mechanism can be adapted to assign agents to multiple size-limited subsets, which may represent tasks to distribute or committees to form. In this setting, we lose only a factor of $1/2$ in the performance guarantee, independent of the number of subsets.

1.2 Related Work

Impartiality as a desirable axiom in multi-agent problems was introduced by De Clippel et al. [11] and was first studied in the context of peer selection in parallel by both Holzman and Moulin [15] and Alon et al. [1]: The work by Holzman and Moulin studied the existence of impartial mechanisms satisfying further axioms such as unanimity and notions of monotonicity, while the research by Alon et al. showed that no deterministic impartial mechanism aiming to select exactly k agents can achieve any constant approximation ratio. In response, Bjelde et al. [4] showed that when fewer than k agents may be selected, $1/k$-optimality is guaranteed by the *bidirectional permutation* mechanism, which picks either one or two agents, depending on the instance. The authors further proved an upper bound of $(k-1)/k$ for any deterministic impartial mechanism.

Continuing the axiomatic line, Tamura and Ohseto [24] studied k-selection in the single-nomination setting and showed that impartiality is compatible with two natural notions of unanimity. Their mechanism was extended to the case of a higher, but constant, maximum number of nominations by Cembrano et al. [9]. Further, Aziz et al. [2] proposed a mechanism satisfying certain monotonicity properties and confirmed its performance in a computational study.

Several works have focused on randomized impartial selection. Alon et al. proposed a family of mechanisms based on a random partition of the agents that yield the first lower bounds on the approximation ratio for this setting, namely $1/4$ for $k = 1$ and $1 - O(1/\sqrt[3]{k})$ for general k. They also provided respective upper bounds of $1/2$ and $1 - \Omega(1/k^2)$. Fischer and Klimm [14] closed the gap for $k = 1$ by giving a $1/2$-approximation algorithm. Bousquet et al. [5] designed a mechanism with an approximation guarantee that goes to one as the maximum score of an agent goes to infinity. A restricted variant of particular importance, first studied in the work of Holzman and Moulin, arises when each agent can vote for exactly one other agent. Here, Fischer and Klimm provided both lower and upper bounds which were later improved by Cembrano et al. [10].

A setting closely related to the impartial selection of k agents is that of *peer review* in which, in contrast to the classic k-selection problem, the votes are weighted and represent a score assigned to a submission. Kurokawa et al. [18] studied a model where first a limited number of weighted votes is sampled and then the selection is performed. The authors proposed an impartial randomized mechanism providing a constant approximation ratio with respect to the (non-impartial) mechanism that randomly samples the votes and selects the best possible set of k agents given these votes. Mattei et al. [21] studied this problem from an axiomatic and experimental point of view, while Lev et al. [19] extended this work to the setting with noisy assessments. Dhull et al. [12] explored the scope and limitations of partition-based mechanisms for peer review in terms of approximating the selection of the best k papers.

Beyond multiplicative approximation, some works have studied the scope and limitations of impartial mechanisms in terms of additive guarantees [6–8] and additional economic axioms [13,20]. Impartiality has also been considered for the selection of agents where preferences come from correlated types [22], for the selection of vertices in graphs with maximal progeny [3,26,27], and for generating social rankings of agents who rank each other [16]. For a survey on incentive handling in peer mechanisms, see Olckers and Walsh [23].

2   Preliminaries

For $n \in \mathbb{N} := \mathbb{Z}_{\geq 1}$, we define the ranges $[n] := \{1, \ldots, n\}$ and $[n]_0 := \{0, \ldots, n-1\}$ and we write $\mathcal{A}_n$ for the set of non-negative $n \times n$ matrices with zero diagonal. An instance of the weighted selection problem is fully described by an integer k and a weight matrix $A \in \mathcal{A}_n$, where k is the number of agents to be selected and A_{ij} corresponds to the weight of the vote that agent i casts for agent j. For $A \in \mathcal{A}_n$, we write A_{-i} for the matrix obtained when removing the i-th row of A. Given $A \in \mathcal{A}_n$ and $R, S \subseteq [n]$, we write

$$\sigma_R(S; A) := \sum_{i \in R,\ j \in S} A_{ij}$$

for the score of the agents in S limited to R, and $\sigma(S; A)$ short for $\sigma_{[n]}(S; A)$. We omit the weight matrix A whenever it is clear from the context and we write j short for $S = \{j\}$ in the above definitions.

Let $n, k \in \mathbb{N}$ with $k < n$ in the following. For $A \in \mathcal{A}_n$, we let

$$\text{OPT}_k(A) := \arg\max_{S \subseteq [n]:\ |S|=k} \sigma(S; A)$$

denote an arbitrary set with the largest score among vertex subsets of size k. We write just OPT_k when the weight matrix is clear.

An (n, k)-selection mechanism is a function $f : \mathcal{A}_n \to 2^{[n]}$ such that $|f(A)| \leq k$ for every $A \in \mathcal{A}_n$. Such a mechanism is *impartial* if, for every pair of instances

$A, A' \in \mathcal{A}_n$ and for all $i \in [n]$ such that $A_{-i} = A'_{-i}$, it holds that $f(A) \cap \{i\} = f(A') \cap \{i\}$. We further call an (n, k)-selection mechanism α-*optimal* if

$$\frac{\sigma(f(A); A)}{\sigma(\mathrm{OPT}_k(A); A)} \geq \alpha$$

holds for all $A \in \mathcal{A}_n$ and some $\alpha \in [0, 1]$.

We write $E \mathbin{\dot{\cup}} F$ for the disjoint union of sets E and F. For a multiset E, we write $\mu_E(e)$ for the multiplicity of $e \in E$ and $\mu(E)$ for the cardinality of E.

A hypergraph is a pair $H = (V, E)$ where V is a finite set of *vertices* and where $E \subseteq 2^V$ is a multiset of *(hyper-)edges*. We say that H is *d-regular* if each vertex is contained in exactly d edges, i.e., $\mu(\{e \in E \mid v \in e\}) = d$ for all $v \in V$; *b-uniform* if each edge contains exactly b vertices, i.e., $|e| = b$ for all $e \in E$; and *linear* if two distinct edges intersect in at most one vertex, i.e., $|e_1 \cap e_2| \leq 1$ for all $e_1, e_2 \in E$ with $\mu_E(e_1) > 1$ or $e_1 \neq e_2$. The *dual* of H is $H^* = (E, X)$ where $X := \{\{e \in E \mid v \in e\} \mid v \in V\}$ is a multiset of sets. One may think of the dual graph in terms of the vertex–edge incidence matrix, which is transposed when taking the dual graph. Note that the dual graph may have repeated edges and loops even if the original graph does not have either.

We call a 2-uniform hypergraph without repeated edges a (simple) graph. For a graph $G = (V, E)$, an edge b-coloring is a mapping $\pi \colon E \to [b]$. It is *feasible* if $\pi(e_1) \neq \pi(e_2)$ for all $e_1, e_2 \in E$ with $e_1 \cap e_2 \neq \emptyset$. Likewise, a vertex b-coloring is a mapping $\pi \colon V \to [b]$ that we call feasible if $\pi(u) \neq \pi(v)$ for all $u, v \in V$ such that $u, v \in e$ for some $e \in E$.

3   Partition Systems

The present work takes inspiration from the *partition mechanism*. This mechanism was first proposed by Alon et al. [1] for the setting of randomized $(n, 1)$-selection, and variants for selecting more than one agent have been studied by Bjelde et al. [4], Aziz et al. [2], and Xu et al. [25]. In its original formulation due to Alon et al., the partition mechanism assigns each agent into a *voter set* S_1 and a *candidate set* S_2 uniformly at random. It then considers only votes from agents in S_1 to agents in S_2 and selects an agent from S_2 with maximum score. This mechanism is impartial as it considers only votes of agents with no chance of being selected and it is $1/4$-optimal, intuitively, as we see every fourth vote in expectation. The (n, k)-selection variant by Bjelde et al. [4] partitions the agents into k sets instead of two and selects one agent from each set that has the highest score from all other sets, additionally considering internal votes that are directed from left to right according to a random permutation of the agents. This variant preserves impartiality and provides a guarantee that varies from $1/2$ to $1 - 1/e$ as k grows from 1 to infinity.

The partition mechanism, although achieving a good ratio when randomization is possible, performs poorly in the deterministic setting. If agents are assigned in any fixed way, votes may be adversarially placed between agents in the same set (and opposite to the order given by the permutation of the agents if such a step is considered), so that the mechanism cannot do any better, in the worst case, than selecting agents with no votes, while the maximum score may be arbitrarily high.

In the following, we build the foundation for a partition-based (n, k)-selection mechanism that is robust against such adversarial placement of votes. To achieve this, agents appear in the candidate set of more than one partition and with a disjoint set of contenders each time. This way, votes not seen for a candidate agent in one partition will be seen in another partition wherein that agent reappears as a candidate. Of course, repeated candidacy may lead to the same agent being selected multiple times, at the expense of contenders with a high number of votes. To minimize this possibility, we let every agent contest just twice and we remove duplicate votes. As our goal is to select up to k agents, we define k such partitions. For now, we make also the simplifying assumption that n and k allow the candidate sets to have equal size b. This is without loss of generality as we may fill smaller partitions with dummy agents who cast and receive no votes and are disfavored when breaking ties. We call a collection of partitions meeting these requirements a *balanced partition system*.

A partition into voters and candidates is fully described by either set. A balanced partition system may thus be written as a family E of candidate subsets of the set of agents V or, in other words, as a hypergraph $H = (V, E)$ without repeated edges, where each $e \in E$ is the candidate set of a single partition. To fulfill the requirements of a balanced partition system, H has to be 2-regular, so that every agent appears in exactly two candidate sets, and b-uniform, so that all candidate sets $e \in E$ have the same size $|e| = b$. The remaining requirement that no two agents compete twice against each other, formally $|e_1 \cap e_2| \leq 1$ for all $e_1, e_2 \in E$ with $e_1 \neq e_2$, translates to H being linear. The following lemma, whose proof is omitted due to space constraints, implies that we can represent a partition system further by a simple graph.

Lemma 1. *A hypergraph is 2-regular and linear if and only if its dual is a simple graph.*

By Lemma 1 and the fact that order and size as well as degree and rank are dual for hypergraphs, there is a one-to-one correspondence between balanced partition systems where n agents are distributed among k candidate sets of size b on the one hand, and b-regular simple graphs of order k and size n on the other hand. In the simple graph representation, edges correspond to agents while incident vertices correspond to candidate sets that the agents appear in.

In the analysis of the mechanism, we will bound the weight selected by it by that of a subset U of top-voted agents that pairwise do not compete. More precisely, U will be a set of maximum weight among a partition of the k top-voted agents into b many subsets with this property. If the mechanism does not select some agent i from U, then only because it makes up for the agent's score in the two partitions that agent i appears in, and which are pairwise disjoint for the agents in U. This leads to a lower bound of $(k/b)/k = 1/b$, stated in Lemma 4. To ensure the existence of b such sets, we require that any subgraph of H induced by

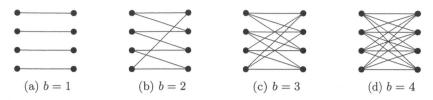

(a) $b = 1$ (b) $b = 2$ (c) $b = 3$ (d) $b = 4$

Fig. 1. The construction of Lemma 2 for $k = 8$ vertices and degree $b \in [4]$: (a) the $4P_2$ ($n = 4$ edges), (b) the cycle C_8 ($n = 8$), (c) the cube graph Q_3 ($n = 12$), and (d) the complete bipartite graph $K_{4,4}$ ($n = 16$). Every edge represents an agent and every vertex corresponds to a partition. A vertex and an edge are incident if the corresponding agent is in the corresponding candidate set.

k vertices can be partitioned into b many (internally) independent sets. We call a balanced partition system whose corresponding hypergraph has this property *robust*. In terms of the b-regular dual graph $G := H^*$, the condition is equivalent to the existence of an edge coloring with b colors for every subgraph induced by k edges: the edges of any one color do not share a vertex, which corresponds to vertices not sharing a hyperedge in H. By Kőnig's line coloring theorem [17], a sufficient condition for such a coloring to exist is that G is bipartite. The proofs of Lemma 3 and 4 will formalize these ideas.

Bipartite and b-regular graphs of even order k and size n exist for all $b = 2n/k$ with $b \leq k/2$. A simple construction is depicted in Fig. 1 and described by the following lemma, whose proof is omitted due to space constraints.

Lemma 2. *Let $b, k, n \in \mathbb{N}$ with $k' := k/2 \in \mathbb{N}$ and $b = 2n/k \leq k'$. Then, $G = (V, E)$ with $V := [k]_0$ and $E := \{\{i, k' + ((i + \ell) \bmod k')\} \mid i \in [k']_0, \ell \in [b]_0\}$ is a b-regular bipartite graph of order k and size n.*

We condense the findings of this section in the following lemma.

Lemma 3. *Let $n, k \in \mathbb{N}$ with $k < n$ be such that $b := 2n/k \in \mathbb{N}$ and $b \leq k/2 \in \mathbb{N}$. Let further V with $|V| = n$ denote a set of agents. Then, one may form k partitions $S_1^p \,\dot\cup\, S_2^p = V$, $p \in [k]$, such that*

(i) $|S_2^p| = b$ for all $p \in [k]$,
(ii) $|S_2^p \cap S_2^q| \leq 1$ for all $p, q \in [k]$ with $p \neq q$,
(iii) $|\{p \in [k] \mid v \in S_2^p\}| = 2$ for all $v \in V$, and
(iv) for every $U \subseteq V$, there is a partition $\bigcup_{t \in [b]} U_t = U$ with $u \in S_2^p \Rightarrow v \notin S_2^p$ for all $t \in [b]$, $u, v \in U_t$ with $u \neq v$, and $p \in [k]$.

Proof. For n, k, and b as in the statement, Lemma 2 guarantees the existence of a b-regular bipartite graph $G = (X, V)$ of order $|X| = k$ and size $|V| = n$. Let $H := G^* = (V, E)$ be its dual graph. Note that H is b-uniform and has order n and size k. By Lemma 1, H is further 2-regular and linear. As $b \geq 2$ by definition, it follows from linearity that H has no repeated edges, i.e., E is a set.

We use H to form a system of partitions of V. First, enumerate E by an arbitrary but fixed bijection $\phi \colon [k] \to E$. Then, for every $p \in [k]$, define a candidate set $S_2^p := \phi(p)$ and the associated voter set $S_1^p := V \setminus \phi(p)$. As H is b-uniform, we have (i) by construction. As it is linear, (ii) follows. Since H is 2-regular, also (iii) holds.

It remains to show property (iv). By Kőnig's line coloring theorem [17], there exists a feasible edge b-coloring $\pi\colon V \to [b]$ of G. Let G' be the subgraph of G induced by an edge set $U \subseteq V$. Clearly, π restricted to U remains a feasible edge b-coloring. The dual $H' := (G')^*$ is the subgraph of H induced by the vertex set U. In terms of H', π assigns colors to vertices. Since π restricted to U is feasible for G', it follows from vertex–edge duality that vertices in H' are colored differently if they appear in a hyperedge together, i.e., π is a feasible vertex coloring for H. Define thus $U_t := \{v \in U \mid \pi(v) = t\}$ for each color $t \in [b]$. Then, the sets U_t are disjoint by definition and $\dot\bigcup_{t \in [b]} U_t = U$ as $\pi(U) \subseteq \pi(V) \subseteq [b]$. Let finally $t \in [b]$ and $u, v \in U_t$ with $u \neq v$ and assume towards a contradiction that $u, v \in S_2^p$ for some $p \in [k]$. Then, $u, v \in \phi(p) \in E$ and $\pi(u) = t = \pi(v)$ by construction of S_2^p and U_t, contradicting that π is a feasible vertex coloring for $H = (V, E)$. $\qquad\square$

Formally, we write $\mathcal{S}(n, k)$ for an arbitrary but fixed sequence $((S_1^p, S_2^p))_{p \in [k]}$ with $S_1^p \dot\cup S_2^p = [n]$ for every $p \in [k]$ that fulfills the conditions of Lemma 3. We assume for technical reasons that $S_2^1 = [b]$.

4   Impartial Selection

We are prepared to construct a mechanism that provides the first approximation guarantee for deterministic impartial selection with weighted votes. Our main result is the following.

Theorem 1. *Let $n, k \in \mathbb{N}$ with $1 < k < n$ and $k - k \bmod 2 \geq 2\sqrt{n}$. Then, there exists an (n, k)-selection mechanism that is impartial and α-optimal with*

$$\alpha = \frac{k - k \bmod 2}{k \left\lceil \frac{2n}{k - k \bmod 2} \right\rceil}.$$

The performance guarantee of Theorem 1 is shown in Fig. 2. It starts from $2/k$ for $k - k \bmod 2 = 2\sqrt{n}$ and grows up to $1/3$ for $k - k \bmod 2 \in [2n/3, n-1]$.

The main idea of the algorithm is as follows. We construct a robust partition system of the set of agents, i.e., a set of k many partitions of the agents into voters and candidates such that each agent appears as a candidate twice and with disjoint sets of contenders. For the second candidacy, we remove votes that are already present in the first candidacy to avoid double-counting. Then, the mechanism selects the top scoring candidate from each partition, possibly selecting some agents twice. This mechanism is impartial as voters and candidates are disjoint in each partition. The performance guarantee stems mainly from the fact that every vote is counted exactly once.

In Sect. 3, we showed that a robust partition system is guaranteed to exist as long as n and k satisfy $k < n$, $b := 2n/k \in \mathbb{N}$ and $b \leq k/2 \in \mathbb{N}$. In the following,

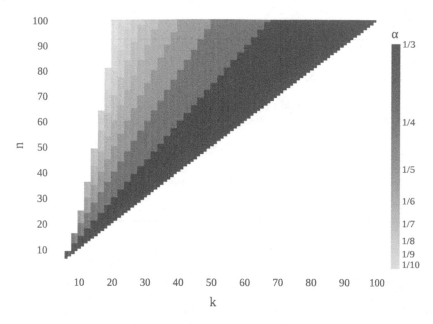

Fig. 2. The performance guarantee of Theorem 1 for permissible n and k.

we assume these conditions in order to define and analyze our mechanism; we lift them in the end to obtain the general result stated in Theorem 1.

Given n and k as in Lemma 3, our selection mechanism is formally described by Algorithm 1; we refer to it as SELECT_k and denote its output by $\text{SELECT}_k(A)$ for a given input matrix $A \in \mathcal{A}_n$. The procedure considers a partition system with the properties stated in Lemma 3 and performs two main steps. Recall that each agent $j \in [n]$ appears in two candidate sets; we denote their indices by $l(j) < r(j) \in [k]$ such that $j \in S_2^{l(j)} \cap S_2^{r(j)}$. The mechanism first computes the *modified score* $\hat{\sigma}_{S_1^p}(j)$ for each $j \in [n]$ and each $p \in \{l(j), r(j)\}$, which is simply the actual score $\sigma_{S_1^{l(j)}}(j)$ for $p = l(j)$. For $p = r(j)$, however, we omit the votes from agents $i \in S_1^{l(j)}$ in order to avoid double counting. The mechanism then selects the vertex with the highest modified score out of each candidate set, breaking ties in favor of the largest index.[1] Figure 3 illustrates a possible execution of SELECT_6 on an instance $A \in \mathcal{A}_9$.

Throughout this section, whenever n, k, and $A \in \mathcal{A}_n$ are fixed, we write $((S_1^1, S_2^1), \ldots, (S_1^k, S_2^k))$, $l(j)$, $r(j)$, $\hat{\sigma}_{S_1^p}(j)$, i^p, and X for each $p \in [k]$ and $j \in [n]$ to refer to the objects defined in SELECT_k. We only specify the input matrix A as an argument when it is not clear from the context. The following lemma constitutes the main technical ingredient for the proof of Theorem 1.

---

[1] We sometimes compare tuples, for example $(\sigma(j), j)$, in lexicographical order. We use standard inequality signs as well as the min and max operators for this purpose.

Algorithm 1: $\text{SELECT}_k(A)$

Input: weight matrix $A \in \mathcal{A}_n$
Output: set $X \subseteq [n]$ with $|X| \leq k$
let $((S_1^1, S_2^1), \dots, (S_1^k, S_2^k)) = \mathcal{S}(n, k)$;
for $j \in [n]$ **do**
    let $\{l(j), r(j)\} = \{p \in [k] : j \in S_2^p\}$ with $l(j) < r(j)$;
    define $\hat{\sigma}_{S_1^{l(j)}}(j) \leftarrow \sigma_{S_1^{l(j)}}(j)$ and $\hat{\sigma}_{S_1^{r(j)}}(j) \leftarrow \sigma_{S_1^{r(j)} \setminus S_1^{l(j)}}(j)$;
end
initialize $X \leftarrow \emptyset$;
for $p \in [k]$ **do**
    take $i^p = \arg\max_{j \in S_2^p}(\hat{\sigma}_{S_1^p}(j), j)$ and update $X \leftarrow X \cup \{i^p\}$
end
return X

Lemma 4. *Let $n, k \in \mathbb{N}$ with $k < n$ be such that $b := 2n/k \in \mathbb{N}$ and $b \leq k/2 \in \mathbb{N}$. Then, SELECT_k is an impartial and $1/b$-optimal (n, k)-selection mechanism.*

Proof. We consider n and k as in the statement. We first note that SELECT_k returns a subset of $[n]$ of size at most k and is well-defined as we have $|\{p \in [k] : j \in S_2^p\}| = 2$ for every $j \in [n]$. The former holds since i^p is a single vertex for every $p \in [k]$ and $X = \bigcup_{p \in [k]} \{i^p\}$; the latter follows from property (iii) of Lemma 3 since $b := 2n/k \in \mathbb{N}$ and $b \leq k/2 \in \mathbb{N}$.

To see that SELECT_k is impartial, let $A, A' \in \mathcal{A}_n$ and $j \in [n]$ such that $A_{-j} = A'_{-j}$. Suppose $j \in \text{SELECT}_k(A)$. From the definition of the mechanism, we have that there is $p \in [k]$ such that $j = \arg\max_{i \in S_2^p}(\hat{\sigma}_{S_1^p}(i; A), i)$. Since $j \in S_2^p$ and $A_{-j} = A'_{-j}$, we have both that $\hat{\sigma}_{S_1^p}(j; A) = \hat{\sigma}_{S_1^p}(j; A')$ and, for every $i \in S_2^p \setminus \{j\}$, that $\hat{\sigma}_{S_1^p}(i; A) = \hat{\sigma}_{S_1^p}(i; A')$. This yields $j = \arg\max_{i \in S_2^p}(\hat{\sigma}_{S_1^p}(i; A'), i)$. Thus, we obtain from the definition of the mechanism that $j \in \text{SELECT}_k(A')$. We conclude that $\text{SELECT}_k(A) \cap \{j\} = \text{SELECT}_k(A') \cap \{j\}$.

It remains to show that SELECT_k has an approximation ratio of $1/b$. To this end, we let $A \in \mathcal{A}_n$ be an arbitrary weight matrix. First, observe that

$$\hat{\sigma}_{S_1^{l(j)}}(j) + \hat{\sigma}_{S_1^{r(j)}}(j) = \sigma_{S_1^{l(j)}}(j) + \sigma_{S_1^{r(j)} \setminus S_1^{l(j)}}(j) = \sigma(j) \tag{1}$$

for every $j \in [n]$, since property (ii) of Lemma 3 implies $S_1^{l(j)} \cup S_1^{r(j)} = [n] \setminus \{j\}$. Furthermore, the definition of i^p yields that

$$\hat{\sigma}_{S_1^p}(i^p) \geq \hat{\sigma}_{S_1^p}(j) \tag{2}$$

for every $p \in [k]$ and $j \in S_2^p$. Given these two facts, we claim that

$$\hat{\sigma}_{S_1^{l(j)}}(i^{l(j)}) + \hat{\sigma}_{S_1^{r(j)}}(i^{r(j)}) \geq \sigma(j) \tag{3}$$

for every $j \in [n]$. To see this, we fix $j \in [n]$. If $i^p = j$ for each $p \in \{l(j), r(j)\}$, inequality (3) follows immediately from equality (1). If $|\{j\} \cap \{i^p : p \in$

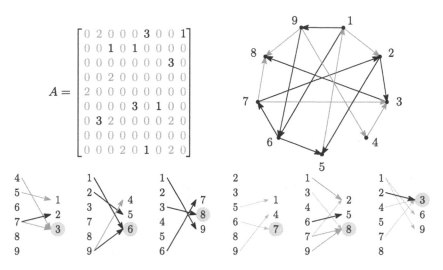

Fig. 3. Example of $\text{SELECT}_6(A)$ for $A \in \mathcal{A}_9$. The weight matrix A is shown alongside its graph representation, where edges of weight 1 are in blue, weight 2 are in orange, weight 3 are in red, and edges of weight 0 are not included. The partition system is given below, where omitted edges are shown in gray. For each partition, the selected vertex is highlighted in light blue. Observe that $\sigma(\text{SELECT}_6(A)) = 17$ and $\sigma(\text{OPT}_6(A)) = 27$; the multiplicative guarantee provided by Lemma 4 for this instance is 1/3. (Color figure online)

$\{l(j), r(j)\}\}| = 1$, say w.l.o.g. $i^{l(j)} = j$ and $i^{r(j)} = h \neq j$, we have that

$$\hat{\sigma}_{S_1^{r(j)}}(h) \geq \hat{\sigma}_{S_1^{r(j)}}(j) = \sigma(j) - \hat{\sigma}_{S_1^{l(j)}}(j),$$

where the inequality follows from (2) and the equality from (1). In this case, inequality (3) follows from $j = i^{l(j)}$ and $h = i^{r(j)}$. Finally, if $j \notin \{i^p : p \in \{l(j), r(j)\}\}$, we have from (2) that

$$\hat{\sigma}_{S_1^{l(j)}}(i^{l(j)}) \geq \hat{\sigma}_{S_1^{l(j)}}(j) \quad \text{and} \quad \hat{\sigma}_{S_1^{r(j)}}(i^{r(j)}) \geq \hat{\sigma}_{S_1^{r(j)}}(j)$$

so that inequality (3) follows from summing up these two inequalities and applying equality (1). This concludes the proof of inequality (3).

Letting χ denote the indicator function for logical propositions, we note that

$$\sigma(\text{SELECT}_k(A)) = \sum_{j \in \text{SELECT}_k(A)} \sigma(j) = \sum_{j \in \text{SELECT}_k(A)} \left(\hat{\sigma}_{S_1^{l(j)}}(j) + \hat{\sigma}_{S_1^{r(j)}}(j) \right)$$

$$\geq \sum_{j \in \text{SELECT}_k(A)} \left(\hat{\sigma}_{S_1^{l(j)}}(j)\chi(j = i^{l(j)}) + \hat{\sigma}_{S_1^{r(j)}}(j)\chi(j = i^{r(j)}) \right)$$

$$= \sum_{p \in [k]} \hat{\sigma}_{S_1^p}(i^p). \tag{4}$$

Indeed, the first equality follows from the definition of $\text{SELECT}_k(A)$, the second one from equality (1), the inequality simply from $\chi(\cdot) \leq 1$, and the last equal-

---

Algorithm 2: $\text{GEN}_{\text{ALG},k}(A)$

Input: weight matrix $A \in \mathcal{A}_n$
Output: set $X \subseteq [n]$ with $|X| \leq k$
define $\tilde{A} \in \mathcal{A}_{\tilde{n}}$ as

$$\tilde{A}_{ij} = \begin{cases} A_{ij} & \text{if } i, j \in [n], \\ 0 & \text{otherwise.} \end{cases}$$

return $\text{ALG}(\tilde{A})$

---

ity follows from the definition of i^p for each $p \in [k]$. We next use inequalities (3) and (4) to conclude the bound stated in the lemma.

For $b := 2n/k$, we know from property (iv) of Lemma 3 that there is a partition $\bigcup_{t \in [b]} U_t = \text{OPT}_k(A)$ such that $i \in S_2^p$ implies $j \notin S_2^p$ for all $t \in [b]$, $i, j \in U_t$ with $i \neq j$, and $p \in [k]$. We obtain that, for every $t \in [b]$,

$$\sigma(\text{SELECT}_k(A)) \geq \sum_{p \in [k]} \hat{\sigma}_{S_1^p}(i^p) \geq \sum_{j \in U_t} \left(\hat{\sigma}_{S_1^{l(j)}}(i^{l(j)}) + \hat{\sigma}_{S_1^{r(j)}}(i^{r(j)}) \right) \geq \sigma(U_t), \tag{5}$$

where the first inequality follows from inequality (4), the second one from the fact that $\{l(i), r(i)\} \cap \{l(j), r(j)\} = \emptyset$ for every $t \in [b]$ and every $i, j \in U_t$ with $i \neq j$, and the last one from inequality (3). This yields

$$\sigma(\text{SELECT}_k(A)) \geq \max_{t \in [b]} \sigma(U_t) \geq \frac{1}{b} \sum_{t \in [b]} \sigma(U_t) = \frac{1}{b}\sigma(\text{OPT}_k(A)).$$

Here, the first inequality follows from (5), the second one from the observation that the maximum of a set of values is at least as large as their average, and the equality from the fact that $\{U_t\}_{t \in [b]}$ is a partition of $\text{OPT}_k(A)$. Therefore, we obtain that SELECT_k is α-optimal for

$$\frac{\sigma(\text{SELECT}_k(A))}{\sigma(\text{OPT}_k(A))} \geq \frac{1}{b} = \alpha. \qquad \square$$

In order to conclude our main result, it only remains to extend the bound given by Lemma 4 to the case where at least one of the conditions $b := 2n/k \in \mathbb{N}$ or $b \leq k/2 \in \mathbb{N}$ is not satisfied. To this end, we show a general way to extend bounds on the approximation ratio for given values of $\tilde{n}$ and $\tilde{k}$ to other values n and k: whenever $n \leq \tilde{n}$ and $k \geq \tilde{k}$, we can do so preserving impartiality and only losing a factor of $\tilde{k}/k$.

Given $k, \tilde{k}, \tilde{n}, n \in \mathbb{N}$ with $k \leq \tilde{k} < \tilde{n} \leq n$, and an $(\tilde{n}, \tilde{k})$-selection mechanism ALG, we can generalize ALG to the (n, k)-selection mechanism $\text{GEN}_{\text{ALG},k}$. This is formally described by Algorithm 2, whose output is denoted by $\text{GEN}_{\text{ALG},k}(A)$ for an input matrix $A \in \mathcal{A}_n$. This algorithm simply extends A to the $\tilde{n} \times \tilde{n}$ matrix $\tilde{A}$ by adding $\tilde{n} - n$ many all-zero rows and columns to it, and then applies ALG on $\tilde{A}$. As before, whenever $\tilde{n}$, n, k, ALG, and $A \in \mathcal{A}_n$ are fixed, we use $\tilde{A}$ to refer to the object defined in Algorithm 2 for this input. In a slight overload of notation, when we consider $A' \in \mathcal{A}_n$ as an input, we write simply $\tilde{A}'$ for the matrix defined in Algorithm 2 on input A'. We obtain the following lemma.

Lemma 5. *Let $\tilde{k}, k, n, \tilde{n} \in \mathbb{N}$ with $\tilde{k} \leq k < n \leq \tilde{n}$ be such that there exists an impartial and $\tilde{\alpha}$-optimal $(\tilde{n}, \tilde{k})$-selection mechanism* ALG. *Then* GEN$_{\text{ALG},k}$ *is an impartial and α-optimal (n, k)-selection mechanism with $\alpha = (\tilde{k}/k)\tilde{\alpha}$.*

Proof. Let n, k, $\tilde{n}$, and $\tilde{k}$ be as in the statement. Let also ALG denote the impartial and $\tilde{\alpha}$-optimal $(\tilde{n}, \tilde{k})$-selection mechanism.

In order to see that GEN$_{\text{ALG},k}$ is impartial, let $A, A' \in \mathcal{A}_n$ and $i \in [n]$ such that $A_{-i} = A'_{-i}$. This implies $\tilde{A}_{-i} = \tilde{A}'_{-i}$, thus the impartiality of ALG yields

$$\text{GEN}_{\text{ALG},k}(A) \cap \{i\} = \text{ALG}(\tilde{A}) \cap \{i\} = \text{ALG}(\tilde{A}') \cap \{i\} = \text{GEN}_{\text{ALG},k}(A') \cap \{i\}.$$

To prove the approximation guarantee, we let $A \in \mathcal{A}_n$ be an arbitrary weight matrix and observe that

$$\frac{\sigma(\text{GEN}_{\text{ALG},k}(A))}{\sigma(\text{OPT}_{\tilde{k}}(\tilde{A}))} = \frac{\sigma(\text{ALG}(\tilde{A}))}{\sigma(\text{OPT}_{\tilde{k}}(\tilde{A}))} \geq \tilde{\alpha}, \tag{6}$$

where the equality follows from the definition of GEN$_{\text{ALG},k}$ and the inequality follows from the $\tilde{\alpha}$-optimality of ALG. On the other hand, as $\tilde{k} \leq k$ and $\sigma(j, \tilde{A}) = 0$ for every $j \notin [n]$, we know that

$$\frac{\sigma(\text{OPT}_k(A))}{k} = \frac{1}{k} \max_{S \subseteq [n]:\, |S|=k} \sigma(S; A) \leq \frac{1}{\tilde{k}} \max_{S \subseteq [n]:\, |S|=\tilde{k}} \sigma(S; A) = \frac{\sigma(\text{OPT}_{\tilde{k}}(\tilde{A}))}{\tilde{k}},$$

i.e., the average score of the k top-voted agents of input A can be no larger than the average score of the $\tilde{k}$ top-voted agents of input $\tilde{A}$. Plugging this inequality into (6) concludes the proof as

$$\frac{\sigma(\text{GEN}_{\text{ALG},k}(A))}{\sigma(\text{OPT}_k(A))} \geq \frac{\tilde{k}}{k} \frac{\sigma(\text{GEN}_{\text{ALG},k}(A))}{\sigma(\text{OPT}_{\tilde{k}}(\tilde{A}))} \geq \frac{\tilde{k}}{k}\tilde{\alpha}. \qquad \square$$

Our main result now follows from the last two lemmas.

Proof of Theorem 1. Let n and k be as in the statement. We define

$$\tilde{k} := k - k \bmod 2 \quad \text{and} \quad \tilde{n} := \frac{k - k \bmod 2}{2} \left\lceil \frac{2n}{k - k \bmod 2} \right\rceil.$$

It is clear that $\tilde{n}, \tilde{k}$ are natural numbers with $\tilde{k} \leq k < n \leq \tilde{n}$ and that

$$b := \frac{2\tilde{n}}{\tilde{k}} = \left\lceil \frac{2n}{k - k \bmod 2} \right\rceil \in \mathbb{N}.$$

Moreover, we have that

$$\tilde{n} = \frac{k - k \bmod 2}{2} \left\lceil \frac{2n}{k - k \bmod 2} \right\rceil \leq \frac{k - k \bmod 2}{2} \left\lceil 2\frac{(k - k \bmod 2)^2}{4}}{k - k \bmod 2} \right\rceil = \frac{\tilde{k}^2}{4},$$

where the inequality follows from the condition $k - k \bmod 2 \geq 2\sqrt{n}$ in the statement. This yields $b = 2\tilde{n}/\tilde{k} \leq \tilde{k}/2 \in \mathbb{N}$. By Lemma 4, this implies that $\text{SELECT}_{\tilde{k}}$ is an impartial and $\tilde{\alpha}$-optimal $(\tilde{n}, \tilde{k})$-selection mechanism with

$$\tilde{\alpha} = \frac{1}{b} = \frac{1}{\left\lceil \frac{2n}{k - k \bmod 2} \right\rceil}.$$

Since $\tilde{n}, \tilde{k} \in \mathbb{N}$ are such that $\tilde{k} \leq k$ and $\tilde{n} \geq n$, Lemma 5 implies that $\text{GEN}_{\text{SELECT}_{\tilde{k}}, k}$ is an impartial and α-optimal (n, k)-selection mechanism with

$$\alpha = \frac{\tilde{k}}{k} \tilde{\alpha} = \frac{k - k \bmod 2}{k \left\lceil \frac{2n}{k - k \bmod 2} \right\rceil}. \qquad \square$$

The mechanism and its approximation ratio naturally extend to the widely studied unweighted setting, where one restricts to matrices $A \in \mathcal{A}_n$ with $A_{ij} \in \{0, 1\}$ for every $i, j \in [n]$. This improves on the previous best lower bound of $1/k$ whenever the number of agents to select is high enough compared to n for Theorem 1 to be applicable: if $k - k \bmod 2 \geq 2\sqrt{n}$, the theorem guarantees the existence of an (n, k)-selection mechanism that is impartial and α-optimal with

$$\alpha = \frac{k - k \bmod 2}{k \left\lceil \frac{2n}{k - k \bmod 2} \right\rceil} \geq \frac{k - k \bmod 2}{k \left\lceil \frac{2(k - k \bmod 2)^2}{4(k - k \bmod 2)} \right\rceil} = \frac{2}{k}.$$

We end this section by showing that the analysis of our (n, k)-selection mechanism SELECT_k for n and k satisfying the conditions of Lemma 4 is tight.

Theorem 2. *Let $n, k \in \mathbb{N}$ with $k < n$ be such that $b := 2n/k \in \mathbb{N}$ and $b \leq k/2 \in \mathbb{N}$. Then, for every $\varepsilon > 0$ we have that SELECT_k is not $(1/b + \varepsilon)$-optimal.*

Proof. Let n and k be as in the statement and consider the partition system $((S_1^1, S_2^1), \ldots, (S_1^k, S_2^k)) = \mathcal{S}(n, k)$. Recall that we defined $\mathcal{S}(n, k)$ such that $S_2^1 = [b]$. Considering $l(j)$ and $r(j)$ as defined in Algorithm 1 for every $j \in [n]$, we note that for each $j \in S_2^1$ we have $l(j) = 1$. For each $j \in S_2^1$, we let $h(j)$ be an arbitrary agent in S_1^1 such that $h(j) \in S_2^{r(j)}$. Such vertex is guaranteed to exist, since from property (ii) of Lemma 3 we know that $S_2^{l(j)} \cap S_2^{r(j)} = \{j\}$, and from property (i) we have that $|S_2^{r(j)}| = b > 1$.

We consider the instance given by $A \in \mathcal{A}_n$ with $A_{ij} = 1$, if $j \in S_2^1$ and $i = h(j)$, and $A_{ij} = 0$, otherwise. Intuitively, this construction aims to have $A_{ij} > 0$ for some $i \in S_1^p$ and $j \in S_2^p$ only if $p = 1$, so that the only agent with a strictly positive score selected by the mechanism, among b agents with a strictly positive score, is i^1. An example of this construction and the corresponding outcome of the mechanism is illustrated in Fig. 4. It is clear that $\text{OPT}_k(A) = [b]$ and $\sigma(\text{OPT}_k(A)) = b$. On the other hand, we have that $\sigma(i^1) = 1$ and, for every $p \in \{2, 3, \ldots, k\}$, that $\hat{\sigma}_{S_1^p}(j) = 0$ for every $j \in S_2^p$. This is because we have $\sigma(j) = 0$ for every $j \notin [b]$ and, whenever there is a $j \in [b] \cap S_p^2$, we also

```
4                 1           1           2           1           1
5                 2           2           3           3           2
    1       4     3     7     5     1     4     2     4       3
6                                               2
    2       5     7     8     6     4     6     5     5       6
7                                               5
    3       6     8     9     8     7     6     8     7       9
8                 8     5     8     7     7     8     9
9                 9     6     9     9     8
```

Fig. 4. Example of the construction of the proof of Theorem 2 for $n = 9$ and $k = 6$ with 3 votes of weight 1: agent 4 votes for agent 1, agent 5 votes for agent 2, and agent 6 votes for agent 3. All votes are only seen in the first partition. Since agents with positive scores have the smallest indices, they are not selected in their second candidate set.

have $h(j) \in S_p^2$. Moreover, for every $p \in \{2, 3, \ldots, k\}$ such that there exists a $j \in [b] \cap S_p^2$, we have that $j \neq \max S_2^p$ since $h(j) \in S_2^p$ and $h(j) > j$. This yields $\sigma(i^p) = 0$ for every $p \in \{2, 3, \ldots, k\}$, thus $\sigma(\text{SELECT}_k(A)) = 1$. This concludes the proof as

$$\frac{\sigma(\text{SELECT}_k(A))}{\sigma(\text{OPT}_k(A))} = \frac{1}{b}. \qquad \square$$

In terms of general upper bounds on the approximation ratio that an impartial mechanism can achieve, the best known is $(k-1)/k$ [4]. Even for the regime $k - k \bmod 2 \geq 2n/3$, in which our mechanism provides a lower bound of $1/3$ and considerably improves the previously best bound of $1/k$ [4], the gap remains large. Further improvements in either lower or upper bounds arise as the main direction for future work.

5 Impartial Assignment

In this section, we consider a generalization of the impartial selection problem in which agents are not selected into one but *assigned* to at most one of m many sets, which we refer to as *jobs*. Each job $\ell \in [m]$ can be assigned at most k agents, so that we obtain the impartial selection problem as the special case where $m = 1$. We first extend the notation from Sect. 2 to this new setting.

For $n, m \in \mathbb{N}$ with $m \leq n$, we consider m-tuples of weight matrices $\mathbf{A} = (A_1, A_2, \ldots, A_m) \in \mathcal{A}_n^m$, each of them representing the weighted votes for one job. Let further $k < n$ in the following; an instance of the assignment problem is then given by the tuple $\mathbf{A}$ and the value k. We let

$$\mathcal{X}_k := \big\{ \mathbf{X} = (X_1, X_2, \ldots, X_m) \in \big(2^{[n]}\big)^m : |X_i| \leq k \text{ and } X_i \cap X_j = \emptyset$$
$$\text{for every } i, j \in [m] \text{ with } i \neq j \big\}$$

denote the set of feasible assignments, i.e., the set of tuples $\mathbf{X}$ containing m pairwise disjoint subsets of agents, each with cardinality at most k. In a slight overload of notation, for $\mathbf{X} \in \mathcal{X}_k$ and $\mathbf{A} \in \mathcal{A}_n^m$, we write

$$\sigma(\mathbf{X}; \mathbf{A}) := \sum_{\ell \in [m]} \sigma(X_\ell; A_\ell)$$

to refer to the sum, over the jobs, of the score of the set assigned to each job according to $\mathbf{X}$, and we simply write $\sigma(\mathbf{X})$ when the instance is clear from the context. Finally, for $\mathbf{A} \in \mathcal{A}_n^m$, we let

$$\text{OPT}_k(\mathbf{A}) := \arg \max_{\mathbf{X} \in \mathcal{X}_k} \sigma(\mathbf{X}; \mathbf{A})$$

denote an arbitrary assignment with the largest score among feasible assignments. We write just OPT_k when the instance is clear.

An (n, m, k)-assignment mechanism is a function $f \colon \mathcal{A}_n^m \to \left(2^{[n]}\right)^m$ such that $f(\mathbf{A}) \in \mathcal{X}_k$ for every $\mathbf{A} \in \mathcal{A}_n^m$. Such a mechanism is *impartial* if, for every pair of instances $\mathbf{A} \in \mathcal{A}_n^m$ and $\mathbf{A}' \in \mathcal{A}_n^m$ and for all agents $i \in [n]$ such that $(A_\ell)_{-i} = (A_\ell')_{-i}$ holds for each job $\ell \in [m]$, it also holds that $(f(\mathbf{A}))_\ell \cap \{i\} = (f(\mathbf{A}'))_\ell \cap \{i\}$ for every $\ell \in [m]$. We further call an (n, m, k)-assignment mechanism α-*optimal* if

$$\frac{\sigma(f(\mathbf{A}); \mathbf{A})}{\sigma(\text{OPT}_k(\mathbf{A}); \mathbf{A})} \geq \alpha$$

holds for all $\mathbf{A} \in \mathcal{A}_n^m$ and some $\alpha \in [0, 1]$.

We are prepared to state the main theorem of this section.

Theorem 3. *Let $n, m, k \in \mathbb{N}$ with $1 < k < n$, $mk \leq n$, and $k - k \bmod 2 \geq 2\sqrt{n}$. Then, there exists an (n, m, k)-assignment mechanism that is impartial and α-optimal with*

$$\alpha = \frac{k - k \bmod 2}{2k \left\lceil \frac{2n}{k - k \bmod 2} \right\rceil}.$$

The proof of this result is omitted due to space constraints. The main ingredient is an adaptation of our mechanism from Sect. 4 that selects from each partition not one but m many agents: one for each set $\ell \in [m]$. We leave the partitioning step unchanged and, for the second step, assign m agents from each candidate set to different jobs in a way that the score obtained for each partition is maximized. In case an agent is assigned to two different jobs, we assign it to the one for which it receives the highest number of votes.

Impartiality of this mechanism follows from a similar reasoning as in the proof of Theorem 1: whenever the vote of an agent is taken into account, the agent is not part of the candidate set. The approximation guarantee makes use of a detailed analysis of the case $b := 2n/k \in \mathbb{N}$ and $b \leq k/2 \in \mathbb{N}$, which is somewhat more intricate than the analysis in Sect. 4. We consider subsets of agents that are assigned to any job in the optimal assignment and are not mutual contenders. We then use the key fact that, when considering the two partitions in which some agent i is in the candidate set, the mechanism assigns agents in a way that the sum of votes of the assigned agents in both partitions is at least the number of votes that i receives for any job. Exploiting the robust partitioning structure as before allows us to take the best of these subsets and conclude via an averaging argument. Here we lose an additional factor of $1/2$ due to the possibility that an agent is initially assigned to two jobs. The extension to general values n, m, and k is then analogous to that of Sect. 4.

References

1. Alon, N., Fischer, F., Procaccia, A., Tennenholtz, M.: Sum of us: strategyproof selection from the selectors. In: Proceedings of the 13th Conference on Theoretical Aspects of Rationality and Knowledge, pp. 101–110 (2011)
2. Aziz, H., Lev, O., Mattei, N., Rosenschein, J.S., Walsh, T.: Strategyproof peer selection using randomization, partitioning, and apportionment. Artif. Intell. **275**, 295–309 (2019)
3. Babichenko, Y., Dean, O., Tennenholtz, M.: Incentive-compatible selection mechanisms for forests. In: Proceedings of the 21st ACM Conference on Economics and Computation, pp. 111–131 (2020)
4. Bjelde, A., Fischer, F., Klimm, M.: Impartial selection and the power of up to two choices. ACM Trans. Econ. Comput. **50**(4), 1–20 (2017)
5. Bousquet, N., Norin, S., Vetta, A.: A near-optimal mechanism for impartial selection. In: Liu, T.-Y., Qi, Q., Ye, Y. (eds.) WINE 2014. LNCS, vol. 8877, pp. 133–146. Springer, Cham (2014). https://doi.org/10.1007/978-3-319-13129-0_10
6. Caragiannis, I., Christodoulou, G., Protopapas, N.: Impartial selection with additive approximation guarantees. Theory Comput. Syst. **66**, 721–742 (2022)
7. Caragiannis, I., Christodoulou, G., Protopapas, N.: Impartial selection with prior information. In: Proceedings of the ACM Web Conference, pp. 3614–3624 (2023)
8. Cembrano, J., Fischer, F., Hannon, D., Klimm, M.: Impartial selection with additive guarantees via iterated deletion. In: Pennock, D.M., Segal, I., Seuken, S. (eds.) Proceedings of the 23rd ACM Conference on Economics and Computation, pp. 1104–1105 (2022)
9. Cembrano, J., Fischer, F., Klimm, M.: Optimal impartial correspondences. In: Hansen, K.A., Liu, T.X., Malekian, A. (eds.) WINE 2022. LNCS, vol. 13778, pp. 187–203. Springer, Cham (2022). https://doi.org/10.1007/978-3-031-22832-2_11
10. Cembrano, J., Fischer, F., Klimm, M.: Improved bounds for single-nomination impartial selection. In: Proceedings of the 24th ACM Conference on Economics and Computation (2023)
11. De Clippel, G., Moulin, H., Tideman, N.: Impartial division of a dollar. J. Econ. Theory **1390**(1), 176–191 (2008)
12. Dhull, K., Jecmen, S., Kothari, P., Shah, N.B.: Strategyproofing peer assessment via partitioning: the price in terms of evaluators' expertise. In: Proceedings of the AAAI Conference on Human Computation and Crowdsourcing, vol. 10, pp. 53–63 (2022)
13. Edelman, P.H., Por, A.: A new axiomatic approach to the impartial nomination problem. Games Econom. Behav. **130**, 443–451 (2021)
14. Fischer, F., Klimm, M.: Optimal impartial selection. SIAM J. Comput. **440**(5), 1263–1285 (2015)
15. Holzman, R., Moulin, H.: Impartial nominations for a prize. Econometrica **810**(1), 173–196 (2013)
16. Kahng, A., Kotturi, Y., Kulkarni, C., Kurokawa, D., Procaccia, A.D.: Ranking wily people who rank each other. In: Proceedings of the 32nd AAAI Conference on Artificial Intelligence, pp. 1087–1094 (2018)
17. Kőnig, D.: Gráfok és alkalmazásuk a determinánsok Žs a halmazok elméletére. Mat. Természettudományi Értesítő **34**, 104–119 (1916)
18. Kurokawa, D., Lev, O., Morgenstern, J., Procaccia, A.D.: Impartial peer review. In: Proceedings of the 24th International Joint Conference on Artificial Intelligence, pp. 582–588 (2015)

19. Lev, O., Mattei, N., Turrini, P., Zhydkov, S.: Peer selection with noisy assessments. arXiv preprint arXiv:2107.10121 (2021)
20. Mackenzie, A.: An axiomatic analysis of the papal conclave. Econ. Theor. **69**, 713–743 (2020)
21. Mattei, N., Turrini, P., Zhydkov, S.: PeerNomination: relaxing exactness for increased accuracy in peer selection. In: Proceedings of the Twenty-Ninth International Conference on International Joint Conferences on Artificial Intelligence, pp. 393–399 (2021)
22. Niemeyer, A., Preusser, A.: Simple allocation with correlated types. Technical report, Working Paper (2022)
23. Olckers, M., Walsh, T.: Manipulation and peer mechanisms: a survey. arXiv preprint arXiv:2210.01984 (2022)
24. Tamura, S., Ohseto, S.: Impartial nomination correspondences. Soc. Choice Welfare **430**(1), 47–54 (2014)
25. Xu, Y., Zhao, H., Shi, X., Zhang, J., Shah, N.B.: On strategyproof conference peer review. In: Proceedings of the 28th International Joint Conference on Artificial Intelligence, pp. 616–622 (2019)
26. Zhang, X., Zhang, Y., Zhao, D.: Incentive compatible mechanism for influential agent selection. In: Caragiannis, I., Hansen, K.A. (eds.) SAGT 2021. LNCS, vol. 12885, pp. 79–93. Springer, Cham (2021). https://doi.org/10.1007/978-3-030-85947-3_6
27. Zhao, Y., Zhang, Y., Zhao, D.: Incentive-compatible selection for one or two influentials. arXiv preprint arXiv:2306.07707 (2023)

Blockchain Participation Games

Pyrros Chaidos[1,2], Aggelos Kiayias[1,3], and Evangelos Markakis[1,4(✉)]

[1] Input Output Global (IOG), Athens, Greece
markakis@gmail.com
[2] National and Kapodistrian University of Athens,
Panepistimioupolis, Ilisia, 16122 Athens, Greece
[3] Athens University of Economics and Business, Patision 76, 10434 Athens, Greece
[4] University of Edinburgh, 10 Crichton St, Newington, Edinburgh EH8 9AB, UK

Abstract. We study game-theoretic models for capturing participation in blockchain systems. Existing blockchains can be naturally viewed as games, where a set of potentially interested users is faced with the dilemma of whether to engage with the protocol or not. Engagement here implies that the user will be asked to complete certain tasks, whenever she is selected to contribute, according to some stochastic process. Apart from the basic dilemma of engaging or not, even more strategic considerations arise in systems where users may be able to declare participation and then retract (while still being able to receive rewards). We propose two models for studying such games, with the first one focusing on the basic dilemma of engaging or not, whereas the latter focuses on the retraction effects. In both models we provide characterization results or necessary conditions on the structure of Nash equilibria. Our findings reveal that appropriate reward mechanisms can be used to stimulate participation and avoid negative effects of free riding, results that are in line with real world blockchain system deployments.

1 Introduction

The main goal of our work is to study aspects of participation and engagement in some of the major blockchain systems from a game-theoretic angle. Blockchain protocols are typified by so called "permissionless participation", where the agents get to decide whether they wish to engage in the protocol or not and if they choose to, they can do so *unilaterally*. This means that the system is capable of making the necessary adjustments to accommodate for increased or decreased participation, while there is no authority that whitelists the agents who participate — for any user, merely downloading the software and running it is sufficient to become a part of the network.

Based on the above, we observe that every running blockchain defines a *participation game*. A simple version of this game, can be described as follows: consider a protocol that operates in distinct units of time that we will call *epochs*. Imagine now a population of potentially interested agents, with a binary action space, who need to decide whether to engage (participate) or not. There are

J. Garg et al. (Eds.): WINE 2023, LNCS 14413, pp. 169–187, 2024.
https://doi.org/10.1007/978-3-031-48974-7_10

two prominent features that we are interested in studying in this work. First, blockchain protocols such as Bitcoin, Algorand, Ouroboros, and Ethereum incorporate a stochastic process, where only some of the agents who chose to participate are *eligible* to contribute within an epoch, and the others are not. This is an essential component that is either achieved via so called, proof of work, or via proof of stake and ensures that the resulting complexity of the protocol will be *sublinear* in the number of participating agents. The second feature is the *threshold* behavior of such systems, where in order for the blockchain protocol to make progress, the number of contributing users (among the eligible ones in each epoch) should exceed a certain *threshold k*. In some protocols, this threshold can be merely 1 (e.g., in Bitcoin it is sufficient that one agent produces a block for the blockchain to advance), while in other protocols, a larger k is required (e.g., Ethereum currently needs a 2/3-majority voting among its randomly selected committees of at least 128 block validators in each epoch).

With respect to the utility of the agents, in most cases of interest, the protocol issues an (expected) reward to those who were both eligible and actually contributed (i.e., completing whatever task was dictated by the protocol for the eligible users) within an epoch, while at the same time, participating incurs a cost, incorporating effort, time and equipment. The utility then clearly depends on the stochastic process that determines the agents' eligibility. Finally, on top of rewards and costs, the protocol also induces a non-negative public benefit which is expressed as an additive "bonus" parameter, enjoyed by all the agents (including those who abstain) as long as the blockchain makes progress. We stress that the game described so far is applicable not only for the process of producing new blocks from one epoch to another, but also for other applications within blockchain systems, where a certain group task needs to be completed, such as producing SNARKs (e.g., [10]) for bootstrapping new users in the system or for building bridges between blockchains.

The model described so far already gives rise to some interesting consequences, as elaborated in Sect. 2. If we delve into the implementations of such reward mechanisms however, there can be a significant burden imposed by keeping track of all participants who contributed in order to issue rewards. This may come in conflict with efficiency considerations and the use of cryptographic primitives which compress the participation information in order to offer complexities sublinear in the number of engaging parties. A concrete example of this behavior in the context of blockchains is compact certificates [17]), which carefully select what signatures to include when composing a multi-user certificate so that the certificate's size is kept small. In systems that utilize such more efficient primitives, it is impossible to reward exactly those who were eligible and participated and thus one has to resort to rewarding even those that may not be fully participating.

From the perspective of our work, using such performance improvements for the reward mechanism opens up a new strategic choice: agents can declare to participate, but afterwards refrain from completing their assigned tasks. We term this richer game, a blockchain participation game with *retraction*. The retraction setting is particularly relevant, when there is a non-negligible cost of completing all the necessary tasks (which can be viewed as an additional cost parameter than the day-to-day running of the protocol). Retraction opens up the possibility that some participants can become *free-riders*, reaping the rewards of an advancing blockchain without incurring costs to themselves.

1.1 Contribution

Based on the previous discussion, we propose a formal framework to study participation games that focus on the aspects of engagement and free riding in some of the major blockchain systems. We start in Sect. 2 with introducing a simple model, where each user is simply faced with the basic dilemma of participating or not. In Sect. 3, we then consider a richer model with the possibility of retraction. Our first observation is that in any attempt to define such games, the trivial profile where nobody participates is an equilibrium. We view this more as an artifact of the definition, and certainly far from what is observed in practice. We are therefore interested in the following questions.

> *Q1: Do these games possess other non-trivial Nash equilibria? If so, how big is the percentage of users that chooses to contribute at an equilibrium? Q2: How should we set the reward to the users so as to incentivize participation for a sufficiently large fraction of users?*

For the model of Sect. 2, we consider different variations of the game based on the selection probability, with an emphasis on the homogeneous case that treats all players equally. Our findings show that we can have equilibria with a very high level of engagement, as long as the reward parameter is set within appropriate ranges, dependent on the game parameters. We also extend our analysis to the non-homogeneous case, and even though the participation level may not always be as high as before, we can still show that it is large enough that the blockchain makes progress with high probability in every epoch.

Moving on, the model of Sect. 3 requires a more involved analysis, since users now have an additional choice of declaring participation and then retracting. Even so, we are able to show that non-trivial equilibria will have a relatively high number of contributors. Overall, we believe our findings reveal a positive picture, confirming that simple reward mechanisms can stimulate engagement; this is consistent with what is observed in various actual blockchains, having reasonably large and active user populations, even with the possibility of retraction.

1.2 Related Work

The games that we introduce in Sect. 2 bear similarities with classic discrete public good games, also referred to as step-level games, which have been extensively studied, both theoretically (see e.g. [9,18] and even more recent variants in [8]), and experimentally, within behavioral game theory (indicatively see e.g., [4]). The main difference is that we have a randomized process for determining the eligible players per epoch, whereas in public good games, any person is automatically eligible to contribute. We also pay special attention to the case where the public benefit is zero (for users who care only for the monetary reward). Finally, our model of Sect. 3 diverges further from public good games, as it has a richer strategy space.

There are already numerous works that have studied various game-theoretic aspects of blockchain systems. A common theme right from the outset of blockchains has been the study of *mining games*, cf. [13], where participants are facing the dilemma of which version of the ledger to extend. Important results in this context include models for selfish-mining, such as [5], as well as [6,7,13,15], that established both positive and negative results on whether the underlying protocol is an equilibrium or whether parties are incentivized to deviate from "honest" behavior and resort to block withholding.

Another common theme has emerged from game-theoretic studies of *pooling* behavior: given that such protocols are permissionless and they lack any mechanism capable to distinguish the participants as separate entities, it is possible for them to form coalitions — called mining pools or stake pools — and act in tandem as a single entity in the protocol. Prior work established various conditions under which such pools arise and studied their relative size [1,2,12,14,16,19].

We stress that the above studies on selfish mining and stake pools are conditioned on having sufficient participants engaged in the system. Hence, we view these approaches as orthogonal to ours, since our goal is to study the basic dilemma of participating or not. The question we are interested in, is whether the participation game itself has favorable equilibria and under what conditions they occur.

2   A Basic Model for Participation Games in Permissionless Protocols

We start with a vanilla game-theoretic model that intends to capture participation in blockchain systems, as described in the introduction.

Let $N = \{1, 2, \ldots, n\}$ be a population of n users. We consider protocols that run on a per epoch basis, and where the goal is to make "progress" in each epoch. Progress may be related to block production in longest chain protocols or it could even be related to other applications within blockchains, where some task needs to be carried out by the users (like issuing some group certificate, a checkpoint, or validating a rollup). We focus on modeling applications with the following two features.

(i) The use of randomization. As it is impractical to ask all users to contribute, due to throughput and other practical considerations, the protocol selects in each epoch only a subset of users that are eligible to contribute.

(ii) The successful completion of the task is *threshold-based*, i.e., there is a public parameter k, so that progress is made only if at least k users contributed towards carrying out the task.

We introduce now some of the relevant model parameters. In particular, we use the following abstractions:

- Let α be the per-epoch cost of participation. This cost can be seen as the average per-epoch cost of running the protocol software throughout an epoch, possibly updating it when necessary. This cost can capture both actual monetary cost (in electricity, etc.) and perceived human effort cost.
- Let r denote the monetary reward[1] that is given to each player who is eligible (i.e., selected, according to the randomization procedure) in a given epoch, as long as progress is made. We assume here that the reward is given to a user after the protocol checks that indeed the user contributed.
- Let k be the threshold, i.e., the minimum required number of eligible participants that need to contribute in an epoch, so that progress is made.
- Let q be the probability with which a player is eligible in a given epoch. We assume an independent Bernoulli trial for each user, hence we do not insist that we have the same number of eligible players per epoch. Reasonable choices for q is to make it sufficiently high so that in expectation we have enough eligible players per epoch to reach the threshold. Indicatively, we could think of q in the range of $[2k/n, 3k/n]$. See also Remark 1 below on the treatment of players with different stake or hashing power.
- Let v be the inherent value that a player associates with the blockchain making progress. We note that this could be quite small compared to the blockchain rewards but it is intended to model the potential benefit that the existence of a blockchain brings to its users.

Remark 1. Our model considers a homogeneous population, where both the probability of selection and the inherent value are the same for every player. We view this as an initial step on the analysis of such systems. In Sect. 2.4 we also cover asymmetric players in terms of the selection probability.

We assume that every user has two possible pure strategies in every epoch. The first is to simply abstain, and the second choice is for the user to participate. The latter means that the user chooses to be active throughout the epoch and also to proceed with completing the necessary tasks if randomly selected to do so. We analyze here players that when they choose to participate, they will abide by the protocol rules. We discuss in Sect. 3 a richer model with "retractors" where users may also consider skipping their assigned tasks even though they expressed willingness to participate.

For brevity, let $\mathcal{S} = \{\perp, \mathrm{P}\}$, be the binary strategy space of each player, where $\perp$ means abstaining and P stands for choosing to participate. A strategy profile is a tuple $s = (s_1, \ldots, s_n) \in \mathcal{S}^n$, specifying a choice for each player. As usual, given a profile s, and a player i, we denote by s_{-i} the profile s, restricted to all players except i.

---

[1] We can think of r as the expected reward per epoch, conditioned on eligibility and on progress being made. The expectation here is to capture the fact that even when players are eligible, they occasionally lose the race and their block is dropped or they just fail to complete their task.

Our goal is to study the pure Nash equilibria of the static 1-epoch game. To define equilibria, we need first to define the (expected) utility that a player receives in a strategy profile. We outline first, in Table 1, the utility under the possible scenarios that may occur.

Table 1. Possible events that may occur for a player i in a given epoch.

Possible scenarios	Progress is made	No Progress
Abstain	v	0
Participate but not eligible	$v - \alpha$	$-\alpha$
Participate and eligible	$r + v - \alpha$	$-\alpha$

2.1 Equilibrium Constraints

To describe the constraints that need to hold in order to have an equilibrium, it is convenient to use the terminology introduced in the following definition.

Definition 1. *Given a player i, consider a strategy profile s_{-i}, for all players except i. Then, let*

- $p(s_{-i})$ *be the probability that progress is made in a given epoch, without taking into account what player i does. This is equal to the probability that at least k people out of the players who chose to participate under the profile s_{-i}, are selected to be eligible.*
- $p(i, s_{-i})$ *be the probability that progress is made in a given epoch, given that i chose to participate, was eligible, and completed her task. This is equal to the probability that at least $k - 1$ people out of the remaining participating players, except i, are eligible to contribute.*

Table 2. Expected utility under the possible strategies for a player i.

Strategy of pl. i	Expected utility of pl. i, given s_{-i}
Abstain	$p(s_{-i})v$
Participate	$(1 - q)p(s_{-i})v + qp(i, s_{-i})(r + v) - \alpha$

Given a profile s_{-i} for all players except i, the expected utility of player i, for each one of his pure strategies is described in Table 2. We can think of any strategy profile $s = (s_1, \ldots, s_n)$, as partitioning the players into 2 sets, the set of possible contributors C, who are the people choosing to participate, and the set A of abstainers. Therefore, a profile s is a Nash equilibrium if and only if

$$u_i(s) \geq u_i(\bot, s_{-i}) \ \forall i \in C \quad \text{and} \quad u_i(s) \geq u_i(P, s_{-i}) \ \forall i \in A$$

After substituting the expressions of Table 2, the above inequalities are equivalent to:

$$q \cdot [(r + v)p(i, s_{-i}) - vp(s_{-i})] \geq \alpha \quad \forall i \in C \tag{1}$$

$$q \cdot [(r + v)p(i, s_{-i}) - vp(s_{-i})] \leq \alpha \quad \forall i \in A \tag{2}$$

2.2   Pure Nash Equilibria for the Case that $v = 0$

To further simplify the game, suppose that $v = 0$. Apart from serving as a simplification, we also find this case to be quite important from the perspective of a protocol designer. One of the goals in the study of such systems is to identify how large should the monetary rewards be in order to incentivize users to engage with the protocol. When v is large, users are already motivated enough and we would need a lower reward to incentivize them, thus, the worst-case scenario in terms of upper bounding the total monetary rewards needed, is when $v = 0$.

Interestingly, this model falls into the broader class of games with *strategic complements*, where the incentive for a player to take the "desirable" action has a monotonic behavior in terms of how many other people took the same action.

Proposition 1. *The family of games with $v = 0$ exhibits strategic complements, as defined in [11], and hence its set of pure Nash equilibria forms a complete lattice.*

We refer to [11] for further discussion on strategic complements. Proposition 1 reveals a structural property on the set of equilibria, but does not give us any further information on their precise form. Hence, we continue by investigating the possible number of contributors that may arise. Suppose that there was an equilibrium with $|C| = \lambda$ contributors and $n - \lambda$ abstainers. By expanding the probability terms in Eqs. (1) and (2), we get the following inequalities.

$$q \cdot \sum_{j=k-1}^{\lambda-1} \binom{\lambda-1}{j} q^j (1-q)^{\lambda-1-j} \geq \frac{\alpha}{r} \tag{3}$$

$$q \cdot \sum_{j=k-1}^{\lambda} \binom{\lambda}{j} q^j (1-q)^{\lambda-j} \leq \frac{\alpha}{r} \tag{4}$$

We elaborate further on how the above inequalities were derived. For (3), it has come from (1), which is equivalent for a player i, to $rp(i, s_{-i}) \geq \alpha$. Now to calculate $p(i, s_{-i})$, one needs to consider all possible ways that progress is made, given that i has completed her task. This corresponds precisely to all the possible ways of selecting $k - 1$ other players to be eligible, out of the $\lambda - 1$ (excluding i) who have chosen to be in C. In an analogous manner, for $p(i, s_{-i})$ in (4), we need to take into account all possible ways of selecting $k - 1$ other players to be eligible, but now out of the λ available contributors (since, for this case $i \in A$).

The main result of this subsection is the characterization obtained in the following theorem, showing a sharp picture, that we can have at most two pure Nash equilibria.

Theorem 1 (Characterization). *We can have at most two Nash equilibria as follows:*

- *The all-out profile $(\bot, \bot, \bot)$, where nobody contributes, is a pure Nash equilibrium for $k > 1$, or when $k = 1$ and $q \leq \frac{\alpha}{r}$.*
- *There is no equilibrium that has both a positive number of contributors and a positive number of abstainers.*
- *The all-in profile, where everyone contributes, is an equilibrium if and only if:*

$$q \cdot \sum_{j=k-1}^{n-1} \binom{n-1}{j} q^j (1-q)^{n-1-j} \geq \frac{\alpha}{r} \qquad (5)$$

Proof. The fact that the all-out profile is an equilibrium is trivial. For the all-in profile, we have that $C = N$ and hence, we only need to check that Eq. (3) holds when $\lambda = n$. But this is true precisely when the ratio α/r satisfies the stated bound.

The most interesting part of the proof is to show that we cannot have any other pure equilibria. For the sake of contradiction, suppose that there is another equilibrium profile s, with λ contributors and $n - \lambda$ users opting out, where $0 < \lambda < n$. We should show that it is not possible to satisfy Eqs. (3) and (4) simultaneously. This is implied by the following lemma.

Lemma 1. *For every integer λ, with $0 < \lambda < n$ it holds that*

$$\sum_{j=k-1}^{\lambda} \binom{\lambda}{j} q^j (1-q)^{\lambda-j} = \sum_{j=k-1}^{\lambda-1} \binom{\lambda-1}{j} q^j (1-q)^{\lambda-1-j} + \binom{\lambda-1}{k-2} q^{k-1} (1-q)^{\lambda-k+1}$$

The proof of Lemma 1 can be obtained as a special case of a more general result, that we present in the full version of this work. □

We refer to the all-out profile as the trivial equilibrium. Despite the existence of such an undesirable equilibrium, we do not view this as a disastrous property. In particular, the all-out profile is hard to be sustained over time, as it is easy to see that any coalition of at least k users could deviate and gain more, and this also agrees with what we observe in practice. On the contrary, the all-in profile has much more attractive properties against coalitional deviations.

Fact 1. *Whenever the all-in profile is an equilibrium, it is also a strong equilibrium, hence no coalition has a profitable deviation.*

Reflecting upon Theorem 1 and Fact 1, we view these as positive news since they show that as long as the reward is sufficiently high, it is possible to incentivize all users to participate.

2.3   Analysis When $v > 0$

Coming now to the case of a positive inherent value, it is natural to expect that this is typically smaller than the monetary reward r. The presence of the value v introduces some differences with the previous subsection and we can no longer have such a sharp characterization as in Theorem 1. Proposition 1 may not always hold either. Nevertheless, for non-trivial equilibria, we still have a relatively high number of contributors, as described below.

Theorem 2. *When $v \leq r$, then any non-trivial equilibrium must have at least $(2 - q)\frac{k-1}{q}$ contributors.*

Theorem 2 provides a necessary condition for λ, but it does not tell us for which values of λ we do have an equilibrium. This is dependent on the other parameters, such as α and v. However, it is possible to check for every λ if there exists a range for the reward r as a function of α, v and k, so that we can have an equilibrium with λ contributors. Additionally, we point out that we can always set the reward appropriately so as to incentivize all players to participate.

Theorem 3. *There always exists a non-empty range for the reward r so that the all-in profile is an equilibrium. Namely this holds as long as*

$$r \geq \frac{\frac{\alpha}{q} - v\binom{n-1}{k-1}q^{k-1}(1-q)^{n-k}}{\sum_{j=k-1}^{n-1}\binom{n-1}{j}q^j(1-q)^{n-1-j}}$$

Note that the smaller the nominator above, the looser the constraint for r, and as v drops down to 0, this is when we get stricter constraints for r (with $v = 0$, we get exactly the bound implied by Theorem 1). To summarize, when $v > 0$, it is conceivable that we have equilibria other than the all-in and the all-out profile, in contrast to Sect. 2.2. However, all the non-trivial equilibria have a sufficiently high number of participants. For reasonable choices of q in the range of $[2k/n, 3k/n]$, as alluded to in the beginning of Sect. 2, Theorem 2 implies that we can have equilibria with more than $n/2$ participants.

2.4   Equilibria for the Non-symmetric Case W.r.t. Selection Probability

We conclude this section by studying the asymmetric case in terms of the selection probability. In particular, suppose that each player i has a possibly different probability q_i of being selected. We will stick to the case where $v = 0$ for simplicity.

The equilibrium constraints are now more complex due to the assymetry. In an equilibrium with a set C of λ contributors and a set A of $n - \lambda$ abstainers, Eqs. (1) and (2) yield the following inequalities:

$$q_i \cdot \sum_{m=k-1}^{\lambda-1} \sum_{S \subseteq C \setminus \{i\}, |S|=m} \left(\prod_{j \in S} q_j \cdot \prod_{j \in C \setminus S \cup \{i\}} (1 - q_j) \right) \geq \frac{\alpha}{r} \quad \forall i \in C \qquad (6)$$

$$q_i \cdot \sum_{m=k-1}^{\lambda} \sum_{S \subseteq C, |S|=m} \left(\prod_{j \in S} q_j \cdot \prod_{j \in C \setminus S} (1 - q_j) \right) \leq \frac{\alpha}{r} \quad \forall i \in A \qquad (7)$$

Despite the added complexity, it is still feasible to understand the composition of contributors and abstainers at equilibrium. Proposition 1 holds for the asymmetric setting as well, hence the equilibrium set forms a complete lattice. Furthermore, the main theorem of this subsection gives a characterization of the possible equilibrium structures that can arise.

Theorem 4. *Given a game with n players, let $q_1 \geq q_2 \geq \cdots \geq q_\ell$ be the distinct selection probabilities. Then for any non-trivial equilibrium, there must exist a threshold q_t so that all players with $q_i \geq q_t$ are contributors, and they are at least k, and all players with $q_i < q_t$ are abstainers.*

The theorem shows that it is conceivable to have non-trivial equilibria with a relatively low (but at least k) number of contributors in the asymmetric case. Nevertheless, since in an actual blockchain system, the players with higher selection probabilities are expected to be the ones who possess higher amounts of cryptocurrency, it shows us that the richer players will "do their duty" and choose to contribute.

Finally, note that the above theorem is a necessary condition telling us how the candidate equilibria may look like, but it does not identify actual equilibrium profiles nor does it give any information on whether we can have multiple equilibria with a different threshold in each of them. Fortunately, we can efficiently investigate all possible equilibria via a small number of checks.

Theorem 5. *Let $q_1 \geq q_2 \geq \cdots \geq q_\ell$ be the distinct selection probabilities. For every $t \in [\ell]$, there exists an equilibrium with the threshold for the contributors being q_t if and only if (6) is satisfied by using $q_i = q_t$ and (7) is satisfied when we use $q_i = q_{t+1}$. As for the all-in profile, there exists a non-empty range for α/r that makes it an equilibrium.*

The following example provides some further intuition about these results.

Example 1. Suppose we have a 4 player game with $q_1 = q_2 = \frac{1}{2}$ and $q_3 = q_4 = \frac{1}{4}$. Let $\alpha = 1, r = 5$ and $k = 2$. This game has 3 different equilibria (all candidate profiles according to Theorem 4 yield equilibria). The all-out profile is trivially an equilibrium. The "rich-only" profile with $C = \{1, 2\}$ is also an equilibrium. The LHS of (6) is $\frac{1}{4} > \frac{1}{5}$ for the contributors. For the non-contributors, the LHS of (7) is $\frac{1}{4} \cdot \frac{3}{4} = \frac{3}{16}$ which is less than $\frac{1}{5}$; i.e., if one of the poor players abstains, the frequency of reward is not enough for the other poor player to participate. Finally, the all-in profile with $C = \{1, 2, 3, 4\}$ also forms an equilibrium. Suppose we now modify the game so that $q_1 = q_2 = \frac{1}{3}$, $q_3 = q_4 = \frac{1}{4}$ and $r = 8$. Then, the "rich-only" profile with $C = \{1, 2\}$ is no longer an equilibrium. The all-in and the all-out profiles are the only equilibria.

3 A Richer Model: Participation Games with Retraction

In this section we study a richer model motivated by two considerations. First of all, in blockchain systems, it can be inefficient to keep a record of which users among the eligible ones in a given epoch participated, so as to reward only them. For instance, using methods such as compact certificates [17], threshold signatures [3] or succinct non-interactive arguments of knowledge (SNARKs), like [10], it is efficient to verify that a sufficient number of users participated, *without* producing a list of those who actually did: the added size of such a list runs contrary to the goal of efficiency of the underlying primitive. Hence one mechanism to consider in such a setting is to provide the reward r to everybody who was eligible in a given epoch without checking if they actually contributed.

The second consideration is the fact that there can be a non-negligible cost of completing the assigned task in the protocol, if a user is eligible, in addition to the cost of participating and running the protocol software. In Sect. 2 this additional cost is implicitly set to zero. But this may not be realistic in some cryptographic systems. One example would be asking validators to produce computationally expensive SNARK proofs for the benefit of new users joining the system: the cost of contributing to a particular proof is significant and also independent of the cost of day-to-day participation. Hence, it is highly relevant to have two different cost parameters.

The introduction of the two above features, creates further strategic considerations. The users are now able to extend their strategy space by the following option: choose to participate in the system (e.g., to check eligibility, so as to receive a reward whenever eligible) but not complete the task (if they feel enough of the other users will do it). This is an undesirable scenario of free riding that may lead to slow or unreliable operation of the entire system. One simple approach to counteract this is to penalise users who fail to do their job, (e.g., in Ethereum 2.0 there are penalties for lack of participation). We observe that if we indeed strip the reward r from users who try to free ride, their strategic choices fall back to the model of Sect. 2 and hence our previous analysis applies directly. Nevertheless, the downside of a penalty mechanism would be the need to perform bookkeeping of all those who engage - something that counteracts the efficiency benefits of the underlying cryptographic primitive. This motivates the question of whether there is an effective mechanism that does not rely on tracking user behaviour. We are specifically interested in extending the reward r to all eligible users, so that we only need to know whether progress was made or not. The benefit of this is a simpler rewards management: users can be paid automatically (e.g. by means of a smart contract), simply by providing a proof of their eligibility.

We highlight below the similarities and differences between the new model and that of Sect. 2:

- As before, α is the average cost of running the protocol throughout an epoch, including the per-epoch check of whether the user is eligible. The parameters k, q, v, are interpreted as in Sect. 2.

- Let β be the cost of contributing, i.e., of performing the task required for the validation process, in case a player is selected in a given epoch. In Sect. 2, we implicitly had $\beta = 0$.
- Under the new model, the reward r is given to all eligible players of an epoch, as long as progress is made, **regardless** of whether they contributed or not.

Every player has now 3 possible pure strategies: (a) abstain, (b) declare to participate, and whenever eligible, contribute, or (c) declare to participate but if eligible, retract. We note that (c) could have been an available strategy in Sect. 2 as well, but because we assumed a zero cost of contributing once eligible, it is weakly dominated. The possible scenarios that can occur, and the corresponding utility, depending on the other players' decisions as well, are shown in Table 3.

Table 3. Possible events that may occur for a player i.

Possible scenarios	Progress is made	No Progress
Abstain	v	0
Declares participation, not eligible	$v - \alpha$	$-\alpha$
Declares participation, eligible and completes tasks	$r + v - \alpha - \beta$	$-\alpha - \beta$
Declares participation, eligible and retracts	$r + v - \alpha$	$-\alpha$

3.1 Equilibrium Constraints

As was the case in Sect. 2, the all-out profile is again trivially an equilibrium. We proceed to examine what other equilibria may exist.

We start with identifying the conditions that need to hold in order for a profile s to be an equilibrium. Given a profile s_{-i} for all players except i, the expected utility of player i, for each one of her pure strategies is described in Table 4. Any strategy profile $s = (s_1, \ldots, s_n)$ partitions the players into 3 sets, the set of possible contributors C, who are the people choosing to participate and complete their task whenever selected to do so, the set of free-riders F, who choose to participate (so that they can get a reward, whenever eligible), but will retract, and the set A of abstainers.

We group the equilibrium constraints into three groups, based on the possible deviations of the players. First of all, for a player $i \in C$, who decided to contribute whenever eligible, she should not have an incentive to retract and move to F. Symmetrically, a player from F should not have an incentive to contribute. After simplifying these inequalities, based on Table 4, these are equivalent to:

$$p(i, s_{-i}) - p(s_{-i}) \geq \frac{\beta}{r + v} \quad \forall i \in C \tag{8}$$

Table 4. Expected utility under the possible pure strategies for a player i.

Strategies of player i	Expected utility of pl. i, given s_{-i}
Abstain	$p(s_{-i})v$
Participate and do not retract	$(1-q)p(s_{-i})v + q[p(i, s_{-i})(r + v) - \beta] - \alpha$
Participate and retract	$(1-q)p(s_{-i})v + qp(s_{-i})(r + v) - \alpha$

$$p(i, s_{-i}) - p(s_{-i}) \le \frac{\beta}{r + v} \quad \forall i \in F \qquad (9)$$

Intuitively, these constraints imply that a player $i \in C$ should be "critical enough". Hence, there should be a lower bound, expressed in (8), on the difference between the success probabilities, i.e., between the probability that progress is made when i is eligible and contributes, and the probability that the progress is made via the remaining players. This lower bound is independent of α but has to depend on β.

In a similar fashion, the players from C should also not have an incentive to abstain, and move to A, and at the same time, the players from A should not have an incentive to participate and contribute. This yields two more inequalities, which after simplifications are equivalent to:

$$q \cdot [(r + v)p(i, s_{-i}) - vp(s_{-i})] \ge \alpha + \beta \cdot q \quad \forall i \in C \qquad (10)$$

$$q \cdot [(r + v)p(i, s_{-i}) - vp(s_{-i})] \le \alpha + \beta \cdot q \quad \forall i \in A \qquad (11)$$

Finally, for a player $i \in F$, she should not have an incentive to abstain. Its counterpart is that players from A should also have no incentive to become free riders. This yields the following:

$$q \cdot r \cdot p(s_{-i}) \ge \alpha \quad \forall i \in F \qquad (12)$$

$$q \cdot r \cdot p(s_{-i}) \le \alpha \quad \forall i \in A \qquad (13)$$

Summarizing, a strategy profile with non-empty sets of contributors, free riders and abstainers is a Nash equilibrium if and only if it satisfies the inequalities (8) to (13). For equilibrium profiles where at least one of the pure strategies is not chosen by any player, one needs to restrict to the corresponding subset of inequalities among (8) to (13).

3.2   Equilibrium Analysis When $v = 0$

In the remainder of the paper, we focus on the case of zero intrinsic value, i.e., $v = 0$. This case is already technically much more involved than its corresponding counterpart in Sect. 2.2. Furthermore, it also serves as a sufficient illustration of the differences in the type of equilibria that may arise when free riding is present. We comment further on this at the end of the subsection.

We can already draw some initial conclusions by Eqs. (8)–(13). Suppose first that we want to check if there exists a non-trivial equilibrium where each pure strategy is chosen by at least one player. We show that this is impossible.

Theorem 6. *There cannot exist an equilibrium where both $C \neq \emptyset$ and $A \neq \emptyset$.*

By Theorem 6, we have that for an equilibrium with a positive number of contributors, either it is the all-in profile or it can also contain some free riders but no abstainers. Note also that in case $C = \emptyset$, we can only have the trivial all-out equilibrium (when $k > 1$), since it is meaningless to have free riders without contributors. Hence, in the sequel, we will examine the co-existence of free riders and contributors at equilibria.

Equilibria with only Contributors and Free Riders. We rewrite the equilibrium constraints, that need to hold for an equilibrium with λ contributors and $n - \lambda$ free riders, where $\lambda > 0$ and $\lambda < n$. From the equilibrium constraints (8)–(13), we need only (8), (9), (10) and (12), since we have no abstainers, and all we need is to ensure that contributors have no incentive to move to F or A and free riders have no incentive to move to C or A. We need first to calculate the relevant success probabilities $p(i, s_{-i})$ and $p(s_{-i})$ in these inequalities, and we can do it in a similar manner as in Sect. 2.2. Namely, for $p(i, s_{-i})$ and $p(s_{-i})$, we will need to consider all possible ways that progress can be made in each case. After carrying out these calculations, the existence of equilibria is equivalent to the following system of inequalities, corresponding to (8), (9), (10) and (12) respectively.

$$\binom{\lambda - 1}{k - 1} q^{k-1} (1 - q)^{\lambda - k} \geq \frac{\beta}{r} \tag{14}$$

$$\binom{\lambda}{k - 1} q^{k-1} (1 - q)^{\lambda - k + 1} \leq \frac{\beta}{r} \tag{15}$$

$$r \cdot q \cdot \sum_{j=k-1}^{\lambda-1} \binom{\lambda - 1}{j} q^j (1 - q)^{\lambda - 1 - j} \geq \alpha + \beta q \tag{16}$$

$$r \cdot q \cdot \sum_{j=k}^{\lambda} \binom{\lambda}{j} q^j (1 - q)^{\lambda - j} \geq \alpha \tag{17}$$

The above system already yields some positive news regarding participation. By looking closely at (14) and (15), we can lower bound the number of contributors.

Claim 1. *At a non-trivial equilibrium, with λ contributors and $n - \lambda$ free riders, it must hold that*

$$n - 1 \geq \lambda \geq \frac{k - 1}{q}$$

To continue the analysis, we define and analyze the following function $f(\lambda, j, q)$, that we will use repeatedly, and is related to the terms of binomial sums.

Definition 2. *For $\lambda \leq n$, $j \leq \lambda$ and $q \in (0,1)$, let*

$$f(\lambda, j, q) = \binom{\lambda - 1}{j - 1} q^{j-1}(1 - q)^{\lambda - j}$$

The function $f(\lambda, j, q)$ equals the probability that a set of exactly $j - 1$ users are selected out of a set of $\lambda - 1$ users, according to the randomization procedure of the protocol. We identify some useful properties for the function f, which we will exploit in the remaining analysis. The following claim is easy to verify.

Claim 2. *For the function $f(\lambda, j, q)$, where $\lambda \leq n$, and $j \leq \lambda$, we have:*

- $f(\lambda + 1, j, q) = \frac{\lambda(1-q)}{\lambda - j + 1} f(\lambda, j, q)$
- *For every $\lambda > \frac{j-1}{q}$, we have $f(\lambda, j, q) > f(\lambda + 1, j, q)$.*
- *For every $\lambda < \frac{j-1}{q}$, we have $f(\lambda, j, q) < f(\lambda + 1, j, q)$.*
- *For $\lambda = \frac{j-1}{q}$ (if this is an integer), we have $f(\lambda + 1, j, q) = f(\lambda, j, q)$.*

By Definition 2, we can see that the first two equilibrium constraints (14) and (15), can be rewritten, using the function f, as

$$f(\lambda, k, q) \geq \frac{\beta}{r} \quad \text{and} \quad f(\lambda + 1, k, q) \leq \frac{\beta}{r} \quad \text{respectively.}$$

We pay particular attention to these two constraints, as they already allow us to conclude on how many contributors there can be at an equilibrium where both contributors and free riders are present. The next two lemmas highlight that once we are given the parameters n, k, q, we cannot have equilibria with many different values for λ. Namely, with the exception of some corner cases, there can only be a single value of λ for equilibria with both contributors and free riders.

Lemma 2. *Given a participation game with retraction, there can be at most one value for $\lambda \in [\frac{k-1}{q}, n)$ that satisfies the equilibrium constraints (14) and (15) both with strict inequality.*

The next lemma deals with the corner case of exact equality in a constraint.

Lemma 3. *Suppose that there exists $\lambda \in [\frac{k-1}{q}, n)$ with $f(\lambda, k, q) = \beta/r$. Then, the constraints (14) and (15) are either satisfied both for λ and $\lambda - 1$ or for λ and $\lambda + 1$, but for no other values in $[\frac{k-1}{q}, n)$.*

Using now Claim 2 again, we can identify the range of β/r that is necessary for an equilibrium to exist.

Lemma 4. *Given a game, there exists a value for λ that satisfies the constraints (14) and (15), if and only if $\frac{\beta}{r} \in [f(n - 1, k, q), f(\lceil \frac{k-1}{q} \rceil, k, q)]$.*

Proof. Recall that we need $\lambda \geq \frac{k-1}{q}$, for an equilibrium to exist, and we also know that f is decreasing when λ satisfies this lower bound, by Claim 2. We can think of the function f when λ varies from $\lceil \frac{k-1}{q} \rceil$ to $n-1$, as creating the subintervals $[f(n-1, k, q), f(n-2, k, q)], [f(n-2, k, q), f(n-3, k, q)], \ldots,$ $[f(\lceil \frac{k-1}{q} \rceil + 1, k, q), f(\lceil \frac{k-1}{q} \rceil, k, q)]$. Hence, if β/r belongs to the range stated in the lemma, it belongs to one of these subintervals. And this means that there exists λ such that $\beta/r \in [f(\lambda+1, k, q), f(\lambda, k, q)]$. But this precisely means that the constraints (14) and (15) are satisfied with this λ. We note that it is also possible that there are two consecutive values for λ that can satisfy the constraints if β/r is equal to one of the endpoints of the subintervals, as described in Lemma 3. $\square$

The next step is to understand the range of α/r for which there exists an equilibrium. This means that we have to deal now with the constraints (16) and (17). Note that both constraints imply an upper bound on α/r, and (16) implies a dependence of this upper bound on β. Hence, by taking the minimum of these two bounds, and by abbreviating the binomial terms using the function f, we can conclude with the following result:

Theorem 7. *Given a participation game with retraction, there exists a non-trivial equilibrium with λ contributors and $n - \lambda$ free riders if and only if the following conditions are met:*

- $n - 1 \geq \lambda \geq \frac{k-1}{q}$
- $\frac{\beta}{r} \in [f(\lambda+1, k, q), f(\lambda, k, q)]$
- $0 \leq \frac{\alpha}{r} \leq \min\{q \cdot \sum_{j=k-1}^{\lambda-1} f(\lambda, j+1, q) - q\frac{\beta}{r}, \ q \cdot \sum_{j=k}^{\lambda} f(\lambda+1, j+1, q)\}$

Moreover, whenever the above constraints are satisfied, there is either a unique value for λ or two consecutive values that can satisfy them.

We illustrate Theorem 7 with the following example.

Example 2. Suppose $k = 13, q = 0.3$, and $n = 60$. Note that $\frac{k-1}{q} = 40$. Hence, a necessary condition to have non-trivial equilibria is that $\frac{\beta}{r} \in [f(n-1, k, q), f(\frac{k-1}{q}, k, q)] = [0.0355, 0.1366]$. Suppose that we choose β and r so that $\frac{\beta}{r} = 0.1$. Then by looking at the range of f, we can verify that λ should be equal to $\lambda = 48$, i.e., one can see that $\frac{\beta}{r} \in [f(49, k, q), f(48, k, q)]$. Then, by looking at the constraints for α we conclude that if $\frac{\alpha}{r} \leq 0.1382$, there exists a non-trivial equilibrium with 48 contributors among the 60 players. Thus, in this game, if we set the reward appropriately (roughly 10 times more than each of the cost parameters), we have an equilibrium with high participation.

The All-in Profile. To finish with the analysis of this section, we also need to check if we can have an equilibrium where everybody participates, i.e., with $|C| = n$. In this case we have a simpler set of constraints and we can identify again precise bounds on the relation between r and the costs α, β, that should hold. In particular, we only need to utilize Eqs. (14) and (16), since there is no other type of players. This implies the following.

Lemma 5. *The all-in profile is an equilibrium if and only if the costs α, β and the monetary reward r satisfy the conditions below. Furthermore, there always exists a non-empty range for the reward r, dependent on the costs α, β, so that*

$$\beta/r \leq f(n,k,q), \quad \text{and also} \quad \alpha/r \leq q \cdot \left(\sum_{j=k-1}^{n} f(n,j+1,q) - \beta/r \right).$$

For the sake of completeness, we can summarize our findings for the existence of equilibria in the following statement.

Corollary 1. *Consider a participation game with retraction.*

- *The all-out profile is always an equilibrium (as long as $k > 1$).*
- *There exist equilibria with λ contributors and $n - \lambda$ free riders only when $\lambda \in \{\lceil \frac{k-1}{q} \rceil, \ldots, n-1\}$, and as long as the reward r, compared to the cost parameters, α and β, satisfies the conditions described in Theorem 7.*
- *There exists a non-empty range for the reward r, so that the all-in profile is an equilibrium, as described in Lemma 5.*

Discussion and Comparisons with Sect. 2. The equilibrium analysis of games with retraction is significantly different from the simpler games we discussed in Sect. 2. The major difference is that as we saw both in Sects. 2.2 and 2.4, players of the same type, i.e., with the same selection probability have to use the same strategy at equilibrium. In contrast, in the richer games we analyzed in this section, players with the same selection probability can utilize different strategies at equilibrium. On the positive side, for both classes of games, we see that beyond the trivial equilibrium, all other equilibria have a relatively high number of contributors, and hence, participation can be incentivized with the use of appropriate rewards.

Alternative Reward Mechanisms. A natural question is whether even simpler mechanisms than the one presented here allow for good equilibria so as to eliminate further the complexity of reward dispensing. For example, consider the case where we distribute rewards to all players who declared to participate, regardless of eligibility. It turns out that such a model would again admit equilibria where progress is made in every epoch with high probability. At the same time however, this also results to more unfair equilibria (in terms of the reward given to the actual contributors) and also possibly higher total expenditure for the protocol. To get some intuition why this may be the case, consider the scheme of the current section, with $q = 1$; this means that everyone who declares participation will be rewarded (since they are, by definition, eligible). Note though that the average total expenditure of the protocol for a reward per epoch r, is linear in q and specifically equal to $q \cdot n \cdot r$. It is easy to verify that with $q = 1$ we can only have a non-trivial equilibrium with exactly k contributors but we would still have to pay all players (with the remaining $n - k$ choosing simply to retract) with a total cost of at least $n(\alpha + \beta)$, highlighting potential issues of fairness and stability in this approach. We defer further elaboration of these aspects in the full version of our work.

4 Conclusions and Future Research

We have demonstrated that by carefully setting reward levels we can achieve equilibria with high engagement in blockchain participation games. Whilst we believe our results are encouraging, we also wish to highlight a number of avenues for future research. Most importantly, it would be interesting to enrich our model with player asymmetries in terms of their costs and selection probability. We have only initiated this in Sect. 2.4, whereas for the model of Sect. 3, such features seem to require a technically more involved analysis. Furthermore, it is also natural to treat repeated versions of these games, either theoretically or via experimental approaches. If some participants can observe the behaviors and contributions of others before committing to their own decisions, it would be interesting to analyze best response dynamics in this setting. Finally, as blockchain participants are actual people subject to their own complex motivations (that often go beyond just maximizing expected profit), it would be beneficial to leverage behavioral game theory in this setting.

References

1. Arnosti, N., Weinberg, S.M.: Bitcoin: a natural oligopoly. In: Blum, A. (ed.) Innovations in Theoretical Computer Science Conference, ITCS. LIPIcs, vol. 124 (2019)
2. Brünjes, L., Kiayias, A., Koutsoupias, E., Stouka, A.: Reward sharing schemes for stake pools. In: EuroS&P, pp. 256–275. IEEE (2020)
3. Desmedt, Y.: Society and group oriented cryptography: a new concept. In: Pomerance, C. (ed.) CRYPTO 1987. LNCS, vol. 293, pp. 120–127. Springer, Heidelberg (1988). https://doi.org/10.1007/3-540-48184-2_8
4. Erev, I., Rapoport, A.: Provision of step-level public goods: the sequential contribution mechanism. J. Confl. Resolut. **34**, 401–425 (1990)
5. Eyal, I., Sirer, E.G.: Majority is not enough: bitcoin mining is vulnerable. In: Financial Cryptography and Data Security, pp. 436–454. Springer (2014)
6. Ferreira, M.V., Hahn, Y.L.S., Weinberg, S.M., Yu, C.: Optimal strategic mining against cryptographic self-selection in proof-of-stake. In: 23rd ACM Conference on Economics and Computation, pp. 89–114. EC '22, ACM (2022)
7. Fiat, A., Karlin, A., Koutsoupias, E., Papadimitriou, C.H.: Energy equilibria in proof-of-work mining. In: Economics and Computation, EC, pp. 489–502 (2019)
8. Gilboa, M., Nisan, N.: Complexity of public goods games on graphs. In: Kanellopoulos, P., Kyropoulou, M., Voudouris, A. (eds.) Algorithmic Game Theory. SAGT 2022. LNCS, vol. 13584, pp. 151–168. Springer, Cham (2022). https://doi.org/10.1007/978-3-031-15714-1_9
9. Gradstein, M., Nitzan, S.: Binary participation and incremental provision of public goods. Soc. Choice Welf. **7**, 171–192 (1990)
10. Groth, J.: On the size of pairing-based non-interactive arguments. In: Fischlin, M., Coron, J.-S. (eds.) EUROCRYPT 2016. LNCS, vol. 9666, pp. 305–326. Springer, Heidelberg (2016). https://doi.org/10.1007/978-3-662-49896-5_11

11. Jackson, M.: Social and Economics Networks. Princeton University Press, Princeton (2008)
12. Kiayias, A.: Decentralizing information technology: the advent of resource based systems. In: Proceedings of the 2022 Symposium on Algorithmic Game Theory (2022)
13. Kiayias, A., Koutsoupias, E., Kyropoulou, M., Tselekounis, Y.: Blockchain mining games. In: Economics and Computation, pp. 365–382. EC '16, ACM (2016)
14. Kiayias, A., Stouka, A.: Coalition-safe equilibria with virtual payoffs. In: Advances in Financial Technologies, AFT '21 (2021)
15. Kroll, J.A., Davey, I.C., Felten, E.W.: The economics of bitcoin mining, or bitcoin in the presence of adversaries. In: Proceedings of the WEIS (2013)
16. Kwon, Y., Liu, J., Kim, M., Song, D., Kim, Y.: Impossibility of full decentralization in permissionless blockchains. In: Advances in Financial Technologies, AFT, pp. 110–123. ACM (2019)
17. Micali, S., Reyzin, L., Vlachos, G., Wahby, R.S., Zeldovich, N.: Compact certificates of collective knowledge. In: Security and Privacy (2021)
18. Palfrey, T., Rosenthal, H.: Participation and the provision of discrete public goods: a strategic analysis. J. Public Econ. **24**, 171–193 (1984)
19. Schrijvers, O., Bonneau, J., Boneh, D., Roughgarden, T.: Incentive compatibility of bitcoin mining pool reward functions. In: Financial Cryptography, FC (2016)

Recovering Single-Crossing Preferences from Approval Ballots

Andrei Constantinescu[(✉)] and Roger Wattenhofer

ETH Zurich, Rämistrasse 101, 8092, Zurich, Switzerland
{aconstantine,wattenhofer}@ethz.ch

Abstract. An electorate with fully-ranked innate preferences casts approval votes over a finite set of alternatives. As a result, only partial information about the true preferences is revealed to the voting authorities. In an effort to understand the nature of the true preferences given only partial information, one might ask whether the unknown innate preferences could possibly be single-crossing. The existence of a polynomial time algorithm to determine this has been asked as an outstanding problem in the works of Elkind and Lackner [18]. We hereby give a polynomial time algorithm determining a single-crossing collection of fully-ranked preferences that could have induced the elicited approval ballots, or reporting the nonexistence thereof. Moreover, we consider the problem of identifying negative instances with a set of forbidden sub-ballots, showing that any such characterization requires infinitely many forbidden configurations.

Keywords: Approval Voting · Single-crossing · Algorithms · Computational Complexity

1 Introduction

One can express their opinion either in *precise* or in *simple* terms. The scientific world favors precision, but real-world decision, voting and election schemes usually adopt simplicity. Both ways have their disadvantages: simple voting is often limited in expressivity, requiring voters to make compromises ("I would really like to vote for candidate A, but candidate B has an actual chance of getting elected."). Precise voting, on the other hand, is too demanding of the voter ("Do I really have to rank all these candidates? I barely know them!"), and hence often not implemented. Indeed, the vast majority of the preference data repository PREFLIB [30] consists of partially-elicited preferences. A modern solution striking a good balance between simplicity and expressive power stands in approval voting [27]. Here ballots ask voters to decide whether they approve or disapprove of each candidate; i.e., is the candidate good enough or not? Approval voting has been successfully applied to a number of settings, including among others multi-winner elections and participatory budgeting (see excellent survey in [26]).

J. Garg et al. (Eds.): WINE 2023, LNCS 14413, pp. 188–206, 2024.
https://doi.org/10.1007/978-3-031-48974-7_11

Starting with the seminal work of Arrow [1], a long line of research has been dedicated to showing the impossibility of computing fair social outcomes. However, these impossibility proofs typically hinge on pathological instances that are unlikely to occur in real-world elections, where political preferences can, for instance, often be explained using a left-right spectrum. Elections following such a spectrum are known as (one-dimensional) Euclidean. This model has received a number of generalizations enhancing its expressive power, including the single-peaked [1,6] and single-crossing [31,35] models, weakening the requirements of the societal axis in two distinct ways. An election is single-crossing if there is an ordering of the voters, called the single-crossing axis, such that, as we sweep through voters in order, preference between any two candidates changes at most once. We omit the definition of single-peaked preferences as it is not required for our work. While the two notions are incomparable, they share many of their attractive properties, including the existence of Condorcet winners and the polynomial-time computability of winners under election schemes for which this is hard in general, like the rules of Young, Dodgson and Kemeny [8] in the single-winner setting, and those of Chamberlin–Courant and Monroe in the multi-winner one [28,34]. The first three follow from the existence of Condorcet winners, while the latter two use dynamic programming [5,11,13,36,37]. When preferences are neither single-peaked nor single-crossing, one can still hope for the removal of a small fraction of voters and/or candidates to allow one of the models to apply.

The models discussed so far are defined for fully-elicited preferences. For approval ballots, their most natural extensions ask whether there are complete-information Euclidean, single-peaked, or single-crossing preferences refining the elicited approval preferences. If so, one calls the elicited preferences *possibly* Euclidean, single-peaked, or single-crossing (short PE, PSP, PSC). Elkind and Lackner [18] study these notions, among others, showing that PE and PSP surprisingly coincide, and PSC is more general (in fact the most general considered in their paper). However, one so far unanswered question is whether PSC collections of ballots can be recognized in polynomial time. In contrast, this is known to be possible for the other notions.

A polynomial-time algorithm for recognizing PSC and computing an associated societal axis has a few important implications. First of all, such an axis is required to apply the known polynomial-time algorithms for computing winning committees for the approval-based Chamberlin–Courant and Monroe rules under single-crossing preferences [11,13,36]. Moreover, Pierczyński and Skowron [33] show that core-stable committees always exist for approval elections under two definitions more restrictive than PSC (one of which being PE = PSP). It is likely that the same will be shown in the future for PSC, as no counterexample is known even for arbitrary approval elections,[1] and an algorithm computing the PSC axis will likely be a necessary component of an algorithm computing core-stable committees for PSC preferences. Finally, such an algorithm can be

---

[1] This is perhaps the main open question in the literature on proportional representation [26].

used by electoral authorities to study the political landscape emerging from an election.

Our Contribution. We give a polynomial time algorithm that takes as input a collection of approval ballots and outputs a collection of fully-ranked ballots with the single-crossing property that could have produced the observed approval votes, reporting accordingly in case this is not possible. In the process, our approach also leads to a novel FPT algorithm for the non-betweenness problem [23]. Our problem has been posed as open by Elkind and Lackner [18].[2]

Moreover, in the full version of the paper [12], we show that, while single-crossing elections admit a characterization in terms of a set of two small forbidden subelections [9], such a characterization ceases to exist for PSC elections. We show this by exhibiting an infinite set of approval elections that are not PSC, but all their proper subelections are.

Related Work. Checking whether an election (with fully-elicited strictly ranked preferences) is single-crossing can be achieved in polynomial time [9,17], in fact even in near-linear time [19]. Checking whether an election can be embedded on a left-right spectrum (i.e., whether it is 1-Euclidean) can also be done in polynomial time [14,15,25]. In both cases, finding witnessing axes/embeddings is also polynomial. On the other hand, deciding whether an embedding exists in d dimensions becomes NP-hard as soon as $d \geq 2$ [32], and heuristic accounts support the difficulty of fitting such higher-dimensional models in practice [10].

Single-peaked elections can also be recognized in polynomial time [3,14,19, 20]. A natural question related to our work is how difficult it is to recover 1-Euclidean preferences or single-peaked preferences from approval ballots. The two notions coincide (PE = PSP), and polynomial-time algorithms have been proposed [18,21] by reduction to the consecutive ones problem (which admits a linear-time algorithm, although using rather complex machinery [7]). In contrast, our algorithm for PSC proceeds directly, using non-betweenness only as a lens. An interesting model that encompasses both single-crossing and single-peaked elections is that of top-monotonic elections [2]. Such elections can be recognized in polynomial time [29] (also the paper whose techniques are closest to ours).

For the case of multi-valued approval, where voters can choose between more than two judgments when filling in the ballot, Fitzsimmons [22] provides a polynomial time recognition algorithm for possibly single-peaked preferences, working even for the case of an unbounded number of possible judgments; i.e., voters give a score between 1 and the number of candidates to each candidate (can also be thought as voters reporting rankings with ties). Their method proceeds by reducing to the consecutive ones problem. Getting similar results for the single-crossing case might seem desirable, but, as shown in [16], this is NP-hard for an unbounded number of possible judgments.

---

[2] Recently, we learned that a solution has in fact been proposed as early as 1979 in the context of the simple plant location problem, by Beresnev and Davydov [4]. This paper is only available in Russian, and Russian-speaking experts seem to believe that the paper is likely missing steps in the arguments. Beresnev and Davydov [4] is referenced in [24], but without details.

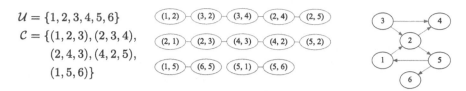

$\mathcal{U} = \{1, 2, 3, 4, 5, 6\}$

$\mathcal{C} = \{(1, 2, 3), (2, 3, 4),$

$(2, 4, 3), (4, 2, 5),$

$(1, 5, 6)\}$

(a) Input instance.   (b) Formula graph, excl. singletons.   (c) Colorful graph.

Fig. 1. Example of the construction of the formula and colorful graphs. Consider the instance of the non-betweenness problem given by $\mathcal{U}$ and $\mathcal{C}$ in Fig. 1a. The corresponding formula graph in Fig. 1b is constructed by adding edges (a, b) — (c, b) and (b, a) — (b, c) for each triple $(a, b, c) \in \mathcal{C}$. Isolated nodes have been omitted. Note that no connected component of the formula graph contains both vertices (a, b) and (b, a) for any $a, b \in \mathcal{U}$. Subsequently, the connected components split into two pairs of complementary components: the two top long red chains and the two bottom short blue chains. The corresponding colorful graph in Fig. 1c is constructed by (arbitrarily) taking the top red component and the right blue component and introducing directed edges corresponding to the ordered pairs in the component. To solve the instance in Fig. 1a one needs to decide for each color whether or not to flip the direction of all edges of that color such that the resulting graph is acyclic. In our case, to avoid the cycle $1 \to 2 \to 5 \to 1$, the possible solutions are to either flip all blue edges or all red edges (but not both, as that leads to cycle $1 \to 5 \to 2 \to 1$ instead). (Color figure online)

Working with approval ballots requires rather different social aggregators to those taking fully-ranked profiles as input. These aggregators have been recently extensively studied in the context of multi-winner elections; see the comprehensive survey of both normative and computational results of Lackner and Skowron [26]. In an effort to translate the known models of voting to approval elections, Elkind and Lackner [18] consider twelve models catered to approval voting, the most general being the domain of possibly single-crossing elections. For all others, their paper shows polynomial-time recognition, while Terzopoulou et al. [38] show that the least restrictive of the others admit finite forbidden substructures characterizations. We resolve the open case of PSC and show that it admits no finite forbidden substructures characterization.

Technical Overview. Our approach for proving the main result proceeds by first translating the input into an instance of the non-betweenness problem, which asks given a ground set $\mathcal{U}$ and a set $\mathcal{C}$ of triples (a, b, c) over $\mathcal{U}$ to compute a linear (i.e., total, irreflexive, antisymmetric, transitive) ordering of the elements of $\mathcal{U}$ such that for none of the triples does b come between a and c. While non-betweenness is NP-hard in general [23], for the type of instances produced by our reduction the problem will turn out to be polynomial. To show this, we begin with an approach to non-betweenness based on boolean satisfiability. In particular, we introduce for every pair of distinct entries $a, b \in \mathcal{U}$ a boolean variable $x_{a,b}$ standing for whether a comes before b in the ordering. Antisymmetry can be captured by adding clauses requiring that $x_{a,b} = \overline{x_{b,a}}$, while for every triple $(a, b, c) \in \mathcal{C}$ one can show that non-betweenness is enforced by adding clauses ensuring that $x_{a,b} = x_{c,b}$. However, transitivity is substantially trickier to model without resorting to 3-literal clauses. To overcome this difficulty, taking a step back, we notice that except for transitivity all constraints are equalities. We take advantage of this fact by constructing the so-called *formula graph*, which has a vertex (a, b) for each variable $x_{a,b}$ in the boolean formula. Edges in the formula graph correspond to the equalities $x_{a,b} = x_{c,b}$ in the boolean formulation above, together with the inferred opposite equalities $x_{b,a} = x_{b,c}$. If some connected component contains both $x_{a,b}$ and $x_{b,a}$, then the instance is not satisfiable. Otherwise, we introduce yet another construct, the *colorful graph*, which is an edge-colored directed graph with vertex set G. One can show that at this point connected components in the formula graph come in complementary pairs $\{S, \overline{S}\}$, where $(a, b) \in S$ whenever $(b, a) \in \overline{S}$. For each such pair we create a new color, the edges of this color corresponding to the ordered pairs in set S. Some colors will only have one edge, which can be safely removed. Our two graph concepts are illustrated in Fig. 1. We then show that satisfying assignments of the boolean formula, this time including transitivity, correspond to acyclic orientations of the colorful graph where edges of each color are either all reversed, or all left in the original orientation. Deciding if such orientations of the colorful graph exist is still NP-hard, but exhausting over all possible orientations already shows that the non-betweenness problem is FPT with respect to the number of colors, a fact which might be of independent interest. At this point, we start studying the structure of the formula and colorful graphs induced by approval elections, showing the surprising result that, unless there is a monochromatic cycle, an acyclic orientation exists and can be computed in polynomial time, from which the election is possibly single-crossing. Proving this result constitutes the most demanding part of our work, and relies on a mixture of combinatorial and computer-aided techniques. Most notably, along the way we prove that if some orientation of the colorful graph is not acyclic, then the induced cycle is either monochromatic or of length three and not repeating colors. With further work, we then prove a strong structural result about colors taking part in three-colored triangles, allowing them to be consistently oriented to complete the proof.

2   Preliminaries

For a non-negative integer n, we write $[n]$ for the set $\{1, 2, \ldots, n\}$. Given a statement S, we write $[S]$ for the Iverson bracket: $[S] = 1$ if S holds, and 0 otherwise.

Given a finite set A, a *partial* order $\succ$ over A is an irreflexive, antisymmetric, transitive relation over A. Two values $a, b \in A$ are *comparable* under $\succ$ if either $a \succ b$ or $b \succ a$; otherwise, they are *incomparable*, in which case we write $a \approx_\succ b$. When any two distinct elements are comparable (i.e., the order is total), order $\succ$ is called a *linear order*. A partial order $\succ$ is called a *weak order* if $\approx_\succ$ forms an equivalence relation (i.e., it is transitive). One can think of weak orders as linear orders allowing for ties, corresponding to incomparable elements. For this reason, when $\succ$ is a weak order, whenever $a \approx_\succ b$ we say that $\succ$ is *indifferent* between a and b. We call *indifference classes* the equivalence classes induced by $a \approx_\succ b$. We call a weak order with at most k indifference classes a *k-weak order*. Note that an order $\succ$ is k-weak if and only if $\succ$ can be induced by a *scoring function* $s : A \to [k]$; i.e., $a \succ b$ if and only if $s(a) > s(b)$. Hence, 2-weak orders are precisely the approval ballots. Given partial orders $\succ$ and $\succ'$, we say that $\succ'$ *extends* $\succ$ if $a \succ' b$ whenever $a \succ b$.

In our setting, an assembly of voters $V = [n]$ expresses their preferences over a set of candidates (alternatives) $C = [m]$. The expressed preferences of voter $i \in V$ consist of a weak order $\succ_i$ over the set of candidates C. The collection of all voter preferences $\mathcal{P} = (\succ_i)_{i \in [n]}$ is known as the *preference profile*. When $\mathcal{P}$ consists solely of linear orders, we say that the electorate has *fully-ranked* preferences. When $\mathcal{P}$ consists solely of 2-weak orders, we say the electorate has *approval* preferences, in which case we will often see $\mathcal{P}$ as an $m \times n$ *approval matrix*, where $\mathcal{P}[c, v] = 1$ if voter v approves of candidate c, and 0 otherwise; i.e., column v contains the approval ballot cast by voter v. Given two profiles $\mathcal{P}$ and $\mathcal{P}'$, we say that $\mathcal{P}'$ is a *subprofile* of $\mathcal{P}$ if $\mathcal{P}'$ can be obtained from $\mathcal{P}$ by removing voters and/or candidates. For approval ballots, subprofiles correspond to removing rows and columns from $\mathcal{P}$. If $\mathcal{P}' \neq \mathcal{P}$ in the previous, then $\mathcal{P}$ is a *proper* subprofile of $\mathcal{P}$.

A profile of weak orders $\mathcal{P}$ is said to be *seemingly single-crossing* (SSC) with respect to a linear order $\lhd$ over the set of voters V, called the *axis*, if there are no three voters i, j, k such that $i \lhd j \lhd k$ and candidates a, b such that $a \succ_i b, b \succ_j a$ and $a \succ_k b$. If additionally profile $\mathcal{P}$ consists solely of linear orders, then under the same conditions $\mathcal{P}$ is *single-crossing* (SC) with respect to $\lhd$. A profile $\mathcal{P}$ of weak orders is said to be *possibly single-crossing* (PSC) with respect to a linear order $\lhd$ of the voters if for all voters $i \in V$ the preference order $\succ_i$ can be extended to a linear order $\succ'_i$ such that $(\succ'_i)_{i \in [n]}$ is single-crossing with respect to $\lhd$. A profile $\mathcal{P}$ is (seemingly/possibly) single-crossing if it is (seemingly/possibly) single-crossing with respect to some linear order $\lhd$ over V. For linear orders, the three notions coincide. More surprisingly, Elkind et al. [16] show that for weak orders the notions of SSC and PSC coincide.[3] Moreover, they show that given a linear order $\lhd$ witnessing SSC for profile $\mathcal{P}$, one can compute in polynomial time for each voter $i \in V$ a linear extension $\succ'_i$ of $\succ_i$ such that $(\succ'_i)_{i \in [n]}$ is single-crossing with respect to $\lhd$. In other words, knowing a seemingly single-crossing axis is enough to compute fully-ranked ballots that could have produced the cast ballots in polynomial time. For completeness, we give a streamlined version of their argument in the full version. Therefore, our focus will be on computing, given a profile $\mathcal{P}$ of weak orders, an SSC axis, or deciding the nonexistence thereof. For general weak orders, Elkind et al. [16] show that deciding axis existence is NP-hard, but they leave it open for k-weak orders for $k \geq 2$, the case $k = 2$ corresponding with the approval voting case.

---

[3] For general partial orders, they show that PSC implies SSC, but not conversely.

Given a finite ground set $\mathcal{U}$, a *non-betweenness* (NB) constraint over $\mathcal{U}$ is a triple of distinct elements $(i, j, k) \in \mathcal{U}^3$. A linear order $\lhd$ over $\mathcal{U}$ satisfies a collection $\mathcal{C}$ of NB constraints if for every $(i, j, k) \in \mathcal{C}$ it does not hold that $i \lhd j \lhd k$ or $k \lhd j \lhd i$; i.e., j is not between i and k in $\lhd$. The problem of deciding whether a linear order $\lhd$ satisfying $\mathcal{C}$ exists is NP-hard [23].[4] Checking for the SSC property reduces to verifying whether a set of NB constraints is satisfiable: given a profile $\mathcal{P}$, construct a set of NB constraints $\mathcal{C}_\mathcal{P}$ over V such that $(i, j, k) \in \mathcal{C}_\mathcal{P}$ iff there are two alternatives $a, b \in C$ such that $a \succ_i b, b \succ_j a$ and $a \succ_k b$. Then, we have the following, which is intuitively enabled by the fact that SSC is invariant with respect to reversing the axis:

Lemma 1. *$\mathcal{P}$ is SSC with respect to a linear order $\lhd$ iff $\mathcal{C}_\mathcal{P}$ is satisfied by $\lhd$.*

3   A Boolean Encoding of Non-Betweenness Constraints

In this section, we develop a general technique for deciding the satisfiability of an arbitrary set $\mathcal{C} \subseteq V^3$ of NB constraints over V. Naturally, our approach does not lead to a polynomial time algorithm for checking satisfiability, as the problem is NP-hard, but analysis in later sections will show that when applied to NB constraint sets $\mathcal{C}_\mathcal{P}$ originating from approval ballots, it leads to a polynomial algorithm for deciding satisfiability.

First, note that a NB constraint (i, j, k) is satisfied by a linear order $\lhd$ if and only if $[i \lhd j] = [k \lhd j]$. Moreover, order $\lhd$ is uniquely determined by the values $[i \lhd j]$ for $i \neq j$. These facts suggest a reformulation of the problem of satisfying NB constraints in terms of the values $[i \lhd j]$, which we do in the following. Given $\mathcal{C}$, we construct a boolean formula $\Phi_\mathcal{C}$ with a variable $x_{i,j}$ for every ordered pair of distinct elements $(i, j) \in V^2$. The clauses of $\Phi_\mathcal{C}$ comprise of the following three constraint sets, corresponding to antisymmetry, agreement with $\mathcal{C}$ and transitivity, respectively:

$\langle 1 \rangle$ For every pair of distinct elements $(i, j) \in V^2$ add the constraint $x_{i,j} = \overline{x_{j,i}}$.
$\langle 2 \rangle$ For every triple $(i, j, k) \in \mathcal{C}$, add the constraint $x_{i,j} = x_{k,j}$.
$\langle 3 \rangle$ For every triple of distinct $(i, j, k) \in V^3$ add the constraint $(x_{i,j} \wedge x_{j,k}) \rightarrow x_{i,k}$.

In the above we used $=$ for the bi-implication operator $\leftrightarrow$, and we wrote $\overline{x}$ for logical negation, also commonly denoted by $\neg x$. Note that, under the presence of set $\langle 1 \rangle$, set $\langle 2 \rangle$ can be reformulated as below since $x_{j,i} = \overline{x_{i,j}} = \overline{x_{k,j}} = x_{j,k}$:

$\langle 2' \rangle$ For every triple $(i, j, k) \in \mathcal{C}$, add the constraints $x_{i,j} = x_{k,j}$ and $x_{j,i} = x_{j,k}$.

---

[4] This is used to prove the hardness of our problem for general weak orders, but the argument requires an unbounded number of indifference classes, so it does not work for bounded k..

Henceforth, we will mostly use $\langle 2' \rangle$, but $\langle 2 \rangle$ will still be more convenient in a few cases. The following lemma formally establishes the correspondence between linear orders satisfying $\mathcal{C}$ and satisfying assignments of $\Phi_\mathcal{C}$. The proof is essentially straightforward by following the definitions, so we leave it for the full version.

Lemma 2. *Linear orders* $\lhd$ *satisfying the NB constraints in* $\mathcal{C}$ *correspond bijectively to satisfying assignments of* $\Phi_\mathcal{C}$ *by* $x_{i,j} = [i \lhd j]$. *Hence,* $\mathcal{C}$ *is satisfiable iff* $\Phi_\mathcal{C}$ *is satisfiable.*

Is this the end of the story? Not quite, because constraint set $\langle 3 \rangle$ encodes transitivity as $(x_{i,j} \wedge x_{j,k}) \rightarrow x_{i,k} \equiv (\overline{x_{i,j}} \vee \overline{x_{j,k}} \vee x_{i,k})$, which is a three-literal clause. However, constraint sets $\langle 1 \rangle$ and $\langle 2 \rangle$ only require two-literal clauses, so there is hope. The approach presented so far is similar in spirit to the approach of Magiera and Faliszewski [29] for recognizing top-monotonic elections. The main difficulty in their case also stems from enforcing transitivity, which similarly seems to require 3-literal clauses at a first glance. The essential observation that they make to progress is that whenever $(i, j, k) \in \mathcal{C}$, then the six entries in $\langle 3 \rangle$ enforcing transitivity for the unordered triple $\{i, j, k\}$ are no longer required, a fact which also holds true in our case. If for every unordered triple we would have at least one NB constraint involving its elements, then none of the transitivity constraints would be required, and the problem could simply be solved by any polynomial 2-CNF solver. This is the main idea of the approach taken in [29]. In our case, on the other hand, it turns out that it is surprisingly tricky to completely eliminate the transitivity constraints, so we will need more insight to deal with them gracefully.

4   Resolving Constraint Sets $\langle 1 \rangle$ and $\langle 2 \rangle$ Using Connected Components, The Formula Graph

Before tackling transitivity, we first turn our attention to constraint sets $\langle 1 \rangle$ and $\langle 2 \rangle$ in isolation, providing a precise characterization of the satisfying assignments of $\Phi_\mathcal{C}$ in the absence of constraint set $\langle 3 \rangle$. This characterization will be instrumental later on when reasoning about the transitivity constraints. In contrast, here Magiera and Faliszewski [29] directly employ a 2-CNF solver in order to tell whether sets $\langle 1 \rangle$ and $\langle 2 \rangle$ are satisfiable in isolation, but doing so only produces one satisfying assignment, rather than a compact representation of all of them, as will be the case for us. Our approach will hinge on the observation that constraint set $\langle 2 \rangle$ only consists of equalities. In what follows, we will work with constraint set $\langle 2' \rangle$ instead of $\langle 2 \rangle$.

To begin, we define the *formula graph* of $\mathcal{C}$, denoted by $G_\mathcal{C}$. The vertex set $V(G_\mathcal{C})$ of $G_\mathcal{C}$ consists of all ordered pairs $(i, j) \in V^2$ with $i \neq j$; i.e., one vertex per variable in the formula. For a vertex (i, j) in $G_\mathcal{C}$ we define the *complementary* vertex $\overline{(i,j)}$ to be (j, i). Note that this is a syntactic notation, and is not to be confused with $\overline{x_{i,j}}$, which signifies the logical negation of a variable. For every constraint $x_u = x_v$ in $\langle 2' \rangle$, where u and v are pairs of distinct elements in V, we add an undirected edge $u - v$ to $G_{\mathcal{C}}$. Intuitively, each edge in $G_{\mathcal{C}}$ signifies an equality that needs to hold in the satisfying assignments of $\langle 1 \rangle$ and $\langle 2' \rangle$. We begin with two preliminary lemmas concerning the connected components of the formula graph.

Lemma 3. *An assignment $(x_u)_{u \in V(G_{\mathcal{C}})}$ satisfies constraint set $\langle 2' \rangle$ if and only if for each connected component S of $G_{\mathcal{C}}$ it holds that $x_u = x_v$ for all $u, v \in S$.*

Lemma 4. *Let S be a connected component in $G_{\mathcal{C}}$, then $\overline{S} := \{\overline{s} : s \in S\}$ is also a connected component in $G_{\mathcal{C}}$. Note moreover that $S = \overline{S}$ or $S \cap \overline{S} = \varnothing$.*

Hence, in light of Lemma 3, if for any connected component S it holds that $S = \overline{S}$, then $\langle 1 \rangle$ and $\langle 2' \rangle$ cannot be simultaneously satisfied. Otherwise, the connected components of $G_{\mathcal{C}}$ can be paired up into "complementary" pairs of distinct connected components $\{S, \overline{S}\}$. In particular, $V(G_{\mathcal{C}}) = \cup_{i=1}^k (S_i \cup \overline{S_i})$ is a *complementary pairs partition* of the graph, where $(S_i)_{i \in [k]}$ are connected components of $G_{\mathcal{C}}$, one per complementary pair. Note that this partition is unique up to changing the roles of S_i and $\overline{S_i}$ for any i, so we will for simplicity refer to it as "the" complementary pairs partition. With this in mind, one can notice that solutions to sets $\langle 1 \rangle$ and $\langle 2' \rangle$ correspond to choosing for each complementary pair $\{S, \overline{S}\}$ one of the two components and setting to true variables corresponding to it and to false variables corresponding to the other component. This is formalized in the following proposition.

Proposition 5. *If $S = \overline{S}$ for any connected component of $G_{\mathcal{C}}$, then there are no assignments satisfying $\langle 1 \rangle$ and $\langle 2' \rangle$. Otherwise, let $V(G_{\mathcal{C}}) = \cup_{i=1}^k (S_i \cup \overline{S_i})$ be the complementary pairs partition. Then, an assignment satisfies $\langle 1 \rangle$ and $\langle 2' \rangle$ iff it is obtained from a tuple $(s_1, \ldots, s_k) \in \{0, 1\}^k$ by setting all variables x_u to s_i for $u \in S_i$ and to $\overline{s_i}$ for $u \in \overline{S_i}$.*

Note that the condition $S \neq \overline{S}$ for all connected components S is equivalent to checking that for all $u \in V(G_{\mathcal{C}})$ vertices u and $\overline{u}$ are in different connected components. Hence, checking whether solutions to $\langle 1 \rangle$ and $\langle 2' \rangle$ exist is equivalent to checking for the latter, which the reader might recognize as the condition used for checking the satisfiability of general 2-CNFs.[5] However, for general 2-CNFs no characterization of the solution space is known since even counting solutions is #P-complete (while in our case we know that there are exactly 2^k solutions).

Henceforth, we will assume that $V(G_{\mathcal{C}}) = \cup_{i=1}^k (S_i \cup \overline{S_i})$ is the complementary pairs partition (otherwise, just report that constraints in $\mathcal{C}$ cannot be satisfied). Then, Proposition 5 gives that the assignments satisfying $\langle 1 \rangle$ and $\langle 2' \rangle$ are precisely those for which a single value x_{S_i} is assigned to each connected component S_i and, similarly, for the complementary component $x_{\overline{S_i}} = \overline{x_{S_i}}$ holds. Note that this already shows that checking the satisfiability of a set of NB constraints is FPT with respect to the number $2k$ of connected components, since one could just try out all 2^k satisfying assignments for $\langle 1 \rangle$ and $\langle 2' \rangle$ and check for transitivity in each case.

---

[5] If we replace our undirected graph by the directed implications graph and the word "connected" by "strongly-connected.".

5 Transitivity by Acyclic Orientations, the Colorful Graph

In this section, we provide an equivalent view of the satisfying assignments of sets $\langle 1 \rangle$ and $\langle 2' \rangle$ as edge-orientations of a certain edge-colored graph. In particular, all edges have a base orientation, and the orientations satisfying $\langle 1 \rangle$ and $\langle 2' \rangle$ correspond to selecting for each color whether to leave all edges as they are, or reverse the direction of all edges of that color. Edge orientations additionally satisfying constraint set $\langle 3 \rangle$ will be those inducing a directed acyclic graph.

To begin, recall that $V(G_C) = \cup_{i=1}^k \left(S_i \cup \overline{S_i} \right)$ is the complementary pairs partition. Consider some connected component S_i. This component consists of a set of ordered pairs $S_i = \{(i_1, j_1), (i_2, j_2), \ldots, (i_\ell, j_\ell)\}$. To have a satisfying assignment of $\langle 1 \rangle$ and $\langle 2' \rangle$, it has to be the case that either $x_{i_1,j_1} = \ldots = x_{i_\ell,j_\ell} = 1$, or $x_{i_1,j_1} = \ldots = x_{i_\ell,j_\ell} = 0$. Both choices lead to satisfying assignments, independently of the choices made for the other components. By Lemma 2, the previous gives that either $[i_1 \lhd j_1] = \ldots = [i_\ell \lhd j_\ell] = 1$, or $[i_1 \lhd j_1] = \ldots = [i_\ell \lhd j_\ell] = 0$, and the choice is independent across S_i's.

Any linear order $\lhd$ over V can be seen as a directed tournament graph with vertex set V, where we draw a directed edge $i \to j$ if $i \lhd j$ and a directed edge $j \to i$ if $j \lhd i$. Consequently, the constraint $[i_1 \lhd j_1] = \ldots = [i_\ell \lhd j_\ell]$ induced by some connected component S_i of the formula graph is equivalent in the tournament of $\lhd$ to the fact that we either have edges $i_1 \to j_1, \ldots, i_\ell \to j_\ell$, or edges $j_1 \to i_1, \ldots, j_\ell \to i_\ell$. In other words, every set S_i is a directed set of edges that in the tournament of $\lhd$ either has to have the given orientation, or exactly the reverse orientation. We call these two options for S_i its two possible *orientations*. We call a choice between the two options for each of the sets $S_1, \ldots, S_k$ an *orientation* of all edge sets. The following shows that out of those orientations, the ones that additionally satisfy set $\langle 3 \rangle$ are those where the resulting directed tournament graph is acyclic. For brevity, we call such orientations *acyclic*.

Proposition 6. *A linear order $\lhd$ over V satisfies the NB constraints in C if and only if it is given by an acyclic orientation of the edge sets $S_1, \ldots, S_k$.*

Some edge sets (connected components in the formula graph) S_i will be singletons; i.e., $|S_i| = 1$. Such sets are not relevant for deciding whether an acyclic orientation exists, because singletons can just be oriented freely after everything else has been oriented. More formally, without loss of generality, assume that edge sets $S_1, \ldots, S_\ell$ are non-singletons; i.e., $|S_i| > 1$ for $i \in [\ell]$; and that edge sets $S_{\ell+1}, \ldots, S_k$ are singletons; i.e., $|S_i| = 1$ for $\ell < i \leq k$. Then, we have the following:

Lemma 7. *An acyclic orientation of edge sets $S_1, \ldots, S_k$ exists if and only if an acyclic orientation of edge sets $S_1, \ldots, S_\ell$ exists.*

Armed as such, we now introduce the *colorful graph* corresponding to the NB constraint set C. The colorful graph is a directed graph with vertex set V and

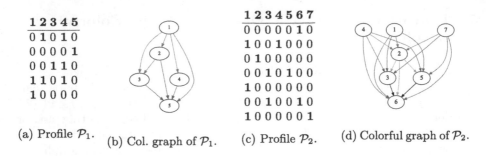

(a) Profile $\mathcal{P}_1$.

(b) Col. graph of $\mathcal{P}_1$.

(c) Profile $\mathcal{P}_2$.

(d) Colorful graph of $\mathcal{P}_2$.

Fig. 2. Two example preference profiles $\mathcal{P}_1, \mathcal{P}_2$, together with the colorful graphs they induce. In Fig. 2d all colors are biclique colors. Namely, the red edges are given by $\{1,4,7\} \times \{2\}$, the green edges by $\{2\} \times \{3,5,6\}$, the black edges by $\{3,5\} \times \{6\}$ and the blue edges by $\{1,4,7\} \times \{3,5,6\}$. Out of these, red, green and black are star biclique. In general, not all colors will be biclique, e.g., in Fig. 2b, only the blue color is biclique. (Color figure online)

edges colored in ℓ colors, identified by the numbers in $[\ell]$. For each $i \in [\ell]$ the edges colored with color i are those in edge set S_i.

Note how for any $u, v \in V$ with $u \neq v$ at most one of the edges $u \to v$ and $v \to u$ can appear in $S_1, \dots, S_\ell$. Moreover, an edge cannot occur in more than one of $S_1, \dots, S_\ell$. As a result, the colorful graph has neither parallel nor anti-parallel edges, and no self-loops. When either edge $a \to b$ or $b \to a$ exists in the colorful graph, we say that $a - b$ exists in the colorful graph. In this case, we write $c(a, b) = c(b, a)$ to denote the color of the respective oriented edge. When the "undirected" edge $a - b$ exists in the graph, and the corresponding directed edge is $a \to b$, we say that $a - b$ is *oriented* from a to b. Finally, when edge $a \to b$ ($a - b$) bears color c, we write $a \xrightarrow{c} b$ ($a \overset{c}{-} b$). Note that the colorful graph is only defined for NB constraint sets $\mathcal{C}$ for which the complementary pairs partition exists.

We can now reformulate our task in terms of the colorful graph. In particular, we want to check whether the edges of the colorful graph can be reoriented such that the resulting graph is acyclic, under the constraint that for each color we choose between flipping (reversing) the direction of all edges of that color or doing nothing. When flipping the direction of edges of a certain color, we say that we "flipped" that color.

6   The Case of Approval Ballots

We now switch our attention to colorful graphs induced by approval preference profiles $\mathcal{P}$; i.e., corresponding to the set of NB constraints $\mathcal{C}_\mathcal{P}$. The structure of these graphs is still combinatorially rich, making reasoning about acyclic orientations challenging, but at the same time significantly less general than for arbitrary sets of NB constraints. Examples of such induced colorful graphs can be found in Fig. 2. For brevity, we henceforth assume that we have already checked in polynomial time whether there are any monochromatic cycles and rejected the input if so. In this section we will prove that if the colorful graph has passed this test, then an acyclic orientation is guaranteed to exist and can be determined in polynomial time. To show this, we prove a number of structural results about inducible colorful graphs, centering around the concept of *biclique colors*, which are colors whose corresponding edges can be described by a cartesian product $A \times B$ for two disjoint sets $A, B \subseteq V$. A color is *star biclique* if it is a biclique color with either $|A| = 1$ or $|B| = 1$. The two concepts are illustrated in Fig. 2. Not all colors in the inducible colorful graphs are biclique, as shown in Fig. 2b. However, our result concerning them will be that whenever edges $a - b - c - a$ exist in the colorful graph and bear different colors, these three colors are biclique, and, moreover, the three cartesian products describing them are of the form $A \times B$, $B \times C$ and $C \times A$. In Fig. 2d this can be observed for the triangle $1 - 2 - 3 - 1$. Using this result, together with an analogue for star biclique colors, we then prove that if some orientation of the colorful graph has a cycle, then the smallest such cycle is of length three, showing that, intuitively, triangles are all one needs to worry about. To complete the proof of the main result, we then show that non-monochromatic triangles neccesarily consist of three different colors, implying that we can ignore all colors taking part in no three-colored triangles and focus on the rest. Afterward, a direct argument reusing our theorem about three-colored triangles can be used to construct an acyclic orientation of the colorful graph in polynomial time, proving the main claim of the paper.

6.1   Biclique Colors

From now on, all colorful graphs we consider are induced by approval preferences, unless stated otherwise. In this section, the goal is the prove that whenever $a - b - c - a$ is a three-colored triangle in the colorful graph, then these three colors are biclique and their edges are given by $A \times B$, $B \times C$ and $C \times A$ for some disjoint sets $A, B, C \subseteq V$ where $a \in A, b \in B, c \in C$. Following up, we then also give an instrumental analog for star biclique colors, namely that if $a - b - c$ are different-colored edges in the colorful graph, and edge $a - c$ does not exist in the colorful graph, then there is a set $B \subseteq V \backslash \{a, c\}$ with $b \in B$ such that the two colors consist of edges $\{a\} \times B$ and $B \times \{c\}$ respectively. This can be intuitively thought of as the case $|A| = |C| = 1$ of the result for three-colored triangles. Henceforth, unless stated otherwise, the preference profiles $\mathcal{P}$ that we consider consist of approval ballots, and are represented through approval matrices.

Our high-level argument will hinge on a number of lower-level results concerning the effect that small local structures in the approval matrix have on the formula and colorful graphs. Below is the first such result that we will need. Its proof is a combination of direct reasoning with case analysis over 4×6 partially filled-in approval matrices. While we have considerably simplified the proof over naive case exhaustion, we found the details inessential, so left them for the full version.

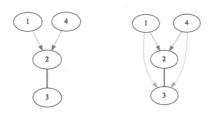

(a) Assumed colors.   (b) Implied colors.

Fig. 3. Illustration of Lemma 8. Figure 3a depicts the assumptions: edge $(1, 2)$ — $(4, 2)$ exists in the formula graph (red) and edge $(2, 3)$ exists in the colorful graph and bears a different color (blue). Additionally, edge $(1, 3)$ is neither red nor blue. Figure 3b shows the edge colors implied by the setup, where green is a color different from red and blue. (Color figure online)

Lemma 8. *Consider four voters, say $1, 2, 3, 4$, such that the formula graph contains the edge $(1, 2)$ — $(4, 2)$ and that edge $(2, 3)$ exists in the colorful graph and has a different color than $(1, 2)$. If edge $(1, 3)$ is either not present in the colorful graph, or it has a different color than $(1, 2)$ and $(2, 3)$, then edge $(1, 3)$ — $(4, 3)$ is guaranteed to be in the formula graph. The scenario is illustrated in Fig. 3.*

It is crucial to notice at this point that Lemma 8, as well as all other results of a similar flavor that we will prove, continue to hold if some of the involved voters coincide. This can be seen by imagining cloning the recurring voters and applying the result with distinct voters. Armed with the previous lower-level result, we are now ready to prove the main result of the section, concerning the biclique structure of colors involved in three-colored triangles.

Theorem 9. *Consider three voters, say $1, 2$ and 3, such that edges $(1, 2)$, $(2, 3)$ and $(3, 1)$ exist in the colorful graph and have different colors, then there exist disjoint sets of voters S_1, S_2 and S_3 such that $1 \in S_1$, $2 \in S_2$ and $3 \in S_3$ and the connected components of $(1, 2)$, $(2, 3)$ and $(3, 1)$ in the formula graph are $S_1 \times S_2$, $S_2 \times S_3$ and $S_3 \times S_1$ respectively.*

Proof. Our proof strategy is summarized in Fig. 4. Consider an inclusion-maximal triple of disjoint sets (S_1, S_2, S_3) such that $1 \in S_1$, $2 \in S_2$, $3 \in S_3$ and the subgraphs of the formula graph induced by $S_1 \times S_2$, $S_2 \times S_3$ and $S_3 \times S_1$ are connected. By inclusion-maximal we mean that for every $v \notin (S_1 \cup S_2 \cup S_3)$ the triples $(S_1 \cup \{v\}, S_2, S_3)$, $(S_1, S_2 \cup \{v\}, S_3)$ and $(S_1, S_2, S_3 \cup \{v\})$ do not satisfy the property. Note that the connectivity assumption implies that in the colorful graph edges in $S_1 \times S_2$ have the color of $(1, 2)$, edges in $S_2 \times S_3$ have the color of $(2, 3)$ and edges in $S_3 \times S_1$ have the color of $(3, 1)$. Denote by C_{12}, C_{23} and C_{31} the connected components of $(1, 2)$, $(2, 3)$ and $(3, 1)$ in the formula graph. We will now show that $C_{12} = S_1 \times S_2$, $C_{23} = S_2 \times S_3$ and $C_{31} = S_3 \times S_1$. Assume for a contradiction that this was not the case, then, without loss of generality, $C_{12} \neq S_1 \times S_2$. Together with the above, this means

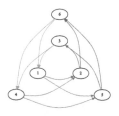 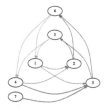 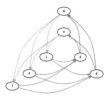

(a) Col. graph for (S_1, S_2, S_3).   (b) Assume $(4, 5, 7) \in \mathcal{C}_{\mathcal{P}}$.   (c) Contradicts maximality.

Fig. 4. Illustration of the proof of Theorem 9. We consider an inclusion-maximal triple (S_1, S_2, S_3) such that $1 \in S_1, 2 \in S_2, 3 \in S_3$ and $S_1 \times S_2$, $S_2 \times S_3$ and $S_3 \times S_1$ are connected in the formula graph, in this case $S_1 = \{1, 4\}, S_2 = \{2, 5\}$ and $S_3 = \{3, 6\}$ (Fig. 4a). Then, we assume for a contradiction that these connected sets are not connected components, in this case because of NB constraint $(4, 5, 7)$ (Fig. 4b). Finally, this implies that $(S_1 \cup \{7\}, S_2, S_3)$ also has the property, contradicting maximality (Fig. 4c).

that $S_1 \times S_2 \subsetneq C_{12}$. Because C_{12} induces a connected subgraph in the formula graph, and $\varnothing \neq S_1 \times S_2 \subsetneq C_{12}$, it follows that there is an edge in the formula graph crossing the cut $(S_1 \times S_2, C_{12} \setminus (S_1 \times S_2))$. Without loss of generality, this edge is of the form $(v_1, v_2) — (x, v_2)$, where $v_1 \in S_1, v_2 \in S_2$ and $x \notin S_1$.

Now, we show that for every $v_3 \in S_3$ there is an edge $(v_1, v_3) — (x, v_3)$ in the formula graph. Note that this means that all edges of the form (x, v_3) for $v_3 \in S_3$ have the color of $(1, 3)$. Consider an arbitrary $v_3 \in S_3$, and note that we can instantiate Lemma 8 with $1 \mapsto v_1, 2 \mapsto v_2, 3 \mapsto v_3$ and $4 \mapsto x$ because $(v_1, v_2) — (x, v_2)$ is in the formula graph and $(v_1, v_2), (v_2, v_3)$ and (v_3, v_1) bear the colors of $(1, 2), (2, 3)$ and $(3, 1)$, which are different. The instantiation gives us that there is an edge $(v_1, v_3) — (x, v_3)$ in the formula graph, as required.

From the above, we know that the edge $(v_1, 3) — (x, 3)$ exists in the formula graph. Similarly, we now show that for every $v_2' \in S_2$ there is an edge $(v_1, v_2') — (x, v_2')$ in the formula graph. This is done analogously, by invoking Lemma 8 with $1 \mapsto v_1, 2 \mapsto 3, 3 \mapsto v_2'$ and $4 \mapsto x$, which is sound because $(v_1, 3) — (x, 3)$ is in the formula graph and $(v_1, 3), (3, v_2')$ and (v_2', v_1) bear the colors of $(1, 3), (3, 2)$ and $(2, 1)$, which are different.

At this point, we have shown that $(S_1 \cup \{x\}) \times S_2$ and $S_3 \times (S_1 \cup \{x\})$ both induce connected subgraphs in the formula graph; in particular, all edges from x to S_2 have the color of $(1, 2)$ and all edges from x to S_3 have the color of $(1, 3)$. We will now additionally show that $x \notin (S_2 \cup S_3)$, from which the triple $(S_1 \cup \{x\}, S_2, S_3)$ also satisfies the property we started with, contradicting the hypothesis that (S_1, S_2, S_3) was inclusion-maximal. First, if $x \in S_3$, then $(1, 2)$ and $(2, 3)$ immediately have the same color, which cannot be the case. Otherwise, if $x \in S_2$, then from our argument $(x, 3)$ has the same color as $(1, 3)$, while the assumptions tell us that it has the same color as $(2, 3)$, meaning that $(1, 3)$ and $(2, 3)$ share the same color, which is again not possible.

The following result can be seen as a star biclique analogue of Theorem 9. Intuitively, one can think of this as the orthogonal case where the edge $1 - 3$ does not exist in the colorful graph and $S_1 = \{1\}, S_3 = \{3\}$. While the two proofs are very similar in spirit, we could not find a transparent way of unifying the two arguments, partly because of less symmetry in the scenario. In particular, this lack of symmetry will require a second low-level result in the spirit of Lemma 8, stated and proven in the full version. See the full version for the relevant proofs.

Theorem 10. *Consider three voters, say $1, 2$ and 3, such that edges $(1,2)$ and $(2,3)$ exist in the colorful graph and have different colors, while edge $(1,3)$ does not exist in the colorful graph. Then, there exists a set of voters S such that $2 \in S$ and $1, 3 \notin S$ with the property that the connected components of $(1,2)$ and $(2,3)$ in the formula graph are $\{1\} \times S$ and $S \times \{3\}$ respectively.*

6.2 Only Three-Colored Triangles Matter, the Main Result

In this section we prove that if some orientation of the colorful graph has a cycle, then the smallest such cycle is a triangle. Then, we show that non-monochromatic triangles neccesarily consist of three different colors. Finally, we prove that any colorful graph with no monochromatic cycles has an acyclic orientation, hinging on Theorem 9. Due to space considerations, the proofs of Lemmas 11 and 13 below are deferred to the full version of the paper.

Lemma 11. *An orientation of the colors in the colorful graph is acyclic if and only if it induces no directed cycles of length three.*

Next, we prove that only three-colored triangles need to be considered; e.g., the case in Fig. 1c cannot occur. For this we will need the following stronger form of Lemma 8:

Lemma 12 (Strengthened Lemma 8). *Consider four voters, say $1, 2, 3, 4$, such that the formula graph contains the edge $(1,2) - (4,2)$ and that edge $(2,3)$ exists in the colorful graph. If the formula graph does not contain edges $(3,2) - (1,2)$ and $(3,2) - (4,2)$, then the formula graph is guaranteed to contain the edge $(1,3) - (4,3)$.*

The proof closely follows that of Lemma 8, but the weaker assumptions make it significantly more difficult to proceed as before in a principled manner. Instead, we resort to a computer program to perform the case analysis (see full version for details).

Lemma 13. *Consider three voters, say $1, 2, 3$ such that edges $1 - 2$ and $2 - 3$ are the same color and oriented $1 \to 2$ and $2 \to 3$ in the colorful graph. If edge $1 - 3$ appears in the colorful graph, then it is of the same color as $1 - 2 - 3$.*

Corollary 14. *An orientation of the colors in the colorful graph is acyclic if and only if it induces no directed cycles of length three consisting of three different colors.*

Theorem 15. *An acyclic orientation of the colorful graph exists and can be computed in poly-time.*

Proof. By Corollary 14, only colors participating in three-colored triangles have to be considered (all other colors we can orient arbitrarily). Let $C = \{c_1, \ldots, c_k\}$ be the set of such colors. For each color $c_i \in C$, applying Theorem 9 to an arbitrary three-colored triangle involving c_i gives that c_i is a biclique color, so the set of edges colored with color c_i in the colored graph is $A_i \times B_i$ for some disjoint sets of vertices A_i, B_i. For a finite set of numbers S, we write $\min S$ for the minimum number in S. We now construct an orientation of the colors in C as follows: for color $c_i \in C$ we orient the edges of color c_i to go $A_i \rightarrow B_i$ if $\min A_i < \min B_i$ and $B_i \rightarrow A_i$ otherwise. Note that $\min A_i \neq \min B_i$ as $A_i \cap B_i = \varnothing$. We now prove that this orientation is acyclic to complete the proof. Assume for a contradiction that the orientation induced a cycle $a \rightarrow b \rightarrow c \rightarrow a$. Then, by Corollary 14, the three edges involved in the cycle bear different colors in the colorful graph, so, by Theorem 9, there exist three disjoint vertex sets A, B, C such that $a \in A, b \in B, c \in C$ and the connected components of $a \rightarrow b$, $b \rightarrow c$ and $c \rightarrow a$ in the formula graph are given by $A \times B$, $B \times C$ and $C \times A$. However, by construction of the orientation, this would mean that $\min A < \min B < \min C < \min A$, a contradiction. $\qed$

Armed as such, we can now prove the main result of our paper.

Theorem 16. *There is a polynomial time algorithm taking as input a collection of approval ballots that outputs a profile of single-crossing preferences that could have lead to the given approval votes. If this is not possible, the algorithm reports accordingly.*

Proof. Start with the approval ballots and build the formula graph. If there are pairs of complementary nodes in the same connected component, report impossibility. Otherwise, build the colorful graph. If a monochromatic cycle exists, report impossibility. Otherwise, use Theorem 15 to compute an acyclic orientation, and extend it to a linear order $\lhd$. Finally, use the approach of Elkind et al. [16], also explained in the full version of our paper, to compute a single-crossing profile with respect to the known axis $\lhd$. $\qed$

7 Conclusions and Future Work

We gave a polynomial time algorithm computing a collection of fully-ranked ballots that could have been induced by the given approval ballots. Furthermore, we showed that there is no finite substructures characterization of possibly single-crossing elections. To prove our main result, we developed a new algorithm for the non-betweenness problem, showing that it is FPT with respect to the number of colors in a certain colorful graph. We expect this FPT algorithm to lead to polynomial algorithms for other ordering problems reducing to the non-betweenness problem by combinatorial analysis of the induced colorful graphs. We note that betweenness reduces to non-betweenness since the constraint "*b* has to be between *a* and *c*" can be modeled as the conjunction of two non-betweenness constraints, namely: "*a* should not be between *b* and *c*" and "*c* should not be between *a* and *b*", so our framework can also be used for betweenness.

As steps for future research, the most natural next question to ask is the complexity of recognizing possibly single-crossing for multi-valued approval, the case of three indifference classes being a natural first candidate to consider. We have empirically found that our proof of correctness no longer applies in this case, since for instance Lemma 8 fails to hold for 3-valued approval. However, we could not find instances where the algorithm as a whole fails the recognition task. We leave it open to settle this question. Another promising avenue for further investigation stands in the case when our algorithm reports that voters' innate preferences are not single-crossing. In this case, one might hope that they are at least somewhat close to being single-crossing, so it would be interesting to investigate how to recover preferences that are nearly single-crossing from approval ballots (see [19, Section 4.8] for a discussion of related results).

References

1. Arrow, K.: Social Choice and Individual Values. John Wiley and Sons, Hoboken (1951)
2. Barberà, S., Moreno, B.: Top monotonicity: a common root for single peakedness, single crossing and the median voter result. Games Econ. Behav. **73**, 345–359 (2011)
3. Bartholdi, J., Trick, M.A.: Stable matching with preferences derived from a psychological model. Oper. Res. Lett. **5**(4), 165–169 (1986)
4. Beresnev, V.L., Davydov, A.I.: On matrices with connectedness property. Upravlyaemye Sistemy **19** (1979). in Russian
5. Betzler, N., Slinko, A., Uhlmann, J.: On the computation of fully proportional representation. J. Artif. Int. Res. **47**(1), 475–519 (2013)
6. Black, D.: On the rationale of group decision-making. J. Polit. Econ. **56**(1), 23–34 (1948)
7. Booth, K.S., Lueker, G.S.: Testing for the consecutive ones property, interval graphs, and graph planarity using PQ-tree algorithms. JCSS **13**(3), 335–379 (1976)
8. Brandt, F., Conitzer, V., Endriss, U., Lang, J., Procaccia, A.D. (eds.): Handbook of Computational Social Choice, 1st edn. Cambridge University Press, USA (2016)
9. Bredereck, R., Chen, J., Woeginger, G.J.: A characterization of the single-crossing domain. Soc. Choice Welf. **41**(4), 989–998 (2013)
10. Busing, F.M.T.A.: Advances in multidimensional unfolding. Ph.D. thesis (2010)
11. Chen, J., Hatschka, C., Simola, S.: Efficient algorithms for monroe and cc rules in multi-winner elections with (nearly) structured preferences. In: ECAI'23, pp. 397–404 (2023)
12. Constantinescu, A., Wattenhofer, R.: Recovering single-crossing preferences from approval ballots. arXiv:2310.03736 (2023)
13. Constantinescu, A.C., Elkind, E.: Proportional representation under single-crossing preferences revisited. In: AAAI'21, vol. 35, no. 6, pp. 5286–5293 (2021)

14. Doignon, J.P., Falmagne, J.C.: A polynomial time algorithm for unidimensional unfolding representations. J. Algorithms **16**(2), 218–233 (1994)
15. Elkind, E., Faliszewski, P.: Recognizing 1-Euclidean preferences: an alternative approach. In: Lavi, R. (ed.) SAGT 2014. LNCS, vol. 8768, pp. 146–157. Springer, Heidelberg (2014). https://doi.org/10.1007/978-3-662-44803-8_13
16. Elkind, E., Faliszewski, P., Lackner, M., Obraztsova, S.: The complexity of recognizing incomplete single-crossing preferences. In: AAAI'15, vol. 29, no. 1, February 2015
17. Elkind, E., Faliszewski, P., Slinko, A.: Clone structures in voters' preferences. In: Proceedings of the ACM Conference on Electronic Commerce, pp. 496–513 (2012)
18. Elkind, E., Lackner, M.: Structure in dichotomous preferences. In: IJCAI'15, pp. 2019–2025 (2015)
19. Elkind, E., Lackner, M., Peters, D.: Preference restrictions in computational social choice: a survey (2022)
20. Escoffier, B., Lang, J., Öztürk, M.: Single-peaked consistency and its complexity, pp. 366–370. IOS Press, NLD (2008)
21. Faliszewski, P., Hemaspaandra, E., Hemaspaandra, L.A., Rothe, J.: The shield that never was: societies with single-peaked preferences are more open to manipulation and control. In: TARK'09, pp. 118–127. ACM, New York, NY, USA (2009)
22. Fitzsimmons, Z.: Single-peaked consistency for weak orders is easy. In: Ramanujam, R. (ed.) TARK'15, vol. 215, pp. 127–140 (2016)
23. Guttmann, W., Maucher, M.: Variations on an ordering theme with constraints. In: IFIP TCS (2006)
24. Klinz, B., Rudolf, R., Woeginger, G.J.: Permuting matrices to avoid forbidden submatrices. Discret. Appl. Math. **60**(1), 223–248 (1995)
25. Knoblauch, V.: Recognizing one-dimensional Euclidean preference profiles. J. Math. Econ. **46**, 1–5 (2010)
26. Lackner, M., Skowron, P.: Multi-Winner Voting with Approval Preferences. Springer International Publishing, Cham (2023). https://doi.org/10.1007/978-3-031-09016-5
27. Laslier, J.F., Sanver, M.R. (eds.): Handbook on Approval Voting. Springer, Berlin, Heidelberg (2010). https://doi.org/10.1007/978-3-642-02839-7
28. Lu, T., Boutilier, C.: Budgeted social choice: from consensus to personalized decision making. In: IJCAI'11, pp. 280–286, January 2011
29. Magiera, K., Faliszewski, P.: Recognizing top-monotonic preference profiles in polynomial time. In: IJCAI'17, pp. 324–330 (2017)
30. Mattei, N., Walsh, T.: Preflib: a library for preferences http://www.preflib.org. In: Perny, P., Pirlot, M., Tsoukiàs, A. (eds.) ADT 2013. LNCS (LNAI), vol. 8176, pp. 259–270. Springer, Heidelberg (2013). https://doi.org/10.1007/978-3-642-41575-3_20
31. Mirrlees, J.: An exploration in the theory of optimum income taxation. Rev. Econ. Stud. **38**(2), 175–208 (1971)
32. Peters, D.: Recognising multidimensional Euclidean preferences. In: AAAI'17, pp. 642–648 (2017)
33. Pierczyński, G., Skowron, P.: Core-stable committees under restricted domains. In: Hansen, K.A., Liu, T.X., Malekian, A. (eds.) Web and Internet Economics. WINE 2022. LNCS, vol. 13778, pp. 311–329. Springer, Cham (2022). https://doi.org/10.1007/978-3-031-22832-2_18
34. Procaccia, A., Rosenschein, J., Zohar, A.: On the complexity of achieving proportional representation. Soc. Choice Welf. **30**, 353–362 (2008)

35. Roberts, K.: Voting over income tax schedules. J. Public Econ. **8**(3), 329–340 (1977)
36. Skowron, P., Yu, L., Faliszewski, P., Elkind, E.: The complexity of fully proportional representation for single-crossing electorates. Theor. Comput. Sci. **569**, 43–57 (2015)
37. Sornat, K., Williams, V.V., Xu, Y.: Near-tight algorithms for the Chamberlin-courant and Thiele voting rules. In: De Raedt, L. (ed.) IJCAI'22, pp. 482–488 (2022)
38. Terzopoulou, Z., Karpov, A., Obraztsova, S.: Restricted domains of dichotomous preferences with possibly incomplete information, pp. 2023–2025. AAMAS'20 (2020)

The Good, the Bad and the Submodular: Fairly Allocating Mixed Manna Under Order-Neutral Submodular Preferences

Cyrus Cousins$^{(\boxtimes)}$ ⓘ, Vignesh Viswanathan ⓘ, and Yair Zick ⓘ

University of Massachusetts, Amherst, MA 01002, USA
{cbcousins,vviswanathan,yzick}@umass.edu

Abstract. We study the problem of fairly allocating indivisible goods (positively valued items) and chores (negatively valued items) among agents with decreasing marginal utilities over items. Our focus is on instances where all the agents have *simple* preferences; specifically, we assume the marginal value of an item can be either -1, 0 or some positive integer c. Under this assumption, we present an efficient algorithm to compute leximin allocations for a broad class of valuation functions we call *order-neutral* submodular valuations. Order-neutral submodular valuations strictly contain the well-studied class of additive valuations but are a strict subset of the class of submodular valuations. We show that these leximin allocations are Lorenz dominating and approximately proportional. We also show that, under further restriction to additive valuations, these leximin allocations are approximately envy-free and guarantee each agent their maximin share. We complement this algorithmic result with a lower bound showing that the problem of computing leximin allocations is NP-hard when c is a rational number.

Keywords: Fair Allocation · Indivisible Items · Mixed Manna

1 Introduction

Fair allocation is a fundamental problem in computational economics. The problem asks how to divide a set of indivisible items among agents with subjective preferences (or valuations) over the items. Most of the literature focuses on the problem of dividing *goods* — items with positive value. However, in several practical applications, such as dividing a set of tasks or allocating shifts to employees, items can be *chores* which provide a negative value to the agents they are allocated to.

Fair allocation with mixed manna (instances containing both goods and chores) is, unsurprisingly, a harder problem than the case with only goods. Several questions which have been answered positively in the only goods setting are either still open or face a negative (impossibility or intractability) result in the mixed goods and chores setting. In particular, very little is known about the computability of *leximin allocations*. A leximin allocation is one that maximizes

J. Garg et al. (Eds.): WINE 2023, LNCS 14413, pp. 207–224, 2024.
https://doi.org/10.1007/978-3-031-48974-7_12

the utility of the agent with the least utility; subject to that, it maximizes the second-least utility, and so on.

In the only goods setting, maximizing Nash welfare is arguably one of the most popular fairness objectives. Unfortunately, the Nash welfare of an allocation, defined as the product of agent utilities, loses its meaning in settings where agent utilities can be negative. In such settings, the leximin objective is a natural substitute. The leximin objective is easy to understand, and it implies an appealing egalitarian notion of fairness. Therefore, designing algorithms that efficiently compute exact (or approximate) leximin allocations in the mixed manna setting is an important research problem.

In this work, we present the first non-trivial results for this problem.

1.1 Our Results

Contributions to the Theory of Fair Allocation. We take a systematic approach towards solving this problem and start with a simple class of valuation functions. We assume all goods are symmetric and provide value c (where c is a positive integer), and all chores are symmetric and provide value -1. We also allow items to provide a value of 0. We assume there are decreasing marginal gains over items; that is, after receiving many items, a good may provide value 0 or even turn into a chore and provide a value of -1. We refer to this class of valuations as $\{-1, 0, c\}$-submodular valuations. With such valuations, there is no clear demarcation between goods and chores; an item may provide positive marginal value when added to an empty bundle but provide negative marginal value when added to a non-empty bundle.

We present an algorithm to compute a leximin allocation for a broad subclass of $\{-1, 0, c\}$ submodular valuations. More specifically, we show that leximin allocations can be computed efficiently when agents have $\{-1, 0, c\}$ submodular valuation functions that satisfy an additional property we call *order-neutrality*. Order-neutrality can be very loosely thought of as a property that requires the number of c-valued items in a bundle to be a monotonically non-decreasing function of the bundle. We analyze these leximin allocations in further detail showing that they are Lorenz dominating and proportional up to one item. We also show that under further restriction to the case where agents have $\{-1, 0, c\}$ additive (or linear) valuations, leximin allocations are approximately envy-free as well as offer each agent their maximin share. We complement this result with lower bounds showing that the problem of computing leximin allocations becomes computationally intractable when c is relaxed to being an arbitrary rational number as opposed to a positive integer.

Technical Contributions. We put forward several interesting combinatorial and algorithmic contributions. We introduce an interesting function class — Order-Neutral Submodular (ONSUB) Valuations. This is a subclass of submodular valuations for which items' marginal contributions takes on a regular structure. While we only analyze this class within the context of fair allocation of indivisible goods, we suspect that the function class may serve as a useful object of study in other domains featuring agents with combinatorial valuations such as matching markets, committee election and participatory budgeting.

In addition, we introduce the weighted item exchange graph. This is an extension of the exchange graph used in matroid theory [24], and more recently in the fair allocation literature [4, 7, 25, 27] to analyze the specific case where all agents have binary submodular valuations. The weighted exchange graph is a more general theoretical tool which allows us to carefully manipulate allocations when agents do not necessarily have binary submodular valuations.

Finally, our key algorithmic contribution is Algorithm 1 which computes leximin allocations. Algorithm 1 operates in three phases: first, it computes a utility maximizing partial allocation. Next, it uses transfer paths to obtain a partial leximin allocation. These partial allocations do not allocate any items that provide a negative marginal value to the agents they are allocated to. To make the allocation complete and allocate the final set of items, the algorithm proceeds to greedily allocate items, whose marginal gain is -1, to the highest utility agents.

1.2   Related Work

Fair allocation with mixed goods and chores has recently gained popularity in the literature. [2] presents definitions of envy-free up to one item (EF1) and proportionality up to one item (PROP1) for the mixed goods and chores setting; [2] also presents an algorithm to efficiently compute EF1 allocations. [11] and [9] further study the existence and computation of approximate envy-free allocations. There have also been a couple of papers studying maximin share fairness with mixed goods and chores. [18] shows that there exists a PTAS to compute a maximin share fair allocation under certain assumptions. On the other hand, [19] shows that the problem of approximating the maximin share of each agent is computationally intractable under additive valuations.

To the best of our knowledge, there are only two works [9, 11] on the fair allocation of mixed goods and chores which present results for non-additive valuation classes. [11] presents several positive results for very specific cases, such as identical valuations, Boolean valuations and settings with two agents. [9] presents an algorithm to compute EF1 allocations under *doubly monotone valuations*. Doubly monotone valuations assume each item is a good (always has positive marginal gain) or a chore (always has negative marginal gain) but otherwise, do not restrict the valuations. Order-neutral submodular valuations (discussed in this paper) relax the assumption that each item must be classified as a good or a chore, but come with the stronger restriction of submodularity. We also note that [2] presents an algorithm for computing an EF1 allocation for doubly monotone valuations; however, this algorithm's correctness was disproved by [9].

The domain restrictions described above are not uncommon in fair allocation. In the mixed goods and chores setting, [17] studies the problem of fair allocation with lexicographic valuations — a restricted subclass of additive valuations. In the goods setting, binary valuations [4, 8, 16, 25, 27] and bivalued valuations [1, 14] have been extensively studied. In the chores setting, bivalued additive valuations

[3,13,15] and binary submodular valuations [5,26] have been well studied. Our results are a natural extension of this line of work.

2 Preliminaries

We use $[k]$ to denote the set $\{1, 2, \ldots, k\}$. Given a set S and an element o, we use $S + o$ and $S - o$ to denote the sets $S \cup \{o\}$ and $S \setminus \{o\}$ respectively.

We have a set of n *agents* $N = [n]$ and a set of m *items* $O = \{o_1, o_2, \ldots, o_m\}$. Each agent $i \in N$ has a *valuation function* $v_i : 2^O \to \mathbb{R}$; $v_i(S)$ denotes the value of the set of items S according to agent i. Given a valuation function v, we let $\Delta_v(S, o) = v(S + o) - v(S)$ denote the marginal utility of adding the item o to the bundle S under v. When clear from context, we sometimes write $\Delta_i(S, o)$ instead of $\Delta_{v_i}(S, o)$ to denote the marginal utility of giving the item o to agent i given that they have already been assigned the bundle S.

An *allocation* $X = (X_0, X_1, \ldots, X_n)$ is an $(n+1)$-partition of the set of items O. X_i denotes the set of items allocated to agent i and X_0 denotes the set of unallocated items. Our goal is to compute *complete* fair allocations — allocations where $X_0 = \emptyset$. When we construct an allocation, we sometimes only define the allocation to each agent $i \in N$; the bundle X_0 is implicitly assumed to contain all the unallocated items. Given an allocation X, we refer to $v_i(X_i)$ as the utility of agent i under the allocation X. We also define the utility vector of an allocation X as the vector $\boldsymbol{u}^X = (v_1(X_1), v_2(X_2), \ldots, v_n(X_n))$.

We define two common methods to compare vectors. We will use these methods extensively in our analysis when comparing allocations.

Definition 2.1 (Lexicographic Dominance). *A vector $\boldsymbol{y} \in \mathbb{R}^n$ lexicographically dominates a vector $\boldsymbol{z} \in \mathbb{R}^n$ (written $\boldsymbol{y} \succ_{lex} \boldsymbol{z}$) if there exists a $k \in [n]$ such that for all $j \in [k-1]$, $y_j = z_j$ and $y_k > z_k$. We sometimes say an allocation X lexicographically dominates an allocation Y if $\boldsymbol{u}^X \succ_{lex} \boldsymbol{u}^Y$.*

Definition 2.2 (Pareto Dominance). *A vector $\boldsymbol{y} \in \mathbb{R}^n$ Pareto dominates a vector $\boldsymbol{z} \in \mathbb{R}^n$ if for all $j \in [n]$, $y_j \geq z_j$ with the inequality being strict for at least one $j \in [n]$. An allocation X Pareto dominates an allocation Y if $\boldsymbol{u}^X$ Pareto dominates $\boldsymbol{u}^Y$.*

2.1 Valuation Functions

In this paper, we will be dealing with the popular class of submodular valuations.

Definition 2.3 (Submodular functions). *A function $v : 2^O \to \mathbb{R}$ is a submodular function if (a) $v(\emptyset) = 0$, and (b) for any $S \subseteq T \subseteq O$ and $o \in O \setminus T$, $\Delta_v(S, o) \geq \Delta_v(T, o)$.*

We also define restricted submodular valuations to formally capture instances where the function has a limited set of marginal values. Throughout this paper, we use the set A to denote an arbitrary set of real numbers.

Definition 2.4 (A-SUB functions). *Given a set of real numbers A, a function $v : 2^O \rightarrow \mathbb{R}$ is an A-SUB function if it is submodular, and every item's marginal contribution is in A. That is, for any set $S \subseteq O$ and an item $o \in O \setminus S$, $\Delta_v(S, o) \in A$.*

Our analysis of submodular functions requires that they satisfy an additional property we call *order-neutrality*. Given a submodular valuation function $v : 2^O \rightarrow \mathbb{R}$, the value of a set of items S can be computed by adding items from S one by one into an empty set and adding up the $|S|$ marginal gains. More formally, given a bijective mapping $\pi : [|S|] \rightarrow S$ which defines the order in which items are added, $v(S)$ can be written as the following telescoping sum:

$$v(S) = \sum_{j \in [|S|]} \Delta_v\left(\bigcup_{\ell \in [j-1]} \pi(\ell), \pi(j)\right).$$

While the value $v(S)$ does not depend on π, the values of the marginal gains in the telescoping sum may depend on π. Given a set S and a bijective mapping $\pi : [|S|] \rightarrow S$ we define the vector $\boldsymbol{v}(S, \pi)$ as the vector of marginal gains (given below) *sorted in ascending order*

$$\left(\Delta_v\big(\emptyset, \pi(1)\big), \Delta_v\big(\pi(1), \pi(2)\big), \ldots, \Delta_v\big(S - \pi(|S|), \pi(|S|)\big)\right).$$

We refer to this vector $\boldsymbol{v}(S, \pi)$ as a *sorted telescoping sum vector* . For any ordering π and set S, the sum of the elements of $\boldsymbol{v}(S, \pi)$ is equal to $v(S)$. A submodular function v is said to be *order-neutral* if for all bundles $S \subseteq O$ and any two orderings $\pi, \pi' : [|S|] \rightarrow S$ of items in the bundle S, we have $\boldsymbol{v}(S, \pi) = \boldsymbol{v}(S, \pi')$; that is, any sorted telescoping sum vector is independent of the order π. For order-neutral submodular valuations, we sometimes drop the π and refer to any sorted telescoping sum vector using $\boldsymbol{v}(S)$. This definition can be similarly extended to A-SUB functions. For readability, we refer to order-neutral A-SUB functions as A-ONSUB functions.

Definition 2.5 (A-ONSUB functions). *Given a set of real numbers A, a function $v : 2^O \rightarrow \mathbb{R}$ is an A-ONSUB function if it is both order-neutral and an A-SUB function.*

We will focus on instances with $\{-1, 0, c\}$-ONSUB valuations where c is a positive integer. Throughout this paper, the only use of c will be to denote an arbitrary positive integer. We also present some results for the restricted setting where agents have additive valuations.

Definition 2.6 (A-ADD functions). *Given a set of real numbers A, a function $v : 2^O \rightarrow \mathbb{R}$ is an A-additive (or simply A-ADD) function if $v(\{o\}) \in A$ for all $o \in O$ and $v(S) = \sum_{o \in S} v(\{o\})$ for all $S \subseteq O$.*

To build intuition for the class of $\{-1, 0, c\}$-ONSUB valuations, we present a few simple examples below.

Example 2.7. Let $O = \{o_1, o_2, o_3, o_4\}$, the following functions $v_1, v_2, v_3 : 2^O \to \mathbb{R}$ are $\{-1, 0, c\}$-ONSUB:

$$v_1(S) = c \min\{|S|, 2\},$$
$$v_2(S) = c\mathbb{I}\{o_1 \in S\} - \mathbb{I}\{o_2 \in S\},$$
$$v_3(S) = c \min\{|S \cap \{o_1, o_2\}|, 1\} - |S \cap \{o_3, o_4\}|.$$

Agent 1's valuation v_1 describes a function where any item provides a value of c but the marginal utility of any item drops to 0 after two items are added to the bundle. v_2 describes a simple additive function where o_1 provides a value of c and o_2 provides a value of -1. v_3 describes a slightly more complex function where o_1 and o_2 are goods but at most one of them can a provide a marginal value of c; o_3 and o_4 are chores providing a marginal value of -1 each.

Note that additive valuations are trivially order-neutral submodular valuations. Unsurprisingly, not all submodular valuations are order-neutral — consider a function v over two items $\{o_1, o_2\}$ such that $v(\{o_1\}) = 0$, $v(\{o_2\}) = 1$ and $v(\{o_1, o_2\}) = 0$. This function is submodular, but not order-neutral, since $v(\{o_1, o_2\})$ has two different sorted telescoping sum vectors. However, it is worth noting that there are many interesting non-additive order-neutral submodular functions. For example, any binary submodular function ($\{0, 1\}$-SUB) is order-neutral.

Proposition 2.8. *When $|A| = 2$, any A-SUB function is order-neutral.*

This proposition implies that the class of $\{0, c\}$-SUB, $\{-1, c\}$-SUB and $\{-1, 0\}$-SUB valuations are all contained in the class of $\{-1, 0, c\}$ order-neutral submodular valuations. It is also worth noting that capped additive valuations, where agents can only receive a positive marginal utility from a fixed number of items also falls under the class of $\{-1, 0, c\}$ order-neutral submodular valuations.

2.2 Fairness and Efficiency Objectives

There are several reasonable fairness objectives used in the fair allocation literature. We discuss most of them in this paper. However, to avoid an overload of definitions, we only define the following two fairness and efficiency objectives in this section.

Utilitarian Social Welfare (USW): The utilitarian social welfare of an allocation X is $\sum_{i \in N} v_i(X_i)$. An allocation X is said to be MAX-USW if it maximizes the utilitarian social welfare.

Leximin: An allocation is said to be leximin if it maximizes the utility of the least valued agent, and subject to that, the utility of the second-least valued agent, and so on [20]. This is usually formalized using the sorted utility vector. The *sorted utility vector* of an allocation X (denoted by s^X) is defined as the utility vector u^X sorted in ascending order. An allocation X is leximin if there is no allocation Y such that $s^Y \succ_{\text{lex}} s^X$.

Other objectives like maximin share, proportionality and envy-freeness are defined in Sect. 6.

2.3   Exchange Graphs and Path Augmentations

In this section, we describe the classic technique of path augmentations. A modified version of these path augmentations is used extensively in our algorithm design. Path augmentations have been used to carefully manipulate allocations when agents have binary submodular valuations ($\{0,1\}$-SUB functions). However, they only work with *clean* allocations.

Definition 2.9 (Clean Allocation). *For any agent $i \in N$, a bundle S is clean with respect to the binary submodualar valuation β_i if $\beta_i(S) = |S|$. An allocation X is said to be clean (w.r.t .$\{\beta_h\}_{h \in N}$) if for all agents $i \in N$, $\beta_i(X_i) = |X_i|$.*

When agents have binary submodular valuations $\{\beta_h\}_{h \in N}$, given a clean allocation X (w.r.t. $\{\beta_h\}_{h \in N}$), we define the *exchange graph* $\mathcal{G}(X, \beta)$ as a directed graph over the set of items O, where an edge exists from o to o' in the exchange graph if $o \in X_j$ and $\beta_j(X_j - o + o') = \beta_j(X_j)$ for some $j \in N$. In other words, an edge exists if agent j can swap item o with the item o' and still retain the same utility level. There is no outgoing edge from any item in X_0.

Let $P = (o_1, o_2, \ldots, o_t)$ be a path in the exchange graph $\mathcal{G}(X, \beta)$ for a clean allocation X. We define a transfer of items along the path P in the allocation X as the operation where o_t is given to the agent who has o_{t-1}, o_{t-1} is given to the agent who has o_{t-2}, and so on until finally o_1 is discarded and becomes freely available. This transfer is called *path augmentation*; the bundle X_i after path augmentation with the path P is denoted by $X_i \wedge P$ and defined as $X_i \wedge P = (X_i - o_t) \oplus \{o_j, o_{j+1} : o_j \in X_i\}$, where $\oplus$ denotes the symmetric set difference operation.

For any clean allocation X and agent i, we define $F_{\beta_i}(X, i) = \{o \in O : \Delta_{\beta_i}(X_i, o) = 1\}$ as the set of items which give agent i a marginal gain of 1 under the valuation β_i. For any agent i, let $P = (o_1, \ldots, o_t)$ be a *shortest* path from $F_{\beta_i}(X, i)$ to X_j for some $j \neq i$. Then path augmentation with the path P and giving o_1 to i results in a clean allocation where the size of i's bundle $|X_i|$ goes up by 1, the size of j's bundle goes down by 1 and all the other agents do not see any change in size. This is formalized below and exists in [6, Lemma 1] and [27, Lemma 3.7].

Lemma 2.10 ([6], [27]). *Let X be a clean allocation with respect to the binary submodular valuations $\{\beta_h\}_{h \in N}$. Let $P = (o_1, \ldots, o_t)$ be a shortest path in the exchange graph $\mathcal{G}(X, \beta)$ from $F_{\beta_i}(X, i)$ to X_j for some $i \in N$ and $j \in N + 0 - i$. Then, the following allocation Y is clean with respect to $\{\beta_h\}_{h \in N}$.*

$$Y_k = \begin{cases} X_k \wedge P & (k \in N + 0 - i) \\ X_i \wedge P + o_1 & (k = i) \end{cases}$$

Moreover, for all $k \in N + 0 - i - j$, $|Y_k| = |X_k|$, $|Y_i| = |X_i| + 1$ and $|Y_j| = |X_j| - 1$.

We also present sufficient conditions for a path to exist. A slight variant of the following lemma appears in [27, Theorem 3.8].

Lemma 2.11 ([27]). *Let X and Y be two clean allocations with respect to the binary submodular valuations $\{\beta_h\}_{h\in N}$. For any agent $i \in N$ such that $|X_i| < |Y_i|$, there is a path from $F_{\beta_i}(X, i)$ to either*

(i) some item in X_k for some $k \in N$ in the exchange graph $\mathcal{G}(X, \beta)$ such that $|X_k| > |Y_k|$, or
(ii) some item in X_0 in $\mathcal{G}(X, \beta)$.

This technique of path augmentations has been extensively exploited in the design of the Yankee Swap algorithm [27]. Given an instance with binary submodular valuations, Yankee Swap computes a clean `MAX-USW` leximin allocation in polynomial time. We use this procedure as a subroutine in our algorithm to compute a partial allocation.

Theorem 2.12 (Yankee Swap [27]). *When agents have binary submodular valuations $\{\beta_h\}_{h\in N}$, there exists an efficient algorithm that computes a clean* `MAX-USW` *leximin allocation.*

We note that Yankee Swap is not the only algorithm to compute clean leximin allocations; [4] also presents an efficient algorithm to do so.

3   Understanding A-ONSUB Valuations

We now present a result about A-ONSUB valuations exploring its connection to binary submodular functions. This connection allows us to adapt path augmentations for our setting. Our arguments generalize the arguments presented by [12, Section 3] about bivalued submodular valuations.

Our result shows that given a threshold value $\tau \in \mathbb{R}$ and an A-ONSUB function v_i, the number of values in the sorted telescoping sum vector greater than or equal to the threshold τ corresponds to a binary submodular function. More formally, for any bundle $S \subseteq O$ and agent $i \in N$, let $\beta_i^\tau(S)$ denote the number of values in the sorted telescoping sum vector $v_i(S)$ greater than or equal to τ. We show that the function β_i^τ is a binary submodular function.

Lemma 3.1. *For any $i \in N$, the function β_i^τ is a binary submodular function.*

This lemma is particularly useful: if any allocation is clean with respect to the valuations $\{\beta_h^\tau\}_{h\in N}$, we can use path augmentations to modify the allocation. In our analysis, we will extensively use the valuations $\{\beta_h^0\}_{h\in N}$ and $\{\beta_h^c\}_{h\in N}$ corresponding to the cases where $\tau = 0$ and $\tau = c$ respectively.

4   Understanding $\{-1, 0, c\}$-ONSUB Valuations

We turn our attention to the specific set of valuations assumed in this paper — $\{-1, 0, c\}$-ONSUB valuations for some positive integer c. We establish a few important technical lemmas for fair allocation instances when all agents have $\{-1, 0, c\}$-ONSUB valuations.

We first show that any allocation X can be decomposed into *three* allocations X^{-1}, X^0 and X^c such that the items with marginal value c are in X^c, items with marginal value 0 are in X^0 and items with marginal value -1 are in X^{-1}.

Lemma 4.1. *When agents have* $\{-1, 0, c\}$*-ONSUB valuations, for any allocation* X*, there exist three allocations* X^{-1}*,* X^0*, and* X^c *such that for each agent* $i \in N$ *(a)* $X_i^{-1} \cup X_i^0 \cup X_i^c = X_i$*, (b)* X_i^{-1}*,* X_i^0*, and* X_i^c *are pairwise disjoint, (c)* $v_i(X_i^c \cup X_i^0) = v_i(X_i^c) = c|X_i^c|$*, and (d)* $v_i(X_i) = c|X_i^c| - |X_i^{-1}|$*.*

We use $Y = Y^c \cup Y^0 \cup Y^{-1}$ to denote the decomposition of any allocation Y into three allocations satisfying the conditions of Lemma 4.1. More generally, given any two allocations X and Y, we refer to the allocation $X \cup Y$ as the allocation where each agent $i \in N$ receives the bundle $X_i \cup Y_i$.

Example 4.2. Consider an instance with two agents $\{1, 2\}$ and four items $\{o_1, o_2, o_3, o_4\}$. Agent valuations are defined as follows:

$$v_1(S) = c \min\{|S \cap \{o_1, o_2, \}|, 1\}, \qquad v_2(S) = -|S \cap \{o_3, o_4\}|.$$

Consider an allocation X where $X_1 = \{o_1, o_2\}$ and $X_2 = \{o_3, o_4\}$. The following is a valid decomposition of X:

$$
\begin{array}{lll}
X_0^c = \{o_2, o_3, o_4\} & X_0^0 = \{o_1, o_3, o_4\} & X_0^{-1} = \{o_1, o_2\} \\
X_1^c = \{o_1\} & X_1^0 = \{o_2\} & X_1^{-1} = \emptyset \\
X_2^c = \emptyset & X_2^0 = \emptyset & X_1^{-1} = \{o_3, o_4\}
\end{array}
$$

There may be other decompositions as well. Specifically, if we swap o_1 and o_2 in the above decomposition, the new set of allocations still corresponds to a valid decomposition.

Note that while order-neutral submodular valuations can have multiple decompositions, $\{-1, 0, c\}$-ADD valuations have a unique decomposition — for any allocation $X = X^c \cup X^0 \cup X^{-1}$ and agent i, X_i^c consists of all items in X_i that agent i values at c, X_i^0 consists of the items that i values at 0 and X_i^{-1} consists of the items that i values at -1.

4.1 Weighted Exchange Graphs

Given an allocation $X = X^c \cup X^0 \cup X^{-1}$, the path augmentation technique introduced in Sect. 2 can be used to manipulate X^c (using the exchange graph $\mathcal{G}(X^c, \beta^c)$) or $X^c \cup X^0$ (using the exchange graph $\mathcal{G}(X^c \cup X^0, \beta^0)$). However, when we use path augmentation with the exchange graph $\mathcal{G}(X^c, \beta^c)$ to manipulate the allocation X^c, we may affect the cleanness of $X^c \cup X^0$ (w.r.t. the valuations $\{\beta_h^0\}_{h \in N}$). To see why, consider the following simple example.

Example 4.3. Consider an example with one agent $\{1\}$ and two items $\{o_1, o_2\}$. The agent's valuation function is defined as follows:

$$v_1(S) = c \min\{|S|, 1\} - \max\{|S| - 1, 0\}.$$

In simple words, the first item in the bundle gives agent 1 a value of c but the second item gives agent 1 a marginal value of -1. Consider the allocations X^c and X^0, where $X_1^c = \emptyset$ and $X_1^0 = \{o_1\}$.

Note that X^c is clean with respect to $\{\beta_h^c\}_{h \in N}$ and $X^c \cup X^0$ is clean with respect to $\{\beta_h^0\}_{h \in N}$. The singleton path (o_2) is one of the shortest paths from $F_{\beta_i^c}(X^c)$ to X_0^c in the exchange graph $\mathcal{G}(X^c, \beta^c)$. Augmenting along this path creates an allocation Y where $Y_1^c = \{o_2\}$ and $X_1^0 = \{o_1\}$.

Validating the correctness of Lemma 2.10, Y^c is indeed clean with respect to $\{\beta_h^c\}_{h \in N}$. However, $Y^c \cup X^0$ is not clean with respect to $\{\beta_h^0\}_{h \in N}$. Note that if we instead chose to augment along the singleton path (o_1) instead of (o_2), we would have not faced this issue.

In the above example, note that X^c and X^0 do not form a valid decomposition of X. This is deliberate; in our algorithm design, we will not assume that X^c, X^0 and X^{-1} form a valid decomposition of X. We will only assume that X^c is clean with respect to $\{\beta_h^c\}_{h \in N}$ and $X^c \cup X^0$ is clean with respect to $\{\beta_h^0\}_{h \in N}$. Our goal is to use path augmentations to modify X^c while retaining both these useful properties. To guarantee that $X^c \cup X^0$ remains clean (w.r.t. $\{\beta_h^0\}_{h \in N}$) even after path augmentation, we present a technique to carefully choose paths in the exchange graph. On a high level, this is done by giving weights to the edges in the exchange graph and choosing the least-weight path as opposed to a shortest path. The way we weigh edges is motivated by Example 4.3 — for all $i \in N$, we give edges from X_i^c to X_i^0 a lower weight than other edges. It turns out that this simple change is sufficient to ensure the cleanness of $X^c \cup X^0$ is clean with respect to $\{\beta_h^0\}_{h \in N}$.

More formally, we define the weighted exchange graph $\mathcal{G}^w(X^c, X^0, \beta^c)$ as a weighted directed graph with the same nodes and edges as $\mathcal{G}(X^c, \beta^c)$. Each edge has a specific weight defined as follows: all edges from $o \in X_i^c$ to $o' \in X_i^0$ for any $i \in N$ are given a weight of $\frac{1}{2}$; the remaining edges are given a weight of 1.

For any agent $i \in N$, we define two paths on this weighted exchange graph. A *Pareto-improving path* is a path from $F_{\beta_i^c}(X^c, i)$ for some $i \in N$ to some item in X_0^c. An *exchange path* is a path from $F_{\beta_i^c}(X^c, i)$ for some $i \in N$ to some item in X_j^c for some $j \in N - i$. We first show that path augmentation along the least-weight Pareto-improving path maintains the cleanness of $X^c \cup X^0$.

Theorem 4.4 (Pareto Improving Paths). *When agents have $\{-1, 0, c\}$-ONSUB valuations, let X^c be a clean allocation with respect to $\{\beta_h^c\}_{h \in N}$ and X^0 be an allocation such that $X^c \cup X^0$ is clean with respect to the valuations $\{\beta_h^0\}_{h \in N}$. Let $X_h^c \cap X_h^0 = \emptyset$ for all $h \in N$. For some agent $i \in N$, if a Pareto-improving path exists from $F_{\beta_i^c}(X^c, i)$ to X_0^c in the weighted exchange graph $\mathcal{G}^w(X^c, X^0, \beta^c)$, then, path augmentation along the least-weight Pareto-improving path $P = (o_1, \ldots, o_t)$ from $F_{\beta_i^c}(X^c, i)$ to X_0^c in the weighted exchange graph $\mathcal{G}^w(X^c, X^0, \beta^c)$ results in the following allocations Y^c and Y^0:*

$$Y_k^c = \begin{cases} X_k^c \wedge P & (k \in N + 0 - i) \\ X_i^c \wedge P + o_1 & (k = i) \end{cases}, \qquad Y_k^0 = \begin{cases} X_k^0 - o_t & (k \in N) \\ X_0^0 + o_t & (k = 0) \end{cases}.$$

Y^c is clean with respect to with respect to $\{\beta_h^c\}_{h \in N}$ and $Y^c \cup Y^0$ is clean with respect to $\{\beta_h^0\}_{h \in N}$. Furthermore, for all $h \in N$, $Y_h^c \cap Y_h^0 = \emptyset$.

In the above Theorem, we also modify Y^0 as part of the path augmentation operation to ensure o_t is not present in two different bundles. We also show that when Pareto-improving paths do not exist, the least-weight exchange path maintains the cleanness of $X^c \cup X^0$.

Theorem 4.5 (Exchange Paths). *When agents have $\{-1, 0, c\}$-ONSUB valuations, let X^c be a clean allocation with respect to $\{\beta_h^c\}_{h \in N}$ and X^0 be an allocation such that $X^c \cup X^0$ is clean with respect to the valuations $\{\beta_h^0\}_{h \in N}$. Let $X_h^c \cap X_h^0 = \emptyset$ for all $h \in N$. If no Pareto-improving path exists from $F_{\beta_i^c}(X, i)$ to X_0^c in the weighted exchange graph $\mathcal{G}^w(X^c, X^0, \beta^c)$, then, path augmentation along the least-weight exchange path $P = (o_1, \ldots, o_t)$ from $F_{\beta_i^c}(X, i)$ to X_j^c (for any $j \in N - i$) in the weighted exchange graph $\mathcal{G}^w(X^c, X^0, \beta^c)$ results in the allocations Y^c and Y^0:*

$$Y_k^c = \begin{cases} X_k^c \wedge P & (k \in N + 0 - i) \\ X_i^c \wedge P + o_1 & (k = i) \end{cases}, \qquad Y_k^0 = X_k^0 \quad (k \in N).$$

Y^c is clean with respect to with respect to $\{\beta_h^c\}_{h \in N}$ and $Y^c \cup Y^0$ is clean with respect to with respect to $\{\beta_h^0\}_{h \in N}$. Furthermore, for all $h \in N$, $Y_h^c \cap Y_h^0 = \emptyset$.

Note that since $\mathcal{G}^w(X^c, X^0, \beta^c)$ and $\mathcal{G}(X^c, \beta^c)$ have the same set of edges, Lemma 2.11 applies to the weighted exchange graph as well.

Lemma 4.6. *When agents have $\{-1, 0, c\}$-ONSUB valuations, let X^c be a clean allocation with respect to $\{\beta_h^c\}_{h \in N}$ and X^0 be an allocation such that $X^c \cup X^0$ is clean with respect to the valuations $\{\beta_h^0\}_{h \in N}$. Let $X_h^c \cap X_h^0 = \emptyset$ for all $h \in N$. For any $i \in N$ and $j \in N+0$, there is a path from $F_{\beta_i^c}(X^c, i)$ to X_j^c in $\mathcal{G}^w(X^c, X^0, \beta^c)$ if and only if there is a path from $F_{\beta_i^c}(X^c, i)$ to X_j^c in $\mathcal{G}(X^c, \beta^c)$.*

Given an allocation $X = X^c \cup X^0 \cup X^{-1}$, these results already hint at a method to modify X^c such that it becomes a leximin allocation with respect to the valuations $\{\beta_h^c\}_{h \in N}$: greedily use path augmentations in the weighted exchange graph until X^c is leximin. This is the high-level approach we use in our algorithm.

5 Leximin Allocations with $\{-1, 0, c\}$-ONSUB Valuations

We are ready to present our algorithm to compute leximin allocations. Our algorithm has three phases. In the first phase, we use Yankee Swap (Theorem 2.12) to compute a MAX-USW allocation X with respect to the valuations $\{\beta_i^0\}_{i \in N}$. We also initialize X^c and X^{-1} to be empty allocations.

In the second phase, we update X^c and X^0 using path augmentations (from Sect. 4.1) until X^c is a leximin partial allocation. This is done by greedily augmenting along min weight Pareto-improving paths and min weight exchange paths until the allocation is leximin.

In the third phase, we update X^{-1} by allocating the remaining items (in $X_0^c \cap X_0^0$). We do so greedily by allocating each item to the agent with the

---

ALGORITHM 1: Leximin Allocations with $\{-1, 0, c\}$-ONSUB Valuations

---

Input : A set of items O and a set of agents N with $\{-1, 0, c\}$-ONSUB
valuations $\{v_h\}_{h \in N}$

Output: A complete leximin allocation

// Phase 1: Make $X^c \cup X^0$ a MAX-USW allocation w.r.t. $\{\beta_h^0\}_{h \in N}$

1   $X^0 \leftarrow$ the output of Yankee Swap with respect to $\{\beta_h^0\}_{h \in N}$

2   $X^c = (X_0^c, \ldots, X_n^c) \leftarrow (O, \emptyset, \ldots, \emptyset)$

              // X^c is clean w.r.t. β^c and $X^c \cup X^0$ is clean w.r.t. β^0

3   $X^{-1} = (X_0^{-1}, \ldots, X_n^{-1}) \leftarrow (O, \emptyset, \ldots, \emptyset)$

     // Phase 2: Make X^c a clean leximin allocation w.r.t. $\{\beta_h^c\}_{h \in N}$

4   **repeat**

5 | **while** *for some* $i \in N$, *there exists a Pareto-improving path from* $F_{\beta_i^c}(X^c, i)$ *in*
       $\mathcal{G}^w(X^c, X^0, \beta^c)$ **do**

6 | | $P = (o_1', \ldots, o_t') \leftarrow$ a min weight Pareto-improving path from $F_{\beta_i^c}(X^c, i)$ to
         X_0^c in $\mathcal{G}^w(X^c, X^0, \beta^c)$

                  /* Augment the allocation with the path P */

7 | | $X_k^c \leftarrow X_k^c \Lambda P$ for all $k \in N + 0 - i$

8 | | $X_i^c \leftarrow X_i^c \Lambda P + o_1'$

9 | | $X_k^0 \leftarrow X_k^0 - o_t'$ for all $k \in N$

10 | | $X_0^0 \leftarrow X_0^0 + o_t'$

11 | **if** *for some* $i \in N$, *there exists an exchange path from* $F_{\beta_i^c}(X^c, i)$ *to some* X_j^c *such
       that either (a)* $|X_i^c| < |X_j^c| + 1$ *or (b)* $|X_i^c| = |X_j^c| + 1$ *and* $i < j$ **then**

12 | | $P = (o_1', \ldots, o_t') \leftarrow$ a min weight exchange path from $F_{\beta_i^c}(X^c, i)$ to X_j^c in
         $\mathcal{G}^w(X^c, X^0, \beta^c)$ /* Augment the allocation with the path P */

13 | | $X_k^c \leftarrow X_k^c \Lambda P$ for all $k \in N + 0 - i$

14 | | $X_i^c \leftarrow X_i^c \Lambda P + o_1'$

15   **while** *at least one path augmentation was done in the iteration*

     // Phase 3: Greedily allocate the items in $X_0^c \cap X_0^0$

16   **while** $|X_0^c \cap X_0^0 \cap X_0^{-1}| > 0$ **do**   /* Unallocated items exist */

17 | $S \leftarrow \underset{h \in N}{\arg\max}\, v_h(X_h^c \cup X_h^0 \cup X_h^{-1})$ /* Set of all max-utility agents */

18 | $i \leftarrow \underset{j \in S}{\max} j$ /* Break ties using index */

19 | $o \leftarrow$ an arbitrary item in $X_0^c \cap X_0^0 \cap X_0^{-1}$

20 | $X_i^{-1} \leftarrow X_i^{-1} + o$

21 | $X_0^{-1} \leftarrow X_0^{-1} - o$

22   **return** $X^c \cup X^0 \cup X^{-1}$

---

highest utility, under the assumption that these items have a marginal utility of
-1. The exact steps are described in Algorithm 1.

We analyze each phase separately and establish key properties that the allocations X^c, X^0 and X^{-1} have at the end of each phase. In order to show computational efficiency, we use the value oracle model where we have oracle access
to each agent's valuation function. A computationally efficient algorithm runs in
polynomial time (in n and m) and only uses a polynomial number of queries to
each value oracle.

5.1 Phase 1

The first phase is a simple setup where we allocate as many non-negative valued items as possible. X^0 is initialized as a clean MAX-USW allocation with respect to the valuations $\{\beta_h^0\}_{h\in N}$. X^c and X^{-1} are initialized as the empty allocation. Note that X^c is trivially clean with respect to the valuations $\{\beta_h^c\}_{h\in N}$.

Lemma 5.1. *At the end of Phase 1, X^c is clean with respect to $\{\beta_h^c\}_{h\in N}$ and $X^c \cup X^0$ is a MAX-USW clean allocation with respect to $\{\beta_h^0\}_{h\in N}$.*

The computational efficiency of Phase 1 relies on the computational efficiency of Yankee Swap. Yankee Swap uses a polynomial number of queries to $\{\beta_h^0\}_{h\in N}$. We can easily construct an efficient oracle for each β_h^0 to ensure Phase 1 runs in polynomial time and a polynomial number of valuation queries.

5.2 Phase 2

In this phase, we use path augmentations to manipulate X^c into a partial leximin allocation. There are three types of paths we check for and augment if it exists:

(a) A Pareto-improving path from $F_{\beta_i^c}(X^c, i)$ for some $i \in N$ to X_0^c in $\mathcal{G}^w(X^c, X^0, \beta^c)$.

(b) An exchange path from $F_{\beta_i^c}(X^c, i)$ for some $i \in N$ to some X_j^c such that $|X_i^c| < |X_j^c| + 1$.

(c) An exchange path from $F_{\beta_i^c}(X^c, i)$ for some $i \in N$ to some X_j^c such that $|X_i^c| = |X_j^c| + 1$ and $i < j$.

Note that these path augmentations work as intended since we maintain the invariant that $X^c \cup X^0$ is clean with respect to $\{\beta_h^0\}_{h\in N}$. Since exchange paths ((b) and (c)) are only guaranteed to work when there are no Pareto-improving paths, we ensure that we augment Pareto-improving paths (a) first before we check and augment exchange paths. This is clear in the steps of Algorithm 1.

Formally, let Z^c be a clean leximin allocation with respect to the valuations $\{\beta_h^c\}_{h\in N}$. If there are multiple, let Z^c be an allocation that is not lexicographically dominated (w.r.t. $\{\beta_h^c\}_{h\in N}$) by any other leximin allocation. Phase 2 ensures that for each agent $i \in N$, $|X_i^c| = |Z_i^c|$. We have the following Lemma.

Lemma 5.2. *Let X^c be a clean allocation with respect to $\{\beta_h^c\}_{h\in N}$ and X^0 be an allocation such that $X^c \cup X^0$ is clean with respect to $\{\beta_h^0\}_{h\in N}$. Let Z^c be a clean leximin allocation with respect to the valuations $\{\beta_h^c\}_{h\in N}$. If there are multiple, let Z^c be an allocation that is not lexicographically dominated (w.r.t. $\{\beta_h^c\}_{h\in N}$) by any other leximin allocation. Then there exists an agent $\ell \in N$ such that $|X_\ell^c| \neq |Z_\ell^c|$ if and only if at least one of the following conditions hold:*

(a) *There exists a Pareto-improving path from $F_{\beta_i^c}(X^c, i)$ for some $i \in N$ to X_0^c in $\mathcal{G}^w(X^c, X^0, \beta^c)$.*

(b) *There exists an exchange path from $F_{\beta_i^c}(X^c, i)$ for some $i \in N$ to some X_j^c such that $|X_i^c| < |X_j^c| + 1$.*

(c) *There exists an exchange path from $F_{\beta_i^c}(X^c, i)$ for some $i \in N$ to some X_j^c such that $|X_i^c| = |X_j^c| + 1$ and $i < j$.*

Lemma 5.2 immediately implies that *if* Phase 2 terminates, it terminates when $|X_h^c| = |Z_h^c|$ for all $h \in N$. Our next Lemma shows that Phase 2 indeed terminates in polynomial time. This result follows from bounding the number of path augmentations our algorithm computes.

Lemma 5.3. *Phase 2 terminates in polynomial time and polynomial valuation queries.*

Note that after each path transfer the number of allocated items in $X^c \cup X^0$ weakly increases. Therefore, $X^c \cup X^0$ remains a clean MAX-USW allocation at the end of Phase 2. Formally, we can make the following observation.

Lemma 5.4. *At the end of Phase 2, X^c is a clean lexicographically dominating leximin allocation with respect to the valuations $\{\beta_h^c\}_{h \in N}$ and $X^c \cup X^0$ is a MAX-USW clean allocation with respect to $\{\beta_h^0\}_{h \in N}$.*

5.3   Phase 3

At this point, $X^c \cup X^0$ is a MAX-USW clean allocation with respect to $\{\beta_h^0\}_{h \in N}$. Thus, all of the remaining items have a marginal value of -1 to every agent. It remains to assign these items in as equitable a manner as possible. In Phase 3, we sequentially allocate the remaining items, giving a "bad" item to the agent with the highest utility.

To carefully compare allocations, we adapt the comparison method of *domination* introduced in [12]. To compare two allocations $Y = Y^c \cup Y^0 \cup Y^{-1}$ and $\hat{Y} = \hat{Y}^c \cup \hat{Y}^0 \cup \hat{Y}^{-1}$, we first check the sorted utility vectors s^{Y^c} and $s^{\hat{Y}^c}$; if s^{Y^c} lexicographically dominates $s^{\hat{Y}^c}$, then we say Y dominates $\hat{Y}$. If s^{Y^c} and $s^{\hat{Y}^c}$ are equal, we check the utility vectors u^{Y^c} and $u^{\hat{Y}^c}$; if u^{Y^c} lexicographically dominates $u^{\hat{Y}^c}$, then we say Y dominates $\hat{Y}$. If u^{Y^c} and $u^{\hat{Y}^c}$ are equal as well, we check u^Y and $u^{\hat{Y}}$; if u^Y lexicographically dominates $u^{\hat{Y}}$, then we say Y dominates $\hat{Y}$.

Definition 5.5 (Domination). *An allocation $Y = Y^c \cup Y^0 \cup Y^{-1}$ dominates the allocation $\hat{Y} = \hat{Y}^c \cup \hat{Y}^0 \cup \hat{Y}^{-1}$ if any of the following conditions hold: (a) $s^{Y^c} \succ_{lex} s^{\hat{Y}^c}$, (b) $s^{Y^c} = s^{\hat{Y}^c}$ and $u^{Y^c} \succ_{lex} u^{\hat{Y}^c}$, or (c) $u^{Y^c} = u^{\hat{Y}^c}$ and $u^Y \succ_{lex} u^{\hat{Y}}$.*

Let $Y = Y^c \cup Y^0 \cup Y^{-1}$ be a *complete* leximin allocation for the original instance with valuations $\{v_h\}_{h \in N}$. If there are multiple leximin allocations, pick one which is not dominated by any other leximin allocation $\hat{Y}$. Due to this specific choice, we sometimes refer to Y as a dominating leximin allocation.

We first show that, just like X^c, Y^c is a lexicographically dominating leximin allocation with respect to the valuations $\{\beta_h^c\}_{h \in N}$.

Lemma 5.6. *After the end of Phase 2, $|X_h^c| = |Y_h^c|$ for all $h \in N$.*

Next, we show that X^c, X^0 and X^{-1} form a valid decomposition of $X^c \cup X^0 \cup X^{-1}$.

Lemma 5.7. *At every iteration in Phase 3, for any agent $i \in N$, $v_i(X_i) = c|X_i^c| - |X_i^{-1}|$.*

Combining these Lemmas, we can show the correctness of our algorithm.

Theorem 5.8. *When agents have $\{-1, 0, c\}$-ONSUB valuations, Algorithm 1 computes a leximin allocation efficiently.*

A useful corollary of this analysis is that Algorithm 1 outputs a MAX-USW allocation.

Corollary 5.9. *When agents have $\{-1, 0, c\}$-ONSUB valuations, Algorithm 1 outputs a MAX-USW allocation.*

6 Properties of the Leximin Allocation

Now that we have shown how a leximin allocation can be computed, we explore its connection to other fairness notions. Specifically, we study the following fairness notions.

Proportionality: An allocation X is said to be proportional if each agent receives at least an n-th fraction of their value for the entire set of items. This is not always possible — consider an instance with two agents and one high valued item. The fair allocation literature has therefore, instead, studied a relaxation of proportionality called proportionality up to one item [2]. An allocation X is *proportional up to one item* (PROP1) if any of the three following conditions hold for every agent $i \in N$: (a) $v_i(X_i) \geq \frac{1}{n}v_i(O)$, (b) $v_i(X_i + o) \geq \frac{1}{n}v_i(O)$ for some $o \in O \setminus X_i$, or(c) $v_i(X_i - o) \geq \frac{1}{n}v_i(O)$ for some $o \in X_i$.

Envy-Freeness: An allocation is *envy-free* if no agent prefers another agent's bundle to their own. This, again, is impossible to guarantee when all items are allocated. We therefore, instead, look at the notion of *envy-freeness up to one item* (EF1) [2,10,22]. When there are both goods and chores, an allocation X is EF1 if for any two agents i, j, there exists either some $o \in X_i$ such that $v_i(X_i - o) \geq v_i(X_j)$, or some $o \in X_j$ such that $v_i(X_i) \geq v_i(X_j - o)$.

Maximin Share: The *maximin share* (MMS) of an agent i is defined as the value they would obtain had they divided the items into n bundles themselves and picked the worst of these bundles. More formally,

$$\text{MMS}_i = \max_{X=(X_1, X_2, \ldots, X_n)} \min_{j \in [n]} v_i(X_j).$$

[23] show that agents cannot always be guaranteed their maximin share; past works [21] instead focus on guaranteeing that every agent receives a fraction of

their maximin share. For some $\varepsilon \in (0,1]$, an allocation X is ε-MMS if for every agent $i \in N$, $v_i(X_i) \geq \varepsilon \cdot \text{MMS}_i$.

Lorenz Dominance: An allocation X is *Lorenz dominating* [4] if for all other allocations Y and all $k \in [n]$, it holds that $\sum_{j \in [k]} s_j^X \geq \sum_{j \in [k]} s_j^Y$ where s^X is the sorted utility vector of X (defined in Sect. 2.2).

Our main result is as follows,

Theorem 6.1. *When agents have $\{-1, 0, c\}$-ONSUB valuations, leximin allocations are guaranteed to be PROP1 and Lorenz dominating. Additionally, when agents have $\{-1, 0, c\}$-ADD valuations, leximin allocations are guaranteed to be EF1 and 1-MMS.*

7   NP-Hardness When c Is Not an Integer

In this section, we show that the problem of computing leximin allocations is NP-hard even for $\{-p, q\}$-ADD valuations for any co-prime integers p and q such that $p \geq 3$. This proof is very similar to the hardness result in [1]. Note that the assumption that p and q are co-prime is necessary since if p divides q, the problem reduces to computing a leximin allocation for agents with $\{-1, \frac{q}{p}\}$-ADD valuations and admits a polynomial time algorithm (Theorem 5.8). More generally, any common divisor of p and q can be eliminated by scaling agent valuations.

Theorem 7.1. *The problem of computing leximin allocations is NP-hard even when agents have $\{-p, q\}$-ADD valuations for any co-prime positive integers p and q such that $p \geq 3$.*

While Theorem 7.1 shows that the problem of computing leximin allocations is NP-hard for most values of p and q, there are two special cases which still remain unresloved — $\{-c, 0, 1\}$-ONSUB valuations and $\{-2, 0, c\}$-ONSUB valuations. We leave these two cases for future work.

8   Conclusions and Future Work

In this work, we study the computation of leximin allocations in instances with mixed goods and chores. We show that when agents have $\{-1, 0, c\}$-ONSUB valuations, leximin allocations can be computed efficiently. We also show that these allocations are Lorenz dominating and approximately proportional.

On a higher level, our work is the first to generalize the path augmentation technique to tri-valued valuation functions. We are hopeful that the tools of weighted exchange graphs and decompositions can be applied to even more general valuation classes. We are also excited by the class of order-neutral submodular valuations. Much like Rado and OXS valuations, we believe order-neutral submodular valuations are an appealing sub-class of submodular valuation functions that warrant further study.

Acknowledgements. We thank anonymous WINE 2023 reviewers and Justin Payan for their useful comments. Viswanathan and Zick are funded by the National Science Foundation grant IIS-2327057.

References

1. Akrami, H., et al.: Maximizing Nash social welfare in 2-value instances. In: Proceedings of the 36th AAAI Conference on Artificial Intelligence (AAAI) (2022)
2. Aziz, H., Caragiannis, I., Igarashi, A., Walsh, T.: Fair allocation of indivisible goods and chores. Auton. Agent. Multi-Agent Syst. **36**, 1–21 (2022)
3. Aziz, H., Lindsay, J., Ritossa, A., Suzuki, M.: Fair allocation of two types of chores. In: Proceedings of the 22nd International Conference on Autonomous Agents and Multi-Agent Systems (AAMAS), pp. 143–151 (2023)
4. Babaioff, M., Ezra, T., Feige, U.: Fair and truthful mechanisms for dichotomous valuations. In: Proceedings of the 35th AAAI Conference on Artificial Intelligence (AAAI), pp. 5119–5126 (2021)
5. Barman, S., Narayan, V., Verma, P.: Fair chore division under binary supermodular costs. In: Proceedings of the 22nd International Conference on Autonomous Agents and Multi-Agent Systems (AAMAS) (2023)
6. Barman, S., Verma, P.: Existence and computation of maximin fair allocations under matroid-rank valuations. In: Proceedings of the 20th International Conference on Autonomous Agents and Multi-Agent Systems (AAMAS), pp. 169–177 (2021)
7. Barman, S., Verma, P.: Truthful and fair mechanisms for matroid-rank valuations. In: Proceedings of the 36th AAAI Conference on Artificial Intelligence (AAAI) (2022)
8. Benabbou, N., Chakraborty, M., Igarashi, A., Zick, Y.: Finding fair and efficient allocations for matroid rank valuations. ACM Trans. Econ. Comput. **9**(4) (2021)
9. Bhaskar, U., Sricharan, A.R., Vaish, R.: On approximate envy-freeness for indivisible chores and mixed resources. In: Proceedings of the Approximation, Randomization, and Combinatorial Optimization. Algorithms and Techniques, APPROX/RANDOM 2021, pp. 1–23 (2021)
10. Budish, E.: The combinatorial assignment problem: approximate competitive equilibrium from equal incomes. J. Polit. Econ. **119**(6), 1061–1103 (2011)
11. Bérczi, K., et al.: Envy-free relaxations for goods, chores, and mixed items (2020)
12. Cousins, C., Viswanathan, V., Zick, Y.: Dividing good and better items among agents with submodular valuations (WINE) (2023). https://doi.org/10.48550/ARXIV.2302.03087, https://arxiv.org/abs/2302.03087
13. Ebadian, S., Peters, D., Shah, N.: How to fairly allocate easy and difficult chores. In: Proceedings of the 21st International Conference on Autonomous Agents and Multi-Agent Systems (AAMAS), pp. 372–380 (2022)
14. Garg, J., Murhekar, A.: Computing fair and efficient allocations with few utility values. In: Caragiannis, I., Hansen, K.A. (eds.) SAGT 2021. LNCS, vol. 12885, pp. 345–359. Springer, Cham (2021). https://doi.org/10.1007/978-3-030-85947-3_23
15. Garg, J., Murhekar, A., Qin, J.: Fair and efficient allocations of chores under bivalued preferences. In: Proceedings of the 36th AAAI Conference on Artificial Intelligence (AAAI) (2022)
16. Halpern, D., Procaccia, A.D., Psomas, A., Shah, N.: Fair division with binary valuations: one rule to rule them all. In: Proceedings of the 16th Conference on Web and Internet Economics (WINE), pp. 370–383 (2020)
17. Hosseini, H., Sikdar, S., Vaish, R., Xia, L.: Fairly dividing mixtures of goods and chores under lexicographic preferences. In: Proceedings of the 22nd International Conference on Autonomous Agents and Multi-Agent Systems (AAMAS) (2023)

18. Kulkarni, R., Mehta, R., Taki, S.: Approximating maximin shares with mixed manna. In: Proceedings of the 21st ACM Conference on Economics and Computation (EC) (2021)
19. Kulkarni, R., Mehta, R., Taki, S.: On the PTAs for maximin shares in an indivisible mixed manna. In: Proceedings of the 35th AAAI Conference on Artificial Intelligence (AAAI), pp. 5523–5530 (2021)
20. Kurokawa, D., Procaccia, A.D., Shah, N.: Leximin allocations in the real world. ACM Trans. Econ. Comput. **6**(34), 1–24 (2018)
21. Kurokawa, D., Procaccia, A.D., Wang, J.: Fair enough: guaranteeing approximate maximin shares. J. ACM **65**(2), 1–27 (2018)
22. Lipton, R.J., Markakis, E., Mossel, E., Saberi, A.: On approximately fair allocations of indivisible goods. In: Proceedings of the 5th ACM Conference on Economics and Computation (EC), pp. 125–131 (2004)
23. Procaccia, A.D., Wang, J.: Fair enough: guaranteeing approximate maximin shares. In: Proceedings of the 15th ACM Conference on Economics and Computation (EC), pp. 675–692 (2014)
24. Cook, W.J., Cunningham, W.H., Pulleyblank, W.R., Schrijver, A.: Combinatorial Optimization. Wiley, Inc. USA (1998). ISBN 047155894X
25. Viswanathan, V., Zick, Y.: A general framework for fair allocation under matroid rank valuations. In: Proceedings of the 24th ACM Conference on Economics and Computation (EC) (2023)
26. Viswanathan, V., Zick, Y.: Weighted notions of fairness with binary supermodular chores. arXiv preprint arXiv:2303.06212 (2023)
27. Viswanathan, V., Zick, Y.: Yankee swap: a fast and simple fair allocation mechanism for matroid rank valuations. In: Proceedings of the 22nd International Conference on Autonomous Agents and Multi-Agent Systems (AAMAS) (2023)

Dividing Good and Great Items Among Agents with Bivalued Submodular Valuations

Cyrus Cousins[ID], Vignesh Viswanathan[(✉)][ID], and Yair Zick[ID]

University of Massachusetts, Amherst, MA 01002, USA
{cbcousins,vviswanathan,yzick}@umass.edu

Abstract. We study the problem of fairly allocating a set of indivisible goods among agents with *bivalued submodular valuations*—each good provides a marginal gain of either a or b ($a < b$) and goods have decreasing marginal gains. This is a natural generalization of two well-studied valuation classes—bivalued additive valuations and binary submodular valuations. We present a simple sequential algorithmic framework, based on the recently introduced Yankee Swap mechanism, that can be adapted to compute a variety of solution concepts, including max Nash welfare (MNW), leximin and p-mean welfare maximizing allocations when a divides b. This result is complemented by an existing result on the computational intractability of MNW and leximin allocations when a does not divide b. We show that MNW and leximin allocations guarantee each agent at least $\frac{2}{5}$ and $\frac{a}{b+2a}$ of their maximin share, respectively, when a divides b. We also show that neither the leximin nor the MNW allocation is guaranteed to be envy free up to one good (EF1). This is surprising since for the simpler classes of bivalued additive valuations and binary submodular valuations, MNW allocations are known to be envy free up to *any* good (EFX).

Keywords: Fair Allocation · Indivisible Goods · Submodular Valuations

1 Introduction

Fair allocation of indivisible goods has gained significant attention in recent years. The problem is simple: we need to assign a set of indivisible *goods* to a set of *agents*. Each agent has a subjective preference over the bundle of goods they receive. Our objective is to find an allocation that satisfies certain *fairness* and *efficiency* (or more generally *justice*) criteria. For example, some allocations maximize the product of agents' utilities, whereas others guarantee that no agent prefers another agent's bundle to their own. The fair allocation literature focuses on the existence and computation of allocations that satisfy a set of *justice criteria* (see [4] for a recent survey). For example, [26] focuses on computing *envy free up to any good* (EFX) allocations while [12] focuses on computing

J. Garg et al. (Eds.): WINE 2023, LNCS 14413, pp. 225–241, 2024.
https://doi.org/10.1007/978-3-031-48974-7_13

Table 1. The computational complexity of computing various justice criteria under various valuation classes. a, b and c are positive integers. NPh is short for NP-hard. SUB is short for submodular and ADD is short for additive.

Justice Criterion	{0,1}-SUB	{1,c}-ADD	{1,c}-SUB	{a,b}-ADD
Nash Welfare	**P** [5]	**P** [1]	**P** (Theorem 2)	**NPh** [1]
Leximin	**P** [5]	**P** (Theorem 3)	**P** (Theorem 3)	**NPh** [1]
p-mean Welfare	**P** [5]	**P** (Theorem 4)	**P** (Theorem 4)	**NPh** [1]

max Nash welfare allocations. Indeed, as these papers show, computing "fair" allocations without any constraint on agent valuations is an intractable problem for most justice criteria [12, 23, 26].

This has led to a recent systematic attempt to "push the computational envelope"—identify simpler classes of agent valuations for which there exist efficiently computable allocations satisfying multiple fairness and efficiency criteria. Several positive results are known when agents have *binary submodular* valuations; that is, their valuations exhibit diminishing returns to scale, and the added benefit of any good is either 0 or 1 [5, 7, 8, 10, 28, 29]. Other works offer positive results when agents have *bivalued* additive preferences, i.e., each good has a value of either a or b, and agents' values for a bundle of goods is the sum of their utility for individual goods [1, 3, 18, 19]. These results are encouraging in the face of known intractability barriers. Moreover, they offer practical benefits: binary submodular valuations naturally arise in settings such as course allocation [29], shift allocation [5] and in public housing assignment [9]. We take this line of work a step further and study a generalization of both bivalued additive valuations and binary submodular valuations.

1.1   Our Contributions

When agents have binary submodular valuations, *leximin* allocations are known to satisfy several desirable criteria. An allocation is leximin (or more precisely, leximin dominant) if it maximizes the welfare of the worst-off agent, then subject to that the welfare of the second worst-off agent and so on. When agents have binary submodular valuations, a leximin allocation maximizes utilitarian and Nash social welfare, is envy-free up to any good (EFX), and offers each agent at least $\frac{1}{2}$ of their maximin share [5]. Furthermore, there exists a simple sequential allocation mechanism that computes a leximin allocation [29].

We consider settings where agents have *bivalued submodular* valuations. That is, the marginal contribution of any good is either a or b, and marginal gains decrease as agents gain more goods. We assume that a divides b, or w.l.o.g. (by rescaling) that $a = 1$ and b is a positive integer. When agents have bivalued submodular valuations, unlike binary submodular valuations, leximin allocations no longer satisfy multiple fairness and efficiency guarantees (see Example 2). Thus, different justice criteria cannot be satisfied by a single allocation. To address this, we present a general, yet surprisingly simple algorithm (called Bivalued

Yankee Swap) that efficiently computes allocations for a broad number of justice criteria. The algorithm (Algorithm 1) is based on the Yankee Swap protocol proposed by [29]: we start with all goods unassigned. At every round we select an agent (based on a selection criterion ϕ). If the selected agent can pick an unassigned good that offers them a high marginal benefit, they do so and we move on. Otherwise, we check whether they can *steal* a high-value good from another agent. We allow an agent to steal a high-value good if the agent who they want to steal from can recover their utility by either taking an unassigned good or stealing a good from another agent. This results in a *transfer path* where the initiating agent increases their utility by b, every other agent retains the same utility, and one good is removed from the pile of unassigned goods. If no such path exists, the agent is no longer allowed to "play" for high-value goods, and can either take low-value goods or none at all.

By simply modifying the agent selection criterion ϕ, this protocol outputs allocations that satisfy a number of "acceptable" justice criteria. We think of justice criteria as ways of comparing allocations; for example, an allocation X is better than an allocation Y according to the Nash-Welfare criterion if it either has fewer agents with zero utilities, or if the product of agent utilities under X is greater than the product of agent utilities under Y. A justice criterion Ψ is acceptable if (informally), it satisfies a notion of Pareto dominance, and if it admits a selection criterion (also referred to as a *gain function*) ϕ that decides which agent should receive a good in a manner consistent with Ψ (see Sect. 4.1 for details).

Several well-known justice criteria are acceptable (see Table 1). For every acceptable justice criterion, our result immediately implies a simple algorithmic framework that computes an allocation maximizing that justice criterion: one needs to simply implement Algorithm 1 with the appropriate gain function ϕ.

To complement our algorithmic results, we further analyze Nash welfare maximizing and leximin allocations. We show that neither are guaranteed to even be envy free up to one good (EF1). This result shows that bivalued submodular valuations signify a departure from both binary submodular and bivalued additive valuations, where the max Nash welfare allocation is guaranteed to be envy free up to any good (EFX). While envy-freeness is not guaranteed under bivalued submodular valuations, we do show that max Nash welfare and leximin allocations offer approximate MMS (maximin share) guarantees to agents. Specifically, we show that max Nash welfare allocations and leximin allocations are $\frac{2}{5}$-MMS and $\frac{a}{b+2a}$-MMS respectively.

1.2 Related Work

Our work is closely related to works on fair allocation under matroid rank (binary submodular) valuations. The problem of fair allocation under matroid rank valuations is reasonably well studied and has seen a surprising number of positive results. [10] shows that utilitarian welfare maximizing envy free up to one good (EF1) allocations always exist and can be computed efficiently. [7] shows that an MMS allocation is guaranteed to exist and can be computed efficiently. [5] shows that a Lorenz dominating allocation (which is both leximin and maximizes Nash welfare) is guaranteed to exist and can be computed efficiently. More recently, [28] presents a general framework (called General Yankee Swap) that can be used to compute weighted notions of fairness such as weighted max Nash welfare and weighted leximin efficiently, in addition to several others. Almost all of the papers in this field use path transfers in their algorithms [5,7,28,29]; a technique which we exploit as well. Our algorithm (called Bivalued Yankee Swap) is closely related to the General Yankee Swap framework [28]. However, due to the complexity of bivalued submodular valuations, the analysis in our setting is more involved.

Our paper also builds on results for fair allocation under bivalued additive valuations. [3] shows that the max Nash welfare allocation is envy free up to any good (EFX) when agents have bivalued additive valuations. [19] showed that an EFX and Pareto optimal allocation can be computed efficiently under bivalued preferences. [1] presents an algorithm to compute a max Nash welfare allocation efficiently when a divides b. This work is arguably the closest to ours. While their algorithm is different, the essential technical ingredients are the same; their algorithm uses transfer paths and decompositions as well. However, their analysis is restricted to bivalued additive valuations and the max Nash welfare allocation. Our results generalize their work both in terms of the class of valuations and the justice criteria considered. In a recent follow up work, [2] presents a polynomial time algorithm to compute a max Nash welfare allocation for the case where a divides $2b$. This case turns out to be surprisingly more complicated than the case where a divides b, requiring a different set of techniques to manipulate the high and low valued goods.

Bivalued additive valuations have also been studied in the realm of chores. [20] and [18] present efficient algorithms to compute an EF1 and Pareto optimal allocation when agents have bivalued preferences.

2   Preliminaries

We use $[t]$ to denote the set $\{1, 2, \ldots, t\}$. For ease of readability, we replace $A \cup \{g\}$ and $A \setminus \{g\}$ with $A + g$ and $A - g$ respectively.

We have a set of n *agents* $N = [n]$ and m *goods* $G = \{g_1, g_2, \ldots, g_m\}$. Each agent $i \in N$ has a *valuation function* $v_i : 2^G \mapsto \mathbb{R}_{\geq 0}$; $v_i(S)$ specifies agent i's value for the bundle S. Given a valuation function v, we let $\Delta_v(S, g) = v(S + g) - v(S)$ denote the marginal utility of the good g to the bundle S under v. When clear from context, we write $\Delta_i(S, g)$ instead of $\Delta_{v_i}(S, g)$ to denote the marginal utility of giving the good g to agent i given that they have already been assigned the bundle S.

Given $a, b \in \mathbb{R}_{\geq 0}$ such that $a < b$, we say that v_i is an (a, b)-*bivalued submodular valuation* if v_i satisfies the following three properties: (a) $v_i(\emptyset) = 0$, (b) $\Delta_i(S, g) \in \{a, b\}$ for all $S \subset G$ and $g \in G \setminus S$, and (c) $\Delta_i(S, g) \geq \Delta_i(T, g)$ for all $S \subseteq T \subset G$ and $g \in G \setminus T$. We use $\{a, b\}$-SUB to denote (a, b)-bivalued submodular valuation functions.

A lot of our analysis uses the class of $\{0,1\}$-SUB valuations. $\{0,1\}$-SUB valuations are also known as *binary submodular valuations*, and have been extensively studied (see Sect. 1.2 for a discussion). When $a = 0$, the valuation function v_i is essentially a scaled binary submodular valuation; specifically, $\frac{1}{b}v_i(\cdot)$ is a $\{0,1\}$-SUB valuation. Existing results under $\{0,1\}$-SUB valuations trivially extend to $\{0,b\}$-SUB valuations as well, and generally offer stronger guarantees than the ones offered in this work. Thus, we focus our attention on the case where $b > a > 0$.

For ease of readability, we scale valuations by a. Under this scaling, all agents have $\{1, \frac{b}{a}\}$-SUB valuations. We further simplify notation and replace the value $\frac{b}{a}$ by c; under this notation, all agents have $\{1, c\}$-SUB valuations where $c > 1$. The value of c is the same for all agents. Note that when a divides b, c is a *natural number*.

An *allocation* $X = (X_0, X_1, X_2, \ldots, X_n)$ is a partition of G into $n+1$ bundles. Each X_i denotes the bundle allocated to agent i; X_0 denotes the unallocated goods. Our goal is to compute *complete* allocations—allocations where $X_0 = \emptyset$. We refer to the value $v_i(X_i)$ as the *utility* of agent i under the allocation X and we define the *utility vector* of an allocation X (denoted by $\boldsymbol{u}^X$) as the vector $(v_1(X_1), v_2(X_2), \ldots, v_n(X_n))$.

For ease of analysis, we sometimes refer to 0 as a dummy agent with valuation function $v_0(S) = c|S|$ and allocated bundle X_0. This is trivially a $\{1, c\}$-SUB function. None of our justice criteria consider agent 0. Our analysis will also use the following definition of lexicographic dominance.

Definition 1 (Lexicographic Dominance). *A vector $\boldsymbol{y} \in \mathbb{R}^n_{\geq 0}$ is said to lexicographically dominate another vector $\boldsymbol{z} \in \mathbb{R}^n_{\geq 0}$ if there exists some $k \in [n]$ such that for all $j \in [k-1]$, $y_j = z_j$ and $y_k > z_k$. This is denoted by $\boldsymbol{y} \succ_{lex} \boldsymbol{z}$. An allocation X is said to lexicographically dominate another allocation Y if $\boldsymbol{u}^X \succ_{lex} \boldsymbol{u}^Y$.*

2.1 Justice Criteria

We consider three central justice criteria.

Leximin: An allocation X is *leximin* if it maximizes the least utility in the allocation and subject to that, maximizes the second lowest utility and so on. This can be formalized using the *sorted utility vector*. The sorted utility vector of an allocation X (denoted by $\boldsymbol{s}^X$) is defined as the utility vector $(v_1(X_1), v_2(X_2), \ldots, v_n(X_n))$ *sorted in ascending order*. An allocation X is leximin if, for no other allocation Y, we have $\boldsymbol{s}^Y \succ_{\text{lex}} \boldsymbol{s}^X$.

Max Nash Welfare (MNW): Let the set of agents who receive a positive utility under an allocation X be denoted by P_X. An allocation X maximizes Nash welfare if it first maximizes the number of agents who receive a positive utility $|P_X|$ and subject to that, maximizes the value $\prod_{i \in P_X} v_i(X_i)$ [12].

p-Mean Welfare: The p-mean welfare of an allocation X is defined as $(\frac{1}{n} \sum_{i \in N} v_i(X_i)^p)^{1/p}$ for $p \leq 1$. Since this value is undefined when $v_i(X_i) = 0$

for any $i \in N$ and $p < 0$, we modify the definition slightly when defining a max p-mean welfare allocation. We again denote P_X as the set of agents who receive a positive utility under the allocation X. An allocation X is said to be a max p-mean welfare allocation for any $p \leq 1$ if the allocation maximizes the size of P_X and subject to that, maximizes $\left(\frac{1}{n} \sum_{i \in P_X} v_i(X_i)^p\right)^{1/p}$.

The p-mean welfare functions have been extensively studied in economics [25], fair machine learning [14–17,22], and, more recently, fair allocation [6]. When p approaches $-\infty$, the p-mean welfare corresponds to the leximin objective and when p approaches 0, the p-mean welfare corresponds to the max Nash welfare objective.

Leximin dominance, Nash welfare and p-mean welfare are three ways of comparing allocations. More generally, we can think of a *justice criterion* Ψ as a way of comparing two allocations X and Y. Notice that we are specifically comparing the utility vectors of allocations. Therefore, we say allocation X is better than Y according to Ψ if $\boldsymbol{u}^X \succ_\Psi \boldsymbol{u}^Y$. To ensure that there indeed exists a Ψ maximizing allocation, we require that $\succeq_\Psi$ be a total ordering over all allocations. An allocation X is maximal with respect to Ψ if it is not Ψ-dominated by any other allocation, i.e., for any other allocation Y, it is not the case that $\boldsymbol{u}^Y \succ_\Psi \boldsymbol{u}^X$. For ease of readability, we sometimes abuse notation and use $X \succ_\Psi Y$ to denote $\boldsymbol{u}^X \succ_\Psi \boldsymbol{u}^Y$.

3   Bivalued Submodular Valuations

We now present some useful properties of bivalued submodular valuations that we use in our algorithms. Informally, we show that the number of high valued goods in any agent's bundle corresponds to a binary submodular function. This allows us to use existing techniques from the fair allocation under binary submodular functions literature in our analysis.

More formally, for each agent $i \in N + 0$, we define a valuation function $\beta_i : 2^G \mapsto \mathbb{R}_{\geq 0}$ as follows: for any $S \subseteq G$, $\beta_i(S)$ is equal to the size of the largest subset T of S such that $v_i(T) = c|T|$; that is, $\beta_i(S) = \max\{|T| : T \subseteq S, v_i(T) = c|T|\}$. In other words, all the goods in T provide a marginal gain of c to i. $\beta_i(S)$ captures the number of goods in S that provide a value of c to i. We show that $\beta_i \in \{0, 1\}$-SUB.

Lemma 1. *For each agent $i \in N + 0$ when v_i is a $\{1, c\}$-SUB valuation, β_i is a binary submodular valuation.*

Using β_i, we can leverage the properties of binary submodular valuations used in the fair allocation literature. However, these properties require bundles to be *clean* as well, as given by the following definition.

Definition 2 (Clean). *For any agent $i \in N + 0$, a bundle S is clean with respect to the binary submodular valuation β_i if $\beta_i(S) = |S|$. By definition, this is equivalent to saying $v_i(S) = c|S|$. An allocation X is said to be clean if for all agents $i \in N + 0$, $\beta_i(X_i) = |X_i|$. By our definition of v_0, any bundle S is clean with respect to the dummy agent 0.*

This is similar to the notion of non-wastefulness used in [1] and cleanness used in [10].

When agents have binary submodular valuations, given a clean allocation X, we define the *exchange graph* $\mathcal{G}(X)$ as a directed graph over the set of goods. An edge exists from g to g' in the exchange graph if $g \in X_j$ and $\beta_j(X_j - g + g') = \beta_j(X_j)$ for some $j \in N+0$. Since our problem instances have *bivalued* submodular valuations, whenever we refer to the exchange graph of an allocation, we refer to it with respect to the *binary* submodular valuations $\{\beta_i\}_{i \in N}$.

Let $P = (g_1, g_2, \ldots, g_t)$ be a path in the exchange graph for a clean allocation X. We define a transfer of goods along the path P in the allocation X as the operation where g_t is given to the agent who has g_{t-1}, g_{t-1} is given to the agent who has g_{t-2} and so on till finally g_1 is discarded. We call this transfer *path augmentation*; the bundle X_i after path augmentation with the path P is denoted by $X_i \wedge P$ and defined as $X_i \wedge P = (X_i - g_t) \oplus \{g_j, g_{j+1} : g_j \in X_i\}$, where $\oplus$ denotes the symmetric set difference operation.

For any clean allocation X and agent i, we define $F_i(X) = \{g \in G : \Delta_{\beta_i}(X_i, g) = 1\}$ as the set of goods which give agent i a marginal gain of 1 under the valuation β_i, i.e., the set of all goods that give agent i a marginal gain of c under v_i. For any agent i, let $P = (g_1, \ldots, g_t)$ be the shortest path from $F_i(X)$ to X_j for some $j \neq i$. Then path augmentation with the path P and giving g_1 to i results in an allocation where i's value for their bundle goes up by c, j's value for their bundle goes down by c and all the other agents do not see any change in value. This is formalized below and exists in [7, Lemma 1] and [29, Lemma 3.7].

Lemma 2 ([7,29]). *Let X be a clean allocation with respect to the binary submodular valuations $\{\beta_i\}_{i \in N+0}$. Let $P = (g_1, \ldots, g_t)$ be the shortest path in the exchange graph $\mathcal{G}(X)$ from $F_i(X)$ to X_j for some $i \in N + 0$ and $j \in N + 0 - i$. Define the allocation Y as follows*

$$Y_k = \begin{cases} X_k \wedge P & k \neq i \\ X_i \wedge P + g_1 & k = i. \end{cases}$$

Then, we have for all $k \in N - i - j$, $\beta_k(Y_k) = \beta_k(X_k)$, $\beta_i(Y_i) = \beta_i(X_i) + 1$ and $\beta_j(Y_j) = \beta_j(X_j) - 1$. Furthermore, the new allocation Y is clean.

We now establish sufficient conditions for a path to exist. We say there is a path from some agent i to some agent j in an allocation X if there is a path from $F_i(X)$ to X_j in the exchange graph $\mathcal{G}(X)$. The following lemma appears in [29, Theorem 3.8].

Lemma 3 ([29]). *Let X and Y be two clean allocations with respect to the binary submodular valuations $\{\beta_i\}_{i \in N+0}$. For any agent i such that $|X_i| < |Y_i|$, there is a path in the exchange graph $\mathcal{G}(X)$ from $F_i(X)$ to some good in X_k for some $k \in N + 0$ such that $|X_k| > |Y_k|$.*

Unfortunately, unlike binary submodular valuations [10], we cannot make any allocation clean without causing a loss in utility to some agents. We can however, efficiently *decompose* any allocation into a clean allocation and a *supplementary* allocation. For any allocation X, the clean part is denoted using X^c (since each good provides a value of c in a clean allocation) and the supplementary allocation is denoted using X^1. The supplementary allocation is a tuple of n disjoint bundles of G, that is $X^1 = (X_1^1, X_2^1, \ldots, X_n^1)$.

The supplementary allocation is not technically an allocation since it is not an $n+1$ partition of G; we call it an allocation nevertheless to improve readability.

Example 1. Consider an example with two agents $\{1,2\}$ and four goods $\{g_1, \ldots, g_4\}$. Agent valuations are given as follows

$$v_1(S) = c|S| \qquad\qquad v_2(S) = c\min\{|S|, 1\} + \max\{|S| - 1, 0\}$$

Consider an allocation X where $X_1 = \{g_1, g_2\}$ and $X_2 = \{g_3, g_4\}$. Note that both goods in X_1 give agent 1 a value of c but only one good in X_2 gives agent 2 a value of c. A natural decomposition in this case would be

$$X_0^c = \{g_4\} \qquad\qquad X_1^c = \{g_1, g_2\} \qquad\qquad X_2^c = \{g_3\}$$
$$X_1^1 = \emptyset \qquad\qquad\qquad X_2^1 = \{g_4\}$$

X^c is the clean part and X^1 is the supplementary part. Note that, decompositions need not be unique. Swapping g_4 and g_3 in the above allocation would result in another decomposition.

Formally, X^c and X^1 is a decomposition of X if for all agents $i \in N$, we have (a) $X_i^1, X_i^c \subseteq X_i$, (b) $X_i^c \cap X_i^1 = \emptyset$, (c) $X_i^c \cup X_i^1 = X_i$, and (d) $|X_i^c| = \beta_i(X_i)$. We show that a decomposition always exists.

Lemma 4. *For any allocation X, there exists a decomposition into a clean allocation X^c and a supplementary allocation X^1 such that for all $i \in N$, we have (a) $X_i^1, X_i^c \subseteq X_i$, (b) $X_i^c \cap X_i^1 = \emptyset$, (c) $X_i^c \cup X_i^1 = X_i$, and (d) $|X_i^c| = \beta_i(X_i)$.*

Note that by design, X_0^c contains all the goods in X^1. This is done to ensure X^c is an $n+1$ partition of G. We do not need to do this for X^1.

We define the union of a clean allocation X^c and a supplementary allocation X^1 (denoted $X^c \cup X^1$) as follows: for each agent $i \in N$, $X_i = X_i^c \cup X_i^1$ and $X_0 = G \backslash \bigcup_{i \in N} X_i$. This definition holds for any pair of clean and supplementary allocations and not just decompositions via Lemma 4. It is easy to see that if an allocation X was decomposed into X^c and X^1 (satisfying the properties of Lemma 4), then $X = X^c \cup X^1$. We will refer to an allocation X as $X = X^c \cup X^1$ to denote a decomposition of the allocation (via Lemma 4) into a clean allocation X^c and a supplementary allocation X^1. As we saw in Example 1, decompositions need not be unique. When we use $X = X^c \cup X^1$, we will refer to any one of the possible decompositions. The exact one will not matter.

Finally, we present a useful metric to compare allocations using their decompositions. We refer to this metric as *domination*. To compare two allocations X and Y, we first compare the sorted utility vectors of X^c and Y^c. If the sorted utility vector X^c lexicographically dominates that of Y^c, then X dominates Y;

if the two allocations X^c and Y^c have the same sorted utility vectors, we compare their utility vectors. If X^c is lexicographically greater than Y^c, then X dominates Y. If X^c and Y^c have the same utility vectors, we compare X and Y lexicographically. This definition is formalized below. Recall that u^X and s^X denote the utility vector and sorted utility vector of the allocation X respectively.

Definition 3 (Domination). *We say an allocation* $X = X^c \cup X^1$ *dominates an allocation* $Y = Y^c \cup Y^1$ *if any of the following three conditions hold:*

(a) $s^{X^c} \succ_{lex} s^{Y^c}$
(b) $s^{X^c} = s^{Y^c}$ *and* $u^{X^c} \succ_{lex} u^{Y^c}$
(c) $u^{X^c} = u^{Y^c}$ *and* $u^X \succ_{lex} u^Y$

An allocation X *is a* dominating Ψ *maximizing allocation if no other* Ψ *maximizing allocation* Y *dominates* X.

4   Bivalued Yankee Swap

We now present *Bivalued Yankee Swap*—a flexible framework for fair allocation under $\{1, c\}$-SUB valuations. The results in this section assume that c is a natural number; in other words, we are interested in $\{a, b\}$-SUB valuations where a divides b.

In the original Yankee Swap algorithm [29], all goods start off initially unallocated. The algorithm proceeds in rounds; at each round, we select an agent based on some criteria. This agent can either take an unallocated good, or initiate a transfer path by stealing a good from another agent, who then steals a good from another agent and so on until some agent steals a good from the pool of unallocated goods. [29] shows that these transfer paths can be efficiently computed and are equivalent to paths on the exchange graph. If there is no transfer path from the agent to an unassigned good, the agent is removed from the game. We continue until all goods have been assigned. This algorithm can also be thought of as a variant of the classical matroid path augmentation algorithm where paths are chosen more carefully so as to maximize a specific justice criterion.

The Bivalued Yankee Swap algorithm is a modified version of this approach. We start by letting agents run Yankee Swap, but require that whenever an agent receives a good (by either taking an unassigned good or by stealing a good from another agent), that good must offer them a marginal gain of c. In addition, we require that every agent who had a good stolen from them fully recovers their utility, i.e., an agent who lost a good of marginal value c, must receive a good of marginal value c in exchange; an agent who lost a good of marginal value 1 must receive a good of marginal value 1 in exchange. Thus, whenever an agent initiates a transfer path, that path results in them receiving an additional utility of c, while all other agents' utilities remain the same. More formally, every agent starts in the set U. If the agent is able to find a path in the exchange graph to an unallocated good, we augment the allocation using this path. If the agent is unable find a path, we remove them from U. Once the agent is removed from

U, no transfer path exists that can give the agent a value of c. Therefore, if an agent outside of U is chosen, we *provisionally* give the agent an arbitrary unassigned good, offering them a marginal gain of 1. Since all unassigned goods offer a marginal gain of 1 for every agent not in U, the choice of which good to allocate to guarantee a value of 1 does not matter. Therefore, we treat this provisionally allocated good as an unallocated good as well; thereby allowing transfer paths to steal this good. If this provisionally allocated good gets stolen, we replace it with another low-value good. The algorithm stops when there are no unallocated goods left. These steps are described in Algorithm 1.

We use a *gain function* ϕ to pick the next agent to invoke a transfer path. Informally, ϕ computes the change in the justice criterion Ψ when a good is assigned to an agent. The gain function concept is used to compute allocations satisfying a diverse set of justice criteria in [28], when agents have binary submodular valuations.

More formally, ϕ maps a tuple $(\boldsymbol{u}^X, i, d)$ consisting of the utility vector of an allocation X, an agent i and a value $d \in \{1, c\}$ to a finite real value. The value $\phi(\boldsymbol{u}^X, i, d)$ quantifies the gain in the justice criterion Ψ of adding a value of d to the agent i given the allocation X. We abuse notation and use $\phi(X, i, d)$ to denote $\phi(\boldsymbol{u}^X, i, d)$.

At each round, we find the agent $i \in U$ who maximizes $\phi(X, i, c)$ (implicitly assuming that i can find a good with a marginal gain of c), and the agent $j \in N \setminus U$ who maximizes $\phi(X, j, 1)$ (implicitly assuming that all goods offer j a marginal gain of 1). We then choose the agent, among i and j, who maximizes ϕ; in the algorithm, these values are denoted by $\texttt{Gain}_c$ and $\texttt{Gain}_1$. Ties are broken first in favor of agents in U and second, in favor of agents with a lower index. Let us present an example of how Algorithm 1 works, with two candidate gain functions: Example 2 shows how a leximin allocation and an MNW allocation are computed for a simple two-agent instance.

Example 2. Consider a setting with six goods $G = \{g_1, \ldots, g_6\}$, and two additive agents 1 and 2. Agent 1's valuation function is $v_1(S) = |S|$, and Agent 2's valuation is $v_2(S) = 5|S|$. In other words, Agent 1 values every good at 1, whereas Agent 2 values every good at 5. We set the gain function to be $\phi_{\texttt{leximin}}(X, i, d) = -6 v_i(X_i) + d$, to compute a leximin allocation (as per Theorem 3), and run Algorithm 1. All agents are initially in U, and we compute $\texttt{Gain}_c, \texttt{Gain}_1$. In the first iteration, $\texttt{Gain}_1 = -\infty$ since all agents are in U and $\texttt{Gain}_c = -6 \times v_1(\emptyset) + 5 = 5$. Both agents 1 and 2 have equal ϕ values so we break ties and choose Agent 1. However, there is no way to give Agent 1 a good which gives them a marginal gain of 5. We therefore remove Agent 1 from U.

In the next iteration, $\texttt{Gain}_c = -6 \times v_2(\emptyset) + 5 = -6 \times 0 + 5 = 5$, and $\texttt{Gain}_1 = -6 \times v_1(\emptyset) + 1 = 1$. Since $\texttt{Gain}_c > \texttt{Gain}_1$, we choose Agent 2 to receive a good. Since they value all the goods at 5, we pick an arbitrary good (say g_1) and allocate it to them.

ALGORITHM 1: Bivalued Yankee Swap

Input : A set of goods G, a set of agents N with $\{1, c\}$-SUB valuations $\{v_h\}_{h \in N}$ and a gain function ϕ.

Output: A dominating Ψ maximizing allocation.

1 $X^c = (X_0^c, X_1^c, \ldots, X_n^c) \leftarrow (G, \emptyset, \ldots, \emptyset)$ `/* X^c has no goods allocated. */`
   `// Invariant: X_0^c stores the provisionally allocated goods as well as`
   `the unallocated goods.`

2 $X^1 = (X_1^1, \ldots, X_n^1) \leftarrow (\emptyset, \ldots, \emptyset)$ `/* Stores the provisional allocation. */`

3 $U \leftarrow N$ `/* Set of agents still in play for c valued goods. */`

4 **while** $\sum_{h \in N} |X_h^1| < |X_0^c|$ **do**   `/* While unallocated goods exist. */`

5      $\text{Gain}_c = \begin{cases} \max_{k \in U} \phi(X^c \cup X^1, k, c) & U \neq \emptyset \\ -\infty & U = \emptyset \end{cases}$

6      $\text{Gain}_1 = \begin{cases} \max_{k \in N \setminus U} \phi(X^c \cup X^1, k, 1) & N \setminus U \neq \emptyset \\ -\infty & N \setminus U = \emptyset \end{cases}$

7   **if** $\text{Gain}_c \geq \text{Gain}_1$ **then**
            `// Try to give an agent from U a value of c.`

8          $S \leftarrow \arg\max_{k \in U} \phi(X^c \cup X^1, k, c)$

9          $i \leftarrow \min_{j \in S} j$

10   **if** *a path in the exchange graph* $\mathcal{G}(X^c)$ *from* $F_i(X^c)$ *to* X_0^c *exists* **then**

11             $P = (g_1', g_2', \ldots, g_k') \leftarrow$ the shortest path from $F_i(X^c)$ to X^0 in $\mathcal{G}(X^c)$
               `// Augment the allocation with the path P and give g_1' to i`

12             $X_k^c \leftarrow X_k^c \wedge P$ for all $k \in N + 0 - i$

13             $X_i^c \leftarrow X_i^c \wedge P + g_1'$

14   **if** $g_k' \in X_j^1$ *for some* $j \in N$ **then**
                   `// Replace good stolen from X_j^1 with an arbitrary`
                   `unallocated good`

15                 $X_j^1 \leftarrow X_j^1 - g_k' + g$ for some $g \in X_0^c \setminus \bigcup_{h \in N} X_h^1$

16   **else**

17             $U \leftarrow U - i$ `/* i is no longer in play for c valued goods. */`

18   **else if** $\text{Gain}_c < \text{Gain}_1$ **then**
            `// Give an agent from N \ U a value of 1.`

19          $S \leftarrow \arg\max_{k \in N \setminus U} \phi(X^c \cup X^1, k, 1)$

20          $i \leftarrow \min_{j \in S} j$

21          $X_i^1 \leftarrow X_i^1 + g$ for some $g \in X_0^c \setminus \bigcup_{h \in N} X_h^1$

22 **return** $X^c \cup X^1$

In the next iteration, $\text{Gain}_c = -6 \times v_2(g_1) + 5 = -6 \times 5 + 5 = -25$ but $\text{Gain}_1 = -6 \times v_1(\emptyset) + 1 = 1$ still. So we have $\text{Gain}_1 > \text{Gain}_c$ and we choose agent 1 to provisionally receive an arbitrary good (say g_2). We will have $\text{Gain}_1 > \text{Gain}_c$ for the remaining iterations as well, yielding the allocation $X_1 = \{g_2, \ldots, g_6\}$, $X_2 = \{g_1\}$, which is indeed leximin.

By modifying the gain function ϕ to be

$$\phi_{\text{MNW}}(X, i, d) = \begin{cases} \frac{v_i(X_i)+d}{v_i(X_i)} & v_i(X_i) > 0 \\ Md & v_i(X_i) = 0 \end{cases},$$

as per Theorem 2 (where M is a very large number), Algorithm 1 outputs an MNW allocation. When no goods are assigned, the first iteration proceeds the exact same way as that of the leximin allocation and Agent 1 is removed from U.

In the second iteration, we have $\text{Gain}_c = 5 \times M$ and $\text{Gain}_1 = M$. Thus, Agent 2 receives a good (say g_1). Next, we have that $\text{Gain}_c = \phi(X, 2, c) = \frac{v_2(g_1)+c}{v_2(g_1)} = \frac{2c}{c} = 2$, and $\text{Gain}_1 = \phi(X, 1, 1) = M \times 1 = M$. Now we have $\text{Gain}_1 > \text{Gain}_c$, so Agent 1 gets g_2. In the next iteration, we still have $\text{Gain}_c = \phi(X, 2, c) = \frac{2c}{c} = 2$, but $\text{Gain}_1 = \phi(X, 1, 1) = \frac{v_1(g_2)+1}{v_1(g_2)} = \frac{2}{1} = 2$ as well. Thus, according to our tiebreaking scheme, we let agent 2 pick a good next, and they get g_3. We continue in a similar manner, and end up with the allocation $X_1 = \{g_2, g_4, g_6\}, X_2 = \{g_1, g_3, g_5\}$, which is indeed MNW.

This example also shows how, unlike the binary submodular valuations case, max Nash welfare and leximin allocations can have significantly different sorted utility vectors.

4.1   When Does Bivalued Yankee Swap Work?

Bivalued Yankee Swap computes a Ψ-maximizing allocation when the following conditions are satisfied. The conditions are defined for any general vector x but it would help to think of x, y and z as utility vectors.

(C1) – Symmetric Pareto Dominance Let $x, y \in \mathbb{Z}_{\geq 0}^n$ be two vectors. Let $s(x)$ denote the vector x sorted in increasing order. If for all $i \in N$, $s(x)_i \geq s(y)_i$, then $x \succeq_\Psi y$. Equality holds if and only if $s(x) = s(y)$.

(C2) – Gain Function Ψ admits a finite valued gain function ϕ that satisfies the following properties:

(G1) Let $x \in \mathbb{Z}_{\geq 0}^n$ be some vector, let $i, j \in N$ be two agents, and d_1, d_2 be two values in $\{1, c\}$. Let $y \in \mathbb{Z}_{\geq 0}^n$ be the vector resulting from adding a value of d_1 to x_i. Let z be the vector resulting from adding a value of d_2 to x_j. If $\phi(x, i, d_1) \geq \phi(x, j, d_2)$, we must have $y \succeq_\Psi z$. Equality holds if and only if $\phi(x, i, d_1) = \phi(x, j, d_2)$.

(G2) For any two vectors $x, y \in \mathbb{Z}_{\geq 0}^n$, an agent $i \in N$ such that $x_i \leq y_i$ and any $d \in \{1, c\}$, we must have $\phi(x, i, d) \geq \phi(y, i, d)$. Equality holds if $x_i = y_i$.

(G3) For any vector $x \in \mathbb{Z}_{\geq 0}^n$ and any two agents $i, j \in N$, if $x_i \leq x_j$, then $\phi(x, i, d) \geq \phi(x, j, d)$ for any $d \in \{1, c\}$. Equality holds if and only if $x_i = x_j$.

There are two differences between our conditions and the conditions of the General Yankee Swap algorithm [28]. First, we strengthen Pareto Dominance to Symmetric Pareto Dominance (C1). Symmetric Pareto Dominance is not biased towards any agent and therefore, two allocations with the same sorted utility vector have the same objective value. As an immediate corollary, weighted notions of fairness like the max weighted Nash welfare objective [13] do not satisfy Symmetric Pareto Dominance (C1); that is, when agents have weights, two allocations with the same sorted utility vector may not have the same objective value. Second, we introduce (G3) which further strengthens our conditions; (G3) states that all things being equal, it is better to increase the utility of lower utility agents than higher utility agents. We conjecture that the justice criteria satisfying the conditions (C1) and (C2) correspond exactly to the set of generalized welfare functions [25, Chapter 3] but are unable to provide a proof. We leave this question to future work.

Our main result is the following.

Theorem 1. *Suppose that all agents in N have $\{1, c\}$-SUB valuations. Let Ψ be a justice criterion that satisfies (C1) and (C2), with a gain function ϕ. Bivalued Yankee Swap with the gain function ϕ outputs a dominating Ψ-maximizing allocation.*

To prove this, we first show an important Lemma stating that for any agent $i \in N$, if there is a path in the exchange graph $\mathcal{G}(Y^c)$ from $F_i(Y^c)$ to Y_j^c such that $|Y_i^c| < |Y_j^c| - 1$, then Y is not a dominating Ψ maximizing allocation. We then use this Lemma to show that when an agent i is removed from U, it happens at the right time. The main idea is somewhat similar to that of General Yankee Swap [28] but the arguments require a significantly more careful analysis.

5 Applying Bivalued Yankee Swap

We now turn to applying Theorem 1 to well known fairness objectives. While we do not prove this explicitly, in all cases, the gain function ϕ can be trivially computed in time $O(\tau)$ (where τ is an upper bound on the time to compute $v_i(S)$ for any i and any S).

5.1 Max Nash Welfare

Recall that a max Nash Welfare allocation X is one that maximizes the number of agents $|P_X|$ who receive a non-zero utility, and subject to that maximizes the product $\prod_{i \in P_X} v_i(X_i)$.

Theorem 2. *When Ψ corresponds to the Nash welfare, Bivalued Yankee Swap run with the following gain function ϕ_{MNW} computes a Nash welfare maximizing allocation.*

$$\phi_{MNW}(X, i, d) = \begin{cases} \frac{v_i(X_i) + d}{v_i(X_i)} & v_i(X_i) > 0 \\ Md & v_i(X_i) = 0 \end{cases}$$

where M is a very large positive number.

5.2 Leximin

Recall that a leximin allocation is one that maximizes the utility of the worst off agent, subject to that, maximizes the utility of the second worst off agent, and so on.

Theorem 3. *When Ψ corresponds to the leximin fairness objective, Bivalued Yankee Swap run with the gain function $\phi_{leximin}(X, i, d) = -(c+1)v_i(X_i) + d$ computes a leximin allocation.*

5.3 p-Mean Welfare

Recall that the max p-mean welfare allocation X first maximizes the number of agents who receive a positive utility $|P_X|$ and subject to that, maximizes

$$M_p(X) = \left(\frac{1}{n} \sum_{i \in P_X} v_i(X_i)^p \right)^{1/p} \quad \text{where } p \leq 1. \text{ We have already shown how}$$

to compute this justice criterion for certain p values: M_0 corresponds to Nash welfare (Sect. 5.1) and $M_{-\infty}$ corresponds to leximin (Sect. 5.2). We now show how to compute a p-mean welfare maximizing allocation for all the other p-values.

Theorem 4. *When Ψ corresponds to the p-mean welfare objective with finite $p < 1$ and $p \neq 0$, Bivalued Yankee Swap run with the following gain function computes a p-mean welfare allocation.*

$$\phi_{p\text{-}w}(X, i, d) = \begin{cases} (v_i(X_i) + d)^p - v_i(X_i)^p & p \in (0,1) \text{ and } v_i(X_i) > 0 \\ v_i(X_i)^p - (v_i(X_i) + d)^p & p < 0 \text{ and } v_i(X_i) > 0 \\ Md & v_i(X_i) = 0 \end{cases}$$

where M is a very large number.

Note that Theorem 4 does not include the case where $p = 1$. This is because the gain function $\phi(X, i, d) = d$ does not satisfy (G3). The case where $p = 1$ corresponds to the utilitarian social welfare (USW) of an allocation. While we can construct a valid gain function to compute a USW optimal allocation, we do not need to. There exists an efficient algorithm for computing a USW optimal allocation without using Bivalued Yankee Swap; that is, compute a clean utilitarian welfare maximizing allocation with respect to the binary submodular valuations $\{\beta_i\}_{i \in N}$ and allocate the remaining goods arbitrarily.

5.4 When c is not a Natural Number

If c is not a natural number (or equivalently, a does not divide b), the complexity of the problem increases significantly and Bivalued Yankee Swap no longer works. This complexity is captured by [1], who show that computing MNW allocations under *bivalued additive valuations* when $a \in \mathbb{N}_{\geq 1}$ and $b \in \mathbb{N}_{\geq 1}$ are coprime is NP-hard. More formally, they show that for every coprime $a \geq 3$ and $b > a$, the problem of computing an MNW allocation is NP-hard. Note that this hardness result does not cover the case where $a = 2$; this case has been recently shown to be in P for additive valuations [2] but remains open for submodular valuations.

6   Maximin Share Guarantees of MNW and Leximin Allocations

We explore the maximin share guarantees of leximin and max Nash welfare allocations. The maximin share of an agent $i \in N$ (denoted by MMS_i) is defined as the utility agent i would receive if they divided the set of goods G into n bundles and picked the worst one (according to their preferences). More formally, $\mathrm{MMS}_i = \max_X \min_{j \in N} v_i(X_j)$. An allocation X is defined as ε-MMS for some $\varepsilon > 0$ if for all agents $i \in N$, $v_i(X_i) \geq \varepsilon \mathrm{MMS}_i$ [11,27]. When agents have binary submodular valuations, both the max Nash welfare and the leximin allocation are guaranteed to be $\frac{1}{2}$-MMS. We prove the following two results about bivalued submodular valuations.

Theorem 5. *Let c be an integer ≥ 2. When agents have $\{1, c\}$-SUB valuations, then any max Nash welfare allocation X is $\frac{2}{5}$-MMS.*

Theorem 6. *Let c be an integer ≥ 2. When agents have $\{1, c\}$-SUB valuations, then any leximin allocation X is $\frac{1}{c+2}$-MMS.*

It is worth noting that the best known MMS-guarantee for submodular valuations is $\frac{1}{3}$ [21]. Theorem 5 shows that the max Nash welfare allocation offers better MMS guarantees, albeit for a restricted subclass of submodular valuations.

7   Envy-Freeness of MNW and Leximin Allocations

Our final technical section deals with the envy-freeness of max Nash welfare and leximin allocations. An allocation X is *envy free up to one good (EF1)* if for all agents $i, j \in N$, $v_i(X_i) \geq v_i(X_j)$ or there exists a good $g \in X_j$ such that $v_i(X_i) \geq v_i(X_j - g)$ [11,24]. An allocation is *envy free up to any good (EFX)* if for all agents $i, j \in N$, and for all goods $g \in X_j$, we have $v_i(X_i) \geq v_i(X_j - g)$ [12].

Under binary submodular valuations, both the leximin and MNW allocations are known to be EFX [5]. Under bivalued additive valuations, MNW allocations are known to be EFX [3]. However, we now show that neither MNW nor leximin allocations are EF1 under bivalued submodular valuations.

Proposition 1. *For every integer $c \geq 2$, neither the max Nash welfare nor the leximin allocation is guaranteed to be EF1 under $\{1, c\}$-SUB valuations.*

8   Conclusions and Future Work

In this work, we study fair allocation under bivalued submodular valuations. Our insights about this class of valuation functions enable us to use path transfers to compute allocations which satisfy strong fairness and efficiency guarantees.

As the first work to study bivalued submodular valuations, we believe our results are merely the tip of this iceberg. We believe that several other positive results can be shown, specifically with respect to MMS guarantees. It is unknown whether MMS allocations exist for this class of valuations. Prior results show that, while they may not exist under general submodular valuations [21], they indeed exist for the simpler classes of bivalued additive valuations [18] and binary submodular valuations [7]. Resolving this problem for bivalued submodular valuations is a natural next step.

We also believe extending these results and the technique of path transfers beyond bivalued submodular valuations is a worthy pursuit. It is unlikely that we will be able to compute optimal MNW or leximin allocations due to several known intractability results. However, we conjecture that it is possible to use a Yankee Swap based method to compute approximate max Nash welfare allocations for more general classes of submodular valuations. One specific class of interest is that of trivalued submodular valuations, e.g. the marginal gain of each good is either 0, 1 or $c > 1$. We intend to explore this class in future work.

Acknowledgements. The authors would like to thank anonymous WINE 2023 reviewers for useful comments. Viswanathan and Zick are funded by the National Science Foundation grant IIS-2327057.

References

1. Akrami, H., et al.: Maximizing Nash social welfare in 2-value instances. In: Proceedings of the 36th AAAI Conference on Artificial Intelligence (AAAI) (2022)
2. Akrami, H., et al.: Maximizing Nash social welfare in 2-value instances: the half-integer case. CoRR abs/2207.10949 (2022)
3. Amanatidis, G., Birmpas, G., Filos-Ratsikas, A., Hollender, A., Voudouris, A.A.: Maximum Nash welfare and other stories about EFX. Theor. Comput. Sci. **863**, 69–85 (2021)
4. Aziz, H., Li, B., Moulin, H., Wu, X.: Algorithmic fair allocation of indivisible items: a survey and new questions. CoRR abs/2202.08713 (2022)
5. Babaioff, M., Ezra, T., Feige, U.: Fair and truthful mechanisms for dichotomous valuations. In: Proceedings of the 35th AAAI Conference on Artificial Intelligence (AAAI), pp. 5119–5126 (2021)
6. Barman, S., Bhaskar, U., Krishna, A., Sundaram, R.G.: Tight approximation algorithms for p-mean welfare under subadditive valuations. In: Proceedings of the 28th Annual European Symposium on Algorithms (ESA), pp. 11:1–11:17 (2020)
7. Barman, S., Verma, P.: Existence and computation of maximin fair allocations under matroid-rank valuations. In: Proceedings of the 20th International Conference on Autonomous Agents and Multi-Agent Systems (AAMAS), pp. 169–177 (2021)
8. Barman, S., Verma, P.: Truthful and fair mechanisms for matroid-rank valuations. In: Proceedings of the 36th AAAI Conference on Artificial Intelligence (AAAI) (2022)
9. Benabbou, N., Chakraborty, M., Elkind, E., Zick, Y.: Fairness towards groups of agents in the allocation of indivisible items. In: Proceedings of the 28th International Joint Conference on Artificial Intelligence (IJCAI), pp. 95–101 (2019)
10. Benabbou, N., Chakraborty, M., Igarashi, A., Zick, Y.: Finding fair and efficient allocations for matroid rank valuations. ACM Trans. Econ. Comput. **9**(4) (2021)

11. Budish, E.: The combinatorial assignment problem: approximate competitive equilibrium from equal incomes. J. Polit. Econ. **119**(6), 1061–1103 (2011)
12. Caragiannis, I., Kurokawa, D., Moulin, H., Procaccia, A.D., Shah, N., Wang, J.: The unreasonable fairness of maximum Nash welfare. In: Proceedings of the 17th ACM Conference on Economics and Computation (EC), pp. 305–322 (2016)
13. Chakraborty, M., Igarashi, A., Suksompong, W., Zick, Y.: Weighted envy-freeness in indivisible item allocation. ACM Trans. Econ. Comput. **9** (2021)
14. Cousins, C.: An axiomatic theory of provably-fair welfare-centric machine learning. In: Proceedings of the 35th Annual Conference on Neural Information Processing Systems (NeurIPS), pp. 16610–16621 (2021)
15. Cousins, C.: Bounds and applications of concentration of measure in fair machine learning and data science. Ph.D. thesis, Brown University (2021)
16. Cousins, C.: Uncertainty and the social planners problem: why sample complexity matters. In: Proceedings of the 2022 ACM Conference on Fairness, Accountability, and Transparency (2022)
17. Cousins, C.: Revisiting fair-PAC learning and the axioms of cardinal welfare. In: Artificial Intelligence and Statistics (AISTATS) (2023)
18. Ebadian, S., Peters, D., Shah, N.: How to fairly allocate easy and difficult chores. In: Proceedings of the 21st International Conference on Autonomous Agents and Multi-Agent Systems (AAMAS), pp. 372–380 (2022)
19. Garg, J., Murhekar, A.: Computing fair and efficient allocations with few utility values. In: Caragiannis, I., Hansen, K.A. (eds.) SAGT 2021. LNCS, vol. 12885, pp. 345–359. Springer, Cham (2021). https://doi.org/10.1007/978-3-030-85947-3_23
20. Garg, J., Murhekar, A., Qin, J.: Fair and efficient allocations of chores under bivalued preferences. In: Proceedings of the 36th AAAI Conference on Artificial Intelligence (AAAI) (2022)
21. Ghodsi, M., HajiAghayi, M., Seddighin, M., Seddighin, S., Yami, H.: Fair allocation of indivisible goods: improvements and generalizations. In: Proceedings of the 19th ACM Conference on Economics and Computation (EC), pp. 539–556 (2018)
22. Heidari, H., Ferrari, C., Gummadi, K., Krause, A.: Fairness behind a veil of ignorance: a welfare analysis for automated decision making. In: Advances in Neural Information Processing Systems, pp. 1265–1276 (2018)
23. Kurokawa, D., Procaccia, A.D., Shah, N.: Leximin allocations in the real world. ACM Trans. Econ. Comput. **6**(34) (2018)
24. Lipton, R.J., Markakis, E., Mossel, E., Saberi, A.: On approximately fair allocations of indivisible goods. In: Proceedings of the 5th ACM Conference on Economics and Computation (EC), pp. 125–131 (2004)
25. Moulin, H.: Fair Division and Collective Welfare. MIT Press, Cambridge (2004)
26. Plaut, B., Roughgarden, T.: Almost envy-freeness with general valuations. SIAM J. Discret. Math. **34**(2), 1039–1068 (2020)
27. Procaccia, A.D., Wang, J.: Fair enough: guaranteeing approximate maximin shares. In: Proceedings of the 15th ACM Conference on Economics and Computation (EC), pp. 675–692 (2014)
28. Viswanathan, V., Zick, Y.: A general framework for fair allocation with matroid rank valuations. In: Proceedings of the 24th ACM Conference on Economics and Computation (EC) (2023)
29. Viswanathan, V., Zick, Y.: Yankee swap: a fast and simple fair allocation mechanism for matroid rank valuations. In: Proceedings of the 22nd International Conference on Autonomous Agents and Multi-Agent Systems (AAMAS) (2023)

Equilibrium Analysis of Customer Attraction Games

Xiaotie Deng[1], Ningyuan Li[1], Weian Li[1], and Qi Qi[2(✉)]

[1] Center on Frontiers of Computing Studies, School of Computer Science, Peking University, Beijing, China
{xiaotie,liningyuan,weian_li}@pku.edu.cn
[2] Gaoling School of Artificial Intelligence, Renmin University of China, Beijing, China
qi.qi@ruc.edu.cn

Abstract. We introduce a game model called "customer attraction game" to demonstrate the competition among online content providers. In this model, customers exhibit interest in various topics. Each content provider selects one topic and benefits from the attracted customers. We investigate both symmetric and asymmetric settings involving agents and customers. In the symmetric setting, the existence of pure Nash equilibrium (PNE) is guaranteed, but finding a PNE is PLS-complete. To address this, we propose a fully polynomial time approximation scheme to identify an approximate PNE. Moreover, the tight Price of Anarchy (PoA) is established. In the asymmetric setting, we show the nonexistence of PNE in certain instances and establish that determining its existence is NP-hard. Nevertheless, we prove the existence of an approximate PNE. Additionally, when agents select topics sequentially, we demonstrate that finding a subgame-perfect equilibrium is PSPACE-hard. Furthermore, we present the sequential PoA for the two-agent setting.

Keywords: Customer Attraction Game · Pure Nash Equilibrium · Complexity · Price of Anarchy · Asymmetry · Sequential Game

1 Introduction

The widespread adoption of the Internet has prompted an increasing number of companies to shift their focus towards online marketplaces. The global digital content market, for instance, has expanded significantly, reaching a staggering 169.2 billion in 2022 and estimated to surpass 173.2 billion in 2023[1]. Concurrently, the rise of social media

---

[1] https://www.marketresearchfuture.com/reports/digital-content-market-11516.

This research was partially supported by the National Natural Science Foundation of China (NSFC) (No. 62172012), and Beijing Outstanding Young Scientist Program No. BJJWZYJH012019100020098, the Fundamental Research Funds for the Central Universities, and the Research Funds of Renmin University of China No. 22XNKJ07, and Major Innovation & Planning Interdisciplinary Platform for the "Double-First Class" Initiative, Renmin University of China.

J. Garg et al. (Eds.): WINE 2023, LNCS 14413, pp. 242–255, 2024.
https://doi.org/10.1007/978-3-031-48974-7_14

platforms has resulted in a surge of individual content creators and users engaging in these platforms. YouTube, as of June 2023, boasts 2.68 billion active users, while TikTok accumulates 1.67 billion active users[2]. Given the vast number of users with diverse needs and interests, both companies and content creators must adopt suitable marketing strategies and select relevant content topics to attract their target customers effectively. As video producers aim to generate more traffic and subscribers, they often tailor their content topics based on the preferences of their target audience. However, with users exhibiting multiple interests and limited internet usage time, producers face stiff competition not only from other producers sharing the same topic but also from any producer targeting overlapping users. Similarly, on digital advertising platforms, online business owners try to choose the right keywords or tags to attract a specific group of potential customers, while facing competition from other advertisers.

Upon examining the present competitive landscape of online customer acquisition, two noteworthy phenomena emerge. On the one hand, in the red ocean market (representing popular topics), excessive competition arises for a limited customer base, resulting in a waste of social resources. On the other hand, in the blue ocean market (representing unpopular topics), existing platforms fail to satisfy the demand of certain customers, leading to a loss of platform users. Inspired by these phenomena, to explore the underlying reasons and investigate the social welfare loss caused by competition, it is necessary to build a behavioral model for the customer attraction scenarios.

Traditionally, the problem of attracting customers is modeled as a location game in the offline markets. Retailers open physical stores with the aim of drawing nearby residents. The store location is determined by considering the number of potential customers within the service range and the competition from neighboring retailers. However, attracting customers online differ significantly. In a location game, the competition typically occurs in a two-dimensional plane or a one-dimensional interval, primarily relying on distance to attract customers. Conversely, attracting online customers involves significantly more complexity as their preferences correspond to a high-dimensional latent feature vector. Consequently, the outcomes of location games are not directly transferrable to online scenarios.

In this work, we present a game model, called "customer attraction game", which captures the common features of online scenarios. In this model, each customer expresses interest in specific topics, while each content producer (acting as an agent) chooses a topic and receives utility based on the customers they attract. However, if a customer is attracted by multiple agents, her utility is divided proportionally to reflect each agent's competitiveness. Particularly, when agents are symmetric, customers randomly select an agent to whom they are attracted. Moreover, when agents are asymmetric, the selection probability aligns with each agent's weight.

We delve into the simultaneous and sequential topic selection behaviors of agents, observing scenarios where agents select topics simultaneously or in a predetermined order. The simultaneous model can capture that the film studios conceive the themes of their entries for a film festival, or advertisers who can change the targeted audience at a relatively low cost. Conversely, the sequential model applies to content creators on

---

[2] https://www.demandsage.com/youtube-stats/, https://www.demandsage.com/tiktok-user-stati stics/.

social media platforms who are cautious about changing topics to avoid losing follow-
ers. To understand the strategic behavior of competing agents, we focus on the stable
state and raise the following questions: Does such a stable state exist? Can this stable
state be reached efficiently? What is the performance of the stable state?

1.1 Main Results

Our model is representative and has wide application. Different from previous papers
of location games, where customers usually live in one or two dimensions, our model
is highly abstract and not restricted by dimensions, which can be regarded as a general
version of traditional location games. This means that our results can be applied to other
more abstract scenarios beyond facility location or political election contexts, opening
up possibilities for broader applications.

In a static symmetric game where all agents have equal weight and strategy space,
we establish the existence of Pure Nash Equilibrium (PNE) and prove that finding a
PNE is PLS-complete. To address this intractability, we propose a Fully Polynomial
Time Approximation Scheme (FPTAS) that computes an ϵ-approximate PNE. Additionally, we provide a tight Price of Anarchy (PoA) for the symmetric setting.

In static asymmetric games, we generalize the results obtained in the symmetric
setting when only the strategy spaces differ among agents. However, when the weights
assigned to agents are different, we demonstrate the nonexistence of PNE and establish the NP-hardness of determining PNE existence. Nonetheless, we show that there
always exist a $O(\log w_{max})$-approximate PNE which can also achieve the $O(\log W)$-
approximately optimal social welfare, where w_{max} and W are the maximum and sum of
all agents' weights.

For sequential games where agents make decisions in order, we reveal the PSPACE-
hardness of finding a subgame-perfect equilibrium (SPE), even in symmetric settings.
Additionally, we analyze the sequential Price of Anarchy (sPoA) and show that it equals
3/2 for the two-agent case. Furthermore, we establish a lower bound of sPoA that
approximately approaches to 2 for the general n-agent case.

We briefly introduce our technical highlights as follows:

- For the static symmetric game, we demonstrate the PLS-completeness of finding
  a PNE through a reduction from the local max-cut problem. Notably, this result
  remains applicable even for the simplest case where all agents have equal weights
  and strategy spaces. Additionally, since the total number of nodes is polynomial
  in the instance of our symmetric game, but the maximum edge weight in a local
  max-cut instance should be exponentially large, we technically encode exponentially
  large edge weights with polynomial number of nodes by utilizing the proportional
  allocation rule.
- For the static asymmetric game, we construct a concise example to show the nonexistence of PNE in the asymmetric game, and utilize it as a gadget in our reduction to
  prove the NP-hardness to determine the existence of PNE. Moreover, we introduce a
  technical lemma that guarantees the PNE equivalence between the asymmetric case
  and the case with symmetric strategy spaces, and establishes the non-existence of
  PNE and NP-hardness results under a nearly-symmetric case where only one agent
  has a different weight.

1.2   Related Work

Our paper is mainly related to two topics: location games and congestion games. The real-world location game is commonly modeled by the classic Hotelling-Downs model [15,24]. Some comprehensive surveys of this topic are presented in [7,17,18]. In the literature, two concepts in the recent papers are similar to the description of customers' preferences in our model: the attraction interval of agents and the tolerance interval of customers. For the former, an agent attracts all customers within her attraction interval. Feldman et al. [19] first introduce the concept of limited attraction and analyze the existence of equilibria as well as the PoA and PoS. Shen and Wang [37] generalize the above model to the case of arbitrary customer distribution. For the latter, an customer is only attracted by agents in her tolerance interval. Ben-Porat and Tennenholtz [5] investigate a model where a customer make a probabilistic choice based on her location's similarity to the facilities. Cohen and Peleg [12] concentrate on the model where each customers has a tolerance interval and visits the closest shop within this range. Recently, there have been studies on two-stage facility location games with strategic agents and clients [29,30]. Compared with these models, the attraction ranges in our model simultaneously reflect the above features, and can be viewed as a highly abstract version of the attraction interval and tolerance interval.

Location games with discrete location spaces are considered by some literature. [38] is the first to propose the game formulation with a finite number of possible locations. Huang [25] studies the mixed strategy equilibrium in a game of three firms with discrete location. Núñez and Scarsini [34] consider a model where each customer has a strict preference over the shop locations. Iimura and von Mouche [26] study the case with general non-increasing demand functions.

A sequential version of the location game, called Voronoi game, where two players compete for customers in specific area by alternately locating points, is proposed by [1], and the winning strategy of the latter player is provided. A series of subsequent papers study Voronoi game on different graphs, such as cycles [32] and general networks [4,16].

In the field of algorithmic game theory, certain concepts and techniques employed in location games have a connection to the research on congestion games, which is first introduced by [35]. Monderer and Shapley [33] prove that any finite potential game is equivalent to a congestion game. Furthermore, the extensive research of congestion games has yielded fruitful results and classic concepts in game theory, for example the existence of PNE [3,21,23,27], PoA or PoS [2,8,10,13,36], and sequential congestion games [9,11,14,31]. Gairing [20] proposes a variant of congestion game, called covering game. It shows that in a covering game with specially designed utility sharing function, every PNE is a constant approximation of the maximum covering problem. While the covering game coincides with the special case of our model where agents are unweighted, but it does not consider the weighted agents. Besides, Gairing [20] focuses more on the perspective of mechanism design, whereas our work focuses on equilibrium analysis under a proportional allocation rule.

The works of Goemans et al. [22] and Bilò et al. [6] are the most relevant to our model. Goemans et al. propose a market sharing game that shares similarities with our symmetric static game. However, their research primarily focuses on unweighted scenarios without considering the weighted case, and they do not study the complexity of finding an equilibrium. In the study conducted by Bilò et al. [6], a project game is introduced where each agent selects a project to contribute to. Similar to our asymmetric model, the agents have different sets of available projects and different weights, with rewards distributed proportionally to the weights. However, an important distinction is that our model allows an agent to attract multiple customers, whereas in their model, each agent can only select one single project.

1.3 Roadmap

In Sect. 2, we formally introduce our customer attraction game. In Sect. 3, we mainly focus on the symmetric setting and show a series of results about PNE. In Sect. 4, the asymmetric setting is investigated. We discuss the sequential version of our game in Sect. 5. In Sect. 6, we give a summary of the whole paper and propose the directions for future work.

2 Model and Preliminaries

In this section, we formally introduce our customer attraction game (CAG). Assume that there are n types of customers represented by n discrete nodes[3] and define the set of nodes as $N = \{1, 2, \cdots, n\}$. For each $j \in N$, its value v_j denotes the number or importance of customers represented by node j. Without loss of generality, we assume that $v_j \in \mathbb{Z}_{\geq 1}$ for any $j \in N$.

There are L candidate topics for agents. For each topic l, its attraction range is defined as $s_l \subseteq N$, i.e., the set of customers who are interested in this topic. Define $\mathcal{S} = \{s_1, s_2, \cdots, s_L\}$ as the collection of attraction ranges of all candidate topics. We assume that $\bigcup_{l=1}^{L} s_l = N$, that is, each customer is attracted by at least one topic.

There are m agents and the set of agents is denoted by $A = \{1, \cdots, m\}$. Each agent $i \in A$ has a weight w_i and we suppose that $w_i \in \mathbb{Z}_{\geq 1}$. The strategy space of agent i is defined as $\mathcal{S}_i \subseteq \mathcal{S}$ representing the available topics provided for agent i. Let the joint strategy space of all agents be $\vec{\mathcal{S}} = \times_{i=1}^{m} \mathcal{S}_i$. When each agent chooses a pure strategy $S_i \in \mathcal{S}_i$, a pure strategy profile is denoted by $\vec{S} = (S_1, \cdots, S_m) \in \vec{\mathcal{S}}$.

Given a pure strategy profile $\vec{S}$, we say that agent i attracts the node j if and only if $j \in S_i$, that is, customers on node j are interested in the topic which agent i selects. For each node $j \in N$, define the load function $c(j, \vec{S})$ as the total weight of the agents attracting j, formally,

$$c(j, \vec{S}) = \sum_{i \in A: j \in S_i} w_i.$$

When node j is attracted by at least one agent, each customer on it selects an agent attracting node j with the probability proportional to each agent's weight. Therefore, in expectation, node j distributes the v_j customers to all agents attracting j in proportion

---

[3] In the following, we sometimes use "node" to represent customers in language.

to their weights. The utility of each agent $i \in A$ is defined as the expected number of attracted customers, that is,

$$U_i(\vec{S}) = \sum_{j \in S_i} \frac{w_i}{c(j, \vec{S})} v_j.$$

Now we formally define an instance of the customer attraction game.

Definition 1. *An instance of customer attraction game is defined by a tuple* $\mathcal{I} = (N, A, \vec{S}, \vec{w}, \vec{v})$, *where*

- N *is the set of all nodes;*
- A *is the set of all agents;*
- $\vec{S} = \times_{i=1}^m S_i$, *where* S_i *is the strategy space of agent i;*
- $\vec{w} = (w_1, w_2, \cdots, w_m) \in \mathbb{Z}_{\geq 1}^m$: *the weights of all agents;*
- $\vec{v} = (v_1, v_2, \cdots, v_n) \in \mathbb{Z}_{\geq 1}^n$: *the values of all nodes.*

We also consider a special case called the symmetric CAG, where all agents are symmetric and all nodes are unit-value. Formally,

- All agents share the same strategy space, that is, for all $i \in A$, $S_i = S$.
- All agents have unit weight, that is, for all $i \in A$, $w_i = 1$.
- All nodes have unit value, that is, for all $j \in N$, $v_j = 1$.

To distinguish, we call the general model defined in Definition 1 as the asymmetric CAG. Sometimes we also discuss the settings in which only part of the above three components are restricted to be symmetric. For convenience, we denote such settings by listing the asymmetric components in the model as the prefix. For example, the notion $(\vec{S}, \vec{v})$-asymmetric CAG means a CAG where the strategy spaces $\vec{S}$ and the node values $\vec{v}$ can be different, while the agents' weights $\vec{w}$ are restricted to be the same.

Next we give the formal definition of pure Nash equilibria (PNE) in the CAG.

Definition 2. *Given an instance* $\mathcal{I}$ *of customer attraction game, a strategy profile* $\vec{S}$ *is a pure Nash equilibrium if, for any* $i \in A$ *and* $S_i' \in S_i$,

$$U_i(S_i, \vec{S}_{-i}) \geq U_i(S_i', \vec{S}_{-i}).$$

We define PNE$(\mathcal{I})$ *as the set containing all pure Nash equilibria of instance* $(\mathcal{I})$.

Sometimes, PNE may not exist for some instances of CAG. For these instances, we focus on the approximate pure Nash equilibrium which is defined formally as

Definition 3. *Given an instance of customer attraction game* $\mathcal{I} = (N, A, \vec{S}, \vec{w}, \vec{v})$, *for any* $\alpha \geq 1$, *a strategy profile* $\vec{S}$ *is an* α-*approximate pure Nash equilibrium if, for any* $i \in A$ *and* $S_i' \in S_i$,

$$U_i(S_i', \vec{S}_{-i}) \leq \alpha U_i(\vec{S}).$$

In this model, for any instance of game $\mathcal{I}$, we define the social welfare with respect to strategy profile, $\vec{S}$, as the sum of the utilities of all m agents, which is also equal to the number of attracted customers under profile $\vec{S}$,

$$\text{SW}(\vec{S}) = \sum_{i=1}^{m} U_i(\vec{S}) = \sum_{\{j : \exists i, \text{such that} j \in S_i | \vec{S}\}} v_j.$$

Let $\vec{S}^*$ be the strategy profile that achieves the optimal social welfare. We define the price of anarchy (PoA) [28] for PNE of an instance as

$$\text{PoA}(\mathcal{I}) = \max_{\vec{S}^{NE} \in \text{PNE}(\mathcal{I})} \frac{\text{SW}(\vec{S}^*)}{\text{SW}(\vec{S}^{NE})}.$$

Consequently, the PoA for CAG among all instances is defined as

$$\text{PoA} = \sup_{\mathcal{I}} \max_{\vec{S}^{NE} \in \text{PNE}(\mathcal{I})} \frac{\text{SW}(\vec{S}^*)}{\text{SW}(\vec{S}^{NE})}.$$

3   Warm-Up: The Symmetric Static Game

As a warm-up, we focus on the symmetric CAG, where all agents are symmetric and all nodes have unit value, as described in Sect. 2 previously. We first prove the existence of PNE by showing that any instance of symmetric CAG is an exact potential game [33]. However, we demonstrate that finding a PNE is PLS-complete. Thus, we propose an FPTAS to output a $(1 + \epsilon)$-approximate PNE. Finally, we give a tight PoA for the symmetric CAG.

Since all agents have identical weights of 1 in the symmetric setting, the load function of one node degenerates to the number of agents attracting this node. That is,

$$c(j, \vec{S}) = |\{i \in A : j \in S_i\}|$$

for any $j \in N$ and any $\vec{S} \in \mathcal{S}$. To show the existence of PNE, we can construct a Rosenthal's potential function [35],

$$\Phi(\vec{S}) = \sum_{j \in N} \sum_{k=1}^{c(j, \vec{S})} \frac{1}{k},$$

for any $\vec{S} \in \mathcal{S}$ and show that any instance $\mathcal{I}$ is an exact potential game. In an exact potential game, whenever an agent changes its strategy to get a better utility, the potential function will also be increased. It means that any PNE is corresponding to a local maximum of $\Phi(\vec{S})$. Since the joint strategy space is finite, there must exist a joint strategy profile such that the potential function cannot be improved by changing the strategy of any agent, which guarantees the existence of PNE.

Theorem 1. *For any instance $\mathcal{I}$ of symmetric customer attraction game, $\mathcal{I}$ is an exact potential game with respect to the potential function $\Phi(\vec{S})$ and the pure Nash equilibrium always exists.*

Note that the proof of Theorem 1 only requires the symmetry of agents' weights. Generally, for any instance $\mathcal{I}$ of $(\vec{S}, \vec{v})$-asymmetric CAG, we can similarly extend the definition of Φ to

$$\Phi(\vec{S}) = \sum_{j \in N} v_j \sum_{k=1}^{c(j, \vec{S})} \frac{1}{k},$$

and one can check that $\Phi(\vec{S})$ is still a potential function.

Corollary 1. *For any instance $\mathcal{I}$ of $(\vec{S}, \vec{v})$-asymmetric customer attraction game, the pure Nash equilibrium always exists.*

Knowing the existence of PNE, a natural way to find a PNE is the best-response dynamics, roughly speaking, which starts with some strategy profile, and iteratively picks up one agent and let this agent deviate to the most beneficial strategy. By the existence of the potential function, it guarantees that the best-response dynamics can find a PNE in finite steps. However, this may take exponential number of steps. We prove that, finding a PNE in the symmetric CAG is PLS-complete. Note that this is a strong hardness result (analogous to the concept of strong NP-hardness), since it do not require the node values to be exponentially large, and the result even holds for the simplest CAG where all agents have the same weight and strategy space, and all nodes have unit value.

Theorem 2. *Finding a Pure Nash Equilibrium of symmetric CAG is PLS-complete.*

Since the problem of finding a PNE is PLS-complete, we turn to designing the efficient algorithm to compute the approximate PNE. Fortunately, for any $\epsilon > 0$, a $(1 + \epsilon)$-approximate pure Nash equilibrium can be found in polynomial time by the best-response dynamics. That is, in each step, if the current strategy profile $\vec{S}$ is not a $(1 + \epsilon)$-approximate PNE, an agent i is chosen and deviates her strategy to S_i', such that

$$U_i(S_i', \vec{S}_{-i}) - U_i(\vec{S}) = \max_{i' \in A, S_{i'}' \in \mathcal{S}_{i'}} (U_{i'}(S_{i'}', \vec{S}_{-i'}) - U_{i'}(\vec{S})).$$

Theorem 3. *Given $0 < \epsilon < 1$, for any instance $\mathcal{I} = (N, A, \vec{S}, \vec{w}, \vec{v})$ of symmetric CAG, the best-response dynamics finds an ϵ-pure Nash equilibrium in $O(\epsilon^{-1}|N||A|\log|A|)$ steps, which implies that it is an FPTAS to compute an $(1 + \epsilon)$-approximate PNE.*

In the rest of this section, we concentrate on the efficiency of PNE, i.e., the PoA in the symmetric setting. Based on the results of literature [39], it is not difficult to check that customer attraction games (even in asymmetric case) belong to the valid utility system introduced by [39], which implies that the upper bound of PoA is 2 directly. On the other hand, we also use another technique to prove a more detailed upper bound, $\min\{2, \max\{n/m, 1\}\}$, on any instance of symmetric CAG. Combining with the examples of lower bound, we finally show that the tight PoA is 2 for the customer attraction games.

Theorem 4. *For any instance* $\mathcal{I} = (N, A, \vec{S}, \vec{w}, \vec{v})$ *of symmetric CAG, given* $|N| = n$ *and* $|A| = m$, *the price of anarchy is* $\min\{2, \max\{n/m, 1\}\}$. *Generally, the price of anarchy for CAG is tight and equal to* 2.

In reality, some instances assuring $n/m < 2$ can achieve a PoA better than 2. When $n/m \geq 2$, the worst PNE of above example leads to one half loss of social welfare, which reflects the competition phenomena where agents pursue the hot topics in the red and blue ocean markets, introduced in Sect. 1.

4   The Asymmetric Static Game

In this section, we begin to investigate the asymmetric CAG. Compared to some elegant results in the symmetric CAG (e.g., the existence of PNE is guaranteed under the symmetric CAG, as well as any $(\vec{S}, \vec{v})$-asymmetric CAG where the weights of agents are restricted to be symmetric, yet), the general asymmetric case becomes trickier, which is due to that the potential function is no longer applicable. In fact, a $(\vec{w})$-asymmetric CAG is generally not an exact potential game, as shown in the following lemma.

Lemma 1. *There is an instance* $\mathcal{I} = (N, A, \vec{S}, \vec{w}, \vec{v})$ *of* $(\vec{w})$-*asymmetric CAG, such that* $\mathcal{I}$ *is not an exact potential game.*

A natural question is, whether the PNE is still guaranteed to exist when the agents have asymmetric weights. Interestingly, we find that when there are only two agents, PNE still exists by Theorem 5.

Theorem 5. *For any instance* $\mathcal{I}$ *of asymmetric customer attraction game with two agents, a pure Nash equilibrium always exists.*

However, once that the number of agents increases to three, we can construct a counterexample to demonstrate that the nonexistence of PNE can be caused by a single weighted agent, i.e., the PNE may not exist, even when all but one of the agents are identically weighted 1, and the strategy spaces and node values are symmetric. To build such example, we first construct an instance of $(\vec{S}, \vec{w}, \vec{v})$-asymmetric CAG (the fully asymmetric model) in Example 1, which has no pure Nash equilibrium. Then we convert it to an instance of $(\vec{w})$-asymmetric CAG, preserving the nonexistence of PNE.

Intuitively, to construct a fully asymmetric model without PNE, we start with two active agents with different weights to form a two-agent 2×2 matrix game. Then, by adding some dummy agents, we can adjust the payoff matrix of these two agents so that the beneficial deviations form a cycle (This action is feasible because the current game is no longer a potential game), which avoids the existence of a PNE.

Example 1 (Nonexistence of PNE).
    Under the $(\vec{S}, \vec{w}, \vec{v})$-asymmetric CAG, we construct an instance $\mathcal{I} = (N, A, \vec{S}, \vec{w}, \vec{v})$. There are 4 nodes labeled as $N = \{q_1, q_2, q_3, q_4\}$. Let $v_{q_1} = 2, v_{q_2} = 1$, $v_{q_3} = 1$ and $v_{q_4} = 2$. There are 3 agents, labeled as $A = \{1, 2, 3\}$, including two active agents 1,2 and one dummy agent 3. The weights of agents are $\vec{w} = (4, 1, 1)$. Both active agents 1 and 2 have two strategies: $s_{1,1} = \{q_1, q_2\}, s_{1,2} = \{q_3, q_4\}, s_{2,1} =$

$\{q_1, q_3\}, s_{2,2} = \{q_2, q_4\}$, i.e., the strategy spaces of the active agents are $\mathcal{S}_1 = \{s_{1,1}, s_{1,2}\}$ and $\mathcal{S}_2 = \{s_{2,1}, s_{2,2}\}$. Dummy agent 3 has one strategy $\mathcal{S}_3 = \{\{q_1, q_4\}\}$. Since agent 3 is dummy, the game can be viewed as a two-player game between agents 1 and 2, each of whom has two strategies. We calculate the utilities of agents 1 and 2, presented as a payoff matrix in Table 1. The weights and values are properly chosen so that agent 1 has incentive to deviate in states $(s_{1,1}, s_{2,1})$ and $(s_{1,2}, s_{2,2})$, while agent 2 has incentive to deviate in states $(s_{1,1}, s_{2,2})$ and $(s_{1,2}, s_{2,1})$. Therefore, there is no PNE in this instance.

Table 1. The payoff matrix of the two-player game between the two active agents

U_1, U_2 \ S_2 S_1	$s_{2,1}$	$s_{2,2}$
$s_{1,1}$	(7/3, 4/3)	(2.4, 1.2)
$s_{1,2}$	(2.4, 1.2)	(7/3, 4/3)

Then, we convert the instance in Example 1 to an instance of $(\vec{w})$-asymmetric CAG, while preserving the nonexistence of PNE. Intuitively, to obtain symmetric strategy spaces, we can modify the original strategy space for each agent by creating a large group of new nodes and adding them to every strategy of this agent, and then join all resulting strategy spaces together to get a common strategy space. When the total value of each large group is designed properly, each agent in the constructed instance will play the role of one agent in the original instance, and thus the constructed instance will be equivalent to the original instance with respect to the existence of PNE. To obtain symmetric node values, for any node j whose value is greater than one, we can simply split node j into a group of nodes with unit value. The summarized results are in the following lemma.

Lemma 2. *Given any instance $\mathcal{I} = (N, A, \vec{\mathcal{S}}, \vec{w}, \vec{v})$ of $(\vec{\mathcal{S}}, \vec{w}, \vec{v})$-asymmetric CAG, if there is one agent $i \in A$ such that $w_i \geq 1$ and $w_{i'} = 1$ for all $i' \neq i$, then an instance $\mathcal{I}' = (N', A', \vec{\mathcal{S}'}, \vec{w}', \vec{v}')$ of $(\vec{w})$-asymmetric CAG can be computed, such that $\mathcal{I}$ has a PNE if and only if $\mathcal{I}'$ has a PNE. The computation time is polynomial in $|N|$, $|A|$, $\sum_{i \in A} |\mathcal{S}_i|$, $\max_{i \in A} w_i$, and $\sum_{j \in N} v_j$. Moreover, the construction can preserve the weights of agents, in other words, $A' = A$ and $\vec{w}' = \vec{w}$.*

With the help of technical Lemma 2, we can modify Example 1 to obtain a counterexample showing that PNE may not exist even when there are only three agents, sharing a common strategy space, and only one agent is not weighted 1.

Theorem 6. *There exists an instance of $(\vec{w})$-asymmetric CAG which has no pure Nash equilibrium, even if there are only three agents and only one agent is not weighted 1.*

The next question is, can we distinguish the instances of asymmetric CAG which have a PNE? We show that it is NP-complete to judge the existence of PNE for any asymmetric CAG, as stated in Theorem 7.

Theorem 7. *The decision problem asking whether a PNE exists for an instance $\mathcal{I}$ of $(\vec{w})$-asymmetric CAG is NP-complete, even if only one agent is not weighted 1.*

As a corollary, deciding the existence of a PNE for an instance under $(\vec{S}, \vec{w}, \vec{v})$-asymmetric CAG model is also NP-complete.

So far, we have shown that there exist instances which do not have a PNE and deciding whether the PNE exists is also NP-hard. However, if we focus on the scope of approximate PNE, we can make sure that a $O(log w_{\max})$-approximate PNE always exists for any $(\vec{S}, \vec{w}, \vec{v})$-asymmetric CAG.

Theorem 8. *For any instance of $(\vec{S}, \vec{w}, \vec{v})$-asymmetric CAG, there exists a $O(\log w_{\max})$-approximate PNE, which also achieves $O(\log W)$-approximately optimal social welfare, where $w_{\max} = \max_{i \in A} w_i$ is the maximum of agents' weights and $W = \sum_{i \in A} w_i$ is the sum of agents' weights.*

Lastly, we discuss the PoA of asymmetric CAG. One can observe that the proof of upper bound 2 on PoA in Theorem 4 actually doesn't require any symmetry of agents and nodes. Therefore, as a corollary, over all instances of asymmetric CAG admitting PNEs, we have PoA = 2. Nevertheless, we also know that approximate PNE always exists for suitable approximation ratios. If we consider the PoA defined on the range of α-approximate PNE, we can get a general conclusion as an extension of Theorem 4.

Theorem 9. *For any instance of $(\vec{S}, \vec{w}, \vec{v})$-asymmetric CAG, the PoA for α-approximate PNE is upper bounded by $1 + \alpha$.*

5   The Sequential Game

In reality, the agents may take actions sequentially in most of the situations. In these scenarios, the agents who act later may be advantaged, since they can see the actions of the former ones and optimize their own strategies. In this section, we study the CAG in sequential setting.

In the sequential CAG, an instance is still defined as the tuple $(N, A, \vec{S}, \vec{w}, \vec{v})$. Without loss of generality, we assume that the agents are labeled as $A = \{1, \cdots, m\}$ and move in the order of $1, \cdots, m$. In round $i = 1, 2, \cdots, m$, the agent i observes the strategies selected by agents $1, \cdots, i-1$, denoted by $\vec{S}_{<i} = (S_1, \cdots, S_{i-1}) \in \times_{i'=1}^{i-1} S_{i'}$ and decides its own strategy $S_i = \sigma_i(\vec{S}_{<i})$. We call $\sigma_i : \times_{i'=1}^{i-1} S_{i'} \to S_i$ as the strategy list of agent i. Given the strategy lists profile $\vec{\sigma} = (\sigma_1, \cdots, \sigma_m)$, the outcome of $\vec{\sigma}$ is the strategy profile formed in the end, formally defined as $\vec{\alpha}(\vec{\sigma}) = (\alpha_1(\vec{\sigma}), \cdots, \alpha_m(\vec{\sigma})) \in \vec{S}$, such that $\alpha_1(\vec{\sigma}) = \sigma_1(\emptyset)$ and $\alpha_i(\vec{\sigma}) = \sigma_i(\alpha_1(\vec{\sigma}), \cdots, \alpha_{i-1}(\vec{\sigma}))$ for any $2 \leq i \leq m$.

When any prefix $\vec{S}_{<i}$ is fixed for some agent $i \in A$, a subgame over agents $i, \cdots, m$ is induced naturally, and the outcome in this subgame is denoted by $\vec{\alpha}(\vec{S}_{<i}, \sigma_i, \cdots, \sigma_m) \in \vec{S}$. We introduce the subgame perfect equilibrium (SPE), defined as follows.

Definition 4. *A strategy lists profile $\vec{\sigma}$ is a subgame perfect equilibrium, if and only if for any agent $i \in A$, any prefix $\vec{S}_{<i} \in \times_{i'=1}^{i-1} S_{i'}$ and any $S_i' \in S_i$, it holds that*

$$U_i \left(\vec{\alpha}(\vec{S}_{<i}, \sigma_i, \cdots, \sigma_m) \right) \geq U_i \left(\vec{\alpha} \left((\vec{S}_{<i}, S_i'), \sigma_{i+1}, \cdots, \sigma_m \right) \right).$$

Let $SPE(\mathcal{I})$ denote the set of all subgame perfect equilibria of instance $\mathcal{I}$.

It is well-known that the SPE always exists and can be found by backward induction: when $\sigma_{i+1}, \cdots, \sigma_m$ is given, agent i can select $\sigma(\vec{S}_{<i}) \in \arg\max_{S_i \in \mathcal{S}_i} U_i$ $\left(\vec{\alpha}(\vec{S}_{<i}, S_i, \sigma_{i+1}, \cdots, \sigma_m) \right)$ for each $\vec{S}_{<i} \in \times_{i'=1}^{i-1} \mathcal{S}_{i'}$. Now we come to discuss the computational complexity of finding an SPE. Since the description length of an SPE is exponentially large, we only care about finding an SPE outcome. However, this is still computationally difficult, as we can prove that a decision version of it is PSPACE-hard.

Theorem 10. *Given an instance $\mathcal{I}$ of sequential CAG, an agent $i \in A$ and an rational number X, the problem to decide whether there exists an SPE $\vec{\sigma} \in \mathrm{SPE}(\mathcal{I})$ such that $U_i(\vec{\alpha}(\vec{\sigma})) \geq X$ is PSPACE-hard.*

Even though it is PSPACE-hard to compute an SPE, we are still curious about the efficiency of the SPE. Similar to the concept of PoA, for the sequential CAG, we define the sequential price of anarchy (sPoA) [31] as

$$
\mathrm{sPoA} = \sup_{\mathcal{I}} \max_{\vec{\sigma} \in \mathrm{SPE}(\mathcal{I})} \frac{\mathrm{SW}(\vec{S}^*)}{\mathrm{SW}(\vec{\alpha}(\vec{\sigma}))},
$$

where $\vec{S}^*$ is the optimal strategy profile in term of the social welfare. In this section, we discuss the tight sPoA for two-agent setting and provide a lower bound of sPoA for general n-agent setting.

Theorem 11. *For any instance $\mathcal{I}$ of sequential CAG with two symmetric agents, i.e., $|A| = 2$ and $w_1 = w_2 = 1$, the sequential price of anarchy is $\frac{3}{2}$. When $|A| > 2$, the sequential price of anarchy is not less than $2 - 1/|A|$.*

6 Conclusions and Future Work

In this paper, we focus on the customer attraction game in three settings. In each setting, we study the existence of pure strategy equilibrium and hardness of finding an equilibrium. Concerned about loss by competition, we also give the (s)PoA of each case.

We propose some research directions: firstly, for the asymmetric static game, we can continue exploring the sufficient conditions to guarantee the existence of PNE. Secondly, the result on the sPoA of the sequential game for more than two agents is still unknown. We conjecture that the sPoA is upper-bounded by 2 when the number of agents is more than two. In addition, the weighted sequential game is also a direction for future research.

References

1. Ahn, H.K., Cheng, S.W., Cheong, O., Golin, M., Van Oostrum, R.: Competitive facility location: the Voronoi game. Theor. Comput. Sci. **310**(1–3), 457–467 (2004)
2. Aland, S., Dumrauf, D., Gairing, M., Monien, B., Schoppmann, F.: Exact price of anarchy for polynomial congestion games. SIAM J. Comput. **40**(5), 1211–1233 (2011)

3. Anshelevich, E., Dasgupta, A., Kleinberg, J., Tardos, É., Wexler, T., Roughgarden, T.: The price of stability for network design with fair cost allocation. SIAM J. Comput. **38**(4), 1602–1623 (2008)
4. Bandyapadhyay, S., Banik, A., Das, S., Sarkar, H.: Voronoi game on graphs. Theor. Comput. Sci. **562**, 270–282 (2015)
5. Ben-Porat, O., Tennenholtz, M.: Shapley facility location games. In: Devanur, N.R., Lu, P. (eds.) WINE 2017. LNCS, vol. 10660, pp. 58–73. Springer, Cham (2017). https://doi.org/10.1007/978-3-319-71924-5_5
6. Bilò, V., Gourvès, L., Monnot, J.: Project games. Theor. Comput. Sci. **940**, 97–111 (2023)
7. Brenner, S.: Location (hotelling) games and applications. Wiley Encyclopedia of Operations Research and Management Science (2010)
8. Chawla, S., Niu, F.: The price of anarchy in Bertrand games. In: Proceedings of the 10th ACM conference on Electronic Commerce, pp. 305–314 (2009)
9. Chen, C., Giessler, P., Mamageishvili, A., Mihalák, M., Penna, P.: Sequential solutions in machine scheduling games. In: Chen, X., Gravin, N., Hoefer, M., Mehta, R. (eds.) WINE 2020. LNCS, vol. 12495, pp. 309–322. Springer, Cham (2020). https://doi.org/10.1007/978-3-030-64946-3_22
10. Chen, H.L., Roughgarden, T.: Network design with weighted players. Theory Comput. Syst. **45**(2), 302–324 (2009)
11. Christodoulou, G., Koutsoupias, E.: On the price of anarchy and stability of correlated equilibria of linear congestion games. In: Brodal, G.S., Leonardi, S. (eds.) ESA 2005. LNCS, vol. 3669, pp. 59–70. Springer, Heidelberg (2005). https://doi.org/10.1007/11561071_8
12. Cohen, A., Peleg, D.: Hotelling games with random tolerance intervals. In: Caragiannis, I., Mirrokni, V., Nikolova, E. (eds.) WINE 2019. LNCS, vol. 11920, pp. 114–128. Springer, Cham (2019). https://doi.org/10.1007/978-3-030-35389-6_9
13. Czumaj, A., Vöcking, B.: Tight bounds for worst-case equilibria. ACM Trans. Algorithms (TALG) **3**(1), 1–17 (2007)
14. de Jong, J., Uetz, M.: The sequential price of anarchy for atomic congestion games. In: Liu, T.-Y., Qi, Q., Ye, Y. (eds.) WINE 2014. LNCS, vol. 8877, pp. 429–434. Springer, Cham (2014). https://doi.org/10.1007/978-3-319-13129-0_35
15. Downs, A.: An economic theory of political action in a democracy. J. Polit. Econ. **65**(2), 135–150 (1957)
16. Dürr, C., Thang, N.K.: Nash equilibria in Voronoi games on graphs. In: Arge, L., Hoffmann, M., Welzl, E. (eds.) ESA 2007. LNCS, vol. 4698, pp. 17–28. Springer, Heidelberg (2007). https://doi.org/10.1007/978-3-540-75520-3_4
17. Eiselt, H.A.: Equilibria in competitive location models. In: Eiselt, H.A., Marianov, V. (eds.) Foundations of Location Analysis. ISORMS, vol. 155, pp. 139–162. Springer, New York (2011). https://doi.org/10.1007/978-1-4419-7572-0_7
18. Eiselt, H.A., Laporte, G., Thisse, J.F.: Competitive location models: a framework and bibliography. Transp. Sci. **27**(1), 44–54 (1993)
19. Feldman, M., Fiat, A., Obraztsova, S.: Variations on the hotelling-downs model. In: Thirtieth AAAI Conference on Artificial Intelligence (2016)
20. Gairing, M.: Covering games: approximation through non-cooperation. In: Leonardi, S. (ed.) WINE 2009. LNCS, vol. 5929, pp. 184–195. Springer, Heidelberg (2009). https://doi.org/10.1007/978-3-642-10841-9_18
21. Gairing, M., Kollias, K., Kotsialou, G.: Existence and efficiency of equilibria for cost-sharing in generalized weighted congestion games. ACM Trans. Econ. Comput. (TEAC) **8**(2), 1–28 (2020)
22. Goemans, M., Li, L.E., Mirrokni, V.S., Thottan, M.: Market sharing games applied to content distribution in ad-hoc networks. In: Proceedings of the 5th ACM International Symposium on Mobile Ad Hoc Networking and Computing, pp. 55–66 (2004)

23. Harks, T., Klimm, M.: On the existence of pure Nash equilibria in weighted congestion games. Math. Oper. Res. **37**(3), 419–436 (2012)
24. Hotelling, H.: Stability in competition. Econ. J. **39**(153), 41–57 (1929)
25. Huang, Z.: The mixed strategy equilibrium of the three-firm location game with discrete location choices. Econ. Bull. **31**(3), 2109–2116 (2011)
26. Iimura, T., von Mouche, P.: Discrete hotelling pure location games: potentials and equilibria. ESAIM Proc. Surv. **71**, 163 (2021)
27. Kollias, K., Roughgarden, T.: Restoring pure equilibria to weighted congestion games. ACM Trans. Econ. Comput. (TEAC) **3**(4), 1–24 (2015)
28. Koutsoupias, E., Papadimitriou, C.: Worst-case equilibria. In: Meinel, C., Tison, S. (eds.) STACS 1999. LNCS, vol. 1563, pp. 404–413. Springer, Heidelberg (1999). https://doi.org/10.1007/3-540-49116-3_38
29. Krogmann, S., Lenzner, P., Molitor, L., Skopalik, A.: Two-stage facility location games with strategic clients and facilities. arXiv preprint arXiv:2105.01425 (2021)
30. Krogmann, S., Lenzner, P., Skopalik, A.: Strategic facility location with clients that minimize total waiting time. In: Proceedings of the AAAI Conference on Artificial Intelligence, vol. 37, no. 5, pp. 5714–5721 (2023)
31. Leme, R.P., Syrgkanis, V., Tardos, É.: The curse of simultaneity. In: Proceedings of the 3rd Innovations in Theoretical Computer Science Conference, pp. 60–67 (2012)
32. Mavronicolas, M., Monien, B., Papadopoulou, V.G., Schoppmann, F.: Voronoi games on cycle graphs. In: Ochmański, E., Tyszkiewicz, J. (eds.) MFCS 2008. LNCS, vol. 5162, pp. 503–514. Springer, Heidelberg (2008). https://doi.org/10.1007/978-3-540-85238-4_41
33. Monderer, D., Shapley, L.S.: Potential games. Games Econ. Behav. **14**(1), 124–143 (1996)
34. Núñez, M., Scarsini, M.: Competing over a finite number of locations. Econ. Theory Bull. **4**(2), 125–136 (2016)
35. Rosenthal, R.W.: A class of games possessing pure-strategy Nash equilibria. Int. J. Game Theory **2**(1), 65–67 (1973)
36. Roughgarden, T., Tardos, É.: How bad is selfish routing? J. ACM (JACM) **49**(2), 236–259 (2002)
37. Shen, W., Wang, Z.: Hotelling-downs model with limited attraction. In: Proceedings of the 16th Conference on Autonomous Agents and MultiAgent Systems, pp. 660–668 (2017)
38. Stevens, B.H.: An application of game theory to a problem in location strategy. Pap. Reg. Sci. Assoc. **7**, 143–157 (1961). https://doi.org/10.1007/BF01969077
39. Vetta, A.: Nash equilibria in competitive societies, with applications to facility location, traffic routing and auctions. In: Proceedings of the 43rd Annual IEEE Symposium on Foundations of Computer Science, pp. 416–425. IEEE (2002)

The Importance of Knowing the Arrival Order in Combinatorial Bayesian Settings

Tomer Ezra[1]([envelope])([ORCID]) and Tamar Garbuz[2]([ORCID])

[1] Simons Laufer Mathematical Sciences Institute, Berkeley, CA 94720, USA
tomer.ezra@gmail.com
[2] Tel Aviv University, Tel Aviv 61390, Israel
garbuz@mail.tau.ac.il
https://tomer-ezra.github.io/

Abstract. We study the measure of order-competitive ratio introduced by Ezra et al. [16] for online algorithms in Bayesian combinatorial settings. In our setting, a decision-maker observes a sequence of elements that are associated with stochastic rewards that are drawn from known priors, but revealed one by one in an online fashion. The decision-maker needs to decide upon the arrival of each element whether to select it or discard it (according to some feasibility constraint), and receives the associated rewards of the selected elements. The order-competitive ratio is defined as the worst-case ratio (over all distribution sequences) between the performance of the best order-unaware and order-aware algorithms, and quantifies the loss incurred due to the lack of knowledge of the arrival order.

Ezra et al. [16] showed how to design algorithms that achieve better approximations with respect to the new benchmark (order-competitive ratio) in the single-choice setting, which raises the natural question of whether the same can be achieved in combinatorial settings. In particular, whether it is possible to achieve a constant approximation with respect to the best online algorithm for downward-closed feasibility constraints, whether $\omega(1/n)$-approximation is achievable for general (non-downward-closed) feasibility constraints, or whether a convergence rate of $1 - o(1/\sqrt{k})$ is achievable for the multi-unit setting. We show, by devising novel constructions that may be of independent interest, that for all three scenarios, the asymptotic lower bounds with respect to the old benchmark, also hold with respect to the new benchmark.

Keywords: Prophet inequality · Order-competitive ratio · Optimal stopping

Tomer Ezra is partially supported by ERC Advanced Grant 788893 AMDROMA "Algorithmic and Mechanism Design Research in Online Markets" and MIUR PRIN project ALGADIMAR "Algorithms, Games, and Digital Markets", by the National Science Foundation under Grant No. DMS-1928930 and by the Alfred P. Sloan Foundation under grant G-2021-16778, while the author was in residence at the Simons Laufer Mathematical Sciences Institute (formerly MSRI) in Berkeley, California, during the Fall 2023 semester.

J. Garg et al. (Eds.): WINE 2023, LNCS 14413, pp. 256–271, 2024.
https://doi.org/10.1007/978-3-031-48974-7_15

1   Introduction

We revisit the prophet inequality problem in combinatorial settings. In the prophet inequality setting [26, 27, 33], there is a sequence of boxes, each contains a stochastic reward drawn from a known distribution. The rewards are revealed one by one to a decision-maker, that needs to decide whether to take the current reward, or continue to the next box. The decision-maker needs to make the decisions in an immediate and irrevocable way, where her goal is to maximize her expected selected reward. The most common performance measure for the analysis of the decision-maker policy is the competitive-ratio, which is the ratio between the expectation of the selected reward and the expected maximum reward. That is, the decision-maker is evaluated by comparison to a "prophet" who can see into the future and select the maximal reward. This framework has been extended to combinatorial settings, where the decision-maker is allowed to select a set of boxes (instead of only one) under some predefined feasibility constraints, such as multi-unit [2, 20], matroids [25], matching [15, 18], and downward-closed (or even general) feasibility constraints [31].

A recent line of work studied the (combinatorial) prophet setting when instead of comparing to the best offline optimum (or the "prophet"), they compare against the best online algorithm [8, 28, 29, 34], and showed how to achieve tighter approximations compared to the best online algorithms.

Recently, Ezra et al. [16] suggested the benchmark termed "order-competitive ratio" defined as the worst-case ratio (over all distribution sequences) between the expectations of the best *order-unaware* algorithm and the best *order-aware* algorithm. Thus, the order-competitive ratio quantifies the loss that is incurred to the algorithm due to an unknown arrival order. Ezra et al. [16] showed that for the single-choice prophet inequality setting, it is possible to achieve $1/\phi$-approximation with respect to the new benchmark (where ϕ is the golden ratio). In particular, they showed a separation between what adaptive and static algorithms can achieve with respect to the new benchmark, while with respect to the optimum offline, there is no such separation as a static threshold can achieve the tight approximation of $1/2$.

The question that motivates this paper is whether one can achieve improved approximations for the new benchmark in combinatorial settings. In particular, whether it is possible to achieve a constant approximation with respect to the best online algorithm for downward-closed feasibility constraints, whether $\omega(1/n)$-approximation is achievable for general (non-downward-closed) feasibility constraints, or whether a convergence rate of $1 - o(1/\sqrt{k})$ is achievable for the multi-unit setting.

1.1   Our Contribution, Techniques, and Challenges

We study this question in three natural and generic combinatorial structures: k-uniform matroid (also known as multi-unit), downward-closed, and arbitrary (not downward-closed) feasibility constraints.

The first scenario we consider is downward-closed feasibility constraints. We first revisit the example in [31] that is based on the upper bound by Babaioff et al. [6] for a different setting, that shows that no algorithm can achieve an approximation of $\omega\left(\frac{\log\log(n)}{\log(n)}\right)$:

Example 1 ([6]). Consider a set of $n = 2^{2^k}$ elements, that are partitioned into $2^{2^k - k}$ parts, each of size 2^k. The reward of each element is 1 with probability 2^{-k} and 0 otherwise. The feasibility constraint is such that the decision-maker is allowed to select elements from at most one part of the partition. The elements arrive in an arbitrary order. It is easy to verify that the expected value of the prophet is $\Omega(2^k)$, since it is a maximum of $2^{2^k - k}$ random variables that are distributed according to $Bin(2^k, 2^{-k})$. On the other hand, no online algorithm can have an expected reward of more than 2, since once the algorithm decides to select an element (with a value at most 1), then the expectation of the sum of the remaining feasible elements is bounded by 1.

As can be observed in Example 1, the instance is constructed in a way that no online algorithm (order aware or unaware) can achieve an expected reward of more than 2, while achieving an expected reward of 1 is trivial. Thus, it fails to show a gap between what order-aware and order-unaware algorithms can achieve. This leads us to our first result.

Result A (Theorem 2): No order-unaware algorithm can achieve an approximation of $\omega\left(\frac{\log\log(n)}{\log(n)}\right)$ with respect to the best order-aware online algorithm for downward-closed feasibility constraints.

To show Result A, we need to develop an entirely different construction than the one used in [6]. Their construction is such that once the online algorithm selects an arbitrary element, it eliminates all the flexibility that the algorithm had in choosing elements due to the feasibility constraint. All attempts that are only based on the construction of the feasibility constraint, are destined to fail since the feasibility constraint will influence both the order-aware and order-unaware algorithms in the same way. Thus, we construct a pair of a feasibility constraint and a distribution over arrival orders. Our elements are partitioned into k layers, and within each layer, the elements are symmetric (with respect to the feasibility constraint). An algorithm needs to select at most one element of each layer. The difference between the elements within the layers, is the role with respect to the arrival order, which draws half of them to be "good", and half of them to be "bad". "Good" elements, are such that the best order-aware algorithm does not lose a lot by choosing them, and "bad" elements, are such that the best order-aware algorithm does lose a lot by choosing them. An order-aware algorithm can distinguish between "good" and "bad" elements and can always choose the "good" ones, while an order-unaware algorithm cannot distinguish between them, therefore cannot do better than guessing and thus it will guess a "bad" one after a constant number of layers in expectation.

The second scenario that we consider is of arbitrary feasibility constraints. For this problem with respect to the best offline algorithm as a benchmark,

Rubinstein [31] showed that no online algorithm can achieve a competitive-ratio of $\omega\left(\frac{1}{n}\right)$. Achieving a competitive-ratio of $\frac{1}{n}$ can be done trivially by selecting the feasible set with the maximal expectation. We next revisit the example in [31] that shows that no online algorithm can achieve an approximation of $\omega\left(\frac{1}{n}\right)$.

Example 2 ([31]). Consider an instance with $n = 2k$ elements, where the collection of feasible sets is $\{\{i, i + k\} \mid i \in [k]\}$. The elements arrive according to the order $1, \ldots, n$, and the value of each element in $[k]$ is deterministically 0, while the value of each element in $\{k + 1, \ldots, n\}$ is 1 with probability $\frac{1}{n}$, and 0 otherwise. The prophet receives a value of 1 if one of the elements of the second type has a non-zero value, which happens with a constant probability. Every online algorithm must select exactly one element among the elements of the first type, which restricts the algorithm to select a specific element of the second type, therefore every online algorithm has an expected value of $\frac{1}{n}$.

As can be observed in Example 2, the instance is constructed with a fixed order, and the optimal algorithm for this feasibility constraint (even for every arrival order), is to discard all zero-value elements and select all elements with a value of 1 as long as there is a way to complete the chosen set to a feasible set. This algorithm is an order-unaware algorithm, and therefore this construction does not induce a separation between what order-unaware and order-aware algorithms can achieve. This leads us to our second result.

Result B (Theorem 3): No order-unaware algorithm can achieve an approximation of $\frac{1+\Omega(1)}{n}$ with respect to the best order-aware online algorithm for general feasibility constraints.

Our result improves upon the result in [31] in two dimensions. First, our result is with respect to the tighter benchmark of the best online algorithm rather than the best offline algorithm. Second, our upper bound matches the lower bound, up to low-order terms (and not just up to a constant).

To show Result B, we create three types of elements: The first type of elements is of elements with a value of 1 with a small probability. Almost all elements are of this type, and the utility of the instance comes from these elements. The feasibility constraint requires selecting exactly one of these elements. The elements of the other two types have a deterministic value of 0, and their role is to limit the ability of the algorithm to select elements of the first type. The feasibility constraint is such that for each subset of elements of type 2, and each element of type 1, there is exactly one subset of elements of type 3 such that their union is feasible. The order of arrival is such that in Phase 1, the elements of type 2 arrive, in Phase 2, most of the elements of type 3, in Phase 3, the elements of type 1 arrive, and in Phase 4, the remaining (few) elements of type 3 arrive. For exactly one subset X of the elements of type 2, it holds that: for each element e of type 1, there is a subset X_e of elements of type 3 that arrive in Phase 4, such that $X \cup \{e\} \cup X_e$ is a feasible set. For all other choices of X, there are at most a few feasible elements of type 3 that arrive at Phase 4, which restricts the algorithm to choose only among a few elements of type 1. The only way to "catch" the value of all the elements of type 1, is to correctly guess the unique good subset X of type 2 with this special property. An order-aware algorithm can always guess it correctly as this information can be derived from the arrival order (since it knows the partition of elements of type 3 between Phase 2 and Phase 4), while an order-unaware cannot guess the correct subset with high enough probability, and therefore it loses a factor of $\frac{1}{n}$ in the approximation.

The third scenario that we consider is of k-capacity feasibility constraints. For this problem with respect to the best offline algorithm as a benchmark, Hajiaghayi et al. [20] showed that no online algorithm can achieve a competitive-ratio of $1 - o\left(\frac{1}{\sqrt{k}}\right)$. Achieving a competitive-ratio of $1 - \Theta\left(\frac{1}{\sqrt{k}}\right)$ with respect to the best-offline is achieved by Alaei et al. [2].

Our last result shows, that one cannot achieve an order-competitive ratio that converges to 1 in a faster rate (up to a constant).

Result C (See Full Version [17]): No order-unaware algorithm can achieve an approximation of $1 - o\left(\frac{1}{\sqrt{k}}\right)$ with respect to the best order-aware online algorithm for k-uniform feasibility constraint.

To show Result C, we construct an instance with three types of elements. The first type is with a deterministic low value, the second type is with a deterministic mid-value, and the third type is randomized, with a probability half of being high, and a probability half of being zero. The order of arrival is such that all the type 2 elements arrive first, and then either all elements with type 1 arrive before all elements of type 3 which is considered the "bad" order, or vice versa which is the "good" order. An algorithm that knows whether it is a good order or a bad order, can adapt the number of elements of type 2 to choose in an optimal way, while an algorithm that does not know the order needs to commit to selecting elements of type 2 before any information regarding the order is revealed. Our analysis then follows by balancing the low, mid, and high values in a way that an order-unaware algorithm that commits to selecting a certain amount of elements of type 2, will be far from the optimal order-aware algorithm for one of the two arrival orders[1].

1.2 Further Related Work

Comparing to the Best Online. Our work is largely related to a line of research that examines alternative benchmarks for the best offline benchmark, and in particular, comparing its performance to the best online algorithm [8,16,24,28,29,32,34]. One example, Niazadeh et al. [28] showed that the original tight prophet inequality bounds comparing the single-pricing with the optimum offline are tight even when comparing to the optimum online as a benchmark (both for the identical and non-identical distributions). Another example is that Papademitriou et al. [29] studied the online stochastic maximum-weight matching problem under vertex arrivals, and presented a polynomial-time algorithm which approximates the optimal online algorithm within a factor of 0.51, which

---

[1] Due to space limit, this result is omitted and can be found in the full version in [17].

was later improved by Saberi and Wajc [32] to 0.526, to $1 - 1/e$ by Braverman et al. [8], and to 0.652 by [34]. Kessel et al. [24] studied a continuous and infinite time horizon counterpart to the classic prophet inequality, term the stationary prophet inequality problem. They showed how to design pricing-based policies which achieve a tight $\frac{1}{2}$-approximation to the optimal offline policy, and a better than $(1 - 1/e)$-approximation of the optimal online policy.

Prophet in Combinatorial Settings. Another line of work, initiated by Kennedy [21,22], and Kertz [23], extends the single-choice optimal stopping problem to multiple-choice settings. Later work extended it to additional combinatorial settings, including multi-unit [2,20] matroids [4,25], polymatroids [12], matching [15,19], combinatorial auctions [10,11,18], and downward-closed (and beyond) feasibility constraints [31].

Different Arrival Models. A related line of work studied different assumptions on the arrival order besides the adversarial order [26,27,33]. Examples for such assumptions are random arrival order (also known as the prophet secretary) [5,9,13,14], and free-order settings, where the algorithm may dictate the arrival order [1,7,30]. Another recent study related to the arrival order has shown that for any arrival order π, the better of π and the reverse order of π achieves a competitive-ratio of at least the inverse of the golden ratio [3].

2   Model

An instance $\mathcal{I}$ of our setting is defined by a triplet $\mathcal{I} = (E, \mathcal{D}, \mathcal{F})$ where E is the ground set of elements, each element $e \in E$ is associated with a distribution $D_e \in \mathcal{D}$, and a feasibility constraint $\mathcal{F} \subseteq 2^E$ over the set of elements (where $\mathcal{F} \neq \emptyset$). The elements arrive one by one. Upon the arrival of element e, its identity is revealed, and a value v_e is drawn independently from the underlying distribution D_e. We call an instance $\mathcal{I}$ binary if for every element $e \in E$, the support of D_e is $\{0, 1\}$.

A decision-maker, who observes the sequence of elements and their values, needs to decide upon the arrival of each element whether to select it or not subject to the feasibility constraint $\mathcal{F}$, which asserts that the set that is chosen at the end of the process (after all elements arrive) must belong to $\mathcal{F}$. Another interpretation of the feasibility constraint, is that the decision-maker must select (respectively discard) element e if all feasible sets that agree with all previous decisions before the arrival of element e, contain (respectively do not contain) element e. A feasibility constraint $\mathcal{F}$ is called *downward-closed* if for every set $S \in \mathcal{F}$, and a subset $T \subseteq S$, then T must be in $\mathcal{F}$. For downward-closed feasibility constraints, discarding elements is always feasible. The decision-maker's utility is the sum of the values of the selected elements. A well-studied special case of downward-closed feasibility constraints is the case of *k-uniform matroid* (also known as multi-unit) for which $\mathcal{F} = \{S \subseteq E \mid |S| \leq k\}$.

We say that a decision-maker (or algorithm) is *order-unaware* if she does not know the arrival order of the elements in advance, and needs to make decisions with uncertainty regarding the order of the future arriving elements. We say that a decision-maker (or algorithm) is *order-aware*, if she knows the order of arrival of the elements in advance, and can base her decisions on this information. Given an instance $\mathcal{I}$, an order of arrival of the elements π, and an algorithm $ALG_{\mathcal{I}}$ (that might be order-unaware, or order-aware), we will denote the expected utility of $ALG_{\mathcal{I}}$ for arrival order π, by $ALG_{\mathcal{I}}(\pi)$. Given an instance $\mathcal{I}$ and an arrival order π, we will denote the order-aware algorithm with the maximal expected utility by $OPT_{\mathcal{I},\pi}$, i.e., $OPT_{\mathcal{I},\pi} \stackrel{\text{def}}{=} \arg\max_{ALG_{\mathcal{I}}} ALG_{\mathcal{I}}(\pi)$.

We want to quantify the importance of knowing the order in advance, and to do so, we use the measure of *order-competitive ratio* proposed by Ezra et al. [16] for the case of choosing a single element (i.e., $\mathcal{F} = \{S \subseteq E \mid |S| \leq 1\}$). Given an instance $\mathcal{I}$, the order-competitive ratio of an order-unaware algorithm $ALG_{\mathcal{I}}$, denoted by $\rho(\mathcal{I}, ALG_{\mathcal{I}})$ is

$$\rho(\mathcal{I}, ALG_{\mathcal{I}}) \stackrel{\text{def}}{=} \min_{\pi} \frac{ALG_{\mathcal{I}}(\pi)}{OPT_{\mathcal{I},\pi}(\pi)}. \tag{1}$$

We use $[j]$ to denote the set $\{1, \ldots, j\}$. Given two partial orders $\pi^1 = (e_1^1, \ldots, e_{k_1}^1)$, and $\pi^2 = (e_1^2, \ldots, e_{k_2}^2)$ over two disjoint subsets of elements $E_1, E_2 \subseteq E$, we define the order $\pi^1 \cdot \pi^2 \stackrel{\text{def}}{=} (e_1^1, \ldots, e_{k_1}^1, e_1^2, \ldots, e_{k_2}^2)$.

In this paper, we use the following forms of Chernoff bound:

Theorem 1 (Chernoff bound). *For a series of n independent Bernoulli random variables $X_1, \ldots, X_n$, and for $X = \sum_{i=1}^{n} X_i$ it holds:*

- *For all $0 \leq \delta \leq 1$, $\Pr\left[|X - E[X]| \geq \delta E[X]\right] \leq 2e^{-\delta^2 \cdot E[X]/3}$.*
- *For all $\delta \geq 0$, $\Pr\left[X \geq (1 + \delta)E[X]\right] \leq e^{-\delta^2 \cdot E[X]/(2+\delta)}$.*

Lastly, for an instance $\mathcal{I} = (E, \mathcal{D}, \mathcal{F})$, and an algorithm $ALG_{\mathcal{I}}$ we denote by $\xi(\mathcal{I}, ALG_{\mathcal{I}})$ the traditional competitive ratio which is

$$\xi(\mathcal{I}, ALG_{\mathcal{I}}) \stackrel{\text{def}}{=} \min_{\pi} \frac{ALG_{\mathcal{I}}(\pi)}{E[\max_{S \in \mathcal{F}} \sum_{e \in S} v_e]}. \tag{2}$$

It is easy to observe, that for every instance $\mathcal{I}$, and an algorithm $ALG_{\mathcal{I}}$,

$$\xi(\mathcal{I}, ALG_{\mathcal{I}}) \leq \rho(\mathcal{I}, ALG_{\mathcal{I}}),$$

thus, every lower bound on the competitive-ratio also applies to the order-competitive ratio (but not vice versa), and any upper bound on the order-competitive ratio also applies to the order-competitive ratio (but not vice versa).

3   Downward-Closed Feasibility Constraints

In this section, we show an upper bound on the order-competitive ratio for the family of downward-closed feasibility constraints. This upper bound also holds with respect to binary instances and matches the best-known upper bound on the competitive-ratio. The current best-known lower bound for the competitive-ratio for downward-closed feasibility constraints of $O\left(\frac{1}{\log^2(n)}\right)$ was proved by Rubinstein [31], and closing this gap is an open question.

Theorem 2. *There exists a constant $\mu > 0$ such that for every $n > 2$ there is a (binary) instance $\mathcal{I} = (E, \mathcal{D}, \mathcal{F})$ with $n = |E|$ and a downward-closed feasibility constraint $\mathcal{F}$ in which for every order-unaware algorithm (deterministic or randomized) $ALG_{\mathcal{I}}$, it holds that*

$$\rho(\mathcal{I}, ALG_{\mathcal{I}}) \leq \frac{\mu \cdot \log \log n}{\log n}.$$

Proof: We assume that $n = \sum_{i=1}^{k} k^i$ for some even k. (Otherwise, we can reduce to the largest $n' \leq n$ that is of this form, by having $n - n'$ redundant elements.) Notice that

$$k \in \Theta \left(\frac{\log n}{\log \log n} \right), \tag{3}$$

since for $k = \frac{\log n}{2 \log \log n}$, it holds that $\sum_{i=1}^{k} k^i \leq k^{k+1} \leq n$, while for $k = \frac{2 \log n}{\log \log n}$, it holds that $\sum_{i=1}^{k} k^i \geq k^k \geq n$ for large enough n. For every string s of length between 1 and k where each character is in $[k]$, we define an element e_s. We denote by s_j for $j \in [|s|]$ the j-th character of the string s, moreover, we denote by $s_{[j]}$ the prefix of s of the first j characters. The set of elements E is defined to be $\{e_s \mid s \in \bigcup_{i=1}^{k} [k]^i\}$. Given a string s and a character j (respectively, another string s'), we denote by sj (respectively, ss') the string-concatenation of j (respectively, s') at the end of string s. We say that an element e_{sj} for a string s and $j \in [k]$ is a child of element e_s, and that e_s is the parent of e_{sj}. (Note that an element can have only one parent, but may have multiple children.) The value of all elements are drawn i.i.d. from the distribution D in which $v = 1$ with probability $\frac{1}{k}$, and $v = 0$ otherwise. Let $\mathcal{D} \stackrel{\text{def}}{=} \{D\}_{e \in E}$. The feasibility constraint $\mathcal{F} \stackrel{\text{def}}{=} \{S \subseteq E \mid$ for every $e_{s_1}, e_{s_2} \in S$, if $|s_1| \leq |s_2|$, then s_1 is a prefix of $s_2\}$ (in other words, only subsets of a single path from the root to one of the leafs are feasible). The instance is then $\mathcal{I} = (E, \mathcal{D}, \mathcal{F})$. It is sufficient to show that for some constant $c > 0$, there is a distribution F over the arrival orders, in which the expected utility of every order-unaware algorithm $ALG_{\mathcal{I}}$ is at most c/k of the expected utility of the optimal order-aware algorithm. I.e.,

$$\exists c \quad \forall ALG_{\mathcal{I}} \qquad E_{\pi \sim F}[ALG_{\mathcal{I}}(\pi)] \leq \frac{c}{k} \cdot E_{\pi \sim F}[OPT_{\mathcal{I}, \pi}(\pi)]. \tag{4}$$

Equation (4) is sufficient since it shows that for every algorithm $ALG_{\mathcal{I}}$ there exists an order π^* (in the support of F) in which $ALG_{\mathcal{I}}(\pi^*) \leq c/k \cdot OPT_{\mathcal{I}, \pi^*}(\pi^*)$, which together with Eq. (3) concludes the proof.

We now define the distribution F over the arrival orders. We first draw independently for every string s of size between 0 and $k - 3$, a random subset of $[k]$ of size $k/2$, which we will denote by r_s. Then, the elements arrive in an arrival order defined by the following recursive formulas. We first define for every string s of size between 0 and $k - 1$ and a parameter $i \in [k - |s|]$:

$$\pi_0(s, i) \overset{\text{def}}{=} (ss')_{s' \in [k]^i},$$

and

$$\pi_0(s) \overset{\text{def}}{=} \pi_0(s, k - |s|) \cdot \ldots \cdot \pi_0(s, 2) \cdot \pi_0(s, 1).$$

We also denote given the random realizations $\{r_s\}_s$, for every string s of size between 0 to $k - 3$ the arrival order

$$\pi_1(s) \overset{\text{def}}{=} (s1, \ldots, sk) \cdot \pi_{\mathbb{1}_{1 \in r_s}}(s1) \cdot \ldots \cdot \pi_{\mathbb{1}_{j \in r_s}}(sj) \cdot \ldots \cdot \pi_{\mathbb{1}_{k \in r_s}}(sk),$$

and for s such that $|s| = k - 2$,

$$\pi_1(s) \overset{\text{def}}{=} (s1, \ldots, sk) \cdot \pi_0(s1) \cdot \ldots \cdot \pi_0(sk).$$

The arrival order is then $\pi_1(\epsilon)$.

For every element e_s, we say that e_s is *good*, if for every $j \in [\min(k - 2, |s|)]$, it holds that $s_j \in r_{s_{[j-1]}}$, and *bad* otherwise. The order of arrival is illustrated in Fig. 1.

We first bound from below the RHS of Eq. (4).

Lemma 1. *For $c' = \frac{\sqrt{e}}{\sqrt{e}-1}$, it holds that $E_{\pi \sim F}[OPT_{\mathcal{I},\pi}(\pi)] \geq k/c'$.*

Proof: We prove this lemma by showing that for every order π in the support of F, it holds that $OPT_{\mathcal{I},\pi}(\pi) \geq \frac{k}{c'}$. Consider an order-aware algorithm (not necessarily $OPT_{\mathcal{I},\pi}$) that selects an element e_s if (1) e_s is feasible, (2) e_s is good, and (3) $v_{e_s} = 1$ or e_s is the last good element to arrive in the set $\{s_{[|s|-1]}j \mid j \in [k]\}$.

By the description of the algorithm we know we will only select good elements, and we will select exactly one element from each layer (elements of strings with the same length). The algorithm receives a utility of 1 from layer $j \in [k]$ if one of the good elements that are the children of the element chosen from layer $j - 1$, has a value of 1. (For elements of layer 1, it is sufficient that one of the good elements, has a value of 1.) Thus, the expected utility of the algorithm is at least the number of layers, times the probability that one of the (at least) $k/2$ elements has a value of 1. Therefore

$$OPT_{\mathcal{I},\pi}(\pi) \geq k \cdot (1 - (1 - \frac{1}{k})^{k/2}) \geq k \cdot (1 - \frac{1}{\sqrt{e}}) = \frac{k}{c'},$$

which concludes the proof. □

We next bound from above the LHS of Eq. (4).

Lemma 2. *For every (deterministic or randomized) order-unaware algorithm $ALG_{\mathcal{I}}$, it holds that $E_{\pi \sim F}[ALG_{\mathcal{I}}(\pi)] \leq 5$.*

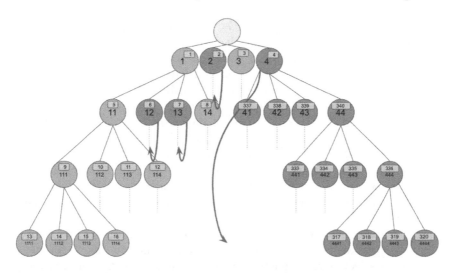

Fig. 1. An example of an instance with $k = 4$. In this example, there are $n = 4 + 4^2 + 4^3 + 4^4 = 340$ elements. The elements are partitioned into layers according to the structure of the feasibility constraint. A feasible set under this constraint is a subset of a path from the root to some leaf in the tree (excluding the root which is not an element). The numbers in the center of the circles represent the corresponding string-identity of the elements. Green circles represent good elements, and red circles represent bad elements. For each layer up to the last two layers, all the children of bad elements are bad, and half of the children of good elements are good (and half of them are bad). For the last two layers, all the children of good elements are good, and all the children of bad elements are bad. In this example, the realizations of the random variables $\{r_s\}_s$ are such that $r_\epsilon = \{1,3\}$, and $r_1 = \{1,4\}$. The numbers in the blue rectangles represent the arrival time of the element according to the arrival order. The outgoing red arrows from bad elements that aren't children of bad elements, represent that after the arrival of this element, the next descendant among the sub-tree rooted at this element that is arriving according to the arrival order, is not a child of the element (as happens with good elements) but rather is a leaf, and the order of arrival of this sub-tree is bottom up. (Color figure online)

Proof: We analyze the performance of $ALG_\mathcal{I}$ by partitioning into three types of contributions: (1) good elements, (2) bad elements that are either children of good elements or in the first layer, and (3) bad elements that are children of bad elements.

We first claim that the expected number of elements of type (1) that $ALG_\mathcal{I}$ selects is at most 2. To show this we can first observe that once a bad element is chosen, then good elements cannot be chosen anymore. After a bad element is chosen, the only elements that can be chosen are the offspring of this element (which are also bad by definition) and the ancestors of the element that haven't arrive yet (which all must be bad). We next observe, that the algorithm can only select good elements in a strictly increasing order (in the length of their corresponding strings). Moreover, for every element e_s from layer j for $j \in [k-2]$, that is a child of a good element or is of layer 1, given the information that the algorithm has up to the arrival of element e_s, the probability of being good is exactly $1/2$. This is since being good, by definition requires that (1) e_s is not a child of a bad element (which the algorithm knows upon the arrival of e_s), and (2) $s_{|s|} \in r_{|s|-1}$, which happens with probability $1/2$. Thus, each time the algorithm tries to select a good element from the first $k-2$ layers, it can no longer select additional good elements with probability $1/2$. If the algorithm reaches layer $k-1$ without selecting a bad element, the algorithm can select at most two more good elements. Therefore if the algorithm tries to "gamble" and select ℓ good elements from the first $k-2$ layers, it selects in expectation at most $\frac{\ell+2}{2^\ell} + \sum_{i=1}^{\ell} \frac{i-1}{2^i} \le 2$ good elements from all k layers[2].

Second, $ALG_\mathcal{I}$ can choose at most one element of type (2). This is since in every feasible set, there is at most one such element. (For every feasible set, only the element that corresponds to the shortest string among the bad ones can be of this type).

Last, the expected utility of $ALG_\mathcal{I}$ from elements of type (3) is at most 2. This is true since we can observe that once a bad element e that is a child of a bad element is selected, the algorithm can only select elements that are ancestors of e. Since there are less than k such elements, and each can contribute a utility of at most $1/k$ in expectation, the expected utility of elements of this type is less than 2. (Element e contributes 1, and its ancestors contributes less than $k \cdot \frac{1}{k}$.) This concludes the proof. □

The theorem follows by combining Lemmata 1 and 2, with Eq. (3). □

4 Non-downward Closed Feasibility Constraints

In this section, we present an upper-bound on the order-competitive ratio of arbitrary (non-downward closed) feasibility constraints. This upper-bound holds even with respect to binary instances. This result is tight since achieving an order-competitive ratio of $\frac{1}{n}$ can be done trivially, by an algorithm that selects the set of elements with the maximum expected sum of values among all feasible sets. Our result also improves the best-known upper bound of the competitive-ratio shown in [31] of $O\left(\frac{1}{n}\right)$ to $\frac{1}{n} + o\left(\frac{1}{n}\right)$.

Theorem 3. *For every constant $\mu > 1$, there exists n_0 such that for every $n \ge n_0$, there exists an instance $\mathcal{I} = (E, \mathcal{D}, \mathcal{F})$ with n elements (i.e., $n = |E|$), in which for every order-unaware algorithm (deterministic or randomized) $ALG_\mathcal{I}$, it holds that $\rho(\mathcal{I}, ALG_\mathcal{I}) \le \frac{\mu}{n}$.*

Proof: We prove that theorem by presenting a construction with n elements, for which no order-unaware algorithm can have an order-competitive ratio of more than $\frac{1}{n} + o(\frac{1}{n})$. We assume for simplicity that $n = 2^{2x}$ for some integer x. Consider an instance $\mathcal{I} = (E, \mathcal{D}, \mathcal{F})$ in which $E = A \cup B \cup C$, where $A = \{a_1, \ldots, a_{k_1}\}$, $B = \{b_1, \ldots, b_{k_2}\}$, and $C = \{c_1 \ldots, c_{k_3}\}$ where $k_1 = 4x$, $k_2 = n - \sqrt{n} - 4x$, and $k_3 = \sqrt{n}$ which sum up to n. The values of all elements in $A \cup C$ are

---

[2] This argument also holds for randomized ℓ.

deterministically 0. The values of all elements in B are 1 with probability $\frac{1}{n^2}$ and 0 otherwise. Let $U_1, \ldots, U_{2^{k_1}}$ be subsets of C which satisfy the conditions from the following lemma:

Lemma 3. *There exists n_0' such that for every $n \geq n_0'$, there exists a collection of sets $U_1, \ldots, U_{2^{k_1}}$ such that:*

- *For all $i \in [2^{k_1}]$, $\log(n) \leq |U_i| \leq 21 \cdot \log(n)$.*
- *For each $j \in [k_3]$, it holds that $|\{i \mid c_j \in U_i\}| \leq \frac{22 \cdot 2^{k_1} \cdot \log(n)}{k_3}$.*
- *For all $i_1, i_2 \in [2^{k_1}]$ such that $i_1 \neq i_2$, it holds that $|U_{i_1} \cap U_{i_2}| \leq 10$.*

Proof: We prove existence by the probabilistic method. For simplicity of presentation, let $\alpha = 10$. Consider a series of random variables X_{ij} that indicate whether $c_j \in U_i$, which are drawn independently according to $Ber\left(\frac{(\alpha+1) \cdot \log(n)}{k_3}\right)$. Note that for the parameter α and for $n \geq 2^{16}$ this probability is guaranteed to be in $[0, 1]$. Let E_i^1 be the event that $|U_i| < \log(n)$ or $|U_i| > (2\alpha + 1) \cdot \log(n)$ (which is equivalent to $|\sum_j X_{ij} - (\alpha + 1) \cdot \log(n)| > \alpha \cdot \log(n)$), let E_j^2 be the event that $|\{i \mid c_j \in U_i\}| > \frac{2(\alpha+1) \cdot 2^{k_1} \cdot \log(n)}{k_3}$ (which is equivalent to $\sum_i X_{ij} > \frac{2 \cdot (\alpha+1) \cdot 2^{k_1} \cdot \log(n)}{k_3}$), let E_{i_1,i_2}^3 be the event that $|U_{i_1} \cap U_{i_2}| > \alpha$ (which is equivalent to $\sum_j X_{i_1 j} \cdot X_{i_2 j} > \alpha$), and let E be the event that one of the formerly defined events occurs, i.e., $E = \left(\bigvee_i E_i^1 \vee \bigvee_j E_j^2 \vee \bigvee_{i_1 \neq i_2} E_{i_1,i_2}^3\right)$. For every $i \in [2^{k_1}]$, it holds that

$$\Pr[E_i^1] = \Pr\left[\left|Bin\left(k_3, \frac{(\alpha+1) \cdot \log(n)}{k_3}\right) - (\alpha+1) \cdot \log(n)\right| > \alpha \cdot \log(n)\right] \leq \frac{2}{n^3},$$

where the inequality is by Chernoff bound. For every $j \in [k_3]$ it holds that

$$\Pr[E_j^2] = \Pr\left[Bin\left(2^{k_1}, \frac{(\alpha+1) \cdot \log(n)}{k_3}\right) > \frac{2(\alpha+1) \cdot 2^{k_1} \cdot \log(n)}{k_3}\right] \leq \frac{1}{n^3},$$

where the inequality is by Chernoff Bound. For all $i_1, i_2 \in [2^{k_1}]$, such that $i_1 \neq i_2$ it holds that

$$\Pr[E_{i_1,i_2}^3] = \Pr\left[Bin\left(k_3, \frac{(\alpha+1)^2 \cdot \log^2(n)}{k_3^2}\right) > \alpha\right]$$

$$\leq \binom{k_3}{\alpha+1} \cdot \left(\frac{(\alpha+1)^2 \cdot \log^2(n)}{k_3^2}\right)^{\alpha+1} \leq \frac{1}{n^5},$$

where the first inequality holds by the union bound, and the second inequality holds for large enough n (for $n > 2^{1000}$). Thus, by the union bound, the probability that one of the events occurs is

$$\Pr[E] \leq 2^{k_1} \cdot \frac{2}{n^3} + k_3 \cdot \frac{1}{n^3} + 2^{2k_1} \cdot \frac{1}{n^5} < 1.$$

Thus, there exist realizations of all X_{ij} in which event E does not occur, which concludes the proof.   □

Next, we name the subsets of A as $V_1, \ldots, V_{2^{k_1}}$, and we define 2^{k_1} corresponding functions. For each $i \in [2^{k_1}]$, we define an arbitrary injective function $f_i : [k_2] \to 2^{U_i}$ (such a function exists since $|U_i| \geq \log(n) \geq \log(k_2)$). We now define the feasibility constraint

$$\mathcal{F} \overset{\text{def}}{=} \{S \mid \exists i, j \text{ such that } S \cap A = V_i \wedge S \cap B = \{b_j\} \wedge S \cap C = f_i(j)\}.$$

For every i, let π_i be the arrival order in which the elements arrive in four phases (within each phase, the order can be arbitrary but during the first phase the order should be consistent for all π_i). Phase 1 is composed of all elements of A. Phase 2 is composed of all elements of $C \setminus U_i$. Phase 3 is composed of all elements in B, and Phase 4 is composed of all elements in U_i, i.e.,

$$\pi_i = \left(\underbrace{A}_{\text{Phase 1}}, \underbrace{C \setminus U_i}_{\text{Phase 2}}, \underbrace{B}_{\text{Phase 3}}, \underbrace{U_i}_{\text{Phase 4}} \right).$$

We next bound from below for every π_i the performance of $OPT_{\mathcal{I}, \pi_i}$ on π_i.

Lemma 4. *For every π_i, it holds that $OPT_{\mathcal{I}, \pi_i}(\pi_i) \geq \frac{1}{n} - o\left(\frac{1}{n}\right)$.*

Proof: Consider the order-aware algorithm, that selects in Phase 1 the subset V_i of A. Then in Phase 2 it selects nothing. In Phase 3 it selects the first element b_j of B that its value is 1 (or the last element of Phase 3, if all of them have values of 0). In Phase 4, the algorithm selects the subset $f_i(j)$ of C. This is always a feasible set. The value of this set is 1, if one of the elements in B has a non-zero value. The lemma then holds since this happens with probability $1 - (1 - \frac{1}{n^2})^{k_2} = \frac{1}{n} - o\left(\frac{1}{n}\right)$. □

In order to bound the performance of a randomized algorithm $ALG_{\mathcal{I}}$, it is sufficient by Yao's principle to define a distribution D_π over arrival orders, and bound the performance of the best deterministic algorithm on the randomized distribution. Consider the distribution D_π, where the order $\pi \sim D_\pi$ is π_i with probability $\frac{1}{2^{k_1}}$ for every $i \in [2^{k_1}]$. We next bound from above the performance of any deterministic algorithm $ALG_{\mathcal{I}}$.

Lemma 5. *For every deterministic algorithm $ALG_{\mathcal{I}}$, it holds that*

$$E_{\pi \sim D_\pi}[ALG_{\mathcal{I}}(\pi)] \leq \frac{1}{n^2} + o\left(\frac{1}{n^2}\right).$$

Proof: Let $ALG_{\mathcal{I}}$ be an arbitrary deterministic algorithm, then since in Phase 1, the order is constant, $ALG_{\mathcal{I}}$ selects deterministically a set $V_{i'} \subseteq A$. We next analyze the performance of $ALG_{\mathcal{I}}$ depending on the realized arrival order π_i. Let $G_{i'} = \{\pi_i \mid U_i \cap U_{i'} \neq \emptyset \wedge \pi_i \neq \pi_{i'}\}$. For every $\pi_i \in G_{i'}$, by Lemma 3 it holds that $|U_i \cap U_{i'}| \leq 10$, then by the end of Phase 2, $ALG_{\mathcal{I}}$ selected a subset of $U_{i'} \setminus U_i$. Since there are at most 10 elements in $U_{i'} \cap U_i$ that didn't arrive by the end of Phase 2, there are at most 2^{10} elements in B that $ALG_{\mathcal{I}}$ can select

that lead to a subset of a feasible set. Thus, it holds that $ALG_{\mathcal{I}}(\pi_i) \leq \frac{2^{10}}{n^2}$.
For the order of arrival $\pi_{i'}$, it holds that $ALG_{\mathcal{I}}(\pi_{i'}) \leq 1 - (1 - \frac{1}{n^2})^{k_2} \leq \frac{1}{n}$.
Otherwise (for every $\pi_i \neq \pi_{i'}$ such that $\pi_i \notin G_{i'}$), it holds that $U_i \cap U_{i'} = \emptyset$,
and therefore by the end of Phase 2, there is only one element that $ALG_{\mathcal{I}}$ can
select which leads to a subset of a feasible set. Thus, $ALG_{\mathcal{I}}(\pi_i) = \frac{1}{n^2}$. The set
$G_{i'}$ is at most of size $\sum_{c_j \in U_{i'}} |\{i \mid c_j \in U_i\}| \leq 21 \cdot \log(n) \cdot \frac{22 \cdot 2^{k_1} \cdot \log(n)}{k_3} = o\left(n^2\right)$,
where the inequality is by Lemma 3. Thus, it holds that $E_{\pi \sim D_\pi}[ALG_{\mathcal{I}}(\pi)] \leq$
$\frac{1}{2^{k_1}} \cdot \frac{1}{n} + \frac{|G_{i'}|}{2^{k_1}} \cdot \frac{2^{10}}{n^2} + \frac{2^{k_1}-1-|G_{i'}|}{2^{k_1}} \cdot \frac{1}{n^2} = \frac{1}{n^2} + o\left(\frac{1}{n^2}\right)$. $\square$

Thus, by combining Lemmata 4, and 5 with Yao's principle, we get that
for every (deterministic or randomized) algorithm $ALG_{\mathcal{I}}$, there exists an arrival
order π_i such that

$$\frac{ALG_{\mathcal{I}}(\pi_i)}{OPT_{\mathcal{I}.\pi_i}(\pi_i)} \leq \frac{1}{n} + o\left(\frac{1}{n}\right),$$

which concludes the proof. $\square$

5   Open Problems

Our goal in this paper was to ask whether, with respect to the new benchmark
of the order-competitive ratio, it is possible to achieve better asymptotic results
than with respect to the traditional competitive-ratio.

One natural Open question is whether in settings where the best competitive-
ratio is half, it is possible to achieve a better than half order-competitive ratio.
Ezra et al. [16] showed that this is possible for single-choice prophet inequality,
but for many other feasibility constraints (e.g., matching, matroids, knapsack,
etc.), this is still an open question. Another open question is what is the best
order-competitive ratio or competitive-ratio for the family downward-closed feasibility constraints, and whether they are the same. The best known lower bound
on the competitive-ratio (and also the order-competitive ratio) is $O\left(\frac{1}{\log^2(n)}\right)$ by
Rubinstein [31].

References

1. Agrawal, S., Sethuraman, J., Zhang, X.: On optimal ordering in the optimal stopping problem. In: EC 2020: The 21st ACM Conference on Economics and Computation, Virtual Event, Hungary, 13–17 July 2020, pp. 187–188. ACM (2020)
2. Alaei, S.: Bayesian combinatorial auctions: expanding single buyer mechanisms to
   many buyers. SIAM J. Comput. 43(2), 930–972 (2014)
3. Arsenis, M., Drosis, O., Kleinberg, R.: Constrained-order prophet inequalities. In:
   Proceedings of the 2021 ACM-SIAM Symposium on Discrete Algorithms (SODA),
   pp. 2034–2046. SIAM (2021)
4. Azar, P.D., Kleinberg, R., Weinberg, S.M.: Prophet inequalities with limited information. In: Proceedings of the Twenty-Fifth Annual ACM-SIAM Symposium on
   Discrete Algorithms, pp. 1358–1377. SIAM (2014)

5. Azar, Y., Chiplunkar, A., Kaplan, H.: Prophet secretary: surpassing the 1-1/e barrier. In: Proceedings of the 2018 ACM Conference on Economics and Computation, pp. 303–318 (2018)
6. Babaioff, M., Immorlica, N., Kleinberg, R.: Matroids, secretary problems, and online mechanisms. In: Symposium on Discrete Algorithms (SODA 2007), pp. 434–443 (2007)
7. Beyhaghi, H., Golrezaei, N., Leme, R.P., Pal, M., Sivan, B.: Improved approximations for free-order prophets and second-price auctions. arXiv preprint arXiv:1807.03435 (2018)
8. Braverman, M., Derakhshan, M., Lovett, A.M.: Max-weight online stochastic matching: improved approximations against the online benchmark. In: EC 2022: The 23rd ACM Conference on Economics and Computation, Boulder, CO, USA, 11–15 July 2022, pp, 967–985. ACM (2022)
9. Correa, J.R., Saona, R., Ziliotto, B.: Prophet secretary through blind strategies. Math. Program. **190**(1), 483–521 (2021)
10. Dutting, P., Feldman, M., Kesselheim, T., Lucier, B.: Prophet inequalities made easy: Stochastic optimization by pricing nonstochastic inputs. SIAM J. Comput. **49**(3), 540–582 (2020)
11. Dütting, P., Kesselheim, T., Lucier, B.: An o(log log m) prophet inequality for subadditive combinatorial auctions. In: 61st IEEE Annual Symposium on Foundations of Computer Science, FOCS 2020, Durham, NC, USA, 16–19 November 2020, pp. 306–317 (2020)
12. Dütting, P., Kleinberg, R.: Polymatroid prophet inequalities. In: Bansal, N., Finocchi, I. (eds.) ESA 2015. LNCS, vol. 9294, pp. 437–449. Springer, Heidelberg (2015). https://doi.org/10.1007/978-3-662-48350-3_37
13. Ehsani, S., Hajiaghayi, M., Kesselheim, T., Singla, S.: Prophet secretary for combinatorial auctions and matroids. In: Proceedings of the Twenty-Ninth Annual ACM-SIAM Symposium on Discrete Algorithms, pp. 700–714. SIAM (2018)
14. Esfandiari, H., Hajiaghayi, M., Liaghat, V., Monemizadeh, M.: Prophet secretary. SIAM J. Discrete Math. **31**(3), 1685–1701 (2017)
15. Ezra, T., Feldman, M., Gravin, N., Tang, Z.G.: Prophet matching with general arrivals. Math. Oper. Res. **47**(2), 878–898 (2022)
16. Ezra, T., Feldman, M., Gravin, N., Tang, Z.G.: "Who is next in line?" On the significance of knowing the arrival order in Bayesian online settings. In: Proceedings of the 2023 Annual ACM-SIAM Symposium on Discrete Algorithms (SODA), pp. 3759–3776. Society for Industrial and Applied Mathematics (2023)
17. Ezra, T., Garbuz, T.: The importance of knowing the arrival order in combinatorial Bayesian settings. arXiv preprint arXiv:2307.02610 (2023)
18. Feldman, M., Gravin, N., Lucier, B.: Combinatorial auctions via posted prices. In: Proceedings of the Twenty-Sixth Annual ACM-SIAM Symposium on Discrete Algorithms, pp. 123–135. SIAM (2014)
19. Gravin, N., Wang, H.: Prophet inequality for bipartite matching: merits of being simple and non adaptive. In: EC, pp. 93–109. ACM (2019)
20. Hajiaghayi, M.T., Kleinberg, R., Sandholm, T.: Automated online mechanism design and prophet inequalities. In: AAAI, vol. 7, pp. 58–65 (2007)
21. Kennedy, D.P.: Optimal stopping of independent random variables and maximizing prophets. Ann. Probab. 566–571 (1985)
22. Kennedy, D.P.: Prophet-type inequalities for multi-choice optimal stopping. Stoch. Process. Their Appl. **24**(1), 77–88 (1987)
23. Kertz, R.P.: Comparison of optimal value and constrained maxima expectations for independent random variables. Adv. Appl. Probab. **18**(2), 311–340 (1986)

24. Kessel, K., Shameli, A., Saberi, A., Wajc, D.: The stationary prophet inequality problem. In: Proceedings of the 23rd ACM Conference on Economics and Computation, pp. 243–244 (2022)
25. Kleinberg, R., Weinberg, S.M.: Matroid prophet inequalities and applications to multi-dimensional mechanism design. Games Econ. Behav. **113**, 97–115 (2019)
26. Krengel, U., Sucheston, L.: Semiamarts and finite values. Bull. Am. Math. Soc. **83**(4), 745–747 (1977)
27. Krengel, U., Sucheston, L.: On semiamarts, amarts, and processes with finite value. Probab. Banach Spaces **4**, 197–266 (1978)
28. Niazadeh, R., Saberi, A., Shameli, A.: Prophet inequalities vs. approximating optimum online. In: Christodoulou, G., Harks, T. (eds.) WINE 2018. LNCS, vol. 11316, pp. 356–374. Springer, Cham (2018). https://doi.org/10.1007/978-3-030-04612-5_24
29. Papadimitriou, C., Pollner, T., Saberi, A., Wajc, D.: Online stochastic max-weight bipartite matching: beyond prophet inequalities. In: Proceedings of the 22nd ACM Conference on Economics and Computation, pp. 763–764 (2021)
30. Peng, B., Tang, Z.G.: Order selection prophet inequality: from threshold optimization to arrival time design. In: FOCS (2022, to appear)
31. Rubinstein, A.: Beyond matroids: secretary problem and prophet inequality with general constraints. In Wichs, D., Mansour, Y. (eds.) Proceedings of the 48th Annual ACM SIGACT Symposium on Theory of Computing, STOC 2016,Cambridge, MA, USA, 18–21 June 2016, pp. 324–332. ACM (2016)
32. Saberi, A., Wajc, D.: The greedy algorithm is not optimal for on-line edge coloring. In: 48th International Colloquium on Automata, Languages, and Programming (ICALP 2021). Schloss Dagstuhl-Leibniz-Zentrum für Informatik (2021)
33. Samuel-Cahn, E.: Comparison of threshold stop rules and maximum for independent nonnegative random variables. Ann. Probab. 1213–1216 (1984)
34. Srinivasan, A., Wajc, D., et al.: Online dependent rounding schemes. arXiv preprint arXiv:2301.08680 (2023)

Prophet Inequalities via the Expected Competitive Ratio

Tomer Ezra[1] , Stefano Leonardi[2] , Rebecca Reiffenhäuser[3] ,
Matteo Russo[2](✉) , and Alexandros Tsigonias-Dimitriadis[4]

[1] Simons Laufer Mathematical Sciences Institute, Berkeley, CA 94720, USA
[2] Department of Computer, Control and Management Engineering Antonio Ruberti,
Sapienza University Rome, Via Ariosto 25, 00185 Rome, Italy
`{leonardi,mrusso}@diag.uniroma1.it`
[3] University of Amsterdam, Amsterdam, The Netherlands
`r.e.m.reiffenhauser@uva.nl`
[4] European Central Bank, Sonnemannstraße 20, 60314 Frankfurt am Main, Germany

Abstract. We consider prophet inequalities under downward-closed constraints. In this problem, a decision-maker makes immediate and irrevocable choices on arriving elements, subject to constraints. Traditionally, performance is compared to the expected offline optimum, called the *Ratio of Expectations* (RoE). However, RoE has limitations as it only guarantees the average performance compared to the optimum, and might perform poorly against the realized ex-post optimal value. We study an alternative performance measure, the *Expected Ratio* (EoR), namely the expectation of the ratio between algorithm's and prophet's value. EoR offers robust guarantees, e.g., a constant EoR implies achieving a constant fraction of the offline optimum with constant probability. For the special case of single-choice problems the EoR coincides with the well-studied notion of probability of selecting the maximum. However, the EoR naturally generalizes the probability of selecting the maximum for combinatorial constraints, which are the main focus of this paper. Specifically, we establish two reductions: for every constraint, RoE and the EoR are at most a constant factor apart. Additionally, we show that the EoR is a stronger benchmark than the RoE in that, for every instance (constraint and distribution), the RoE is at least a constant fraction of the EoR, but not vice versa. Both these reductions imply a wealth of EoR results in multiple settings where RoE results are known.

Keywords: Prophet Inequalities · Online Decision-Making · Downward-closed Feasibility Constraints

1 Introduction

Prophet Inequalities are one of optimal stopping theory's most prominent problem classes. In the classic prophet inequality, a decision-maker must select an element e from an online sequence of elements E immediately and irrevocably.

The sequence is revealed one by one in an online fashion, and the decision-maker wants to maximize the weight of the chosen element, where each element's weight is drawn from some distribution D_e. The decision-maker knows the distributions and is compared to a *prophet*, who knows all the realizations of the weights in advance. A classic result of Krengel, Sucheston and Garling [28,29], and Samuel-Cahn [33] asserts that the decision-maker can extract at least half of the prophet's expected reward and that this result is tight.

A vast body of research has studied the classic prophet inequality and its variants, where the objective function is to maximize the ratio between what the algorithm gets in expectation and the expected weight of the ex-post optimum. We use the shorthand RoE to signify this *ratio of expectations*. However, this benchmark has shortcomings for many applications of prophet inequalities. Oftentimes, the decision-maker is not only concerned about the expected value, but she also wants to have some guarantees with respect to the ex-post outcome. The concept of risk aversion has been defined in various ways in the literature: a common underlying principle is that the involved parties often want to avoid the possibility of *extremely bad* outcomes.

As our first example shows, such risk-averse decision-makers might prefer to select a box with a deterministic weight of 1, even though the second box's expected weight is slightly larger. This is because the weight of the second box has a high probability of having a value of 0, and it is much riskier to choose this option for just a marginal improvement in the expected utility.

Example 1. Consider a setting with two boxes. The first box's weight w_1 is deterministically 1, and the second box's weight w_2 is 0 with probability $1 - \varepsilon$ and $\frac{1+2\varepsilon}{\varepsilon}$ with probability ε, for $\varepsilon \in (0,1]$ (Fig. 1).

The decision-maker's expected utility would be 1 if she selects the first box, and is $1+2\varepsilon$ if she selects the second box. While picking the second box maximizes the RoE, this is a much riskier choice that brings only a negligible improvement.

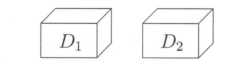

Fig. 1. Two boxes: $w_1 = 1$, $w_2 \sim D_2 = \begin{cases} 0, \text{ w.p. } 1 - \varepsilon \\ \frac{1+2\varepsilon}{\varepsilon}, \text{ w.p. } \varepsilon \end{cases}$.

Since maximizing the RoE does not capture the phenomenon of risk aversion, we would like to define a benchmark that does. A first suggestion is the *probability of selecting the maximum* (PbM), introduced by Gilbert and Mosteller [21] for the case of i.i.d. elements, for which the PbM = 0.5801. In the non-i.i.d. case, in worst-case order, Esfandiari et al. [15] show a tight bound on PbM of $1/e$.

A decision-maker that maximizes the PbM selects the first box in Example 1, and thus picks the maximum with probability close to 1. Another different approach from the PbM is the *expected ratio*, EoR, between the algorithm's weight and the weight of the ex-post optimum (originally suggested in [34] for other domains). In the full version of the paper [17], we establish that PbM and EoR are essentially identical measures of performance in single-choice settings. This no longer holds for richer variants of prophet inequalities beyond single-choice, which is the main focus of our paper.

We study the natural extension of classic prophet inequalities, termed prophet inequalities with combinatorial constraints [31], where the decision-maker is allowed to select more than a single element according to a prede-fined (downward-closed) feasibility constraint. These types of constraints cap-ture the idea that if a given set is feasible, so are all its subsets: examples include knapsack, matchings, and general matroids, as well as their intersection. In the next example, one can observe that, for any online algorithm, the probability of selecting the maximum is exponentially small in the number of elements; put another way, the probability of selecting the exact optimum offline is negligible and, thus, the guarantees of the PbM measure do not extend to combinatorial prophet inequalities.

Example 2. Consider a setting with n pairs of boxes, such that, for each pair i, one box has weight $w_{1,i} = 1$ deterministically, and the second has weight $w_{2,i}$, equal to 0 with probability 1/2 and to 2 with probability 1/2. The (downward-closed) feasibility constraint is that at most one box from each pair can be selected (a partition matroid), and the decision-maker gets the sum over the selected set (Fig. 2).

Fig. 2. n pairs of boxes: for each pair, $w_{1,i} = 1$, $w_{2,i} \sim D_2 = \begin{cases} 0, & \text{w.p. } 1/2 \\ 2, & \text{w.p. } 1/2 \end{cases}$.

The probability of selecting the maximum is the probability that the algo-rithm chooses the larger realized value for each pair of boxes. An online algorithm cannot select the maximum of each pair with probability greater than 1/2. Since all realizations are independent, we have an upper bound of PbM = $1/2^n$ for every online algorithm (see the full version of the paper [17]). This motivates choosing a different measure of performance in combinatorial settings. In par-ticular, for this example, the algorithm that always selects the first box for each pair guarantees good expected ratio. Indeed, by Jensen's Inequality (see the full version of the paper [17]), we have

$$\mathbf{E}\left[\frac{\mathsf{ALG}}{\mathsf{OPT}}\right] \geq \frac{n}{\mathbf{E}\left[\sum_{i \in [n]} \max\{w_{1,i}, w_{2,i}\}\right]} = \frac{n}{\frac{3}{2}n} = \frac{2}{3}.$$

Note that this algorithm also guarantees $\mathsf{ALG} \geq \left(\frac{2}{3} - \varepsilon\right) \cdot \mathsf{OPT}$ with high prob-ability, which is a type of guarantee a risk-averse decision-maker would desire.

Combining that EoR and PbM are equivalent for single-choice settings and that PbM is unachievable beyond single-choice, we believe that the EoR is the right alternative to the PbM in combinatorial settings. In the full version of the paper [17], we discuss other possible extensions of PbM and show their shortcomings in combinatorial settings.

It is important to note that there are some instances where optimizing EoR leads to bad guarantees for risk-averse decision-makers. This is the case for both the EoR and the RoE (see the full version of the paper [17]). However, there are cases where having to average over many runs to obtain a good ratio does not meet the problem requirements. Consider a platform or marketplace that faces a frequently repeated (e.g., daily) resource allocation problem. The task is to allocate limited resources to a stream of customers, subject to any underlying downward-closed constraint about what can be allocated and to whom. The platform wants to maximize an objective, such as social welfare or revenue. From the perspective of both the platform and the customers, it is often desirable to know that some ex-post guarantees will be satisfied. Specifically, the customers on a given day might want to know that if they have a high value for some subset of the resources, they will have a fair chance at getting it, and the platform also wants to ensure that on every instance, it will allocate a good fraction of the resources to the customers who value them highly. In such scenarios, designing a strategy that maximizes the EoR, rather than the RoE or some other performance measure, will guarantee that.

In the full version of the paper [17], we provide a series of claims, which establish that our definition of EoR (in contrast to the RoE) guarantees the best-we-can-hope-for when minimizing the risk compared to the ex-post value. More specifically, in the full version of the paper [17] we show that a "good" (i.e., relatively high constant) EoR directly implies that we attain a constant fraction of the optimum with constant probability. Moreover, as shown in the full version of the paper [17], no (qualitatively) better bi-criteria approximation can be achieved; there exist simple feasibility constraint-distribution pairs for which either no constant approximation to the maximum is possible with high probability, or no near-optimal approximation to the maximum can be attained with constant probability. Therefore, settling for the ex-post guarantee of the EoR is best possible in combinatorial prophet inequalities. Moreover, instead of aiming directly for such bi-criteria results, our main goal in this paper is to suggest a natural alternative performance measure and present its properties and insights it provides. Thus, we believe that, apart from its simplicity, the two main reasons that make the EoR an interesting objective function are (1) that is the "right" generalization of PbM beyond the single-choice setting, and, (2) that it captures well one of the natural ways to think of risk-aversion in online decision-making.

We further investigate how the notions of the ratio of expectations and the expected ratio are connected to each other. As a first step, the following examples show why a constant RoE algorithm does not guarantee a constant EoR, and vice versa.

First, consider Example 1. The canonical 1/2-competitive (and tight) RoE algorithm for the single-choice problem is that of setting a single threshold $\tau :=$ $\mathbf{E}\left[\text{OPT}\right]/2 = \mathbf{E}\left[\max\{w_1, w_2\}\right]/2 = (1 - \varepsilon + 1 + 2\varepsilon)/2 = 1 + \varepsilon/2$, and accepting the first box whose weight exceeds τ. We now analyze the performance of such algorithm, measured according to EoR:

$$\text{EoR} := \mathbf{E}\left[\frac{\text{ALG}}{\text{OPT}}\right] = (1 - \varepsilon) \cdot \frac{0}{1} + \varepsilon \cdot \frac{(1 + 2\varepsilon)/\varepsilon}{(1 + 2\varepsilon)/\varepsilon} = \varepsilon,$$

since the algorithm would only accept if the value is at least $1 + \varepsilon/2$, which only happens if the second box realization is $(1 + 2\varepsilon)/\varepsilon$. This algorithm has no EoR guarantee since ε can be arbitrarily small.

Second, the next example shows that a constant EoR algorithm is not necessarily constant competitive in the RoE sense.

Example 3. Consider a setting with two boxes, one with a weight $w_1 = 1$ deterministically, and the second with a weight w_2, which is ε^2 with probability $1 - \varepsilon$ and $1/\varepsilon^2$ with probability ε, for $\varepsilon \in (0, 1]$ (Fig. 3).

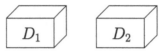

Fig. 3. Two boxes: $w_1 = 1$, $w_2 \sim D_2 = \begin{cases} \varepsilon^2, & \text{w.p. } 1 - \varepsilon \\ 1/\varepsilon^2, & \text{w.p. } \varepsilon \end{cases}$.

The algorithm that always selects the first box achieves $\mathbf{E}\left[\frac{\text{ALG}}{\text{OPT}}\right] = (1 - \varepsilon) \cdot \frac{1}{1} + \varepsilon \cdot \frac{1}{1/\varepsilon^2} > 1 - \varepsilon$. On the contrary, $\mathbf{E}\left[\text{ALG}\right] = 1$ and $\mathbf{E}\left[\text{OPT}\right] = 1 - \varepsilon + \frac{1}{\varepsilon} > \frac{1}{\varepsilon}$. Thus,

$$\text{RoE} := \frac{\mathbf{E}\left[\text{ALG}\right]}{\mathbf{E}\left[\text{OPT}\right]} < \frac{1}{1/\varepsilon} = \varepsilon.$$

As ε can be arbitrarily small, this algorithm does not guarantee a constant RoE.

The aforementioned examples demonstrate that algorithms exhibiting a constant guarantee for one performance measure (such as the optimal RoE algorithm, which gives a guarantee of 1/2 for any instance) might fail miserably in some instances for the other performance measure. This motivates us to deeper understand whether, and for which settings, a good algorithm for RoE can be transformed to a good algorithm for EoR, and vice versa. The **guiding question** of this paper is, therefore,

What is the relation between RoE and EoR?

1.1   Our Contributions

As a motivation for our study, in the full version of the paper [17], we show the equivalence between PbM and EoR in the single-choice setting. We present two proofs for this equivalence; one is an adaptation of the worst-case example for the PbM measure by Esfandiari et al. [15]. The second proof is based on the observation that for each product distribution, we can construct a new product distribution for which the EoR is arbitrarily close to the PbM of the original distribution.

Our main results establish two reductions between EoR and RoE. In particular, we show that for every downward-closed feasibility constraint, RoE and EoR are at most a multiplicative constant factor apart (see Sect. 3 and Sect. 4). For the following informal statements of the three main results, we first introduce some basic notation. We use $\mathsf{RoE}(\mathcal{F})$ (similarly $\mathsf{EoR}(\mathcal{F})$) to denote the ratio between the performance of the *best* algorithm against the offline optimum on the *worst-case* distribution, given a family of feasibility constraints $\mathcal{F}$. Note that, in principle, we expect the worst-case distributions to be different for the two measures. In the second and third statement, we use the stronger notions of $\mathsf{RoE}(\mathcal{F}, D)$ and $\mathsf{EoR}(\mathcal{F}, D)$. Here, the ratio expresses the guarantee of the best algorithm against the offline optimum on the worst *constraint-distribution pair* $(\mathcal{F}, D)$. This means that the input now consists not only of a family $\mathcal{F}$, but also of a product (i.e., the distributions of the elements' weights are independent) distribution D.

Theorem (Equivalence between RoE and EoR, Corollary 2). *For every downward-closed family of feasibility constraints $\mathcal{F}$ it holds that*

$$\frac{\mathsf{RoE}(\mathcal{F})}{\mathsf{EoR}(\mathcal{F})} \in \Theta(1).$$

In the next result, we show that the EoR is a stronger benchmark than the RoE in the sense that for every instance composed of a feasibility constraint and a product distribution the RoE is at least a constant of the EoR.

Theorem (EoR to RoE reduction, *Theorem 3*). *For every downward-closed family of feasibility constraints $\mathcal{F}$, and a product distribution D it holds that*

$$\mathsf{RoE}(\mathcal{F}, D) \geq \frac{\mathsf{EoR}(\mathcal{F}, D)}{68}.$$

We complement this by showing that the parallel result cannot be achieved in the other direction (i.e., from RoE to EoR).

Theorem (RoE to EoR impossibility, *Corollary 3*). *For every $\varepsilon > 0$, there exist a feasibility constraint $\mathcal{F}$ and a product distribution D in which $\mathsf{EoR}(\mathcal{F}, D) \leq \varepsilon$ and $\mathsf{RoE}(\mathcal{F}, D) \in \Omega(1)$.*

1.2 Our Techniques

A key ingredient of our proof is a distinction between cases where the contribution to the value of the prophet comes from a large number of boxes and cases where the contribution mainly comes from a small set of boxes. If we are in the latter case, one can just run a simple threshold strategy and have a good guarantee. Otherwise, we use our second key ingredient which is analyzing the structure of the offline optimum function (value of the prophet) in the event that the threshold algorithm does not have a good enough guarantee. In particular, we show that, under such event, the normalized offline optimum function is *self-bounding* (see Definition 3), and therefore well-concentrated [6]. Our Proposition 1 and its extended-XOS counterpart (see the full version of the paper [17]) generalize claims about the self-bounding property of normalized offline optima shown in [4, 35].

Feasibility-Based Reduction: RoE($\mathcal{F}$) vs. EoR($\mathcal{F}$). To prove that the EoR($\mathcal{F}$) is at least a constant times RoE($\mathcal{F}$), we first calculate the threshold for which the maximal value exceeds it with probability of half. We use this to perform a *tail-core* split: Intuitively, if the expected offline optimum is not too large (i.e., close to the threshold), then our algorithm tries to catch a "superstar" (i.e., the first element with a weight above the threshold). To simplify our analysis, we count only the gain in the cases where exactly one such element is realized (the tail event). This happens with constant probability, and since the expected offline optimum is relatively small, we always get a good fraction of it by picking this unique element.

When instead, the expected offline optimum is large (i.e., far from the threshold), we run a constant competitive RoE algorithm in a black-box fashion. As already pointed out, an algorithm with constant RoE does not necessarily achieve any guarantee for EoR. We overcome this obstacle through our case distinction and the self-bounding properties we show for the optimum. In particular, we upper bound the value of the offline optimum with high enough probability and lower bound the RoE algorithm expected value conditioned on the optimum not being too large.

An immediate corollary of our result and [31] is that, for downward-closed feasibility constraints, the EoR is in $\Omega\left(\frac{1}{\log^2 n}\right)$. However, we prove a much stronger result: For every specific feasibility constraint, the EoR is a constant away from the RoE (which implies trivially the former assertion). In particular, if for some feasibility constraint the RoE is $\Omega\left(\frac{1}{\log\log n}\right)$, then our result shows that the EoR is approximately the same (and not that it is just bounded by $\Omega\left(\frac{1}{\log^2 n}\right)$).

Instance-Based Reduction: RoE($\mathcal{F}, D$) vs. EoR($\mathcal{F}, D$). For the other direction, we show a stronger result, in that for every instance (a feasibility constraint, and a product distribution) the RoE is at least a constant fraction of the EoR.

To achieve this result, we show that either the original EoR algorithm achieves up to a constant the same RoE guarantee, or that the simple threshold algorithm that achieves half of the expectation of the maximal element, guarantees a good

RoE. We remark that both our reductions are constructive: We could interpret them as using a RoE algorithm black-box to design an EoR one, and vice versa. Further extensions and implications (XOS objectives, unknown prior, different assumptions on the arrival order, etc.) are discussed in Sect. 6.

1.3   Related Work

For early work on prophet inequalities, starting from the classic model and some of its most important variants, we refer the reader to the comprehensive survey of Hill and Kertz [25]. The topic of prophet inequalities has recently regained strong interest, primarily among researchers in theoretical computer science, due to its connections to (algorithmic) mechanism design and, in particular, posted price mechanisms [7,10,23,27]. The surveys of Correa et al. [9] and Lucier [30] provide detailed overviews of recent results in prophet inequalities and their connections to mechanism design, respectively.

This recent surge of interest has given rise to a stream of work, extending the classic prophet inequality to more general objective functions beyond single-choice (including submodular [8], XOS [18], and monotone subadditive functions [32]), different assumptions on the arrival order, and rich combinatorial feasibility constraints. Among the latter, some notable results include k-uniform matroid [1, 23,26], matching [2,16,18,22], general matroid or knapsack [7,11,13,19,27], and polymatroid constraints [12]. Among the most general environments considered (in which non-trivial positive results can be achieved) are arbitrary downward-closed feasibility constraints [31,32]. Combined with our framework, these results immediately give corresponding (lower and upper) bounds on the EoR. Note that [31] also considers non-downward-closed feasibility constraints, but shows that it is impossible to achieve an RoE larger than $O(1/n)$.

One of our goals in this paper is to go beyond the traditional measure of performance in prophet inequalities, i.e., the ratio of expectations, and understand how natural alternatives perform in a wide range of scenarios. While the EoR measure has not been studied before in the context of prophet inequalities, Garg et al. [20] considered it for Bayesian cost minimization problems, such as the Online Stochastic Steiner tree problem, where they show an upper bound of $O(\log \log n)$ on the gap between RoE and EoR: whether this gap is constant is an open question up to this day. Furthermore, Hartline et al. [24, Appendix A.1] study a similar notion to the EoR and compare it to the RoE in the context of prior independent mechanism design. Their goal is to measure the performance of an algorithm without knowledge of the input distribution against the best algorithm with full distributional knowledge. We defer the reader to the full version of the paper for more details on past literature [17].

2   Preliminaries

2.1   Model and Notation

We consider a setting where there is a ground set of elements E, and each element $e \in E$ is associated with a non-negative weight $w_e \sim D_e$. We assume that the distributions have no point masses[1], and we denote by $D = \times_{e \in E} D_e$ the product distribution. The elements are presented with their weights in an online fashion to a decision-maker who needs to decide immediately and irrevocably whether to accept the current element or not. The decision-maker must ensure that the set of selected elements belongs to a predefined family of downward-closed feasibility constraints $\mathcal{F}$ at all times. The goal of the decision-maker is to maximize the weight of the selected set. We make use of the following definitions and notations.

Definition 1. *Let $w \in \mathbb{R}_{\geq 0}^{|E|}$ be a non-negative weight vector. We define OPT: $\mathbb{R}_{\geq 0}^{|E|} \to \mathcal{F}$ to be the function mapping a vector of weights to a maximum-weight set in family $\mathcal{F}$. Namely,*

$$OPT(w) = \arg\max_{S \in \mathcal{F}} \sum_{e \in S} w_e.$$

Moreover, we abuse notation of vector w and use $w(S)$ to denote the sum of weights in set S, i.e., $w(S) := \sum_{e \in S} w_e$.

Definition 2. *Given a downward-closed family of feasible sets $\mathcal{F}$, we define $f_{\mathcal{F}}$: $\mathbb{R}_{\geq 0}^{|E|} \to \mathbb{R}$ to be the function that, given a weight vector w, returns the maximal weight of a feasible set in $\mathcal{F}$, i.e., $f_{\mathcal{F}}(w) := w(OPT(w))$. When clear from context, we omit $\mathcal{F}$ and use f instead of $f_{\mathcal{F}}$.*

Given an online algorithm ALG, we denote the (possibly random) set chosen by it given an input w by $\mathsf{ALG}(w)$. We will denote by $a_{\mathsf{ALG}}(w) := w(\mathsf{ALG}(w))$ the weight of the feasible set chosen by the algorithm for a specific realization w, and when clear from context, we omit ALG from the notation and use $a(w)$ instead of $a_{\mathsf{ALG}}(w)$. Our objective is to design algorithms that maximize the expected ratio between what the online algorithm gets, and the weight of the offline optimum. To measure our performance, given a downward-closed family $\mathcal{F}$, a product distribution D, and an algorithm ALG, we define

$$\mathsf{EoR}(\mathcal{F}, D, \mathsf{ALG}) := \mathbf{E}\left[\frac{w(\mathsf{ALG}(w))}{f(w)}\right],$$

where the expectation runs over the stochastic generation of the input, as well as the (possible) randomness of the algorithm. Similarly, we define

$$\mathsf{EoR}(\mathcal{F}, D) := \sup_{\mathsf{ALG}} \mathsf{EoR}(\mathcal{F}, D, \mathsf{ALG}), \tag{1}$$

and

$$\mathsf{EoR}(\mathcal{F}) := \inf_{D} \mathsf{EoR}(\mathcal{F}, D). \tag{2}$$

---

[1] We assume that there are no point masses for simplicity of presentation. All of our theorems can be adjusted to the case where there are point masses.

We will compare our results to the standard objective of maximizing the ratio of expectations between the algorithm and the offline optimum. Accordingly, we denote

$$\mathsf{RoE}(\mathcal{F}, D, \mathsf{ALG}) := \frac{\mathbf{E}\left[w(\mathsf{ALG}(\boldsymbol{w}))\right]}{\mathbf{E}\left[f(\boldsymbol{w})\right]}.$$

The final benchmark we will compare our results to is the probability of selecting an optimal (offline) set, defined by

$$\mathsf{PbM}(\mathcal{F}, D, \mathsf{ALG}) := \Pr\left[w(\mathsf{ALG}(\boldsymbol{w})) = f(\boldsymbol{w})\right].$$

Analogously to Eqs. (1), and (2), we define $\mathsf{RoE}(\mathcal{F}, D)$, $\mathsf{RoE}(\mathcal{F})$, $\mathsf{PbM}(\mathcal{F}, D)$, and $\mathsf{PbM}(\mathcal{F})$.

2.2 Structural Properties

As a first step before stating and proving the main results, we derive several properties of f that may be of interest beyond this paper. The main technical tool that we use throughout to guarantee only a constant-factor loss in the reduction is the self-bounding property of the (normalized) offline optimum. Since this property resembles a "smoothness" condition when removing one of the coordinates of the input vector, we can only prove it if we restrict the support of the weights.

Definition 3 (Self-bounding functions). *Let $\boldsymbol{x} := (x_1, \ldots, x_n)$ be a vector of independent random variables, and $\mathcal{X}$ the corresponding product space. Similarly, let $\boldsymbol{x}^{(i)} := (x_1, \ldots, x_{i-1}, x_{i+1}, \ldots, x_n)$ be the same vector deprived of the ith coordinate, and $\mathcal{X}^{(i)}$ the corresponding product space. A function $g : \mathcal{X} \to \mathbb{R}$ is said to be self-bounding if there exists a series of functions $\{g_i\}_{1 \leq i \leq n}$, such that each $g_i : \mathcal{X}^{(i)} \to \mathbb{R}$ satisfies*

$$0 \leq g(\boldsymbol{x}) - g_i(\boldsymbol{x}^{(i)}) \leq 1,$$

$$\sum_{i \in [n]} \left(g(\boldsymbol{x}) - g_i(\boldsymbol{x}^{(i)})\right) \leq g(\boldsymbol{x}).$$

Proposition 1 (Properties of $f_{\mathcal{F}}$). *For every downward-closed family of sets $\mathcal{F}$, the function $f\ (= f_{\mathcal{F}})$ satisfies the following properties:*

1. *f is 1-Lipschitz.*
2. *f is monotone, i.e., if $\boldsymbol{u} \geq \boldsymbol{v}$ point-wise, then $f(\boldsymbol{u}) \geq f(\boldsymbol{v})$.*
3. *For every $\tau > 0$, the function f/τ restricted to the domain $[0, \tau]^{|E|}$ is self-bounding.*

In the full version of the paper [17], we generalize the above proposition to arbitrary (extended) XOS functions.

The main attribute of self-bounding functions that we use is the following:

Theorem 1 (BLM Inequality[5]). *For a self-bounding function $g : \mathcal{X} \to \mathbb{R}$, it holds that:*

$$\text{For every } z > 0, \qquad \Pr\left[g(\boldsymbol{x}) \geq \mathbf{E}[g(\boldsymbol{x})] + z\right] \leq e^{-\frac{3z^2}{6\mathbf{E}[g(\boldsymbol{x})]+2z}}.$$

$$\text{For every } z < \mathbf{E}[g(\boldsymbol{x})], \Pr\left[g(\boldsymbol{x}) \leq \mathbf{E}[g(\boldsymbol{x})] - z\right] \leq e^{-\frac{z^2}{2\mathbf{E}[g(\boldsymbol{x})]}}.$$

3   From RoE to EoR: Feasibility-Based Reduction

Before presenting our reduction from RoE to EoR, we start with a few definitions and observations. We defer their proofs, as well as other auxiliary claims, to the full version of the paper [17]. Fixing a parameter $\gamma \in (0, 1)$, we define the threshold τ given a set of elements E with corresponding distributions $\{D_e\}_{e \in E}$, to be such that

$$\Pr\left[\tau \geq \max_{e \in E} w_e\right] = \gamma. \tag{3}$$

Such a τ exists and is unique for every γ since we assume that there are no point masses. For every $e \in E$, we denote by $\overline{D}_e$ the distribution $D_{e|w_e \leq \tau}$, as per the τ defined in Eq. (3). This is well defined since $\gamma > 0$, and, therefore, the probability that $w_e \leq \tau$ is at least $\gamma > 0$. Given a realization of $\boldsymbol{w}$, let $\overline{\boldsymbol{w}} \in \mathbb{R}_{\geq 0}^{|E|}$ be the weight vector determined by the following process: For each e, if $w_e \leq \tau$ then $\overline{w}_e = w_e$; otherwise, let $\overline{w}_e$ be a fresh (independent) draw from $\overline{D}_e$. Note that the distribution of $\overline{\boldsymbol{w}}$ is a product distribution, where for each $e \in E$, $\overline{w}_e$ is distributed according to $\overline{D}_e$.

We next define the two events that we will use in our analysis.

Definition 4. *Let us define the following events.*

1. ***Core.*** $\mathcal{E}_0 := \{\forall e \in E : w_e \leq \tau\}$.
2. ***Tail.*** $\mathcal{E}_1 := \{\exists! e \in E : w_e > \tau\}$,

where the symbol "$\exists!$" signifies the existence of a unique such element.

The next observation enables us to flexibly change whenever needed from conditioning on $\mathcal{E}_0$ to working directly with the truncated distribution, and vice versa.

Observation 1. The distribution of $\overline{D}$ is identical to the distribution of D conditioned on event $\mathcal{E}_0$.

We are now ready to present our reduction from RoE to EoR.

Theorem 2. *For every downward-closed family of feasibility constraints $\mathcal{F}$, it holds that*

$$\mathsf{EoR}(\mathcal{F}) \geq \frac{\mathsf{RoE}(\mathcal{F})}{12}. \tag{4}$$

Note that in the reduction of Algorithm 1, part of the input is a subroutine $\mathsf{ALG_{RoE}}$ that has an RoE at least as large as α. The condition on the event $\mathcal{E}_0$ just means that all we need to know is that this α-guarantee holds when all weights are below the chosen threshold τ. Starting from $\mathsf{ALG_{RoE}}$, we design an algorithm that uses $\mathsf{ALG_{RoE}}$ as a black box in one of the two cases and achieves an EoR which is at most a multiplicative constant factor away from the RoE.

ALGORITHM 1: RoE-to-EoR

Data: Ground set E, distributions D_e, feasibility family $\mathcal{F}$, a subroutine $\mathsf{ALG_{RoE}}$

Parameters: $\gamma \in (0,1)$, $c > 0$

Assumption: $\mathsf{ALG_{RoE}}$ satisfies $\mathbf{E}\left[a_{\mathsf{ALG_{RoE}}}(w) \mid \mathcal{E}_0\right] \geq \alpha \cdot \mathbf{E}\left[f(w) \mid \mathcal{E}_0\right]$

Result: Subset $\mathsf{ALG}(w) \subseteq E$ such that $\mathsf{ALG}(w) \in \mathcal{F}$

Calculate τ according to Equation (3);

Calculate $W := \mathbf{E}\left[f(\overline{w})\right]$;

if $W \leq c \cdot \tau$ **then**

　　Return the first element $e^* \in E$ such that $w_{e^*} \geq \tau$;

　　If no such element exists, return $\emptyset$;

else

　　As long as $w_e \leq \tau$ run $\mathsf{ALG_{RoE}}(w)$;

　　If $w_e > \tau$, reject all remaining elements;

end

In order to prove Theorem 2, by the definition of $\mathsf{RoE}(\mathcal{F})$, we will assume the existence of an algorithm $\mathsf{ALG_{RoE}}$ that satisfies:

$$
\mathbf{E}_{w \sim D}\left[a_{\mathsf{ALG_{RoE}}}(w) \mid \mathcal{E}_0\right] = \mathbf{E}_{\overline{w} \sim \overline{D}}\left[a_{\mathsf{ALG_{RoE}}}(w)\right]
$$
$$
\geq \alpha \cdot \mathbf{E}_{\overline{w} \sim \overline{D}}\left[f(\overline{w})\right] = \alpha \cdot \mathbf{E}_{w \sim D}\left[f(w) \mid \mathcal{E}_0\right], \quad (5)
$$

where the equalities hold by Observation 1. Our analysis distinguishes between two cases according to whether $W := \mathbf{E}\left[f(\overline{w})\right] \leq c \cdot \tau$. Lemma 1 analyzes the case where $W \leq c \cdot \tau$, and Lemma 2 the case where $W > c \cdot \tau$. In the remainder, for ease of notation, we use $a(w)$ instead of $a_{\mathsf{ALG_{RoE}}}(w)$.

Lemma 1 ("Catch the superstar"). *For all constants $\delta > 1, k \geq 1$ and $c \geq \frac{4+2\delta}{3(\delta-1)^2} \log \frac{k}{\alpha}$, if $W \leq c \cdot \tau$, then Algorithm 1 satisfies*

$$
\mathbf{E}\left[\frac{a(w)}{f(w)}\right] \geq \frac{\gamma \log(1/\gamma)}{c+1}. \quad (6)
$$

Lemma 2 ("Run the Combinatorial Algorithm"). *For all constants $\delta > 1, k \geq 1$ and $c \geq \frac{4+2\delta}{3(\delta-1)^2} \log \frac{k}{\alpha}$, if $W > c \cdot \tau$, then Algorithm 1 satisfies*

$$
\mathbf{E}\left[\frac{a(w)}{f(w)}\right] \geq \frac{\gamma}{\delta} \frac{k-\delta}{k} \alpha. \quad (7)
$$

4   From **EoR** to **RoE**: Instance-Based Reduction

In this section, we show an instance-based reduction from EoR to RoE. Unlike Theorem 2, our next result shows that the RoE is always at least a constant fraction of the EoR, for every pair of (downward-closed) feasibility constraint $\mathcal{F}$ and product distribution D. We will assume the existence of an algorithm $\mathsf{ALG}_{\mathsf{EoR}}(\boldsymbol{w})$ that satisfies $\mathbf{E}_{\boldsymbol{w} \sim D}\left[a_{\mathsf{ALG}_{\mathsf{EoR}}}(\boldsymbol{w})/f(\boldsymbol{w})\right] \geq \alpha$. For ease of notation, we denote the value of Algorithm 2 on $\boldsymbol{w}$ by $a(\boldsymbol{w})$.

ALGORITHM 2: EoR-to-RoE

Data: Ground set E, distributions D_e, feasibility $\mathcal{F}$, a subroutine $\mathsf{ALG}_{\mathsf{EoR}}$
Assumption: $\mathsf{ALG}_{\mathsf{EoR}}(\boldsymbol{w})$ satisfies that $\mathbf{E}_{\boldsymbol{w} \sim D}\left[a_{\mathsf{ALG}_{\mathsf{EoR}}}(\boldsymbol{w})/f(\boldsymbol{w})\right] = \alpha$
Result: Subset $\mathsf{ALG}(\boldsymbol{w}) \subseteq E$ such that $\mathsf{ALG}(\boldsymbol{w}) \in \mathcal{F}$
Let $A := \mathbf{E}_{\boldsymbol{w} \sim D}[\max_{e \in E} w_e]$;
if $A \geq \alpha \cdot \mathbf{E}_{\boldsymbol{w} \sim D}[f(\boldsymbol{w})]/34$ then
$\quad$| $\quad$ Return the first element $e^* \in E$ such that $w_{e^*} \geq \frac{A}{2}$;
$\quad$| $\quad$ If no such element exists, return $\emptyset$;
else
$\quad$| $\quad$ Run $\mathsf{ALG}_{\mathsf{EoR}}(\boldsymbol{w})$;
end

Theorem 3. *For every downward-closed family of feasibility constraints $\mathcal{F}$, and every product distribution D it holds that*

$$\mathsf{RoE}(\mathcal{F}, D) \geq \frac{\mathsf{EoR}(\mathcal{F}, D)}{68}. \tag{8}$$

An immediate corollary of Theorem 3 is:

Corollary 1. *For every downward-closed feasibility constraint $\mathcal{F}$ it holds that*

$$\mathsf{RoE}(\mathcal{F}) \geq \frac{\mathsf{EoR}(\mathcal{F})}{68}.$$

The above corollary and Theorem 2 imply

Corollary 2. *For every downward-closed feasibility constraint $\mathcal{F}$ it holds that*

$$\frac{\mathsf{RoE}(\mathcal{F})}{\mathsf{EoR}(\mathcal{F})} \in \Theta(1).$$

5   From **RoE** to **EoR**: An Impossibility of Instance-Based Reduction

In this section, we show that an instance-based reduction from RoE to EoR is unachievable. This is in contrast to the reduction of Theorem 3 from EoR to RoE. In particular, we show that there are a feasibility constraint $\mathcal{F}$ and a product distribution D for which $\mathsf{RoE}(\mathcal{F}, D)$ is constant but $\mathsf{EoR}(\mathcal{F}, D)$ is sub-constant. We show this using the following stronger claim:

Proposition 2. *For every feasibility constraint $\mathcal{F}$ there exists a product distribution D such that $\mathsf{EoR}(\mathcal{F}, D) \in O\left(\mathsf{RoE}(\mathcal{F})\right)$ while $\mathsf{RoE}(\mathcal{F}, D) \in \Omega(1)$.*

We know by the example presented in Appendix B of [31] that is based on an example from [3] for a different setting, that there exists a feasibility constraint with n elements such that $\mathsf{RoE}(\mathcal{F}) \in O\left(\frac{\log \log(n)}{\log(n)}\right)$. Combining Proposition 2 with this example for large enough n implies that:

Corollary 3. *For every $\epsilon > 0$, there exist a feasibility constraint $\mathcal{F}$ and a product distribution D in which $\mathsf{EoR}(\mathcal{F}, D) \leq \epsilon$ and $\mathsf{RoE}(\mathcal{F}, D) \in \Omega(1)$.*

6 Discussion

In this paper, we studied the performance of combinatorial prophet inequalities via the expected ratio (EoR). We focus on its connections to the standard measure of performance in the literature, i.e., the ratio of expectations (RoE). We establish that, for every downward-closed feasibility constraint, the gap between $\mathsf{RoE}(\mathcal{F})$ and $\mathsf{EoR}(\mathcal{F})$ is at most a constant. Moreover, we show that the EoR is an even stronger benchmark in the sense that $\mathsf{RoE}(\mathcal{F}, D)$ is at least a constant of $\mathsf{EoR}(\mathcal{F}, D)$, but not vice versa.

We want to remark that Algorithm 1 and Algorithm 2 are constructive ways to prove Theorem 2 Theorem 3. For example, the purpose of Algorithm 1 is to show that for every family of feasibility constraints $\mathcal{F}$, $\mathsf{EoR}(\mathcal{F}) \geq \mathsf{RoE}(\mathcal{F})/12$. For $\alpha = \mathsf{RoE}(\mathcal{F})$, by definition of $\mathsf{RoE}(\mathcal{F})$, there exists an algorithm that satisfies the assumption made in the algorithm, and for the proof of the theorem, this can be used to show an existence of an algorithm with $\mathsf{EoR}(\mathcal{F}) \geq \alpha/12$. Therefore, the assumption simply restates the starting point of the reduction of Theorem 2.

In the remainder of this section, we state some remarks and discuss extensions that follow from our proofs and techniques.

Arrival Order. We first note that our results hold for any arrival order of the elements, such as random [14], free [36], or batch arrival order [16].

Single-Sample. We can also consider scenarios in which the decision-maker does not have full knowledge of the distributions of the elements' weights. In fact, our reductions can be adjusted (with slightly worse constants) to scenarios in which the decision-maker has only a single sample from each distribution (see the full version of the paper for a formal discussion [17]).

Extension to XOS Functions. Our results extend beyond additive functions over downward-closed feasibility constraints, namely to extended-XOS functions (see the full version of the paper for a formal discussion [17]). We generalize the optimum function (Definition 1) to be $\mathsf{OPT}(\boldsymbol{w}) := \arg\max_{S \in \mathcal{F}} \max_{i \in [\ell]} \langle b_i, \boldsymbol{w}_S \rangle$, where $\boldsymbol{w}_S \in \mathbb{R}^{|E|}$ is the vector of elements weights in S (and 0 for elements not in

S), while each $b_i \in \mathbb{R}^{|E|}$ is a vector of nonnegative coefficients. Similarly to the additive case, we have that $f(\boldsymbol{w}) = \max_{\substack{i \in [\ell] \\ S \in \mathcal{F}}} \langle b_i, \boldsymbol{w}_S \rangle$. We can show that, by setting $a_{ij}^S := b_{ij} \cdot \mathbf{1}_{j \in S}$, $f(\boldsymbol{w})$ can be expressed as $f(\boldsymbol{w}) = \max_{\substack{i \in [\ell] \\ S \in \mathcal{F}}} \langle a_i^S, \boldsymbol{w}_S \rangle$. Moreover, the functions resulting from projecting all such f's onto $\{0,1\}^{|E|}$ not only are XOS but describe all XOS functions. We can now run Algorithm 1, and perform the case distinction with the modified threshold τ being such that $\Pr\left[\exists j \in [[E]] : w_j > \frac{\tau}{\max_{i \in [\ell]} a_{ij}}\right] = \gamma$. With this at hand, the "catch the superstar" subroutine becomes selecting the first element j with $w_j > \frac{\tau}{\max_{i \in [\ell]} a_{ij}}$, while the "run the combinatorial algorithm" remains unaltered (up to slight modifications, described in the full version of the paper [17]). For Algorithm 2, we redefine $A := \mathbf{E}[\max_{i,j} a_{ij} \cdot w_j]$ and repeat a similar analysis to the one above. All in all, we lose an additional $\max_{i,j} a_{ij}$ factor in the expected ratio of Theorem 2 and Theorem 3, and we get that:

$$\mathsf{EoR}(\mathcal{F}) \geq \frac{\mathsf{RoE}(\mathcal{F})}{12 \cdot \max_{i,j} a_{ij}}, \quad \text{and} \quad \mathsf{RoE}(\mathcal{F}, D) \geq \frac{\mathsf{EoR}(\mathcal{F}, D)}{68 \cdot \max_{i,j} a_{ij}}.$$

Other Measures of Performance. In this work, we study the expected ratio as our measure of performance. Another natural performance measure is the expected inverse ratio, i.e., $\mathsf{EoIR}(\mathcal{F}) := \mathbf{E}\left[f(\boldsymbol{w})/a(\boldsymbol{w})\right]$. We now show that such a measure may be unbounded even for the single-choice case. Let us consider again Example 3. Fix a (randomized) algorithm that selects the first element with probability p, and let $a_p(\boldsymbol{w})$ be its performance on input $\boldsymbol{w}$. Then we have

$$\mathsf{EoIR}(\mathcal{F}) \geq \min_p \mathbf{E}\left[\frac{f(\boldsymbol{w})}{a_p(\boldsymbol{w})}\right]$$

$$\geq \min_p (1-\varepsilon) \cdot \left(p \cdot \frac{1}{1} + (1-p) \cdot \frac{1}{\varepsilon^2}\right) + \varepsilon \cdot \left(p \cdot \frac{1/\varepsilon^2}{1} + (1-p) \cdot \frac{1/\varepsilon^2}{1/\varepsilon^2}\right)$$

$$= \min_p \frac{1-\varepsilon}{\varepsilon^2} + \varepsilon - \left(\frac{1-\varepsilon}{\varepsilon^2} - \frac{1}{\varepsilon} - 1 + 2\varepsilon\right) p \geq \frac{1}{\varepsilon} + 1 - \varepsilon > \frac{1}{\varepsilon}.$$

Hereby, the second inequality holds since the algorithm selects the second box (given it is realized) with probability at most $1-p$, while the third inequality follows from setting $p = 1$ to minimize the expression. As ε can be made arbitrarily small, we have an unbounded EoIR.

Despite the above simple impossibility result in maximization problems, the same measure of performance could be of use when the decision-maker seeks to minimize a function subject to, e.g., covering constraints with stochastic inputs. As mentioned in Sect. 1.3, Garg et al. [20] study the relation between RoE and EoR for minimization problems, such as Online Steiner Tree and Traveling Salesman Problem with stochastic inputs. On the other hand, the EoIR measure in this setting remains unexplored. It would be interesting to understand whether the reductions provided in Sects. 3 and 4 are generalizable to the minimization setting (both for the EoR and the EoIR).

Gap Between EoR *and* RoE. Despite the similarity between the benchmarks, it is not at all obvious whether the maximal gap between RoE and EoR for prophet settings and every downward-closed feasibility constraint would be constant (it is an open question whether this gap is constant in other Bayesian settings [20]). In our feasibility-based reduction from RoE to EoR, we lose a constant of $\frac{1}{12}$. Losing a constant is unavoidable already from the single-choice setting, where there is a (tight) gap of $\frac{2}{e}$. It would be interesting to study whether this is the worst gap possible. In the other direction (i.e., from EoR to RoE), the gap in the reduction is also not tight; it is even possible that the RoE($\mathcal{F}$) is at least the EoR($\mathcal{F}$) for every feasibility constraint $\mathcal{F}$. However, we know that RoE($\mathcal{F}, D$) can be smaller than EoR($\mathcal{F}, D$) by a factor of 2 by Example 1 and can be unboundedly larger than EoR($\mathcal{F}, D$). Additionally, finding the exact value of EoR($\mathcal{F}$) for specific downward-closed constraints (e.g., matching, matroid, knapsack, etc.) is an interesting open question.

Acknowledgements. Partially supported by the ERC Advanced Grant 788893 AMDROMA "Algorithmic and Mechanism Design Research in Online Markets" and MIUR PRIN project ALGADIMAR "Algorithms, Games, and Digital Markets", FAIR (Future Artificial Intelligence Research) project, funded by the NextGenerationEU program within the PNRR-PE-AI scheme (M4C2, investment 1.3, line on Artificial Intelligence). The last author further acknowledges the support of the Alexander von Humboldt Foundation with funds from the German Federal Ministry of Education and Research (BMBF), the Deutsche Forschungsgemeinschaft (DFG, German Research Foundation) - Projektnummer 277991500, the COST Action CA16228 "European Network for Game Theory" (GAMENET), and ANID, Chile, grant ACT210005. Most of this work was done while the author was at TU Munich and Universidad de Chile. The views expressed in this paper are the author's and do not necessarily reflect those of the European Central Bank or the Eurosystem. Tomer Ezra was also supported by the National Science Foundation under Grant No. DMS-1928930 and by the Alfred P. Sloan Foundation under grant G-2021-16778, while the author was in residence at the Simons Laufer Mathematical Sciences Institute (formerly MSRI) in Berkeley, California, during the Fall 2023 semester.

References

1. Alaei, S.: Bayesian combinatorial auctions: expanding single buyer mechanisms to many buyers. In: 2011 IEEE 52nd Annual Symposium on Foundations of Computer Science, pp. 512–521 (2011)
2. Alaei, S., Hajiaghayi, M., Liaghat, V.: Online prophet-inequality matching with applications to ad allocation. In: Proceedings of the 13th ACM Conference on Electronic Commerce, EC 2012. ACM Press (2012)
3. Babaioff, M., Immorlica, N., Kleinberg, R.: Matroids, secretary problems, and online mechanisms. In: Proceedings of the Eighteenth Annual ACM-SIAM Symposium on Discrete Algorithms, SODA 2007, pp. 434–443. SIAM (2007)
4. Blum, A., Caragiannis, I., Haghtalab, N., Procaccia, A.D., Procaccia, E.B., Vaish, R.: Opting into optimal matchings. In: Klein, P.N. (ed.) Proceedings of the Twenty-Eighth Annual ACM-SIAM Symposium on Discrete Algorithms, SODA 2017, pp. 2351–2363. SIAM (2017)

5. Boucheron, S., Lugosi, G., Massart, P.: A sharp concentration inequality with application. Random Struct. Algorithms **16**(3), 277–292 (2000)
6. Boucheron, S., Lugosi, G., Massart, P.: On concentration of self-bounding functions. Electron. J. Probab. **14** (2009)
7. Chawla, S., Hartline, J.D., Malec, D.L., Sivan, B.: Multi-parameter mechanism design and sequential posted pricing. In: Schulman, L.J. (ed.) Proceedings of the 42nd ACM Symposium on Theory of Computing, STOC 2010. ACM (2010)
8. Chekuri, C., Livanos, V.: On submodular prophet inequalities and correlation gap. In: Algorithmic Game Theory - 14th International Symposium, SAGT 2021. Lecture Notes in Computer Science, vol. 12885, p. 410. Springer, Cham (2021)
9. Correa, J., Foncea, P., Hoeksma, R., Oosterwijk, T., Vredeveld, T.: Recent developments in prophet inequalities. SIGecom Exch. **17**(1), 61–70 (2019)
10. Correa, J., Foncea, P., Pizarro, D., Verdugo, V.: From pricing to prophets, and back! Oper. Res. Lett. **47**(1), 25–29 (2019)
11. Dütting, P., Feldman, M., Kesselheim, T., Lucier, B.: Prophet inequalities made easy: stochastic optimization by pricing nonstochastic inputs. SIAM J. Comput. **49**(3), 540–582 (2020)
12. Dütting, P., Kleinberg, R.: Polymatroid prophet inequalities. In: Bansal, N., Finocchi, I. (eds.) ESA 2015. LNCS, vol. 9294, pp. 437–449. Springer, Heidelberg (2015). https://doi.org/10.1007/978-3-662-48350-3_37
13. Ehsani, S., Hajiaghayi, M., Kesselheim, T., Singla, S.: Prophet secretary for combinatorial auctions and matroids. In: Proceedings of the Twenty-Ninth Annual ACM-SIAM Symposium on Discrete Algorithms, SODA 2018. SIAM (2018)
14. Esfandiari, H., Hajiaghayi, M., Liaghat, V., Monemizadeh, M.: Prophet secretary. SIAM J. Discret. Math. **31**(3), 1685–1701 (2017)
15. Esfandiari, H., Hajiaghayi, M., Lucier, B., Mitzenmacher, M.: Prophets, secretaries, and maximizing the probability of choosing the best. In: The 23rd International Conference on Artificial Intelligence and Statistics, AISTATS 2020. Proceedings of Machine Learning Research, vol. 108, pp. 3717–3727. PMLR (2020)
16. Ezra, T., Feldman, M., Gravin, N., Tang, Z.G.: Online stochastic max-weight matching: prophet inequality for vertex and edge arrival models. In: EC 2020: The 21st ACM Conference on Economics and Computation, pp. 769–787. ACM (2020)
17. Ezra, T., Leonardi, S., Reiffenhäuser, R., Russo, M., Tsigonias-Dimitriadis, A.: Prophet inequalities via the expected competitive ratio. CoRR abs/2207.03361 (2022)
18. Feldman, M., Gravin, N., Lucier, B.: Combinatorial auctions via posted prices. In: Indyk, P. (ed.) Proceedings of the Twenty-Sixth Annual ACM-SIAM Symposium on Discrete Algorithms, SODA 2015, pp. 123–135. SIAM (2015)
19. Feldman, M., Svensson, O., Zenklusen, R.: Online contention resolution schemes with applications to Bayesian selection problems. SIAM J. Comput. **50**(2), 255–300 (2021)
20. Garg, N., Gupta, A., Leonardi, S., Sankowski, P.: Stochastic analyses for online combinatorial optimization problems. In: Proceedings of the Nineteenth Annual ACM-SIAM Symposium on Discrete Algorithms, SODA 2008, pp. 942–951 (2008)
21. Gilbert, J.P., Mosteller, F.: Recognizing the maximum of a sequence. J. Am. Stat. Assoc. **61**(313), 35–73 (1966)
22. Gravin, N., Wang, H.: Prophet inequality for bipartite matching: merits of being simple and non adaptive. In: Proceedings of the 2019 ACM Conference on Economics and Computation, EC 2019, pp. 93–109. ACM (2019)

23. Hajiaghayi, M.T., Kleinberg, R., Sandholm, T.: Automated online mechanism design and prophet inequalities. In: Proceedings of the 22nd National Conference on Artificial Intelligence, AAAI 2007, vol. 1, pp. 58–65. AAAI Press (2007)
24. Hartline, J., Johnsen, A.: Lower bounds for prior independent algorithms (2021)
25. Hill, T.P., Kertz, R.P.: A survey of prophet inequalities in optimal stopping theory. Contemp. Math. **125** (1992)
26. Jiang, J., Ma, W., Zhang, J.: Tight guarantees for multi-unit prophet inequalities and online stochastic knapsack. In: Proceedings of the 2022 Annual ACM-SIAM Symposium on Discrete Algorithms (SODA) (2022)
27. Kleinberg, R., Weinberg, S.M.: Matroid prophet inequalities and applications to multi-dimensional mechanism design. Games Econ. Behav. **113**, 97–115 (2019)
28. Krengel, U., Sucheston, L.: Semiamarts and finite values. Bull. Am. Math. Soc. **83**(4), 745–747 (1977)
29. Krengel, U., Sucheston, L.: On semiamarts, amarts, and processes with finite value. Adv. Probab. Relat. Top. **4**, 197–266 (1978)
30. Lucier, B.: An economic view of prophet inequalities. ACM SIGecom Exchanges **16**(1), 24–47 (2017)
31. Rubinstein, A.: Beyond matroids: secretary problem and prophet inequality with general constraints. In: Proceedings of the 48th Annual ACM SIGACT Symposium on Theory of Computing, STOC 2016, pp. 324–332. ACM (2016)
32. Rubinstein, A., Singla, S.: Combinatorial prophet inequalities. In: Proceedings of the Twenty-Eighth Annual ACM-SIAM Symposium on Discrete Algorithms, SODA 2017, pp. 1671–1687. SIAM (2017)
33. Samuel-Cahn, E.: Comparison of threshold stop rules and maximum for independent nonnegative random variables. Ann. Probab. **12**(4) (1984)
34. Scharbrodt, M., Schickinger, T., Steger, A.: A new average case analysis for completion time scheduling. J. ACM **53**(1), 121–146 (2006)
35. Vondrák, J.: A note on concentration of submodular functions (2010)
36. Yan, Q.: Mechanism design via correlation gap. In: Proceedings of the Twenty-Second Annual ACM-SIAM Symposium on Discrete Algorithms, SODA 2011, pp. 710–719 (2011)

Smoothed Analysis of Social Choice Revisited

Bailey Flanigan[1], Daniel Halpern[2(✉)], and Alexandros Psomas[3]

[1] Carnegie Mellon University, Pittsburgh, USA
[2] Harvard University, Cambridge, USA
dhalpern@g.harvard.edu
[3] Purdue University, West Lafayette, USA

Abstract. A canonical problem in social choice is how to aggregate ranked votes: that is, given n voters' rankings over m candidates, what *voting rule* f should we use to aggregate these votes and select a single winner? One standard method for comparing voting rules is by their satisfaction of *axioms*—properties that we want a "reasonable" rule to satisfy. Unfortunately, this approach leads to several impossibilities: no voting rule can simultaneously satisfy all the properties we would want, at least in the worst case over all possible inputs.

Motivated by this, we consider a relaxation of this worst case requirement: a "smoothed" model of social choice, where votes are independently perturbed with small amounts of noise. If no matter which input profile we start with, the probability of an axiom being satisfied post-noise becomes large as the number of voters n grows, we take it to be as good as satisfied—called "smoothed-satisfied"—even if it may be violated in the worst case.

Within our smoothed model - a mild restriction of Lirong Xia's - we give a cohesive overview of when smoothed noise is sufficient to overcome axiomatic impossibilities. We give simple sufficient conditions for smoothed-satisfaction or smoothed-violation of several axioms and paradoxes, including most that have been studied plus some previously unstudied. We then observe that in a practically important subclass of noise models, convergence to smoothed satisfaction is prohibitively slow as n grows. Motivated by this, we prove bounds specifically within a canonical noise model from this subclass—the Mallows model. Here, we find a more nuanced picture on exactly when smoothed analysis can help.

Keywords: Social Choice · Smoothed Analysis

1 Introduction

One of the most canonical problems in social choice is how to aggregate votes: that is, given a *preference profile* consisting of n voters' rankings over m alternatives, what *voting rule* f should we use to aggregate this preference profile into a single winner? A predominant way of comparing voting rules is by their satisfaction of *axioms*—logical statements that describe basic, natural properties that voting rules should satisfy.

© The Author(s), under exclusive license to Springer Nature Switzerland AG 2024
J. Garg et al. (Eds.): WINE 2023, LNCS 14413, pp. 290–309, 2024.
https://doi.org/10.1007/978-3-031-48974-7_17

Traditionally, it is said that a voting rule *satisfies* an axiom if the logical statement specified by the axiom holds on *all possible* preference profiles; if there exists even one profile on which it does not hold (a *counterexample*), the rule is said to *violate* the axiom. Unfortunately, under this worst-case notion of satisfaction, *all* voting rules violate at least some common-sense axioms.[1] There is hope, however, because this worst-case approach may cover up an important distinction between voting rules: while some rules may fail to satisfy an axiom on large swaths of preference profiles, others may fail only on precisely contrived instances. This distinction is of practical significance because voting rules of the latter type should usually satisfy the axiom in practice, where the lack of perfect correlation between people's preferences makes extremely contrived counterexamples unlikely to occur.

In 2020, a paper by Lirong Xia aimed to capture this intuition by modeling profiles as *semi-random* — mostly adversarial, but with random perturbations [19]. More precisely, Xia's model assumes that "noisy" (but otherwise adversarial) profiles arise in the following way: first, let an adversary choose voters' "types," with each type corresponding to a distribution over rankings. The ultimate profile is then sampled by selecting each voter's ranking independently from their type distribution.

In this paper, we capture the same intuition using a slightly restricted version of Xia's semi-random model. Our model produces "noisy" profiles in the following way: first, let an adversary fix a *starting profile*. Then, apply a small amount of noise independently to each individual ranking in the profile — that is, for each ranking, draw a new ranking from some distribution (this distribution can be arbitrary, up to regularity conditions). We will refer to this as the *smoothed* model, as it is directly analogous to Spielman and Teng's "smoothed" model in the distinct context of linear programming. We discuss this connection and the precise relationship between our model and Xia's in Sect. 1.2. The key takeaway is that ours is a slight restriction, assumed primarily for ease of exposition.

Because the smoothed model goes so minimally beyond the worst case, resolving axioms in this way is quite meaningful: the small amount of noise it assumes should be sufficient to escape counterexamples only if they are truly isolated among other profiles, i.e., contrived. The model is also well-motivated practically, as people's preferences are established to be susceptible to small shocks, in daily life and even in the preference elicitation process [5,11,12]. The existing work in this area gives some hope of such positive results: it shows that under all standard voting rules considered, the axioms Group-strategyproofness [22], Resolvability [20], and Participation [21] are satisfied post-noise with high probability as n grows large. In contrast, the axiom Condorcet consistency remains violated with high probability, even after semi-random noise, by a popular class of rules [21]. Here we see heterogeneity across axioms: for some, smoothed noise reliably circumvents impossibilities, whereas, for others, it seems insufficient. This motivates our first question:

---

[1] E.g., the Gibbard-Satterthwaite theorem says that any onto, single-winner, non-dictatorial voting rule with $m > 2$ is not strategyproof [9]. For more examples, see the Comparison of electoral systems Wikipedia page.

Question 1: *What properties of axiomatic impossibilities make them resolvable by smoothed noise?*

This question remains to be fully addressed, in part because existing work has mostly analyzed axioms one by one, using technically intricate arguments to get precise asymptotic convergence rates in n. While this approach has yielded detailed insights about specific axioms and voting rules, it remains to distill higher-level patterns across rules and axioms.

Our second research question is a refinement of the first and deals with understanding when semi-random noise can circumvent impossibilities, *absent a key assumption made in the related work so far.* This assumption is called *positivity* in the related work, and it amounts to assuming that when a ranking is perturbed, there is at least some fixed probability, MIN-PROB, of *any other ranking* resulting from this perturbation. Although it is not explicitly written, existing upper bounds on rates of convergence for the axioms we consider implicitly depend multiplicatively on $1/\text{MIN-PROB}^{3/2}$. Although when MIN-PROB is large, this polynomial dependence is not too demanding, when MIN-PROB is small, enormous numbers of voters may be necessary to guarantee a reasonable probability of axiom satisfaction.

Although assuming that MIN-PROB is large enough to avoid such issues may seem innocuous, it is anything but. This is because noise models in which MIN-PROB is very small represent perhaps the most realistic subclass of noise models: they encompass any noise model in which small perturbations are unlikely to *completely reverse* someone's ranking (or make similarly drastic changes). Given that small real-world shocks that may perturb people's preferences are unlikely to qualitatively change their opinions (to, e.g., the extent that their rankings are reversed), a more realistic noise model might be one in which "less" noise corresponds to lower probability of drawing a perturbed ranking that reflects extreme opinion change. This motivates our second research question:

Question 2: *What properties of axiomatic impossibilities make them resolvable by smoothed noise, for noise models in which rankings reflecting extreme opinion change are drawn with low (or zero) probability?*

1.1   Results and Contributions

Question 1. In Sect. 3, we study the general smoothed model. We distill two patterns: one across axioms where smoothed noise *does* help, and another across axioms where it does not. In the process, we do the first smoothed (or semi-random) analysis of several core axioms in social choice. We also re-analyze some axioms studied in past work, but we prove these results anew, with derivations that are standardized across axioms and yield bounds that explicitly depend on MIN-PROB, which will be useful in Question 2. We delineate which analyses are new and which are re-proven in Sect. 1.2.

As summarized in Table 1 (located in Sect. 3.1), we first show negative results for the axioms CONDORCET CONSISTENCY, INDEPENDENCE OF IRRELEVANT ALTERNATIVES, CONSISTENCY, MAJORITY. Across all voting rules we study, we find that these axioms are all smoothed-violated whenever they are violated, i.e., smoothed noise is insufficient to circumvent impossibilities for these axioms. Moreover, across these axioms and rules, smoothed violation occurs for the same reason: there exist large contiguous segments of the profile space consisting of counterexamples. As shown in the left-hand pane of Fig. 1, if such regions exist, then our starting profile can be chosen to be inside one, in which case small perturbations are insufficient to produce profiles outside it. We formalize this underlying pattern into a generic sufficient condition for smoothed-violation in Lemma 2.

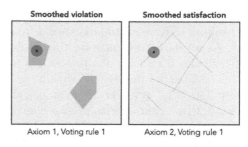

Smoothed violation	Smoothed satisfaction
Axiom 1, Voting rule 1	Axiom 2, Voting rule 1

Fig. 1. Both panes represent (abstractly) the space of all possible profiles. Red profiles are counterexamples in which the voting rule fails to satisfy the axiom. The black dot is the starting profile, and surrounding region is where the resulting profile is likely to end up after applying noise. (Color figure online)

As summarized in Table 2 (located in Sect. 3.2), we next show positive results for the axioms RESOLVABILITY and group notions of STRATEGYPROOFNESS, PARTICIPATION, and MONOTONICITY. In the spirit of generalizing across axioms, we are able to analyze the latter three at once by proving the smoothed-satisfaction of a more general axiom, called $\rho(n)$-GROUP-STABILITY, which requires that the behavior of $\rho(n)$ voters should not be able to affect the outcome of the election (essentially framing *Margin of Victory* as a binary property [17,22]). The common feature of these axioms, which permits their smoothed-satisfaction, is that their counterexamples must occur on or near hyperplanes. As depicted in the right-hand pane of Fig. 1, this means that any "well-behaved" noise, applied to any starting profile, will place very little probability mass on the sliver of counterexamples contained within (potentially some nonzero but shrinking distance of) the measure-zero hyperplane. We formalize this underlying pattern into a generic sufficient condition for smoothed-satisfaction in Lemma 4.

Question 2. In Sect. 4, we pursue the first convergence rates to smoothed satisfaction that are not parameterized by MIN-PROB, and which do not rely on the minimum probability being nonzero. Although the core ideas we use naturally extend to broader classes of noise models, for concreteness, we pursue these bounds specifically in the *Mallows model* — perhaps the most canonical model of noisy rankings. This model has the realistic property we want: the probability of drawing a ranking decreases in its *Kendall tau distance*[2] from the original rank-

---

[2] The Kendall tau distance between two rankings π, π' is the number of pairs of candidates on which they disagree.

ing. More precisely, for a noise parameter $\phi \in [0, 1]$, the probability of drawing a ranking at Kendall tau distance d from the original ranking is proportional to ϕ^d. As a result, MIN-PROB for this model is exponentially small: the probability of reversing a ranking due to Mallows noise is $\phi^{\Omega(m^2)}$.

The Mallows model falls within our smoothed model, so our impossibilities from Sect. 3.1 carry over. From among the remaining axioms of interest (those in Sect. 3.2), we characterize precise convergence rates to smoothed satisfaction of RESOLVABILITY and $\rho(n)$-GROUP-STRATEGYPROOFNESS in the Mallows model.

While MIN-PROB-parameterized upper bounds yielded heterogeneity across axioms, they were almost homogeneous across voting rules per axiom, i.e., for each axiom, either nearly all rules satisfied, or nearly all rules violated them. Interestingly, these new bounds reveal diversity across voting rules as well. As summarized in Table 3, while PLURALITY and BORDA COUNT are smoothed-satisfied at a rate polynomial in ϕ and m, we can lower bound the convergence rates of MAXIMIN and VETO to include a $1/\phi^{\Omega(m)}$ term. This essentially means the number of voters required to guarantee satisfaction is exponential in these parameters, making this property less appealing. Across these voting rules, we can distill a pattern: voting rules whose outcomes are sensitive to changes throughout ranking positions are helped substantially by Mallows noise, while those whose outcomes remain fixed despite potentially many swaps are helped much less. Zooming out, this conceptual finding extends beyond the Mallows model and the studied axioms, suggesting that for reasonable noise models, only voting rules that are truly sensitive to small changes in voters' preferences will have brittle impossibilities.

Bonus Contribution: Smoothed Analysis of Arrow's Theorem. As an extension, in Sect. 5 we do the first smoothed (or semi-random) analysis of *Arrow's Theorem* [1] — perhaps one of the most influential impossibilities in social choice. Surprisingly, we show that this impossibility is resolved with high probability under the smoothed model. However, as we show, this resolution is for a trivial reason related to Arrow's definition of the axiom *Non-dictatorship*. We therefore identify a slight strengthening of Arrow's theorem, and pose the open question of whether smoothed noise is sufficient to resolve it.

1.2 Related Work

Our model is designed to be a close analog to Spielman and Teng's celebrated "smoothed" model, introduced in their analysis of the Simplex algorithm [18]. We give a detailed description of the parallels between our model and theirs in the appendix of the full version.[3] Within social choice, the model most closely related to ours is Lirong Xia's semi-random model [19],[4] of which our model is a mild restriction. To see how Xia's model generalizes ours, in our model

---

[3] For the full version, see https://arxiv.org/abs/2206.14684.

[4] To avoid confusion with our model, we clarity: Xia's model is called "smoothed" in the original paper, but is renamed "semi-random" in most subsequent work.

we always have $m!$ "types", with each type corresponding to a possible starting ranking. Each type's associated distribution is then the noise added to that ranking. The key restriction we make is that, while Xia's model allows types with potentially different "shapes" of noise distributions, we assume that all rankings are perturbed by a common noise distribution[m], just centered at different rankings. This restriction is technically mild, as the assumption critical to analyses in both models is that noise is *independent* across rankings, and many of our results could be expressed in the more general model but in a less concise way. Because of that, we chose to stick to a slightly more restricted model for ease of exposition.

Beyond Xia's model, axiom satisfaction has also been studied in other, less-adversarial randomized models. One popular such model is *Impartial Culture*, where profiles are drawn uniformly and i.i.d. [8,10,14]. Slightly more generally, the axiom Strategyproofness was studied in a *non-uniform* i.i.d model in [15]. Further afield, there is empirical work on the frequency of axiom violations on simulated inputs and real elections [7,16].

Now we situate our results among past work on axiom satisfaction in the semi-random model. First, our results in Sect. 4 are completely distinct from this existing line of work, pursuing bounds that do not rely on the positivity assumption core to the semi-random model. In contrast, our results in Sect. 3 are for generic noise distributions that *are* subject to the positivity assumption, along with all other regularity conditions imposed in Xia's semi-random model. In the general smoothed model, we present the first smoothed (or semi-random) analysis of the axioms CONSISTENCY, INDEPENDENCE OF IRRELEVANT ALTERNATIVES, and MAJORITY, along with Arrow's impossibility. We also introduce and analyze a new axiom, $\rho(n)$-GROUP-STABILITY, which we show generalizes multiple standard axioms. In the interest of identifying patterns, we also re-analyze several axiomatic impossibilities for which there already exist similar bounds,[5] though we show them via different proofs and give bounds that are explicit in their dependence on MIN-PROB. For clarity, we distinguish axioms and impossibilities analyzed for the first time in this paper in Table 1, Table 2, and Table 3 with *. The set of voting rules we study overlaps — but not perfectly — with the existing work on these axioms. In the appendix, we offer a detailed comparison between the techniques we use to prove our bounds versus those used in past work.

2   Model

Rankings. There are m candidates M and n voters N. Voters express their preferences over the candidates as complete rankings. Formally, a ranking π is a bijection from indices to candidates $[m] \to M$, where $\pi(j)$ represents the candidate ranked j-th in π. Let $a \succ_\pi b$ to denote that candidate a is preferred

---

[5] These axiomatic results include RESOLVABILITY [20]; STRATEGYPROOFNESS [22]; CONDORCET and PARTICIPATION [21], plus Moulin's impossibility of satisfying them together [23]; and Condorcet's voting paradox and the ANR impossibility theorem [19].

to b in ranking π, or formally, $\pi^{-1}(a) < \pi^{-1}(b)$. $\mathcal{L}(M)$ is the set of $m!$ possible rankings, or just $\mathcal{L}$ when M is clear from context. Letting the rankings in $\mathcal{L}$ be implicitly ordered, the last ($m!$-th) ranking in $\mathcal{L}$ is π_{-1}, and the set of rankings without this element as $\mathcal{L}_{-1} = \mathcal{L} \setminus \{\pi_{-1}\}$. For reasons to be clarified next, we will often work with $\mathcal{L}_{-1}$ instead of $\mathcal{L}$.

Profiles and Profile Histograms. A profile $\boldsymbol{\pi} = (\pi_i | i \in N)$ is a vector of n rankings, where π_i is voter i's ranking. Let $\Pi_n = \prod_{i=1}^{n} \mathcal{L}$ be the set of all profiles on n voters, and let $\Pi = \bigcup_{n \in \mathbb{Z}^+} \Pi_n$ be the set of all profiles overall. We define addition over profiles in the natural way: $(\boldsymbol{\pi} + \boldsymbol{\pi}') = (\pi_i | i \in N) \| (\pi'_i | i \in N')$.[6]

Instead of working directly with profiles, we will work primarily with their *histograms*. A *histogram* $\mathbf{h}$ is an $|\mathcal{L}_{-1}|$-length vector whose π-th entry h_π is the proportion of voters with ranking π in a profile ($\mathbf{h}$ is indexed only up to $\mathcal{L}_{-1}$ because the histogram must add to 1, so the $m!$-th index is redundant). The histogram associated with a particular profile $\boldsymbol{\pi}$ is $\mathbf{h}^\pi$, with π-th entry $\mathbf{h}_\pi^\pi := 1/n |\{i : \pi_i = \pi\}|$. We define the simplex of all possible histograms as $\Delta_{\mathbf{h}} := \{\mathbf{h} \in [0,1]^{|\mathcal{L}_{-1}|} : \sum_{\pi \in \mathcal{L}_{-1}} h_\pi \leq 1\}$. Of course, H includes vectors with irrational entries that could never correspond to a well-defined profile; nonetheless, it will be useful to consider the completed space. In order to talk about only the histograms that are realizable from well-defined profiles, we also define $H^{\mathbb{Q}} = H \cap \mathbb{Q}^{|\mathcal{L}_{-1}|}$ to be the subset of H of vectors with rational components.

Finally, we will use *histogram operator* $\mathcal{H}(\cdot)$ to transform more general, profile-based objects into their histogram-based analogs. For example, $\mathcal{H}(\boldsymbol{\pi}) = \mathbf{h}^\pi$, and $\mathcal{H}(\Pi) = H^{\mathbb{Q}}$. For a single ranking, $\mathcal{H}(\pi)$ is a $|\mathcal{L}_{-1}|$-length basis vector with a 1 at the π-th index ($\mathcal{H}(\pi_{-1})$ is the 0s vector). We will also apply this operator to *distributions* over profiles and rankings in the natural way, first drawing a profile or ranking from the distribution, and then considering its corresponding histogram.

Voting Rules. A *voting rule* is a function $R : \Pi \mapsto 2^M$ mapping a given profile to a set of winning candidates. Then, $R(\boldsymbol{\pi})$ is the set of winners chosen by the voting rule on $\boldsymbol{\pi}$. If a voting rule by its standard definition results in a tie, rather than specifying a tie-breaking rule, we assume it returns all such winners (this assumption is for ease of exposition only). Let $\mathcal{R}$ be the set of all voting rules. We will study several specific rules, defined colloquially below and formally in the appendix.

Positional Scoring Rules (PSRs) are represented by m-length vectors of weakly decreasing scores $s_1 \geq s_2 \geq \cdots \geq s_m$, where $s_1 = 1$ and $s_m = 0$. Candidate a receives s_i points for each voter that ranks it i-th; the candidate with the most points wins. We consider three specific PSRs, defined by their score vectors: PLURALITY: $(1, 0, \ldots, 0, 0)$, BORDA: $(1, 1 - \frac{1}{m-1}, \ldots, \frac{1}{m-1}, 0)$, and VETO: $(1, 1, \ldots, 1, 0)$. Beyond PSRs, we study the rules MINIMAX, KEMENY-YOUNG, and COPELAND. MINIMAX selects the candidate with the smallest *maximum*

---

[6] Here $\|$ represents concatenation, and thus the resulting profile $\boldsymbol{\pi} + \boldsymbol{\pi}'$ contains $|N| + |N'|$ voters. We also extend addition to permit positive integer multiples of profiles: $z\boldsymbol{\pi}$ means adding a profile together z times.

pairwise domination.[7] KEMENY-YOUNG selects the candidate ranked first in the ranking with the minimal sum of *Kendall tau* distances from voters' rankings.[8] COPELAND selects the alternative with the most points, giving a 1 point for each candidate b it pairwise dominates, and $1/2$ point for each b with which it pairwise ties. Finally, in our analysis, we will often study a general class of voting rules called *hyperplane rules*. This class is known to be equivalent to *generalized scoring rules* [24] and encompasses essentially every standard voting rule, including all those listed above.

Definition 1 (Hyperplane rules [15]). *Note that given a set of ℓ affine-hyperplanes, these hyperplanes partition the space of histograms into at most 3^ℓ regions, as every point is either on a hyperplane or on one of two sides. We say that R is a hyperplane rule if there exists a finite set of affine hyperplanes $H_1, \ldots, H_\ell$ such that R is constant on each such region.*

We will sometimes subdivide this class of rules further: *decisive* hyperplane rules are those which output a single winner on profiles that are not on any hyperplane; *non-decisive* hyperplane rules are all others.

Axioms. Formally, an *axiom* is a function $A : \mathcal{R} \to (\Pi \to \{\text{True}, \text{False}\})$, mapping a voting rule R to another mapping describing whether A is satisfied by R on any given profile. We can think of $A(R)$ as representing the true/false statement "R is consistent with A on π". We will study several standard axioms, defined formally in the appendix and described colloquially below.

R satisfies RESOLVABILITY on π if it selects a single winner. R satisfies CONDORCET CONSISTENCY (abbreviated as CONDORCET) on π if it selects the *Condorcet winner* (the candidate that pairwise dominates all other candidates), or by default if π has no Condorcet winner. R satisfies MAJORITY on π if it selects the *majority winner* (the candidate ranked first by a majority of voters), or by default if π does not have a majority winner. R satisfies CONSISTENCY on π if there is no partition of π into subprofiles such that a unique candidate b, which is *not* the winner in π, is chosen as the winner on all subprofiles in that partition. R satisfies INDEPENDENCE OF IRRELEVANT ALTERNATIVES (IIA) on π if the winner, a, cannot be made to lose to b by adjusting votes in a way that does not change the relative positions of a and b.

We also use $\rho(n)$-GROUP-STABILITY, which colloquially requires that the outcome of voting rules be *stable* to a change in the behavior of up to $\rho(n)$ voters. This is essentially framing *margin of victory* as a binary condition.

Definition 2 ($\rho(n)$-GROUP-STABILITY). *For a given rule R, $\rho(n)$-GROUP-STABILITY(R) is satisfied if, for every pair of profiles π, π' that differ on at most $\rho(n)$ of voters' rankings, $R(\pi) = R(\pi')$.*

---

[7] Useful definitions: a *pairwise-dominates* b in π when over half of voters rank a ahead of b: $\left|\{\pi_i | a \succ_{\pi_i} b\}\right| > n/2$. a and b *pairwise tie* when $\left|\{\pi_i | a \succ_{\pi_i} b\}\right| = n/2$. a's *maximum pairwise domination* is equal to $\max_{b \neq a} \left|\{\pi_i | a \succ_\pi b\}\right|$.

[8] The Kendall tau distance between rankings π, π' is the total number of swaps required to transform π into π'.

This axiom implies strong, $\rho(n)$-parameterized group-level versions of three axioms of common interest: $\rho(n)$-GROUP-STRATEGYPROOFNESS — no group of up to $\rho(n)$ voters can strategically misreport their preferences in concert and cause R to output an alternative they all weakly prefer, and at least one of them strongly prefers; $\rho(n)$-GROUP-PARTICIPATION — no group of up to $\rho(n)$ voters can leave the election and cause R to output an alternative they all weakly prefer, and at least one of them strongly prefers); and $\rho(n)$-GROUP-MONOTONICITY — no group of up to $\rho(n)$ voters can weakly decrease the position of an alternative c, which is not currently a winner by R, in their rankings, and cause c to become a winner by R.[9] We formally define these axioms in the appendix.

2.1   Smoothed Model of Profiles

Noise Distributions Over Rankings. A *noise distribution* is effectively a distribution over rankings; formally, it is a distribution over permutations σ : $[m] \rightarrow [m]$. When we "apply noise" to a ranking π via a noise distribution $\mathcal{S}_\phi$ (with parameter ϕ to be defined later), we are drawing a random permutation $\sigma \sim \mathcal{S}_\phi$, and then permuting π according to σ. Abusing notation slightly, we represent the distribution over rankings induced by perturbing π with noise model $\mathcal{S}_\phi$ as $\mathcal{S}_\phi(\pi)$.[10]

In this paper, we consider the class of noise distributions $\mathcal{S}$, encompassing any distribution over rankings satisfying the minimal regularity conditions in Assumptions 1 and 2 below. All $\mathcal{S}_\phi \in \mathcal{S}$ are parameterized by a dispersion parameter $\phi \in [0, 1]$, which captures the "level" of noise applied by $\mathcal{S}_\phi$. We will use ϕ to implement multiple different such measures throughout our results. Assumptions 1 and 2 impose the minimal requirements to ensure that ϕ reasonably measures the amount of noise, and that the noise distribution is practical to work with. As we implement specific uses of ϕ, we will impose additional assumptions as needed in later sections.

Assumption 1 (Extremal values). *The distribution $\mathcal{S}_0$ is the point mass on the identity (so that $\phi = 0$ corresponds to no noise added). The distribution $\mathcal{S}_1$ is uniform over all permutations (so that $\phi = 1$ corresponds maximum noise). Note that this implies that for all profiles $\boldsymbol{\pi} \in \Pi$, $\mathcal{S}_1(\boldsymbol{\pi})$ is equivalent to the impartial culture model (see, e.g., [6]).*

Assumption 2 (Continuity). *For all σ, the probability $\mathcal{S}_\phi$ places on σ is continuous in ϕ.*

---

[9] The fact that $\rho(n)$-GROUP-STABILITY implies the first and third of these axioms is by definition. For the second, $\rho(n)$-GROUP-STABILITY technically implies $\frac{\rho(n)}{1-\rho(n)}$-GROUP-PARTICIPATION, since this axiom involves $\rho(n)$ voters leaving the electorate; this does not change any of the asymptotic results, and we handle it in our proofs.

[10] More formally, $\mathcal{S}_\phi(\pi) = \pi + \sigma$ with $\sigma \sim \mathcal{S}_\phi$. Here, the $+$ operator represents composition: if $\sigma(i) = j$, then the i-th ranked candidate in the perturbed ranking will be $\pi(j)$.

Sampling Noisy Profiles. Applying noise via $\mathcal{S}_\phi$ to an entire *profile* means applying it to every ranking within it *independently* (an assumption also made in the semi-random model [19]). Formally, if π is our starting profile, then our noisy profile is drawn from the distribution $\prod_{i=1}^{n} \mathcal{S}_\phi(\pi_i)$, where each $\mathcal{S}_\phi(\pi_i)$ is independent. The resulting noisy profile is denoted $\mathcal{S}_\phi(\pi)$. We will treat $\mathcal{S}_\phi$, $\mathcal{S}_\phi(\pi)$, and $\mathcal{S}_\phi(\pi)$ as distributions and random variables interchangeably.

Since we will work with profiles in histogram form, so we will usually reason about distributions over rankings and profiles *projected into histograms space*. We express the distribution over ranking histograms induced by noise distribution $\mathcal{S}_\phi(\pi)$ as $\mathcal{H}(\mathcal{S}_\phi(\pi))$. Where the former is a distribution over rankings, the latter is a distribution over basis vectors. Then, the corresponding noise distribution over profile histograms is, naturally, $\mathcal{H}(\mathcal{S}_\phi(\pi)) = 1/n \sum_{i=1}^{n} \mathcal{H}(\mathcal{S}_\phi(\pi_i))$.

2.2   Smoothed-Satisfaction and Smoothed-Violation of Axioms

First, we formally define worst-case notions of axiom satisfaction and violation. Recall that $A(R) : \Pi \to \{\text{True}, \text{False}\}$ intakes a given profile and outputs whether rule R satisfies axiom A on that profile. Then, we say that π is a *counterexample* to $A(R)$ iff $(A(R))(\pi) = \text{False}$. Let $\Pi^{\neg A(R)}$ be the set of all profiles that are counterexamples to $A(R)$. Then, in the worst-case sense, R *satisfies* A if $\Pi^{\neg A(R)} = \emptyset$ (i.e., no counterexample exists); otherwise, R *violates* A.

Now, we define what it means for R to satisfy or violate A in the smoothed model. Conceptually, R *smoothed-satisfies* A if the probability that R satisfies A, after applying a noise distribution, converges to 1 as n grows large. In contrast, R *smoothed-violates* A if the probability of R satisfying A converges to 0 as n grows large. Note that worst-case satisfaction implies smoothed-satisfaction, and smoothed-violation implies violation.

Definition 3 (smoothed-satisfied). *Voting rule R $\mathcal{S}$-smoothed-satisfies axiom A at a rate $f(n, \phi)$ if, for all $n \in \mathbb{Z}^+$, profiles $\pi \in \Pi_n$, and $\phi \in (0, 1]$,*

$$\sup_{\phi' \in [\phi, 1]} \Pr\left[\mathcal{S}_{\phi'}(\pi) \in \Pi^{\neg A(R)}\right] \leq f(n, \phi).$$

Definition 4 (smoothed-violated). *A voting rule R $\mathcal{S}$-smoothed-violates axiom A at a rate of $f(n)$ if there exists an $n_0 \in \mathbb{Z}^+$, a profile $\pi \in \Pi_{n_0}$ and a $\phi \in (0, 1]$ such that for all $z \in \mathbb{Z}^+$,*

$$\inf_{\phi' \in [0, \phi]} \Pr\left[\mathcal{S}_{\phi'}(z\pi) \in \Pi^{\neg A(R)}\right] \geq 1 - f(z\,n_0).$$

Note that z is used just to let the composition of the electorate remain consistent as n grows without introducing divisibility issues.

Per Definition 3, convergence to smoothed-satisfaction occurs eventually for *all* $\phi \in (0, 1]$, although the rate depends on ϕ. When we show smoothed-violation, in contrast, we are saying there exists a constant amount of noise ϕ such that this amount, or any less, will not be enough to ensure satisfaction of

A by R as n grows large. Note the gap between these two definitions: smoothed-violated is not the negation of smoothed-satisfied, and a claim could therefore be "in-between" these definitions, satisfying neither. This will not end up being the case for any of the voting rules or criteria we study.

3 Patterns Across Smoothed-Violated, Smoothed-Satisfied Axioms

3.1 Condorcet, Majority, Consistency, IIA

This section is dedicated to axioms about which we prove negative results. As summarized in Table 1, we find that for all the voting rules we study, smoothed noise is insufficient to prevent existing violations of any of these axioms. Across these axioms, the reason for this insufficiency is the same: as formalized in our generic sufficient condition in Lemma 2, the histogram space contains contiguous regions of counterexamples.

Table 1. Summary of results stated in Theorem 1. s-violated = smoothed-violated.

	Axioms			
Voting Rules	CONDORCET	MAJORITY*	CONSISTENCY*	IIA*
PLURALITY	s-violated	satisfied	satisfied	s-violated
PSRs \ PLURALITY	s-violated	s-violated	satisfied	s-violated
MINIMAX	satisfied	satisfied	s-violated	s-violated
KEMENY-YOUNG	satisfied	satisfied	s-violated	s-violated
COPELAND	satisfied	satisfied	s-violated	s-violated

Theorem 1. *For all $A \in \{$CONDORCET, MAJORITY, CONSISTENCY, IIA$\}$ and all $R \in$ PSRs$\cup \{$MINIMAX, KEMENY-YOUNG, COPELAND$\}$, if R violates A, then R smoothed-violates A.*

The proof of this theorem is found in the appendix; we explain the intuition here. The core idea is that for small enough ϕ, $\mathcal{H}(\mathcal{S}_\phi(\pi))$ concentrates at an exponential rate very near the starting histogram $\mathbf{h}^\pi$ (Lemma 1). This lemma, proven in the appendix, follows from the fact that noise is applied independently across rankings, allowing the application of a simple Hoeffding bound.

Lemma 1. *Let $\mathcal{S}$ be a noise model. For all $\varepsilon > 0$, there exists a $\phi \in (0, 1]$ such that for all $\phi' \in [0, \phi]$ and profiles $\pi \in \Pi_n$ on n voters,*

$$\Pr\left[\|\mathcal{H}(\mathcal{S}_{\phi'}(\pi)) - \mathbf{h}^\pi\|_1 < \varepsilon\right] > 1 - \exp\left(\varepsilon^2 n/2\right).$$

From this concentration follows a general sufficient condition for smoothed-violation (Lemma 2): that there exists a counterexample π such that $\mathbf{h}^\pi$ is contained within a ball of counterexamples. Then, as the noise distribution concentrates around $\mathbf{h}^\pi$, most of its probability mass is contained in the ball, and the probability of a counterexample converges to 1.

Lemma 2. *Fix a noise model S, axiom A, and rule R. Suppose there exists a profile π and radius $r > 0$ such that for all profiles $\pi' \in \Pi$ satisfying (1) $|\pi'| = z|\pi|$ for some $z \in \mathbb{Z}^+$ and (2) $\|\mathbf{h}^{\pi'} - \mathbf{h}^{\pi}\|_1 < r$, it is the case that $\pi' \in \Pi^{\neg A(R)}$. Then, R smoothed-violates A at a rate of*

$$f(n) = \exp\left(-r^2 n/2\right).$$

Proof. Fix a noise model S, an axiom A, and a voting rule R. Fix a profile π and radius $r > 0$ satisfying the preconditions of the theorem. Choose $\varepsilon = r$ and let ϕ be the one from Lemma 1 corresponding to ε and S. Fix an arbitrary $z \in \mathbb{Z}^+$ and $\phi' \in [0, \phi]$. Notice that $\mathbf{h}^{\pi} = \mathbf{h}^{z\pi}$. Hence, Lemma 1 guarantees that the post-noise histogram lies in an r-radius ball around the original histogram $\mathbf{h}^{\pi}$ with high probability—that is, $\mathcal{H}(S_{\phi'}(z\pi)) \in B_r^{L_1}(\mathbf{h}^{\pi})$ with probability at least $1 - f(z|\pi|) = 1 - \exp\left(-r^2 z|\pi|/2\right)$. Note also that every profile in the support of $S_{\phi'}(z\pi)$ is guaranteed to have $z|\pi|$ voters. These two facts, taken with both preconditions of the theorem, imply that the post-noise profile histogram is a counterexample with high probability—that is, $S_{\phi'}(z\pi) \in \Pi^{\neg A(R)}$ with probability at least $1 - f(z|\pi|)$. It then follows, by the definition of smoothed violation (Definition 4), that R smoothed-violates A at a rate of $f(z|\pi|)$, as needed.

Finally, we conclude Theorem 1 by finding counterexamples for each A, R that exist within balls of other counterexamples, and then applying Lemma 2. Across rules and axioms, simple counterexamples suffice, supporting the intuition that the insufficiency of smoothed noise is not a quirk of these rules and axioms. Rather, smoothed noise may be insufficient across many interpretable rules and axioms, because — as a natural consequence of their interpretability — they behave similarly on similar profiles.

3.2 RESOLVABILITY, STRATEGYPROOFNESS, PARTICIPATION, and MONOTONICITY

This section is dedicated to axioms about which we prove (mostly) positive results, summarized in Table 2. Across these axioms, our proofs use a common property that makes smoothed-noise *sufficient* for circumventing impossibilities: that across voting rules, their counterexamples are restricted to regions on or near a limited number of hyperplanes.

As are upper bounds in Xia's semi-random model [19], our upper bounds here will be parameterized by MIN-PROB$(S_\phi) := \min_\sigma \Pr[S_\phi = \sigma]$, the smallest probability S_ϕ assigns to any permutation. This parameterization motivates two additional assumptions: ASSUMPTION 3 ensures that $1/$MIN-PROB(S_ϕ) is well-defined, and ASSUMPTION 4 describes how MIN-PROB implements ϕ as a measurement of the noisiness of S_ϕ.

Assumption 3 (Positivity). *For all $\phi \in (0, 1]$, MIN-PROB$(S_\phi) > 0$. That is, for any nonzero amount of noise, the resulting distribution assigns positive probability to all permutations.*

Table 2. Summary of results stated in Theorem 2 and Theorem 3. s-satisfied = smoothed-satisfied. Positive results for $o(\sqrt{n})$-GROUP-STABILITY directly imply positive results for $o(\sqrt{n})$-GROUP-STRATEGYPROOFNESS, $o(\sqrt{n})$-GROUP-PARTICIPATION, and $o(\sqrt{n})$-GROUP-MONOTONICITY.

	Voting Rules	Axioms	
		RESOLVABILITY	$o(\sqrt{n})$-GROUP STABILITY*
	PLURALITY	s-satisfied	s-satisfied
(Decisive	PSRs \ PLURALITY	s-satisfied	s-satisfied
hyperplane	MINIMAX	s-satisfied	s-satisfied
rules)	KEMENY-YOUNG	s-satisfied	s-satisfied
	COPELAND	s-violated	s-satisfied

Assumption 4 (Weak Monotonicity). *The value* MIN-PROB$(\mathcal{S}_\phi)$ *is non-decreasing in* ϕ.[11]

First, Theorem 2 shows that RESOLVABILITY is smoothed-satisfied for all decisive hyperplane rules. Notably, these rules exclude the known rule COPELAND, due to its lack of sensitivity to changes in rankings in regions surrounding ties.

Theorem 2. *All decisive hyperplane rules smoothed-satisfy* RESOLVABILITY *at rate* $O_m(1/\sqrt{n}\cdot\text{MIN-PROB}(\mathcal{S}_\phi)^{3/2})$. *All non-decisive hyperplane rules smoothed-violate* RESOLVABILITY.

The formal proof is found in the appendix. This result corely relies on the fact that, essentially regardless of the noise distribution over *rankings* $\mathcal{S}_\phi$, the resulting noise distribution over *entire profile histograms* $\mathcal{H}(\mathcal{S}_\phi(\pi))$ must converge uniformly[12] to a multi-dimensional Gaussian distribution at a rate of $O(1/\sqrt{n})$ (Lemma 3):

Lemma 3. *Let* $\mathcal{S}$ *be a noise model,* $\phi \in [0,1]$ *a parameter, and* $\pi \in \Pi_n$ *a profile on* n *voters. Then, for all convex sets* $X \subseteq \mathbb{R}^{m!-1}$,

$$\left| \Pr\left[\mathcal{H}(\mathcal{S}_\phi(\pi)) \in X\right] - \Pr\left[\mathcal{N}\left(\mathbb{E}[\mathcal{H}(\mathcal{S}_\phi(\pi))], \text{Cov}[\mathcal{H}(\mathcal{S}_\phi(\pi))]\right) \in X\right] \right|$$
$$\leq O((m!)^{7/4})/\sqrt{n} \cdot \text{MIN-PROB}(\mathcal{S}_\phi)^{3/2}.$$

Intuitively, $\mathcal{H}(\mathcal{S}_\phi(\pi))$ converges to a Gaussian because it is akin to the sum of independent indicators (literally, independently-drawn rankings represented in

---

[11] We note that our results don't centrally depend on Assumption 4—Assumptions 2 and 3 are sufficient to give the same high-level results. We include Assumption 4 because it is not prohibitive, it simplifies the exposition, and allows us to give more useful parameterized bounds.

[12] "Uniform" (over profiles) convergence means that for all profiles π of any fixed n, $\mathcal{H}(\mathcal{S}_\phi(\pi))$ is at most $O(1/\sqrt{n})$ "distance away" from the the Gaussian distribution with expectation and variance corresponding to that of $\mathcal{H}(\mathcal{S}_\phi(\pi))$.

histogram space as basis vectors). We prove this convergence in the appendix using a general form of the Berry Esseen bound [3].

This convergence allows us to prove a general sufficient condition for the smoothed-satisfaction: that the set of counterexamples are contained with a measure-zero subset of histograms (Lemma 4).

Lemma 4. *Fix a noise model S, voting rule R, and axiom A. If there exists some set X such that (1) $X = \bigcup_{j=1}^{\ell} X_j$ where each $X_j \subseteq \Delta_{\mathbf{h}}$ is convex, (2) $\mathcal{H}\left(\Pi^{\neg A(R)}\right) \subseteq X$, and (3) X is measure zero (noting that X is necessarily measurable), then R **smoothed-satisfies** A at a rate of*

$$f(n, \phi) = \ell \cdot O((m!)^{7/4}) / \left(\sqrt{n} \cdot \text{MIN-PROB}(S_\phi)^{3/2}\right).$$

The formal proof is found in the appendix, and uses that the Gaussian places zero probability mass over measure-zero regions of its support. This concludes our analysis of RESOLVABILITY.

Now, we show that $\rho(n)$-GROUP-STABILITY (Definition 2) is smoothed-satisfied by all hyperplane rules for $\rho(n) \in o(\sqrt{n})$. Because $\rho(n)$-GROUP-STABILITY implies $\rho(n)$-GROUP-STRATEGYPROOFNESS, $\rho(n)$-GROUP-PARTICIPATION, and $\rho(n)$-GROUP-MONOTONICITY, these axioms must also be smoothed-satisfied by all hyperplane rules.

Theorem 3. *$o(\sqrt{n})$-GROUP-STABILITY is smoothed-satisfied by all hyperplane rules at a rate of $O_m(1/\sqrt{n}\cdot\text{MIN-PROB}(S_\phi)^{3/2}) + o(1)$.*

We will now prove Theorem 3 via the following anti-concentration lemma, which states that for any π, $\mathcal{H}(S_\phi(\pi))$ places limited probability mass within distance $\delta(n)$ of any specific hyperplane as n grows large, so long as $\delta(n)$ is decreasing sufficiently quickly in n.

Lemma 5. *Let $\mathcal{G}$ be the set of all hyperplanes in $\mathbb{R}^{m!-1}$. For all noise models S, parameters $\phi \in [0, 1]$, and $\delta(n) \in o(1/\sqrt{n})$, we have the following, where d is the L_1 distance.*

$$\sup_{G \in \mathcal{G}} \sup_{\phi' \in [\phi, 1]} \sup_{\pi \in \Pi_n} \Pr\left[d(\mathcal{H}(S_{\phi'}(\pi)), G) \leq \delta(n)\right]$$

$$\in O\left(\delta(n)\sqrt{n}/\sqrt{\text{MIN-PROB}(S_\phi)} + 1/\text{MIN-PROB}(S_\phi)^{3/2}\sqrt{n}\right) \in o(1).$$

This lemma, proven in the appendix, shows that even if $\mathbb{E}[\mathcal{H}(S_\phi(\pi))]$ falls within $\delta(n)$ distance of the hyperplane (we can think of this as it falling within a "thick" hyperplane), the width of this thick hyperplane is shrinking faster than the distribution over histograms concentrates as n grows large. To make the convergence rate in Lemma 5 $o(1)$, $\delta(n)$ must be in $o(1/\sqrt{n})$.

To apply this lemma to show the smoothed-satisfaction of $\rho(n)$-GROUP-STABILITY, first observe that for a group of size $\rho(n)$ to be able to impact the outcome of the election in π (i.e., for π to be a counterexample), that group must be *pivotal* in π — that is, $\mathbf{h}^\pi$ must lie within some $\rho(n)$-dependent distance from a profile on which the winner changes. Because the set of such profiles are defined by finitely-many hyperplanes (by definition of hyperplane rules), counterexamples are then restricted to "thick" hyperplanes. To apply Lemma 5 for each such hyperplane, we need their width to be $o(1/\sqrt{n})$ in histogram space, corresponding to a coalition of size $\rho(n) \in o(\sqrt{n})$. We conclude by union bounding over the finite number of hyperplanes referred to in Definition 1.   □

4   Beyond Dependence on the Minimum Probability

All convergence rates to smoothed-satisfaction proven so far — both in Sect. 3.2 and in the related work [19] — depend on $1/\sqrt{n} \cdot \text{MIN-PROB}^3$. As a result, when MIN-PROB is very small, extremely large n is required to get reasonably low probability of an axiom violation: examining the relative n and MIN-PROB dependency above, if we decrease MIN-PROB by factor ℓ, we need to increase the number of voters n by a factor of ℓ^3 in order to recover the same bound.[13]

Motivated by the lack of good bounds for the important class of noise models with small MIN-PROB, we now pursue the first MIN-PROB-independent bounds on convergence rates. For concreteness, we specifically pursue these bounds within the well-established *Mallows model*, as defined below; however, as we will illustrate in detail throughout this section, our results rely only weakly on the properties of this model, and should apply to more general models as well.

Definition 5 (Mallows noise model [13]). *Let* $d : \mathcal{L} \times \mathcal{L} \to \mathbb{N}$ *be the Kendall tau distance, and let* $\phi \in [0, 1]$. *Then, the Mallows model* $\mathcal{S}_\phi^{Mallows}$ *is defined as follows, where* $Z = \sum_{\pi' \in \mathcal{L}} \phi^{d(\pi', \pi)}$ *is a normalizing term.*

$$\Pr\left[\mathcal{S}_\phi^{Mallows}(\pi) = \pi'\right] = \phi^{d(\pi, \pi')}/Z.$$

The Mallows model is an attractive case study for proving MIN-PROB-independent bounds for two reasons. First, it precisely captures the intuition motivating this analysis: that long-range swaps should be rare under less noise, making MIN-PROB low. Longer range swaps are *so* rare in this model, in fact, that MIN-PROB($\mathcal{S}_\phi^{Mallows}$) approaches zero exponentially fast as m grows: the maximum Kendall tau distance between rankings is $\binom{m}{2}$, so MIN-PROB($\mathcal{S}_\phi^{Mallows}$) $\in$ $\phi^{\Omega(m^2)}$. This motivates our second reason: that despite its importance in social choice, existing convergence rates grow poorer at an exponential rate for Mallows noise as m gets large and/or ϕ gets small.

---

[13] One may wonder if different techniques could potentially improve this dependence. Although we do not have a tight cubic lower bound and the exact bound should depend on the exact voting rule/axiom, we can at least get a linear one. Indeed, consider the noise distribution that switches to any ranking other than the starting one with probability ε, but stays on the starting one with probability $1 - (m! - 1)\varepsilon$. The minimum probability here is ε, but we need $n \in \omega(\varepsilon/m!)$ to ensure that the profile changes at all with high probability, a necessary condition to smoothed-satisfy any axiom that fails on even a single profile.

Within the Mallows model, we characterize the precise m, ϕ, n-dependent rates at which RESOLVABILITY and $o(\sqrt{n})$-STRATEGY-PROOFNESS are smoothed-satisfied by four diverse voting rules: PLURALITY, BORDA, VETO, and MINIMAX. Recall that these voting rules are already known to smoothed-satisfy both axioms by our results in Sect. 3.2. The key difference in this analysis is that, while before n was being treated as the only variable (with m and the noise level treated as constants), we now consider more closely the convergence depends on m and ϕ. In particular, we are interested in how large n must be (as a function of m and ϕ) for the rate to be $o(1)$ (i.e., satisfaction occurring with high probability). We summarize our results in Table 3, framed to directly answer this question. The formal statements and proofs are below, in Sect. 4.1.

Table 3. Bounds on rates of smoothed-satisfaction under Mallows noise model.

Voting Rules	Axioms		
	RESOLVABILITY	n^p-GROUP-STRATEGYPROOF	
PLURALITY	$\omega(m^5/\phi)$ is sufficient	$\omega((m^5/\phi)^{1/(1-p)})$ is sufficient	(Proposition 1)
BORDA	$\omega\left(m^6/\phi\right)$ is sufficient	$\omega((m^8/\phi)^{1/(1-p)})$ is sufficient	(Proposition 2)
VETO	$\Omega(1/\phi^{m-2})$ is necessary	$\Omega(1/\phi^{m-2})$ is necessary	(Proposition 3)
MINIMAX	$\Omega\left(1/m\phi^{\lfloor m/2 \rfloor}\right)$ is necessary	$\Omega\left(1/m\phi^{\lfloor m/4 \rfloor}\right)$ is necessary	(Proposition 4)

What is striking in these results is a clear separation between voting rules, which was not visible in our MIN-PROB-parameterized bounds. The convergence rates of PLURALITY and BORDA do not get dramatically worse as ϕ (and thus MIN-PROB) gets small; put another way, as ϕ scales down, n must scale up proportionally to maintain roughly the same probability of satisfaction. In contrast, MINIMAX and VETO require exponentially (in m) large n.[14]

We can, as before, distill a pattern explaining this gap: the voting rules that achieve the best rates (PLURALITY, BORDA are those whose outcomes are more sensitive to local swaps across the support (or, in critical areas of the support, in the case of PLURALITY). The outcomes of VETO and MAXIMIN, in contrast, are fairly insensitive to local swaps, allowing us to show that local swaps are not enough to overcome impossibilities. While this pattern is perhaps unsurprising in retrospect, it may be important in informing the choice of voting rules.

4.1   Formal Statements and Proofs

We defer the proofs on PLURALITY, BORDA, VETO and MINIMAX to the appendix. Our arguments will rely only very weakly on the properties of the

---

[14] To put this in perspective, suppose $m = 6$ and we decrease ϕ from $1/5$ to $1/10$. To maintain a similar probability of violating RESOLVABILITY, if we are using PLURALITY or BORDA it is sufficient to double the number of voters; if we are using VETO, one needs at least $2^4 = 16$ times as many voters, and for MINIMAX, one needs at least $2^3 = 8$ times as many voters. This gap only gets steeper as m gets larger.

Mallows model, essentially requiring just that the noise rarely induces swaps between distantly-ranked candidates. To emphasize the kinds of noise models to which our arguments can generalize, we now recap the precise properties of the noise model required for the proof corresponding to each voting rule. For PLU-RALITY, the key property of the noise distribution we use is that no alternative is ranked first post-noise with probability near 1. For BORDA, the key property is that there is no ℓ for which two candidates will be exactly ℓ positions apart post-noise with probability near 1. For VETO, the key property is that there is an exponentially small probability of a given voter moving one of their top two candidates to last place. For MINIMAX, the key property is that it is unlikely for two candidates a distance of $m/2$ apart to swap. Before presenting our results, we establish a few useful properties of Mallows noise.

Lemma 6 (First- and last-place probability [2]). *From starting ranking π, the probability that $\pi(j)$ is ranked first in a ranking drawn from $\mathcal{S}_\phi^{Mallows}(\pi)$ is proportional to ϕ^j, i.e., $\phi^j / \sum_{j'=1}^m \phi^{j'}$. Symmetrically, the probability $\pi(j)$ is ranked last is proportional to ϕ^{m-j}.*

From [13], if two candidates i, j are k positions apart in π (i.e., $\pi(i)$ and $\pi(j)$ with $i - j = k$), the probability that they retain their relative order post-Mallows noise is increasing in k, and is always at least $1/2$. This was refined by [4] to an exact value of their swap probability:

Lemma 7 (swap probability [4]). *For candidates i, j such that $\pi(i)$ and $\pi(j)$ with $i - j = k$, their probability of swapping post-Mallows noise is*

$$q(k) := \frac{1}{1-\phi^{k+1}} \left(1 - \frac{(1-\phi)k\phi^k}{1-\phi^k} \right).$$

We will often parameterize our bounds by the probability of the opposite event — that i and j at distance k *do* swap, which we denote by $\bar{q}(k) := 1 - q(k)$. We will use the following upper bounds on $\bar{q}(k)$, derived in the appendix.

Lemma 8. *For all $k \in [m]$ and $\phi \in [0, 1]$, $\bar{q}(k) \leq \min\left\{ k\phi^k, \phi^{k/2}, \frac{\phi}{1+\phi} \right\}$.*

Proposition 1 (PLURALITY). PLURALITY *smoothed-satisfies* RESOLVABIL-ITY *at a rate of at most*

$$O\left(m^{5/2}/\sqrt{n\phi} + m\exp(-n/6m) \right)$$

and, for all $\rho(n) \leq \frac{n}{6m}$, smoothed-satisfies $\rho(n)$-GROUP-STABILITY at a rate of

$$O\left(\rho(n)m^{5/2}/\sqrt{n\phi} + m\exp(-n/9m) \right).$$

Proposition 2 (BORDA). BORDA *smoothed-satisfies* RESOLVABILITY *at a rate of $O\left(m^3/\sqrt{n\phi} \right)$, and smoothed-satisfies $\rho(n)$–GROUP-STABILITY at a rate of $O\left(\rho(n)m^4/\sqrt{n\phi} \right)$.*

Proposition 3 (VETO). *The rate* VETO *smoothed-satisfies* RESOLVABILITY *is lower bounded by* $1 - n\phi^{m-2}$, *and, as long as* $n \geq m$, *the rate* 1-STRATEGY-PROOFNESS *is smoothed-satisfied is lower bounded by* $1 - n\phi^{m-2} - m\phi^{\lfloor \frac{n}{m} \rfloor}$.

Proposition 4 (MINIMAX). *The rate at which* MINIMAX *smoothed-satisfies* RESOLVABILITY *is lower bounded by*

$$\Omega\left(1/\sqrt{n\bar{q}(\lfloor m/2 \rfloor)} - m \exp\left(-n(1 - 2\bar{q}(\lfloor m/2 \rfloor))^2/32\right)\right),$$

at least for n *divisible by* 4, *and the rate for* 1-GROUP-STRATEGYPROOFNESS *is lower bounded by*

$$1 - 3nq(\lfloor m/4 \rfloor) - O\left(m \exp\left(-(n-2)(1 - 2\bar{q}(\lfloor m/2 \rfloor))^2/32\right)\right).$$

Note that using the fact that $\bar{q}(\lfloor m/2 \rfloor) \in O(1/\log m)$, a necessary condition for a high probability bound is $n \in \omega(\bar{q}(\lfloor m/2 \rfloor))$. We can plug in our various upper bounds from Lemma 8 on $\bar{q}(\lfloor m/2 \rfloor)$. For the bound in Table 3, the second of these implies that $n \in \omega(1/(m\phi^{\lfloor m/2 \rfloor}))$ is necessary.

5 Discussion

Since the motivation for the smoothed model is fundamentally practical, we first outline some key practical takeaways from our work. These takeaways address the question: *how much can a bit of smoothed noise give us, when, and why?*

Although the related work so far has been largely optimistic about smoothed analysis in social choice, our analysis in Sect. 3.1 illustrates that for many rules and axioms, smoothed noise may be insufficient to circumvent impossibilities. Our sufficient condition for smoothed violation, moreover, suggests that the insufficiency of smoothed noise may be tied fundamentally to axioms' and voting rules' simplicity and interpretability, suggesting that these negative results are difficult to get around without making other compromises.

On the other hand, our results in Sect. 3.2 show that for certain kinds of axioms — i.e., those whose counterexamples must lie near hyperplanes or other small-measure structures — a small amount of noise *is* enough to circumvent impossibilities. However, our results in Sect. 4 paint a more nuanced picture: if we believe that in practice, any perturbations are truly unlikely to cause drastic opinion changes, then our choice of voting rule is far more important than past work suggests. In particular, while past MIN-PROB-parameterized upper bounds yield similar convergence rates across voting rules (for any given axiom), our results in the Mallows model — for which existing bounds do not usefully apply — show that actually, certain voting rules converge to smoothed-satisfaction far faster than others. The voting rules that converge faster are, perhaps unsurprisingly, those whose outcomes are more sensitive to local swaps.

With these takeaways in hand, in the appendix, we provide extensions and lay out future work. The key extension we provide there shows that **Arrow's impossibility is resolved under smoothed noise** — but only because trivial rules can satisfy the axiom NON-DICTATORSHIP. We propose a more meaningful strengthening of Arrow's impossibility; determining whether it is resolved under smoothed noise remains an open question. We also discuss opportunities for generalizing our results from Sect. 4 to generic models in which drastic opinion change is impossible under low noise.

References

1. Arrow, K.: Individual values and social choice, vol. 24. Wiley, Nueva York (1951)
2. Awasthi, P., Blum, A., Sheffet, O., Vijayaraghavan, A.: Learning mixtures of ranking models. In: Proceedings of the 27th Annual Conference on Neural Information Processing Systems (NeurIPS), pp. 2609–2617 (2014)
3. Bentkus, V.: A Lyapunov-type bound in rd. Theory Probab. Appl. **49**(2), 311–323 (2005)
4. Boehmer, N., Faliszewski, P., Kraiczy, S.: Properties of the mallows model depending on the number of alternatives: a warning for an experimentalist. In: Proceedings of the 40th International Conference on Machine Learning (ICML) (2023)
5. Dillman, D.A., Christian, L.M.: Survey mode as a source of instability in responses across surveys. Field Methods **17**(1), 30–52 (2005)
6. Eğecioğlu, Ö., Giritligil, A.E.: The impartial, anonymous, and neutral culture model: a probability model for sampling public preference structures. J. Math. Sociol. **37**(4), 203–222 (2013)
7. Felsenthal, D.S., Maoz, Z., Rapoport, A.: An empirical evaluation of six voting procedures: do they really make any difference? Br. J. Polit. Sci. **23**(1), 1–27 (1993)
8. Gehrlein, W.V.: Condorcet's paradox and the likelihood of its occurrence: different perspectives on balanced preferences. Theor. Decis. **52**(2), 171–199 (2002)
9. Gibbard, A.: Manipulation of voting schemes. Econometrica **41**, 587–602 (1973)
10. Green-Armytage, J., Tideman, T.N., Cosman, R.: Statistical evaluation of voting rules. Soc. Choice Welf. **46**(1), 183–212 (2016)
11. Kahneman, D., Tversky, A.: Prospect theory: an analysis of decision under risk. In: Handbook of the Fundamentals of Financial Decision Making: Part I, pp. 99–127. World Scientific (2013)
12. Lee, L., Amir, O., Ariely, D.: In search of homo economicus: cognitive noise and the role of emotion in preference consistency. J. Consum. Res. **36**(2), 173–187 (2009)
13. Mallows, C.L.: Non-null ranking models. I. Biometrika **44**(1/2), 114–130 (1957)
14. Mossel, E.: Probabilistic view of voting, paradoxes, and manipulation. Bull. Am. Math. Soc. **59**(3), 297–330 (2022)
15. Mossel, E., Procaccia, A.D., Rácz, M.Z.: A smooth transition from powerlessness to absolute power. J. Artif. Intell. Res. **48**, 923–951 (2013)
16. Plassmann, F., Tideman, T.N.: How frequently do different voting rules encounter voting paradoxes in three-candidate elections? Soc. Choice Welf. **42**(1), 31–75 (2014)
17. Pritchard, G., Slinko, A.: On the average minimum size of a manipulating coalition. Soc. Choice Welf. **27**, 263–277 (2006)
18. Spielman, D.A., Teng, S.H.: Smoothed analysis of algorithms: why the simplex algorithm usually takes polynomial time. J. ACM **51**(3), 385–463 (2004)
19. Xia, L.: The smoothed possibility of social choice. In: Proceedings of the 33rd Annual Conference on Neural Information Processing Systems (NeurIPS), pp. 11044–11055 (2020)

20. Xia, L.: How likely are large elections tied? In: Proceedings of the 22nd ACM Conference on Economics and Computation (EC), pp. 884–885 (2021)
21. Xia, L.: The semi-random satisfaction of voting axioms. In: Proceedings of the 34th Annual Conference on Neural Information Processing Systems (NeurIPS), pp. 6075–6086 (2021)
22. Xia, L.: The impact of a coalition: assessing the likelihood of voter influence in large elections. In: Proceedings of the 24th ACM Conference on Economics and Computation (EC) (2023). forthcoming
23. Xia, L.: Semi-random impossibilities of condorcet criterion. In: Proceedings of the 37th AAAI Conference on Artificial Intelligence (AAAI), pp. 5867–5875 (2023)
24. Xia, L., Conitzer, V.: Generalized scoring rules and the frequency of coalitional manipulability. In: Proceedings of the 9th ACM Conference on Economics and Computation (EC), pp. 109–118 (2008)

A Discrete and Bounded Locally Envy-Free Cake Cutting Protocol on Trees

Ganesh Ghalme[1]($\boxtimes$), Xin Huang[2], Yuka Machino[3], and Nidhi Rathi[4]

[1] Indian Institute of Technology Hyderabad, Sangareddy, India
`ganeshghalme@ai.iith.ac.in`
[2] Kyushu University, Fukuoka, Japan
[3] Massachusetts Institute of Technology, Cambridge, USA
`yukam997@mit.edu`
[4] Max Planck Institute for Informatics, University of Saarland, Saarbrücken, Germany
`nrathi@mpi-inf.mpg.de`

Abstract. We study the classic problem of *fairly* allocating a divisible resource modeled as a unit interval $[0, 1]$ and referred to as a *cake*. In a landmark result, Aziz and Mackenzie [4] gave the first discrete and bounded protocol for computing an *envy-free cake division*, but with a huge query complexity consisting of six towers of exponent in the number of agents, n. However, the best-known lower bound for the same is $\Omega(n^2)$, leaving a massive gap in our understanding of the complexity of the problem.

In this work, we study an important variant of the problem where agents are embedded on a graph whose edges determine agent relations. Given a graph, the goal is to find a *locally envy-free* allocation where every agent values her share of the cake at least as much as that of any of her *neighbors'* share. We identify a *non-trivial* graph structure, namely a tree having depth at most 2 (DEPTH2TREE), that admits a query efficient protocol to find locally envy-free allocations using $O(n^4 \log n)$ queries under the standard Robertson-Webb (RW) query model. To the best of our knowledge, this is the first such non-trivial graph structure. In our second result, we develop a novel cake-division protocol that finds a locally envy-free allocation among n agents on *any* TREE graph using $O(n^{2n})$ RW queries. Though exponential, our protocol for TREE graphs achieves a significant improvement over the best-known query complexity of six-towers-of-n for complete graphs.

1 Introduction

The problem of fairly dividing a set of resources among a set of participating agents is one of the most fundamental problems in distributive justice, with roots

N. Rathi—A part of this work was done when N. Rathi was at Aarhus University, Denmark.

J. Garg et al. (Eds.): WINE 2023, LNCS 14413, pp. 310–328, 2024.
https://doi.org/10.1007/978-3-031-48974-7_18

dating back to biblical time. However, arguably the first formal mathematical approach towards this problem was initiated by Steinhaus [24] (see also [14]). Over time, this problem has not only found interest in the academic communities of various disciplines such as social sciences, economics, mathematics, and computer science (see [9,10,23] for excellent expositions) but has also found relevance in a wide-range of real-world applications [12,16,21]. The problem of *cake division* provides an elegant abstraction to several situations where a divisible resource is to be allocated among agents with heterogeneous preferences such as division of land, allocation of radio and television spectrum, rent division, to name a few (see [19] for implementations of cake-cutting methods).

Formally, a cake-division instance consists of n agents, each having a cardinal preference over the cake that is modeled as a unit interval $[0, 1]$. These preferences are specified by a valuation function v_i's, and we write $v_i(I)$ to denote agent i's value for the piece $I \subseteq [0, 1]$. The goal is to partition the cake into n bundles (may consist of finitely many intervals) and assign them to the n agents such that this assignment is consider *fair*. A central notion of fairness in resource-allocation settings is *envy-freeness* that deems an allocation *fair* if every agent prefers her allocated share over that of any other agent [17]. The appeal of envy-freeness can be rightfully perceived from its strong existential result: under very mild assumptions, an envy-free cake division is guaranteed to always exist [15,25].

The algorithmic results for the problem, however, remain elusive. In fact, Stromquist [26] proved that there does not exist a finite protocol[1] for computing an envy-free cake division (with contiguous pieces) in an adversarial model. Furthermore, Deng et al. [13] showed that this problem is PPAD-hard when agents have ordinal valuations. The best-known protocol of Aziz and Mackenzie [4] for finding envy-free cake divisions with non-contiguous pieces has a super-exponential query complexity bound of $O(n^{n^{n^{n^{n^n}}}})$. In contrast, the best known lower bound for the problem is $\Omega(n^2)$ leaving a massive gap in our understanding of the query complexity of the problem [22].

Current techniques do not seem well suited to yield an immediate answer to developing better protocols for finding envy-free cake divisions. Hence, one of the promising approach is to find efficient protocols for instances satisfying certain properties. In this work, we study one such interesting variant where we assume that envy comparisons between agents are dictated by an underlying graph over agents. The goal here is to find a *locally envy-free* allocation of the cake such that no agent envies her neighbor(s) in the given graph G. Note that when G is a complete graph, we retrieve the classical setting of envy-free cake division.

The above-described graphical framework opens various interesting directions for understanding the problem of fairness in cake division (see [1,7,8,11,27]). From a practical point of view, it captures many situations in which global knowledge is unavailable or unrealistic. For instance, when the graph represents social connections between a group of people, it is reasonable to assume that agents only envy the agents whom they know (i.e., friends or friends of friends).

---

[1] We consider the protocol under the standard Robertson-Webb query model.

Similarly, when a graph represents rank hierarchy in an organization, it is reasonable to assume that agents only envy their immediate neighbors (i.e. colleagues).

The primary objective of this paper, however, is to study *local envy-freeness* from a theoretical standpoint. Given that the state-of-the-art protocol for finding envy-free allocation for a complete graph requires super-exponential queries, a natural line of research—and the focus of this work—is to identify interesting graph structures that admit faster protocols for computing locally envy-free allocations.

Our Results and Techniques: Our work focuses on cake-division instances where the underlying graph over the agents is either a TREE or a DEPTH2TREE. A TREE graph represents a setting where agents are embedded on a rooted tree (see Fig. 1), while a DEPTH2TREE graph is a special case of tree graph with a depth at most two. Our protocols operate under the standard Robertson-Webb query model [23] (defined in Sect. 2), and hence are discrete in nature.

We begin in Sect. 3 by addressing an open problem listed in [1] and [7] that asks for identifying interesting classes of graph structures that admit polynomial-query algorithms for local envy-freeness. Our result identifies the first *non-trivial* graph structure over n agents, namely a DEPTH2TREE, for which we develop an efficient protocol for computing locally envy-free allocations. As a warm-up example, a simpler protocol for 4-agents-on-a-line graph and 5-agents-on-a-line graph can be found in the full version of the paper [18].

Theorem 1. *For cake-division instances with n agents on a DEPTH2TREE, there exists a discrete protocol that finds a locally envy-free allocation using at most $O(n^3 \log(n))$* **cut** *and $O(n^4 \log(n))$* **eval** *queries under RW model.*

Interestingly, the techniques developed for DEPTH2TREE graph do not extend trivially to the TREE graphs. Nonetheless, inspired by its main ideas, in Sect. 4, we develop a recursive protocol that computes a locally envy-free allocation among n agents on *any* TREE graph with a query complexity of $O(n^{2n})$. The idea of recursion imparts notable simplicity to our protocol, even though the analysis is somewhat intricate. Our next result addresses the open problem of designing a discrete and bounded protocol for local envy-freeness on trees mentioned in [1].

Theorem 2. *For cake-division instances with n agents on a TREE, there exists a discrete protocol that computes a locally envy-free allocation using at most $O(n^{2n})$ Robertson-Webb queries.*

1.1   Related Literature

The celebrated Selfridge-Conway protocol [23] finds an envy-free allocation among three agents using 5 queries. Aziz and Mackenzie [5] proposed a cake-cutting protocol that finds an envy-free allocation among four agents in (close to) 600 queries. This bound was then improved by Amanatidis et al. [2] to 171 queries. As discussed above, despite significant efforts, the problem of developing efficient envy-free cake-cutting (discrete) protocols for $n \geq 5$ agents remains largely open. Additionally, multiple hardness results have motivated various interesting settings. For example, efficient fair cake-cutting protocols for interesting classes of valuations have been developed in [6,20]. The work of Arunachaleswaran et al. [3] developed an approximation algorithm that efficiently finds a cake division with contiguous pieces wherein the envy is multiplicatively bounded within a factor of $2 + O(1/n)$.

The problem of fair cake division with graphical (envy) constraints was first introduced by Abebe et al. [1]. They characterize the set of directed graphs for which an *oblivious* single-cutter protocol—a protocol that uses a single agent to cut the cake into pieces—admits a bounded query complexity for locally envy-free allocations in the Robertson-Webb query model. In contrast, our work studies a class of *undirected* graphs that are significantly harder to analyze, and surprisingly develop comparable upper bounds. In another closely related paper, Bei et al. [7] develops *a moving-knife protocol*[2] that outputs an envy-free allocation on tree graphs. In more recent work, Bei et al. [8] develop a discrete and bounded *locally proportional* protocol for any given graph. Tucker [27] compliments this result by providing a lower bound of $\Omega(n^2)$ on the query complexity of obtaining locally proportional allocation in the Robertson-Webb model. In contrast, our work addresses a stronger guarantee of local envy-freeness. We address the open questions raised in [1,8] by (a) developing a discrete and bounded protocol for TREE graphs with single-exponential query complexity, and (b) constructing a query-efficient discrete protocol that finds locally envy-free allocations among n agents on DEPTH2TREE.

2   The Setting

We write $[k]$ to denote the set $\{1, 2, \ldots, k\}$ for a positive integer k.

This work considers the problem of fairly allocating a divisible resource—modeled as a unit interval $[0, 1]$ and referred to as a *cake*—among n agents denoted by $(a_1, a_2, \ldots, a_n)$. For an agent a_i for $i \in [n]$, we write v_i to specify her cardinal valuations over the intervals in $[0, 1]$. In particular, $v_i(I) \in \mathbb{R}^+ \cup \{0\}$ represents the valuation of agent a_i for the interval $I \subseteq [0, 1]$. For brevity, we will write $v_i(x, y)$ instead of $v_i([x, y])$ to denote agent a_i's value for an interval $[x, y] \subseteq [0, 1]$. Following the standard convention, we assume that v_is are non-negative, additive,[3] and non-atomic.[4] Additionally, without loss of generality, we assume that the valuations are normalized i.e., $v_i(0, 1) = 1$ for all $i \in [n]$.

We write G to denote the underlying graph structure over the agents, often written as *agents are on G*. Here, the nodes of G represent agents and the edges among them correspond to *envy constraints*. We will write $\mathcal{I}$ to refer to a cake-division instance with graph constraints.

---

[2] A moving-knife protocol may not be implementable in discrete steps under standard RW query model, and hence our result for TREE graphs is stronger than that of [7].

[3] For any two disjoint intervals $I_1, I_2 \subseteq [0, 1]$, we have $v_i(I_1 \cup I_2) = v_i(I_1) + v_i(I_2)$.

[4] For any interval $[x, y]$ and any $\lambda \in [0, 1]$, $\exists y' \in [x, y]$ such that $v_i(x, y') = \lambda \cdot v_i(x, y)$.

2.1   Preliminaries

For cake-division instance with n agents we define an *allocation* $\mathcal{A} = (A_1, A_2, \ldots, A_n)$ of the cake $[0, 1]$ to be a collection of n pair-wise disjoint pieces such that $\cup_{i \in [n]} A_i = [0, 1]$. Here, a piece or a bundle A_i (a finite union of intervals of the cake $[0, 1]$) is assigned to agent a_i for $i \in [n]$. We say $\mathcal{A}$ is a *partial* allocation if the union of A_is forms a strict subset of $[0, 1]$. In this work, we study the fairness notion of *local envy-freeness* defined below.

Definition 1 (Local Envy-freeness). *Given a cake-division instance with a graph G, an allocation $\mathcal{A} = (A_1, A_2, \cdots, A_n)$ is said to be* locally envy-free (on G) *if for any agent a_i for $i \in [n]$ and any piece $A_j \in \mathcal{A}$ such that a_i and a_j have an edge in G, we have $v_i(A_i) \geq v_i(A_j)$. When G is a complete graph, a locally envy-free allocation is called as an* envy-free *allocation.*

Definition 2 (Robertson-Webb (RW) query model). *Our protocols operate under the standard Robertson-Webb query model [23] that allows access agents' valuations via the following two types of queries:*[5]

1. *Cut query: Given a point $x \in [0, 1]$ and a target value $\tau \in [0, 1]$, $\mathbf{cut}_i(x, \tau)$ asks agent a_i to report the smallest $y \in [x, 1]$ such that $v_i(x, y) = \tau$. If such a y does not exist, then the response is some arbitrary number, say -1.*
2. *Evaluation query: Given $0 \leq x < y \leq 1$, $\mathbf{eval}_i(x, y)$ asks agent a_i to report her value $v_i(x, y)$ for the interval $[x, y]$ of the cake.*

Our protocols use the following three standard procedures (formal descriptions are deferred to the full version of the paper [18]).

Select: Given a collection of pieces $\mathcal{X}$, $m \leq |\mathcal{X}|$, and an agent a_i, SELECT$(\mathcal{X}, a_i, m)$ returns the m highest-valued pieces in $\mathcal{X}$ according to v_i. It is easy to see that SELECT requires zero $\mathbf{cut}$ queries and at most $|\mathcal{X}|$ $\mathbf{eval}$ queries.

Trim: Given a collection of pieces $\mathcal{X}$ and an agent a_i, TRIM$(\mathcal{X}, a_i)$ returns a collection of $|\mathcal{X}|$-many piece, each of value equal to her smallest-valued piece in $\mathcal{X}$ along with some *residue R*. It first finds the lowest valued piece according to v_i and makes the remaining pieces of value equal to it by trimming. The *residue* is the collection of all the trimmings formed in the procedure. The TRIM procedure requires $|\mathcal{X}| - 1$ $\mathbf{cut}$ queries and $|\mathcal{X}|$ $\mathbf{eval}$ queries.

Equal: Given a collection of pieces $\mathcal{X}$ and an agent a_i, EQUAL$(\mathcal{X}, a_i)$ redistributes among the pieces in $\mathcal{X}$ (creating no residue) such that each piece is equally valued by a_i. It also identifies a bundle in the original collection that has a value larger than the average value of the bundles (according to v_i). Note that while both EQUAL and TRIM procedures return an allocation where all the pieces are equally valued by a_i, TRIM may generate a residue whereas EQUAL redistributes the pieces of the cake without leaving any residue. The EQUAL procedure requires $|\mathcal{X}| - 1$ $\mathbf{cut}$ queries and $|\mathcal{X}|$ $\mathbf{eval}$ queries.

---

[5] See a remark in the full version of the paper [18].

3   Depth2Tree: A Tractable Instance

In this section, we consider cake-division instances where the underlying graph over the agents is a DEPTH2TREE and develop a novel protocol to compute locally envy-free allocations among n agents using poynomially-many queries (Theorem 1). We assume that the graph is rooted at agent a_r and we write D to denote the set of neighbours/children of agent a_r. Each agent $a_i \in D$ has $\ell_i + 1$ neighbours, i.e., she is connected to $\ell_i \geq 0$ leaf agents. In addition, we write $L(i)$ to denote the set of children of agent $a_i \in D$.

Overview of the* D2Tree *Protocol: For cake-division instances with n agents on a DEPTH2TREE, we develop a protocol D2TREE that progressively builds an allocation $\mathcal{A} = (A_1, \ldots, A_n)$ among n agents. D2TREE primarily consists of a while-loop that has three phases: Selection, Trimming, and Equaling. The protocol maintains a set $\text{Tr} \subseteq D$ of agents who perform the TRIM operation during the execution of our protocol.

We initialize the set $\text{Tr} = D$ to contain all the neighbour agents of a_r. The key goal of the while-loop is to create *domination* (see Definition 3) for agent a_r over her each neighbour in D one by one. An agent $a_i \in \text{Tr}$ gets removed from this set as soon as a_r *dominates* her; we show that the set Tr shrinks as the algorithm progresses (Lemma 1). And finally, the while-loop terminates when $\text{Tr} = \emptyset$.

We will call the unallocated part of the cake obtained at the end of each round of the while-loop as *residue*, denoted by R. In the beginning, the residue $R = [0, 1]$ and it keeps decreasing with subsequent rounds of the while-loop. Throughout the algorithm, each agent $a_i \in D$ maintains a set $\mathcal{A}^{(i)} = (A_0^{(i)}, \ldots, A_{\ell_i}^{(i)})$ of $\ell_i + 1$ bundles of equal value to her. Note that the leaf agents do not perform any operation throughout the entire execution of the while-loop.

Definition 3 (D2Tree Domination Condition). *At any round of the while-loop with residue R, we say that agent a_r with bundle A_r dominates her neighbour agent $a_i \in D$ with a collection $\mathcal{A}^{(i)} = (A_0^{(i)}, \ldots, A_{\ell_i}^{(i)})$ of $\ell_i + 1$ bundles if*

1. *For bundle $A_0^{(i)} \in \mathcal{A}^{(i)}$, we have $v_r(A_r) = v_r(A_0^{(i)})$*
2. *For all the remaining bundles in the set $\mathcal{A}^{(i)}$, we have $v_r(A_r) - v_r(A_k^{(i)}) \geq v_r(R)$ for all $k \in [\ell_i]$*

The domination condition says that bundle A_r has become sufficiently more valuable than the bundles $A_k^{(i)} \in \mathcal{A}^{(i)}$ for $k \in [\ell_i]$ according to agent a_r such that even after the whole of residue (of that round) is added to any bundle in $\mathcal{A}^{(i)}$, a_r will not envy the recipient of that bundle.

Any round t of the while-loop with residue $R = R^t$ and the current set Tr of the trimmer agents consists of the following three phases:

Selection Phase: In the beginning, agent a_r divides the residue R^t into n equal pieces, each of value $v_r(R^t)/n$ to her; we denote this set by $\mathcal{X}$ (in Step 3). Then, one by one, each agent $a_i \in D$ selects her $\ell_i + 1$ most favorite (available) pieces from $\mathcal{X}$ (see Step 5). We denote the set of these selected pieces by $\mathcal{X}^{(i)} = \{X_0^{(i)}, \ldots, X_{\ell_i}^{(i)}\} \subseteq \mathcal{X}$, where $X_0^{(i)}$ is a_i's least valued piece in $\mathcal{X}^{(i)}$. Note that, we have $v_r(\mathcal{X}^{(i)}) = (\ell_i + 1) \cdot v_r(R^t)/n$.

After every neighbour agent of a_r in D has made her selection, the remaining single piece (of value $v_r(R^t)/n$) from $\mathcal{X}$ is added to the bundle A_r (in Step 6).

Trimming Phase: This phase begins with every agent $a_i \in Tr$ adding her least-valued piece, $X_0^{(i)} \in \mathcal{X}^{(i)}$ to bundle $A_0^{(i)}$ (in Step 12). This implies that

$$v_r(A_r) = v_r(A_0^{(i)}) \tag{1}$$

and hence the first condition of domination will be satisfied. This is due to the fact that both the bundles A_r and $A_0^{(i)}$ get a piece of value $v_r(R^t)/n$ in each round t. For the remaining ℓ_i bundles, agent $a_i \in Tr$ performs a TRIM procedure, making these ℓ_i bundles of value equal to $v_i(X_0^{(i)})$, and forming the set $\mathcal{Y}^{(i)}$ (see Step 10). The residue due to this operation is added to the residue R^{t+1} for the next round $t+1$ (Step 11). All the bundles in $\mathcal{Y}^{(i)}$ obtained from the Trimming phase are added one to each bundle in $\mathcal{A}^{(i)}$ in a way that helps us achieve the desired dominance (see Steps 12–14). Due to the TRIM operation, we have

$$v_r(A_r) \geq v_r(A_k^{(i)}) \text{ for all } k \in [\ell_i] \tag{2}$$

We show that after repeated applications of the TRIM operation, agent a_r achieves domination over agent a_i. In particular, we prove (in Lemma 1) that after every $O(n \log n)$ rounds of the while-loop, there exists some agent $a_i \in Tr$ over whom dominance is achieved.

Equaling Phase: Let agent a_r achieves dominance over some agent $a_i \in D$ in round $t - 1$, after which she is removed from the set Tr. That is, we have $v_r(A_r) - v_r(A_k^{(i)}) \geq v_r(R^t)$ for all $k \in [\ell_i]$, where[6] R^t is the residue formed at the end of round $t - 1$, and hence is the residue at the beginning of round t.

In round t, agent a_i still begins with the Selection phase as before. Her bundle $A_0^{(i)}$ receives a trimmed piece from her set $\mathcal{Y}^{(i)}$; see Steps 18–19. Therefore, for all the subsequent rounds, we obtain

$$v_r(A_r) \geq v_r(A_0^{(i)}) \tag{3}$$

She next performs the EQUAL operation (in Step 17) which does not produce any residue. We will show that due to the second condition of the dominance (Definition 3), no matter how the residue of future rounds is distributed among the bundles of $\mathcal{A}^{(i)}$, agent a_r will have no envy towards any of the bundles formed during this procedure.

Termination of the While-Loop: Once the set Tr becomes empty and agent a_r dominates every agent $a_i \in D$, the while-loop terminates. The final allocation

---

[6] Here, the bundles A_r and $A_k^{(i)}$ for $k \in [\ell_i]$ are the ones that were formed till the end of round $t - 1$. For brevity, we do not add the notation $t - 1$ in these bundles.

is formed in Steps 24–27: agent a_r receives bundle A_r, each leaf agent $a \in L_i$ corresponding to agent $a_i \in D$ chooses her favorite bundle from the set $\mathcal{A}^{(i)}$ formed by her parent agent a_i, while a_i receives the last remaining bundle in the above set. This creates a complete allocation $\mathcal{A}$ of the cake that we will show is locally envy-free on DEPTH2TREE.

Theorem 1. *For cake-division instances with n agents on a* DEPTH2TREE, *there exists a discrete protocol that finds a locally envy-free allocation using at most $O(n^3 \log(n))$* **cut** *and $O(n^4 \log(n))$* **eval** *queries under RW model.*

Proof. We begin by establishing three important properties of D2TREE (in Lemma 1) that are crucial in establishing the desired polynomial upper bound on its query complexity. Their proofs appear in the full version of the paper [18].

Lemma 1. *The following properties hold true for D2TREE protocol:*

1. *In every round of the while-loop, D2TREE protocol makes $O(n)$ cuts on the cake using $O(n)$* **cut** *and $O(n^2)$* **eval** *queries.*
2. *Agent a_r's valuation for the residues in two consecutive rounds t and $t + 1$ of the while-loop in D2TREE satisfies $v_r(R^{t+1}) \leq (1 - (|D| + 1)/n)v_r(R^t)$.*
3. *Agent a_r starts dominating at least one agent from the set Tr in every $O(n \log n)$ rounds of the while-loop in D2TREE, after which it is removed from the set Tr. That is, the size of the set $|Tr|$ decreases by one in every $O(n \log n)$ rounds of the while-loop.*

For a given cake-division instance, we will prove that the output allocation $\mathcal{A} = (A_1, \ldots, A_n)$ of D2TREE is locally envy-free by splitting the analysis into three following cases.

(a) **Root agent:** Consider an arbitrary agent $a_i \in D$ and the set $\mathcal{A}^{(i)}$ of ℓ_i many bundles formed after the termination of the while-loop in D2TREE. Note that, agent a_i is assigned one bundle from the set $\mathcal{A}^{(i)}$. Therefore, to prove local envy-freeness for agent a_r, it suffices to show that agent a_r prefers her own bundle A_r over any bundle in the set $\mathcal{A}^{(i)}$. Note that, Eqs. (1) and (2) imply $v_r(A_r) \geq v_r(A_k^{(i)})$ for all $k \in \{0\} \cup [\ell_i]$ throughout the trimming phase of agent a_i. Once agent a_r dominates agent a_i, say with respect to residue R^t of round t, we must have $v_r(A_r) \geq v_r(A_k^{(i)}) + v_r(R^t)$ for all $k \in [\ell_i]$. Therefore, agent a_r will not envy any of the bundles in the set $\mathcal{A}^{(i)}$ in future rounds $t' \geq t + 1$ irrespective of how residue $R^{t'}$ is divided into these bundles in the equaling phase of agent a_i. Furthermore, recall that Eq. (3) ensures that agent a_r does not envy the bundle $A_0^{(i)}$ as well. Overall, agent a_r has no local envy in the final allocation.

(b) **Neighbour agents:** Consider an arbitrary agent $a_i \in D$. Throughout the execution of our algorithm, every bundle $A_k^{(i)} \in \mathcal{A}^{(i)}$ is of equal value to agent a_i. This follows due to the properties of TRIM and EQUAL operations. Therefore, when the leaf agents of agent a_i (in the set L_i) selects her bundle from $\mathcal{A}^{(i)}$ in Step 25, agent a_i will have no envy towards them.

Next, we will show that a_i has no envy towards the root agent as well. Recall that in the selection phase, a_i chooses her favourite $\ell_i + 1$ pieces from $\mathcal{X}$ (in Step

---

D2Tree: Local Envy-freeness for n agents on DEPTH2TREE

Input: A cake-division instance $\mathcal{I}$ on DEPTH2TREE $=(n, a_r, D, \{\ell_i\}_{a_i \in D})$
Output: A locally envy-free allocation.

1   Initialize $R \leftarrow [0, 1]$, set of trimmer agents $\mathrm{Tr} \leftarrow D$, bundles $A_0^{(i)}, \ldots, A_{\ell_i}^{(i)} \leftarrow \emptyset$
     for each $a_i \in D$ and a bundle $A_r \leftarrow \emptyset$ for the root agent.

2   **while** $\mathrm{Tr} \neq \emptyset$ **do**

3 | Agent a_r divides R into n equally-valued pieces that are kept in the set $\mathcal{X}$
       —— Selection——

4 | **for** $a_i \in D$ **do**

5 | | $\mathcal{X}^{(i)} \leftarrow$ SELECT$(a_i, \mathcal{X}, \ell_i + 1)$

6 | Set $A_r \leftarrow A_r \cup (\mathcal{X} \setminus \cup_{a_i \in D}(\mathcal{X}^{(i)}))$ /* The remaining single piece from $\mathcal{X}$ */

       ——Trimming——

7 | Set $R \leftarrow \emptyset$

8 | **for** $a_i \in \mathrm{Tr}$ **do**

9 | | Let $X_0^{(i)} = \arg\min_{X \in \mathcal{X}^{(i)}} v_i(X)$ /* This piece won't be trimmed   */

10 | | $(\mathcal{Y}^{(i)}, R') \leftarrow$ TRIM$(a_i, \mathcal{X}^{(i)})$

11 | | Set $R \leftarrow R \cup R'$

12 | | $A_0^{(i)} \leftarrow A_0^{(i)} \cup X_0^{(i)}$ and $\mathcal{Y}^{(i)} \leftarrow \mathcal{Y}^{(i)} \setminus X_0^{(i)}$

13 | | Let $A_w^{(i)} = \arg\max_{1 \leq k \leq \ell_i} v_r(A_k^{(i)})$ and $Y_g^{(i)} = \arg\min_{1 \leq k \leq \ell_i} v_r(Y_k^{(i)})$

14 | | $A_w^{(i)} \leftarrow A_w^{(i)} \cup Y_g^{(i)}$ and $\mathcal{Y}^{(i)} \leftarrow \mathcal{Y}^{(i)} \setminus Y_g^{(i)}$ /* Trying to achieve
           domination on $A_w^{(i)}$ for the root agent   */

15 | | For each $k \neq 0, w$, add one arbitrary piece from $\mathcal{Y}^{(i)}$ to $A_k^{(i)}$

       ——Equaling——

16 | **for** $a_i \in D \setminus \mathrm{Tr}$ **do**

17 | | $(\mathcal{Y}^{(i)}, Y_*) \leftarrow$ EQUAL$(a_i, \mathcal{X}^{(i)})$

18 | | Let $Y_k^{(i)} \in \mathcal{Y}^{(i)}$ be the piece such that $Y_k^{(i)} \subseteq Y_*$ /* There is only one
           piece satisfying this condition.   */

19 | | $A_0^i \leftarrow A_0^i \cup Y_k^{(i)}$ /* This ensures that root will not envy the
           bundle A_0^i */

20 | | For each $k \neq 0$, add one arbitrary piece from $\mathcal{Y}^{(i)}$ to the bundle $A_k^{(i)}$

       ——Checking Domination——

21 | **for** $a_i \in \mathrm{Tr}$ **do**

22 | | **if** *agent a_r dominates agent a_i (see Definition 3)*

23 | | | $\mathrm{Tr} \leftarrow \mathrm{Tr} \setminus \{i\}$

      ——Choose after while loop——

24   **for** $a_i \in D$ **do**

25 | **for** $a_j \in L(i)$ **do**

26 | | a_j is allocated her favorite (available) bundle from $(A_k^{(i)})_{k \in [\ell_i]}$

27 | a_i is allocated the remaining bundle

28   **return** The allocation $A_r \cup (A_k^{(i)})_{k \in [\ell_i], a_i \in D}$

---

5). The remaining single piece is added to agent a_r's bundle A_r. Therefore, in each round of the while-loop, the increment for each bundle $A_k^{(i)}$ for $k \geq 0$ is as large as the increment in bundle A_r in the view of agent a_i. That is, we have $v_i(A_k^{(i)}) \geq v_i(A_r)$ for all $k \in \{0\} \cup [\ell_i]$, and hence no local envy for agent a_i.

(c) **Leaf agents:** It is trivial to observe that any leaf agent will have no local envy since she chooses her favourite bundle before her neighbour agent.

Overall, D2TREE outputs a locally envy-free allocation among n agents on a DEPTH2TREE.

Counting Queries: Lemma 1 ensures that after $O(n \log n)$ rounds, the number of agents $a_i \in D$ who are in the set Tr decreases at least by one. Hence, the while-loop terminates (i.e. when Tr $= \emptyset$) after at most $O(n^2 \log n)$ many rounds. By Lemma 1, we know that each round of the while-loop requires $O(n)$ cut and $O(n^2)$ eval queries. Hence, D2TREE requires $O(n^3 \log n)$ cut queries and $O(n^4 \log n)$ eval queries. This completes our proof.   □

4   Local Envy-Freeness on a Tree

In this section, we develop a recursive protocol DOMINATION(R, k) that finds a locally envy-free allocation for n agents on a TREE graph using at most $O(n^{2n})$ RW queries (Theorem 2). Without loss of generality, we *always* assume that the agents $a_1, a_2, \ldots, a_n$ are indexed according to some arbitrary topological ordering; making agent a_n as the root node and a_1 a leaf node in the graph. The topological order over the agents ensures that any descendant of agent a_j for $j \in [n]$ must have an index smaller than j.

Terminology: For a given cake-division instance with n agents on a TREE, we define the following sets for any $j \in [n]$.

1. $D_j := \{j\} \cup \{i \in [j-1] : a_i \text{ is a descendant of } a_j\}$ is the set containing index j and the indices of descendants of a_j in the underlying TREE. We write $d_j = |D_j|$ to denote the size of the set D_j.
2. C_j denotes the set of indices of the immediate descendants (or children) of a_j in the underlying TREE. Furthermore, we write p_j to denote the index of the parent of agent a_j.

For a given fixed index $k \in [n]$, we say an agent a_j is *active* if $j \geq k+1$, otherwise she is *inactive*. Next we define important sets used in the analysis of the protocol.

1. For an active agent a_j, Inchild$(k, a_j) := \{i \mid i \leq k \text{ and } i \in C_j\}$ is the set consisting the indices of all her inactive children. This set is empty for inactive agents.
2. For an active agent a_j, Inact$(k, a_j) := \{j\} \cup \{t \in D_i \mid i \in \text{Inchild}(k, a_j)\}$ is the set consisting of the indices of all the descendants of her inactive children. This set is empty for inactive agents.
3. For an allocation $\mathcal{B} = (B_1, \ldots, B_n)$, we define Storage$(k, a_j) := \{B_i \mid i \in \text{Inact}(k, a_j)\}$ as the set that stores the bundles assigned to agents whose indices are in Inact(k, a_j).

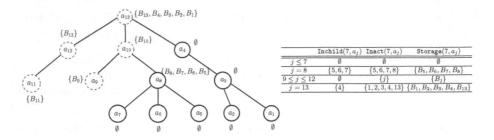

Fig. 1. The left figure depicts a representative example with 13 agents on a TREE. The agent corresponding to every node is written inside the circle. For a fixed index $k = 8$, the red dashed nodes and the black nodes represent the active and inactive agents respectively. The adjoining table details the sets $\texttt{Inchild}(k-1, a_j)$, $\texttt{Inact}(k-1, a_j)$, and $\texttt{Storage}(k-1, a_j)$ for an allocation $\mathcal{B} = (B_1, \ldots, B_n)$ with respect to $k = 8$. The set written next to a_j in the figure is her $\texttt{Storage}(k-1, a_j)$. (Color figure online)

For an allocation $\mathcal{B}$, the collection of storage sets $\{\texttt{Storage}(k, a_j)$ for $j \in [n]\}$ creates a partition of its bundles. When the value of k is obvious from the context, we will use the phrase *storage of* a_j to refer to her $\texttt{Storage}(k, a_j)$ set.

Definition 4 (k-Fair allocation for Trees). *Consider a cake-division instance with n agents on a* TREE. *For a given piece $R \subseteq [0,1]$, we say an allocation $\mathcal{B} = (B_1, \ldots, B_n)$ (of R) is k-FAIR if for each agent a_j with $j \geq k$ the following conditions hold.*

C1. Agent a_j does not envy her neighbours.
C2. $v_j(B_j) = v_j(B)$ for all $B \in \texttt{Storage}(k-1, a_j)$, and
C3. $v_j(B_j) \geq v_j(B)$ for all $B \in \texttt{Storage}(k-1, a_\ell)$ such that a_ℓ is an active child of a_j (with respect to index $k-1$).

The above-defined notion of k-FAIRness forms the crux of our technical ideas. We explain this notion with the following example (Example 1).

Example 1. Consider an instance of a TREE graph with 13 agents depicted in Fig. 1.

An 8-FAIR allocation $\mathcal{B} = \{B_1, \ldots, B_{13}\}$ of some $R \subseteq [0,1]$ satisfies the following conditions.

- *C1:* Agents $a_8, a_9, \ldots, a_{13}$ do not envy their neighbors.
- *C2:* We have $v_8(B_8) = v_8(B)$ for all $B \in \{B_5, B_6, B_7, B_8\}$ and $v_{13}(B_{13}) = v_{13}(B)$ for all $B \in \{B_1, B_2, B_3, B_4, B_{13}\}$.
- *C3:* We detail the condition for $j = 10$, and other cases can be dealt similarly. Agent a_{10} has two active children (with respect to index $k-1 = 7$): a_9 and a_8 with storage sets $\{B_9\}$ and $\{B_5, B_6, B_7, B_8\}$ respectively. Hence, for $j = 10$ we have $v_{10}(B_{10}) \geq v_{10}(B)$ for $B \in \{B_5, B_6, B_7, B_8, B_9\}$.

Note that any 1-FAIR allocation is locally envy-free for agents on a TREE. For any piece $R \subseteq [0,1]$ and index $k \geq 1$, we will develop a recursive protocol DOMINATION(R, k) that is always k-FAIR (see Lemma 2). Hence, DOMINATION($[0,1], 1$) will output the desired locally envy-free allocation among n agents on a TREE.

Overview of Domination(R, k): For $k = n$ and any piece $R \subseteq [0,1]$, protocol DOMINATION(R, n) is defined in a straight-forward manner: agent a_n cuts R into n equal pieces according to her. It is easy to see that this allocation is indeed n-FAIR. For $k \geq 1$, our protocol DOMINATION(R, k) successively constructs a k-FAIR allocation $\mathcal{A} = (A_1, \ldots, A_n)$ of R among n agents in multiple rounds. It does so by repeatedly invoking DOMINATION($R, k+1$) and using its $(k+1)$-FAIR output allocations. We will refer to R as *residue* and it keeps on shrinking as the algorithm proceeds. DOMINATION(R, k) primarily consists of a while-loop that terminates when the residue reduces so much that the parent agent a_{p_k} (of agent a_k) satisfies a certain *domination condition* (as stated in Step 8 of DOMINATION(R, k)) over the bundles of agents in D_k with respect to the current residue. This *domination* serves as a crucial property for creating the desired k-FAIR allocation (without creating any envy for agent a_{p_k}). We prove that the domination is achieved in polynomial many rounds of the while-loop (see Lemma 2) that becomes the backbone argument to establish the desired query complexity of our protocol DOMINATION($[0,1], 2$).

Let us consider an arbitrary round t of the while-loop (Step 3–17) during the execution of DOMINATION(R, k), and denote the residue at its beginning as R^t; where $R^1 = R$. The round begins with invoking DOMINATION($R^t, k+1$) to obtain a $(k+1)$-FAIR allocation $\mathcal{B}^t = (B_1^t, B_2^t, \ldots, B_n^t)$ of R^t. Throughout round t, we focus (and modify *some* of) the bundles in the Storage(k, a_{p_k}) set corresponding to $\mathcal{B}^t$. Recall that, due to $(k+1)$-FAIRness of $\mathcal{B}^t$, agent a_{p_k} values each bundle in her Storage(k, a_{p_k}) set equally. Also, note that Storage(k, a_k) corresponding to $\mathcal{B}^t$ is empty and that is exactly what agent a_k is trying to build during this recursive step. The challenge is to form the Storage($k-1, a_k$) set corresponding to the output allocation $\mathcal{A}$ while ensuring its k-FAIRness. Observe that, this allocation $\mathcal{A}$ (and its Storage($k-1, a_k$) set) is then used by DOMINATION($R, k-1$) in the next recursive step.

The while-loop of our protocol consists of three phases: Selection, Trimming, and Equaling where only the agent a_k performs the associated operations.
- In the first phase of SELECTION, as the name suggests, agent a_k *selects* her $|D_k| = d_k$-many favorite pieces from the Storage(k, a_{p_k}) set (see Step 4), denoted by the set $\mathcal{X}^t$.[7] Since agent a_{p_k} values each bundle in her storage set equally, we can re-index all the selected bundles in $\mathcal{X}^t$ to match the indices in the set D_k (and accordingly re-index the remaining non-selected bundles as well).

---

[7] Recall that the set D_k consists of the indices of the descendants of agent a_k and her own index. Since the indices of the agents follow topological ordering, $d_k \leq k$ for any $k \in [n]$. Moreover, since $p_k > k$, the set Storage(k, a_{p_k}) set cannot be empty.

Now, for every $j \notin D_k$, the *intact*[8] bundle B_j is added to the bundle A_j of agent a_j. We use this fact to establish Condition C1 of k-FAIRness for output allocation $\mathcal{A}$ (in Lemma 2).

- If agent a_{p_k} has not yet achieved the *domination* (as stated in Step 8), then agent a_k enters into the Trimming phase and performs a TRIM operation (Step 11) on the set $\mathcal{X}^t$ of bundles chosen in Step 5 to obtain a trimmed set $\mathcal{Y}^t$ of d_k-many bundles. The residue obtained due to this trimming process becomes the residue R^{t+1} for the next round. The bundles in $\mathcal{Y}^t$ are allocated (in Steps 11–14) among agents in D_k in a way that expedites the desired domination and ensures Condition C3 of k-FAIRness for allocation $\mathcal{A}$.

- The key observation here is that the value of the residue according to agent a_{p_k} decreases exponentially fast. This ensures that the said domination for agent a_{p_k} is achieved in at most $d_k + d_k \log d_k$ iterations of the while-loop (Lemma 2). As soon as the domination condition is satisfied, agent a_k performs an EQUAL operation (instead of TRIM) on the output allocation of DOMINATION($R^{d_k + d_k \log d_k}, k+1$). This process produces no residue and the while-loop terminates. Towards this end, agent a_k produces d_k-many equally valued bundles due to Trim and Equal operations that forms her Storage($k-1, a_k$) set. This helps in establishing Condition C2 of k-FAIRness for allocation $\mathcal{A}$.

Finally, the count of $d_k + d_k \log d_k$ on the number of rounds of while-loop leads to the desired runtime for our protocol. We will prove that DOMINATION(R, k) outputs a k-FAIR allocation, by showing that all three conditions (in Definition 4) are satisfied (see Lemma 2).

Overall, the final output of DOMINATION($[0, 1], 1$) is a 1-FAIR allocation of the cake $[0, 1]$. The run-time of DOMINATION(R, k) and its recursive nature leads to the query complexity n^{2n} for the DOMINATION($[0, 1], 2$).

Notation Guide for Domination(R, k): We say any round t of the while-loop has residue R^t at its beginning. We write $\mathcal{B}^t$ to be the output of DOMINATION($R^t, k+1$) in Step 3. Furthermore, $\mathcal{X}^t$ denotes the output of the SELECT procedure (Step 4) and $\mathcal{Y}^t$ denotes the output of the TRIM and EQUAL procedures performed by agent a_k (Steps 10 and 16). We will drop the superscript t whenever it is clear from the context. Finally, we write $\mathcal{A} = (A_1, \ldots, A_n)$ to be the output allocation of DOMINATION(R, k).

Theorem 2. *For cake-division instances with n agents on a* TREE, *there exists a discrete protocol that computes a locally envy-free allocation using at most $O(n^{2n})$ Robertson-Webb queries.*

We begin with a crucial lemma (Lemma 2) which proves that DOMINATION(R, k) returns a k-FAIR allocation after at most $d_k + d_k \log(d_k)$ many runs of the while loop. This property forms the crux of our recursive protocol DOMINATION(R, k).

---

[8] A bundle is said to be *intact* if it is in the form as present in $\mathcal{B}^t$ obtained from DOMINATION($R^t, k+1$) and has not been modified.

Recursion step: DOMINATION(R, k) for TREES

Input: A cake-division instance $\mathcal{I}$ on a TREE, a piece $R \subseteq [0,1]$, and an index $k \in \{1, \ldots, n-1\}$.

Output: A k-FAIR allocation of R.

1 Initialize bundles $A_i \leftarrow \emptyset$ for $i \in [n]$ and set a counter $c \leftarrow 0$
2 **while** $R \neq \emptyset$ **do**
3       $\mathcal{B} \leftarrow$ DOMINATION$(R, k+1)$
       —— Selection——
4       $\mathcal{X} \leftarrow$ SELECT$(a_k, \texttt{Storage}(k, a_{p_k}), d_k)$ /* Storage set for $\mathcal{B}$ */
5   Re-index the bundles in $\mathcal{X}$ and $\texttt{Storage}(k, a_{p_k}) \setminus \mathcal{X}$ so that they bear the
       indices in D_k and $\texttt{Inact}(k, a_{p_k}) \setminus D_k$ respectively
       /* We can do this because Step 3 ensures that agent a_{p_k} is
         indifferent towards the bundles in $\texttt{Storage}(k, a_{p_k})$ */
6   **for** $j \notin D_k$ **do**
7         $A_j \leftarrow A_j \cup B_j$ **if** $\exists\, \ell \in D_k$ such that $v_{p_k}(A_{p_k}) - v_{p_k}(A_\ell) \leq v_{p_k}(R)$
         /* Checking the domination condition for agent a_{p_k} */
         ——Trimming——
8   Set $R \leftarrow \emptyset$
9       $(\mathcal{Y}, R) \leftarrow$ TRIM$(a_k, \mathcal{X})$
10   Let $Y_g = \arg\min_{\ell \in D(k)} v_{p_k}(Y_\ell)$ and $w = c \mod d_k + 1$
11     $A_w \leftarrow A_w \cup Y_g$ and $\mathcal{Y} \leftarrow \mathcal{Y} \setminus Y_g$
       /* Trying to achieve domination on A_w for the agent a_{p_k} */
12   For each $\ell \in D_k \setminus \{w\}$, add one arbitrary piece from $\mathcal{Y}$ to A_ℓ
13     $c \to c + 1$
14   **else**
         ——Equaling——
15        $\mathcal{Y} \leftarrow$ EQUAL$(a_k, \mathcal{X})$
16   For each $i \in D_k$, add one arbitrary piece from $\mathcal{Y}$ to A_i
17 **return** Allocation $\mathcal{A} = (A_1, \ldots, A_n)$

Lemma 2. *Consider any cake-division instance with n agents on a TREE graph. For any $R \subseteq [0,1]$ and $k \in [n]$, DOMINATION(R, k) computes a k-FAIR allocation in $d_k + d_k \log d_k$ rounds of the while-loop.*

Next, we state three properties in Lemma 3 that are instrumental in proving the above lemma. Its proof is deferred to the full version of the paper [18].

Lemma 3. DOMINATION(R, k) *has following three properties:*

1. *For any $j \notin D(k) \cup p_k$, $\texttt{Storage}(k, a_j) = \texttt{Storage}(k-1, a_j)$. Furthermore, this set remains intact during the entire execution of DOMINATION(R, k). Moreover, for $j = p_k$ we have $\texttt{Storage}(k-1, a_{p_k}) \subseteq \texttt{Storage}(k, a_{p_k})$.*
2. *For any round t of the while-loop during the protocol DOMINATION(R, k). Then, after $t + d_k \log d_k$ rounds of the while-loop, we obtain $v_{p_k}(R^{t+d_k \log d_k}) \leq c_t$. Here, $c_t := \max_{\ell \in D_k} \{v_{p_k}(B_{p_k}^t) - v_{p_k}(Y_\ell^t)\}$ and bundles Y_k^t for $k \in [\ell_i]$ are obtained after the TRIM procedure in Step 10 of DOMINATION(R, k) protocol performed by agent a_k in round t.*

3. *During the execution of* DOMINATION(R, k), *the difference between the value of agent* a_{p_k} *for her bundle and for the bundle of any agent in the set* $D(k)$ *increases with each round of the while-loop i.e., for any round* t, *we have*

$$v_{p_k}(A^t_{p_k}) - v_{p_k}(A^t_\ell) \leq v_{p_k}(A^{t+1}_{p_k}) - v_{p_k}(A^{t+1}_\ell) \quad \textit{for all } \ell \in D_k$$

where $\mathcal{A}^t$ *and* $\mathcal{A}^{t+1}$ *are the allocations at the end of rounds* t *and* $t+1$.

Proof of Lemma 2: Given any piece $R \subseteq [0, 1]$ and $k \in [n]$, we begin by proving that the output allocation, $\mathcal{A} = (A_1, A_2, \ldots, A_n)$ of the DOMINATION(R, k) is k-FAIR. Towards the end, we will prove the desired count on the number of while-loops that suffices to achieve so.

We will proceed via recursion on $k \in [n]$. Recall that DOMINATION(R, n) asks agent a_n to simply divide R into n equal pieces, making it n-FAIR trivially. Now, let us assume that the claim holds true for $k + 1$, and we will prove it for k. That is, in every round of the while-loop, the output allocation $\mathcal{B}$ of DOMINATION$(R, k+1)$ is $(k+1)$-FAIR. We will show that DOMINATION(R, k) is k-FAIR by proving that its output allocation $\mathcal{A}$ satisfies Conditions C2 and C3 and then finally Condition C1 will follow.

-**Condition C2**: For each round of the while-loop, since $\mathcal{B}$ is $(k + 1)$-FAIR, we have $v_j(B_j) = v_j(B)$ for all $B \in \mathtt{Storage}(k, a_j)$ for all $j \geq k + 1$ from Condition C2. For $j \geq k + 1$ and $j \neq p_k$, note that Lemma 3 implies that the bundles in the $\mathtt{Storage}(k, a_j)$ set remains intact during DOMINATION(R, k). Since $\mathtt{Storage}(k, a_j) = \mathtt{Storage}(k - 1, a_j)$, the desired condition is satisfied for these agents.

Now, for $j = p_k$, the induction hypothesis implies that $v_{p_k}(B_{p_k}) = v_{p_k}(B)$ for all $B \in \mathtt{Storage}(k, a_{p_k})$. And Lemma 3 implies that $\mathtt{Storage}(k - 1, a_{p_k}) \subseteq \mathtt{Storage}(k, a_{p_k})$. Hence, we obtain the desired relation.

Finally, let us consider agent a_k. We know that agent a_k selects d_k many bundles from $\mathtt{Storage}(k, a_{p_k})$ in each round of the while-loop and performs a Trim procedure on this set to make them all of equal value to her. The equaling phase maintains this property, hence establishing Condition C2 of k-FAIRness.

- **Condition C3**: Let us consider agents a_k and a_{p_k}. We show that $v_{p_k}(A_{p_k}) \geq v_{p_k}(A)$ for all $A \in \mathtt{Storage}(k - 1, a_k)$. Towards the end of the protocol DOMINATION(R, k), we know that agent a_k creates her $\mathtt{Storage}(k - 1, a_k)$ set containing d_k many equally-valued bundles. In each round of the while-loop, a_k selects her d_k-many favorite pieces from the set $\mathtt{Storage}(k, a_{p_k})$ (of $(k + 1)$-FAIR allocation of that round) in Step 4 and performs TRIM until a_{p_k} starts dominating her.

At the termination round, say T, of the while-loop, we have (by the domination condition stated in Step 8)

$$v_{p_k}(A^T_{p_k}) - v_{p_k}(A^T_\ell) \geq v_{k+1}(R^T) \text{ for all } \ell \in D(k),$$

where A^T_ℓ denotes the bundle of agent a_ℓ formed at the end of round T of the while-loop. The domination condition implies that the residue R^T is small enough that it does not induce any envy for a_{p_k} even if R^T is fully allocated to any bundle A_ℓ^T for $\ell \in D(k)$. The Equaling phase maintains the similar relation, and hence we obtain $v_{p_k}(A_{p_k}) \geq v_{p_k}(A_j)$ for all $j \in D_k$. Since, these A_j's form the Storage$(k-1, a_k)$ set, we obtain the desired relation.

Finally, for any other active child a_ℓ of a_{p_k}, note that Lemma 3 says that the Storage(k, a_ℓ) remains intact, i.e., we have Storage$(k-1, a_\ell) =$ Storage(k, a_ℓ). Since bundle $B_{p_k} \in$ Storage(k, a_{p_k}) is present in Storage$(k-1, a_{p_k})$, Condition C3 follows from the induction hypothesis.

- **Condition C1**: Let us first consider agent a_k. In any round t of the while-loop, she selects her d_k most preferred pieces from Storage(k, a_{p_k}) set (corresponding to $\mathcal{B}$). We re-index the bundles such that B_{p_k} is one of the remaining pieces, and that is allocated to agent a_{p_k}. The Trimming and Equaling phases ensure that we maintain $v_k(A_k) \geq v_k(A_{p_k})$. The fact that a_k does not envy any of her children follows from Condition C2, proved above. Condition C3 can be used to prove that a_{p_k} does not envy non-children agent a_k as well.

Consider an agent a_j such that $j \notin (D_k \cup p_k)$. All that is left is to prove that agent a_j does not envy her neighbours in the output allocation $\mathcal{A}$. This is true since the topological ordering ensures that any agent in the set $D(k)$ has an index that is less than k. Since agent a_j is allocated an intact piece from $\mathcal{B}$ in Steps 6–7, using the induction hypothesis, we obtain that she does not envy her neighbours in the allocation $\mathcal{A}$. This proves that $\mathcal{A}$ satisfies Condition C1 of k-FAIRness.

Runtime Analysis: To establish the runtime of DOMINATION(R, k), we will consider the first d_k rounds of the while-loop during its execution. If our protocol terminates before d_k rounds, we are done. If not, observe that Steps 11–12 allocate the smallest piece after trimming (i.e. $\arg\min_{\ell \in D(k)} v_{p_k}(Y_\ell)$) to different bundles in the first d_k rounds. Assume, without loss of generality, bundle A_h is allocated the smallest trimmed piece in round $h \in [d_k]$. Recall the definition of the maximum trimmed value $c_h := \max_{\ell \in D_k} \{v_{p_k}(B_{p_k}^h) - v_{p_k}(X_\ell^h)\}$ for round h. Since agent a_{p_k} gets bundle $B_{p_k}^h$ (which is not trimmed) in this round, we know that $v_{p_k}(A_{p_k}^h) - v_{p_k}(A_\ell^h) \geq c_h$ for all $\ell \in D(k)$. Using this inequality for all $h \leq d_k$ we have,

$$v_{p_k}(A_{p_k}^{d_k + d_k \log d_k}) - v_{p_k}(A_\ell^{d_k + d_k \log d_k}) \geq v_{p_k}(A_{p_k}^h) - v_{p_k}(A_h^h) \quad \text{(by Lemma 3)}$$
$$\geq c_m \quad \text{(by Step 12)}$$
$$\geq v_{p_k}(R^{d_k + d_k \log d_k}) \quad \text{(by Lemma 3)}$$

Therefore, after at most $d_k + d_k \log d_k$ rounds, the agent a_k enters the Equaling phase and the while-loop terminates to output the final k-FAIR allocation. □

Proof of Theorem 2: Note that the k-FAIRness of DOMINATION(R, k) ensures that the final output allocation $\mathcal{A}^*$ of DOMINATION$([0, 1], 2)$ is locally envy-free.

We denote the query complexity of DOMINATION(R, k) by T_k for $k \in [n]$; we will prove $T_1 = O(n^{2n})$. Note that, each round of the while-loop during the execution of the protocol DOMINATION(R, k) requires d_k eval queries and $d_k - 1$ cut queries. Lemma 2 proves that DOMINATION(R, k) outputs a k-FAIR allocation in $d_k + d_k \log d_k$ rounds of the while-loop. Now, let us first observe the execution of DOMINATION$(R, 1)$ protocol. It invokes DOMINATION$(R, 2)$ which makes T_2 many queries to output $\mathcal{C}$. The corresponding set (in $\mathcal{C}$) Storage$(1, a_i) = \{C_i\}$ for every agent a_i, except the parent agent of a_1. We have Storage$(1, a_{p_1}) = \{C_1, C_{p_1}\}$, out of which agent a_1 selects her favorite bundle (by making two eval queries) and the remaining bundle. That is, we can write $T_1 = T_2 + 2$.

Since the agents are indexed according to the topological order, we have $d_k \leq k$ for all $k \in [n]$. Hence, we will derive the query complexity in terms of k instead of d_k. We prove that $T_k \leq \sum_{j=k}^{n} j \prod_{i=k}^{j} (3i \log i)$ using induction on k. For the base case $k = n$, we know that $T_n = n \leq 3n^2 \log(n)$. Assuming that the bound holds for T_{k+1} and writing T_k in terms of T_{k+1} we have,

$$T_k \leq (d_k + d_k \log d_k) T_{k+1} + d_k(d_k + d_k \log d_k) \leq (k + k \log k) T_{k+1} + 2k^2 \log k$$

$$\leq 3k \log k \sum_{j=k+1}^{n} j \prod_{i=k+1}^{j} (3i \log i) + 2k^2 \log k = \sum_{j=k+1}^{n} j \prod_{i=k}^{j} (3i \log i) + 2k^2 \log k$$

$$< \sum_{j=k+1}^{n} j \prod_{i=k}^{j} (3i \log i) + k(3k \log k) \leq \sum_{j=k}^{n} j \prod_{i=k}^{j} (3i \log i). \tag{4}$$

Let us now finally bound T_2 using the bound described in Eq. (4). We obtain $T_2 \leq \sum_{j=}^{n} j \prod_{i=2}^{j} (3i \log i)$. Let us now denote $h_j = j \prod_{i=2}^{j} (3i \log(i)) = j \cdot 3^j \cdot j! \prod_{i=2}^{j} \log i$. Note that, for all $2 \leq j < n$ we have $2h_j < h_{j+1}$, and hence, $\sum_{j=2}^{n-1} h_j < h_n$. We have $T_2 \leq \sum_{j=2}^{n} j \prod_{i=2}^{j} (3i \log i) = \sum_{j=2}^{n} h_j \leq 2h_n = 2n \cdot 3^n \cdot n! (\log n)^n$. Using Stirling's approximation, we obtain $T_2 = O(n^{2n})$. $\qquad\square$

5   Discussion and Future Directions

In this paper, we studied the open problems stated in [1,7] by (a) developing a discrete and bounded protocol for local envy-freeness for n agents on TREE graphs with a single-exponential query complexity, and (b) constructing a query-efficient protocol for computing locally envy-free allocations among n agents on DEPTH2TREE. We believe that exploring the complexity of envy-free cake division with graphical constraints will give us novel insights and help us understand the hidden bottlenecks in the query complexity of the general problem.

Our work raises an interesting question of developing query efficient algorithms for finding locally envy-free allocations for fixed parameters such as arboricity or the tree-width of the graph. A second interesting research direction is to study trees with a constant depth and check if can we develop query-efficient protocols for these graphs. Developing efficient protocols for graphs such as a cycle or a bipartite graph is also an interesting future direction.

Acknowledgments. Part of work was done when Xin was at Technion–Israel Institute of Technology and supported by the Aly Kaufman Fellowship. Ganesh was supported by Department of Science and Technology, India under grant CRG/2022/007927. The authors thank Siddharth Barman, Ioannis Caragiannis, and Amik Raj Behera for their helpful comments.

References

1. Abebe, R., Kleinberg, J., Parkes, D.C.: Fair division via social comparison. In: Proceedings of the 16th Conference on Autonomous Agents and MultiAgent Systems (AAMAS), pp. 281–289 (2017)
2. Amanatidis, G., Christodoulou, G., Fearnley, J., Markakis, E., Psomas, C.-A., Vakaliou, E.: An improved envy-free cake cutting protocol for four agents. In: Deng, X. (ed.) SAGT 2018. LNCS, vol. 11059, pp. 87–99. Springer, Cham (2018). https://doi.org/10.1007/978-3-319-99660-8_9
3. Arunachaleswaran, E.R., Barman, S., Kumar, R., Rathi, N.: Fair and efficient cake division with connected pieces. In: Caragiannis, I., Mirrokni, V., Nikolova, E. (eds.) WINE 2019. LNCS, vol. 11920, pp. 57–70. Springer, Cham (2019). https://doi.org/10.1007/978-3-030-35389-6_5
4. Aziz, H., Mackenzie, S.: A discrete and bounded envy-free cake cutting protocol for any number of agents. In: Proceedings of the 57th Annual Symposium on Foundations of Computer Science (FOCS), pp. 416–427 (2016)
5. Aziz, H., Mackenzie, S.: A discrete and bounded envy-free cake cutting protocol for four agents. In: Proceedings of the 48th Annual ACM Symposium on Theory of Computing, pp. 454–464. (STOC) (2016)
6. Barman, S., Rathi, N.: Fair cake division under monotone likelihood ratios. In: Mathematics of Operations Research, pp. 1875–1903 (2022)
7. Bei, X., Qiao, Y., Zhang, S.: Networked fairness in cake cutting. In: Proceedings of the 26th International Joint Conference on Artificial Intelligence, (IJCAI), pp. 3632–3638 (2017)
8. Bei, X., Sun, X., Wu, H., Zhang, J., Zhang, Z., Zi, W.: Cake cutting on graphs: a discrete and bounded proportional protocol. In: Proceedings of the Fourteenth Annual ACM-SIAM Symposium on Discrete Algorithms (SODA), pp. 2114–2123 (2020)
9. Brams, S.J., Taylor, A.D.: Fair Division: From Cake-Cutting to Dispute Resolution. Cambridge University Press, Cambridge (1996)
10. Brandt, F., Conitzer, V., Endriss, U., Lang, J., Procaccia, A.D.: Handbook of Computational Social Choice. Cambridge University Press, Cambridge (2016)
11. Bredereck, R., Kaczmarczyk, A., Niedermeier, R.: Envy-free allocations respecting social networks. Artif. Intell. **305**, 103664 (2022)
12. Budish, E.: The combinatorial assignment problem: approximate competitive equilibrium from equal incomes. J. Polit. Econ. **119**, 1061–1103 (2011)
13. Deng, X., Qi, Q., Saberi, A.: Algorithmic solutions for envy-free cake cutting. Oper. Res. **60**(6), 1461–1476 (2012)
14. Dubins, L.E., Spanier, E.H.: How to cut a cake fairly. Am. Math. Monthly **68**(1P1), 1–17 (1961)
15. Edward Su, F.: Rental harmony: Sperner's lemma in fair division. Am. Math. Monthly **106**(10), 930–942 (1999)
16. Etkin, R., Parekh, A., Tse, D.: Spectrum sharing for unlicensed bands. IEEE J. Sel. Areas Commun. **25**(3), 517–528 (2007)

17. Foley, D.K.: Resource Allocation and the Public Sector, vol. 7:45–98. Yale Economic Essays (1966)
18. Ghalme, G., Huang, X., Machino, Y., Rathi, N.: A discrete and bounded locally envy-free cake cutting protocol on trees. arXiv preprint arXiv:2211.06458 (2022)
19. Goldman, J., Procaccia, A.D.: Spliddit: unleashing fair division algorithms. ACM SIGecom Exchanges **13**(2), 41–46 (2015)
20. Kurokawa, D., Lai, J., Procaccia, A.: How to cut a cake before the party ends. In: Proceedings of the AAAI Conference on Artificial Intelligence (AAAI), pp. 555–561 (2013)
21. Moulin, H.: Fair Division and Collective Welfare. MIT press, Cambridge (2004)
22. Procaccia, A.D.: Thou shalt covet thy neighbor's cake. In: Proceedings of the 21st International Joint Conference on Artificial Intelligence (IJCAI), pp. 239–244 (2009)
23. Robertson, J., Webb, W.: Cake-Cutting Algorithms: Be Fair if You Can. AK Peters/CRC Press, Natick (1998)
24. Steinhaus, H.: The problem of fair division. Econometrica **16**, 101–104 (1948)
25. Stromquist, W.: How to cut a cake fairly. Am. Math. Mon. **87**(8), 640–644 (1980)
26. Stromquist, W.: Envy-free cake divisions cannot be found by finite protocols. Electron. J. Combinatorics **15**(1), R11 (2008)
27. Tucker-Foltz, J.: Thou shalt covet the average of thy neighbors' cakes. Inf. Process. Lett. **180**, 106341 (2023)

A Mechanism for Participatory Budgeting with Funding Constraints and Project Interactions

Mohak Goyal[(✉)] [iD], Sahasrajit Sarmasarkar [iD], and Ashish Goel [iD]

Stanford University, Stanford, CA, USA
{mohakg,sahasras,ashishg}@stanford.edu

Abstract. Participatory budgeting (PB) has been widely adopted and has attracted significant research efforts; however, there is a lack of mechanisms for PB which elicit project interactions, such as substitution and complementarity, from voters. Also, the outcomes of PB in practice are subject to various minimum/maximum funding constraints on 'types' of projects. There is an insufficient understanding of how these funding constraints affect PB's strategic and computational complexities. We propose a novel preference elicitation scheme for PB which allows voters to express how their utilities from projects within 'groups' interact. We consider preference aggregation done under minimum and maximum funding constraints on 'types' of projects, where a project can have multiple type labels as long as this classification can be defined by a 1-laminar structure (henceforth called 1-laminar funding constraints). Overall, we extend the Knapsack voting model of Goel et al. [23] in two ways – enriching the preference elicitation scheme to include project interactions and generalizing the preference aggregation scheme to include 1-laminar funding constraints. We show that the strategyproofness results of Goel et al. [23] for Knapsack voting continue to hold under 1-laminar funding constraints. Moreover, when the funding constraints cannot be described by a 1-laminar structure, strategyproofness does not hold. Although project interactions often break the strategyproofness, we study a special case of vote profiles where truthful voting is a Nash equilibrium under substitution project interactions. We then turn to the study of the computational complexity of preference aggregation. Social welfare maximization under project interactions is NP-hard. As a workaround for practical instances, we give a fixed parameter tractable (FPT) algorithm for social welfare maximization with respect to the maximum number of projects in a group when the overall budget is specified in a fixed number of bits. We also give an FPT algorithm with respect to the number of distinct votes.

Keywords: Participatory budgeting · Fixed parameter tractability · Nash equilibrium

M. Goyal and S. Sarmasarkar—We thank Lodewijk Gelauff and Sanath Kumar Krishnamurthy for their insightful comments on this work. The omitted proofs and more detailed discussions are in the full paper [24] available online.

J. Garg et al. (Eds.): WINE 2023, LNCS 14413, pp. 329–347, 2024.
https://doi.org/10.1007/978-3-031-48974-7_19

1   Introduction

Participatory Budgeting (PB) is a process through which residents can vote on a city government's use of public funds [10,18,39]. Residents might, for example, vote on allocating funds between projects like street repairs or enhancing public safety. PB has been shown to promote citizen engagement, government transparency, and good governance [20,29,40]. Projects in PB have a fixed cost, and there is an overall budget B of funds that the city can spend.

Several voting methods have been used in PB [5]. The most widely used methods in practice are *K-Approval* [7], in which voters approve up to K projects on the ballot and *Approval* [6], in which voters approve any number of projects that they like. These methods are preferred for their simplicity for voters. In *Ranking* [1], voters rank all the projects in order of value or value-for-money. PB organizers often want voters to understand the budgetary constraints they face. For this, a popular choice is *Knapsack voting* [23], in which voters select any number of projects, subject to their costs satisfying the budget constraint.

However, all these methods ignore utility interactions from different projects. Most existing work in PB (with the notable exceptions of [14,26]) assume that the utilities of voters are additively separable over different projects. This fails to capture many real-world complexities of voter preferences [2]. For example, consider the following two projects proposed to enhance public safety.

Example 1. Project 1: Install more streetlights in area A.
Project 2: Hire additional police officers for area A.

For some voters, either of these projects is sufficient to solve the problem, but doing both would be excessive. For these voters, the two projects are *substitutes*. However, some other voters may think that both projects are necessary to make area A safe – and doing only one project would be a waste of money since, in that case, they will continue to avoid the area. For these voters, the two projects are *complements*. Another group of voters may think that additional police officers are not required and only streetlights are required for the area. For these voters, these two projects are *independent* (and they like only Project 1.)

Example 1, with only two projects, illustrates the complexity of eliciting different voters' preferences in PB in real-world scenarios. All the methods we discussed earlier, i.e., K-Approval, Approval, Ranking, and Knapsack voting, fail to capture the different preferences of the voters in Example 1.

There is another natural type of project interaction, which occurs when some projects contradict each other, as in the following example for road development.

Example 2. Project 1: Widen the car lane at street X.
Project 2: Build a bike lane on street X.

Often due to physical constraints, at most one of these projects can be done. In such cases, these projects are *contradictory*. Ideally, the PB ballot *must* inform the voters of such a constraint and restrict the space of possible votes accordingly.

Previous works [14,26] have modelled project interactions in PB via various utility functions. However, their preference elicitation schemes do not enable voters to express their opinions on various project interactions. For example, Jain et al. [26] use Approval voting and assume that the PB organizer knows the project interactions. Fairstein et al. [14] allow voters to express their own groups of projects with interacting utilities; however, they only consider substitution project interactions. This paper aims to fill this gap and design an intuitive preference elicitation scheme under which voters can express a broader range of project interactions than in the existing literature.

Another contribution of this paper is the study of the implications of funding constraints on PB's strategic and computational aspects. From a fairness point of view in budgetary tasks, imposing maximum and/or minimum funding constraints on types of projects is often desirable. Moreover, a project can have multiple 'types'. For this, we consider a 1-laminar structure of type labels of projects. As we show in Observation 2, when the funding constraints are defined on this 1-laminar structure of labelling, these can be represented as a rooted tree – which, in turn, implies a hierarchical ordering of these project labellings.

Such hierarchically defined funding constraints are natural to study for PB[1]. The highest order constraint is the overall budget constraint B. The second-order constraint may, for example, be on the funding to different districts in the city. Further down, for each district, we may have constraints on department and sub-department levels. Each sub-department may be proposing several projects on the ballot. Therefore, the 1-laminar structure of funding constraints closely captures the situations that may arise in real-world PB[2].

1.1 Overview of Our Proposed Model

Same as Jain et al. [26], we adopt a framework wherein the projects on the ballot are partitioned into *groups* by the PB organizer. These groups would typically correspond to a theme – for example, public safety, road development, food support, etc. We assume that projects in different groups have non-interacting utilities for all voters; however, for projects within a group, there can be one of four possibilities regarding project interactions: the projects are *1) substitutes, 2) complementary, 3) contradictory, or 4) independent (no interaction)*. We formally describe the class of preferences expressible in our proposed method in Observation 1 in Sect. 2.2. Crucially, in our model, voters need not agree on these project interactions except when the projects are contradictory (in this case, the ballot is designed to reflect the interaction).

---

[1] See, e.g., here and here that the US federal discretionary and mandatory fundings are described hierarchically.

[2] Note that the funding constraints do not restrict the space of possible votes – these constraints are imposed only on the PB outcome. It is conceivable that a PB organizer may impose these constraints on the votes too. However, we believe that this is not a good idea. It makes voting complicated and restricts voters' freedom to express their opinions. For example, in a PB election in Austin, TX, USA, in 2020, certain groups expressed dissatisfaction with the limitations of the budget input tool as to how much funding they could deduct from the police department [15]. The city then conducted a follow-up election removing those constraints from the ballot.

Votes in our scheme have three simple parts, including distributing funds to groups, approving projects within groups of interest, and answering yes/no *complementarity* questions for each group of interest. The formal details are given in Definition 1. We explain in Sect. 2.2 how these three simple parts of the voting scheme come together and provide an intuitive language for the voters to express project interactions.

We adopt a natural generalization of the overlap utility function, which was first given by Goel et al. [23] for Knapsack voting, where it captures, in dollar terms, the agreement between the voter's 'ideal outcome' and the PB outcome. Our generalization of the overlap utility (Definition 2) captures project interactions; that is, for a bundle of complementary projects, a voter receives utility only if the entire bundle is funded, and for a bundle of substitute projects, the voter's utility saturates at a point that they specify via their vote.

We take a utilitarian approach and consider maximizing the sum of the utilities of all voters under 1-laminar funding constraints. We call our mechanism "Participatory budgeting with project interactions" (PBPI).

1.2   Our Results

We study the incentives of strategic voting in PBPI (Sect. 3). For singleton groups, PBPI is same as Knapsack voting. We show that the strategyproofness[3] result of Goel et al.[4] [23] continues to hold under 1-laminar funding constraints. This result has an important implication: the 1-laminar funding constraints do not make the voting mechanism more complex for a voter. These constraints may, for example, be shared in a separate document from the actual ballot, and the voter may or may not review it to make an 'informed' decision. Moreover, this result is tight in the sense that if the funding constraints do not have a 1-laminar structure, then the mechanism is not strategyproof.

With project interactions, PBPI is often not strategyproof; we study an interesting special case of vote profiles for which truthful voting is a Nash equilibrium [33] under substitution project interactions. The arguments used in the proof are subtle and require the construction of a 2-layer algorithm and careful analysis of voter strategies and potential benefits from various possible strategic deviations.

We then study the computational complexity of preference aggregation in PBPI. Due to the complex project interactions captured in our model, identifying a social welfare maximizing set of projects under budget constraints is NP-hard. We show that when the project groups have a fixed maximum size, and the budget is specified in a fixed number of bits, the problem is fixed-parameter tractable (FPT). For this, we provide a recursive algorithm (Sect. 4). Notably, this result holds also with 1-laminar funding constraints. This result implies that

---

[3] A voting mechanism is strategyproof [17] if it is a weakly-dominant strategy for all voters to vote truthfully.

[4] Goel et al. [23] showed that Knapsack voting is strategyproof for unit-cost projects under the overlap utility function. Further, their model does not consider project interactions, equivalent to PBPI with singleton groups.

for most real-world instances of PB (where not too many projects are expected to have interacting utilities together), computational complexity is not a worry for PB designers, even if the voter turnout is large. We also study the case where the number of distinct votes is small, and give an FPT algorithm for it. While not applicable to general PB elections where all voters submit independent votes, this result is important for cases where a small number of elected representatives engage in a budgetary tasks. Each voter, in this case, may also have a 'weight' denoting how many people they represent.

1.3 Related Work

On Project Interactions in PB. Jain et al. [26] first proposed a model of PB where projects are divided into groups, and only the projects within a group can have interacting utilities. Our model differs in the following ways:

- They use approval voting. This choice of the preference elicitation method implies that the PB organizers need to assume the knowledge of project interactions. This further implies that in the eyes of the mechanism, all voters have the same project interactions. Our preference elicitation scheme allows different voters to have different project interactions, which they can express in the vote. Overall, in the trade-off between expressiveness and simplicity, we adopt a more expressive method, whereas theirs is simpler.
- Their utility function does not consider project costs – it depends only on the number of projects funded from a voter's approval set. Our generalization of the overlap utility function accounts for project costs. This is important when projects have vastly different scales, costs, and utilities.
- PBPI models contradictory projects explicitly, unlike their model.
- In addition to an overall budget constraint (as considered in Jain et al. [26]), we also consider 1-laminar funding constraints on the PB outcome.

Fairstein et al. [14] take an egalitarian approach and give a mechanism which gives a *proportional* outcome while accounting for substitute projects, where voters can express which projects are substitutes.

On Strategic Voting. Goel et al. [23] study the Knapsack voting model with overlap utilities and show that their mechanism is strategyproof if one of the two assumptions holds: 1) partial funding of projects is allowed, 2) projects have unit cost.

Freeman et al. [16] study a class of "moving-phantom" strategyproof mechanisms for PB. They show that the social welfare-maximizing mechanism is the unique Pareto-optimal one in this class of mechanisms.

Yang and Wang [41] study strategyproofness in MW elections for various types of restrictions on the outcome, represented by a graph of the alternatives. For example, they consider cases where the outcome has to be a connected subgraph, an independent set, or a subgraph of bounded diameter. Their results do not cover the class of constraints we study, 1-laminar funding constraints.

On Funding Constraints in PB. Patel et al. [34] study a more general model than PB called *group fair knapsack*. They consider minimum and maximum constraints on the total weight of items from a type in the outcome and provide approximation algorithms for several problem variants.

Jain et al. [27] consider a model where each group of projects has a budget limit in addition to the overall PB budget limit. While social welfare maximization is NP-hard in their setup, they give efficient algorithms for several special cases. Chen et al. [11] consider funding constraints in a setup where the overall budget depends on donations from citizens.

Constraints on the outcome similar in spirit to our funding constraints have been studied in MW elections [3,8,35,41]. Closest to our work, Bredereck et al. [8] study the computational aspects of identifying an outcome under various scoring rules and diversity constraints defined over a 1-laminar classification.

2   Model

In this section, we describe PBPI. Figure 1 is an example of a simple ballot.

2.1   Ballot Design and Preference Elicitation Scheme

There are n voters $\{1, 2, \ldots, n\}$ denoted by $[n]$ and m projects given by $P = \{p_1, p_2, \ldots, p_m\}$. Project p_j has a fixed positive rational cost c_j. B denotes the total budget of funds. An *outcome* of an instance of PB is a bundle of projects Q that satisfies the budget constraint $\sum_{j \in Q} c_j \leq B$. The PB organizer partitions the set of projects P into r *groups* – this partition is $Z = \{z_1, z_2, \ldots, z_r\}$ such that $\cup_{j \in [r]} z_j = P$ and $z_j \cap z_k = \emptyset$ for all $j \neq k$. Groups of projects on the ballot can be of one of two forms: *contradictory* and *non-contradictory*.

In a group of contradictory projects, there is a real-world constraint that at most one of the projects can be implemented (recall Example 2 on road widening). PBPI imposes this constraint on all voters.

Definition 1 (Vote in PBPI).   *Voter $i's$ vote v^i has three components: (f^i, s^i, t^i).*

- *The* fund allocation *for group g, $f_g^i \geq 0$ is the amount of funds that voter i allocates to g. f^i must satisfy the budget constraint $\sum_{g \in [r]} f_g^i \leq B$.*
- *The* approval set *in group g, s_g^i, is the subset of z_g that voter i approve.*
- *Finally, t_g^i is the binary answer to the* complementarity question *in group g, such that $t_g^i = 1$ if voter i considers s_g^i to be complementary, and $t_g^i = 0$ if voter i considers s_g^i to be substitutes or independent.*

The complementarity question is not asked for groups of contradictory projects (see, for example, Group 3 in Fig. 1) where at most one project can be done – here t_g^i is fixed to 0 for notational convenience. We refer to the set of all votes $\{v^i | i \in [n]\}$ as the *vote profile* V. We further assume that the project

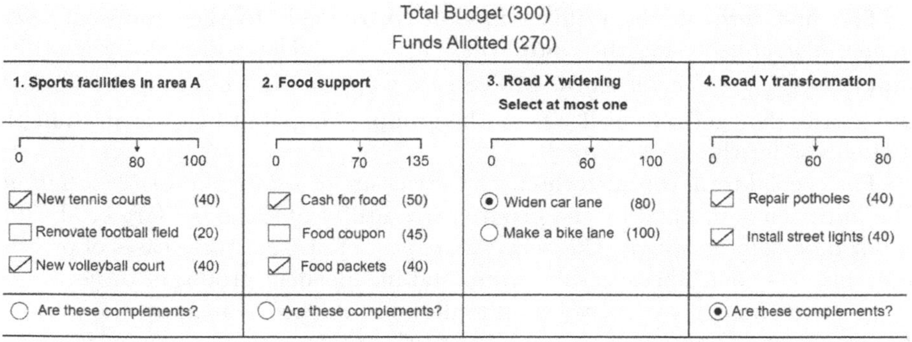

Total Budget (300)

Funds Allotted (270)

1. Sports facilities in area A	2. Food support	3. Road X widening Select at most one	4. Road Y transformation
⌐ ↓ ⌐ 0 80 100	⌐ ↓ ⌐ 0 70 135	⌐ ↓ ⌐ 0 60 100	⌐ ↓ ⌐ 0 60 80
☑ New tennis courts (40) ☐ Renovate football field (20) ☑ New volleyball court (40)	☑ Cash for food (50) ☐ Food coupon (45) ☑ Food packets (40)	◉ Widen car lane (80) ○ Make a bike lane (100)	☑ Repair potholes (40) ☑ Install street lights (40)
○ Are these complements?	○ Are these complements?		◉ Are these complements?

Fig. 1. A simple example of a PBPI ballot with a partial vote marked. There are four groups of projects. Groups 1, 2, and 4 have non-contradictory projects. Group 3 is of contradictory projects; a voter can approve at most one project from this group. The total budget is $B = 300$. The costs of projects are given in parentheses next to its name. The fund allocation f_g^i to each group is represented by a slider. The total funds allocated by the voter are represented by 'Funds Allotted' and are constrained to be at most B. Finally, there is a complementarity question, via which the voter can indicate if their approved projects in the group are complementary for them.

costs c_j, total budget B, and fund allocations f_g^i $\forall i \in [n], g \in [r]$ are positive integers. This is reasonable since the parameters C and B can be rescaled to be integers as long as these are rational without changing the outcome of PB. Typically in real-world PB, costs and budgets are not specified at high precision.

2.2   Utility Function and the Class of Expressible Preferences

We adopt a generalization of the overlap utility [23]. In this subsection, we describe how our preference elicitation method enables voters to express their project interactions and also define the class of expressible preferences in PBPI.

Definition 2 (Utility). *The utility of voter i with vote $v^i = (f^i, s^i, t^i)$ from a bundle of projects Q is:*

$$u_i(Q) = \sum_{g=1}^{r} \left(t_g^i \cdot \mathbb{I}(s_g^i \subseteq Q) \cdot \min(f_g^i, \sum_{j \in \{Q \cap s_g^i\}} c_j) + (1 - t_g^i) \cdot \min(f_g^i, \sum_{j \in \{Q \cap s_g^i\}} c_j) \right)$$

Here $\mathbb{I}(\cdot)$ denotes the indicator function.

Definition 3 (Social Welfare). *The sum of all voters' utilities is social welfare $U(Q) = \sum_{i \in [n]} u_i(Q)$. We use social welfare and social utility interchangeably.*

The first term in the utility function corresponds to the groups of *complementary projects* for the voter (i.e., $t_g^i = 1$). The voter receives utility $\min(f_g^i, \sum_{j \in \{Q \cap s_g^i\}} c_j)$ if all the projects they approve in this group are funded. Otherwise, they get zero utility from the group. Group 4 in Fig. 1 is an example of this case for the marked vote.

The second term captures the utility structure of *substitution with satiation*. This function is inspired by the Leontief-free utility function of Garg et al. [19]. In our model, interestingly, this term can capture the other three cases of project interactions - contradictory, substitute, and independent groups of projects.

In *contradictory groups*, voters are allowed to approve at most one project. If voter i approves project $j^* \in z_g$, their utility from group g is $\min(f_g^i, c_{j^*})$ if $j^* \in Q$ and 0 otherwise. See, e.g., Group 3 in Fig. 1.

For non-contradictory groups, if a voter believes that the projects they approve are *substitutes*, they allocate as many funds f_g^i at which their utility is capped (i.e., their point of satiation). Their utility from the projects in s_g^i adds up linearly up to the point where it saturates. An example is Group 2 in Fig. 1.

However, if the projects are *independent* for them, voters can express this by allocating funds f_g^i that cover the cost of all the projects they approve in the group. An example is Group 1 in Fig. 1 for the marked vote.

Note that PBPI is a generalization of the knapsack voting framework of Goel et al. [23]. If all projects are *independent*, then a voter can allocate f_g^i equal to the sum of costs of the projects they approve in the group and set $t_g^i = 0$. PBPI can, in fact, elicit a much richer class of preferences from individual voters.

We summarize the class of preferences expressible in PBPI below.

Observation 1. *For a group of contradictory projects, a voter can express their:*

– **A. Top choice**, *and the utility they get from it, which can be up to its cost.*

For non-contradictory groups, a voter can express the following preferences:

– **B1. Perfect complements:** *The voter views the set of projects s_g^i as one unit, and their utility from this unit is f_g^i (or the cost of the unit, whichever is lower) if this unit is implemented entirely and 0 otherwise.*
– **B2. Perfect substitutes with satiation:** *The utility from each project is equal to its cost, but their total utility from the group saturates at f_g^i.*
– **B3. Independent:** *The voter views all projects in s_g^i as independent, and the utility from each equals its cost (as in the Knapsack model [23]).*

The cases B2 and B3 are differentiated in the vote by the satiation level expressed in f_g^i. For $t_g^i = 0$, when f_g^i is at least equal to the sum of the costs of all the projects approved by the voter, the group is of independent projects for them, and substitute projects otherwise.

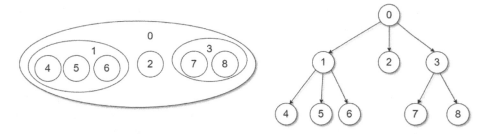

Fig. 2. Each node corresponds to a label or 'type' of projects. The set of projects with label 1 is the union of the set of projects with labels 4, 5, and 6. All projects have the default label 0. The 1-laminar structure of the labels enables the representation of the label relations as a rooted tree as described in Observation 2.

2.3   Preference Aggregation and Funding Constraints

A line of work on PB [27,34] has studied the problem of imposing minimum and/or maximum *funding constraints* on types of projects to ensure the diversity of the outcome. This is sometimes done with the idea that a particular type of projects may be more beneficial to certain demographics in society. Since fairness-motivated interventions based on demographics are often hard to implement in PB [21] or are disallowed by law [38], imposing funding constraints on types of projects is a reasonable proxy to obtain a diverse and equitable outcome.

Let L be a set of labels. Denote the set of projects with the label $l \in L$ by P_l. We refer to the set of projects P_l as 'type' l. Each project can have any number of labels from set L. These labels could, for example, indicate the project's geographic location or theme (e.g. infrastructure, education, etc.). However, this general system of project labels is difficult to study formally for computational complexity and strategic voting. Therefore we make the following assumptions (Fig. 2).

Assumption 1 (1-laminar labelling). *The labelling is 1-laminar . That is, for any two distinct labels x and y, the sets P_x and P_y satisfy either a) $P_x \cap P_y = \emptyset$ or b) $P_x \subset P_y$ or c) $P_y \subset P_x$.*

Observation 2. *Any 1-laminar labelling can be represented as a rooted tree, denoted by $\mathcal{T}_L$. There is a node corresponding to each label. The root node corresponds to a default label (which we construct for notational convenience) that applies to all projects. Nodes of all labels y such that $P_y \subset P_x$ are children of the node representing label x if there is no z satisfying $P_y \subset P_z \subset P_x$.*

Assumption 2. *All projects in a group on the ballot have the same labels.*

This assumption implies that we can introduce a new level at the bottom of the rooted tree of the labels wherein the 'group' represents a new type label and forms the leaves of the tree.

Definition 4 (1-laminar funding constraints). *A valid outcome Q of the PB election must satisfy the following constraints:*

$$B_l^{\min} \leq \sum_{j \in Q \cap P_l} c_j \leq B_l^{\max} \qquad \text{for all } l \in L. \qquad (1)$$

Note that the overall budget constraint can be seen as being associated with the default label, which applies to all projects and corresponds to the root of the tree representation of the labelling. Later in the paper, we will study if including the 1-laminar funding constraints imposes additional challenges for the computational complexity of preference aggregation or have consequences regarding strategic voting. We have positive results on both fronts - Theorems 3, 4, and 5.

We consider mechanisms for PB that produce *non-fractional* and *deterministic* outcomes. In the special case where 1-laminar funding constraints are not imposed on the outcome, we denote the problem of identifying an social welfare-maximizing bundle of projects under budget constraints by SWM-PB; with 1-laminar funding constraints, we call this problem FC-SWM-PB.

2.4   Limitations of the Model

Same as the previous works in this line [14,26], PBPI does not capture the case where both substitutes and complementary projects are present in a group. For example, in a group of projects $\{p_1, p_2, p_3\}$, it is possible that a voter finds p_1 and p_2 to be complementary, but p_3 to be a substitute for the bundle $\{p_1, p_2\}$. PBPI does not enable voters to express such project interactions.

PBPI also does not model *ranked preferences over substitute projects*, which could have been elicited via a ranking-based voting scheme within groups instead of approval in PBPI. Further, our utility function does not model *soft complements*, i.e., per our utility function, a voter cannot get partial utility if only a part of their approval set in a group of complementary projects is funded.

For a group $\{p_1, p_2\}$, voters in PBPI cannot express preferences of the following type: "p_1 and p_2 are independent for me, but my utility from each is only half as much as their respective costs." This is because our utility function cannot distinguish between this case and the case where p_1 and p_2 are substitutes for the voter. A voter could express this type of preference if p_1 and p_2 were in singleton groups on the ballot.

We can mathematically model the above-stated and even more complex project interactions with cost-based utility functions. Our positives result on computational complexity in Theorems 4 and 5 (FPT with respect to the maximum group size) will hold for any utility function as long as all project interactions are within their own groups. However, we do not adopt a more complex utility function since it would require an equally expressive preference elicitation method which may not be intuitive for the voters in the real world.

3   Incentives for Strategic Voting in PBPI

Goel et al. [23] showed that Knapsack voting is strategyproof under overlap utility if one of the two assumptions holds: 1) partial funding of projects is allowed, 2) projects have unit cost. Technically, these assumptions provide the same leverage, ensuring that a greedy algorithm is optimal for preference aggregation. In this section[5], we work with the assumption that projects are unit-cost[6].

Assumption 3 (Unit-Cost Projects). *All projects have cost $c_j = 1 \; \forall j \in [m]$.*

Therefore, our results in this section are for the MW election setting. First, we consider the special case of PBPI without project interaction, i.e., the same as Knapsack voting. In this case, we show that even with 1-laminar funding constraints on the outcome, Knapsack voting is strategyproof, i.e., voters are incentivised to vote truthfully, disregarding the 1-laminar funding constraints.

Theorem 1. *With singleton groups, unit-cost projects (Assumption 3), and the 1-laminar labelling (Assumption 1) for the funding constraints (Definition 4), PBPI is strategyproof.*

The proof uses Theorem 3 (given later). Per Theorem 3, truth-telling is a weakly dominant strategy for a voter i when the other voters vote truthfully. Suppose some other voters $i' \in I$ did not vote truthfully. In the case of singleton groups, the untruthful votes can be seen as the truthful votes of an alternative set of voters. The result follows.

Note that the result also extends to the fractional knapsack considered by Goel et al. [23], since each project can then be considered a collection of unit-cost projects.

The 1-laminar labelling (Assumption 1) is a fairly general setup for defining the funding constraints in practice. In fact, technically it is *necessary* for Theorem 1 to hold. Towards this, we have the following result.

Theorem 2. *PBPI with singleton groups and unit-cost projects is not strategyproof if the labelling for the funding constraints does not satisfy the 1-laminar structure (Assumption 1).*

We now study the effect of considering project interactions on the incentives to vote truthfully. Unfortunately, as is seen often in social choice theory [22,36], voters in PBPI may be incentivised to deviate from their truthful votes to attain a better outcome for themselves when project interactions are considered.

---

[5] We do not need this assumption in the section on computational complexity .

[6] It is easy to show that under overlap utility, the unit-cost assumption is crucial for strategyproofness to hold in Knapsack voting and PBPI. We sketch an example here. Consider a case with four projects whose costs are $c_1 = 1$, $c_2 = 2$, $c_3 = 2$, and $c_4 = 3$. Budget B is 4 units. There are no project interactions. A voter who only likes project 4 is incentivised to vote for project 1 too when all projects have equal approvals otherwise.

Observation 3. *PBPI is not strategyproof even with Assumption 3.*

Proof (with complementarity project interaction). Consider this example.
There are $n = 3$ voters and $m = 6$ projects, each with cost $c_j = 1$. The budget is $B = 3$. There are no 1-laminar funding constraints. There are two groups of projects $z_1 = \{p_1, p_2, p_3\}$, and $z_2 = \{p_4, p_5, p_6\}$. In the truthful votes, all projects are independent for all three voters. Voters 1 and 2 approve projects p_1, p_2, and p_4 and set funds $f_1^1 = f_1^2 = 2$ and $f_2^1 = f_2^2 = 1$. Whereas voter 3 approves projects p_3, p_5, and p_6, and sets $f_1^3 = 1$ and $f_2^3 = 2$. In this case, the outcome of SWM-PB is $\{p_1, p_2, p_4\}$.

Voter 3 has the incentive to modify their vote and set $f_1^3 = 3$, approve all of z_1, and report that these projects are complements, i.e., $t_1^3 = 1$. Now, $\{p_1, p_2, p_3\}$ is the outcome of SWM-PB – it increases the utility of voter 3 by 1 unit.

The example above shows how the complementarity project interaction can potentially lead to profitable strategic deviations for voters. However, this type of project interaction is not required to render PBPI not strategyproof.

Observation 4. *PBPI is not strategyproof under Assumption 3 when there are only substitutes allowed.*

Despite these negative observations, it is interesting to study special classes of vote profiles where voters do not have incentives to deviate from truthful voting even with project interactions. We now describe such a class of vote profiles.

3.1   Case with a Strict Total Order in Subgroups of Substitutes

Here we give a class of vote profiles for which truthful voting is a Nash equilibrium. *Subgroups* of a group are defined by an arbitrary partition of the projects in the group. In this class of vote profiles, there is no complementarity protect interaction, voters agree on the type of project interaction in a group, and for substitutes, it entails *some* partition of groups into subgroups such that there is a strict total order[7] of projects within each subgroup. We further need that all voters who approve project p also approve all projects of a higher order than p.

Definition 5 (Special Vote Profile). *For each non-contradictory group g, one of the following holds:*

1. *(Independents). All voters $i \in [n]$ consider the group to be of independent projects and set $f_g^i = |s_g^i|$.*
2. *(Substitutes). There exists a partition of z_g into "sub-groups" and there is a strict total order in each sub-group such that every voter who approves a project $p \in z_g$ also approves all projects of a higher order than p. No voter approves projects from multiple subgroups in a group.*

---

[7] A strict total order on a set S is a relation on S that is irreflexive, anti-symmetric, transitive, and every pair of elements of S is comparable.

Observe that we do not make any assumption on fund allocations f_g^i for substitute projects. Also, this structure does not assume which groups any voter funds. However, it does assume that voters who allocate funds to a group agree on the type of project interaction in any group. Also, note that having a strict total order on some partition of a group into subgroups is a strictly weaker condition than having a strict total order over entire groups of projects. As a clarification, while studying the vote profile of Definition 5, we do not make any assumption on the space of expressible votes, other than that there are no complementarity project interactions considered.

The class of vote profiles in Definition 5 is important to study since voters who are residents of an area may agree on the relative usefulness of projects in a group but may not agree on how many projects must be done from any group.

Having explained the case of vote profiles of Definition 5, we now discuss our results on it. First, when the vote profile V is per Definition 5, a greedy aggregation algorithm (Algortihm 1) produces a social welfare-maximizing outcome under an overall budget constraint and 1-laminar funding constraints. This is not particularly surprising since in contradictory and independent groups, utilities of individual projects are additively separable, and in substitute groups, the strict total order in subgroups provides an optimal ordering to fund projects. The 1-laminar funding constraints dictate which projects are eligible for funding at any step of the algorithm.

The surprising result is that the same thing also holds when one voter's vote doesn't follow Definition 5 (Lemma 1). This one vote could disagree with others on which groups are independents and which are substitutes, could approve projects across subgroups, or violate the strict total order assumption within subgroups. The fact that the greedy algorithm still produces a social welfare-maximizing outcome when one vote does not follow Definition 5 has important implications for the study of strategic voting in our model.

We can leverage Lemma 1 to show that when the truthful votes (underlying, not observed) are per Definition 5, truthful voting is a Nash equilibrium[8] (Theorem 3). This result has important practical consequences, which we discuss later in the section after giving the results formally.

Greedy Algorithm. We first describe the Greedy Algorithm 1, which we use in Lemma 1 and Theorem 3. First, without the 1-laminar funding constraints, a greedy algorithm is simple; it will construct the outcome Q in B steps, adding a project in each step which brings the best improvement in social welfare. With 1-laminar funding constraints, our Greedy Algorithm 1 is run in two passes. We assume that the funding constraints $B^{\min}$ and $B^{\max}$ are such that at least one valid outcome of PB exists. In the first pass, the algorithm fulfils the minimum funding constraints given by $B^{\min}$. Recall from Observation 2 that the 1-laminar labelling can be represented as a rooted tree denoted by $\mathcal{T}_L$, and the levels of the

---

[8] We do not obtain strategyproofness here; if multiple votes violate Definition 5, then there can arise opportunities for profitable strategic deviations for voters.

Greedy Algorithm 1: Input: $(Z, B, V, L, P_l \; \forall \, l \in L, B^{\max}, B^{\min})$, Output: Q

Let $\mathbb{Q}$ be the set of outcomes satisfying the 1-laminar 'maximum' funding constraints

Initialize $Q \leftarrow \emptyset$

1. First Pass   ▷ Satisfy minimum funding constraints

for *Traverse* $\mathcal{T}_L$ *in reverse order[9], index node by l* **do**

$\quad$ **while** $|Q \cap P_l| < B_l^{\min}$ **do**

$$Q \leftarrow Q \cup \Big\{ \underset{\rho \in \{j \mid j \in P_l \setminus Q, \; j \cup Q \in \mathbb{Q}\}}{\arg\max} U(Q \cup \{\rho\}) \Big\}$$

$\quad$ **end**

end

2. Second Pass   ▷ Complete Q while respecting funding constraints

while $|Q| < B$ **do**

$$Q \leftarrow Q \cup \Big\{ \underset{\rho \in \{j \mid j \notin Q, \; j \cup Q \in \mathbb{Q}\}}{\arg\max} U(Q \cup \{\rho\}) \Big\}$$

end

---

tree form a hierarchy of the labellings. Within this pass, the algorithm proceeds in the reverse hierarchy order for the type labelling.

In the second pass, the algorithm produces the outcome Q while respecting all the maximum funding constraints while adding one project to the outcome at a time. The algorithm pseudocode is given formally in Greedy Algorithm 1. We now state and prove the main results of this section.

In this subsection, we assume that there are no complementarity project interactions, and in fact, the complementarity question is disabled and we have t_g^i for all voters $i \in [n]$ and all groups $g \in [r]$.

Lemma 1. *Under Assumptions 1, 2, and 3, when at most one vote deviates from Definition 5, the outcome of Greedy Algorithm 1 is a solution to FC-SWM-PB.*

Proof Sketch. We first show that when every vote follows Definition 5, Greedy Algorithm 1 gives a solution to FC-SWP-PB. This trivially follows for contradictory and independent groups and follows from strict total ordering on the set of approved projects within every subgroup of substitute projects.

When one voter violates Definition 5, we break down the problem to each group of projects – we first show that the set of projects selected from any group g for any social welfare maximising outcome (up to tie-breaking order) from group g is monotonically increasing with the funding allocated to the group g. Then, we show that the greedy Algorithm 1 always yields an optimal solution using the tree representation of the 1-laminar funding constraints.

We now use Lemma 1 to study the incentives of strategic voting.

---

[9] A traversal of a rooted tree in the reverse order is defined as the traversal which starts with the leaves in arbitrary order, followed by the deletion of all the leaves, and repeating the process till the entire graph is traversed.

Theorem 3. *Under Assumptions 1, 2, and 3, when the true vote profile is per Definition 5, truthful voting is a Nash equilibrium in PBPI with FC-SWM-PB.*

The proof is technical. It crucially uses Lemma 1. For a voter i deviating from their truthful vote, we break down their vote into that for each group $g \in [r]$ and then show that for any state of their vote in the other groups, truthful voting in a group g is a weakly dominant strategy as long as the other voters' votes are per Definition 5.

Implications of the Results. Lemma 1 and Theorem 3 signify that for the class of vote profiles given in Definition 5, PBPI satisfies two key desiderata of voting mechanisms – efficient computation of outcome (via the Greedy Algorithm 1) and an incentive for voters to vote truthfully, thereby making voting simpler.

Theorem 3 has important consequences for PB. First, under the setup of this subsection, the funding constraints do not present additional complexities for voting. Therefore PB organizers and policymakers need not worry about the cognitive complexity of the mechanism when deciding whether they must impose 1-laminar funding constraints on the PB outcome. Second, since the 1-laminar funding constraints have no role in deciding the voting strategy, it can be justified to release this information as a separate document and unclutter the actual ballot. It may be possible to run the PB election even when the parameters of the funding constraints are yet to be decided by a process which doesn't depend on the PB election.

4   Computational Complexity of Preference Aggregation

Since our mechanism for PB can model relatively complex preferences for all voters, the amount of information to be processed by the preference aggregation algorithm can be substantial. As for all voting schemes, it is important to study the computational complexity of aggregating the votes for any objective function of the social planner. We first make a negative observation. Due to the complex project interactions in our model, SWM-PB cannot be solved in polynomial time unless $P = NP$.

Observation 5. SWM-PB *is* NP-hard.

The proof is via a reduction from the maximum set coverage problem. This result is unsurprising since many models with project interactions in PB elections face this issue [26, 27]. FC-SWM-PB is also NP-hard since it is at least as hard as SWM-PB.

For real-world voting problems, we often deal with scenarios where some parameters are small. On a positive note, we show that FC-SWM-PB can be solved in polynomial time for reasonable real-world parameters of the model.

We first consider the case where the number of projects that must be grouped together for project interactions is small. This is expected to be the case for most real-world PB elections. We also need the technical condition that the number of bits required to specify the budget B is a fixed parameter. This can naturally be true for real-world ballots where the required precision of costs and funds is not very high. For example, in a case where the total budget is 10^6 currency units, all costs and votes can be reasonably specified in units of 10^3 currency units. In this example, we will have $B = 1000$.

4.1   FPT with Respect to the Maximum Size of a Group and log(B)

Let s_{max} denote the maximum size of a group of projects. We will show that if s_{max} is fixed, SWM-PB and FC-SWM-PB are computationally tractable. This suggests that PB organizers must design the ballot with reasonably small groups if computational complexity is a concern.

Theorem 4. *Under Assumption 2, FC-SWM-PB is FPT with respect to* $(\log(B), s_{max})$.

Proof Sketch. Observe that for any possible fund allocation to group g, we can compute the social-welfare maximizing or *best* subset of z_g in time $O(n2^{|z_g|})$ by doing a brute-force search over all subsets of z_g. We run a recursion to find the optimal fund allocation to each group. The overall computational complexity of the recursion is therefore $O(nrB \cdot 2^{s_{max}})$. To satisfy the funding constraints, we set the utility of the terms of the recursion violating them to $-\infty$.

Note that we do not need the assumption on the 1-laminar structure of the funding constraints (Assumption 1) for this result to hold. It will hold for any arbitrary set of type labels on projects and associated minimum and maximum funding constraints as long as Assumption 2 holds.

4.2   FPT with Respect to the Number of Distinct Votes

In this subsection, we deviate from the general framework of PB and study the case where the number of distinct votes is small. Often, budgetary tasks are undertaken by a small number of elected representatives, and each representative may be voting on behalf of a different number of voters. This framework may also be relevant in the realm of delegation voting [25]. Another use case would be a PB format where (an unrestricted number of) voters are asked to choose one out of a fixed set of 'prototype outcomes'. Yet another use case would be when a small number of voters are queried randomly to get a 'quick pulse' of the people's opinions.

For simplicity, we will overload the notation and use n for the number of distinct voters and w_i for the frequency or weight of the vote of voter i for each $i \in [n]$. In this notation, the social utility of an outcome Q is given by $\sum_{i \in [n]} w_i u_i(Q)$, which is the objective function of FC-SWM-PB here.

Theorem 5. *Under Assumptions 2,   3, FC-SWM-PB is FPT with respect to* n.[10]

---

[10] For unit cost projects, B is bounded by m, and we don't need it as a fixed parameter.

Proof Sketch. We divide the projects into partitions approved by unique subsets of voters. We then solve a mixed integer program (MIP) with variables corresponding to the number of projects funded from each partition. By the result of Bredereck et al. [9], concave utility functions can also be incorporated in MIP, and the runtime is exponential in the number of variables.

Discussion: The runtime is doubly exponential in n, resulting from having project interaction in the objective. Due to this, the techniques of [30], who gave singly exponential time algorithms for several combinatorial voting problems, do not directly apply to our setup. We can, however, use the approximation scheme of [37] to get an arbitrarily close approximation in singly exponential time (our model is 'p-subseparable' per their terminology).

5 Conclusions

While the use of *categories* or *groups* of projects in PB ballots is now standard in theory [14,26,28], experimental studies (e.g. the study on Amazon Mechanical Turk by Fairstein et al. [13]), and practice (e.g. the 2020 Long Beach, USA PB election at https://budget.pbstanford.org/longBeach2020), there is no existing work on leveraging this partition to design a preference elicitation method which can enable voters to express a wide variety of project interactions with only a reasonable amount of cognitive effort[11]. We fill this gap by providing a mechanism with this property, which can also naturally integrate 1-laminar funding constraints without creating additional strategic or computational complexities, is deterministic[12], is computationally tractable in reasonable parameter regimes (Theorem 4), and is also robust to unilateral strategic deviations for a class of vote profiles (Theorem 3). Therefore, our proposed design, PBPI, is a strong candidate for PB in the real world.

Empirically studying the expressiveness and simplicity of our PB mechanism is important. This may be done similar to how Fairstein et al. [13] study other common PB mechanisms. Including more complex project interactions in an intuitive voting scheme for PB continues to be an important avenue for future research. For our positive results on strategic voting, we need to make assumptions on the cost of projects and drop the complementarity project interactions – designing strategyproof mechanisms for PB without these assumptions would be a major contribution. Studying the properties of preference aggregation methods that maximize the Nash welfare [32] or characterizing the core of PB [12,31] under project interactions are also interesting research directions.

---

[11] A line of work in PB studies the amount of cognitive effort a voter must apply in a voting mechanism (see, for example, [4,13]). While there is a lack of consensus in the literature on the relative effort PB mechanisms require from voters, and we have not evaluated PBPI empirically, we believe that our scheme needs only a comparable amount of effort from voters as existing methods. This is because each part of a vote (fund allocation to groups, approval, and complementarity questions) is intuitive.

[12] Being deterministic is often desirable for voting mechanisms. One of the reasons is the difficulty of verifying implementation correctness in randomized schemes.

References

1. Arrow, K.J., Sen, A., Suzumura, K.: Handbook of Social Choice and Welfare, vol. 2. Elsevier, Amsterdam (2010)
2. Aziz, H., Shah, N.: Participatory budgeting: models and approaches. In: Rudas, T., Péli, G. (eds.) Pathways Between Social Science and Computational Social Science. CSS, pp. 215–236. Springer, Cham (2021). https://doi.org/10.1007/978-3-030-54936-7_10
3. Bei, X., Liu, S., Poon, C.K., Wang, H.: Candidate selections with proportional fairness constraints. Auton. Agent. Multi-Agent Syst. **36**, 1–32 (2022)
4. Benade, G., Itzhak, N., Shah, N., Procaccia, A.D., Gal, Y.: Efficiency and usability of participatory budgeting methods (2018)
5. Benade, G., Nath, S., Procaccia, A.D., Shah, N.: Preference elicitation for participatory budgeting. Manag. Sci. **67**(5), 2813–2827 (2021)
6. Brams, S.J., Fishburn, P.C.: Approval voting. Am. Polit. Sci. Rev. **72**(3), 831–847 (1978)
7. Brams, S.J., Fishburn, P.C.: Voting procedures. Handb. Soc. Choice Welfare **1**, 173–236 (2002)
8. Bredereck, R., Faliszewski, P., Igarashi, A., Lackner, M., Skowron, P.: Multiwinner elections with diversity constraints. In: Proceedings of the AAAI Conference on Artificial Intelligence, vol. 32 (2018)
9. Bredereck, R., Faliszewski, P., Niedermeier, R., Skowron, P., Talmon, N.: Mixed integer programming with convex/concave constraints: fixed-parameter tractability and applications to multicovering and voting. Theoret. Comput. Sci. **814**, 86–105 (2020)
10. Cabannes, Y.: Participatory budgeting: a significant contribution to participatory democracy. Environ. Urban. **16**(1), 27–46 (2004)
11. Chen, J., Lackner, M., Maly, J.: Participatory budgeting with donations and diversity constraints. In: Proceedings of the AAAI Conference on Artificial Intelligence, vol. 36, pp. 9323–9330 (2022)
12. Fain, B., Goel, A., Munagala, K.: The core of the participatory budgeting problem. In: Cai, Y., Vetta, A. (eds.) WINE 2016. LNCS, vol. 10123, pp. 384–399. Springer, Heidelberg (2016). https://doi.org/10.1007/978-3-662-54110-4_27
13. Fairstein, R., Benadè, G., Gal, K.: Participatory budgeting design for the real world. arXiv:2302.13316 (2023)
14. Fairstein, R., Meir, R., Gal, K.: Proportional participatory budgeting with substitute projects. arXiv preprint arXiv:2106.05360 (2021)
15. Falcon, R.: Austin justice coalition survey tool finds majority of local participants want to cut APD budget by USD 236M (2020). https://www.kxan.com/news/local/austin/austin-justice-coalition-survey-tool-finds-majority-of-local-participants-want-to-cut-apd-budget-by-236m/
16. Freeman, R., Pennock, D.M., Peters, D., Vaughan, J.W.: Truthful aggregation of budget proposals. J. Econ. Theory **193**, 105234 (2021)
17. Fudenberg, D., Tirole, J.: Game Theory. MIT press, Cambridge (1991)
18. Ganuza, E., Baiocchi, G.: The power of ambiguity: How participatory budgeting travels the globe. J. Public Deliberation **8**, (2012)
19. Garg, J., Mehta, R., Vazirani, V.V.: Substitution with satiation: a new class of utility functions and a complementary pivot algorithm. Math. Oper. Res. **43**(3), 996–1024 (2018)

20. Gatto, A., Sadik-Zada, E.R.: Governance matters. fieldwork on participatory budgeting, voting, and development from Campania, Italy. J. Public Affairs **22**, e2769 (2021)
21. Gelauff, L., Goel, A., Munagala, K., Yandamuri, S.: Advertising for demographically fair outcomes. arXiv:2006.03983 (2020)
22. Gibbard, A.: Manipulation of voting schemes: a general result. Econometrica, 587–601 (1973)
23. Goel, A., Krishnaswamy, A.K., Sakshuwong, S., Aitamurto, T.: Knapsack voting for participatory budgeting. ACM Trans. Econ. Comput. **7**(2), 1–27 (2019)
24. Goyal, M., Sarmasarkar, S., Goel, A.: A mechanism for participatory budgeting with funding constraints and project interactions (2023)
25. Green-Armytage, J.: Direct voting and proxy voting. Const. Polit. Econ. **26**, 190–220 (2015)
26. Jain, P., Sornat, K., Talmon, N.: Participatory budgeting with project interactions. In: Proceedings of the Twenty-Ninth International Joint Conference on Artificial Intelligence, IJCAI-20, pp. 386–392 (2020)
27. Jain, P., Sornat, K., Talmon, N., Zehavi, M.: Participatory budgeting with project groups. arXiv:2012.05213 (2020)
28. Jain, P., Talmon, N., Bulteau, L.: Partition aggregation for participatory budgeting. In: Proceedings of the 20th International Conference on Autonomous Agents and MultiAgent Systems, pp. 665–673 (2021)
29. Johnson, C., Carlson, H.J., Reynolds, S.: Testing the participation hypothesis: evidence from participatory budgeting. Polit. Behav. **45**, 1–30 (2021)
30. Knop, D., Koutecký, M., Mnich, M.: Voting and bribing in single-exponential time. ACM Trans. Econ. Comput. (TEAC) **8**(3), 1–28 (2020)
31. Munagala, K., Shen, Y., Wang, K.: Auditing for core stability in participatory budgeting. In: Hansen, K.A., Liu, T.X., Malekian, A. (eds.) WINE 2022. LNCS, vol. 13778, pp. 292–310. Springer, Cham (2022). https://doi.org/10.1007/978-3-031-22832-2_17
32. Nash Jr, J.F.: The bargaining problem. Econometrica: J. Econometric Soc. 155–162 (1950)
33. Nash, J.F., Jr.: Equilibrium points in n-person games. Proc. Natl. Acad. Sci. **36**(1), 48–49 (1950)
34. Patel, D., Khan, A., Louis, A.: Group fairness for knapsack problems. arXiv preprint arXiv:2006.07832 (2020)
35. Rey, S., Endriss, U., de Haan, R.: Designing participatory budgeting mechanisms grounded in judgment aggregation. In: Proceedings of the International Conference on Principles of Knowledge Representation and Reasoning, vol. 17, pp. 692–702 (2020)
36. Satterthwaite, M.A.: Strategy-proofness and arrow's conditions: existence and correspondence theorems for voting procedures and social welfare functions. J. Econ. Theory **10**(2), 187–217 (1975)
37. Skowron, P.: Fpt approximation schemes for maximizing submodular functions. Inf. Comput. **257**, 65–78 (2017)
38. Sowell, T.: Affirmative Action Around the World: An Empirical Study. Yale University Press, New Haven (2004)
39. Wainwright, H.: Making a people's budget in Porto Alegre. NACLA Rep. Am. **36**, 37–42 (2003)
40. Wampler, B.: A guide to participatory budgeting. In: Participatory Budgeting. World Bank (2007)
41. Yang, Y., Wang, J.: Multiwinner voting with restricted admissible sets: Complexity and strategyproofness. In: IJCAI, pp. 576–582 (2018)

Randomized Algorithm for MPMD on Two Sources

Kun He[1], Sizhe Li[1], Enze Sun[2], Yuyi Wang[3(✉)], Roger Wattenhofer[4],
and Weihao Zhu[5]

[1] Huazhong University of Science and Technology, Wuhan, China
{brooklet60,sizheree}@hust.edu.cn
[2] The University of Hong Kong, Hong Kong, China
sunenze@connect.hku.hk
[3] CRRC Zhuzhou Institute, Zhuzhou, China
yuyiw920@163.com
[4] ETH Zurich, Zurich, Switzerland
wattenhofer@ethz.ch
[5] University of Illinois, Urbana-Champaign, Urbana, USA
weihaoz3@illinois.edu

Abstract. A 3-competitive deterministic algorithm for the problem of
min-cost perfect matching with delays on two sources (2-MPMD) was
proposed years ago. However, whether randomness leads to a more competitive algorithm remains open. 2-MPMD is similar to the famous ski
rental problem. Indeed, for both problems, we must choose between continuing to pay a repeating cost or a one-time fee. There is a memoryless
randomized algorithm for ski rental that is more competitive than its
best deterministic algorithm. But, surprisingly, memoryless randomized
algorithms for 2-MPMD cannot do better than 3-competitive. In this
paper, we devise a 2-competitive randomized algorithm for 2-MPMD.
Moreover, we prove that 2 is also the lower bound.

Keywords: Online matching with delay · Randomized algorithms ·
Competitive analysis · Lower bound

1 Introduction

"Patience is the best remedy for every trouble," said Plautus, a Roman play-
wright of the Old Latin period. Even though one might disagree with his opinion,
recently, patience has shown its power in many online problems. These problems,
called online problems with delay, allow waiting (at a cost) before any major
decision is made and attract much attention. In particular, for the problem of
online matching with delays, in which new arrival requests are not required to
be matched immediately, due to its wide range of applications such as ridesharing, game player matching, and so on, several interesting results have been
obtained [2,3,11,12,18].

Part of this work was done when W. Zhu was at Shanghai Jiao Tong University.

J. Garg et al. (Eds.): WINE 2023, LNCS 14413, pp. 348–365, 2024.
https://doi.org/10.1007/978-3-031-48974-7_20

In this paper, we study online matching with delays restricted to two sources, which is the most straightforward metric, so-called the min-cost perfect matching with delays on two sources (2-MPMD) problem [12]. Under the assumption of paired request sequences (see Sect. 2.2), for every arrival request pair, there is a choice between continuing to pay a repeating delaying cost or paying a one-time (larger) matching cost to eliminate the repeating cost, which is similar to the famous ski rental problem.

There are competitive deterministic algorithms for ski rental and 2-MPMD problems. The best for ski rental, the break-even algorithm, is $(2 - 1/b)$-competitive, where b is the ratio between buying and renting costs. The best deterministic algorithm for 2-MPMD borrows ideas from the break-even algorithm and achieves 3-competitiveness [12]. It is well-known for the ski rental problem that there exists a $\approx \frac{e}{e-1}$-competitive randomized algorithm, which is strictly more competitive than the best deterministic algorithm [15]. It is natural to ask whether this is the case for the 2-MPMD problem, i.e., whether there exists a randomized algorithm that is more competitive than the best deterministic algorithm.

The optimal randomized algorithm for the ski rental problem is memoryless [15]. Hence, the first attempt to design a randomized 2-MPMD algorithm also explored memoryless algorithms. However, it turned out that no memoryless algorithm for 2-MPMD can be better than 3-competitive [12]. It remains a challenge whether there is a more competitive randomized algorithm for 2-MPMD.

In this paper, we answer the above question by devising a randomized algorithm and showing that its competitive ratio is 2, which is significantly smaller than the ratio of the best deterministic algorithm. Besides, we also prove that no other randomized algorithm for 2-MPMD can further reduce the competitive ratio.

1.1   Related Work

Matching has long been a classic problem in combinatorial optimization ever since the seminal work of Edmonds [9,10]. Starting with the famous paper of Karp et al. [16], online matching problems attract much attention where many different versions have been investigated in [1,6–8,13,14,16,17,20–24]. In these versions, requests on one side of a bipartite graph are given previously, while requests on the other side are given in an online fashion, and the goal is usually to maximize the matching size.

The min-cost perfect matching with delays problem (MPMD) was first introduced by Emek et al. [11]. In their settings, requests arrive in a continuous online fashion in a metric space, and the algorithm should match all requests to each other with some delays when necessary. The goal is to minimize the total cost: the sum of metric distances between matched requests plus the sum of waiting times of each request. They presented an online randomized algorithm with competitive ratio $O\left(\log^2 n + \log \Delta\right)$, where n is the number of points in the metric space and Δ is the aspect ratio. Azar et al. [3] later showed another online algorithm with competitive ratio $O(\log n)$, while they also proved that no online

MPMD algorithm could be better than $\Omega(\sqrt{\log n})$-competitive. Azar et al. [4] devised a deterministic algorithm with a competitive ratio of $O(m^{\log(\frac{3}{2}+\varepsilon)})$ for any fixed $\varepsilon > 0$ and m being the number of requests, which is the first deterministic algorithm achieving a sub-linear competitive ratio. Ashlagi et al. [2] improved the lower bound to $\Omega\left(\frac{\log n}{\log \log n}\right)$. They also investigated a bipartite variant of the MPMD problem, where they obtained an upper bound $O(\log n)$ and a lower bound $\Omega\left(\sqrt{\frac{\log n}{\log \log n}}\right)$. Mari et al. [19] studied a stochastic version of the MPMD problem where the input requests follow a Poisson arrival process. They presented two online deterministic algorithms which are constantly competitive.

Problems of online min-cost perfect matching with non-linear delays have also been studied. Liu et al. [18] presented an online algorithm with competitive ratio $O(k)$ on any k-point uniform metric space for some convex time cost functions. Later, a constant competitive ratio algorithm was devised by Azar et al. [5] for some concave time cost functions.

For the special case of the online 2-MPMD problem, Emek et al. [12] showed a deterministic variant of the online algorithm in [11], which is 3-competitive. They also proved that any deterministic online algorithm has a competitive ratio of at least 3.

1.2 Proof Overview

In the online 2-MPMD problem, the main challenge is to make decisions between continuing to pay a repeating cost or a one-time fee. As no memoryless algorithm for 2-MPMD can be better than 3-competitive [12], we need to memorize some useful information for future decisions. Inspired by the optimal algorithm, we give a new concept "state", which helps us to make decisions with a competitive ratio guaranteed. The algorithm we devise is actually "Markovian". That is, at any time, the distribution of future decisions is only related to the current state value. When proving the lower bound, we construct a distribution of randomized sequences with the help of "state", which is also "Markovian". More precisely, at any moment, if all previous requests are fixed, the distribution of the future sequence is only based on the current state value. This helps us to make the expected cost ratio of any online deterministic algorithm be at least 2, which finishes the whole proof.

1.3 Paper Organization

The rest of the paper is organized as follows. We introduce necessary notations and background knowledge in Sect. 2. In Sect. 3, we present an optimal offline algorithm and give a detailed analysis of the structure of optimal solutions, which leads to a core concept of "state". In Sect. 4, we show a 2-competitive randomized online algorithm. Finally, we prove that 2 is the tight lower bound in Sect. 5.

2   Preliminaries

2.1   Model

In an instance of the 2-MPMD problem, there are two sources a and b, and a request sequence R. Each request r belongs to one source, denoted $x(r) \in \{a, b\}$, and arrives at time $t(r)$. We consider the request set R in a continuous-time online fashion so that request $r \in R$ is unknown until its arriving time $t(r)$.

Assume throughout that $|R|$ is even. We need to match all requests into pairs, so the result should be a perfect matching of R. In this online setting, the algorithm is allowed to delay the matching of any request. A request is said to be *open* if it has not been matched yet. Otherwise, it is called *matched*.

Formally, for any two requests r_1, r_2, we denote an operation $M(r_1, r_2, t)$ as matching r_1 and r_2 at time $t \geq \max\{t(r_1), t(r_2)\}$. The cost of the match M consists of *space cost* and *time cost*. If requests r_1 and r_2 have different sources, the space cost is 1, and otherwise, 0. The time costs incurred by match M on requests r_1 and r_2 are $t - t(r_1)$ and $t - t(r_2)$, respectively. The cost of M, denoted as $cost(M)$, is defined to be the sum of its space cost and the time costs of r_1 and r_2.

The set of matches produced by an algorithm ALG is denoted as M_{ALG}. The total cost incurred by algorithm ALG for request sequence R is defined as $cost_{ALG}(R) = \sum_{M \in M_{ALG}} cost(M)$. The *cost ratio* of algorithm ALG on request sequence R is the ratio of $cost_{ALG}(R)$ to $cost_{OPT}(R)$, where OPT is an optimal offline algorithm.

For a deterministic algorithm ALG, it is said to be α-competitive if there exists a constant β such that $cost_{ALG}(R) \leq \alpha \cdot cost_{OPT}(R) + \beta$ for any request sequence R. For a randomized algorithm ALG, it is said to be α-competitive if there exists a constant β such that $\mathbf{E}[cost_{ALG}(R)] \leq \alpha \cdot cost_{OPT}(R) + \beta$ for any request sequence R.

An algorithm is called *smart* if the following property holds: Once a request r arrives at its source where there exists an open request r', the algorithm matches r and r' at once. Note that during the execution of a smart algorithm, there can never be more than one open request at a single source. According to [12], only focusing on smart online algorithms is not harmful at all.

2.2   Paired Sequences

For simplicity, we first restrict the given request sequence R to a *paired sequence*. That is, it can be partitioned into request pairs consisting of two requests with distinct sources arriving at the same time.

Our smart algorithm in Sect. 4.1 for paired sequences can be generalized in an online way to cope with general sequences with a competitive ratio guaranteed. This generalization is introduced in Sect. 4.2.

With the above assumption, we can characterize R by a sequence of request pairs. For a request pair p, define its arrival time $t(p)$ as the arrival time of its requests. A pair is said to be *open* if its requests have not been matched yet;

otherwise, it is said to be *matched*. Note that two requests in the same pair get matched at the same time in a smart algorithm. We say a smart algorithm ALG is *waiting* at time t if there is an open pair unmatched; otherwise, we say it is *idle*. The key point of the problem is to find the right moment to turn the algorithm from waiting to idle.

If ALG matches two requests in an open pair p, we call it an *internal match*, denoted $\texttt{Internal}(p)$. If ALG keeps waiting until a new pair p' arrives and matches these two pairs without any space cost, we call it an *external match*, denoted $\texttt{External}(p, p')$.

Assume that the paired sequence $R = (p_1, p_2, \ldots, p_{|R|})$ where $|R|$ is the number of pairs. We call consecutive pairs in the sequence a *segment*, e.g., $A = (p_4, p_5, \cdots, p_9)$ is a segment. For two consecutive segments A and B, $A \cup B$ denotes the concatenated segment of A and B. If a sequence can be partitioned into disjoint segments $A_1, A_2, \ldots, A_k$ and the time gap between any two consecutive segments A_i and A_{i+1}, which is the gap between the arrival time of the last request in A_i and the first request in A_{i+1}, is large enough for any $1 \leq i \leq k-1$, then $cost_{OPT}(\cup_{i=1}^{k} A_i)$ equals to $\sum_{i=1}^{k} cost_{OPT}(A_i)$. In this case, we call these segments *cost-independent*. For the sake of convenience, we define $l_i := t(p_{i+1}) - t(p_i)$ for $1 \leq i < |R|$. The parity of a segment is defined to be the parity of the number of pairs it contains.

3   Structure of the Optimal Solution

In this section, we propose an offline algorithm TOPI (the optimal phase identification algorithm) which finds the optimal solution when given R is a paired sequence. The algorithm first assigns a *state value* S to each pair and divides R into several special segments, which are called *phases*. This subroutine is called Phase Identification (**PI**). Then, TOPI uses two other subroutines Match_{even} and Match_{odd}, to deal with each segment independently, according to its parity. Although TOPI is offline, the subroutine **PI** is online and this gives us the intuition to partition the whole sequence into cost-independent phases in our randomized algorithm in Sect. 4. The state value in TOPI can also help us analyze the competitive ratio of the randomized algorithm.

We start from the Phase Identification, which obtains the state value S_i of the i-th pair and a set of phases F that forms a partition of the sequence R. Initially, F only has one phase containing the first pair, of which the state value is defined to be $S_1 = 1$. For every other pair p_i where $i > 1$, its state value is assigned based on the previous state value S_{i-1} and the time gap l_{i-1}. If $l_{i-1} \geq S_{i-1}$, **PI** starts a new phase containing only one pair p_i with its state value $S_i = 1$. Otherwise, **PI** adds pair p_i to the end of the last phase and assigns its state value to be $S_i = 1 - S_{i-1} + l_{i-1}$. The subroutine is as shown in Algorithm 1. For a better understanding, see Fig. 1 for an example.

Note that in the implementation, for each phase, we only record indices of pairs it contains. As each phase is a special segment, the indices of its pairs are

Algorithm 1. Phase Identification (PI)

1: $S_1 \leftarrow 1, f \leftarrow (1), F = \emptyset$.
2: **for** i from 2 to $|R_p|/2$ **do**
3:     **if** $l_{i-1} \geq S_{i-1}$ **then**
4:         $S_i \leftarrow 1$
5:         add f to F
6:         reset f as a phase which only contains i
7:     **else**
8:         $S_i \leftarrow 1 - S_{i-1} + l_{i-1}$
9:         attach i to the end of f
10:     **end if**
11: **end for**
12: add f to F

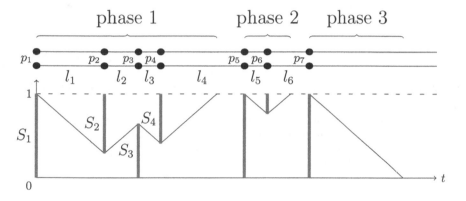

Fig. 1. The phase partition of a sequence with seven pairs. For each pair in the first phase, the length of a red line segment equals the state value of the corresponding pair. (Color figure online)

consecutive. For simplicity, we denote a phase f with k pairs as $(f_1, \cdots, f_k)$, where f_i indicates the index of the i-th pair and $f_{i+1} - f_i = 1$ for $1 \leq i < k$.

Now we introduce the other two subroutines Match_{even} and Match_{odd}. Match_{even} externally matches every two adjacent pairs, while Match_{odd} internally matches the first pair and externally matches every two adjacent pairs. Precisely, assume that we have a phase with k pairs $f = (f_1, \cdots, f_k)$. Match_{even} performs $\text{External}(p_{f_{2i-1}}, p_{f_{2i}})$ for $1 \leq i \leq \lfloor k/2 \rfloor$. Match_{odd} performs $\text{Internal}(p_{f_1})$ and $\text{External}(p_{f_{2i}}, p_{f_{2i+1}})$ for $1 \leq i \leq \lfloor (k-1)/2 \rfloor$. Note that Match_{even} leaves the last pair in an odd phase unmatched, while Match_{odd} leaves the last pair in an even phase unmatched. In our algorithm TOPI, we never apply Match_{even} to odd phases or Match_{odd} to even phases.

For every phase, two cost functions $odd(t)$ and $even(t)$ are defined as the accumulated cost till time t if we run Match_{odd} and Match_{even} for this phase. We denote $odd_f(t)$ and $even_f(t)$ as the cost functions of phase f. The *state function* $s(t)$ is defined as follows: Suppose d is the number of arrived pairs till time t. If d is odd, then $s(t)$ is defined as $\frac{odd(t)-even(t)+1}{2}$, otherwise it is defined as $\frac{even(t)-odd(t)+1}{2}$. See Fig. 2 for an example.

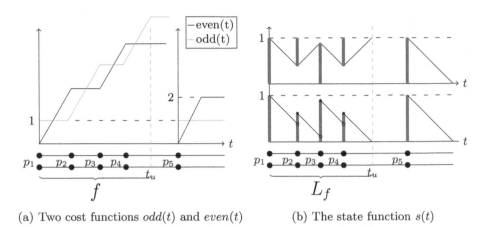

(a) Two cost functions $odd(t)$ and $even(t)$ (b) The state function $s(t)$

Fig. 2. Cost functions $odd(t)$ and $even(t)$ of a sequence with 5 pairs and the corresponding state value. Here the sequence is partitioned into two phases, For the first phase f, L_f is its length, and t_u is its ending time. In subfigure (a), the cost function holds when the corresponding subroutine is idle, and the cost increases at a speed of 2 when the subroutine is waiting. Subfigure (b) shows that the state function $s(t)$ can be expressed by a continuous fold line.

Recall that we utilize the state values to construct phases in Algorithm 1. The following lemma shows a connection between state values and state functions.

Lemma 1. *For a pair sequence $R = (p_1, p_2, \ldots, p_{|R|})$ where $|R|$ is the number of pairs, assume that $f = (f_1, \cdots, f_k)$ is a phase with k pairs. For every $f_1 \leq i \leq f_k$, $s(t(p_i)) = S_i$.*

Proof. When $i = f_1$, it is obvious that $s(t(p_i)) = S_i = 1$. Now we assume that $i > f_1$.

If $i - f_1$ is odd, then $odd(t(p_i)) = odd(t(p_{i-1}))$ and $even(t(p_i)) - even(t(p_{i-1})) = 2l_{i-1}$. Thus,

$$s(t(p_i)) = (even(t(p_i)) - odd(t(p_i)) + 1)/2$$
$$= (even(t(p_{i-1})) + 2l_{i-1} - odd(t(p_{i-1})) + 1)/2$$
$$= 1 - s(t(p_{i-1})) + l_{i-1}.$$

If $i - f_1$ is even, we can analogously show that $s(t(p_i)) = 1 - s(t(p_{i-1})) + l_{i-1}$. According to Algorithm 1, $S_i = 1 - S_{i-1} + l_{i-1}$, which implies that $s(t(p_i)) = S_i$ for any $f_1 \leq i \leq f_k$. This finishes the proof.

We can see that, for each phase $f = (f_1, \cdots, f_k)$, the state function decreases at unit rate during the time interval $[t(p_i), t(p_{i+1}))$ for any $f_1 \le i < f_k$. After the last pair p_{f_k} arrives, the state function also keeps decreasing at the unit rate until it becomes 0 and remains 0 until the next phase starts. Let t_u be the ending time of phase f, which is defined to be the earliest time such that $s(t_u) = 0$ while all pairs in the phase have arrived. Note that $t_u = t(p_{f_k}) + S_{f_k}$. Define $L_f = t_u - t(p_{f_1})$ as the length of the phase f.

Now, we introduce the offline algorithm TOPI as follows. Given a paired sequence, we first use Algorithm 1 to obtain phases. For each phase, if it is an even phase, we use Match_{even} to deal with it. Otherwise, we use Match_{odd} instead. The following theorem shows that the algorithm TOPI is optimal.

Theorem 1. *The algorithm* TOPI *is optimal for 2-MPMD problems when the given request sequence R is a paired sequence.*

Proof. Suppose the given paired sequence $R = (p_1, \cdots, p_n)$ where n is the number of pairs. For $1 \le i \le n$, define R_i as the segment containing the first i pairs, i.e., $R_i = (p_1, p_2, \cdots, p_i)$. Specifically, we denote R_0 as an empty sequence. For $0 \le i \le n$, we say R_i is verified if TOPI produces an optimal solution on it, i.e., $cost_{\mathsf{TOPI}}(R_i) = cost_{OPT}(R_i)$. We use inductive methods to verify R_i for all $0 \le i \le n$.

Base Case: When $i = 0$, $cost_{\mathsf{TOPI}}(R_0) = cost_{OPT}(R_0) = 0$. When $i = 1$, $cost_{\mathsf{TOPI}}(R_1) = cost_{OPT}(R_1) = 1$. Both R_0 and R_1 are trivially verified.

Inductive Step: For $i \ge 2$, assume that both R_{i-1} and R_{i-2} are already verified. Now, we verify R_i as follows.

Consider the last pair p_i in R_i. In the optimal solution for R_i, there are two possible cases for pair p_i:

Case 1: OPT performs $\mathtt{External}(p_{i-1}, p_i)$ and $cost_{OPT}(R_i) = cost_{OPT}(R_{i-2}) + 2l_{i-1}$.

Case 2: OPT performs $\mathtt{Internal}(p_i)$ and $cost_{OPT}(R_i) = cost_{OPT}(R_{i-1}) + 1$.

Now we come to TOPI. Let $f = (f_1, \cdots, f_k)$ be the last phase in F when Algorithm 1 receives pair p_i. There are three different options when a new pair arrives as follows:

Option A: Add pair p_i into the current phase, and TOPI performs $\mathtt{External}(p_{i-1}, p_i)$. Thus, $cost_{\mathsf{TOPI}}(R_i) = cost_{\mathsf{TOPI}}(R_{i-2}) + 2l_{i-1}$.

Option B: Start a new phase with pairs p_{i-1} and p_i, and TOPI performs $\mathtt{External}(p_{i-1}, p_i)$. Thus, $cost_{\mathsf{TOPI}}(R_i) = cost_{\mathsf{TOPI}}(R_{i-2}) + 2l_{i-1}$.

Option C: Start a new phase with the single pair p_i, and TOPI preforms $\mathtt{Internal}(p_i)$. Thus, $cost_{\mathsf{TOPI}}(R_i) = cost_{\mathsf{TOPI}}(R_{i-1}) + 1$.

Now, we prove that TOPI is optimal on the decision for pair p_i. Without loss of generality, assume that $i - f_1$ is odd.

Suppose Option A or Option B is selected. According to Algorithm 1, condition $l_{i-1} < s(t(p_{i-1}))$ holds, which implies that $even_f(t(p_{i-1})) + 2l_{i-1} <$

$odd_f(t(p_{i-1})) + 1$. As $i - f_1$ is odd, $even_f(t(p_{i-1})) = even_f(t(p_{i-2}))$, which leads to $even_f(t(p_{i-2})) + 2l_{i-1} < odd_f(t(p_{i-1})) + 1$. All operations performed in other phases (except phase f) by TOPI are the same. Hence, $cost_{\mathsf{TOPI}}(R_{i-2}) - even_f(t(p_{i-2})) = cost_{\mathsf{TOPI}}(R_{i-1}) - odd_f(t(p_{i-1}))$[1]. Plugging it back, we have $cost_{\mathsf{TOPI}}(R_{i-2}) + 2l_{i-1} < cost_{\mathsf{TOPI}}(R_{i-1}) + 1$, which means that Option C is worse than Option A or Option B.

Now assume that Option C is selected. According to Algorithm 1, condition $l_{i-1} \geq s(t(p_{i-1}))$ holds. Similarly, we have $cost_{\mathsf{TOPI}}(R_{i-2}) + 2l_{i-1} \geq cost_{\mathsf{TOPI}}(R_{i-1}) + 1$, which implies that Option C is no worse than Option A or Option B. Thus, segment R_i is verified.

Conclusion: Therefore, algorithm TOPI always selects optimal options and produces an optimal solution.

From above, we see that the phases obtained from Algorithm 1 are cost-independent. The optimal solution for the whole sequence is equivalent to the combination of all solutions for each phase. For an individual phase, the following lemma shows that its length equals the cost of its optimal solution, which is of great use when analyzing the competitive ratio of the randomized algorithm in Sect. 4.

Lemma 2. _For a phase with k pairs $f = (f_1, f_2, \cdots, f_k)$, its length equals to the cost of its optimal solution $OPT(f)$, i.e., $L_f = Cost_{OPT}(f)$._

Proof. Consider the cost functions odd_f and $even_f$ of phase f. Recall that $odd_f(t(p_{f_1})) = 1$ and $even_f(t(p_{f_1})) = 0$. During the time interval $[t(p_{f_1}), t(p_{f_k})]$, exactly one of these two functions increases at the speed of 2, and the other one holds. That is, $t(p_{f_k}) - t(p_{f_1}) = \frac{odd_f(t(p_{f_k})) + even_f(t(p_{f_k})) - 1}{2}$.

If k is odd, then $Cost_{OPT}(f)$ equals to $odd_f(t(p_{f_k}))$. Thus,

$$\begin{aligned}
L_f &= s(t(p_{f_k})) + (t(p_{f_k}) - t(p_{f_1})) \\
&= \frac{odd_f(t(p_{f_k})) - even_f(t(p_{f_k})) + 1}{2} + \frac{odd_f(t(p_{f_k})) + even_f(t(p_{f_k})) - 1}{2} \\
&= odd_f(t(p_{f_k})) = Cost_{OPT}(f).
\end{aligned}$$

Analogously, if k is even, the above proof is still correct after exchanging the $even_f$ and odd_f notations.

4 2-Competitive Randomized Algorithm

In this section, we introduce our 2-competitive randomized algorithm. In Sect. 4.1, we present the Randomized Delaying State-based Algorithm (RDSA) for paired sequences and prove its competitive ratio. In Sect. 4.2, we show a generalized version of RDSA for all request sequences with a competitive ratio guaranteed.

---

[1] If phase f starts with p_{i-1}, then we say $even_f(t(p_{i-2})) = even_f(t(p_{i-1})) = 0$. As $i - f_1$ is odd, the left part of the equality is exactly the cost of TOPI before phase f starts. A similar argument applies for the right part, which implies the correctness of the equality.

4.1 Randomized Algorithm

The randomized 2-competitive algorithm RDSA works as follows.

Algorithm 2. Randomized Delaying State-based Algorithm

1: $S_1 \leftarrow 1$
2: **while** Receive a pair p_i at time $t(p_i)$ when algorithm is idle **do**
3:     $S_i \leftarrow \min\{1 - S_{i-1} + l_{i-1}, 1\}$
4:     Sample $X_i \sim \mathbf{U}(0, S_i)$
5:     **if** There is an open pair p_{i+1} arriving in $(t(p_i), t(p_i) + X_i]$ **then**
6:         $S_{i+1} \leftarrow \min\{1 - S_i + l_i, 1\}$
7:         External(p_i, p_{i+1}) at time $t(p_{i+1})$
8:     **else**
9:         Internal(p_i) at time $t(p_i) + X$
10:     **end if**
11: **end while**

The state value S_i is assigned to each pair p_i in the same way as Algorithm 1. Here, we ensure that the state value never exceeds 1.[2] If RDSA is idle when pair p_i arrives, we uniformly sample X_i from 0 to S_i, then delay pair p_i for X_i seconds. Precisely, if the next pair p_{i+1} arrives no later than $t(p_i) + X_i$, RDSA performs External(p_i, p_{i+1}) at time $t(p_{i+1})$. Otherwise, RDSA performs Internal(p_i) at time $t(p_i) + X_i$.

In Algorithm 2, we cannot determine whether the algorithm is waiting or idle when a new pair arrives due to randomness. Nevertheless, in one single phase with k pairs $f = (f_1, \cdots, f_k)$ and $f_1 \leq i \leq f_k$, the probability that RDSA is idle when pair p_i arrives, which is denoted as P_i, equals to the state value $s(t(p_i))$.

Lemma 3. *For one single phase $f = (f_1, \cdots, f_k)$ and $f_1 \leq i \leq f_k$, $P_i = s(t(p_i))$.*

Proof. We use inductive methods on i to finish the proof. For the first pair p_{f_1}, $P_1 = s(t(p_{f_1})) = 1$. Now, we assume that $f_1 < i \leq f_k$ and $P_{i-1} = s(t(p_{i-1}))$.

Consider two consecutive pairs p_{i-1} and p_i. There are two possible cases that RDSA may be idle when p_i arrives as follows.

Case 1: RDSA performs External(p_{i-2}, p_{i-1}) and is waiting when p_{i-1} arrives. The probability that this case happens is $1 - P_{i-1}$.
Case 2: RDSA performs Internal(p_{i-1}) during interval $[t(p_{i-1}), t(p_i))$, where X_{i-1} sampled is less than l_{i-1}. The probability for this case is $P_{i-1} \cdot \frac{l_{i-1}}{S_{i-1}} = l_{i-1}$.

In summary, for $f_1 < i \leq f_k$, the probability that RDSA is idle when p_i arrives equals to $P_i = 1 - P_{i-1} + l_{i-1} = s(t(p_i))$, which finishes the proof.

---

[2] If l_{i-1} is too large, S_i will be assigned to 1, which implies that a new phase starts. This is similar to Algorithm 1.

Now, we prove that the randomized online algorithm RDSA is 2-competitive on every phase.

Theorem 2. *For any phase* $f = (f_1, f_2, \cdots, f_k)$, *the cost ratio for algorithm* RDSA *on* f *is* 2.

Proof. Let L_f be the length and t_u be the ending time of the phase. Denote $rdsa(t)$ as the expected total costs (space costs and time costs) that RDSA has incurred by time t. According to Lemma 2, the cost of an optimal solution on phase f equals to L_f. We will prove that the expected costs of RDSA on f is exactly $2L_f$. More precisely, we use inductive methods to verify that $rdsa(t(p_i)) = 2(t(p_i) - t(p_{f_1})) + S_i \cdot (1 - S_i)$ for $f_1 \le i \le f_k$.

Base Case: When $i = 1$, $rdsa(t_{f_1}) = 0$ and $S_1 = 1$, which leads to the equality directly.

Inductive Step: For $f_1 \le i < f_k$, assuming that the equality holds for i, we now prove that it is also true for $i + 1$.

Consider the time interval $(t(p_i), t(p_{i+1})]$. Compared with $rdsa(t(p_i))$, some extra costs are included in $rdsa(t(p_{i+1}))$ if and only if RDSA is idle when p_i arrives, of which the probability is P_i based on Lemma 3.

Now, let's discuss the delaying time X_i sampled by RDSA for pair p_i. If $X_i < l_i$, RDSA performs Internal(p_i) at time $t(p_i) + X_i$ with extra cost $2X_i + 1$. Otherwise, $X_i \ge l_i$, and RDSA performs External(p_i, p_{i+1}) with extra cost $2l_i$. According to Lemma 3, $P_i = s(t(p_i)) = S_i$. As X_i is sampled uniformly from $(0, S_i)$, the expected extra cost incurred during the time interval $(t(p_i), t(p_{i+1})]$ equals to

$$P_i \cdot \left(\int_0^{l_i} \frac{2X_i + 1}{S_i} d(X_i) + \int_{l_i}^{S_i} \frac{2l_i}{S_i} d(X_i) \right) = l_i \cdot (1 + 2S_i - l_i).$$

Hence,

$$\begin{aligned} rdsa(t(p_{i+1})) &= rdsa(t(p_i)) + l_i \cdot (1 + 2S_i - l_i) \\ &= 2(t(p_i) - t(p_{f_1})) + S_i \cdot (1 - S_i) + l_i \cdot (1 + 2S_i - l_i) \\ &= 2(t(p_{i+1}) - t(p_{f_1})) + S_{i+1} \cdot (1 - S_{i+1}). \end{aligned}$$

The last equality is correct as $l_i = t(p_{i+1}) - t(p_i)$ and $S_{i+1} = 1 - S_i + l_i$ when $f_1 \le i < f_k$.

Conclusion: From above, the equality holds for any $f_1 \le i \le f_k$.

Afterwards, we consider costs incurred during $(t(p_{f_k}), t_u]$, which can only contain the cost of Internal(p_{f_k}) at time $t(p_{f_k}) + X_{f_k}$. The expected cost is $P_{f_k} \cdot (2\mathbf{E}[X_{f_k}] + 1) = S_{f_k} \cdot (S_{f_k} + 1)$. Therefore, $rdsa(t_u) = rdsa(t(p_{f_k})) + S_{f_k} \cdot (S_{f_k} + 1) = 2L_f$, which completes the proof.

Recall that phases are cost-independent for the optimal algorithm. As the first pair in each phase has state value 1, all phases are also cost-independent for RDSA. Thus, the cost ratio for RDSA on a paired sequence R is

$$\frac{\mathbf{E}[Cost_{RDSA}(R)]}{Cost_{OPT}(R)} = \frac{\sum_f \mathbf{E}[Cost_{RDSA}(f)]}{\sum_f Cost_{OPT}(f)} = \frac{2 \sum_f Cost_{OPT}(f)}{\sum_f Cost_{OPT}(f)} = 2.$$

Corollary 1. *For any paired sequence R, the cost ratio for algorithm* RDSA *on R is 2.*

4.2   The Generalized Version of RDSA

In this section, we introduce the generalized version of RDSA, which works for all general sequences. Suppose the given request sequence R consists of n requests, $r_1, \ldots, r_n$, where $t(r_1) \leq \ldots \leq t(r_n)$. For any smart algorithm ALG, there are no more than two open requests at any moment. In this way, the time axis can be partitioned into *odd zones* and *even zones* according to the parity of the number of requests being open. More precisely, for $1 \leq i \leq n$, the interval $[t(r_i), t(r_{i+1}))$ is called an odd zone when i is odd and an even zone when i is even[3]. Note that every smart algorithm, including the optimal algorithm, has exactly one open request during each odd zone, which leads to inevitable waiting costs. In the generalized version, we still consider some requests in pairs and define an arrival time for each pair. We present the generalized version of RDSA as follows.

Initially, there are no requests, and the algorithm is idle.

When the algorithm is idle, we check the next two requests r_i and r_{i+1} where i is odd. The time interval $[t(r_i), t(r_{i+1}))$ is an odd zone. At time $t(r_{i+1})$, if $x(r_i) = x(r_{i+1})$, the algorithm matches these two requests and remains idle. Otherwise, the algorithm views these two requests as a pair with arrival time $t(r_{i+1})$ and turns to wait. A state value S is also assigned to this pair, similar to Algorithm 1. If it is the first pair, its state value is 1. Otherwise, its state value equals $\min\{1 - S' + l, 1\}$, where S' is the state value of the previous pair, and l the total lengths of even zones from the arrival time of the last pair to $t(r_{i+1})$. Similarly, we sample a delaying time $X \sim \mathbf{U}(0, S)$.

When the algorithm is waiting, let R_1 and R_2 be the two open requests in one pair with state value S where $x(R_1) = a, x(R_2) = b$. Similarly, we check the next two requests r_i, r_{i+1}. The interval $[t(r_i), t(r_{i+1}))$ is an odd zone. Without loss of generality, assume that $x(r_i) = a$. At the time t_i, the algorithm matches requests R_1 and r_i. If $x(r_{i+1}) = a$, requests R_2 and r_{i+1} are viewed as a pair. The state value and the arrival time assigned to the pair are equal to those assigned to the pair R_1 and R_2. Otherwise, when $x(r_{i+1}) = b$, the algorithm matches R_2 and r_{i+1} and turns idle. Meanwhile, requests r_i and r_{i+1} are viewed as a pair, of which the state value is $\min\{1 - S' + l, 1\}$ and the arrival time is $t(r_{i+1})$, which is similar as before. Note that the algorithm will not be waiting for too long. More precisely, once the total lengths of even zones from time $\min\{t(R_1), t(R_2)\}$ reaches X, requests R_1 and R_2 get matched immediately, and the algorithm turns idle.

The following theorem shows that the generalized version of RDSA also has a competitive ratio of 2.

Theorem 3. *The generalized version of algorithm* RDSA *is 2-competitive for online 2-MPMD problems.*

---

[3] Here, we define $t(r_{n+1}) = +\infty$.

Proof. Consider an odd zone $[t(r_i), t(r_{i+1}))$ where i is odd. Let $\Delta t = t(r_{i+1}) - t(r_i)$. For any smart algorithm, including the generalized version of RDSA, there exists exactly one open request during this interval. For all requests that arrive no earlier than $t(r_{i+1})$, we let them arrive Δt earlier.[4] In this way, the cost of the optimal solution decreases by Δt. Simultaneously, the cost of the solution produced by the generalized version of RDSA also decreases by Δt, which implies that its competitive ratio does not increase.

Now, the length of any odd zone is zero, which implies that the input sequence has been changed into a paired sequence. The generalized version of RDSA works the same way as the original version. Based on Corollary 1, the algorithm is 2-competitive.

5   Tight Lower Bound

In this section, we use Yao's principle [25] to derive the lower bound of 2-MPMD problems, which shows that our randomized algorithm is optimal. We construct a distribution of randomized paired sequences to make the expected cost ratio of any online deterministic algorithm at least 2.

5.1   Sequence Construction

Recall that both the optimal algorithm and RDSA are based on the state function, which can be described by a fold line, where each turning point corresponds to a pair in the sequence. This provides us with some intuition to construct the distribution of the randomized paired sequence $\tilde{R}$ as follows.

We place the first pair at time 0 and set its state value as 1. Suppose the last pair we have placed has arrival time t and state value S. We sample a non-negative $X \sim \text{Exp}(1)$[5]. If $X < S$, we place a pair at time $t + X$ with state value $1 - S + X$ and repeat the process. Otherwise, we finish the construction.

The whole paired sequence we obtain is actually one single phase. We can define its state function $s(t)$ in the same way as Sect. 3. Suppose there exists one pair with arrival time t and state value $S = s(t)$. Let $P_t(x)$ be the probability that no placement occurs during the interval $(t, t + x)$. Hence, when $x < s(t)$, $P_t(x) = \int_0^x e^{-y} dy = e^{-x}$, which is unrelated to t. For simplicity, we use $P(x)$ instead of $P_t(x)$ when $x \le s(t)$ and $P(x) = e^{-x}$.

5.2   Cost Ratio Analysis

Now, we are able to compute the expected costs incurred by OPT and any online deterministic algorithm ALG. Note that for a pair with state value S,

---

[4] We do not let them arrive earlier in the algorithm. This operation is only for better analyzing the competitive ratio.

[5] Recall Exp(1) is the exponential distribution, of which the probability density function is $f(x) = e^{-x}$ for $x \ge 0$.

the distribution of the randomized segment after the pair is only related to S. We first discuss the expected cost incurred by OPT, which equals the expected length of the phase.

Lemma 4. *The expected cost incurred by OPT on $\tilde{R}$ is at most 1.*

Proof. For $1 \leq i$ and $0 \leq s \leq 1$, assuming the i-th pair p_i is placed at time t with state value s, let $T(i, s)$ be the expected length of the phase after the i-th pair. Note that $T(1, 1)$ is exactly what we want. There are two possible cases for the $(i + 1)$-th pair as follows.

Case 1: The $(i + 1)$-th pair doesn't exist, which implies that the ending time of the sequence is $t + s$. The probability of this case is $P(s) = e^{-s}$, and the extra cost is s.

Case 2: The $(i + 1)$-th pair has arrival time $t + x$ where $0 \leq x < s$, of which the probability density is $P(x) = e^{-x}$ and the extra cost is $x + T(i + 1, 1 - s + x)$.

Hence, for $1 \leq i$ and $0 \leq s \leq 1$,

$$T(i, s) = s \cdot e^{-s} + \int_0^s (x + T(i + 1, 1 - s + x)) \cdot e^{-x} dx. \tag{1}$$

Let $L > 0$ be a sufficiently large constant and define $T'(i, s)$ as an auxiliary function such that the same equation holds for $T'(i, s)$ and $T'(i, s) = -s^2 + 2s$ when $i \geq L$. Then, for $i = L - 1, \cdots, 1$, we can use inductive methods as follows:

$$T'(i, s) = s \cdot e^{-s} + \int_0^s (x + T'(i + 1, 1 - s + x)) \cdot e^{-x} dx$$

$$= s \cdot e^{-s} + \int_0^s (x - (1 - s + x)^2 + (1 - s + x)) \cdot e^{-x} dx$$

$$= -s^2 + 2s.$$

The second equality above is correct due to the inductive assumptions.

Therefore, $T'(i, s) = -s^2 + 2s$ for all $i \geq 1$ and $0 \leq s \leq 1$. Specifically, $T'(1, 1) = 1$. Now, let's consider the gap between $T(1, 1)$ and $T'(1, 1)$. For $\ell \geq L$, the probability that the phase contains at least ℓ pairs is at most $(1 - P(1))^\ell = (1 - e^{-1})^\ell$. For a phase with ℓ pairs where the L-th pair has state value s, the actual cost incurred after the L-th pair is at most $\ell - L + 1^6$, whereas in T' this cost has been assumed to be $-s^2 + 2s \geq 0$. Using this, we can upper bound the difference between $T(1, 1)$ and $T'(1, 1)$ as follows:

$$T(1, 1) \leq T'(1, 1) + \sum_{\ell=L}^{+\infty} (\ell - L + 1) \cdot (1 - e^{-1})^\ell = 1 + e^2 \cdot (1 - e^{-1})^L,$$

which implies $T(1, 1) \leq 1$ as L can be arbitrarily large. This completes the proof.

---

[6] We can simply internally match all pairs, which will not incur fewer costs than the optimal algorithm does.

Now, we come to the expected cost incurred by any online deterministic algorithm.

Lemma 5. *For any online deterministic algorithm ALG, the expected cost incurred by ALG is at least 2.*

Proof. For $1 \leq i$ and $0 \leq s \leq 1$, assume that the i-th pair p_i is placed at time t with state value s. Let $ALG(i, s, 0)$ and $ALG(i, s, 1)$ be the expected costs incurred by ALG after the i-th pair where ALG is idle or waiting after receiving p_i. Note that $ALG(1, 1, 1)$ is exactly what we want.

Suppose ALG becomes idle when receiving pair p_i. If p_i is the last pair of the phase, then no extra cost is incurred. Otherwise, assume that the first pair after p_i is placed at time $t + x$, of which the probability density is e^{-x}. In this way, ALG turns waiting at time $t + x$, which indicates that

$$ALG(i, s, 0) = \int_0^s ALG(i+1, 1-s+x, 1) \cdot e^{-x} dx. \qquad (2)$$

Now, let's consider the case that ALG becomes waiting when receiving pair p_i. Suppose that ALG delays pair p_i for m seconds, where $m \leq s$. There are three possible cases as follows.

Case 1: The $(i+1)$-th pair doesn't exist, which implies that the ending time of the sequence is $t+s$. The probability of this case is $P(s) = e^{-s}$ and ALG performs `Internal`(p_i) at time $t + m$, of which the extra cost equals to $2m + 1$.

Case 2: The $(i+1)$-th pair p_{i+1} has arrival time $t + x$ and $x \leq m$, of which the probability density is e^{-x}. ALG performs `External`(p_i, p_{i+1}) at time $t + x$ and turns idle, of which the extra cost equals to $2x + ALG(i + 1, 1 - s + x, 0)$.

Case 3: The $(i+1)$-th pair p_{i+1} has arrival time $t + x$ and $x > m$, of which the probability density is e^{-x}. ALG performs `Internal`(p_i) at time $t + m$ and turns idle. At time $t + x$, it turns to wait again. In this case, the extra cost equals $2m + 1 + ALG(i + 1, 1 - s + x, 1)$.

Hence, for $1 \leq i$ and $0 \leq s \leq 1$,

$$ALG(i, s, 1) = (2m + 1) \cdot e^{-s} + \int_0^m (2x + ALG(i + 1, 1 - s + x, 0)) \cdot e^{-x} dx$$

$$+ \int_m^s (2m + 1 + ALG(i + 1, 1 - s + x, 1)) \cdot e^{-x} dx. \qquad (3)$$

Similar to Lemma 4, let $L > 0$ be a sufficiently large constant and define $ALG'(i, s, 0)$ and $ALG'(i, s, 1)$ as two auxiliary functions such that the same equations hold and $ALG'(i, s, j) = j - s^2 + 2s$ when $i \geq L$ and $j \in \{0, 1\}$. Then, for $i = L - 1, \cdots, 1$, we can use inductive methods as follows:

$$ALG'(i, s, 0) = \int_0^s ALG'(i+1, 1-s+x, 1) \cdot e^{-x} dx$$
$$= \int_0^s (1 - (1-s+x)^2 + 2(1-s+x)) \cdot e^{-x} dx$$
$$= -s^2 + 2s,$$
$$ALG'(i, s, 1) = (2m+1) \cdot e^{-s} + \int_0^m (2x + ALG'(i+1, 1-s+x, 0)) \cdot e^{-x} dx$$
$$+ \int_m^s (2m+1 + ALG'(i+1, 1-s+x, 1) \cdot e^{-x} dx$$
$$= (2m+1) \cdot e^{-s} + \int_0^m (2x - (1-s+x)^2 + 2(1-s+x)) \cdot e^{-x} dx$$
$$+ \int_m^s (2m+2 - (1-s+x)^2 + 2(1-s+x)) \cdot e^{-x} dx$$
$$= 1 - s^2 + 2s.$$

Therefore, $ALG'(i, s, j) = j - s^2 + 2s$ when $i \geq L$ and $j \in \{0,1\}$. Specifically, $ALG'(1,1,1) = 2$. Now, let's consider the gap between $ALG(1,1,1)$ and $ALG'(1,1,1)$. For $\ell \geq L$, the probability that the phase contains at least ℓ pairs is at most $(1 - P(1))^\ell = (1 - e^{-1})^\ell$. For a phase with ℓ pairs where the L-th pair has state value s, the actual cost incurred after the L-th pair is no less than 0, whereas in T' this cost has been assumed to be $1 - s^2 + 2s \geq 2$. Therefore, we can upper bound the difference between $ALG(1,1,1)$ and $ALG'(1,1,1)$ as follows:

$$ALG(1,1,1) \geq ALG'(1,1,1) - \sum_{\ell=L}^{+\infty} 2(1 - e^{-1})^\ell = 2 - 2e(1 - e^{-1})^L,$$

which implies $ALG(1,1,1) \geq 2$ as L can be arbitrarily large. This completes the proof.

Finally, we are able to come to the lower bound of 2.

Theorem 4. *The competitive ratio of any randomized online algorithm on 2-MPMD is at least 2.*

Proof. According to Lemma 4 and Lemma 5, $ALG(1,1,1) \geq 2$ and $T(1,1) \leq 1$, which implies that for the randomized data $\tilde{R}$, the expected cost ratio of any online deterministic algorithm is at least 2. Based on Yao's principle [25], the competitive ratio of any randomized online algorithm on 2-MPMD is at least 2, which finishes the proof.

Acknowledgments. K. He and S. Li are supported by the National Natural Science Foundation of China (Grant No. U22B2017). E. Sun is supported by the National Natural Science Foundation of China (Grant No. 6212290003). Y. Wang is supported by the National Key Research and Development Program of China (Grant No. SQ2022YFB4300064).

References

1. Aggarwal, G., Goel, G., Karande, C., Mehta, A.: Online vertex-weighted bipartite matching and single-bid budgeted allocations. In: Proceedings of the Twenty-Second Annual ACM-SIAM Symposium on Discrete Algorithms, pp. 1253–1264. SIAM (2011)
2. Ashlagi, I., et al.: Min-cost bipartite perfect matching with delays. In: 20th International Workshop on Approximation Algorithms for Combinatorial Optimization Problems (APPROX), Berkeley, California, USA, August 2017
3. Azar, Y., Chiplunkar, A., Kaplan, H.: Polylogarithmic bounds on the competitiveness of min-cost perfect matching with delays. In: Proceedings of the Twenty-Eighth Annual ACM-SIAM Symposium on Discrete Algorithms, pp. 1051–1061. SIAM (2017)
4. Azar, Y., Jacob Fanani, A.: Deterministic min-cost matching with delays. In: Epstein, L., Erlebach, T. (eds.) WAOA 2018. LNCS, vol. 11312, pp. 21–35. Springer, Cham (2018). https://doi.org/10.1007/978-3-030-04693-4_2
5. Azar, Y., Ren, R., Vainstein, D.: The min-cost matching with concave delays problem. In: Marx, D. (ed.) Proceedings of the 2021 ACM-SIAM Symposium on Discrete Algorithms, SODA 2021, Virtual Conference, 10–13 January 2021, pp. 301–320. SIAM (2021)
6. Bansal, N., Buchbinder, N., Gupta, A., Naor, J.S.: A randomized o (log 2 k)-competitive algorithm for metric bipartite matching. Algorithmica **68**(2), 390–403 (2014)
7. Birnbaum, B., Mathieu, C.: On-line bipartite matching made simple. ACM SIGACT News **39**(1), 80–87 (2008)
8. Devanur, N.R., Jain, K., Kleinberg, R.D.: Randomized primal-dual analysis of ranking for online bipartite matching. In: Proceedings of the Twenty-Fourth Annual ACM-SIAM Symposium on Discrete Algorithms, pp. 101–107. SIAM (2013)
9. Edmonds, J.: Maximum matching and a polyhedron with 0, 1-vertices. J. Res. Natl. Bureau Stand. B **69**, 125–130 (1965)
10. Edmonds, J.: Paths, trees, and flowers. Can. J. Math. **17**, 449–467 (1965)
11. Emek, Y., Kutten, S., Wattenhofer, R.: Online matching: haste makes waste! In: 48th Annual Symposium on Theory of Computing (STOC), June 2016
12. Emek, Y., Shapiro, Y., Wang, Y.: Minimum cost perfect matching with delays for two sources. In: 10th International Conference on Algorithms and Complexity (CIAC), Athens, Greece, May 2017
13. Goel, G., Mehta, A.: Online budgeted matching in random input models with applications to adwords. In: SODA, vol. 8, pp. 982–991. Citeseer (2008)
14. Kalyanasundaram, B., Pruhs, K.: Online weighted matching. J. Algorithms **14**(3), 478–488 (1993)
15. Karlin, A.R., Manasse, M.S., McGeoch, L.A., Owicki, S.: Competitive randomized algorithms for nonuniform problems. Algorithmica **11**(6), 542–571 (1994)
16. Karp, R.M., Vazirani, U.V., Vazirani, V.V.: An optimal algorithm for on-line bipartite matching. In: Proceedings of the Twenty-Second Annual ACM Symposium on Theory of Computing, pp. 352–358 (1990)
17. Khuller, S., Mitchell, S.G., Vazirani, V.V.: On-line algorithms for weighted bipartite matching and stable marriages. Theor. Comput. Sci. **127**(2), 255–267 (1994)
18. Liu, X., Pan, Z., Wang, Y., Wattenhofer, R.: Impatient online matching. In: 29th International Symposium on Algorithms and Computation (ISAAC 2018), vol. 123, pp. 62–1. Schloss Dagstuhl-Leibniz-Zentrum für Informatik (2018)

19. Mari, M., Pawlowski, M., Ren, R., Sankowski, P.: Online matching with delays and stochastic arrival times. CoRR abs/2210.07018 (2022)
20. Mehta, A.: Online matching and ad allocation. Found. Trends Theor. Comput. Sci. **8**(4), 265–368 (2013)
21. Mehta, A., Saberi, A., Vazirani, U., Vazirani, V.: Adwords and generalized online matching. J. ACM (JACM) **54**(5), 22-es (2007)
22. Meyerson, A., Nanavati, A., Poplawski, L.: Randomized online algorithms for minimum metric bipartite matching. In: Proceedings of the Seventeenth Annual ACM-SIAM Symposium on Discrete Algorithm, pp. 954–959 (2006)
23. Miyazaki, S.: On the advice complexity of online bipartite matching and online stable marriage. Inf. Process. Lett. **114**(12), 714–717 (2014)
24. Naor, J., Wajc, D.: Near-optimum online ad allocation for targeted advertising. ACM Trans. Econ. Comput. (TEAC) **6**(3–4), 1–20 (2018)
25. Yao, A.C.C.: Probabilistic computations: toward a unified measure of complexity. In: 18th Annual Symposium on Foundations of Computer Science (sfcs 1977), pp. 222–227. IEEE (1977)

Polyhedral Clinching Auctions
for Indivisible Goods

Hiroshi Hirai[1] and Ryosuke Sato[2]

[1] Nagoya University, Nagoya, Japan
hirai.hiroshi@math.nagoya-u.ac.jp
[2] University of Tokyo, Tokyo, Japan
ryosuke-sato-517@g.ecc.u-tokyo.ac.jp

Abstract. In this study, we propose the polyhedral clinching auction for indivisible goods, which has so far been studied for divisible goods. As in the divisible setting by Goel et al. (2015), our mechanism enjoys incentive compatibility, individual rationality, and Pareto optimality, and works with polymatroidal environments. A notable feature for the indivisible setting is that the whole procedure can be conducted in time polynomial of the number of buyers and goods. Moreover, we show additional efficiency guarantees, recently established by Sato for the divisible setting: The liquid welfare (LW) of our mechanism achieves more than 1/2 of the optimal LW, and that the social welfare is more than the optimal LW.

Keywords: Auctions · Budget Constraints · Polymatroids · Liquid Welfare

1 Introduction

The theoretical foundation for budget-constrained auctions is an unavoidable step toward further social implementation of auction theory. A representative example of such auctions is ad auctions (e.g., Edelman et al. [9]), where advertisers naturally have budgets for their advertising costs. However, it is well known [6,7] that designing auctions with budget constraints is theoretically difficult: Any budget-feasible mechanism cannot achieve the desirable goal of satisfying all of incentive compatibility (IC), individual rationality (IR), and constant approximation to the optimal social welfare (SW).

When budgets are public, Dobzinski et al. [6] proposed a budget-feasible mechanism that builds on the clinching framework of Ausubel [1]. Their mechanism satisfies IC, IR, and Pareto Optimality (PO), a weaker notion of efficiency than SW. They also showed that their mechanism is the only budget-feasible mechanism satisfying IC, IR, and PO. These results have inspired further research [4,8,10,12–14,16] for extending their mechanism to various settings.

Polyhedral clinching auction by Goel et al. [13] is the most outstanding of these, and can even be applied to complex environments expressed by polymatroids. Their mechanism is a clever fusion of auction theory and polymatroid theory. This has brought further extensions, such as concave budget constraints [12]

J. Garg et al. (Eds.): WINE 2023, LNCS 14413, pp. 366–383, 2024.
https://doi.org/10.1007/978-3-031-48974-7_21

and two-sided markets [16, 22]. Particularly, a recent result by Sato [22] established a new type of efficiency guarantees. Thus, the polyhedral clinching auction is a standard framework for the theory of budget-constrained auctions.

These results of the polyhedral clinching auctions are all restricted to auctions of divisible goods, though many auctions deal with indivisible goods. In this study, we address the polyhedral clinching auction for indivisible goods to enlarge its power of applicability.

Our Contributions. We propose the polyhedral clinching auction for indivisible goods, based on the framework of the one for divisible goods in Goel et al. [13]. Our mechanism exhibits the desirable properties expected from theirs. That is, it satisfies IC, IR, and PO and works with polymatroidal environments. This means that it is applicable to a wide range of auctions, such as multi-unit auctions in Dobzinski et al. [6], matching markets in Fiat et al. [10], ad slot auctions in Colini-Baldeschi et al. [4], and video-on-demand in Bikhchandani et al. [3]. In addition, a promising future research is two-sided extensions of our results, as already proceeded for the divisible settings in Hirai and Sato [16] and Sato [22]. Particularly, such an extension includes the reservation exchange markets in Goel et al. [11]—a setting of two-sided markets for display advertising.

As in the divisible setting in [13], each iteration of our mechanism can be done in polynomial time. A notable feature specific to the indivisible setting is iteration complexity. The total number of the iterations is also polynomially bounded in the number of buyers and the goods. Thus, the whole procedure can be implemented in polynomial time.

In addition to the above PO, we establish two types of efficiency guarantees. The first one is that our mechanism achieves *liquid welfare* (LW) more than $1/2$ of the optimal LW. This is the first LW guarantee for clinching auctions with indivisible goods. Here LW [7, 24] is a payment-based efficiency measure for budget-constrained auctions, and is defined as the sum of the total admissibility-to-pay, which is the minimum of valuation of the allocated goods (willingness-to-pay) and budget (ability-to-pay). Our LW guarantee is understood as an indivisible and one-sided version of the one recently established by Sato [22] for the divisible setting. The notable point is that the LW guarantee holds for such general auctions, even in indivisible setting, while other existing work [7] on LW guarantees for clinching auctions is only limited to simple settings.

The second one is that our mechanism achieves SW more than the optimal LW. This type of efficiency guarantee, which compares the SW of mechanisms with the optimal LW, was initiated by Syrgkanis and Tardos [24] in the Bayesian setting, and was recently obtained by Sato [22] for clinching auctions with divisible goods in the prior-free setting. In budget-constrained auctions, modifying the valuations to make the market non-budgeted is often considered; see, e.g., Lehmann et al. [19]. If each buyer's valuation is modified to the budget-additive valuation, then LW is interpreted as the SW. Then, the optimal LW is used as the target value of SW and can be thought as a reasonable benchmark. Thus, this guarantee provides another evidence for high efficiency of our mechanism.

Our Techniques. *Tight sets lemma* [12,13] characterizes the dropping of buyers and the final allocation, and is a powerful tool for efficiency guarantees of polyhedral clinching auctions. For showing efficiency guarantees (PO, LW, SW) mentioned above, we establish a new and the first tight sets lemma for indivisible setting. Although our (hard-budget) setting is a natural indivisible version of the one in Goel et al. [13], the indivisibility causes complications in various places and prevents straightforward generalization in both its formulation and proof. We utilize the notions of *dropping prices* and *unsaturation* by Goel et al. [12] invented for a more complex divisible setting (concave-budget setting), and formalize and prove our indivisible tight sets lemma.

Even with our new tight sets lemma, the indivisibility still prevents a straightforward adaptation of previous techniques showing efficiency guarantees, especially the LW guarantee. The proof of the above 2-approximation LW guarantee is based on the idea of Sato [22] for divisible setting, and is obtained by establishing the inequality

$$\mathrm{LW}^{\mathrm{M}} \geq p^{\mathrm{f}}(N) \geq \mathrm{LW}^{\mathrm{OPT}} - \mathrm{LW}^{\mathrm{M}},$$

where LW^{M} and $p^{\mathrm{f}}(N)$ are the LW value and the total payments, respectively, in our mechanism, and $\mathrm{LW}^{\mathrm{OPT}}$ is the optimal LW value. In the divisible setting of [22], the second inequality is obtained by using the LW optimal allocation to provide a lower bound on future payments. However, this approach does not fit in our setting due to indivisibility. Instead, we introduce a new technique of lower bounding the future payments via virtual buyers and the associated virtual optimal LW allocation. This new technique is interesting in its own right, and expected to be applied to LW guarantees of other auctions.

Other Related Works. Auctions of indivisible goods are ubiquitous in the real-world. For unbudgeted settings, its theory already has a wealth of knowledge; see, e.g., Krishna [17] and Nisan et al. [21]. Auctions with (poly)matroid constraints were initiated by Bikhchandani et al. [3]. They considered buyers who have concave valuations and no budget limits. Our framework captures a budgeted extension of their framework if all buyers have additive valuations within their budgets. For budgeted auctions, Dobzinski et al. [6] proposed the adaptive clinching auction and showed that it has IC, IR, and PO. Later, Fiat et al. [10] and Colini-Baldeschi et al. [4] extended their mechanism to a market represented by a bipartite graph. Our framework is also viewed as a generalization of theirs to polymatroidal settings.

LW was introduced independently and simultaneously by Dobzinski and Leme [7] and Syrgkanis and Tardos [24]. The existing LW guarantees for auctions (with public budgets) are as follows: For clinching auctions, Dobzinski and Leme [7] showed that the adaptive clinching auction in Dobzinski et al. [6] achieves 2-approximation to the optimal LW. Recently, Sato [22] showed that the polyhedral clinching auction in Hirai and Sato [16] achieves 2-approximation to the optimal LW even under polymatroidal constraints. Our results are viewed as an indivisible and one-sided version of his results. For unit price auctions,

Dobzinski and Leme [7] also showed that their unit price auction achieves 2-approximation to the optimal LW. Later, Lu and Xiao [20] proposed another unit price auction and improved the guarantee to $(1+\sqrt{5})/2$. It is an interesting research direction to incorporate polymatroidal constraints with their mechanisms, for which our results may help.

Organization of This Paper. The rest of the paper is organized as follows. In Sect. 2, we introduce our model. In Sect. 3, we propose our mechanism and provide some basic properties. In Sect. 4, we analyze the structure of our mechanism and obtain the tight sets lemma. In Sect. 5, we provide the efficiency guarantees for our mechanism with respect to PO, LW, and SW. For lack of space, all omitted proofs are given in the full paper [15].

Notation. Let $\mathbb{R}_+$ (resp. $\mathbb{Z}_+$) denote the set of nonnegative real numbers (resp. integers), and let $\mathbb{R}_{++}$ (resp. $\mathbb{Z}_{++}$) denote the set of positive real numbers (resp. integers). For a set N, let $\mathbb{R}_+^N$ (resp. $\mathbb{Z}_+^N$) denote the set of all functions from N to $\mathbb{R}_+$ (resp. $\mathbb{Z}_+$). For $x \in \mathbb{R}_+^N$, we often denote $x(i)$ by x_i, and write it as $x = (x_i)_{i \in N}$. For $S \subseteq N$, let $x(S)$ denote the sum of $x(i)$ over $i \in S$, i.e., $x(S) := \sum_{i \in S} x(i)$. In addition, we often denote a singleton $\{i\}$ by i.

2 Our Model

Consider a market with multiple buyers and one seller who plays the role of the auctioneer. The seller auctions multiple units of a single indivisible good. Let $N := \{1, 2, \ldots, n\}$ be the set of buyers. Each buyer $i \in N$ has three real numbers v_i, v_i', B_i, where $v_i, v_i' \in \mathbb{R}_+$ are a valuation and a bid of buyer i, respectively, for a unit of the good, and $B_i \in \mathbb{R}_{++}$ is a budget of buyer i, i.e., the maximum total payment that i can pay in the auction. The valuation of each buyer is private information unknown to other buyers and the seller, and we assume that the budget is public information available to the seller. The seller determines the allocation based on a predetermined mechanism.

The allocation $\mathcal{A} := (x, p)$ is a pair of $x \in \mathbb{Z}_+^N$ and $p \in \mathbb{R}^N$, where x_i is the number of indivisible goods allocated to buyer i, and p_i is the payment of buyer i for their goods. Then, the budget constraints are described as $p_i \leq B_i$ $(i \in N)$. We are given an integer-valued monotone submodular function $f : 2^N \to \mathbb{Z}_+$ that represents the feasible allocation of goods. Note that an integer-valued function $f : 2^N \to \mathbb{Z}_+$ is monotone submodular if it satisfies (i) $f(\emptyset) = 0$, (ii) $f(S) \leq f(T)$ for each $S \subseteq T$, and (iii) $f(S \cup e) - f(S) \geq f(T \cup e) - f(T)$ for each $S \subseteq T$ and $e \in N \setminus T$. For any set of buyers $S \subseteq N$, the buyers in S can transact at most $f(S)$ amounts of goods through the auction. Note that $f(N)$ means the total goods sold in the auction. This condition is equivalent to $x \in P(f)$ using the polymatroid $P(f) := \{x \in \mathbb{R}_+^N \mid x(S) \leq f(S) \ (S \subseteq N)\}$. We often denote $P(f)$ by P. We also assume $f(N) = f(N \setminus i)$ for each $i \in N$, which implies that competition exists among buyers for each good at the beginning.

The utilities of the buyers are defined by

$$u_i(\mathcal{A}) := \begin{cases} v_i x_i - p_i & \text{if } p_i \leq B_i, \\ -\infty & \text{otherwise.} \end{cases}$$

Thus, the utilities of the buyers are quasi-linear if the budget constraints are satisfied, and otherwise, the utilities go to $-\infty$. The utility of the seller is defined as the revenue of the seller, i.e., $u_s(\mathcal{A}) := \sum_{i \in N} p_i$. The mechanism is a map $\mathcal{M} : \mathcal{I} \to \mathcal{A}$ from information $\mathcal{I}$ to allocation $\mathcal{A}$. Note that $\mathcal{I}$ includes all information that the seller can access, and thus $\mathcal{I} = (N, \{v'_i\}_{i \in N}, \{B_i\}_{i \in N}, f)$. We call a mechanism $\mathcal{M}$ budget feasible if, for any $\mathcal{I}$, mechanism $\mathcal{M}$ outputs an allocation that satisfies the budget constraints.

We consider to design an efficient budget feasible mechanism $\mathcal{M}$ that satisfies incentive compatibility (IC) and individual rationality (IR). A mechanism satisfies IC if for any $(\mathcal{I}, \{v_i\}_{i \in N})$, it holds $u_i(\mathcal{M}(\mathcal{I}_i)) \geq u_i(\mathcal{M}(\mathcal{I}))$ for each $i \in N$, where $\mathcal{I}_i$ denotes the information obtained from $\mathcal{I}$ by replacing v'_i with v_i. Intuitively, IC guarantees that the best strategy for each buyer is to report their true valuation. When the mechanism satisfies IC, it satisfies IR if, for any $(\mathcal{I}, \{v_i\}_{i \in N})$, it holds $u_i(\mathcal{M}(\mathcal{I}_i)) \geq 0$ for each $i \in N$. Intuitively, IR guarantees that each buyer obtains nonnegative utility when the buyer reports the true valuation.

The efficiency of the mechanism can be evaluated by the followings: Social welfare(SW) is defined as the sum of the valuations of the allocated goods for all buyers, and it can be interpreted as the sum of the utilities of all participants. In other words, $\text{SW}(\mathcal{A}) := \sum_{i \in N} v_i x_i = \sum_{i \in N} u_i(\mathcal{A}) + u_s(\mathcal{A})$. This is the standard efficiency measure used for the auctions. As mentioned, however, it is known (e.g., Dobzinski and Leme [7]) that for any $\alpha < n$, there is no budget feasible mechanism that achieves α-approximation to the optimal SW with IC and IR.

The alternative measure for budget-constrained auctions is LW, which is defined by $\text{LW}(\mathcal{A}) := \sum_{i \in N} \min(v_i x_i, B_i)$ for allocation $\mathcal{A}$. LW represents the sum of the possible payments that buyers can pay for their allocated goods. Another type of efficiency guarantee suitable for budget constraints is Pareto optimality (PO). A mechanism satisfies PO if for any $(\mathcal{I}, \{v_i\}_{i \in N})$, there is no other allocation $\mathcal{A}'$ with (i) $u_i(\mathcal{A}') \geq u_i(\mathcal{M}(\mathcal{I}))$ for each $i \in N$, (ii) $u_s(\mathcal{A}') \geq u_s(\mathcal{M}(\mathcal{I}))$, and (iii) at least one inequality holds without equality.

3   Polyhedral Clinching Auctions for Indivisible Goods

In this section, we describe our mechanism. Our mechanism incorporates the polyhedral approach of Goel et al. [13] to the (budgeted) clinching auctions in previous indivisible settings (e.g., [4,6,10]). A full description of our mechanism is presented in Algorithms 1 and 2.

Now we outline the mechanism. The price clock $c \in \mathbb{R}_+$ represents a transaction price for one unit of the good. Our mechanism is an ascending auction, where c gradually increases. For the current price c, the demand vector $d := (d_i)_{i \in N} \in \mathbb{Z}_+^N$ represents the maximum possible amounts of transaction for

---

Algorithm 1. Polyhedral Clinching Auction for Indivisible Goods

1: $x_i = 0$, $p_i := 0$, $d_i := f(i) + 1$ $(i \in N)$ and $c := 0$.
2: **while** Active buyers exist **do**
3:     Increase c until there appears an active buyer j such that $v'_j = c$ or $d_j = \frac{B_j - p_j}{c}$.
4:     **while** $\exists$ active buyer j with $v'_j = c$ **do**
5:         Pick such a buyer j and let $d_j := 0$
6:         Clinching(f, x, p, d, c)
7:     **end while**
8:     **while** $\exists$ active buyer j with $d_i = \frac{B_i - p_i}{c}$ **do**
9:         Pick such a buyer j and let $d_j := d_j - 1$
10:     Clinching(f, x, p, d, c)
11:     **end while**
12: **end while**
13: Output $(x^f, p^f) := (x, p)$.

Algorithm 2. Clinching (f, x, p, c, d)

1: **for** $i = 1, 2, \ldots, n$ **do**
2:     Clinch a maximal increase δ_i satisfying $P^i_{x,d}(\delta_i) = P^i_{x,d}(0)$; see equations (1) and (2)
3:     $x_i := x_i + \delta_i$, $p_i := p_i + c\delta_i$, $d_i := d_i - \delta_i$
4: **end for**

---

buyers. We call buyer i *active* if $d_i > 0$, and *dropping* if d_i just reaches zero. In our mechanism, buyers are dropping by either of (i) demand update in line 5 or 9, or (ii) clinching the goods in Algorithm 2.

Initially, the allocation (x, p) is all zero, the price clock c is set to zero, and the demand d_i is set to $f(i) + 1$ for each buyer $i \in N$. Then, the following procedure is repeated as long as there are active buyers: At the beginning of an iteration, the price clock c is updated to $\min_{i \in N; d_i > 0} \min\{v'_i, (B_i - p_i)/d_i\}$ in line 3. When the price clock c is updated, there exists a set of active buyers i such that $c = v'_i$ or $d_i = (B_i - p_i)/c$. Then, the demands of such buyers are updated: The demand d_i of a buyer i with $c = v'_i$ decreases to zero in line 5, and that of a buyer i with $c < v'_i$ and $d_i = (B_i - p_i)/c$ decreases by one in line 9. After each case of the demand update, if the clinching condition (described below) is satisfied, the buyers clinch some amount of goods. Then, the allocation (x, p) and the demands d are updated. After processing both cases, if there exists an active buyer, the next iteration is performed. Otherwise, Algorithm 1 terminates and outputs the final allocation (x^f, p^f).

The clinching steps in lines 6 and 10 are described in Algorithm 2. Let x and d be the allocation of goods and the demand vector, respectively, just before the execution of Algorithm 2 in an iteration. We consider two polytopes that represent the feasible transactions of buyers and describe the clinching condition. For a polymatroid P, and vectors $x \in P$ and $d \in \mathbb{R}^N_+$, we define the *remnant supply polytope* $P_{x,d}$ by $P_{x,d} := \{y \in \mathbb{R}^N_+ \mid x + y \in P, \ y_i \le d_i \ (i \in N)\}$, which indicates the feasible transaction of buyers from the current iteration. In addition, when buyer i clinches $w_i \in \mathbb{Z}_+$ units,

$$P^i_{x,d}(w_i) := \{u|_{N\backslash i} \mid u \in P_{x,d} \text{ and } u_i = w_i\} \tag{1}$$

represents a feasible transaction of buyers $N\backslash i$. The clinching amount δ_i of buyer i is then described by the following:

$$\delta_i = \sup\{w_i \geq 0 \mid P^i_{x,d}(w_i) = P^i_{x,d}(0)\}. \tag{2}$$

Thus, each buyer i clinches the maximal possible amount δ_i, not affecting the feasible transactions of other buyers $N\backslash i$. This intuition of clinching is consistent with the ones in previous works [1,2,4,6,8,10,12–14] on clinching auctions.

The polytopes $P_{x,d}$ and $P^i_{x,d}(w_i)$ are known to be polymatroids. Let $f_{x,d}$ denote the monotone submodular function for $P_{x,d}$. Sato [22] pointed out that $f_{x,d}$ can be described by a simple formula in his divisible setting. We show that this result also holds in our indivisible setting.

Theorem 1. *Let x and d be the allocation of goods and the demand vector, respectively, in Algorithm 1. Then, it holds $f_{x,d}(S) = \min\limits_{S' \subseteq S}\{f(S') - x(S') + d(S\backslash S')\}$ for any $S \subseteq N$.*

Theorem 1 is useful for analyzing our mechanism. For instance, δ can be computed as follows:

Proposition 1. *In the execution of Algorithm 2, it holds $\delta_i = f_{x,d}(N) - f_{x,d}(N\backslash i) \leq d_i$ for each $i \in N$, where x and d are the allocation of goods and the demand vector, respectively, just before the execution of Algorithm 2.*

Proposition 1 implies that, provided the value oracle of f, the value δ_i can be computed in polynomial time by a submodular minimization algorithm, such as in Lee et al. [18]. In addition, Proposition 1 implies that the amount of goods allocated to each buyer in Algorithm 2 is independent of the order of the buyers.

We investigate the properties of our mechanism in the following. To this end, we fix an input $(\mathcal{I}, v)$. We first consider the properties specific to our indivisible setting: There are two major differences between our mechanism and that in Goel et al. [13]. The first is on buyer's demand: The demands must be an integer vector because our model deals with indivisible goods. Thus, the demand for fewer than one unit is rounded down, which makes the function $f_{x,d}$ integer-valued. Therefore, δ is an integer vector in each iteration.

Proposition 2. *In Algorithm 2, it holds $\delta_i \in \mathbb{Z}_+$ for each $i \in N$.*

The second is on price update: Our mechanism sets a common price c for all buyers and does not use a fixed step size for price increases. This is based on the idea of Bikhchandani et al. [3] and Fiat et al. [10], and it plays an essential role in the iteration bounds; therefore, in the computational complexity. In our mechanism, the total sum of initial demands is $\sum_{i \in N} f(i) + n$, and in each iteration, it is guaranteed that the total sum of the demands is decreased by at least one. Therefore, the following lemma holds:

Lemma 1. *Our mechanism terminates after at most $\sum_{i \in N} f(i) + n$ iterations.*

Note that it holds $\sum_{i \in N} f(i) + n \leq n(f(N) + 1)$ based on the monotonicity of f. Further, as stated above, each iteration can be computed using the submodular minimization algorithm [18], which has runtime polynomial in the number of n, provided the value oracle of f is given. Therefore, our mechanism can be computed in polynomial time.[1]

Moreover, our mechanism inherits several desirable properties from the polyhedral clinching auction by Goel et al. [13]:

Proposition 3. *At the end of the auction, it holds $x^{\mathrm{f}}(N) = f(N)$.*

Theorem 2. *Our mechanism is budget feasible and satisfies IC and IR.*

From Theorem 2, our mechanism satisfies IC, and thus, we assume in the rest of the paper that all buyers bid truthfully, that is, $v'_i = v_i$ for every $i \in N$.

4 Structural Properties of the Mechanism

In this section, we establish the structures of the tight sets lemma (*tight sets lemma*) for our mechanism, which is necessary for our efficiency guarantees in Sect. 5. We call a set $T \subseteq N$ *tight* if $x^{\mathrm{f}}(T) = f(T)$.

4.1 Tight Sets Lemma

We provide the characterization of the dropping of buyers. In our mechanism, the dropping prices are needed to describe the state of such buyers.

Definition 1 (Goel et al. [12]). *In an execution of Algorithm 1, the dropping price ϕ_i of buyer i is defined as the first price for which i had zero demand.*

Obviously, it holds that $v_i \geq \phi_i$ for each $i \in N$. The goal of this section is:

Theorem 3 (Tight sets lemma). *Let $i_1, i_2, \ldots, i_t$ be the buyers dropping by demand update in line 5 or 9, where they are sorted in the reverse order of their dropping, that is, $\phi_{i_1} \geq \cdots \geq \phi_{i_t}$. For each $k = 1, 2, \ldots, t$, let X_k denote the set of active buyers just before the drop of i_k. Then we obtain:*

(i) *It holds $i_k \in X_k \backslash X_{k-1}$. Moreover, it holds $\phi_{i_k} = v_{i_k}$ or $B_{i_k} - p^{\mathrm{f}}_{i_k} = \phi_{i_k}$.*

(ii) *For buyer $i \in X_k \backslash (X_{k-1} \cup i_k)$, it holds that $\phi_i = \phi_{i_k}$ and $B_i - p^{\mathrm{f}}_i \leq \phi_i$. Moreover, if there exists a buyer $\ell \in X_k \backslash (X_{k-1} \cup i_k)$ with $B_\ell - p^{\mathrm{f}}_\ell = \phi_\ell$; then it holds $v_i > \phi_i$ for each $i \in X_k \backslash X_{k-1}$, and $B_{i_k} - p^{\mathrm{f}}_{i_k} = \phi_{i_k}$.*

(iii) *$\emptyset = X_0 \subset X_1 \subset X_2 \subset \cdots \subset X_t = N$ is a chain of tight sets.*

The tight sets lemma by Goel et al. [13] utilizes a simple structure of their mechanism that buyers dropping by clinching exhaust their budget, and other buyers have valuations equal to their dropping prices. However, these are *not* preserved in our setting. Therefore, we need the notions of dropping prices and unsaturation in [12] to illustrate sharper information for dropping of buyers.

---

[1] We assume that the number $f(N)$ is part of the input size.

4.2   Unsaturation

We begin by the definition of unsaturation. Throughout this section, let x and d denote the allocation of goods and demand vector in an iteration. Also, for demand vector $d \in \mathbb{Z}_+^N$, we define d^{-k} by $d^{-k} := (0, d_{-k})$, where $d_{-k} \in \mathbb{Z}_+^{N \setminus k}$ denotes the demands of buyers $N \setminus k$.

Definition 2 (Goel et al. [12]). For buyers $i, k \in N$, buyer i is *k-unsaturated* if for any maximal vector $z \in P_{x, d^{-k}}$, it holds $z_i = d_i^{-k}$. Buyer i is *k-saturated* if buyer i is not *k-unsaturated*.

The binary relation "i is *k*-unsaturated" is denoted by "$i \lesssim k$". Unsaturation is closely related to the demands of buyers and the clinching amount. These relationships are illustrated by the followings:

Lemma 2. *If $i \lesssim k$ just before line 5 or 9 in an iteration, then it holds $d_i \leq d_k$.*

Proposition 4. *Suppose that the demand of buyer i decreases from d_i to d_i' in line 5 or 9 in an iteration. For each buyer $k \neq i$, if $i \lesssim k$ just before the demand update, then in the subsequent execution of Algorithm 2, it holds $\delta_k = d_i - d_i'$. Otherwise, $\delta_k < d_i - d_i'$.*

4.3   Proof of Theorem 3

Our main focus in the proof is the relationship between the dropping prices of buyers in each tight set, where the following two propositions will help.

Proposition 5. *If a buyer drops by clinching some goods in Algorithm 2, then there is a buyer who drops by the demand update in line 5 or 9 just before the execution of Algorithm 2.*

Proposition 5 implies that buyers who drop by clinching have the same dropping prices as the last buyer who drops by the demand update. To prove this, we use the following lemma.

Lemma 3. *Just after the execution of Algorithm 2, it holds $f_{x,d}(N) = f_{x,d}(N \setminus i) = f_{x,d^{-i}}(N)$ for each $i \in N$.*

Proof of Proposition 5. We show that if the demand d_i of buyer i decreases by one in line 9 and still has a positive demand d_i', then no buyer drops in the subsequent execution of Algorithm 2.

After the demand update, the value $f_{x,d}(N \setminus i)$ is unchanged since it is independent of d_i. Then, $f_{x,d}(N)$ is also unchanged by the monotonicity of $f_{x,d}$ and Lemma 3. By Proposition 1 and Lemma 3, we have $\delta_i = f_{x,d}(N) - f_{x,d}(N \setminus i) = 0$, which means i is still active just after the execution of Algorithm 2.

Then, we consider the clinching amount δ_k of buyer $k \neq i$ in Algorithm 2. If $i \lesssim k$ just before the demand update, it follows from Lemma 2 that $d_i \leq d_k$. Thus, it follows from Proposition 4 that $\delta_k = d_i - d_i' \leq d_k - d_i' < d_k$, where the last inequality follows by $d_i' > 0$. Otherwise, it follows from Proposition 4 that

$\delta_k < d_i - d'_i = 1$, and thus, we have $\delta_k = 0$ since δ_k is an integer. Therefore, no buyer drops out of the auction by clinching just after i's demand update. This implies that if a buyer drops by clinching, then there exists a buyer who drops by the demand update just before the execution of Algorithm 2.   □

Proposition 6. *After the execution of Algorithm 2 in any iteration of Algorithm 1, for the set of active buyers S, it holds $x^{\mathrm{f}}(S) = f(S)$.*

This proposition immediately proves Property (iii) in Theorem 3. Combining this with Proposition 5, we have $\phi_i = \phi_j$ for each $k = 1, 2, \ldots, t$ and $i, j \in X_k \backslash X_{k-1}$. Now we prove Theorem 3. We also use the following lemmas.

Lemma 4. *Buyers who drop out of the auction have their demands updated in line 5 or 9 at least once.*

Lemma 5. *Let i be an active buyer. If the demand d_i has never been updated in line 9, then $d_i = f(i) + 1 - x_i \leq \frac{B_i - p_i}{c}$. If the demand d_i has just been updated in line 9 of an iteration, then it holds $d_i = \frac{B_i - p_i}{c} - 1$ for the rest of that iteration. In other cases, it holds $d_i = \left\lfloor \frac{B_i - p_i}{c} \right\rfloor$.*

Proof of Theorem 3. Property (iii) holds from Proposition 6. For Property (i), let i_k ($k \in N$) be a buyer who drops in line 5 or 9 of Algorithm 1. Buyer i_k drops by either of the following:

- The price is equal to their valuation (line 5).
- The price is equal to the remaining budget (line 9).

These two cases correspond to the two cases in Property (i), respectively.

For Property (ii), suppose that $i \in X_k \backslash (X_{k-1} \cup i_k)$. Then, $\phi_i = \phi_{i_k}$ holds by Propositions 5 and 6. By Lemma 4, the demand d_i has been updated in line 9 at least once. After that, d_i changes with keeping the inequality $d_i \geq \frac{B_i - p_i}{c} - 1$ by Lemma 5. Just when the demand d_i decreases to zero by clinching, $\frac{B_i - p_i}{\phi_i}$ decreases by the same amount in Algorithm 2. This implies $B_i - p_i^{\mathrm{f}} \leq \phi_i$.

Suppose that there exists a buyer $\ell \in X_k \backslash (X_{k-1} \cup i_k)$ with $B_\ell - p_\ell^{\mathrm{f}} = \phi_\ell$. This implies just before the execution of Algorithm 2 that ℓ drops, it holds $d_\ell = \frac{B_\ell - p_\ell}{\phi_\ell} - 1$ because d_ℓ and $\frac{B_\ell - p_\ell}{\phi_\ell}$ decrease by the same amount in Algorithm 2. By Lemma 5, ℓ's demand has been updated in line 9 in the same iteration as their dropping. This means i_k drops by the demand update in line 9 (after that of ℓ), and thus $B_{i_k} - p_{i_k}^{\mathrm{f}} = \phi_{i_k}$. Moreover, just before the demand update of i_k, the buyers with valuations equal to ϕ_{i_k} have already dropped in the demand update in line 5 in the same iteration. Therefore, it holds $v_i > \phi_i$ for $i \in X_k \backslash X_{k-1}$.   □

5   Efficiency

In this section, we provide three types of efficiency guarantees for our mechanism. Our tight sets lemma (Theorem 3) plays critical roles in the proofs.

5.1 Pareto Optimality

We first show that our mechanism satisfies Pareto optimality, which has been the efficiency goal in many previous studies for clinching auctions with budgets.

Theorem 4. *Our mechanism satisfies PO.*

The proof is an inductive argument with respect to $\{X_k\}_{k\in\{0,1,\dots,t\}}$ in the tight set lemma (Theorem 4), as in the proof of Goel et al. [12] for their divisible setting. Instead of dropping prices $\{\phi_{i_k}\}_{k\in\{1,2,\dots,t\}}$ (as they used), we use a new non-increasing sequence $\{\theta_k\}_{k\in\{1,2,\dots,t\}}$ defined by

$$\theta_k = \min_{i\in X_k} v_i \quad (k \in \{1,2,\dots,t\}) \tag{3}$$

due to the difference of the tight sets lemma. By construction, the following properties hold.

Lemma 6. *The sequence $\{\theta_k\}_{k\in\{1,\dots,t\}}$ constructed by (3) satisfies the following:*

(i) *It holds $v_i \geq \theta_k$ for each $k \in \{1,2,\dots,t\}$ and $i \in X_k$.*
(ii) *$\{\theta_k\}_{k\in\{1,\dots,t\}}$ is non-increasing on k.*
(iii) *For each $k \in \{1,2,\dots,t\}$, if $\phi_{i_k} = v_{i_k}$, then it holds $\phi_{i_k} = \theta_k = v_{i_k}$. Otherwise, it holds $\phi_{i_k} < \theta_k \leq v_{i_k}$.*
(iv) *It holds $B_i - p_i^f < \theta_k$ for each $k \in \{1,2,\dots,t\}$ and $i \in X_k \backslash X_{k-1}$ with $v_i > \phi_{i_k}$.*

Proof of Theorem 4. Suppose that there exists an allocation $\mathcal{A} := (x',p')$ satisfying (i) $v_i x_i^f - p_i^f \leq v_i x_i' - p_i'$ for each $i \in N$, (ii) $p^f(N) \leq p'(N)$, and (iii) at least one inequality holds without equality. Combining these inequalities, we have $\sum_{i\in N} v_i x_i^f < \sum_{i\in N} v_i x_i'$. Let $\{\theta_k\}_{k\in\{1,2,\dots,t\}}$ be the sequence in Lemma 6. Then, we show that $\theta_k(x_i^f - x_i') \leq p_i^f - p_i'$ for each $k \in \{1,2,\dots,t\}$ and $i \in X_k \backslash X_{k-1}$ by the following case-by-case analysis:

1. Suppose that $x_i^f \geq x_i'$. Then, by Property (i) of Lemma 6, it holds $\theta_k(x_i^f - x_i') \leq v_i(x_i^f - x_i') \leq p_i^f - p_i'$.
2. Suppose that $x_i^f < x_i'$ and $v_i > \phi_{i_k}$. By the indivisibility of the good and $p_i' \leq B_i < p_i^f + \theta_k$ from Property (iv) of Lemma 6, it holds $\theta_k(x_i^f - x_i') \leq -\theta_k < p_i^f - p_i'$.
3. Suppose that $x_i^f < x_i'$ and $v_i = \phi_{i_k}$. This means that buyer i_k drops out of the auction in line 5 of Algorithm 1. Thus, we have $v_{i_k} = \theta_k = \phi_{i_k} = v_i$ by Property (iii) of Lemma 6. Then, it holds $\theta_k(x_i^f - x_i') = v_i(x_i^f - x_i') \leq p_i^f - p_i'$.

From the above, we can also see that if all the inequalities hold in equality, we have $v_i = \theta_k$ or $x_i^f = x_i'$ for each $k \in \{1,2,\dots,t\}$ and $i \in X_k \backslash X_{k-1}$.

Now we prove $p^f(X_k) - p'(X_k) \geq \theta_k(x^f(X_k) - x'(X_k)) \geq 0$ for each $k \in \{0,1,\dots,t\}$, where $\theta_0 := \theta_1$. For $k = 0$, the inequality trivially holds by $X_0 = \emptyset$. Suppose that the inequality holds for $k - 1$. We prove it holds for k. Since

$x'(X_{k-1}) \leq f(X_{k-1}) = x^{\mathrm{f}}(X_{k-1})$ and $x'(X_k) \leq f(X_k) = x^{\mathrm{f}}(X_k)$ by Property (iii) of Theorem 3, we have

$$0 \leq \theta_k(x^{\mathrm{f}}(X_k) - x'(X_k))$$
$$\leq \theta_{k-1}(x^{\mathrm{f}}(X_{k-1}) - x'(X_{k-1})) + \theta_k(x^{\mathrm{f}}(X_k \backslash X_{k-1}) - x'(X_k \backslash X_{k-1}))$$
$$\leq p^{\mathrm{f}}(X_{k-1}) - p'(X_{k-1}) + \theta_k(x^{\mathrm{f}}(X_k \backslash X_{k-1}) - x'(X_k \backslash X_{k-1}))$$
$$\leq p^{\mathrm{f}}(X_{k-1}) - p'(X_{k-1}) + p^{\mathrm{f}}(X_k \backslash X_{k-1}) - p'(X_k \backslash X_{k-1}) = p^{\mathrm{f}}(X_k) - p'(X_k),$$

where the second inequality follows by Property (ii) of Lemma 6, the third inequality follows by the assumption, and the fourth inequality follows by $\theta_k(x_i^{\mathrm{f}} - x_i') \leq p_i^{\mathrm{f}} - p_i'$ for each $k \in \{1, 2, \ldots, t\}$ and $i \in X_k \backslash X_{k-1}$. By substituting k with t, we have $p^{\mathrm{f}}(N) - p'(N) \geq \theta_t(x^{\mathrm{f}}(N) - x'(N)) \geq 0$. Since we assume $p^{\mathrm{f}}(N) \leq p'(N)$, all the inequalities hold in equality. Therefore, we have $v_i = \theta_k$ or $x_i^{\mathrm{f}} = x_i'$ for each $k \in \{1, 2, \ldots, t\}$ and $i \in X_k \backslash X_{k-1}$.

Using this, we prove $\sum_{i \in X_k} v_i(x_i^{\mathrm{f}} - x_i') \geq \theta_k(x^{\mathrm{f}}(X_k) - x'(X_k)) \geq 0$ for each $k \in \{0, 1, \ldots, t\}$. For $k = 0$, the inequality trivially holds by $X_0 = \emptyset$. Suppose that $\sum_{i \in X_{k-1}} v_i(x_i^{\mathrm{f}} - x_i') \geq \theta_{k-1}(x^{\mathrm{f}}(X_{k-1}) - x'(X_{k-1}))$ holds for $k-1$. We prove it holds for k. Then, we have

$$\sum_{i \in X_k} v_i(x_i^{\mathrm{f}} - x_i') \geq \theta_{k-1}(x^{\mathrm{f}}(X_{k-1}) - x'(X_{k-1})) + \sum_{i \in X_k \backslash X_{k-1}} \theta_k(x_i^{\mathrm{f}} - x_i')$$
$$\geq \theta_k(x^{\mathrm{f}}(X_{k-1}) - x'(X_{k-1})) + \sum_{i \in X_k \backslash X_{k-1}} \theta_k(x_i^{\mathrm{f}} - x_i')$$
$$= \theta_k(x^{\mathrm{f}}(X_k) - x'(X_k)) \geq 0,$$

where the first inequality holds by the assumption, and $v_i = \theta_k$ or $x_i^{\mathrm{f}} = x_i'$ for each $i \in X_k \backslash X_{k-1}$, and the second inequality holds by Property (ii) of Lemma 6, and the third inequality holds by $x^{\mathrm{f}}(X_k) = f(X_k) \geq x'(X_k)$ by Property (iii) of Theorem 3. Thus, it holds $\sum_{i \in X_k} v_i(x_i^{\mathrm{f}} - x_i') \geq \theta_k(x^{\mathrm{f}}(X_k) - x'(X_k)) \geq 0$ for any $k \in \{0, 1, \ldots, t\}$. Substitute k with t, we have $\sum_{i \in N} v_i x_i^{\mathrm{f}} \geq \sum_{i \in N} v_i x_i'$, which contradicts the hypothesis. $\square$

5.2   Liquid Welfare

Let LW^{M} and $\mathrm{LW}^{\mathrm{OPT}}$ denote the LW of our mechanism and the optimal LW, respectively. We provide the first LW guarantee for clinching auctions with indivisible goods, which holds even in polymatroidal environments.

Theorem 5. *It holds* $\mathrm{LW}^{\mathrm{M}} \geq \frac{1}{2}\mathrm{LW}^{\mathrm{OPT}}$.

To prove Theorem 5, we establish

$$\mathrm{LW}^{\mathrm{M}} \geq p^{\mathrm{f}}(N) \geq \mathrm{LW}^{\mathrm{OPT}} - \mathrm{LW}^{\mathrm{M}},$$

implying $\mathrm{LW}^{\mathrm{M}} \geq \frac{1}{2}\mathrm{LW}^{\mathrm{OPT}}$ as in Sato [22]. The first inequality is easy by

$$\mathrm{LW}^{\mathrm{M}} = \sum_{i \in N} \min(v_i x_i^{\mathrm{f}}, B_i) \geq p^{\mathrm{f}}(N), \tag{4}$$

where $v_i x_i^{\mathrm{f}} \geq p_i^{\mathrm{f}}$ $(i \in N)$ holds by IR, and $B_i \geq p_i^{\mathrm{f}}$ $(i \in N)$ holds by budget feasibility due to Theorem 2. In the following, we prove the second inequality.

We follow the outline of the proof in Sato [22]: We first provide the formula for an LW optimal allocation, and then show a lower bound on the number of goods remaining at any point in our mechanism. Since the remaining goods are sold at the price equal to or higher than the current one, we can provide a lower bound on future payments. Then, considering the initial step of the auction, we obtain a lower bound on the total payments sufficient for the LW guarantee.

However, several differences due to indivisibility prevent us from making a straightforward extension. One difference is the change in the formula of an LW optimal allocation. The formula for the divisible setting of Sato [22] does not necessarily yield an LW optimal allocation in our indivisible setting. Another difference is the number of remaining goods. Since the demands for fewer than one unit are rounded down due to the indivisibility, the number of remaining goods might be fewer than the lower bound in [22]. These differences are illustrated by the examples in the full paper [15].

To provide the formula for an LW optimal allocation in our setting, we divide each buyer into two virtual buyers in the following way: For each buyer $i \in N$, consider two copies i_a, i_b, where i_a represents a buyer whose valuation is v_i and their budget is $\left\lfloor \frac{B_i}{v_i} \right\rfloor v_i$, and i_b represents a buyer whose valuation and budget are $B_i - \left\lfloor \frac{B_i}{v_i} \right\rfloor v_i$[2]. Let $N' := \cup_{i \in N} \{i_a, i_b\}$. We define a map $\Gamma : 2^{N'} \to 2^N$ that outputs the union of buyers i with $\{i_a, i_b\} \cap S' \neq \emptyset$ for each $S' \subseteq N'$. In addition, we define a new monotone submodular function $f' : 2^{N'} \to \mathbb{Z}_+$ by $f'(S') := f(\Gamma(S'))$ $(S' \subseteq N')$. Note that f' is again a monotone submodular function; see, e.g., Section 44.6g of Schrijver [23]. Then, an allocation for the original market can be constructed simply by summing the allocated goods of the corresponding two buyers from an allocation for N'.

An LW optimal allocation in our setting can be obtained by the greedy procedure described below. Suppose that buyers in N' are ordered in descending order according to their valuations. For each buyer $i' \in N'$, let $H_{i'} := \{1, 2, \ldots, i' - 1\}$ denote the set of buyer $k \in N'$ who has a higher valuation than i', or who has the same valuation as i' and $\Gamma(k)$ is numbered before $\Gamma(i')$ in N.

Proposition 7. *An LW optimal allocation* $\tilde{x}^* := (\tilde{x}_i^*)_{i \in N}$ *is given by* $\tilde{x}_i^* = x_{i_a}^* + x_{i_b}^*$ *for each* $i \in N$, *where* $x_{i'}^* = \min(\frac{B_{i'}}{v_{i'}}, \min_{H \subseteq H_{i'}} \{f'(H \cup i') - x_{i'}^*(H)\})$ *for each* $i' \in N'$.

Using this allocation, we have

$$\mathrm{LW}^{\mathrm{OPT}} = \sum_{i \in N} \min(v_i \tilde{x}_i^*, B_i) = \sum_{i \in N} (v_{i_a} x_{i_a}^* + v_{i_b} x_{i_b}^*), \tag{5}$$

---

[2] If $B_i - \left\lfloor \frac{B_i}{v_i} \right\rfloor v_i = 0$, then we define $v_{i_b} = v_i$ and $B_{i_b} = 0$.

where the second equality holds by the definition of v_{i_b}. Proposition 7 indicates that for each $i \in N$, the buyer i_a (resp. i_b) can obtain at most $\left\lfloor \frac{B_i}{v_i} \right\rfloor$ (resp. 1) units in this optimal allocation.

Using the LW optimal allocation, we then provide a lower bound on the remaining goods in Algorithm 1. Consider an iteration of Algorithm 1, where x and d are the allocation of goods and the demand vector, respectively, and c is the price. Define $X := \{i \in N \mid d_i > 0\}$, $Y := \{i \in X \mid x_i^{\mathrm{f}} < \tilde{x}_i^*\}$ as in Sato [22]. In addition, we define $Y_c := \{i \in Y \mid c < v_{i_b}\}$, and $x_a^*(S) := \sum_{i \in S} x_{i_a}^*$ and $x_b^*(S) := \sum_{i \in S} x_{i_b}^*$ for any $S \subseteq N$.

Proposition 8. It holds $f_{x,d}(Y) - x_a^*(Y) + x(Y) - x_b^*(Y_c) \geq 0$.

If Y_c of Proposition 8 is replaced by Y, the left-hand side is changed to $f_{x,d}(Y) - \tilde{x}^*(Y) + x(Y)$, which is the same as the one in Theorem 3.9 of Sato [22]. However, due to the indivisibility, the number of remaining goods might be fewer than this bound. Thus, to handle with the difference, we use x_a^* and x_b^* instead of $\tilde{x}^*$, and define a new set Y_c to obtain the sharp lower bound.

Using Proposition 8, we provide a lower bound of payments in our mechanism for the set of buyer i with $x_i^{\mathrm{f}} \geq \tilde{x}_i^*$. As in Proposition 8, we use the optimal allocation of virtual buyers instead of that of original buyers.

Theorem 6. It holds $\displaystyle \sum_{i \in N; x_i^{\mathrm{f}} \geq \tilde{x}_i^*} p_i^{\mathrm{f}} \geq \sum_{i \in N; x_i^{\mathrm{f}} < \tilde{x}_i^*} (\phi_i x_{i_a}^* + v_{i_b} x_{i_b}^* - \phi_i x_i^{\mathrm{f}})$.

In the proof, we use the following proposition derived from tight sets lemma (Theorem 3). This proposition illustrates the relationship between the allocation of goods in our mechanism and the optimal allocation for some buyers.

Proposition 9. If $x_i^{\mathrm{f}} < \tilde{x}_i^*$ and $v_i > \phi_i$ for some $i \in N$, then it holds $x_{i_a}^* = \left\lfloor \frac{B_i}{v_i} \right\rfloor$, $x_{i_b}^* = 1$, and $v_{i_b} \leq \phi_i$.

Moreover, we use *backward* mathematical induction as in the proof of Theorem 5.9 in Sato [22]. We show that, throughout the execution of Algorithm 1, it holds

$$\sum_{i \in X \setminus Y} (p_i^{\mathrm{f}} - p_i) \geq \sum_{i \in Y} \phi_i (x_{i_a}^* - x_i^{\mathrm{f}}) + \sum_{i \in Y_c} v_{i_b} x_{i_b}^* + c(f_{x,d}(X) - x_a^*(Y) - x_b^*(Y_c) + x(Y)).$$

(6)

Note that inequality (6) provides a lower bound on future payments of buyers. We show that both sides of (6) are zero at the end of the auction, and that the left-hand side gradually becomes larger than the right-hand side as we go back to the beginning.

Proof of Theorem 6. At the end of Algorithm 1, inequality (6) holds because both sides are equal to 0 by $X = Y = Y_c = \emptyset$. Using the following case-by-case analysis, we prove that if it holds at the end of an iteration, it holds at the beginning of the iteration. Note that in the following, we use

$$f_{x,d}(X) + x(Y) = f_{x,d}(N) + x(Y) = f(N) - x(N) + x(Y) = f(N) - x(N \setminus Y),$$

where the first equality follows from $f_{x,d}(X) \leq f_{x,d}(N)$ by the monotonicity of $f_{x,d}$ and $f_{x,d}(N) \leq f_{x,d}(X) + d(N \backslash X) = f_{x,d}(X)$ due to the definition of $f_{x,d}$ and X, and the second equality follows from Proposition 3.

(i) Execution of Algorithm 2:

For buyer i who belongs to $X \backslash Y$ just before clinching $\delta_i > 0$ amount of goods, the left-hand side of inequality (6) is decreased by $c\delta_i$. The first and the second terms on the right-hand side are unchanged because $i \notin Y$, and the third term on the right side is decreased by $c\delta_i$ by $f_{x,d}(X) + x(Y) = f(N) - x(N \backslash Y)$. Thus, both sides of inequality (6) are decreased by $c\delta_i$. For buyer $i \in Y$ who is still active after clinching, the left-hand side and the first and the second terms on the right-hand side are unchanged. Moreover, the third term on the right side is also unchanged by $f_{x,d}(X) + x(Y) = f(N) - x(N \backslash Y)$. Thus, both sides of inequality (6) are unchanged. For buyer $i \in Y$ whose demand is positive before the execution of Algorithm 2 and becomes zero by clinching, it holds from $d_i > 0$ that $c \leq v_i$. If $c = v_i$, then $v_{i_b} \leq v_i = c = \phi_i$, and thus $i \notin Y_c$. Otherwise, from Proposition 9, we have $v_{i_b} \leq \phi_i = c$, which means $i \notin Y_c$. Therefore, the left-hand side is unchanged by $i \in Y \backslash Y_c$. On the right-hand side, the first term is decreased by $\phi_i(x_{i_a}^* - x_i^f) = c(x_{i_a}^* - x_i^f)$, and the second term is unchanged, and the third term is increased by $c(x_{i_a}^* - x_i^f)$ by $f_{x,d}(X) + x(Y) = f(N) - x(N \backslash Y)$. Then, both sides of inequality (6) are unchanged. Therefore, if (6) holds after the execution of Algorithm 2, it holds before that.

(ii) The demand update:

Suppose that the demand of buyer i is updated in line 5 or 9. If the demand d_i is still positive after the update, both sides of inequality (6) are obviously unchanged because $f_{x,d}(X) + x(Y) = f(N) - x(N \backslash Y)$ is unchanged. In the following, we consider the case where the demand d_i becomes zero. The left-hand side of (6) is unchanged by $p_i = p_i^f$ before the update. Suppose that $i \in X \backslash Y$. Then, the right-hand side is obviously unchanged. Suppose that $i \in Y_c$ before the update. Then, it holds $c = \phi_i < v_{i_b} < v_i$, which contradicts with Proposition 9. Suppose that $i \in Y \backslash Y_c$ before the update. The first term on the right-hand side is reduced by $\phi_i(x_{i_a}^* - x_i^f) = c(x_{i_a}^* - x_i^f)$, the second term remains unchanged, and the third term is increased by $c(x_{i_a}^* - x_i^f)$. Thus, the right-hand side remains unchanged. Therefore, if (6) holds after the price update, it holds before that.

(iii) The price update:

It suffices to consider the change of Y_c in the second and third terms on the right-hand side of (6) because x and d are unchanged by the price update. Let $\tilde{c}$ be the price before the update. By Proposition 8, it holds that $f_{x,d}(Y) - x_a^*(Y) + x(Y) \geq x_b^*(Y_{\tilde{c}}) \geq x_b^*(Y_c)$ by $Y_{\tilde{c}} \supseteq Y_c$. Thus,

$$\sum_{i \in Y_c} v_{i_b} x_{i_b}^* + c(f_{x,d}(X) - x_a^*(Y) - x_b^*(Y_c) + x(Y))$$

$$\geq \sum_{i \in Y_c} v_{i_b} x_{i_b}^* + c(f_{x,d}(X) - x_a^*(Y) - x_b^*(Y_{\tilde{c}}) + x(Y)) + c x_b^*(Y_{\tilde{c}} \backslash Y_c)$$

$$\geq \sum_{i \in Y_{\tilde{c}}} v_{i_b} x_{i_b}^* + \tilde{c}(f_{x,d}(X) - x_a^*(Y) - x_b^*(Y_{\tilde{c}}) + x(Y)),$$

where the last inequality holds for $\tilde{c} < v_{i_b} \leq c$ for each $i \in Y_{\tilde{c}} \backslash Y_c$. Thus, the left-hand side of inequality (6) is unchanged, and the right-hand side is increased by the price update. Therefore, if inequality (6) holds after the price update, it holds before the price update.

Therefore, inequality (6) holds throughout Algorithm 2. At the beginning of the auction, we have $\sum_{i \in N; x_i^f \geq \tilde{x}_i^*} p_i^f \geq \sum_{i \in N; x_i^f < \tilde{x}_i^*} (\phi_i x_{i_a}^* + v_{i_b} x_i^* - \phi_i x_i^f)$. $\quad\Box$

The following lemma shows that our guarantee on payments in Theorem 6 is sufficient to obtain Theorem 5.

Lemma 7. *For each $i \in N$ with $x_i^f < \tilde{x}_i^*$, it holds*

$$\phi_i x_{i_a}^* + v_{i_b} x_{i_b}^* - \phi_i x_i^f = \min(v_i \tilde{x}_i^*, B_i) - \min(v_i x_i^f, B_i).$$

Proof of Theorem 5. For buyer i with $x_i^f < \tilde{x}_i^*$, by Lemma 7, it holds $\phi_i x_{i_a}^* + v_{i_b} x_{i_b}^* - \phi_i x_i^f = \min(v_i \tilde{x}_i^*, B_i) - \min(v_i x_i^f, B_i)$. Then, by $p_i^f \geq 0$ $(i \in N)$ and Theorem 6, we have

$$p^f(N) \geq \sum_{i \in N; x_i^f \geq \tilde{x}_i^*} p_i^f \geq \sum_{i \in N; x_i^f < \tilde{x}_i^*} \left(\phi_i x_{i_a}^* + v_{i_b} x_{i_b}^* - \phi_i x_i^f\right)$$

$$= \sum_{i \in N; x_i^f < \tilde{x}_i^*} \left(\min(v_i \tilde{x}_i^*, B_i) - \min(v_i x_i^f, B_i)\right)$$

$$\geq \sum_{i \in N} \left(\min(v_i \tilde{x}_i^*, B_i) - \min(v_i x_i^f, B_i)\right) = \mathrm{LW}^{\mathrm{OPT}} - \mathrm{LW}^{\mathrm{M}},$$

where the last inequality holds by $\min(v_i \tilde{x}_i^*, B_i) - \min(v_i x_i^f, B_i) \leq 0$ for each $i \in N$ with $x_i^f \geq \tilde{x}_i^*$. By Eq. (4), we have $\mathrm{LW}^{\mathrm{M}} \geq p^f(N) \geq \mathrm{LW}^{\mathrm{OPT}} - \mathrm{LW}^{\mathrm{M}}$. Therefore, we have $\mathrm{LW}^{\mathrm{M}} \geq \frac{1}{2} \mathrm{LW}^{\mathrm{OPT}}$. $\quad\Box$

Remark 1. In the full paper [15], we provide an example to show that the inequality in Theorem 5 is tight. Moreover, we also provide a lower bound of 4/3 for the LW guarantee in our indivisible setting by extending Theorem 5.1 of Dobzinski and Leme [7].

5.3 Social Welfare

Let SW^{M} denote the SW of our mechanism. Then, we have the following:

Theorem 7. *It holds $\mathrm{SW}^{\mathrm{M}} = \sum_{i \in N} v_i x_i^f \geq \mathrm{LW}^{\mathrm{OPT}}$.*

This inequality is tight because the optimal LW is equal to the optimal SW if the budgets of all buyers are sufficiently large. In this case, Theorem 7 implies that our mechanism outputs an allocation that maximizes the SW.

The proof is an inductive argument with respect to $\{X_k\}_{k \in \{0,1,\ldots,t\}}$ in the tight set lemma (Theorem 4) as in Sato [22]. On the other hand, we need to use $\{\phi_{i_k}\}_{k\in\{1,2...,t\}}$ instead of $\{v_{i_k}\}_{k\in\{1,2...,t\}}$ because $\{v_{i_k}\}_{k\in\{1,2...,t\}}$ is not necessarily monotone in our setting. Then, by Eq. (5), we have

$$\text{SW}^{\text{M}} - \text{LW}^{\text{OPT}} = \sum_{i\in N}\left(v_i x_i^{\text{f}} - (v_{i_a} x_{i_a}^* + v_{i_b} x_{i_b}^*)\right).$$

To link the right-hand side to dropping prices, we use the following lemma:

Lemma 8. *For each $i \in N$, it holds $v_i x_i^{\text{f}} - (v_{i_a} x_{i_a}^* + v_{i_b} x_{i_b}^*) \geq \phi_i(x_i^{\text{f}} - \tilde{x}_i^*)$.*

Proof of Theorem 7. Using Theorem 3, we prove

$$\sum_{i\in X_k}\left(v_i x_i^{\text{f}} - (v_{i_a} x_{i_a}^* + v_{i_b} x_{i_b}^*)\right) \geq \phi_{i_k}(x^{\text{f}}(X_k) - \tilde{x}^*(X_k))$$

for each $k \in \{0,1,\ldots,t\}$ by mathematical induction (where we set $\phi_{i_0} := \phi_{i_1}$). For $k = 0$, both sides are equal to 0 by $X_0 = \emptyset$. Suppose that the inequality holds for $k - 1$. By Property (iii) of Theorem 3, it holds $x^{\text{f}}(X_{k-1}) = f(X_{k-1}) \geq \tilde{x}^*(X_{k-1})$. Then, for $i \in X_k \backslash X_{k-1}$, by Lemma 8 and $\phi_i = \phi_{i_k}$ due to Theorem 3, we have $v_i x_i^{\text{f}} - (v_{i_a} x_{i_a}^* + v_{i_b} x_{i_b}^*) \geq \phi_{i_k}(x_i^{\text{f}} - \tilde{x}_i^*)$. Therefore, we have

$$\sum_{i\in X_k}(v_i x_i^{\text{f}} - (v_{i_a} x_{i_a}^* + v_{i_b} x_{i_b}^*))$$

$$\geq \phi_{i_{k-1}}(x^{\text{f}}(X_{k-1}) - \tilde{x}^*(X_{k-1})) + \phi_{i_k}\sum_{i\in X_k\backslash X_{k-1}}(x_i^{\text{f}} - \tilde{x}_i^*)$$

$$\geq \phi_{i_k}(x^{\text{f}}(X_k) - \tilde{x}^*(X_k)),$$

where the second inequality holds by $\phi_{i_{k-1}} \geq \phi_{i_k}$ and $x^{\text{f}}(X_{k-1}) \geq \tilde{x}^*(X_{k-1})$. Substituting k with t, it holds $\sum_{i\in X_t} x_i^{\text{f}} = f(N) \geq \sum_{i\in X_t} \tilde{x}_i^*$ by Proposition 3 and $X_t = N$. Thus, we have

$$\text{SW}^{\text{M}} = \sum_{i\in N} v_i x_i^{\text{f}} \geq \sum_{i\in N}(v_{i_a} x_{i_a}^* + v_{i_b} x_{i_b}^*) = \text{LW}^{\text{OPT}},$$

where the last equality holds by Eq. (5). $\qquad\square$

Remark 2. Another SW guarantee of clinching auctions was given by Devanur et al. [5]. They showed that in their divisible setting, clinching auctions yield an envy-free allocation and achieve a two-approximation of SW to the maximum SW among all envy-free allocations. However, extending their result to our indivisible setting seems unrealistic, see the full paper [15] for more detail.

Acknowledgements. We thank the referees for helpful feedback and suggestions. This work was supported by Grant-in-Aid for JSPS Research Fellow Grant Number JP22KJ1137, Japan and Grant-in-Aid for Challenging Research (Exploratory) Grant Number JP21K19759, Japan.

References

1. Ausubel, L.M.: An efficient ascending-bid auction for multiple objects. Am. Econ. Rev. **94**(5), 1452–1475 (2004)
2. Bhattacharya, S., Conitzer, V., Munagala, K., Xia, L.: Incentive compatible budget elicitation in multi-unit auctions. In: SODA, pp. 554–572 (2010)
3. Bikhchandani, S., de Vries, S., Schummer, J., Vohra, R.V.: An ascending vickrey auction for selling bases of a matroid. Oper. Res. **59**(2), 400–413 (2011)
4. Colini-Baldeschi, R., Leonardi, S., Henzinger, M., Starnberger, M.: On multiple keyword sponsored search auctions with budgets. ACM Trans. Econ. Comput. **4**(1), 2:1–2:34 (2015)
5. Devanur, N.R., Ha, B.Q., Hartline, J.D.: Prior-free auctions for budgeted agents. In: EC, pp. 287–304 (2013)
6. Dobzinski, S., Lavi, R., Nisan, N.: Multi-unit auctions with budget limits. Games Econ. Behav. **74**(2), 486–503 (2012)
7. Dobzinski, S., Leme, R.P.: Efficiency guarantees in auctions with budgets. In: ICALP, pp. 392–404 (2014)
8. Dütting, P., Henzinger, M., Starnberger, M.: Auctions for heterogeneous items and budget limits. ACM Trans. Econ. Comput. **4**(1), 4:1–4:17 (2015)
9. Edelman, B., Ostrovsky, M., Schwarz, M.: Internet advertising and the generalized second-price auction: selling billions of dollars worth of keywords. Am. Econ. Rev. **97**(1), 242–259 (2007)
10. Fiat, A., Leonardi, S., Saia, J., Sankowski, P.: Single valued combinatorial auctions with budgets. In: EC, pp. 223–232 (2011)
11. Goel, G., Leonardi, S., Mirrokni, V., Nikzad, A., Leme, R.P.: Reservation exchange markets for internet advertising. In: ICALP, pp. 142:1–142:13 (2016)
12. Goel, G., Mirrokni, V., Leme, R.P.: Clinching auctions beyond hard budget constraints. In: EC, pp. 167–184 (2014)
13. Goel, G., Mirrokni, V., Leme, R.P.: Polyhedral clinching auctions and the adwords polytope. J. ACM **62**(3), 18:1–18:27 (2015)
14. Goel, G., Mirrokni, V., Leme, R.P.: Clinching auctions with online supply. Games Econ. Behav. **123**, 342–358 (2020)
15. Hirai, H., Sato, R.: Polyhedral clinching auctions for indivisible goods (2023). https://arxiv.org/abs/2303.00231
16. Hirai, H., Sato, R.: Polyhedral clinching auctions for two-sided markets. Math. Oper. Res. **47**(1), 259–285 (2022)
17. Krishna, V.: Auction Theory, 2nd edn. Academic Press, San Diego (2010)
18. Lee, Y.T., Sidford, A., Wong, S.C.W.: A faster cutting plane method and its implications for combinatorial and convex optimization. In: FOCS, pp. 1049–1065 (2015)
19. Lehmann, B., Lehmann, D., Nisan, N.: Combinatorial auctions with decreasing marginal utilities. Games Econ. Behav. **55**(2), 270–296 (2006)
20. Lu, P., Xiao, T.: Improved efficiency guarantees in auctions with budgets. In: EC, pp. 397–413 (2015)
21. Nisan, N., Roughgarden, T., Tardos, E., Vazirani, V.V.: Algorithmic Game Theory. Cambridge University Press, New York (2007)
22. Sato, R.: Polyhedral clinching auctions with a single sample (2023). https://arxiv.org/abs/2302.03458
23. Schrijver, A.: Combinatorial Optimization. Springer, Heidelberg (2003)
24. Syrgkanis, V., Tardos, E.: Composable and efficient mechanisms. In: STOC, pp. 211–220 (2013)

Online Matching with Stochastic Rewards: Advanced Analyses Using Configuration Linear Programs

Zhiyi Huang[1], Hanrui Jiang[2], Aocheng Shen[2], Junkai Song[1], Zhiang Wu[3], and Qiankun Zhang[2(✉)]

[1] The University of Hong Kong, Hong Kong SAR, China
`zhiyi@cs.hku.hk, dsgsjk@connect.hku.hk`
[2] Huazhong University of Science and Technology, Wuhan, China
`{hanry,awsonshen,qiankun}@hust.edu.cn`
[3] Hong Kong University of Science and Technology, Hong Kong SAR, China
`zwube@connect.ust.hk`

Abstract. Mehta and Panigrahi (2012) proposed Online Matching with Stochastic Rewards, which generalizes the Online Bipartite Matching problem of Karp, Vazirani, and Vazirani (1990) by associating the edges with success probabilities. This new feature captures the pay-per-click model in online advertising. Recently, Huang and Zhang (2020) studied this problem under the online primal dual framework using the Configuration Linear Program (LP), and got the best known competitive ratios of the Stochastic Balance algorithm. Their work suggests that the more expressive Configuration LP is more suitable for this problem than the Matching LP.

This paper advances the theory of Configuration LP in two directions. Our technical contribution includes a characterization of the joint matching outcome of an offline vertex and *all its neighbors*. This characterization may be of independent interest, and is aligned with the spirit of Configuration LP. By contrast, previous analyses of Ranking generally focus on only one neighbor. Second, we designed a Stochastic Configuration LP that captures a stochastic benchmark proposed by Goyal and Udwani (2020), who used a Path-based LP. The Stochastic Configuration LP is smaller and simpler than the Path-based LP. Moreover, using the new LP we improved the competitive ratio of Stochastic Balance from 0.596 to 0.611 when the success probabilities are infinitesimal, and to 0.613 when the success probabilities are further equal.

1 Introduction

Suppose that Alice is planning an upcoming trip to Hawaii and she just types "Hawaii resort" in a search engine. Once she presses the return key, the search

Zhiyi Huang is supported by the Research Grants Council of Hong Kong (Grant No. 17203022). Qiankun Zhang is supported by National Natural Science Foundation of China (Grant No. 62302183) and National Science Foundation of Hubei Province (Grant No. 2023AFB258).

J. Garg et al. (Eds.): WINE 2023, LNCS 14413, pp. 384–401, 2024.
https://doi.org/10.1007/978-3-031-48974-7_22

engine will provide a list of related websites, among which several websites have a small "Ad" tag next to them to indicate that they are sponsored results. We will refer to such websites as advertisers. If Alice clicks on a sponsored result, the advertiser would pay the search engine, where the amount depends on its bid for the keywords among other factors. This is one of many scenarios of online advertising, which contributes hundreds of billions of US dollars to the annual revenue of major IT companies.

Online advertising presents many unique challenges, each of which has led to a long line of research. For example, the search engine must decide which advertiser shall be returned to a search query without knowing what search queries may come next. The essence of this online decision-making problem is captured by Online Bipartite Matching, which amazingly was introduced by Karp, Vazirani, and Vazirani [22] well before online advertising even existed. To see the connection, consider the advertisers and search queries as vertices on the two sides of a bipartite graph. The advertisers are known from the beginning and we call them *offline vertices*. The search queries come one by one and we call them *online vertices*. The algorithm needs to decide how to match each online vertex on its arrival, with the goal of maximizing the matching size. We measure an algorithm by the worst-case ratio of its expected matching size to the size of the optimal matching in hindsight, a.k.a., the *competitive ratio*.

Readers may notice a disparity, however, if they compare our example scenario of online advertising with the Online Bipartite Matching problem. In online advertising, the search engine cannot control whether a user clicks on the ad or not. The best that a search engine could do is to estimate how likely the user will click, based on the keywords and the information it gathers about the user. To this end, an attempt to match an online vertex to an offline neighbor only succeeds with some probability, rather than with certainty like in Online Bipartite Matching. Because of this disparity, Mehta and Panigrahi [25] proposed Online Matching with Stochastic Rewards, extending Online Bipartite Matching by associating the edges with success probabilities.

Much progress has been made in Online Matching with Stochastic Rewards in the last decade. Most related to this paper is the work of Huang and Zhang [17], who successfully applied the online primal-dual framework to this problem. Their analysis of the Stochastic Balance algorithm, which matches each online vertex to a neighbor that has been matched the least number of times so far, yielded the best competitive ratios to date. Conceptually, they found that the Matching Linear Program (LP) is not expressive enough for an online primal-dual analysis of this problem, and instead one could consider the Configuration LP.

There are still unanswered questions, however, that cast some doubt on the usefulness of the Configuration LP. For instance, Mehta and Panigrahi [25] showed that the Ranking algorithm, which randomly ranks the offline vertices at the beginning and matches each online vertex to the neighbor with the highest rank, is another competitive algorithm for Online Matching with Stochastic Rewards. However, Huang and Zhang [17] failed to offer any non-trivial competitive analysis for the Ranking algorithm using the online primal-dual analysis with Configuration LP. Moreover, Goyal and Udwani [13] introduced an alternative benchmark, which we will refer to as the stochastic benchmark (see Sect. 2 for details). They proved that the competitive ratios of both Stochastic Balance and Ranking would be strictly larger if they were compared against the stochastic benchmark. To do so, they designed a Path-based LP which is completely different than the Configuration LP. Can we reproduce or even improve these results by further developing the theory of online primal-dual with Configuration LP?

1.1 Our Results

Our first contribution is *an online primal-dual analysis of Ranking based on the Configuration LP*, improving the competitive ratio of Ranking from 0.534 to 0.572. To present the technical novelty that enables this analysis, we need to first explain the intuition behind the usefulness of the Configuration LP in an online primal-dual analysis compared to the simple Matching LP. The essence of an online primal-dual analysis is to design an appropriate gain-splitting rule. When the algorithm matches an online vertex v to an offline neighbor u and increases the expected objective accordingly by success probability p_{uv}, imagine that we further split the gain of p_{uv} between its two vertices u and v. We need the gain splitting rule to satisfy some LP-dependent conditions which essentially say that the total expected gain of neighboring vertices are sufficiently large. More precisely, if we use the Matching LP, the condition is:

$$p_{uv} \cdot \mathbf{E}[\text{gain of } u] + \mathbf{E}[\text{gain of } v] \geq \Gamma \cdot p_{uv}, \tag{1}$$

where Γ is the competitive ratio that we seek to prove for the algorithm.

By contrast, if we use the Configuration LP, the critical condition no longer considers just a single edge (u, v). Instead, it examines an offline vertex u and a *subset of neighbors* S whose success probabilities sum to about 1, and requires that:

$$\mathbf{E}[\text{gain of } u] + \sum_{v \in S} \mathbf{E}[\text{gain of } v] \geq \Gamma. \tag{2}$$

In other words, the configuration LP only needs an amortized version of Eq. (1), which sums over $v \in S$. Even if Eq. (1) fails in the worst-case scenario with respect to (w.r.t.) an offline vertex u and a single neighbor v, Eq. (2) may still hold if such worst-case cannot happen to all $v \in S$ at the same time. Note that, however, to use the power of the weaker condition, we need to characterize the joint matching outcome of u and a subset of neighbors S, rather than doing so for each $v \in S$ separately. Huang and Zhang [17] gave such a characterization for Stochastic Balance.

This paper finds that even the joint matching outcome of u and S is insufficient for getting a useful characterization of the worst-case scenario of Eq. (2) w.r.t. Ranking. Instead, we consider the joint outcome of u and *all its neighbors*, including those outside S. Characterizing the entire neighborhood's matching outcome may seem counter-intuitive *a priori*, which may be why Huang and

Zhang [17] failed to get a non-trivial competitive ratio for Ranking. The characterization and the new approach in general may be useful for other online matching problems whose offline vertices can be matched more than once, e.g., the AdWords problem [26].

Our second contribution is *a new Stochastic Configuration LP*, which we design to bound the stochastic benchmark of Goyal and Udwani [13]. The stochastic benchmark refers to the best objective achievable by an algorithm which knows the graph and the edges' success probabilities but not the realization of edge successes and failures, and which still needs to match online vertices by their arrival order. It is natural to consider LPs whose decision variables depend on the observed edge successes and failures to capture the stochastic benchmark. This is the approach of Goyal and Udwani [13], who referred to the observed edge successes and failures as a sample path and gave a Path-based LP. Their LP is huge, however, because the number of sample paths is exponential in the number of edges by definition. Further, compared to the Matching and Configuration LPs, the Path-based LP has an extra set of consistency constraints: if a sample path is a sub-path of another sample path, then the relevant matching outcome must be consistent. These constraints lead to new dual variables that are outside the scope of existing gain-splitting methods. Goyal and Udwani [13] still found a canonical gain-splitting rule and sufficient conditions for proving competitive ratios for Ranking and Stochastic Balance, but their conditions do not correspond to the constraints of Path-based LP's dual. Therefore, it is unclear how to apply their approach to other problems.

We use a different strategy to design Stochastic Configuration LP for the stochastic benchmark. Instead of using variables x_{uS} for the probability an offline vertex u is matched to a subset of online vertices S as in the Configuration LP, we consider variables y_{uS} that represents the probability an offline vertex u *would be matched to subset S if all matches to u failed*. The size of the Stochastic Configuration LP is therefore exponential only in the number of vertices, like the original Configuration LP. Further, these variables implicitly capture the consistency requirements and therefore no additional consistency constraints are needed. As a result, the online primal dual analysis can directly work with the (approximate) dual constraints. Compared to condition (2) given by the Configuration LP, the condition w.r.t. the stochastic variant scales the gains of online vertices and the competitive ratio on the right-hand-side by some multipliers smaller than 1, with the multiplier of competitive ratio being smaller. Hence, the resulting condition is indeed easier to satisfy.

Using the Stochastic Configuration LP, we improved the competitive ratios of Stochastic Balance w.r.t. the stochastic benchmark from 0.596 to 0.611 when the success probabilities are infinitesimal, and to 0.613 if the success probabilities

Table 1. Summary of Results. The non-stochastic and stochastic benchmarks are denoted as OPT and S-OPT respectively. The results for Stochastic Balance apply to infinitesimal success probabilities. We round the competitive ratios down to the third digit after the decimal point.

	OPT	S-OPT	
Ranking (Equal Prob.)	0.534 [25] → **0.572**	$1 - \frac{1}{e} \approx 0.632$ [13]	
Stochastic Balance (Equal Prob.)	0.576 [17]	0.596 [13]	→ **0.613**
Stochastic Balance (General)	0.572 [17]	0.596 [13]	→ **0.611**

are further equal.[1,2] We also reproduce the optimal $1 - \frac{1}{e}$ competitive ratio of Ranking w.r.t. the stochastic benchmark (Table 1).

1.2   Related Works

Our analyses follow the (randomized) online primal dual framework by Devanur, Jain, and Kleinberg [7], who were inspired by Birnbaum and Mathieu [2] and Goel and Mehta [12].

An online advertising platform's objective is rarely as simple as maximizing the matching size. For instance, advertisers usually have different values for different keywords. Hence, the literature has subsequently introduced many variants of the original Online Bipartite Matching problem by Karp et al. [22]. Aggarwal et al. [1] generalized Ranking and the optimal $1 - \frac{1}{e}$ competitive ratio to the (offline) vertex-weighted problem. Feldman et al. [9] considered the edge-weighted problem with free disposal, and gave a $1 - \frac{1}{e}$ competitive algorithm assuming large capacities of offline vertices; see also Devanur et al. [5] for an online primal dual version of this result. Their algorithm may be viewed as a generalization of the Balance algorithm for unweighted matching by Kalyanasundaram and Pruh [20], which also inspired the Stochastic Balance algorithm by Mehta and Panigrahi [25] for Online Matching with Stochastic Rewards. Recently, Fahrbach et al. [8] gave the first algorithm that is strictly better than $\frac{1}{2}$-competitive for the edge-weighted problem without large-capacity assumption. The ratio was later improved independently to 0.509 by Shin and An [27], to 0.519 by Gao et al. [11], and to 0.536 by Blanc and Charikar [3]. Another important variant is the AdWords problem by Mehta et al. [26], where each offline vertex may be matched multiple times and has a budget-additive valuation. Under a small bid assumption, Mehta et al. [26] gave an optimal $1 - \frac{1}{e}$ competitive algorithm. See also Buchbinder, Jain, and Naor [4] for an online primal dual

---

[1] We remark that the competitive ratio of Goyal and Udwani [13] holds for the optimal value of the Path-based LP, which may be strictly larger than the optimal value of our Stochastic Configuration LP in some cases. Nevertheless, both LPs' optimal values upper bound the stochastic benchmark.

[2] We thank an anonymous reviewer for pointing out that a preliminary arXiv version (v5) of [13] claims a competitive ratio of 0.605 for the general infinitesimal case without proof. Nevertheless, our bound is also larger than that.

analysis of this algorithm, and Devanur and Jain [6] for a generalization called Online Matching with Concave Returns. Huang, Zhang, and Zhang [18] recently broke the $\frac{1}{2}$ barrier for AdWords with general bids and gave a 0.501-competitive algorithm, using an approach similar to Fahrbach et al. [8]. A recent line of results [23,29,30] studied budget-oblivious algorithms for AdWords, whose allocation policies do not depend on the budgets of offline vertices.

Besides the stochastic edge successes and failures in Online Matching with Stochastic Rewards, the literature has also studied other stochastic models of online matching problems. Mahdian and Yan [24] and Karande, Mehta, and Tripathi [21] independently showed that Ranking is strictly better than $1 - \frac{1}{e}$ competitive if online vertices arrive by random order. Huang et al. [16] extended this result to the vertex-weighted problem, which was further improved upon by Jin and Williamson [19]. The problem with stochastically generated graphs was first studied by Feldman et al. [10], who named it Online Stochastic Matching. We also refer readers to the most recent works on this problem by Huang and Shu [14], Huang, Shu, and Yan [15], and Tang, Wu, and Wu [28].

1.3 Paper Outline

Section 2 presents a formal definition of online matching with stochastic rewards, the benchmarks we will compare against, and existing linear programs and algorithms. Section 3 introduces our new stochastic Configuration LPs, and their properties. Section 4 shows the online primal-dual analyses of Ranking. Finally, our results for Stochastic Balance are deferred to full paper because of page limitation and the fact that the improvements mainly come from a combination of the structural lemmas by Huang and Zhang [17] and the new stochastic Configuration LP proposed in this paper. All proofs in this paper are deferred to full paper.

2 Preliminaries

We write x^+ for function $\max\{x, 0\}$. For any vector x and any index i, we write x_{-i} for the vector with the i-th entry removed, and (x_i, x_{-i}) for a vector whose i-th entry is x_i, and whose other entries are x_{-i}.

2.1 Model

Consider a bipartite graph $G = (U, V, E)$. Vertices in U are referred to as *offline vertices* because they are known to the algorithm from the beginning. Vertices in V arrive one by one, and are referred to as *online vertices*. We will write $v \prec v'$ to denote that v arrives before v'. Each edge (u, v) is associated with a success probability $0 \leq p_{uv} \leq 1$; further define $p_{uS} = \sum_{v \in S} p_{uv}$ for any $u \in U$ and any $S \subseteq V$. When an online vertex $v \in V$ arrives, the algorithm sees v's incident edges and the corresponding success probabilities. Then, the algorithm makes an irrevocable matching decision about v. Should the algorithm choose to match v to an offline vertex u, the match would succeed with probability p_{uv}. If a match is unsuccessful, the offline vertex u can still be matched to future online vertices, but the online vertex v will not get a second chance. An offline vertex $u \in U$ is *successful* if there is an edge that is matched to u and is successful; u is *unsuccessful* otherwise. We want to maximize the number of successful offline vertices.

We remark that the results in this paper generalize to the vertex-weighted problem, where offline vertices have positive weights and we want to maximize the total weight of successful offline vertices. Readers familiar with the online matching literature shall not find this surprising since almost all known results for unweighted online matching problems generalize to the vertex-weighted problems. This version of the paper only presents the unweighted case to keep the exposition simple.

By allowing $p_{uv} = 0$, we may assume without loss of generality that $E = U \times V$ which will simplify the exposition in some parts of this paper. We say that u and v are neighbors if $p_{uv} > 0$. We say that the success probabilities are equal if p_{uv} is either 0 and p for some $0 \leq p \leq 1$.

Stochastic Budgets. Mehta and Panigrahi [25] showed that when the success probabilities are infinitesimal, we can consider an equivalent model in which the randomness plays a different role. At the beginning, each offline vertex $u \in U$ independently draws a budget θ_u from the exponential distribution with mean 1. When an online vertex $v \in V$ arrives, the algorithm may match it to any offline neighbor and collect a gain that equals the success probability of the edge. However, the total gain from an offline vertex u is capped by its budget θ_u, i.e., it is either the sum of success probabilities of the edges matched to u, denoted as ℓ_u and referred to as its *load*, or its budget θ_u, whichever is smaller. Further, the budget θ_u is unknown to the algorithm until the moment when the load ℓ_u exceeds the budget, which corresponds to u's succeeding.

Benchmarks and Competitive Analysis. The *offline (non-stochastic) optimum*, denoted as OPT, refers to the optimal objective achievable by a computationally unlimited offline algorithm that knows the graph and success probabilities, and when the budgets are non-stochastically equal to 1. The offline optimum upper bounds the objective achievable by any (online or offline) algorithm in the model with stochastic budgets [25].

The *offline stochastic optimum*, denoted as S-OPT, refers to the optimal objective achievable by a computationally unlimited offline algorithm that knows the graph and success probabilities, but can only match the online vertices one by one by their arrival order, and can only observe the stochastic budget of an offline vertex when its load exceeds the budget like online algorithms.

An online algorithm is Γ-competitive w.r.t. one of these benchmarks if for *any instance* of Online Matching with Stochastic Rewards, the expected objective given by the algorithm is at least a Γ fraction of the benchmark. Here $0 \leq \Gamma \leq 1$ is called the *competitive ratio*.

2.2 Existing Linear Programs

Consider $0 \leq x_{uv} \leq 1$ as the probability that u is matched to v (successful or not) for any $u \in U$ and any $v \in V$. The *Matching LP* is defined as:

$$\max_{x \geq 0} \quad \sum_{u \in U} \sum_{v \in V} p_{uv} x_{uv}$$

$$\text{s.t.} \quad \sum_{v \in V} p_{uv} x_{uv} \leq 1 \qquad \forall u \in U$$

$$\sum_{u \in U} x_{uv} \leq 1 \qquad \forall v \in V$$

The first constraint states that the expected load of an offline vertex u, which is equal to the probability that u succeeds, is at most 1. The second constraint says that each online vertex v can be matched to at most one offline vertex. An offline allocation yields a feasible solution to the Matching LP by the aforementioned interpretation of x_{uv}. Hence, the optimal objective of Matching LP upper bounds the offline optimum OPT.

For any offline vertex $u \in U$ and any subset of online vertices $S \subseteq U$, let $\bar{p}_{uS} = \min\{p_{uS}, 1\}$, where recall that $p_{uS} = \sum_{v \in S} p_{uv}$. Consider $0 \leq x_{uS} \leq 1$ as the probability that S is the subset of online vertices matched to u by the algorithm. Huang and Zhang [17] used the following *Configuration LP* and its dual LP:

$$\max_{x \geq 0} \quad \sum_{u \in U} \sum_{S \subseteq V} \bar{p}_{uS} x_{uS} \qquad \textbf{(Primal)}$$

$$\text{s.t.} \quad \sum_{S \subseteq V} x_{uS} \leq 1 \qquad \forall u \in U$$

$$\sum_{u \in U} \sum_{S \subseteq V : v \in S} x_{uS} \leq 1 \qquad \forall v \in V$$

$$\min_{\alpha, \beta \geq 0} \quad \sum_{u \in U} \alpha_u + \sum_{v \in V} \beta_v \qquad \textbf{(Dual)}$$

$$\text{s.t.} \quad \alpha_u + \sum_{v \in S} \beta_v \geq \bar{p}_{uS} \qquad \forall u \in U, \forall S \subseteq V \qquad (3)$$

The optimal objective of Configuration LP also upper bounds the offline optimum OPT. In fact, it coincides with the optimal objective of Matching LP when the success probabilities are infinitesimal (c.f., Huang and Zhang [17]). Nevertheless, its dual structure is better suited for an online primal dual analysis.

Lemma 1 (c.f., Lemma 3 of Huang and Zhang [17]). *Suppose that an algorithm for Online Matching with Stochastic Rewards can further split p_{uv} between α_u and β_v each time it matches an edge (u, v), such that the dual constraint (3) holds up to factor Γ in expectation:*

$$\mathbf{E}\left[\alpha_u + \sum_{v \in S} \beta_v\right] \geq \Gamma \cdot \bar{p}_{uS}.$$

Then, the algorithm is Γ-competitive w.r.t. OPT.

Finally, Goyal and Udwani [13] introduced the *Path-based LP* which is tailored for the offline stochastic optimum S-OPT. A *sample path* refers to all information of what has happened so far, including the subset of edges matched by the algorithm, and the realization of whether these matches are successful. This LP considers variable $0 \leq x_{uv}^{\omega} \leq 1$ for any edge (u, v) which represents the probability that the algorithm matches v to u conditioned on a sample path ω. We omit the details of Path-based LP because it is not directly related to the LPs in this paper, and its constraints are too complicated to be covered concisely here. We remark that the number of possible sample paths and thus the number of variables of Path-based LP are exponential in the number of *edges*. By contrast, the number of variables of Configuration LP is exponential in the number of *vertices*, which is typically much smaller than the number of edges.

2.3   Existing Algorithms

Ranking. We will consider the *Ranking* algorithm in the case of equal success probabilities. At the beginning, draw a rank $\rho_u \in [0, 1]$ uniformly at random for each offline vertex $u \in U$. Then, on the arrival of each online vertex $v \in V$, match it to the unsuccessful neighbor that has the smallest rank. We can break ties arbitrarily since they occur with zero probability.

Stochastic Balance (Equal Probability). On the arrival of each online vertex $v \in V$, match it to the unsuccessful neighbor that has the smallest load, breaking ties arbitrarily, e.g., by the lexicographical order. Recall that the load of an offline vertex is the sum of success probabilities of the edges that have been matched to it so far.

Stochastic Balance (General). We first describe a fractional algorithm. On the arrival of each online vertex $v \in V$, consider a continuous process that matches v fractionally to the unsuccessful neighbor $u \in U$ with the largest $p_{uv}(1 - g(\ell_u))$ where $g : [0, 1] \to [0, 1]$ is a non-decreasing function chosen by the algorithm designer. If the success probabilities are infinitesimal, one can further convert it to an integral algorithm with no loss in the competitive ratio, through a reduction by Huang and Zhang [17] based on randomized rounding. Hence, it suffices to analyze the fractional algorithm's competitive ratio.

3   Stochastic Configuration Linear Programs

3.1   Stochastic Thresholds

Even when the success probabilities are arbitrary, we may still consider an equivalent model that generalizes the viewpoint of stochastic budgets for infinitesimal success probabilities. We call this the *stochastic thresholds viewpoint*. At the beginning, each offline vertex u independently draws a threshold τ_u uniformly from $[0, 1]$. When an online vertex v arrives, the algorithm may match it to any offline neighbor and collect a gain which equals the success probability of the edge. For each offline vertex u and any subset of online vertices S, define:

$$\tilde{p}_{uS} = 1 - \prod_{v \in S} (1 - p_{uv}).$$

An offline vertex u is successful (and can no longer be matched) if the subset of online vertices S matched to it satisfies $\tilde{p}_{uS} \geq \tau_u$. Further, the threshold τ_u is unknown to the algorithm until the moment that offline vertex u becomes successful.

Observe that when the success probabilities are infinitesimal, we have $\tilde{p}_{uS} = 1 - e^{-p_{uS}}$, and the stochastic budget and threshold of an offline vertex $u \in U$ satisfy $\theta_u = -\ln(1 - \tau_u)$.

3.2   Reduced-Form Stochastic Configuration Linear Program

Consider $0 \leq y_{uS} \leq 1$ as the probability that a subset of online vertices S would be matched to an offline vertex u if it has an infinite stochastic budget $\theta_u = \infty$ (i.e., when $\tau_u = 1$ if the success probabilities are not infinitesimal), over the randomness of the stochastic budgets θ_{-u} of other offline vertices and the randomness of the algorithm. Further, for any $S \subseteq V$ and $v \in V$, let:

$$S(v) = \{v' \in S : v' \prec v\}$$

denote the subset of online vertices in S that arrive before v. We consider the following *Reduced-form Stochastic Configuration LP* and its dual:

$$\max_{y \geq 0} \quad \sum_{u \in U} \sum_{S \subseteq V} \tilde{p}_{uS} \, y_{uS} \qquad \qquad \text{(Primal)}$$

$$\text{s.t.} \quad \sum_{S \subseteq V} y_{uS} \leq 1 \qquad \qquad \forall u \in U \qquad (4)$$

$$\sum_{u \in U} \sum_{S \subseteq V : v \in S} (1 - \tilde{p}_{uS(v)}) y_{uS} \leq 1 \qquad \forall v \in V \qquad (5)$$

$$\min_{\alpha, \beta \geq 0} \quad \sum_{u \in U} \alpha_u + \sum_v \beta_v \qquad \qquad \text{(Dual)}$$

$$\text{s.t.} \quad \alpha_u + \sum_{v \in S} (1 - \tilde{p}_{uS(v)}) \beta_v \geq \tilde{p}_{uS} \qquad \forall u \in U, \forall S \subseteq V \qquad (6)$$

Theorem 1. *The optimal objective of Reduced-form Stochastic Configuration LP is greater than or equal to S-OPT.*

Lemma 2. *Suppose that an algorithm for Online Matching with Stochastic Rewards can further split p_{uv} between α_u and β_v each time it matches an edge (u, v), such that the dual constraint (6) holds up to factor Γ in expectation, i.e.:*

$$\mathbf{E}\left[\alpha_u + \sum_{v \in S} (1 - \tilde{p}_{uS(v)}) \beta_v \right] \geq \Gamma \cdot \tilde{p}_{uS}.$$

Then, the algorithm is Γ-competitive w.r.t. S-OPT.

3.3   Stochastic Configuration Linear Program

Consider $0 \leq y_{uS}(\theta_{-u}) \leq 1$ (respectively, $y_{uS}(\tau_{-u})$ if the success probabilities are *not* infinitesimal) as the probability that a subset of online vertices S would be matched to an offline vertex u if it has an infinite stochastic budget $\theta_u = \infty$ (respectively, if $\tau_u = 1$), *conditioned on the stochastic budgets* θ_{-u} (respectively, the stochastic thresholds τ_{-u}) of other offline vertices, and over the randomness of the algorithm. This paper will further consider an even more expressive Stochastic Configuration LP and its dual in some analyses.

$$
\underset{y \geq 0}{\text{maximize}} \quad \sum_{u \in U} \sum_{S \subseteq V} \mathbf{E}_{\theta_{-u}} \Big[\tilde{p}_{uS} \, y_{uS}(\theta_{-u}) \Big] \qquad \textbf{(Primal)}
$$

$$
\text{subject to} \quad \sum_{S \subseteq V} y_{uS}(\theta_{-u}) \leq 1 \qquad \forall u \in U, \forall \theta_{-u} \qquad (7)
$$

$$
\sum_{u} \sum_{S \subseteq V : v \in S, \theta_u \geq p_{uS}(v)} y_{uS}(\theta_{-u}) \leq 1 \qquad \forall v \in V, \forall \theta \qquad (8)
$$

$$
\underset{\alpha, \beta \geq 0}{\text{minimize}} \quad \mathbf{E}_{\theta} \Big[\sum_{u} \alpha_u(\theta_{-u}) + \sum_{v \in V} \beta_v(\theta) \Big] \qquad \textbf{(Dual)}
$$

$$
\text{subject to} \quad \alpha_u(\theta_{-u}) + \mathbf{E}_{\theta_u} \Big[\sum_{v \in S : \theta_u \geq p_{uS}(v)} \beta_v(\theta_u, \theta_{-u}) \Big] \geq \tilde{p}_{uS} \quad \forall u \in U, \forall S \subseteq V, \forall \theta_{-u}
$$

$$
(9)
$$

Theorem 2. *The optimal objective of Stochastic Configuration LP is at least* S-OPT.

Lemma 3. *Suppose that an algorithm for Online Matching with Stochastic Rewards can further split p_{uv} between α_u and β_v each time it matches an edge (u, v). Let $\alpha_u(\theta_{-u})$ be the expectation of α_u conditioned on any stochastic budgets θ_{-u} of offline vertices other than u, and let $\beta(\theta)$ be the expectation of β_v conditioned on any stochastic budgets θ of all vertices. If the dual constraint (6) holds up to factor Γ, i.e.:*

$$
\alpha_u(\theta_{-u}) + \mathbf{E}_{\theta_u} \Big[\sum_{v \in S : \theta_u \geq p_{uS}(v)} \beta_v(\theta_u, \theta_{-u}) \Big] \geq \Gamma \cdot \tilde{p}_{uS},
$$

then the algorithm is Γ-competitive w.r.t. S-OPT.

4   Ranking

4.1   Basics

We will assume throughout the section that the ranks of offline vertices are distinct, since that happens with probability 1. Below we first develop some basic elements that will be used by the proofs of both results in this section.

Dual Updates. We start by explaining the dual update rule associated with Ranking, which is identical to the existing one in the literature. The dual variables are initially 0. Let $g : [0,1] \to [0,1]$ be a non-decreasing gain splitting function to be determined. We will assume that $g(1) = 1$ which will allow us to handle a boundary case of the analysis under a unified framework. Recall that ρ_u denotes the rank of u which is uniformly drawn between 0 and 1 at the beginning by the Ranking algorithm. When the algorithm matches an online vertex v to an offline vertex u, increase α_u by $p \cdot g(\rho_u)$, and set β_v as $p \cdot \left(1 - g(\rho_u)\right)$, where recall that p denotes the equal success probability of matching neighboring vertices.

Characterization of Matching. All analyses in this section will fix an offline vertex $u \in U$ and the ranks ρ_{-u} and stochastic thresholds τ_{-u} of other offline vertices. A canonical analysis of Ranking, such as those for Online Bipartite Matching (c.f., Devanur et al. [7]), would further fix a neighboring online vertex $v \in V$ and characterize the matching outcome of u and v. For the problem of Online Matching with Stochastic Rewards and in the spirit of Configuration LPs, however, we need to characterize the joint matching outcome of u and *all its neighbors*.

We first define some notations. Consider an imaginary run of Ranking with vertex u removed, while keeping the ranks of other offline vertices as ρ_{-u}. Further consider an online vertex v in u's neighborhood. If v is matched to an offline vertex u' in the imaginary run, define the v's *critical rank* as $\mu_v = \rho_{u'}$. If v is not matched, define its critical rank as $\mu_v = 1$. Finally, for any $0 \le \mu \le 1$, let $N_u(\mu)$ be the set of u's neighbors whose critical ranks are greater than or equal to μ.

Lemma 4. *Suppose that u's rank is ρ_u and u's stochastic threshold is such that u succeeds after i matches. Then, the subset of online vertices matched to u is the first $\min\{i, |N_u(\rho_u)|\}$ vertices in $N_u(\rho_u)$ according to the arrival order.*

Expectation of Dual Variables. The next invariant about dual variable α_u follows from the above gain splitting rule.

Lemma 5. *For any offline vertex $u \in U$, any ranks ρ and stochastic thresholds τ of offline vertices, and the corresponding load ℓ_u of u, we have:*

$$\alpha_u = \ell_u \cdot g(\rho_u).$$

As a corollary, we can further bound the expectation of α_u conditioned on the ranks ρ_{-u} and stochastic thresholds τ_{-u} of other offline vertices.

Lemma 6. *For any offline vertex $u \in U$, any ranks ρ_{-u} and stochastic thresholds τ_{-u} of offline vertices other than u, we have:*

$$\mathbf{E}_{\rho_u, \tau_u}\left[\alpha_u \mid \rho_{-u}, \tau_{-u}\right] = \int_0^1 \left(1 - (1-p)^{|N_u(\rho_u)|}\right) g(\rho_u)\, \mathrm{d}\rho_u.$$

Recall that $N_u(\mu)$ is the subset of u's neighbors whose critical ranks are greater than or equal to μ. Further let $N_u(\mu, v)$ be the subset of vertices in $N_u(\mu)$ that arrive before v.

Lemma 7. *For any offline vertex* $u \in U$, *any* u's *neighbor* v, *any ranks* ρ_{-u} *and stochastic thresholds* τ_{-u} *of offline vertices other than* u, *we have:*

$$\mathbf{E}_{\rho_u, \tau_u}\left[\beta_v \mid \rho_{-u}, \tau_{-u}\right] \geq p\left(1 - g(\mu_v) + \int_0^{\mu_v} (1-p)^{|N_u(\rho_u, v)|}(g(\mu_v) - g(\rho_u)) \, d\rho_u\right).$$

4.2   Non-stochastic Benchmark

Theorem 3. *Ranking is at least* 0.572 *-competitive w.r.t. the offline (non-stochastic) benchmark* OPT *for any instance with equal success probabilities.*

The rest of the subsection is devoted to proving Theorem 3 by an online primal dual analysis with the Configuration LP. By Lemma 1, it suffices to show that $\mathbf{E}\left[\alpha_u + \sum_{v \in S} \beta_v\right] \geq \Gamma \cdot \bar{p}_{uS}$ for any offline vertex $u \in U$ and any subset of online vertices $S \subseteq V$, and with the stated competitive ratio $\Gamma = 0.572$. We will prove it further conditioning on any ranks ρ_{-u} and stochastic thresholds τ_{-u} of other vertices, i.e.:

$$\mathbf{E}_{\rho_u, \tau_u}\left[\alpha_u + \sum_{v \in S} \beta_v \mid \rho_{-u}, \tau_{-u}\right] \geq \Gamma \cdot \bar{p}_{uS}, \tag{10}$$

First we apply the bounds for dual variables' expectation from Lemmas 6 and 7 to the LHS of the inequality:

$$\mathbf{E}_{\rho_u, \tau_u}\left[\alpha_u + \sum_{v \in S} \beta_v \mid \rho_{-u}, \tau_{-u}\right] \geq \int_0^1 \Big(1 - \underbrace{(1-p)^{|N_u(\rho_u)|}}_{(a)}\Big)g(\rho_u) \, d\rho_u \tag{11}$$

$$+ \underbrace{\sum_{v \in S} p}_{(b)}\left(1 - g(\mu_v) + \int_0^{\mu_v} \underbrace{(1-p)^{|N_u(\rho_u, v)|}}_{(c)}(g(\mu_v) - g(\rho_u)) \, d\rho_u\right).$$

Assumptions. The rest of the analysis will treat the above as a pure inequality problem, where the arrival order and critical ranks of u's neighbors may be chosen arbitrarily by an adversary. In particular, when we characterize the worst-case scenario of the inequality we shall not concern about how to construct an instance to get the specified critical ranks. With this treatment, we can make several assumptions that on the one hand are without loss of generality, and on the other hand simplify the subsequent analysis.

First recall the assumption that the offline vertices' ranks are distinct, because the exceptions happen with zero probability. In other words, we may relax and simplify the RHS assuming distinct ranks, which for example implies that u's rank ρ_u does not equal the critical rank μ_v of any v. Although the resulting bound may be violated for some zero measure subset of ranks ρ_u, the bound still holds after the integration.

Further, it suffices to consider the above inequality for infinitesimal success probabilities. For $p = 1$ it is the online bipartite matching problem by Karp et al. [22], for which Ranking is $1 - \frac{1}{e}$ competitive. For any instance with $0 < p < 1$ and any large integer n, we can instead consider a new instance such that (i) the equal success probabilities is $p' = 1 - \sqrt[n]{1-p}$, (ii) each online vertex in the original instance has n copies in the new instance, which arrive consecutively and have the same critical ranks as the original vertex, and (iii) the subset S' of concerned in the new instance contains $m = \lfloor \frac{p}{p'} \rfloor$ copies of each vertex $v \in S$ in the original one. By this construction, the RHS of Eq. (10) is the same in the two instances, up to an error from the rounding of m that diminishes as n tends to infinity. Further, the RHS of Eq. (11) is smaller in new instance because (i) and (ii) ensure that part (a) stays the same and part (c) decreases, and (iii) ensures that the changes to part (b) weakly decreases the expression. Hence, it suffices to establish the inequality in the new instance which satisfy the assumption of infinitesimal success probability. We may therefore rewrite the expression as:

$$\mathbf{E}_{\rho_u, \tau_u}\left[\alpha_u + \sum_{v \in S} \beta_v \mid \rho_{-u}, \tau_{-u} \right] \geq \int_0^1 \left(1 - e^{-p|N_u(\rho_u)|} \right) g(\rho_u) \, \mathrm{d}\rho_u \tag{12}$$

$$+ \sum_{v \in S} p\big(1 - g(\mu_v)\big) + \sum_{v \in S} \int_0^{\mu_v} e^{-p|N_u(\rho_u, v)|} \left(1 - e^{-p} \right) \big(g(\mu_v) - g(\rho_u) \big) \, \mathrm{d}\rho_u.$$

Finally, it is sufficient to establish the stated approximate dual feasibility condition when the critical ranks μ_v are distinct for different v. Otherwise, we may slightly perturb the critical ranks, e.g., by a random noise drawn from $[-\epsilon, \epsilon]$ for a sufficiently small $\epsilon > 0$, to get a new instance satisfying our assumption. The resulting RHS of Eq. (12) converges to the original one when the magnitude of perturbations tends to zero.

For any $0 \leq \mu \leq 1$, let:

$$P(\mu) = p \cdot \big| N_u(\mu) \cap S \big|, \qquad \bar{P}(\mu) = p \cdot \big| N_u(\mu) \backslash S \big|$$

denote the sums of success probabilities of u's neighbors with critical ranks greater than or equal to μ, inside S and outside S respectively.

Characterization of Worst-Case. We next present a series of lemmas that characterize the worst-case of various aspects of the instance, including the arrival order, the size of S, and critical ranks, etc.

Lemma 8 (Worst Arrival Order). *We have:*

$$\mathbf{E}_{\rho_u, \tau_u}\left[\alpha_u + \sum_{v \in S} \beta_v \mid \rho_{-u}, \tau_{-u} \right] \geq \int_0^1 \left(1 - e^{-\bar{P}(\rho_u)} \right) g(\rho_u) \, \mathrm{d}\rho_u$$

$$- \int_0^1 \big(1 - g(\mu_v)\big) \, \mathrm{d}P(\mu_v) - \int_0^1 e^{-P(\rho_u) - \bar{P}(\rho_u)} \int_{\rho_u}^1 g(\mu_v) \, \mathrm{d}e^{P(\mu_v)} \, \mathrm{d}\rho_u. \tag{13}$$

The lemma's inequality would hold with equality if online vertices in S arrive by increasing order of critical ranks, and vertices outside S arrive before those in S. We remark that the negative signs in front of the second and third integration come from the fact that $P(\mu)$ is non-increasing. All three parts make positive contribution to the RHS.

Lemma 9 (Worst Critical Ranks Outside S). *Given any non-increasing function $P : [0,1] \rightarrow [0,\infty)$, the RHS of Eq. (13) is minimized when $\bar{P}(\mu)$ is a step function that equals ∞ for $\mu < \mu_0$ and equals 0 for $\mu > \mu_0$, for some threshold μ_0.*

The lemma indicates that letting all online vertices outside S have the same critical rank μ_0 is the worst-case scenario.

Given Lemma 9, and noting that $p_{uS} = P(0)$ and thus $\bar{p}_{uS} = \min\{P(0), 1\}$, it suffices to find a gain splitting function g such that for any non-increasing function $P : [0,1] \rightarrow [0,\infty)$ we have:

$$\int_0^{\mu_0} g(\rho_u)\,\mathrm{d}\rho_u - \int_0^1 \left(1 - g(\mu_v)\right)\mathrm{d}P(\mu_v) - \int_{\mu_0}^1 e^{-P(\rho_u)} \int_{\rho_u}^1 g(\mu_v)\,\mathrm{d}e^{P(\mu_v)}\,\mathrm{d}\rho_u$$
$$\geq \Gamma \cdot \min\{P(0), 1\}.$$
$$(14)$$

Lemma 10 (Worst Size of S). *If Eq. (14) holds for all non-increasing function P that satisfies $p_{uS} = P(0) = 1$, then it also holds for an arbitrary non-increasing function P.*

Therefore, our task further simplifies to finding a gain splitting function g such that for any non-increasing function $P : [0,1] \rightarrow [0,1]$ with $P(0) = 1$ we have:

$$\int_0^{\mu_0} g(\rho_u)\,\mathrm{d}\rho_u - \int_0^1 \left(1 - g(\mu_v)\right)\mathrm{d}P(\mu_v) - \int_{\mu_0}^1 e^{-P(\rho_u)} \int_{\rho_u}^1 g(\mu_v)\,\mathrm{d}e^{P(\mu_v)}\,\mathrm{d}\rho_u \geq \Gamma.$$
$$(15)$$

This is the moment when we finally specify the gain splitting function g:

$$g(\rho) = \begin{cases} \min\left\{\frac{c}{e-(e-1)\rho}, 1 - \frac{1}{e}\right\} & 0 \leq \rho < 1; \\ 1 & \rho = 1. \end{cases}$$
$$(16)$$

for a constant $c \approx 1.161$ such that $\int_0^1 g(x) = 1 - g(0) > 0.572$.

Lemma 11 (Worst Critical Ranks Inside S). *For the function g in Eq. (15) is minimized when P approaches a step function in the limit, i.e., when for some $\rho_0 > \mu_0$:*

$$P(\rho) = \begin{cases} 1 & 0 \leq \rho < \rho_0 - \epsilon; \\ \frac{\rho_0 - \rho}{\epsilon} & \rho_0 - \epsilon \leq \rho \leq \rho_0; \\ 0 & \rho_0 < \rho \leq 1. \end{cases}$$

and let $\epsilon \rightarrow 0$.

Despite its complex look, the lemma actually gives a simple characterization of the worst-case critical ranks inside S: all online vertices in S have the same critical rank $\rho_0 > \mu_0$. We have the complex form because we write the lower bounds of the LHS of approximate dual feasibility through a function P, which help simplify the proofs of previous lemmas but is inconvenient when we need to represent identical critical ranks.

Finally, we apply the above worst-case function P and focus on optimizing the gain splitting function g w.r.t. the resulting differential inequality.

Lemma 12. *For any $\mu_0 < \rho_0$, and for the function g in Eq. (16), we have:*

$$\int_0^{\mu_0} g(\rho_u)\,\mathrm{d}\rho_u + \left(1 - g(\rho_0)\right) + \left(1 - \frac{1}{e}\right)(\rho_0 - \mu_0)g(\rho_0) \geq \Gamma,$$

for the stated competitive ratio $\Gamma = 0.572$

4.3 Stochastic Benchmark

Theorem 4. *Ranking is $1 - \frac{1}{e}$-competitive w.r.t. the offline stochastic benchmark S-OPT for any instance with equal success probabilities.*

The rest of the subsection is devoted to proving Theorem 4 by an online primal dual analysis with the Reduced-form Stochastic Configuration LP. By Lemma 2, it is sufficient to show that $\mathbf{E}\left[\alpha_u + \sum_{v \in S}(1 - \tilde{p}_{uS(v)})\beta_v\right] \geq \Gamma \tilde{p}_{uS}$ for any offline vertex $u \in U$ and any subset of online vertices $S \subseteq V$, and with the stated competitive ratio $\Gamma = 1 - \frac{1}{e}$. We will prove it further conditioning on any ranks ρ_{-u} and stochastic thresholds τ_{-u} of other vertices, i.e.:

$$\mathbf{E}_{\rho_u, \tau_u}\left[\alpha_u + \sum_{v \in S}(1 - \tilde{p}_{uS(v)})\beta_v \mid \rho_{-u}, \tau_{-u}\right] \geq \left(1 - \frac{1}{e}\right) \cdot \tilde{p}_{uS} \qquad (17)$$

Let $v_1, v_2, \ldots, v_n$ be the online vertices in S, sorted by their critical ranks, which we denote as $\mu_1 \geq \mu_2 \geq \cdots \geq \mu_n$. We next apply the lower bounds for the dual variables' expectation to the LHS of Eq. (17). First we have:

$$\mathbf{E}_{\rho_u, \tau_u}\left[\alpha_u \mid \rho_{-u}, \tau_{-u}\right] \geq \int_0^1 \left(1 - (1 - p)^{|N_u(\rho_u)|}\right)g(\rho_u)\,\mathrm{d}\rho_u \qquad \text{(Lemma 6)}$$

$$\geq \int_0^1 \left(1 - (1 - p)^{|N_u(\rho_u) \cap S|}\right)g(\rho_u)\,\mathrm{d}\rho_u$$

$$= \int_0^1 \sum_{i=1}^{|N_u(\rho_u) \cap S|} p(1 - p)^{i-1}g(\rho_u)\,\mathrm{d}\rho_u$$

$$= \sum_{i=1}^n p(1 - p)^{i-1}\int_0^{\mu_i} g(\rho_u)\,\mathrm{d}\rho_u.$$

To bound the expected contribution by β_v's, we apply a weaker version of Lemma 7, dropping the second part on the RHS of the lemma's inequality. Suppose that the vertices in S arrive by order $v_{\pi(1)} \prec v_{\pi(1)} \prec \ldots \prec v_{pi(n)}$. We get that:

$$\mathbf{E}_{\rho_u, \tau_u}\left[\sum_{v \in S}(1 - \tilde{p}_{uS(v)})\beta_v \mid \rho_{-u}, \tau_{-u}\right]$$

$$\geq \sum_{i=1}^{n} p(1-p)^{i-1}\left(1 - g(\mu_{\pi(i)})\right) \qquad \text{(Lemma 7)}$$

$$\geq \sum_{i=1}^{n} p(1-p)^{i-1}\left(1 - g(\mu_i)\right). \qquad \text{(rearrangement inequality)}$$

Combining the two bounds, we get that the LHS of Eq. (17) is at least:

$$\sum_{i=1}^{n} p(1-p)^{i-1}\underbrace{\left(\int_0^{\mu_i} g(\rho_u)\, d\rho_u + 1 - g(\mu_i)\right)}_{(\star)}$$

Next we choose $g(x) = e^{x-1}$ just like the analysis of Ranking for the original Online Bipartite Matching problem. This choice ensures that $(\star)$ equals $1 - \frac{1}{e}$. The above bound is therefore:

$$\left(1 - \frac{1}{e}\right)\sum_{i=1}^{n} p(1-p)^{i-1} = \left(1 - \frac{1}{e}\right)\left(1 - (1-p)^n\right) = \left(1 - \frac{1}{e}\right) \cdot \tilde{p}_{uS}.$$

References

1. Aggarwal, G., Goel, G., Karande, C., Mehta, A.: Online vertex-weighted bipartite matching and single-bid budgeted allocations. In: SODA, pp. 1253–1264. SIAM (2011)
2. Birnbaum, B., Mathieu, C.: On-line bipartite matching made simple. ACM SIGACT News **39**(1), 80–87 (2008)
3. Blanc, G., Charikar, M.: Multiway online correlated selection. In: FOCS, pp. 1277–1284. IEEE (2022)
4. Buchbinder, N., Jain, K., Naor, J.S.: Online primal-dual algorithms for maximizing ad-auctions revenue. In: Arge, L., Hoffmann, M., Welzl, E. (eds.) ESA 2007. LNCS, vol. 4698, pp. 253–264. Springer, Heidelberg (2007). https://doi.org/10.1007/978-3-540-75520-3_24
5. Devanur, N.R., Huang, Z., Korula, N., Mirrokni, V.S., Yan, Q.: Whole-page optimization and submodular welfare maximization with online bidders. ACM Trans. Econ. Comput. **4**(3), 1–20 (2016)
6. Devanur, N.R., Jain, K.: Online matching with concave returns. In: STOC, pp. 137–144 (2012)
7. Devanur, N.R., Jain, K., Kleinberg, R.D.: Randomized primal-dual analysis of ranking for online bipartite matching. In: SODA, pp. 101–107. SIAM (2013)
8. Fahrbach, M., Huang, Z., Tao, R., Zadimoghaddam, M.: Edge-weighted online bipartite matching. J. ACM **69**(6), 1–35 (2022)

9. Feldman, J., Korula, N., Mirrokni, V., Muthukrishnan, S., Pál, M.: Online ad assignment with free disposal. In: Leonardi, S. (ed.) WINE 2009. LNCS, vol. 5929, pp. 374–385. Springer, Heidelberg (2009). https://doi.org/10.1007/978-3-642-10841-9_34

10. Feldman, J., Mehta, A., Mirrokni, V., Muthukrishnan, S.: Online stochastic matching: beating $1 - \frac{1}{e}$. In: FOCS, pp. 117–126. IEEE (2009)

11. Gao, R., He, Z., Huang, Z., Nie, Z., Yuan, B., Zhong, Y.: Improved online correlated selection. In: FOCS, pp. 1265–1276. IEEE (2021)

12. Goel, G., Mehta, A.: Online budgeted matching in random input models with applications to adwords. In: SODA, pp. 982–991 (2008)

13. Goyal, V., Udwani, R.: Online matching with stochastic rewards: optimal competitive ratio via path-based formulation. Oper. Res. **71**(2), 563–580 (2023)

14. Huang, Z., Shu, X.: Online stochastic matching, Poisson arrivals, and the natural linear program. In: STOC, pp. 682–693 (2021)

15. Huang, Z., Shu, X., Yan, S.: The power of multiple choices in online stochastic matching. In: STOC, pp. 91–103 (2022)

16. Huang, Z., Tang, Z.G., Wu, X., Zhang, Y.: Online vertex-weighted bipartite matching: beating $1-\frac{1}{e}$ with random arrivals. ACM Trans. Algorithms **15**(3), 1–15 (2019)

17. Huang, Z., Zhang, Q.: Online primal dual meets online matching with stochastic rewards: configuration LP to the rescue. In: STOC, pp. 1153–1164. ACM (2020)

18. Huang, Z., Zhang, Q., Zhang, Y.: AdWords in a panorama. In: FOCS, pp. 1416–1426. IEEE (2020)

19. Jin, B., Williamson, D.P.: Improved analysis of RANKING for online vertex-weighted bipartite matching in the random order model. In: Feldman, M., Fu, H., Talgam-Cohen, I. (eds.) WINE 2021. LNCS, vol. 13112, pp. 207–225. Springer, Cham (2022). https://doi.org/10.1007/978-3-030-94676-0_12

20. Kalyanasundaram, B., Pruhs, K.R.: An optimal deterministic algorithm for online b-matching. Theor. Comput. Sci. **233**(1–2), 319–325 (2000)

21. Karande, C., Mehta, A., Tripathi, P.: Online bipartite matching with unknown distributions. In: STOC, pp. 587–596 (2011)

22. Karp, R.M., Vazirani, U.V., Vazirani, V.V.: An optimal algorithm for on-line bipartite matching. In: STOC, pp. 352–358. ACM (1990)

23. Liang, J., Tang, Z.G., Xu, Y., Zhang, Y., Zhou, R.: On the perturbation function of ranking and balance for weighted online bipartite matching. ArXiv abs/2210.10370 (2022)

24. Mahdian, M., Yan, Q.: Online bipartite matching with random arrivals: an approach based on strongly factor-revealing LPs. In: STOC, pp. 597–606 (2011)

25. Mehta, A., Panigrahi, D.: Online matching with stochastic rewards. In: FOCS, pp. 728–737. IEEE (2012)

26. Mehta, A., Saberi, A., Vazirani, U., Vazirani, V.: Adwords and generalized online matching. J. ACM **54**(5), 22-es (2007)

27. Shin, Y., An, H.C.: Making three out of two: three-way online correlated selection. In: ISAAC. Schloss Dagstuhl-Leibniz-Zentrum für Informatik (2021)

28. Tang, Z.G., Wu, J., Wu, H.: (Fractional) online stochastic matching via fine-grained offline statistics. In: STOC, pp. 77–90 (2022)

29. Udwani, R.: Adwords with unknown budgets and beyond. ArXiv abs/2110.00504 (2021)

30. Vazirani, V.V.: Towards a practical, budget-oblivious algorithm for the adwords problem under small bids. ArXiv abs/2107.10777 (2022)

Online Nash Welfare Maximization Without Predictions

Zhiyi Huang[1]([⊠])[iD], Minming Li[2][iD], Xinkai Shu[1][iD], and Tianze Wei[2][iD]

[1] University of Hong Kong, Kowloon, Hong Kong
{zhiyi,xkshu}@cs.hku.hk
[2] City University of Hong Kong, Kowloon, Hong Kong
minming.li@cityu.edu.hk, t.z.wei-8@my.cityu.edu.hk

Abstract. The maximization of Nash welfare, which equals the geometric mean of agents' utilities, is widely studied because it balances efficiency and fairness in resource allocation problems. Banerjee, Gkatzelis, Gorokh, and Jin (2022) recently introduced the model of online Nash welfare maximization for T divisible items and N agents with additive utilities with predictions of each agent's utility for receiving all items. They gave online algorithms whose competitive ratios are logarithmic. We initiate the study of online Nash welfare maximization *without predictions*, assuming either that the agents' utilities for receiving all items differ by a bounded ratio, or that their utilities for the Nash welfare maximizing allocation differ by a bounded ratio. We design online algorithms whose competitive ratios are logarithmic in the aforementioned ratios of agents' utilities and the number of agents.

Keywords: Fair division · Online algorithm · Nash welfare

1 Introduction

Suppose that you run a daycare center. From time to time, some organizations donate candies for children who have different preferences over various kinds of candies: Alice likes chocolates, Bob prefers gummy bears, Charlie favors lollipops, and so forth. The candies come in abundance, so you may consider it as a resource allocation problem of *divisible* items. How would you distribute the candies to the children?

Naturally, we would like to allocate the candies based on the children's values for them. We will assume that the children have *additive utilities* for receiving bundles of candies. Following this idea, it may be tempting to allocate the candies to the children in a way that maximizes the sum of children's utilities for the candies that they received, a.k.a., the social welfare. The example below, however, highlights the potential unfairness of this approach.

Example 1. Consider distributing two packs of chocolates and two packs of gummy bears to two children Alice and Bob. Alice has values of 100 and 15

This work is supported in part by an NSFC grant (No. 6212290003).

J. Garg et al. (Eds.): WINE 2023, LNCS 14413, pp. 402–419, 2024.
https://doi.org/10.1007/978-3-031-48974-7_23

for receiving a pack of chocolates and gummy bears, respectively. Bob, on the other hand, has values of 1 and 10 for chocolates and gummy bears, respectively.

The social welfare maximizing allocation gives everything to Alice for her higher values, and leaves nothing for Bob. This is blatantly unfair, especially at a daycare center.

What if we distribute evenly and let each child have a pack of chocolates and a pack of gummy bears? While this is fair, it is inefficient. The gummy bears contribute little to Alice's utility, while the chocolates do not add much to Bob's utility.

The question then becomes how to allocate the candies if we want to strike a balance between efficiency and fairness. In the above example, allocating both packs of chocolates to Alice and both packs of gummy bears to Bob is a decent option. In general, maximizing the *Nash welfare*, i.e., the geometric mean of the children's utilities, is a fundamental objective designed to satisfy a set of axiomatic properties (see, e.g., Kaneko and Nakamura [24] and Moulin [30]), and, in particular, is a good proxy for balancing efficiency and fairness. Informally, it achieves a natural compromise between maximizing utilitarian social welfare, i.e., maximizing the sum of children's utilities, and maximizing egalitarian social welfare, i.e., maximizing the minimum utility among children, corresponding to extreme efficiency and extreme fairness, respectively. Indeed, the allocation that maximizes Nash welfare also satisfies several standard notions of fairness, such as envy-freeness and proportionality [33]. It also coincides with the aforementioned decent allocation in the above example. Further, the Nash welfare maximization problem is *scale invariant*: if we increase an agent's values for all items by the same multiplicative factor, the Nash welfare maximizing allocation will remain the same. Moreover, maximizing the Nash welfare with divisible items and additive utilities reduces to solving the Eisenberg-Gale convex program [17]. We can therefore compute this allocation in polynomial time.

There is one more complication though. We cannot predict when the donations of candies will be made, what kinds of candies will be donated, and in what quantities. When some organizations make a donation, we need to distribute the candies to the children immediately without much information about future donations.

Besides the toy example above, many real online resource allocation problems for heterogeneous agents require a balance between fairness and efficiency, e.g., allocating food each day to people in need [32], or allocating shared computational resources to users [22]. Hence, our task becomes designing online allocation algorithms for Nash welfare maximization, whose performance is evaluated by *competitive ratio*, that is, the worst ratio between the online algorithm's solution and the offline optimal solution. From now on, we will more generally talk about items and agents.

This problem was first studied by Banerjee et al. [6]. They noticed that under worst-case analysis, no algorithm can have a competitive ratio better than the trivial $O(N)$, where N is the number of agents. They then turned to the model of online algorithms with predictions, in which the online algorithm is further given some predictions of each agent's utility for receiving the set of all items, which they called the *monopolist utility*. If the predictions were accurate, their algorithm would be $O(\log N)$-competitive.[1] They further proved that the competitive ratio is asymptotically optimal for the problem.

1.1   Our Contribution

Online Algorithms Without Predictions. Is it truly hopeless to design good online algorithms *without predictions*? By examining the hard instance by Banerjee et al. [6], which they designed to rule out the existence of good online algorithms without predictions, we noticed that the agents' values therein differ by an exponential factor, a rare phenomenon in real resource allocation problems. Can we design online algorithms for "natural instances" where agents' values are not as extreme?

Motivated by the above observation, this paper introduces the notions of *balanced* and *impartial* instances. An instance is λ-*balanced* if the agents' monopolist utilities differ by, at most, a multiplicative factor λ. For such instances, our main result is a $\log^{1+o(1)}(\lambda N)$-competitive online algorithm. We further show that this competitive ratio is nearly optimal in the sense that no online algorithm can be better than $\log^{1-o(1)}(\lambda N)$ competitive.

On the other hand, an instance is μ-*impartial* if the agents' utilities for the Nash welfare maximizing allocation differ by, at most, a multiplicative factor μ. For μ-impartial instances, our main result is a $\log^{2+o(1)} \mu$-competitive online algorithm. We remark that this competitive ratio is independent of the number of agents or items, and thus, this would be an $O(1)$-competitive algorithm if the instance is $O(1)$-impartial. In the special case when the agents' values for different items are binary, we have an improved competitive ratio $\log^{1+o(1)} \mu$; we further prove that it is nearly optimal. Besides, we also give algorithms when λ and μ are unknown.

Implications for Online Algorithms with Predictions. The algorithms from this paper further imply new results in the model of online algorithms with predictions. If we had accurate predictions of the agents' monopolist utilities, we could then normalize the agents' values accordingly to obtain a 1-balanced instance using the fact that Nash welfare maximization is scale invariant. If we had (α, β)-approximate predictions, i.e., if they were at worst an α-factor larger or a β-factor smaller than the agents' monopolist utilities, the normalized instance would be $\alpha\beta$-balanced. Therefore, the aforementioned competitive online algorithm for balanced instances implies a $\log^{1+o(1)}(\alpha\beta N)$-competitive online algorithm for the model with (α, β)-approximate predictions, which improves the $O(\alpha \log \beta N)$-competitive online algorithm given by Banerjee et al. [6] when α is sufficiently large. Compared to Banerjee et al. [6], the assumption of balanced instances is

---

[1] Banerjee et al. [6] also gave an $O(\log T)$-competitive ratio where T is the number of items, under the assumption that items have unit supplies. Since this paper considers arbitrary supplies, the dependence in T is no longer valid.

Table 1. Summary of results. Here λ^* and μ^* are the smallest numbers for which the instance is λ^*-balanced and μ^*-impartial. We omit the constants and $\log\log$ factors for brevity. We also omit the trivial bound of N.

	Algorithm	Lower Bound
λ^*-Balanced	$\log \lambda^* N$	$\log \lambda^* N$
μ^*-Impartial	$\min\left\{ \log \mu^* N, \ \log^2 \mu^* \right\}$	$\log \mu^*$
(α, β)-Predictions	$\alpha \log \beta N$ [6] $\to \ \log \alpha \beta N$	
(α, β)-Predictions for Nash welfare maximizing utilities	$\min\left\{ \log \alpha \beta N, \ \log^2 \alpha \beta \right\}$	

slightly weaker since an instance with predictions can be transformed to a balanced instance by scaling and the monopolist utilities are known, which means there is more prior knowledge.

Similarly, if we had (α, β)-approximate predictions for the agents' utilities for the Nash welfare maximizing allocation, we can run the aforementioned online algorithm for impartial instances on the normalized $\alpha\beta$-impartial instance to yield an $O(\log^2 \alpha\beta)$ competitive ratio. Note that this is $O(1)$ for $\alpha, \beta = O(1)$, independent of the number of agents N; by contrast, we need a logarithmic dependence in N even with perfect predictions on the monopolist utilities. Conceptually, it indicates that it is more useful to predict agents' utilities for the Nash welfare maximizing allocations, rather than their monopolist utilities.

See Table 1 for a summary of our results.

1.2   Other Related Works

There is a vast literature on online algorithms for resource allocation problems, such as online packing (c.f., Alon et al. [2] and Buchbinder and Naor [11]) and online matching problems (c.f., Karp et al. [25] and Mehta [29]). Most related to this paper is the work by Devanur and Jain [15] on online matching with concave returns. Maximizing the Nash welfare can be equivalently formulated as maximizing the sum of logarithms of the agents' utilities, a special case of the model of Devanur and Jain [15] when the return functions are the logarithm. However, their results do not imply competitive online algorithms for our problem because they studied the multiplicative competitive ratio with respect to (w.r.t.) the sum of log-utilities. By contrast, Banerjee et al. [6] and this paper consider the multiplicative competitive ratio w.r.t. the geometric mean of utilities, which corresponds to the additive approximation factor w.r.t. the sum of log-utilities. Barman et al. [9] studied online algorithms for the more general p-mean welfare maximization, which captures the Nash welfare as a special case when $p = 0$. Their result for Nash welfare can be interpreted as an $O(\log^3 N)$-competitive algorithm for 1-balanced instances; by contrast our algorithm is $O(\log^{1+o(1)}(\lambda N))$-competitive for general λ-balanced instances. Gao et al. [18] and Liao et al. [28] considered the market equilibrium in the online fair allocation with stationary and non-stationary inputs, and they show that a mechanism named PACE can approach the competitive equilibrium in the mean square at the rate of $O(\frac{\log t}{t})$, where t is the time step.

Nash welfare was introduced about 70 years ago [24,31]. Its maximization has been extensively studied in the offline setting. The problem of divisible items and additive utilities is equivalent to the Eisenberg-Gale convex program [17]. Much effort has been devoted to the problem of indivisible items and various kinds of utilities. Even for additive utilities, the problem is NP-hard [30] and APX-hard [26]. Cole and Gkatzelis [14] gave the first constant-approximation algorithm for additive utilities. Subsequently, Anari et al. [4] applied the theory of stable polynomials to get an e-approximation algorithm. Cole et al. [13] further improved the analysis to show a 2-approximation. Finally, Barman et al. [10] gave a new algorithm that achieves the state-of-the-art $e^{1/e} < 1.45$-approximation. The problem with more general utilities has also been studied. Garg et al. [19] designed a constant-approximation algorithm for budget-additive utilities. Anari et al. [3] introduced a constant-approximation algorithm for separable, piecewise-linear concave utilities. For submodular utilities, Garg et al. [20] proposed an $O(n \log n)$-approximation algorithm, and Li and Vondrák [27] recently obtained the first constant-approximation algorithm for submodular utilities. Last but not least, Barman et al. [8] and Chaudhury et al. [12] independently developed $O(n)$-approximation algorithms for the even more general subadditive utilities.

Azar et al. [5] studied an online resource allocation problem whose competitive algorithms are also competitive w.r.t. Nash welfare. However, they assumed that any agent's values for different items could only differ by a bounded ratio; their competitive ratio is logarithmic in the numbers of agents and items as well as this bounded ratio. By contrast, our notions of balance and impartiality compare different agents' utilities. Therefore, their results do not apply to our model. Besides that, there is a significant line of work about online resource allocation, which focuses on different objectives, e.g., Gkatzelis et al. [21] considered maximizing utilitarian social welfare with envy-freeness in online resource allocation. Banerjee et al. [7] studied an online public items allocation problem with predictions, where the objective is to achieve proportional fairness.

2   Preliminaries

2.1   Nash Welfare Maximization

Consider N agents and T divisible items. Denote the supply of an item t by s_t. We represent an allocation of items to agents by a non-negative real vector $\boldsymbol{x} = (x_{it})_{1 \leq i \leq N, 1 \leq t \leq T}$, where x_{it} is the amount of item t allocated to agent i. A feasible allocation must comply with the supply constraints of items; that is, it must satisfy $\sum_{i=1}^{N} x_{it} \leq s_t$ for any item t. For each agent $1 \leq i \leq N$ and each item $1 \leq t \leq T$, $v_{it} \geq 0$ denotes agent i's value for receiving a unit of item t. The agents' utilities are *additive*: If an agent i receives x_{it} amount of each item t, its utility equals $u_i = \sum_{t=1}^{T} x_{it} v_{it}$.

Definition 1. *The Nash welfare of an allocation* $\boldsymbol{x} = (x_{it})_{1\leq i\leq N, 1\leq t\leq T}$ *is the geometric mean of the agents' utilities:*

$$\left(\prod_{i=1}^{N} u_i\right)^{1/N} = \prod_{i=1}^{N}\left(\sum_{t=1}^{T} x_{it}v_{it}\right)^{1/N}.$$

We also consider an equivalent objective, namely, maximizing the sum of logarithms of utilities:

$$\max\ \sum_{i=1}^{N}\log u_i.$$

Let $\boldsymbol{x}^* = (x_{it}^*)_{1\leq i\leq N, 1\leq t\leq T}$ denote the allocation that maximizes the Nash welfare, breaking ties arbitrarily. Let $\boldsymbol{u}^* = (u_i^*)_{1\leq i\leq N}$ denote the corresponding utilities of the agents.

Maximizing the Nash welfare is a good proxy for balancing the efficiency and fairness of the allocation. Concretely, when the items are divisible, it implies that any agent would get at least $\frac{1}{N}$ of its monopolist utility for receiving all items, a notion of fairness known as proportionality.

Lemma 1 (c.f., Vazirani [34]). *The Nash welfare maximizing allocation* $\boldsymbol{x}^*$ *and the corresponding utilities* $\boldsymbol{u}^*$ *satisfy that for any agent* $1 \leq i \leq N$:

$$u_i^* \geq \frac{1}{N}\sum_{t=1}^{T} s_t v_{it}.$$

2.2 Online Algorithms

This paper considers an online setting in which the items arrive one by one, while the agents are known from the beginning. When each item arrives, the algorithm observes the item's supply and the agents' values for the item. It must then decide how to allocate the item immediately and irrevocably. We consider the ratio of the optimal Nash welfare to the expected Nash welfare of the algorithm's allocation. The *competitive ratio* of an algorithm is the maximum of this ratio over all possible instances.

Greedy Algorithms with Anticipated Utilities. A natural strategy to allocate an item t is first to estimate how much utility each agent i could get from the other items, denoted as u'_{it} and referred to as agent i's anticipated utility. We may then allocate item t greedily to maximize the Nash welfare based on the anticipated utilities. For example, a natural yet conservative form of anticipated utilities in our model would be the agents' utilities from the previous items. In other models, the anticipated utilities may depend on some prior knowledge of the agents' values; for instance, the algorithm of Banerjee et al. [6] may be viewed as a greedy algorithm with anticipated utilities that depend on the agents' monopolist utilities.

It is therefore instructive to first examine some basic properties of this family of algorithms. We may formulate the allocation problem conditioned on antici- pated utilities u'_{it} as a convex program:

$$\text{maximize} \quad \sum_{i=1}^{N} \log(u'_{it} + v_{it}x_{it}) \tag{1}$$

$$\text{subject to} \quad \sum_{i=1}^{N} x_{it} \leq s_t \quad \text{and} \quad x_{it} \geq 0 \text{ for any agent } i.$$

Since the objective is monotone in x_{it}'s, the optimal solution to the above program, denoted as $\boldsymbol{z}_t = (z_{it})_{1 \leq i \leq N}$, must satisfy the first constraint with equality, i.e.:

$$\sum_{i=1}^{N} z_{it} = s_t. \tag{2}$$

Further, for the optimal multiplier $\nu^* \geq 0$ of the first constraint and for any agent $1 \leq i \leq N$, the optimality conditions imply that:

$$z_{it} = \max\left\{\frac{1}{\nu^*} - \frac{u'_{it}}{v_{it}}, 0\right\}. \tag{3}$$

We give the following lemma to bound the gain of $\sum_{i=1}^{N} \log u'_{it}$ when we follow the optimal solution from Eq. (3) to allocate an item t.

Lemma 2. *The optimal solution $\boldsymbol{z}$ to convex program (1) satisfies that:*

$$\sum_{i=1}^{N} \log(u'_{it} + v_{it}z_{it}) - \sum_{i=1}^{N} \log u'_{it} \geq s_t \cdot \max_{1 \leq i \leq N} \frac{v_{it}}{u'_{it} + v_{it}z_{it}}.$$

Randomized versus Deterministic Algorithms. For divisible items, the best com- petitive ratios achievable by randomized and deterministic online algorithms are the same. Given any randomized algorithm, we may convert it into a determin- istic one such that for each item t, the amount of item t that the deterministic algorithm allocates to any agent i equals the expected amount that the random- ized algorithm would allocate. Since the Nash welfare is a concave function of the allocation, the deterministic algorithm yields a weakly larger Nash welfare. Therefore, our positive results will describe online algorithms in their random- ized form if that simplifies the presentation. In our negative results, on the other hand, it suffices to consider deterministic algorithms.

2.3 Balanced and Impartial Instances

Banerjee et al. [6] showed that for arbitrary instances, no online algorithm can have a competitive ratio better than the trivial $O(N)$. They then explored the model of online algorithms with predictions, where the algorithms know the agents' monopolist utilities for receiving all items.

We observe that the agents' values in the hard instances of Banerjee et al. [6] are lopsided: the largest value is exponentially larger than the smallest. This is rarely the case in real resource allocation problems. Recall the example of allocating food and computational resources. It is almost impossible to see agents' values towards the same item differ substantially. Hence, this paper aims to design algorithms with good performance on "natural instances" in which different agents are of "similar importance". We define two notions of "natural instances". Motivated by Banerjee et al. [6], our first notion compares the agents' monopolist utilities for receiving all items.

Definition 2. *An instance is* λ-balanced *if for any agents* $1 \leq i, j \leq N$:

$$\sum_{t=1}^{T} s_t v_{it} \leq \lambda \sum_{t=1}^{T} s_t v_{jt}.$$

We shall refer to $\lambda^* = \frac{\max_{1 \leq i \leq N} \sum_{t=1}^{T} s_t v_{it}}{\min_{1 \leq i \leq N} \sum_{t=1}^{T} s_t v_{it}}$ *as the* balance ratio *of the instance.*

We remark again that competitive online algorithms for λ-balanced instances can be transformed into online algorithms with predictions in the model of Banerjee et al. [6]. Given prediction $\tilde{V}_i$ of each agent i's monopolist utility $V_i = \sum_{t=1}^{T} s_t v_{it}$ that is (α, β)-approximate, i.e., $\alpha V_i \geq \tilde{V}_i \geq \frac{1}{\beta} V_i$, we can normalize agent i's values to be $v'_{it} = \frac{v_{it}}{V_i}$ for all items t to get an $\alpha\beta$-balanced instance.

Our second notion examines whether different agents get similar utilities in the Nash welfare maximizing allocation. This notion is motivated by the fairness notion called *equitability* in the fair division literature [1,16], where an allocation is called equitable if all agents receive the same utility.

Definition 3. *Given any instance, an allocation* $\boldsymbol{x}$ *is* μ-impartial *if the corresponding utilities* $\boldsymbol{u}$ *satisfy that for any agents* $1 \leq i, j \leq N$:

$$u_i \leq \mu \cdot u_j.$$

An instance is μ-impartial *if any Nash welfare maximizing allocation* $\boldsymbol{x}^*$ *is* μ-impartial. *Let* μ^* *be the smallest value of* μ *such that the instance is* μ-impartial; *we shall refer to* μ^* *as the* impartiality ratio *of the instance.*

Impartiality ratio can be seen as an approximate equitability ratio of the Nash welfare maximizing allocation and therefore captures whether Nash welfare maximizing allocation is intuitively fair. Therefore, we consider the impartiality ratio to be an indicator of whether Nash welfare is a suitable fairness notion for the given instance. Impartiality has a simple yet important implication: For each item, we may ignore the agents whose values for the item are much smaller than the highest value. We make this precise with the next lemma.

Lemma 3. *For any agent i and item t, if $x_{it}^* > 0$ then $v_{it} \geq \frac{1}{\mu^*} \max_{1 \leq j \leq N} v_{jt}$.*

Finally, the balance ratio and impartiality ratio are within a factor N from each other.

Lemma 4. *For any instance, we have $\lambda^* \leq \mu^* N$ and $\mu^* \leq \lambda^* N$.*

In conclusion, these two notions reflect the "naturality" of an instance, as a replacement of prediction in previous works. In Sect. 3, we study balanced instances and achieve $O(\log \lambda^* N)$-competitive ratio. For impartial instances, besides the $O(\log \mu^* N)$ result as a corollary of Lemma 4, in Sect. 4 we give an online algorithm whose competitive ratio is $O(\log^2 \mu^*)$, only depending on the impartiality ratio. (All log log factors are omitted here.) Therefore, hopefully the impartiality ratio is the better one to use.

3   Balanced Instances

This section studies balanced instances. Section 3.1 introduces our algorithm under an additional assumption that we were given an upper bound of the balance ratio (Sect. 3.1). Then, Sect. 3.2 explains how to remove this assumption by guessing the balance ratio, while losing at most log log factors in the competitive ratio.

3.1   Algorithm with a Known Upper Bound of the Balance Ratio

Suppose that we are given an upper bound λ of the instance's balance ratio λ^*. In other words, we know that the instance is λ-balanced. The algorithm will divide each item t's supply equally into two halves. On the one hand, it allocates the first half equally to all agents, a naïve strategy that is certainly fair but is not efficient enough to approximately maximize the Nash welfare on its own. On the other hand, it greedily allocates the second half of the item to maximize the Nash welfare assuming that each agent would get not only their utilities for the second halves of the previous items allocated to them, but also a fraction of the sum of *all agents'* monopolist utilities for all known items. Concretely, let $z_t = (z_{it})_{1 \leq i \leq N}$ denote the allocation of the second half of item t. For anticipated utilities:

$$
u_{it}' = \underbrace{\frac{1}{2\lambda N^2} \sum_{j=1}^{N} \sum_{t'=1}^{t} v_{jt'} s_{t'}}_{\substack{\text{sum of monopolist utilities} \\ \text{for all known items,} \\ \textbf{including item } t}} + \underbrace{\sum_{t'=1}^{t-1} v_{it'} z_{it'}}_{\substack{i\text{'s utility for the second halves} \\ \text{of previous items allocated to } i, \\ \textbf{excluding item } t}} \tag{4}
$$

the algorithm chooses z to maximize $\sum_{i=1}^{N} \log \left(u_{it}' + v_{it} z_{it} \right)$ subject to $\sum_{i=1}^{N} z_{it} \leq \frac{s_t}{2}$ and $z_{it} \geq 0$ for all agents i. We call this algorithm Half-and-Half. See Algorithm 1 for a formal definition.

Algorithm 1: Half-and-Half (for λ-balanced instances)

1 for each item $1 \leq t \leq T$ **do**
2 $\quad$ Let $y_{it} = \frac{s_t}{2N}$ for all agents $1 \leq i \leq N$.
3 $\quad$ Let z_{it} maximize (for anticipated utilities u'_{it} defined in Equation (4))

$$\sum_{i=1}^{N} \log(u'_{it} + v_{it} z_{it})$$

$\quad$ subject to $\sum_{i=1}^{N} z_{it} \leq \frac{s_t}{2}$ and $z_{it} \geq 0$ for all agents $1 \leq i \leq N$.
4 $\quad$ Allocate $x_{it} = y_{it} + z_{it}$ amount of item t to each agent $1 \leq i \leq N$.

Theorem 1. *Algorithm 1 is $O(\log \lambda N)$-competitive.*

A comparison with the Set-Aside Greedy algorithm of Banerjee et al. [6] is warranted, since readers familiar with the previous algorithm may have noticed the similarity between the two algorithms. Both algorithms divide each item equally into two halves; both allocate the first half equally, and the second half by some greedy algorithm with anticipated utilities. The difference lies in the designs of anticipated utilities. The Set-Aside Greedy algorithm may be viewed as replacing the first part of our anticipated utility in Eq. (4) by a $\frac{1}{2N}$ fraction of the prediction on agent i's monopolist utility. Following the idea of Set-Aside Greedy, a natural attempt is to use each agent i's monopolist utility for all known items as the prediction of its final monopolist utility. In the online setting, however, the items that contribute the most to an agent's monopolist utility may come at the very end. In that case, the algorithm would underestimate the agent's final utility for a long time, and, as a result, might unnecessarily allocate many items to this agent in the early rounds. Our solution is to aggregate the monopolist utilities of *all agents* for the known items into an anticipated utility for *every agent*, an idea driven by the assumption of balanced instances.

3.2 Guessing the Balance Ratio

When we have no prior knowledge of the balance ratio, we can guess an upper bound of the balance ratio by sampling from an appropriate distribution. Since the final ratio depends logarithmically on the upper bound, it suffices to make a good enough guess that is, at most, a polynomial of the true balance ratio λ^*.

Concretely, we shall consider a sequence of numbers starting from 2, such that each sequel number is the square of the previous number. We will sample each number λ with a probability that is inverse polynomial in $\log \log \lambda$; this ensures that the correct guess is made with a sufficiently large probability. Finally, we apply the prior-dependent Half-and-Half algorithm (Algorithm 1) with the guessed upper bound λ. See Algorithm 2 for a formal definition.

Theorem 2. *Algorithm 2 is $O(\log \lambda^* N (\log \log \lambda^*)^2)$-competitive.*

Algorithm 2: Half-and-Half (for instances with unknown balance ratio)

---

1 Sample λ such that it equals 2^{2^k} with probability $\frac{6}{\pi^2} \cdot \frac{1}{(k+1)^2}$ for any non-negative integer k.

2 Run Algorithm 1 with λ to allocate the items to the agents.

Considering the relation of balance and impartiality ratios (Lemma 4), Theorem 2 can directly imply Theorem 3, which means the algorithm is also competitive for impartial instances. The next section will develop algorithms tailored for impartial instances with competitive ratios independent of the number of agents.

Theorem 3. *Algorithm 2 is $O\left(\log \mu^* N (\log \log \mu^* N)^2\right)$-competitive.*

4   Impartial Instances

This section considers impartial instances. Section 4.1 examines a special case of binary values: an agent i's value for an item t is either $v_{it} = 0$, or some value $v_{it} = v_t$ that is the same for all agents. We will show that a simple greedy algorithm, which was referred to as Myopic Greedy by Banerjee et al. [6], is $O(\log \mu^*)$-competitive in this special case. This is nearly tight, as the next section will show an almost matching lower bound. Section 4.2 then explains how to reduce the general case to the binary case under an additional assumption that an upper bound μ of the impartiality ratio μ^* is given, i.e., if we know that the instance is μ-impartial. Finally, Sect. 4.3 applies the same technique as in the previous section to randomly guess an upper bound, removing the additional assumption while losing a factor that depends on $\log \log \mu^*$ in the competitive ratio.

4.1   Myopic Greedy and Binary Values

The Myopic Greedy algorithm simply allocates each item t greedily to maximize the Nash welfare conditioned on the allocation before item t. In other words, it is a greedy algorithm with anticipated utilities, for which the anticipated utility of an agent equals its utility for the items allocated to it so far. See Algorithm 3 for a formal definition.

The rest of this subsection will assume that the agents' values are binary, that is, for any item $1 \leq t \leq T$ and any agent $1 \leq i \leq N$, either $v_{it} = 0$ or $v_{it} = v_t$.

Theorem 4. *Algorithm 3 is $O(\log \mu^*)$-competitive if the agents have binary values for the items.*

Algorithm 3: Myopic Greedy

---

1 **for** each item $1 \leq t \leq T$ **do**
2 $\quad$ Let the allocation $\boldsymbol{x}_t = (x_{it})_{1 \leq i \leq N}$ maximize:

$$\sum_{i=1}^{N} \log \left(\sum_{t'=1}^{t-1} v_{it'} x_{it'} + v_{it} x_{it} \right)$$

$\quad$ subject to $\sum_{i=1}^{N} x_{it} \leq s_t$ and $x_{it} \geq 0$ for all agents $1 \leq i \leq N$.

In fact, we will prove a slightly stronger result so that in the next subsection, we can reduce the general case to the case of binary values. We formulate the stronger claim as the next lemma.

Lemma 5. *For any feasible allocation* $\tilde{\boldsymbol{x}} = (\tilde{x}_{it})_{1 \leq i \leq N, 1 \leq t \leq T}$ *and the agents' corresponding utilities* $\tilde{\boldsymbol{u}} = (\tilde{u}_i)_{1 \leq i \leq N}$, *if allocation* $\tilde{\boldsymbol{x}}$ *is* $\tilde{\mu}$-*impartial, then the Nash welfare of the allocation by Algorithm 3 is at least an* $O(\log \tilde{\mu})$ *approximation to the Nash welfare of allocation* $\tilde{\boldsymbol{x}}$, *i.e.:*

$$\left(\frac{\prod_{i=1}^{N} \tilde{u}_i}{\prod_{i=1}^{N} u_i} \right)^{\frac{1}{N}} = O(\log \tilde{\mu}).$$

Theorem 4 follows as a corollary by letting $\tilde{\boldsymbol{x}}$ be the Nash welfare maximizing allocation $\boldsymbol{x}^*$.

The rest of the subsection focuses on proving Lemma 5. We assume without loss of generality that the agents are sorted by their utilities for the algorithm's allocation, i.e.:

$$u_1 \leq u_2 \leq \cdots \leq u_N.$$

We start with the following lemma, which links the agents' utilities $\boldsymbol{u}$ for the algorithm's allocation, and their utilities $\tilde{\boldsymbol{u}}$ for the benchmark allocation $\tilde{\boldsymbol{x}}$.

Lemma 6. *For any* $1 \leq i \leq N$:

$$\sum_{j=1}^{i} \tilde{u}_j \leq \sum_{j=1}^{N} \min\{u_i, u_j\}. \tag{5}$$

If we view the agents' utilities $\boldsymbol{u}$ for the algorithm's allocation as variables, Lemma 6 offers a set of linear inequalities that relates them with the benchmark utilities $\tilde{\boldsymbol{u}}$. The next lemma shows that subject to these inequalities, the smallest Nash welfare w.r.t. $\boldsymbol{u}$ is achieved when the inequalities all hold with equalities.

Lemma 7. *For any non-negative* $\boldsymbol{u} = (u_i)_{1 \leq i \leq N}$ *and* $\tilde{\boldsymbol{u}} = (\tilde{u}_i)_{1 \leq i \leq N}$ *that satisfy the inequality in Eq. (5), we have:*

$$\sum_{i=1}^{N} \log u_i \geq \sum_{i=1}^{N} \log \left(\sum_{j=1}^{i} \frac{\tilde{u}_j}{N+1-j} \right). \tag{6}$$

We will relax the denominators on the right-hand side of Eq. (6) to be N in the subsequent analysis. We get that:

$$\sum_{i=1}^{N} \log u_i \geq \sum_{i=1}^{N} \log \left(\frac{1}{i} \sum_{j=1}^{i} \tilde{u}_j \right) - \log \frac{N^N}{N!}. \qquad (7)$$

If the benchmark allocation $\tilde{\boldsymbol{x}}$ was 1-impartial, for all agents $1 \leq i \leq N$, we would have $\tilde{u}_i = \left(\prod_{j=1}^{N} \tilde{u}_j \right)^{\frac{1}{N}}$. In that case, the above Inequality (7) would already imply a constant competitive ratio because it can be written as:

$$\log \left(\prod_{i=1}^{N} u_i \right)^{\frac{1}{N}} \geq \log \left(\prod_{i=1}^{N} \tilde{u}_i \right)^{\frac{1}{N}} - \log \frac{N}{(N!)^{\frac{1}{N}}},$$

where the second term on the right-hand side is $O(1)$ by Stirling's approximation.

For μ-impartial instances with an arbitrary $\mu \geq 1$, we need one more inequality given in the next lemma. The form of the inequality is clean, so we suspect that it may have already been proved in the past. To our best effort, however, we only found Hardy's Inequality and its several variants (c.f., Hardy et al. [23]) to have a similar spirit as they also compare the function values of a sequence of non-negative real numbers and the functions values of their prefix averages; but those inequalities do not imply our lemma. It would be interesting to find further applications of this inequality in other problems.

Lemma 8. *Suppose that $a_1, a_2, \ldots, a_N$ are real numbers between 1 and μ. We have:*

$$\frac{1}{N} \sum_{i=1}^{N} \log a_i - \frac{1}{N} \sum_{i=1}^{N} \log \left(\frac{1}{i} \sum_{j=1}^{i} a_j \right) \leq \log(\log \mu + 1) + 1.$$

Lemma 5 then follows by combining Eq. (7), Stirling's approximation, and Lemma 8 with $a_i = \frac{\tilde{u}_i}{\min_{1 \leq j \leq N} \tilde{u}_j}$ and $\mu = \tilde{\mu}$.

4.2 Algorithm with a Known Upper Bound of the Impartiality Ratio

Suppose that we are given an upper bound μ of the instance's impartiality ratio μ^*. In other words, we know that the instance is μ-impartial. For any item t, let $\bar{v}_t = \max_{i \in [N]} v_{it}$ denote the maximum value of any agent for item t. By Lemma 3, we should only allocate item t to the agents whose values for item t is at least $\frac{1}{\mu} \bar{v}_t$. A naïve approach is to treat all such agents as if they had value $\bar{v}_t$ for the item, and then apply the Myopic Greedy algorithm; doing so would lose an extra factor μ in the competitive ratio in the worst case. Instead, we divide the item equally into $\lceil \log \mu \rceil$ sub-items, each with supply $\frac{s_t}{\lceil \log \mu \rceil}$. For the j-th sub-item, which we refer to as sub-item (t, j), the agents' values are rounded

---

Algorithm 4: Greedy with Rounded Values (for μ-impartial instances)

---

1 **for** each item $1 \leq t \leq T$ **do**
2   Let $\bar{v}_t = \max_{1 \leq i \leq N} v_{it}$
3   **for** $j = 1$ to $\lceil \log \mu \rceil$ **do**
4   Let there be a sub-item (t, j) with supply $\frac{s_t}{\lceil \log \mu \rceil}$.
5   Let each agent i's value for the sub-item be:

$$v_{i(t,j)} = \begin{cases} \frac{\bar{v}_t}{2^j} & \text{if } v_{it} \geq \frac{\bar{v}_t}{2^j} \\ 0 & \text{otherwise.} \end{cases}$$

6   Let Algorithm 3 allocate the sub-items and let the allocation be $(x_{i(t,j)})_{1 \leq i \leq N, 1 \leq j \leq \lceil \log \mu \rceil}$.
7   Allocate $x_{it} = \sum_{j=1}^{\lceil \log \mu \rceil} x_{i(t,j)}$ amount of item t to each agent $1 \leq i \leq N$.

Algorithm 5: Greedy with Rounded Values (for instances with unknown impartiality ratio)

---

1 Sample μ such that it equals 2^{2^k} with probability $\frac{6}{\pi^2} \cdot \frac{1}{(k+1)^2}$ for any non-negative integer k.
2 Run Algorithm 4 with μ to allocate the items to the agents.

---

down to either $\frac{\bar{v}_t}{2^j}$ or 0. That is, an agent i has value $\frac{\bar{v}_t}{2^j}$ for sub-item (t, j) if its original value for item t is at least as much, and has value 0 for this sub-item otherwise. Then, we run the Myopic Greedy algorithm to allocate the sub-items. See Algorithm 4 for a formal definition.

Theorem 5. *Algorithm 4 is $O(\log^2 \mu)$-competitive.*

4.3 Guessing the Impartiality Ratio

This is almost verbatim to the counterpart for balanced instances. When we have no prior knowledge of the impartiality ratio, we can guess an upper bound by sampling from an appropriate distribution. Since the final ratio depends logarithmically on the upper bound, it suffices to make a good enough guess that is, at most, a polynomial of the true impartiality ratio μ^*.

We shall again consider a sequence of numbers starting from 2, such that each sequel number is the square of the previous number. We will sample each number μ with a probability that is inverse polynomial in $\log \log \mu$; this ensures that the correct guess is made with a sufficiently large probability. Finally, we apply the prior-dependent Greedy with Rounded Values algorithm (Algorithm 4) with the guessed upper bound μ. See Algorithm 5 for a formal definition.

Theorem 6. *Algorithm 5 is $O\left(\log^2 \mu^* (\log \log \mu^*)^2\right)$-competitive.*

Table 2. Illustration of a hard instance. Rows are items and columns are agents. The number in the intersection of the t-th row and the i-th column is agent i's value for item t.

t	Agent 1	Agent 2	Agent 3	$\cdots$	Agent n
1	n^2	n^2	n^2		n^2
2	0	n^4	n^4		n^4
3	0	0	n^6		n^6
			$\cdots$		
n	0	0	0		n^{2n}

Compared to the results of balanced instances, the results of impartial instances show that if the utilities of agents in Nash welfare maximizing allocation can be predicted, we can achieve a constant competitive ratio, which means the prediction of the impartiality ratio may be more helpful.

5   Lower Bounds

We first restate a lower bound by Banerjee et al. [6] under our model.

Lemma 9 (Banerjee et al. [6], Theorem 3). *No online algorithm can achieve a competitive ratio better than* $\log^{1-o(1)} N$,[2] *even for 1-balanced instances.*

Next, we prove that the logarithmic dependence on the balance ratio or the impartiality ratio is necessary. The main ingredient is the instance given by Table 2, which is a variant of another hard instance by Banerjee et al. [6] (c.f., Theorem 4 therein). The instance has N agents, and $T = N$ items of unit supplies. Agent i's value for item t is N^{2i} if $t \leq i$, and is 0 otherwise. Recall that for lower bounds, it suffices to consider deterministic algorithms. Hence, we may assume without loss of generality that agent i receives the least amount of item $t = i$ among agents i to N in the algorithm's allocation.

Lemma 10. *No algorithm can be better than* $\frac{n-1}{e}$ *competitive for the instance given by Table 2.*

Lemma 11. *The imbalance ratio and impartiality ratio of the instance given by Table 2 are at most* n^{2n}.

We can now derive our lower bounds as corollaries of the above lemmas.

Theorem 7. *No algorithm can be better than* $\min\left\{\log^{1-o(1)} N\lambda^*, \frac{N-1}{e}\right\}$- *competitive.*

---

[2] The original theorem by Banerjee et al. [6] only claimed a weaker bound of $\log^{1-\varepsilon} N$ but their proof actually showed a slightly stronger bound that we restate here.

Theorem 8. *No algorithm can be better than* $\min\{\log^{1-o(1)}\mu^*, \frac{N-1}{e}\}$- *competitive.*

We remark that Theorem 8 still holds even if the agents' values for the items are either 0 or 1. For each item t in Table 2, instead of letting it have unit supply and letting agents t to n have values n^{2t} for it, we can instead let it have supply n^{2t} and let agents t to n have values 1 for it. Scaling the allocation of item t in the original instance by n^{2t} factor would give an allocation for the 0-1 value version, with the same utilities for all agents.

6   Conclusion

In this paper, we initiate the study of online Nash welfare maximization *without predictions*. We define *balance ratio* and *impartiality ratio* for an instance, and design online algorithms whose competitive ratios only depend on the logarithms of the aforementioned ratios of agents' utilities and the number of agents. One possible extension of our work is to close the gap between the lower and upper bounds in general instances. In addition to that, our technique may provide an insight into how to remove predictions for other online problems.

References

1. Alon, N.: Splitting necklaces. Adv. Math. **63**(3), 247–253 (1987)
2. Alon, N., Awerbuch, B., Azar, Y., Buchbinder, N., Naor, J.: A general approach to online network optimization problems. ACM Trans. Algorithms **2**(4), 640–660 (2006)
3. Anari, N., Mai, T., Gharan, S.O., Vazirani, V.V.: Nash social welfare for indivisible items under separable, piecewise-linear concave utilities. In: Proceedings of the 29th Annual ACM-SIAM Symposium on Discrete Algorithms, pp. 2274–2290 (2018)
4. Anari, N., Gharan, S.O., Saberi, A., Singh, M.: Nash social welfare, matrix permanent, and stable polynomials. In: Proceedings of the 8th Innovations in Theoretical Computer Science Conference, p. 36 (2017)
5. Azar, Y., Buchbinder, N., Jain, K.: How to allocate goods in an online market? In: de Berg, M., Meyer, U. (eds.) ESA 2010. LNCS, vol. 6347, pp. 51–62. Springer, Heidelberg (2010). https://doi.org/10.1007/978-3-642-15781-3_5
6. Banerjee, S., Gkatzelis, V., Gorokh, A., Jin, B.: Online Nash social welfare maximization with predictions. In: Proceedings of the 33rd Annual ACM-SIAM Symposium on Discrete Algorithms, pp. 1–19 (2022)
7. Banerjee, S., Gkatzelis, V., Hossain, S., Jin, B., Micha, E., Shah, N.: Proportionally fair online allocation of public goods with predictions. arXiv preprint arXiv:2209.15305 (2022)
8. Barman, S., Bhaskar, U., Krishna, A., Sundaram, R.G.: Tight approximation algorithms for p-mean welfare under subadditive valuations. In: Proceedings of the 28th Annual European Symposium on Algorithms, pp. 11:1–11:17 (2020)
9. Barman, S., Khan, A., Maiti, A.: Universal and tight online algorithms for generalized-mean welfare. In: Proceedings of the AAAI Conference on Artificial Intelligence, pp. 4793–4800 (2022)

10. Barman, S., Krishnamurthy, S.K., Vaish, R.: Finding fair and efficient allocations. In: Proceedings of the 19th ACM Conference on Economics and Computation, pp. 557–574 (2018)
11. Buchbinder, N., Naor, J.: Online primal-dual algorithms for covering and packing. Math. Oper. Res. **34**(2), 270–286 (2009)
12. Chaudhury, B.R., Garg, J., Mehta, R.: Fair and efficient allocations under subadditive valuations. In: Proceedings of the AAAI Conference on Artificial Intelligence, vol. 35, pp. 5269–5276 (2021)
13. Cole, R., et al.: Convex program duality, Fisher markets, and Nash social welfare. In: Proceedings of the 18th ACM Conference on Economics and Computation, pp. 459–460 (2017)
14. Cole, R., Gkatzelis, V.: Approximating the Nash social welfare with indivisible items. In: Proceedings of the 47th Annual ACM Symposium on Theory of Computing, pp. 371–380 (2015)
15. Devanur, N.R., Jain, K.: Online matching with concave returns. In: Proceedings of the 44th Annual ACM Symposium on Theory of Computing, pp. 137–144 (2012)
16. Dubins, L.E., Spanier, E.H.: How to cut a cake fairly. Am. Math. Monthly **68**(1P1), 1–17 (1961)
17. Eisenberg, E., Gale, D.: Consensus of subjective probabilities: the pari-mutuel method. Ann. Math. Stat. **30**(1), 165–168 (1959)
18. Gao, Y., Peysakhovich, A., Kroer, C.: Online market equilibrium with application to fair division. In: Advances in Neural Information Processing Systems, vol. 34, pp. 27305–27318 (2021)
19. Garg, J., Hoefer, M., Mehlhorn, K.: Approximating the Nash social welfare with budget-additive valuations. In: Proceedings of the 29th Annual ACM-SIAM Symposium on Discrete Algorithms, pp. 2326–2340 (2018)
20. Garg, J., Kulkarni, P., Kulkarni, R.: Approximating Nash social welfare under submodular valuations through (un)matchings. In: Proceedings of the 31st Annual ACM-SIAM Symposium on Discrete Algorithms, pp. 2673–2687 (2020)
21. Gkatzelis, V., Psomas, A., Tan, X.: Fair and efficient online allocations with normalized valuations. In: Proceedings of the AAAI Conference on Artificial Intelligence, pp. 5440–5447 (2021)
22. Hao, F., Kodialam, M., Lakshman, T., Mukherjee, S.: Online allocation of virtual machines in a distributed cloud. IEEE/ACM Trans. Networking **25**(1), 238–249 (2016)
23. Hardy, G.H., Littlewood, J.E., Pólya, G.: Inequalities. Cambridge University Press, Cambridge (1952)
24. Kaneko, M., Nakamura, K.: The Nash social welfare function. Econometrica **47**(2), 423–435 (1979)
25. Karp, R.M., Vazirani, U.V., Vazirani, V.V.: An optimal algorithm for on-line bipartite matching. In: Proceedings of the 22nd Annual ACM Symposium on Theory of Computing, pp. 352–358 (1990)
26. Lee, E.: APX-hardness of maximizing Nash social welfare with indivisible items. Inf. Process. Lett. **122**, 17–20 (2017)
27. Li, W., Vondrák, J.: A constant-factor approximation algorithm for Nash social welfare with submodular valuations. In: 2021 IEEE 62nd Annual Symposium on Foundations of Computer Science, pp. 25–36 (2022)
28. Liao, L., Gao, Y., Kroer, C.: Nonstationary dual averaging and online fair allocation. In: Advances in Neural Information Processing Systems, vol. 35, pp. 37159–37172 (2022)

29. Mehta, A.: Online matching and ad allocation. Found. Trends Theor. Comput. Sci. **8**(4), 265–368 (2013)
30. Moulin, H.: Fair Division and Collective Welfare. MIT Press, Cambridge (2004)
31. Nash, J.F., Jr.: The bargaining problem. Econometrica **18**(2), 155–162 (1950)
32. Prendergast, C.: How food banks use markets to feed the poor. J. Econ. Perspect. **31**(4), 145–162 (2017)
33. Varian, H.R.: Equity, envy, and efficiency. J. Econ. Theory **9**(1), 63–91 (1974)
34. Vazirani, V.V.: Combinatorial algorithms for market equilibria. In: Algorithmic Game Theory, pp. 103–134. Cambridge University Press (2007)

The Price of Anarchy of Probabilistic Serial in One-Sided Allocation Problems

Sissi Jiang[ID], Ndiame Ndiaye[✉][ID], Adrian Vetta[ID], and Eggie Wu[ID]

McGill University, 845 Rue Sherbrooke O, Montréal, QC H3A 0G4, Canada
{yuhe.jiang,ndiame.ndiaye,qihan.wu}@mail.mcgill.ca,
adrian.vetta@mcgill.ca

Abstract. We study "fair mechanisms" for the (asymmetric) one-sided allocation problem with m items and n multi-unit demand agents with additive, unit-sum valuations. The symmetric case ($m = n$), the one-sided matching problem, has been studied extensively for the special class of unit demand agents, in particular with respect to the folklore *Random Priority* mechanism and the *Probabilistic Serial* mechanism, introduced by Bogomolnaia and Moulin [6]. These are both fair mechanisms and attention has focused on their structural properties, incentives, and performance with respect to social welfare. Under the standard assumption of unit-sum valuation functions, Christodoulou et al. [10] proved that the price of anarchy is $\Theta(\sqrt{n})$ in the one-sided matching problem for both the Random Priority and Probabilistic Serial mechanisms. Whilst both Random Priority and Probabilistic Serial are *ordinal mechanisms*, these approximation guarantees are the best possible even for the broader class of *cardinal mechanisms*.

To extend these results to the general setting of the one-sided allocation problems there are two technical obstacles. One, asymmetry ($m \neq n$) is problematic especially when the number of items is much greater than the number of agents, $m \gg n$. Two, it is necessary to study multi-unit demand agents rather than simply unit demand agents. For this paper, our focus is on Probabilistic Serial. Our first main result is an upper bound of $O(\sqrt{n} \cdot \log m)$ on the price of anarchy for the asymmetric one-sided allocation problem with multi-unit demand agents. We then present a complementary lower bound of $\Omega(\sqrt{n})$ for any fair mechanism. That lower bound is unsurprising. More intriguing is our second main result: the price of anarchy of Probabilistic Serial degrades with the number of items. Specifically, a logarithmic dependence on the number of items is necessary as we show a lower bound of $\Omega(\min\{n, \log m\})$.

Keywords: One-Sided Allocation Problem · Price of Anarchy · Probabilistic Serial · Fair Division

© The Author(s), under exclusive license to Springer Nature Switzerland AG 2024
J. Garg et al. (Eds.): WINE 2023, LNCS 14413, pp. 420–437, 2024.
https://doi.org/10.1007/978-3-031-48974-7_24

1   Introduction

In the *one-sided matching problem* a set of m items must be matched in a *fair* manner to a set of $n = m$ (symmetry) agents.[1] This is a classical problem in economics and computer science with numerous practical applications, such as assigning children to schools, patients to doctors, workers to tasks, social housing to people, etc. Consequently, there has been a huge amount of research concerning matching mechanisms, their incentive and structural properties, and the social quality of the outcomes they induce. Of course, these mechanisms are restricted by the fact that the allocation must be a matching. Equivalently, this constraint can be viewed as an assumption of *unit demand* valuation functions, where each agent desires at most one good. However, unit demand valuations are very restrictive. Indeed, in mechanism design primary focus is on *multi-unit demand* valuations and Budish et al. [8] highlight the importance of moving beyond unit-demand agents in the field of *fair* mechanism design. Moreover, in most practical applications the number of items differs from the number of agents ($m \neq n$, asymmetry) and/or the agents have *multi-unit demand* valuations. For example, in estate division or the allocation of shifts to employees, university courses to students, landing and hanger slots to airlines, etc. This motivates our work: we study fair allocation mechanisms for the asymmetric one-sided allocation problem with multi-unit demand agents and analyse the quality of the outcomes they produce with respect to social welfare.

1.1   Background

The one-sided matching problem with indivisible items was formally introduced by Hylland and Zeckhauser [13] in 1979, where they studied the competitive equilibrium from equal incomes (CEEI) mechanism. This mechanism is "fair" by the equal incomes assumption. It is also *envy-free* but not strategy-proof and, indeed, early work in the economics community focused on the structural and incentive properties of matching mechanisms. For example, Zhou [18] gave an impossibility result showing the non-existence of a mechanism that is simultaneously strategy-proof, pareto optimal, and symmetric. See [2,16] for surveys on the one-sided matching problem and on matching markets more generally.

Since monetary transfers are typically not allowed in the one-sided matching problem, it belongs to the field of mechanism design without money [15]. A folklore mechanism in this realm is Random Priority (RP). Applied to the one-sided matching problem, this mechanism orders the agents uniformly at random. The agents then, in turn, select their favorite item that has not previously been selected. This mechanism, also popularly known as Random Serial Dictatorship (RSD) [1], is strategy-proof.

---

[1] To avoid confusion, we emphasize that symmetry in this paper refers to an equal number of agents and items. This is a standard definition in one-sided markets. (The term symmetric does have a second meaning in the literature: all the agents have identical valuation functions).

Another prominent mechanism is Probabilistic Serial (PS), introduced by Bogomolnaia and Moulin [6] in 2001. This is a "consumption" mechanism: to begin, every agent *consumes* their favorite item at the same *consumption rate*. When the favorite item of an agent is completely consumed (that is, together all the agents have consumed exactly one unit of that item) then this agent switches to consume its next favorite item, etc. Since its discovery, Probabilistic Serial has become the most well-studied mechanism for the one-sided matching problem. It has many desirable properties such as envy-freeness and ordinal efficiency when the agents are truthful [6]. However, unlike Random Priority, it is not strategy-proof and some of its desirable properties fail to hold when the agents are strategic [11]. Several extensions to the mechanism have been proposed; see, for example, [3,8,14]. Aziz et al. studied the manipulability of Probabilistic Serial [5] and proved the existence of pure strategy Nash equilibria under the mechanism [4].

An important recent line of research in the computer science community has been to quantify the social welfare of allocations induced by a mechanism in comparison to the optimal obtainable social welfare. Two approaches abound in the literature [12,17]. First is the *approximation ratio*, where agents are assumed to report truthfully to the mechanism. Second, and more interestingly from a game-theoretic perspective, is the *price of anarchy*, where agents are assumed to be strategic [10]. However, for mechanism design without money, these measures are of little interest without a normalization assumption. As a result, the standard normalization assumption [7,9,10,12,17] is that the valuation function of each agent is *unit-sum*. Specifically, agent i has a non-negative value $v_i'(j)$ for item j and $\sum_j v_i'(j) = 1$. Under the unit-sum assumption, a breakthrough result of Christodoulou et al. [10] is that price of anarchy is $\Theta(\sqrt{n})$ for both the Random Priority and Probabilistic Serial mechanisms for the one-sided matching problem. From a practical perspective, unit-sum valuation constraints have wide application. One topical example is sports drafts, such as salary caps in US major sports leagues or the Indian Premier League auction in cricket. Here a unit-sum restriction on bids or salaries imposes equity across teams.

We remark that both Random Priority and Probabilistic Serial have the characteristic that they are *ordinal mechanisms*. Specifically, rather than requiring the entire valuation function of each agent, they need only the preference ordering on the items induced by the valuation function. Interestingly, despite being ordinal mechanisms, these bounds are the best possible even for the broader class of *cardinal mechanisms* where agents are required to submit their entire valuation function [10].

1.2   Overview and Results

The aim of this work is to extend the study of one-sided allocation problems beyond matchings to general allocations. Ergo, we consider asymmetric allocation problems and allow for agents with multi-unit demand valuation functions rather than unit demand valuations. After describing the setting, we shift our focus to the *Probabilistic Serial* (PS) mechanism. We describe the model and the known properties of Probabilistic Serial in Sect. 2.

In Sect. 3 we present structural theorems for the Probabilistic Serial mechanism. In Sect. 4 we prove our main results. First, we use our structural theorems to show that the Probabilistic Serial mechanism has a price of anarchy of $O(\sqrt{n} \cdot \log m)$ for the asymmetric one-sided allocation problem and multi-unit demand agents with additive unit-sum valuations.[2] Second, we prove a lower bound of $\Omega(\sqrt{n})$ on the price of anarchy for any "fair" mechanism, where a mechanism is deemed fair if each agent obtains the same number of items in expectation (as is the case for RP and PS). Third, we present a more intriguing lower bound: the price of anarchy degrades with the number of items. Specifically, a lower bound of $\Omega(\min\{n, \log m\})$ for Probabilistic Serial is shown; thus, a logarithmic dependence on the number of items is necessary for the price of anarchy in the asymmetric one-sided allocation problem for Probabilistic Serial.

Finally, we wrap up in Sect. 5 with discussions on (i) the price of stability, and (ii) the performance of (extensions of) the Random Priority mechanism in the asymmetric one-sided allocation problem.

2   The One-Sided Allocation Problem

In this section we present the asymmetric one-sided allocation problem with multi-unit demand agents. There is a set I of n agents and a set J of m items. Each agent $i \in I$ has a non-negative value $v_i'(j)$ for item $j \in J$. The agents have additive *multi-unit demands*: agent i has a value $v_i'(S) = \sum_{j \in S} v_i'(j)$ for any collection $S \subseteq J$ of items.[3] Furthermore, we assume that valuation functions are *unit-sum*, that is, $\sum_{j \in J} v_i'(j) = 1$ for every agent i. Denote by $\mathcal{U}$ the set of unit-sum valuation functions.

Our focus is on direct revelation *random* mechanisms. Given a unit-sum valuation function v_i', agent i can report to the allocation mechanism M a, possibly non-truthful, unit-sum valuation function v_i. We denote the space of feasible reports the mechanism may receive by $\mathcal{V} \equiv \mathcal{U}^n$. Given a set $v = \{v_1, v_2, \ldots v_n\} \in \mathcal{V}$ of reported valuations, let $v_{-i} = \{v_1, \ldots, v_{i-1}, v_{i+1}, \ldots v_n\}$ be the set of reported valuations excluding agent i. We define $M(v) = M(\{v_i, v_{-i}\})$ to be the *random variable* of the bundle of items allocated to agent i by the mechanism M given the reported valuations $v = \{v_i, v_{-i}\}$. We say that $v_i'(M(v))$ is the *expected payoff* to agent i, the expectation of the true value they have for the bundle they are allocated. Further, v_i is a *best response* to v_{-i} if it maximizes the resultant payoff to agent i, that is, $v_i = \mathrm{argmax}_{\hat{v}_i \in \mathcal{U}} v_i'(M(\{\hat{v}_i, v_{-i}\}))$. The reported valuation v is a *Nash equilibrium* if v_i is a best response to v_{-i}, for every agent $i \in I$. Denote by $NE(v')$ the set of valuations which are Nash equilibria with respect to the true valuations $v' = \{v_1', v_2', \ldots, v_n'\}$.

---

[2] We remark that the proof of [10] for Probabilistic Serial does not apply with multi-unit demand agents, even in the simple symmetric ($m = n$) setting.

[3] In contrast, for a *unit demand* agent i, we have $v_i'(S) = \max_{j \in S} v_i'(j)$.

The *social welfare* of the allocation given by the mechanism is $\sum_{i \in I} v_i'(M(v))$. Observe that, for additive valuation functions, the social welfare is maximized by simply assigning each item to the agent that values it the most. Thus the optimal welfare is $OPT(v') = \sum_{j \in J} \max_{i \in I} v_i(j)$. The *price of anarchy* is the worst-case ratio between the optimal welfare and the social welfare of the *worst* Nash equilibrium, namely $\sup_{v'} \sup_{v \in NE(v')} \frac{OPT(v')}{\sum_{i \in I} v_i'(M(v))}$. Similarly, the *price of stability* is the worst-case ratio between the optimal welfare and the social welfare of the *best* Nash equilibrium.

2.1   The Probabilistic Serial Mechanism

Let $v = \{v_1, v_2, \ldots v_n\} \in \mathcal{V}$ be the reported unit-sum valuations. Each item has a size (quantity) of one unit, and each agent i will consume one item at a rate of 1 at any time. At time $t \in \left[0, \frac{m}{n}\right]$,[4] let $r^v(t)$ be the set of *remaining* items, that is, the items that have not yet been entirely consumed. Each agent consumes the remaining item for which it has highest value. Specifically, at time t, agent consumes the highest value item from $r^v(t)$. For our analysis, if there are ties and there are two items that an agent has the same value for, then the tie is broken adversarially.

It is known that Probabilistic Serial is not strategy-proof. This motivates studying strategic agents and Nash equilibria under this mechanism. We are especially interested in calculating the price of anarchy of the mechanism. The price of anarchy is known for the one-sided matching problem due to work by Christodoulou et al. [10].

Theorem 2.1. *[10] For the one-sided matching problem with unit-sum valuations, the price of anarchy of Probabilistic Serial is $\Theta(\sqrt{n})$.*

In fact, this guarantee extends beyond Nash equilibria to coarse correlated equilibria and to Bayesian settings. Furthermore, they show this guarantee is the best possible.

Theorem 2.2. *[10] For the one-sided matching problem with unit-sum valuations, the price of anarchy of any unit-sum mechanism is $\Omega(\sqrt{n})$.*

The aim of this paper is to extend to results of [10] to the asymmetric one-sided allocation problem with multi-unit demand valuations. This we achieve in Sect. 4.1 with our main positive result:

Theorem 2.3. *For the asymmetric one-sided allocation problem with multi-unit demand agents, the price of anarchy of Probabilistic Serial is $O(\sqrt{n} \cdot \log m)$.*[5]

---

[4] At time $\frac{m}{n}$ all the items have been consumed because there are m units and each of the n agents consumes at rate 1.

[5] While we do not show the existence of pure strategy Nash equilibria of probabilistic serial in the multi-unit demand setting, the proof by Aziz et al. for the unit demand case [4] also applies to the multi-unit demand setting.

This result is tight to within the logarithmic factor. In particular, in Sect. 4.2 we show that a lower bound of $\Omega(\sqrt{n})$ holds in this setting for *any* fair mechanism. Furthermore, our main negative result is that a logarithmic dependence on m is necessary in the lower bound for Probabilistic Serial. In particular, the price of anarchy of Probabilistic Serial degrades with the number of items.

Theorem 2.4. *For the asymmetric one-sided allocation problem with multi-unit demand agents, the price of anarchy of Probabilistic Serial is $\Omega(\min\{n, \log m\})$.*

The rest of the paper is dedicated to proving these results.

3   Bounding the Payoff of an Agent

To quantify the price of anarchy we require an understanding of the allocations and payoffs induced at a Nash equilibrium $v \in \mathcal{V}$. This is difficult to do directly. So a standard approach is, for each agent i, to fix the strategies v_{-i} of the other agents and hypothesize about the payoff obtainable if the agent plays an alternative strategy. This lower bounds the payoff obtained by the strategy v_i because it is a best response to v_{-i}. Summing over all agents then gives a lower bound on social welfare.

But what alternative strategies should be considered? In Sects. 3.1 and 3.2, we study two simple strategies for each agent: changing the highest ranked item and changing the k highest ranked items. We prove structural properties of these strategies and then use these properties to prove a technical lemma. This technical lemma can be viewed as a generalization to the asymmetric allocation problem with multi-unit demand agents of the main technical lemma of [10] for the symmetric allocation problem with unit demand agents.

3.1   Bidding for One Item

As stated, a natural approach in trying to quantify the social welfare of a Nash equilibrium v is to consider alternate strategies for the agents. An important starting point to understand alternate strategies is considering what happens if an agent shifts to reporting a specific item j as their favorite item. We denote this report by $\hat{u}_j$.

To analyze this change of strategy, let $t_j(v)$ be the minimum value of t such that j has been entirely consumed; recall this is the *consumption time* of item j under the Nash equilibrium v. Now denote by $\tilde{t}_j = t_j(\hat{u}_j, v_{-i})$, the consumption time of item j when agent i bids single-mindedly for j and the other agents $-i$ report v_{-i}.

Two properties of this alternative report will be useful. First, regardless of the strategies of the other agents $-i$, the consumption time of item j will be minimized if agent i consumes item j first. Second, at a Nash equilibrium, if agent i deviates and consumes item j first, with j being some item whose consumption time was at most 1, then the consumption time of item j can decrease by at most 75%.

Before proving these two properties we remark that whilst the first property may seem self-evident there is a major subtlety due to dynamic knock-on effects. Indeed, when an item $\ell \neq j$ for which many agents have high value has been entirely consumed, the consumption rates get redistributed among the remaining items. It is necessary to show that bidding for ℓ does not decrease the completion time of j despite leading to the other agents consuming more of j. The key to the proof is showing that these indirect knock-on effects do not outweigh the direct effects of consuming j.

Lemma 3.1. *Given any v_{-i}, the consumption time of item j is minimized when agent i reports j as their favorite item. That is, $\min_{\overline{u} \in \mathcal{U}} t_j(\overline{u}, v_{-i}) = \tilde{t}_j$ and $\mathrm{argmin}_{\overline{u} \in \mathcal{U}} t_j(\overline{u}, v_{-i}) = \hat{u}_j$*

Proof. Take any agent $i \in I = [n]$, any item $j \in J = [m]$ and any $v \in \mathcal{V}$. Throughout the proof v_{-i} will be fixed but the strategy of agent i will change from v_i to some $\hat{u}_j$, where i reports j as the highest value item. Assume without loss of generality that the items of J are labelled in non-decreasing consumption time; that is, $(t_{x_j})_{j \in [m]}$ is non-decreasing.

We wish to show that at any time $t \leq \tilde{t}_j$, the time when j is consumed in the alternative strategy, the following properties, which imply that j is consumed faster in the alternative strategy, hold:

1. The set of items which have been consumed under the alternative strategy is contained in the set of items which have been consumed in the original strategy.
2. The consumption rate with the original strategy of any $j' \in [m] \setminus \{j\}$ that has not been consumed is at least its consumption rate with the alternative strategy. The total consumption rate of a set X of items in the original strategy is at most the total consumption rate of X after i switches to reporting j as their favorite item.

Consider the first time t such that either condition is no longer true. Given that the consumption rate and the set of consumed items only changes when some item has been consumed, t must be the consumption time of an item in either the original strategy or the alternative strategy. Now assume that Condition 1 no longer holds. Since the condition held up to time t, the item(s) that has been consumed at time t must have been consumed in the alternative strategy but not the original one. However, by Condition 2, items other than j were being consumed slower in the alternative strategy before time t so they can't be consumed earlier. Hence Condition 1 must still hold up to time t.

Assume that some item j' which has not been consumed in the original strategy is the favorite item of an agent i' in the alternative strategy. By Condition 1 holding up to time τ_k, the set of available items in the alternative strategy contains the set of available items in the original one, so it must also be its favorite available item in the original strategy. Similarly, since no item outside X is consumed before j has been consumed in the original strategy, we can show that no agent has an item $j_1 \notin X$ as their favorite item in the alternative strategy and an item $j_2 \in X$ as their favorite in the original strategy as both j_1 and j_2 must be available in both strategies. Since the consumption rate of an item is the number of agents who have it as their favorite item, the two aforementioned results imply that Condition 2 still holds at time t.

We have shown that both conditions hold throughout the mechanism until j has been consumed with the alternative strategy. So, because X is consumed faster with the alternative strategy than with the original strategy until j has been consumed, j must be consumed faster with the alternative strategy than with the original strategy. As this is true for any strategy, the alternative strategy minimizes the consumption time of j.   □

The consumption time of each item in the Nash equilibrium will be denoted as the time $t_j = t_j(v)$. For convenience we denote $t_0 = 0$. Without loss of generality, we relabel the items so that t_j is increasing with j.

A second property which we require is the following lemma from [10].

Lemma 3.2. *[10] Let v be a pure Nash equilibrium. Take any agent i and any item j whose consumption time is at most 1. The consumption time of item j decreases by at most 75% if agent i switches to the single-minded strategy $\hat{u}_j$, that is if $t_j \leq 1$ then $\tilde{t}_j \geq \frac{1}{4} \cdot t_j$.*

This lemma only concerns items consumed by time 1.[6] Of course, later we must also consider items whose consumption time is after time 1.

3.2   Bidding for a Sequence

Unfortunately, unlike in the matching case, it is not enough to consider changing the reported value for a single item when we have multi-unit demand agents. Indeed, this is intuitively obvious. If an agent wins many items in the optimal solution to the allocation problem then a strategy that targets a single item will likely to do very poorly in comparison.

To circumvent this problem, we now need to assume an agent changes their report to a target set $X = \{x_1, x_2, \ldots, x_k\}$. That is, we assume that the agent reports x_i as their i^{th} favorite item. We denote this strategy as $\hat{u}_X$.

Using the two lemmas we obtained in the previous section, we can analyse the consequences of this deviation. Specifically, we prove the following technical lemma.

Lemma 3.3. *For any agent i, let v'_i be the true value i has for the items and let v be any pure Nash equilibrium with respect to v'. Then, for any sequence of items $X = \{x_1, x_2, \ldots, x_k\}$ whose consumption times are bounded above by 1, it holds that:*

$$v'_i(PS(v)) \geq \frac{1}{4} \cdot \sum_{\ell=1}^{k} (t_{x_\ell} - t_{x_{\ell-1}}) \cdot v'_i(x_\ell).$$

---

[6] We remark that this condition on the consumption time is not present in the original statement of the lemma as it is not needed when there are only n items.

Proof. Recall that we have:

- t_ℓ: consumption time of item x_ℓ in the Nash equilibrium v.
- $\tilde{t}_\ell$: consumption time of item x_ℓ in $(\hat{u}_X, v_{-i})$ where i ranks X as their favorite items.

We additionally denote the consumption time of item j when an agent switches to $\hat{u}_X$ as $\hat{t}_j = t_j(\hat{u}_X, v_{-i})$. Moreover, for convenience we denote $\hat{t}_0 = 0$.

By Lemma 3.1 we have $\tilde{t}_j \geq \hat{t}_j$. By Lemma 3.2 we have $\hat{t}_j \geq \frac{1}{4}t_j$. Therefore, $\tilde{t}_j \geq \frac{1}{4}t_j$. Now recall, under the strategy $\hat{u}_X$, the items of $X = \{x_1, x_2, \ldots, x_k\}$ are ordered in decreasing order of value for agent i. This means that before time $\tilde{t}_{x_\ell}$ the agent consumes an item whose value is at least $v_i'(x_\ell)$. In particular, because $\tilde{t}_{x_\ell} \geq \frac{1}{4}t_{x_\ell}$, if $t_{x_{\ell-1}} \leq t_{x_\ell}$ then during the interval $[\frac{1}{4}t_{x_{\ell-1}}, \frac{1}{4}t_{x_\ell}]$ agent i consumes an item of value at least $v_i'(x_\ell)$. Therefore, agent i has a payoff of

$$v'(PS(\{\hat{u}_X, v_{-i}\})) \geq \sum_{\ell=1}^{k} \max\left(\frac{1}{4}t_{x_\ell} - \frac{1}{4}t_{x_{\ell-1}}, 0\right) \cdot v_i'(x_\ell)$$

$$\geq \sum_{\ell=1}^{k} \left(\frac{1}{4}t_{x_\ell} - \frac{1}{4}t_{x_{\ell-1}}\right) \cdot v_i'(x_\ell)$$

$$\geq \frac{1}{4} \cdot \sum_{\ell=1}^{k} \left(t_{x_\ell} - t_{x_{\ell-1}}\right) \cdot v_i'(x_\ell)$$

As v is a Nash equilibrium, it must also give agent i a payoff of at least $\frac{1}{4} \cdot \sum_{i=1}^{k} \left(t_{x_\ell} - t_{x_{\ell-1}}\right) \cdot v_i'(x_\ell)$. $\qquad\square$

4   The Price of Anarchy

We are now ready to quantify the price of anarchy. We begin with the upper bound in Sect. 4.1. We will then present complementary lower bounds in Sect. 4.2.

4.1   An Upper Bound on the Price of Anarchy

To give an upper bound on the price of anarchy, we proceed in two steps. In the first step, we will assume that the number of agents and items is the same, that is we consider the *symmetric case* where $m = n$ and prove that the price of anarchy is at most $O(\sqrt{n} \cdot \log n)$ with multi-unit demand agents. Then we extend this result to the *asymmetric case*, where m is arbitrary, and show the price of anarchy is $O(\sqrt{n} \cdot \log m)$.

The Symmetric Case. It is clear that in the symmetric case where $m = n$, the price of anarchy for the unit demand setting is better than the price of anarchy for the multi-unit demand case as the same strategies would give the same expected payoff but the optimal solution of the multi-unit demand is at better. Throughout this section, we show that having multi-unit demand agents changes the price of anarchy by a factor of at most $O(\log(n))$.

Observe that with $m = n$, the completion time of each item is at most 1, allowing us to apply Lemma 3.3 for all items.

We now formulate the price of anarchy as an optimization program. This optimization program is very difficult to handle directly. So our basic approach will be to apply a series of relaxations and simplifications until we obtain a program we can solve. The task is to ensure the transformations are consistent with generating upper bounds and that they do not degrade the value of the objective function excessively.

We now show the following result which will be key in proving the main result, when the number of items is at most the number of agents. We give a detailed overview of the proof here, but the full proof is deferred to the full version of the paper due to space constraints.

Theorem 4.1. *In the symmetric one-sided allocation problem with multi-unit demand agents, let OPT be the social welfare of the optimal allocation. The social welfare of any Nash equilibrium is at least $\Omega\left(\frac{OPT}{\sqrt{n}\cdot\log n}\right)$ for Probabilistic Serial.*

Proof. Let $\{X_1, X_2, \ldots, X_n\}$ be the optimal allocation, where each agent i receives the bundle of items $X_i = \{x_1^i, \ldots, x_{k_i}^i\}$. Here we assume the items in X_i are ordered by increasing consumption time in the Nash equilibrium v. We can then use Lemma 3.3 to lower bound the social welfare of the Nash equilibrium v. To do this first note that, whilst the items of X_i are ordered by consumption time they are not ordered by value. In particular, for the lower bound in Lemma 3.3 we use the right-to-left maxima of $\{v_i'(x_1^i), v_i'(x_2^i), \ldots, v_i'(x_{k_i}^i)\}$. This gives a lower bound on the social welfare of the Nash equilibrium v. We may then bound the price of anarchy using the optimization program with the following objective function:

$$\min \frac{1}{4\cdot OPT} \cdot \sum_{i=1}^{n}\sum_{\ell=1}^{k_i} \max_{\ell'=\ell,\ldots,k_i} \left\{v_{i(x_{\ell'})}' \cdot (t_{x_\ell^i}(v) - t_{x_\ell^i}(v))\right\}. \tag{1}$$

Where we are optimizing over v' the true valuations of the agents, v the reported valuations which are a Nash equilibrium with respect to v', t the consumption time of the items, and OPT the social welfare of the optimal allocation given the valuations v'. We then bound the price of anarchy by minimizing the objective function over the following set of constraints:

$$\text{s.t.} \quad \sum_{i=1}^{n}\sum_{\ell=1}^{k_i} v'_{i,x^i_\ell} = OPT \tag{2}$$

$$\bigcup_{i=1}^{n} \{x^i_\ell : \ell \in [k_i]\} = [n] \tag{3}$$

$$\sum_{j=1}^{n} v'_i(j) = 1 \quad \forall i \in [n] \tag{4}$$

$$\sum_{j=1}^{n} v_i(j) = 1 \quad \forall i \in [n] \tag{5}$$

$$t_j(v) \le t_{j+1}(v) \ \forall j \in [n-1] \tag{6}$$

$$v \in NE(v') \tag{7}$$

Let's understand this optimization program. Constraints (4) and (5) state that, for every agent, v'_i and v_i are unit-sum valuation functions. Constraints (2) and (3) ensure $\{X_1, X_2, \ldots, X_n\}$ is a partition of the items with optimal social welfare (with respect to the true valuation functions v'). Next the constraint (6) forces the items to be ordered by increasing consumption time. Finally, the constraint (7) states that v is a Nash equilibrium with respect to the true valuations v'. The objective function (1) then gives a worst-case bound on the price of anarchy using Lemma 3.3.

However, this optimization program is difficult to analyze so our task now is to simplify the program without weakening the resultant price of anarchy guarantee.

To do this, our first step is to fix OPT, the social welfare of the optimal solution. In doing so we may omit the denominator from the objective function (1). Second, observe that the bound can only be worse if we relax or remove constraints from the optimization program. In particular, we omit the Nash equilibrium constraint (7). The third and most important step is to use a change of variables in order to prove that in the worst case an agent who receives a set of size k in the optimal allocation has value $\frac{1}{k}$ for each item in that set. (*The technical details for these steps are given in the full version of the paper*)

Denote $t^{(i)}$ to be the largest consumption time of an item allocated to i in the optimal allocation and ℓ_i to be the number of items allocated to i in the optimal allocation. Then, we can bound the value of the Nash equilibrium below by:

$$\min \quad \sum_{i=1}^{k} \frac{t_{x^i_{\ell_i}}}{\ell_i} \quad \text{s.t.} \quad \bigcup_{i=1}^{n} \{x^i_\ell : \ell \in [k_i]\} = [n]$$

To evaluate this, recall that the items are labelled in increasing order of consumption time. These consumption times then satisfy the following property.

Lemma 4.2. *The consumption time of item j must satisfy $t_j \ge \frac{j}{n}$.*

Proof. Each agent consumes an item at a rate of 1. Consequently, the total consumption rate of all agents is n. Thus at time $t = \frac{j}{n}$ the number of units consumed of all goods is exactly j. But the quantity of each good is each is exactly 1, so at most j goods can have been completely consumed at time $t = \frac{j}{n}$. Hence the consumption time of good j is $t_j \geq \frac{j}{n}$. $\square$

Next we partition the agents into groups depending upon their ℓ_i. Specifically, let $I_\tau = \left\{i \in [k] : \ell_i \in [2^\tau, 2^{\tau+1})\right\}$ for all $0 \leq \tau \leq \lceil \log n \rceil$. Using Lemma 4.2, we can show that the bundles assigned to q agents of I_τ have been consumed at time at least $\frac{2^\tau \cdot q}{n}$. Using this bound, on the consumption time of I_τ with the highest cardinality, we get a bound of $\frac{k^2}{4n \log^2 n + o(n \log n)}$ on the value of the optimization program. (*The technical details of this step are given in teh full version of the paper*)

We are now ready to prove our price of anarchy upper bound. By applying Lemma 3.3, with $X = [n]$, at any Nash equilibrium each agent is guaranteed a payoff of at least $\frac{1}{4n}$. Thus the social welfare of any Nash equilibrium is at least $\frac{1}{4}$. The price of anarchy is then at most

$$\max_{k \in [1,n]} \left\{ \min \left\{ \frac{k}{\frac{1}{4}} , \frac{k}{\frac{1}{4} \cdot \frac{k^2}{4n \log^2 n + o(n \log n)}} \right\} \right\} = 2\sqrt{n} \cdot \log n + o\left(\sqrt{n \log n}\right)$$

So, the price of anarchy of the Probabilistic Serial mechanism is $O\left(\sqrt{n} \cdot \log n\right)$. in the symmetric case, even with multi-unit demand agents. $\square$

The Asymmetric Case. To extend our proof of the upper bound to the asymmetric setting, we proceed by case analysis. The first case is when the items which are consumed before time 1 make the largest contribution to social welfare. In this case, the proof used to show Theorem 4.1 can be used to show an upper bound of $O(\sqrt{n} \cdot \log m)$. In the second case, when the items which are consumed after time 1 make the largest contribution to social welfare, we show there is then a small set of items that make a significant contribution to social welfare; moreover, the agents have a strategy to win these items with constant probability.

To prove our main positive result, we begin with the following result concerning the first case.

Theorem 4.3. *Let X be the set of items whose consumption time is at most 1. If OPT_X is the social welfare of the optimal allocation of these items, then the social welfare of any Nash equilibrium is at least $\Omega\left(\frac{OPT_X}{\sqrt{n} \cdot \log |X|}\right)$.*

Proof. The proof of Theorem 4.1 for the symmetric case applies in bounding the value the agents get in a Nash equilibrium using only items in X. However, we are not guaranteed to have n items in X, instead we have $|X| \leq n$.[7] So, denoting $k = |X|$, when we simplify the optimization program we get the following

$$\min \quad \sum_{i=1}^{k} \frac{t_i}{\ell_i} \quad \text{s.t.} \quad \bigcup_{i=1}^{n} \{x_\ell^i : \ell \in [k_i]\} = [k]$$

We can then mimic the rest of the proof of Theorem 4.1 and split the items into $\log k$ sets containing items 2^τ to $2^{\tau+1} - 1$. Replacing $\log n$ by $\log k$ in the proof, we get a bound of $2\sqrt{n} \cdot \log k + o\left(\sqrt{n \log k}\right)$, as required.   □

Unfortunately, for the case $m > n$, we cannot directly use the proof of the symmetric case. Instead, we apply a different approach based upon the following lemma.

Lemma 4.4. *For any agent i, let v_i' be the true value i has for the items and let v be any pure Nash equilibrium with respect to v'. Then, for any sequence of items $X = \{x_1, x_2, \ldots, x_k\}$ whose consumption times are bounded below by $q \geq 2$, for some $k < \lfloor q \rfloor$ it holds that:*

$$v_i'(PS(v)) \geq \sum_{\ell=1}^{k} v_i'(x_\ell).$$

Proof. If agent i bids for the items of X from time 0 to $\lfloor q \rfloor - 1$, then the remaining items will be consumed more slowly. In particular, it will take longer for the remaining agents to switch from consuming the other items to consuming items from X. But the consumption time of items in X is at least q and the minimum non-zero consumption rate in Probabilistic Serial is 1. Therefore, no agent may consume from the set X before time $\lfloor q \rfloor - 1$. Thus, i is the only agent consuming these items so it consumes all of X for a duration of time k. Hence, i is guaranteed to win all the items in X and the result follows.   □

We can now prove our upper bound for the price of anarchy of Probabilistic Serial.

Theorem 2.3. *For the asymmetric one-sided allocation problem with multi-unit demand agents, the price of anarchy of Probabilistic Serial is $O(\sqrt{n} \cdot \log m)$.*[8]

Proof. First, assume that $OPT \leq 4\sqrt{n} \cdot \log m$. Then a truthful report of the preferences for Probabilistic Serial will guarantee each agent an expected payoff of $\frac{1}{n}$. This implies that the ratio between the optimal value and the value of the worst Nash equilibrium is at most $\frac{1}{4\sqrt{n} \log m}$.

Second, assume $OPT > 4\sqrt{n} \log m$. Take a Nash equilibrium v. Now let X_0 be the set of items with consumption time between 0 and 1 and let X_ℓ be the set

---

[7] We remark that Lemma 4.2 implies that $|X| \leq n$.

[8] While we do not show the existence of pure strategy Nash equilibria of probabilistic serial in the multi-unit demand setting, the proof by Aziz et al. for the unit demand case [4] also applies to the multi-unit demand setting.

of items with consumption time between $2^{\ell-1}$ and 2^{ℓ}, for $\ell = 1, \ldots, \log(m/n)$. Denote by OPT_{ℓ} the contribution of items in X_{ℓ} to the value of OPT.

We have two cases. Either $OPT_0 \geq \frac{OPT}{2}$ or $\sum_{\ell=1}^{\log(m/n)} OPT_{\ell} \geq \frac{OPT}{2}$. If $OPT_0 \geq \frac{OPT}{2}$ then the result follows from Theorem 4.3.

Otherwise, we have $\sum_{\ell=1}^{\log(m/n)} OPT_{\ell} \geq \frac{OPT}{2}$. Now take ℓ maximizing OPT_{ℓ}. We can now find disjoint sets $X_{i,\ell}$ so that $|X_{i,\ell}| \leq 2^{\ell-1}$ and $\sum_{i=1}^{n} \sum_{j \in X_{i,\ell}} v_i'(j) \geq \frac{OPT}{8\sqrt{n}\log m}$. If each agent can be guaranteed a constant proportion of $X_{i,\ell}$ in the Nash equilibrium, then the result holds. Using Lemma 4.4, if $\ell \neq 1$, agent i can obtain all of $X_{i,\ell}$ giving a bound of $\frac{OPT}{8\sqrt{n}\log m}$. On the other hand, if $\ell = 1$, we know that at most half of the item $X_{i,1}$ has been consumed at time $\frac{1}{2}$ in the Nash equilibrium. If an agent were to switch to consuming that item first, they would get a half unit of the unique item in $X_{i,1}$. This gives a lower bound of $\frac{OPT}{16\sqrt{n}\log m}$. Hence the agents can get a constant proportion of the $X_{i,\ell}$ and their contribution to the optimal welfare is $\Omega\left(\frac{OPT}{\sqrt{n}\log m}\right)$. So the price of anarchy is at most $O(\sqrt{n} \cdot \log m)$. □

4.2 A Lower Bound on the Price of Anarchy

We now present two lower bounds on the price of anarchy. For our first lower bound, we verify that Theorem 2.2 extends to the symmetric one-sided allocation problem with multi-unit demand agents for a specific class of mechanisms, fair mechanisms in which all agents have the same expected number of items.

Theorem 4.5. *For the symmetric one-side allocation problem, the pure price of anarchy of any unit-sum fair mechanism is $\Omega(\sqrt{n})$.*

Proof. Consider the example used by Christodoulou et al. [10] to prove Theorem 2.2 for the matching problem. Take the following valuation function:

$$v_i(j) = \begin{cases} \frac{1}{n} + \varepsilon & \text{if } i = j \cdot \sqrt{n} + i' \text{ for } i' = 1, \ldots, \sqrt{n} \\ \frac{1}{n} - \frac{\varepsilon}{n-1} & \text{otherwise} \end{cases}$$

Now consider a Nash equilibrium for v. Let i_j be the index of the agent who has positive value for item j but has the smallest probability of being assigned j in the Nash equilibrium. Next, create a new valuation $v_i'(j)$ which is $v_i(j)$ if $i \neq i_{j'}$ for any j' and which is 1 if $i = i_j$ and 0 if $i = i_{j'} \neq i_j$

Since the agents get the same number of items in expectation, a Nash equilibrium for v is also a Nash equilibrium for v' where the agents maximize their probability of getting their favorite item. The social welfare of the optimal allocation is $\sqrt{n}$. At the Nash equilibrium, since the agents i_j get assigned j with probability at most $\frac{1}{\sqrt{n}}$, the social welfare is at most $\sqrt{n} \cdot \frac{1}{\sqrt{n}} + \sqrt{n} \cdot \left(1 - \frac{1}{\sqrt{n}}\right) \cdot \left(\frac{1}{n} + \frac{1}{n^3}\right) \leq 3$. This gives a lower bound of $\Omega(\sqrt{n})$ on the price of anarchy. □

This bound is not surprising as with multi-unit demand agents the optimal allocation has higher welfare than the optimal matching. However, our second lower bound is more surprising: for Probabilistic Serial in the asymmetric setting, the price of anarchy deteriorates with the number of items!

Theorem 2.4. *For the asymmetric one-sided allocation problem with multi-unit demand agents, the price of anarchy of Probabilistic Serial is $\Omega(\min\{n\,,\log m\})$.*

Proof. Assume we are given k and let $q = o(k)$. Let the number of agents be $n = k+q$ and the number of items be $m = 2^q - 1$. Assume that, for $i = 1,\ldots, k$, agent i has value 1 for item 1 and 0 for the remaining items. For $i = k+1,\ldots, k+q$, agent i has value $\frac{1}{2^i}$ for items 2^{i-k} to $2^{i+1-k} - 1$. Assume, without loss of generality, that when the items they have positive value for have been consumed, agents consume from the lowest indexed item at a rate of 1.

Then, the items will be consumed in order and until item j has been consumed, agents 1 to $k + \lfloor \log j \rfloor$ are only consuming items 1 to j. So item j is consumed at time $\frac{j}{k + \lfloor \log j \rfloor} \geq \frac{2j}{n}$. Thus, agents 1 to k are consuming item 1 which they value at 1 for a time duration of $\frac{1}{k} < \frac{4}{n}$. Then agents $k+1$ to $k+q$ will consume items 2^{i-k} to $2^{i+1-k} - 1$, which they value at $\frac{1}{2^{i-k}}$, at a rate of 1 for a time duration of at most $\frac{2^{i+1-k}-1}{n}$. Hence the value they obtain is at most $\frac{2^{i+1-k}-1}{n} \cdot \frac{1}{2^{i-k}} < \frac{4}{n}$.

Since every agent has value at most $\frac{4}{n}$ from the mechanism, the total social welfare is at most 4. However, the optimal allocation has welfare q so the price of anarchy of Probabilistic Serial Serial is at least $\Omega(q) = \Omega(\log m)$.

Note, it is easy to see that $O(n)$ is always an upper bound on the price of anarchy in any instance. □

We conjecture that the price of anarchy for n agents and m items is, in fact, $\Theta(\min\{n\,,\sqrt{n}\cdot\log m\})$. This would imply our upper bound of $O(\sqrt{n}\cdot\log m)$ is tight and that allowing for agents to get multiple items worsens the price of anarchy.

5   Related Problems

5.1   The Price of Stability and General Equilibria

We obtain similar bounds for the price of stability. Below the upper bound follows immediately from our price of anarchy bound. The lower bound follows from the lower bound on the price of stability of Probabilistic Serial for unit demand agents [10].

Theorem 5.1. *For the one-sided allocation problem with multi-unit demand agents, the price of stability of Probabilistic Serial is at least $\Omega(\sqrt{n})$ and at most $O(\sqrt{n}\cdot\log n)$.*

In their paper, Christodoulou et al. also consider more general equilibrium concepts: *coarse correlated equilibria* and *Bayes-Nash equilibria*. They show that their $O(\sqrt{n})$ bound extends to the coarse correlated price of anarchy and the

Bayesian price of anarchy [10]. Since our results on the bound of an agent's payoff does not depend on the other players acting optimally, we can also extend our results to these equilibria.

Theorem 5.2. *For the one-sided allocation problem with multi-unit demand agents, the coarse correlated price of anarchy of probabilistic serial is $O(\sqrt{n} \cdot \log n)$. The Bayesian price of anarchy of probabilistic serial is $O(\sqrt{n} \cdot \log n)$*

5.2   The Random Priority Mechanism

The focus of this paper has been the Probabilistic Serial mechanism. How does the other classical mechanism, Random Priority, perform in the asymmetric one-sided allocation problem? To answer this question, we remark that there are two natural ways to implement Random Priority in the asymmetric setting:

- *Random Priority.* Each agent is randomly sampled once and, upon selection, picks their favorite $\frac{m}{n}$ items from amongst those that are still available.
- *Repeated Random Priority.* Agents are sampled repeatedly m times in a row (uniformly and independently) and, upon selection, the selected agent picks its favorite available item.

Theorem 5.3. *For the asymmetric one-sided allocation problem with multi-unit demand agents, the price of anarchy of Random Priority is at least n when $m \geq n^2$.*

Proof. The upper bound follows because each agent is guaranteed a payoff of at least $\frac{1}{n}$.

The lower bound follows from assuming we have $m = n^2$ items and agent i has value $1 - \varepsilon$ for item i and $\frac{\varepsilon}{n-1}$ for items $i' \in [n]\backslash\{i\}$ (and 0 for the rest). Then, the first agent to be selected gets value 1 but the remaining agents get value 0. In the optimal allocation, every agent gets value $1 - \varepsilon$, so the optimal welfare is $n \cdot (1 - \varepsilon)$. □

Theorem 5.4. *For the asymmetric one-sided allocation problem with multi-unit demand agents, the price of anarchy of Repeated Random Priority is at least $\Omega(\min\{n, \log m\})$.*

Proof. Consider the example from the proof of Theorem 2.4. The expected value for an agent who get 2^i items is $\frac{1}{2^i}$ multiplied by their expected number of items won from the set of items they are interested in. This is at most the number of times that appear in the first 2^{i+1} rounds, omitting those where an agent with a higher index wins. This can be bounded by the expectation of a binomial $B\left(2^{i+1}, \frac{1}{k+i}\right)$ which is $\frac{2^{i+1}}{k+i} \leq \frac{2^{i+2}}{n}$. This implies that their expected value is at most $\frac{4}{n}$. Thus the welfare of the allocation is at most 4, even though the optimal social welfare is q. Consequently, when there are $2^q - 1$ items the welfare of the best Nash equilibrium is at most $\frac{OPT}{q} = \frac{OPT}{\log m}$. □

6   Conclusion

We studied fair mechanisms for the asymmetric one-sided allocation problem with multi-unit demand agents. A natural open problem is to close the logarithmic gap between the upper and lower bounds in Theorem 2.3 and Theorem 4.5. Another interesting line of research is to study whether our results extend to other classes of valuation function, specifically, non-additive valuation functions. We remark that while the price of anarchy bounds for the unit demand setting extend to *unit-range* valuations this is *not* the case for multi-unit demands. For unit-range valuations the price of anarchy is $\Omega(n)$ for any fair mechanism.[9]

Acknowledgements. The authors are grateful to Hervé Moulin for comments and advice. We also thank the referees for suggestions improving the paper.

References

1. Abdulkadirogl, A., Sonmez, T.: Random serial dictatorship and the core from random endowments in house allocation problems. Econometrica **66**(3), 689–701 (1998). http://www.jstor.org/stable/2998580
2. Abdulkadirogl, A., Sonmez, T.: Matching Markets: Theory and Practice, Econometric Society Monographs, vol. 1, pp. 3–47. Cambridge University Press, Cambridge (2013). https://doi.org/10.1017/CBO9781139060011.002
3. Ashlagi, I., Saberi, A., Shameli, A.: Assignment mechanisms under distributional constraints. Oper. Res. **68**(2), 467–479 (2020). https://doi.org/10.1287/opre.2019.1887
4. Aziz, H., Gaspers, S., Mackenzie, S., Mattei, N., Narodytska, N., Walsh, T.: Equilibria under the probabilistic serial rule. In: Proceedings of 24th International Conference on Artificial Intelligence (AAAI), pp. 1105–1112 (2015). https://doi.org/10.48550/ARXIV.1502.04888, https://arxiv.org/abs/1502.04888
5. Aziz, H., Gaspers, S., Mackenzie, S., Mattei, N., Narodytska, N., Walsh, T.: Manipulating the probabilistic serial rule. In: Proceeding of Autonomous Agents and Multiagent Systems International Conference (AAMAS), pp. 1451–1459 (2015). https://arxiv.org/abs/1502.04888
6. Bogomolnaia, A., Moulin, H.: A new solution to the random assignment problem. J. Econ. Theory **100**(2), 295–328 (2001). https://doi.org/10.1006/jeth.2000.2710, https://www.sciencedirect.com/science/article/pii/S0022053100927108
7. Boutilier, C., Caragiannis, I., Haber, S., Lu, T., Procaccia, A., Sheffet, O.: Optimal social choice functions: a utilitarian view. In: Proceedings of 13th Conference on Electronic Commerce (EC), pp. 197–214 (2013)
8. Budish, E., Che, Y., Kojima, F., Milgrom, P.: Designing random allocation mechanisms: theory and applications. Am. Econ. Rev. **103**(2), 585–623 (2013). http://www.jstor.org/stable/23469677
9. Caragiannis, I., Kaklamanis, C., Kanellopoulos, P., Kyropoulou, M.: The efficiency of fair division. Theory Comput. Syst. **50**(4), 589–610 (2012). https://doi.org/10.1007/s00224-011-9359-y

---

[9] To see this, take a single agent with value 1 for every item and let other agents having value 1 for the first item and ε for the remaining items.

10. Christodoulou, G., Filos-Ratsikas, A., Frederiksen, S.K.S., Goldberg, P.W., Zhang, J., Zhang, J.: Social welfare in one-sided matching mechanisms. In: Osman, N., Sierra, C. (eds.) AAMAS 2016. LNCS (LNAI), vol. 10002, pp. 30–50. Springer, Cham (2016). https://doi.org/10.1007/978-3-319-46882-2_3

11. Ekici, O., Kesten, O.: An equilibrium analysis of the probabilistic serial mechanism. Int. J. Game Theory (2016). https://doi.org/10.1007/s00182-015-0475-9

12. Filos-Ratsikas, A., Frederiksen, S.K.S., Zhang, J.: Social welfare in one-sided matchings: random priority and beyond. In: Lavi, R. (ed.) SAGT 2014. LNCS, vol. 8768, pp. 1–12. Springer, Heidelberg (2014). https://doi.org/10.1007/978-3-662-44803-8_1

13. Hylland, A., Zeckhauser, R.: The efficient allocation of individuals to positions. J. Polit. Econ. **87**(2), 293–314 (1979). http://www.jstor.org/stable/1832088

14. Katta, A., Sethuraman, J.: A solution to the random assignment problem on the full preference domain. J. Econ. Theory **131**(1), 231–250 (2006). https://doi.org/10.1016/j.jet.2005.05.001, https://www.sciencedirect.com/science/article/pii/S0022053105001079

15. Procaccia, A., Tennenholtz, M.: Approximate mechanism design without money. In: Proceedings of 10th Conference on Electronic Commerce (EC), pp. 177–186 (2009)

16. Sonmez, T., Unver, U.: Matching, allocation and exchange of discrete resources. Handbook of Social Economics, vol. 1, pp. 781–852. North-Holland (2011). https://doi.org/10.1016/B978-0-444-53187-2.00017-6, https://www.sciencedirect.com/science/article/pii/B9780444531872000176

17. Zhang, J.: Tight social welfare approximation of probabilistic serial. Theor. Comput. Sci. **934**, 1–6 (2022). https://doi.org/10.1016/j.tcs.2022.08.003, https://www.sciencedirect.com/science/article/pii/S0304397522004807

18. Zhou, L.: On a conjecture by Gale about one-sided matching problems. J. Econ. Theory **52**(1), 123–135 (1990). https://doi.org/10.1016/0022-0531(90)90070-Z, https://www.sciencedirect.com/science/article/pii/002205319090070Z

An Adaptive and Verifiably Proportional Method for Participatory Budgeting

Sonja Kraiczy$^{(\boxtimes)}$ and Edith Elkind

Department of Computer Science, University of Oxford,
7 Parks Rd, Oxford OX1 3QG, UK
{Sonja.Kraiczy,Edith.Ekind}@cs.ox.ac.uk

Abstract. Participatory Budgeting (PB) is a form of participatory democracy in which citizens select a set of projects to be implemented, subject to a budget constraint. The Method of Equal Shares (MES), introduced in [18], is a simple iterative method for this task, which runs in polynomial time and satisfies a demanding proportionality axiom (Extended Justified Representation) in the setting of approval utilities. However, a downside of MES is that it is non-exhaustive: given an MES outcome, it may be possible to expand it by adding new projects without violating the budget constraint. To complete the outcome, the approach currently used in practice (e.g., in Wieliczka in Apr 2023, https://equalshares.net/resources/zielony-milion/) is as follows: given an instance with budget b, one searches for a budget $b' \geq b$ such that when MES is executed with budget b', it produces a maximal feasible solution for b. The search is greedy, i.e., one has to execute MES from scratch for each value of b'. To avoid redundant computation, we introduce a variant of MES, which we call Adaptive Method of Equal Shares (AMES). Our method is budget-adaptive, in the sense that, given an outcome W for a budget b and a new budget $b' > b$, it can compute the outcome W' for budget b' by leveraging similarities between W and W'. This eliminates the need to recompute solutions from scratch when increasing virtual budgets. Furthermore, AMES satisfies EJR in a certifiable way: given the output of our method, one can check in time $O(n \log n + mn)$ that it provides EJR (here, n is the number of voters and m is the number of projects). We evaluate the potential of AMES on real-world PB data, showing that small increases in budget typically require only minor modifications of the outcome.

Keywords: Computational Social Choice · Participatory Budgeting

1 Introduction

Participatory Budgeting (PB) gives residents of a city the power to decide how (part of a) public budget will be spent. Starting with Porto Alegre in Brazil, PB is now used in many cities around the world including locations in France, Iceland, Italy, Poland, Spain and many more [24]. Typically, the city council

© The Author(s), under exclusive license to Springer Nature Switzerland AG 2024
J. Garg et al. (Eds.): WINE 2023, LNCS 14413, pp. 438–455, 2024.
https://doi.org/10.1007/978-3-031-48974-7_25

collects proposals for projects to be funded [10], such as building new cycle lanes or refurbishing a playground, and then the residents vote to indicate which of the proposed projects they would like to see implemented. These votes are then aggregated into a final outcome, which must respect the budget constraint [5]. In practice, it is common to use the *greedy rule*, which asks the residents which projects they approve, sorts the projects by the number of approvals (from highest to lowest), and adds projects one by one in this order until the budget is exhausted. However, the greedy rule is very far from being proportional: if 51% of residents approve one set of projects, while 49% approve a disjoint set of projects, it will consider all projects approved by the 51% majority before those approved by the 49% minority.

In contrast, the recently introduced Method of Equal Shares (MES) [19] gives all voters equal voting power. The outcome of MES is proportional, as formalized by the demanding EJR (Extended Justified Representation) axiom [1]. (This axiom was originally formulated for the simpler setting where all projects have equal cost, known as multiwinner voting, and then extended to the PB setting [18].) Indeed, MES is the first voting rule to satisfy the EJR axiom in the context of participatory budgeting. MES has already been used in practice in Wieliczka (Poland) in April 2023, and in Aarau (Switzerland) in June 2023 [20]. Informally, it proceeds by (virtually) giving each voter an equal part of the budget and selecting projects iteratively one by one; the cost of each selected project is shared (almost) evenly among its supporters. In each iteration, MES selects a project so as to minimize the per-head price paid by its supporters: for instance, it will prioritize a project that costs $10,000 and is supported by 2000 voters over a project that costs $8,000 and and is supported by 1000 voters. It stops when none of the remaining projects can be afforded by its supporters.

Efficient Completion. A downside of MES is that it may fail to exhaust the budget, i.e., there may be projects that receive approvals from some voters and would fit within the remaining budget, but remain unselected. Therefore, a natural question is how to best complete the outcome of MES. In practice, the current approach to the completion problem (for a detailed discussion, see [20]) is to run the method multiple times, increasing the (virtual) budget in each iteration. In more detail, suppose the real budget available is three million dollars, but the output of MES with such a budget is not exhaustive. Then we increase every voter's budget by one dollar and run MES with the increased budget. We proceed in this way until either the outcome is exhaustive or the next budget increase would result in an outcome whose total cost is more than three million dollars.

However, running the Method of Equal Shares from scratch repeatedly may be very inefficient: Intuitively, the outcome is unlikely to change much if the budget is increased by a small amount. Hence, it would be desirable to leverage the overlap between the outcomes of successive iterations, and construct the outcome iteratively, only modifying it to the extent necessary, instead of executing MES from scratch each time we increase the budget.

Efficient Verifiablility. Another key challenge in multiwinner voting and participatory budgeting is efficiently verifying the proportionality of an outcome. In particular, verifiability is important in the context of blockchains, and specifically their applications in Decentralized Autonomous Organizations (DAOs) [6,22]. DAOs are based on creating a large network of nodes belonging to human stakeholders, with each node running a protocol locally. The vision is for such a network to offer services and make democratic decisions about various matters in a decentralized fashion, without the need to trust or rely on a central authority. The protocol should work even if the computational resources of the humans behind nodes are limited, such as off-the-shelf computers or smartphones, so that as many as possible can benefit from the network.

Some blockchain networks use multiwinner voting to appoint validators [8, 9,13]. Validators are special roles that nodes can take on: they have to validate transactions and receive a monetary reward for doing so (or get punished for adversarial behavior). It is desirable to select validators (from the set of candidate nodes) in a proportional manner, both to increase voting nodes' satisfaction [14], and to avoid the centralization of power [12]. Unfortunately, many proportional multiwinner rules are computationally hard [3] or else have prohibitively slow polynomial running time, so recent work by [11] proposes efficient verifiability as a solution. The key observation is that the (expensive) computational task of choosing the validators can be performed by a non-trusted party ("off-chain") as long as the proportionality of the proposed solution can be efficiently checked by any node. Unfortunately, checking whether an arbitrary outcome satisfies the EJR axiom (or even the weaker PJR axiom) is NP-hard even in the setting of multiwinner voting [1,2]. Nevertheless, [11] present a new multiwinner voting rule phragmms, such that the outputs of this rule satisfy PJR and this can be efficiently verified in time linear in the input size (the rule outputs auxiliary information, which can be used as a certificate of PJR). Now, verifiability remains a relevant concern in the broader context of participatory budgeting: it is natural for DAOs to allocate funds for new projects based on stakeholder votes. However, prior to our work it was not known whether there exist voting rules for participatory budgeting that satisfy demanding proportionality axioms (such as EJR) in a verifiable way.

Our Contribution. We present a method for participatory budgeting with approval utilities that is closely related to the Method of Equal Shares. Specifically, we define a weak order $\triangleright$ on feasible solutions (i.e., pairs of the form (W, X), where W is the selected set of projects and X describes how the costs of these projects are shared among the voters) that is inspired by the definition of the MES rule. Our method, which we call the Adaptive Method of Equal Shares (AMES), operates by executing greedy local search with respect to $\triangleright$, i.e., it outputs solutions that cannot be improved with respect to $\triangleright$ by simple transformations. Interestingly, in all solutions output by AMES the cost of each project is shared *exactly* equally among its supporters. Moreover, if a MES solution for a given instance has this 'exact equal sharing' property, it is also in the output of AMES on this instance, i.e., the two methods are indeed similar.

We show that our method has several desirable properties:

1. AMES satisfies the EJR axiom for approval utilities.
2. AMES runs in time $O(mn \log n + mn)$, where m is the number of projects and n is the number of voters. Moreover, it admits a very efficient algorithm for verifying that its output satisfies EJR: by using $O(mn)$ auxiliary data, it can verify EJR in time $O(mn)$, i.e., linear in the input size.
3. AMES is budget-adaptive: To determine its output for budgets $b_1 < b_2 \ldots < b_t$, instead of having to start from scratch for every single budget, we can use the output for budget b_i to obtain the output of the rule for budget b_{i+1}.

This combination of properties in unique: AMES is the first polynomial-time voting method to satisfy EJR in the approval PB setting that is budget-adaptive and verifiably proportional. Indeed, in the PB setting Phragmén's rule [15, 21] is naturally budget-adaptive, but fails EJR even for multiwinner voting (i.e., for PB with unit costs), while Local Search PAV [2] is fully adaptive, but is known to fail EJR in the PB setting [18]. For multiwinner setting, Brill and Peters [7], have recently proposed a new axiom, which they call EJR+. This axiom strengthens EJR, can be verified in time $O(kmn)$ in general (where k is the size of the committee), and is satisfied by a simple greedy rule. However, the analysis in [7] does not show linear-time verifiability; moreover, the verifiability results of [7], just as those of [11], only apply to multiwinner approval voting rather than the more general setting of participatory budgeting.

To gain insights on the practicality of our method, we analyze real-world participatory budgeting data from several cities in Poland, and consider the average number of changes in the outcome as a result of increases in the budget. Our findings confirm that AMES, and the budget-adaptive approach in general, has great potential to make practical implementation much faster.

We also show that, by handling ties carefully, we can ensure that the output of the algorithm run from scratch with budget b is identical to the output of the adaptive algorithm that starts from a feasible solution for a budget $b' < b$ and then (gradually) raises the budget to b. This property is attractive, because it ensures that AMES is as easy to explain to the voters as the standard MES rule: instead of explaining how the algorithm modifies the solution with each budget increase, one can simply demonstrate its execution for the final budget b.

Related Work. The special case of participatory budgeting with approval ballots where all projects have the same cost is known as *approval-based multiwinner voting*, and there is a very substantial literature on approval-based multiwinner voting rules, their axiomatic properties, and algorithmic complexity [16]. The EJR axiom was first introduced in this context in [1], and the first polynomial-time rule shown to satisfy it was Local Search PAV [2], followed by the arguably more natural Method of Equal Shares [19].

In recent years, there has been an explosion of interest in participatory budgeting; by now there is a wide variety of different models and approaches [5]. Proportionality axioms for participatory budgeting were first considered by [4]. In [18]. the authors extend the Method of Equal Shares from the multiwinner voting setting [19] to a general PB model with general additive utilities. They show that MES satisfies EJR up to one project, giving the first such polynomial-time voting method for participatory budgeting. For the special case of participatory budgeting with approval utilities (which is the focus of our work), [18] show that MES satisfies EJR.

Our adaptive method uses a stability notion very similar to that of Peters et al. [17], in that they also consider justifying outcomes of an election by means of a price system that satisfies the stability condition. Peters et al. are concerned with the existence and properties of this stability notion for a given committee size, which may not always exist, while our formulation is not tied to a committee size. Instead, we show that we can use the stability notion to efficiently move through the space of intermediate outcomes of MES when increasing the committee size/ budget.

2 Preliminaries

For each $t \in \mathbb{N}$, we write $[t] = \{1, 2, \ldots, t\}$.

Participatory Budgeting. We first introduce the model of participatory budgeting with approval ballots. An *election* is a tuple $E = (N, P, (A(i))_{i \in N}, b, cost)$, where:

1. $N = [n]$ and $P = \{p_1, \ldots, p_m\}$ are the sets of *voters* and *projects*, respectively;
2. for each $i \in N$ the set $A(i) \subseteq P$ is the *ballot* of voter i, i.e., the set of projects approved by i;
3. $b \in \mathbb{Q}_{>0}$ is the available budget;
4. $cost : P \to \mathbb{Q}_{>0}$ is a function that for each $p \in P$ indicates the cost of selecting p. For each $W \subset P$, we denote the total cost of W by $cost(W) = \sum_{p \in W} cost(p)$.

We assume every project is approved by at least one voter.

An *outcome* is a set of projects $W \subseteq P$ that is *feasible*, i.e., satisfies $cost(W) \leq b$. We denote the set of all outcomes for an election $E = (N, P, (A(i))_{i \in N}, b, cost)$ by $\mathcal{W}(E)$; sometimes, to emphasize the dependence on the budget b, we write $\mathcal{W}(b)$ instead of $\mathcal{W}(E)$. We say that an outcome W is *exhaustive* if it is a maximal feasible set of projects, i.e., $W \cup \{p\} \notin \mathcal{W}(E)$ for each $p \in P \setminus W$. We assume that the voters have approval utilities, i.e., the utility of voter $i \in N$ from an outcome $W \subseteq P$ is given by $|A(i) \cap W|$. Our goal is to select an outcome based on voters' ballots. An *aggregation rule* (or, in short, a *rule*) is a function $\mathcal{R}$ that for each election E selects an outcome $\mathcal{R}(E) \in \mathcal{W}(E)$, called the *winning outcome*.

Load Distribution. Given an election $E = (N, P, (A(i))_{i \in N}, b, cost)$, a *load distribution* for an outcome $W \in \mathcal{W}(E)$ is a collection of rational numbers $X = (x_{i,p})_{i \in N, p \in P}$ that form a feasible solution to the following linear program:

$$\sum_{i \in N} x_{i,p} = cost(p) \qquad \text{for all } p \in W \qquad (1)$$

$$0 \leq x_{i,p} \leq cost(p) \qquad \text{for all } i \in N, p \in W \qquad (2)$$

$$x_{i,p} = 0 \qquad \text{for all } i \in N, p \in P \setminus (W \cap A(i)) \qquad (3)$$

Given a load distribution X and a voter $i \in N$, we write $X_i = \sum_{p \in P} x_{i,p}$ to denote the total load of i, and write $N_p(X)$ to denote the set of voters who pay for p in X:

$$N_p(X) = \{i \in N : x_{i,p} > 0\}.$$

We denote the set of all load distributions for an outcome W by $\mathcal{X}(W)$. Further, given an election $E = (N, P, (A(i))_{i \in N}, b, cost)$, we set

$$\mathcal{WX}(E) = \{(W, X) : W \in \mathcal{W}(E), X \in \mathcal{X}(W)\};$$

just as before, we write $\mathcal{WX}(b)$ instead of $\mathcal{WX}(E)$ to emphasize the dependence on b, and refer to elements of $\mathcal{WX}(b)$ as *solutions for b* (or simply *solutions*). Intuitively, the linear program in Eqs. (1)–(3) describes how voters in N can share the cost of projects in W; each project is associated with a (rational) cost (or, load), and the cost of a project can only be shared by voters who approve that project. Note that this linear program is feasible as long as we assume that every project in W has at least one approval.

We will say that a load distribution $X \in \mathcal{X}(W)$ is *priceable* [18] if $\sum_{p \in W} x_{i,p} \leq \frac{b}{n}$ for all $i \in N$. We say that X is *equal-shares* if it is priceable and for every $p \in W$ and every pair of voters $i, j \in N_p(X)$ we have $x_{i,p} = x_{j,p}$, that is, the voters who pay for p share the cost of p exactly equally. We will say that a solution (W, X) is *equal-shares* (resp., *priceable*) if X is an equal-shares (resp., priceable) load distribution in $\mathcal{X}(W)$; an outcome W is equal-shares (resp., priceable) if there exists an $X \in \mathcal{X}(W)$ such that (W, X) is equal-shares (resp., priceable). We denote the set of all equal-share solutions for an election E by $\mathcal{WX}^=(E)$; again, we will sometimes write $\mathcal{WX}^=(b)$ instead of $\mathcal{WX}^=(E)$ to emphasize the dependence on b.

Example 1. We will use the following instance of participatory budgeting as a running example:

$$E = (N, P, (A(i))_{i \in N}, b, cost) \text{ where}$$
$$N = \{1, 2, 3\}, P = \{p_1, p_2, p_3, p_4, p_5\}, b = 35, A(1) = A(2) = A(3) = P,$$
$$cost(p_j) = 6 \text{ for } j \in \{1, 2, 3\}, cost(p_4) = 7, cost(p_5) = 10.$$

We focus on the outcome $W = P$; note that $cost(W) = 35 = b$.

Consider the load distribution X given by $x_{i,p} = cost(p)/3$ for all $i \in N$, $p \in P$. Note that X is priceable and equal-shares and hence $(W, X) \in \mathcal{WX}^=(E)$.

In contrast, consider the load distribution X' given by $x'_{1,p} = 35/9$ for $p \in \{p_1, p_2, p_3\}$, $x'_{2,p_4} = 7$, $x'_{2,p_1} = x'_{2,p_2} = 19/9$, $x'_{2,p_3} = 4/9$, $x'_{3,p_5} = 10$, $x'_{3,p_3} = 5/3$ (all values not explicitly specified are 0). This load distribution is priceable: we have $X_i = 35/3$ for each $i \in N$. However, it is not equal-shares: all three voters make different contributions towards p_3.

Moreover, the load distribution X'' given by $x''_{1,p} = 6$ for $p \in \{p_1, p_2, p_3\}$, $x''_{2,p_4} = 7$, $x''_{3,p_3} = 10$ (all values not specified are 0) is not priceable: we have $X_1 = 18 > 35/3$. Nevertheless, it is a load distribution for W: all projects in W are fully paid for, and each voter only pays for projects in W she approves.

Extended Justified Representation (EJR). Given a participatory budgeting election $E = (N, P, (A(i))_{i \in N}, b, cost)$ and a subset of projects $T \subseteq P$, we say that a group of voters S is T-cohesive if $\frac{|S|}{n} \geq \frac{cost(T)}{b}$ and $T \subseteq \cap_{i \in S} A(i)$. An outcome $W \in \mathcal{W}(E)$ is said to provide *Extended Justified Representation (EJR)* for approval utilities if for each $T \subset P$ and each T-cohesive group S of voters there exists a voter $i \in S$ such that $|A(i) \cap W| \geq |T|$. A rule $\mathcal{R}$ satisfies *Extended Justified Representation* if for each election E the outcome $\mathcal{R}(E)$ provides EJR.

Method of Equal Shares. The Method of Equal Shares (MES) is defined for the general PB model, in which voters may have real-valued additive utilities [18,19]. In this work we focus on approval utilities only, and hence we describe MES for approval utilities.

Fix an election $E = (N, P, (A(i))_{i \in N}, b, cost)$. Initially, each voter $i \in N$ is allocated a fixed budget of $\frac{b}{n}$. MES builds a solution (W, X) for b. It starts by setting $W = \varnothing$, $x_{i,p} = 0$ for all $i \in N$, $p \in P$. At each iteration, MES selects one project to be added, i.e., it sets $W \leftarrow W \cup \{p\}$ for some project p and updates the load distribution X so as to pay for p. The cost of p is shared equally among its supporters, with the following exception: if, by contributing her share of the cost, the voter would have to exceed her budget $\frac{b}{n}$, she simply contributes her entire remaining budget instead. Formally, given the current outcome W and load distribution X, for each project $p \in P \backslash W$ we compute the quantity

$$\rho(p) = \min \left\{ \rho : \sum_{i : p \in A(i)} \min \left\{ \rho, \frac{b}{n} - X_i \right\} = cost(p) \right\},$$

add a project p with the smallest value of $\rho(p)$ to W, and set $x_{i,p} = \min\{\rho(p), \frac{b}{n} - X_i\}$ if $p \in A(i)$ and $x_{i,p} = 0$ otherwise. If $\rho(p) = +\infty$ for all projects in $P \backslash W$, the algorithm terminates. Importantly, this may happen even if W is not exhaustive.

3   Stable Equal-Shares Solutions Allow for Fast Verification of EJR

We will now define a notion of stability for equal-shares solutions; our definition is designed to ensure that every stable outcome satisfies EJR.

The intuition behind our definition is as follows. If a solution (W, X) satisfies $X_i < \frac{b}{n}$, voter i is willing to contribute her remaining budget to support further projects in $A(i)$. Moreover, even if i does not have enough budget left to contribute to new projects, she may still prefer to spend money on more cost-efficient projects. To achieve this, voter i may want to withdraw her support from a project and reallocate it to a more cost-efficient project.

To formalize this intuition, we first define a weak order on the set of equal-shares solutions for a given election.

Per-voter Price. Given an election $E = (N, P, (A(i))_{i \in N}, b, cost)$, an outcome $W \in \mathcal{W}(E)$ and an equal-shares load distribution $X \in \mathcal{X}(W)$, we define the *per-voter price* of a project $p \in W$ with respect to X as the cost of p divided by the number of voters who pay for p in X:

$$\pi(p, X) = \frac{cost(p)}{|N_p(X)|};$$

note that, since (W, X) is an equal-shares solution, this is exactly the amount that each of the voters who pays for p contributes towards the cost of p. If $p \notin W$, we set $\pi(p, X) = +\infty$. The *per-voter price vector* of a load distribution $X \in \mathcal{X}(W)$ is the list of numbers $\pi(X) = (\pi(p, X))_{p \in P}$, sorted from the smallest to the largest, with ties broken based on a fixed order of projects in P. We write $\pi_j(X)$ to denote the j-th entry of $\pi(X)$. Given two solutions $(W, X), (W', X') \in \mathcal{WX}^=(E)$, we write $(W, X) \rhd (W', X')$ if $\pi(X) \preceq_{lex} \pi(X')$, where $\preceq_{lex}$ is the lexicographic order on m-tuples. Note that for every pair of solutions $(W, X), (W', X') \in \mathcal{WX}^=(E)$ we have $(W, X) \rhd (W', X')$ or $(W', X') \rhd (W, X)$, i.e., $\rhd$ is a weak order on $\mathcal{WX}^=(E)$.

Stability. We are now ready to define our notion of stability. Consider an election $E = (N, P, (A(i))_{i \in N}, b, cost)$ and a solution $(W, X) \in \mathcal{WX}^=(E)$. For each voter $i \in N$, let $z_i = \max\{x_{i,p} : p \in W\}$ so $z_i \geq 0$ if $i \in N_p(X)$ for some $p \in W$ and $z_i = -\infty$ otherwise. Let $K = \{\frac{cost(p)}{i} : p \in P, i \in N\}$ and let $\epsilon = \min_{x,y \in K, x \neq y} |x - y|$ denote a lower bound on the minimum possible positive difference between two payments in X. Let

$$\kappa_i(X) = \max\left\{z_i - \epsilon, \frac{b}{n} - X_i\right\}; \tag{4}$$

we refer to the quantity $\kappa_i(X)$ as the *capacity* of voter i (and omit X from the notation when it is clear from the context). Intuitively, $\kappa_i(X)$ is the amount that i is prepared to contribute towards a new project that she approves, either by using up her unspent budget (as captured by the $\frac{b}{n} - X_i$ term), or by withdrawing her contribution from one of the projects with the highest per-voter price among the ones she is currently paying for, and using these funds to pay for a less expensive project (i.e., one with per-voter price at most $z_i - \epsilon$). Note that this includes the scenario in which voters decrease their contribution to a project by sharing its load with an increased number of voters.

Definition 1. A solution $(W, X) \in \mathcal{WX}^=(E)$ is *unstable* if there is a project $p \in P$ and a positive integer t with $t > |N_p(X)|$ such that $|\{i \in N : p \in A(i), \kappa_i(X) \geq \frac{cost(p)}{t}\}| \geq t$; otherwise we say that (W, X) is *stable*.

We denote the set of all stable equal-shares solutions for an election E by $\mathcal{WX}^*(E)$, and write $\mathcal{WX}^*(b)$ instead of $\mathcal{WX}^*(E)$ when we want to emphasize the dependence on the budget b.

3.1   Properties of Stable Outcomes: Proportionality and Verifiability

Our primary goal in this section is to show that stable outcomes provide EJR. To this end, we establish a property that may be of independent interest (we defer the proof, as well as some of the subsequent proofs, to the full version of the paper).

Lemma 1. *Given an election E, a subset of voters $V \subseteq N$ and a subset of projects $T \subseteq P$, consider the election $E' = (V, T, (A'(i))_{i \in V}, b, cost')$ with $A'(i) = A(i) \cap T$ for all $i \in V$ and $cost'(p) = cost(p)$ for all $p \in T$. Fix a pair of solutions $(W, X) \in \mathcal{WX}^*(E)$, $(W', X') \in \mathcal{WX}^*(E')$ and a voter $i \in V$, and let $r = |\{p : x'_{i,p} > 0\}|$. Then there exists a voter $j \in V$ such that $|(A(i) \cup A(j)) \cap W| \geq r$. In particular, there exists a set of projects P' with $P' \subseteq A(i) \cap W'$ such that $P' \subseteq W$ and voter j approves at least $r - |P'|$ projects in $W \backslash P'$.*

We use Lemma 1 to show that if $(W, X) \in \mathcal{WX}^*(E)$ then W provides EJR.

Theorem 1. *If (W, X) is a stable solution for an election E, then W provides EJR for approval utilities.*

Proof. Consider a set of projects $T \subseteq P$ and a group of voters V of size at least $\frac{cost(T)}{b} \cdot n$ such that $T \subset A(i)$ for all $i \in V$. We will show that $|A(i) \cap W| \geq |T|$ for some $i \in V$. Consider the restricted instance E' where all approvals but those from V to T are removed, and the equal-shares solution (T, X'), where $x'_{i,p} = \frac{cost(p)}{|V|}$ for every $i \in V$ and $p \in T$, and all other entries of X' are zero. Note that every voter $i \in V$ approves $|T|$ projects in T. Further, (T, X') is in $\mathcal{WX}^*(E')$, since all projects that are approved by some voter in E' are selected, and the cost of each project is shared by all $|V|$ voters. Consider an arbitrary voter $i \in V$. By Lemma 1, since (W, X) is a stable solution for the original election E, there is a $T' \subset A(i) \cap T$ with $T' \subseteq W$ such that some voter $j \in V$ approves additional $|T| - |T'|$ projects in $W \backslash T'$. But since j approves all projects in $T' \subset T$ as well, it follows that she approves at least $|T|$ projects in W, as desired.   □

Next, we will show that stability can be verified very efficiently. Since stability implies EJR, it follows that for any PB rule that outputs stable solutions (so in particular for the AMES rule, to be defined in the next section), we can quickly verify that the associated outcome provides EJR. Importantly, this verification procedure may be much faster than evaluating the rule on a given election.

This result may appear counterintuitive, since checking whether a given outcome provides EJR is known to be NP-hard, even in the context of multiwinner voting [1]. This apparent contradiction is resolved by observing that the input to our verification procedure is a solution, i.e., a pair (W, X), rather than an outcome W: the auxiliary information provided by X enables the verification algorithm to run in nearly-linear time.

Theorem 2. *There is a verification process $\mathcal{V}(W, X)$ that, given an election $E = (N, P, (A(i))_{i \in N}, b, cost)$ with $|N| = n$, $|P| = m$ and a solution in $\mathcal{WX}(E)$, decides the stability of (W, X) and runs in time $O(n \log n + mn)$.*

In fact, the proof of Theorem 2 shows that we can achieve linear-time verifiability if we can pass the list of capacities $\mathcal{K} = (\kappa_i(X))_{i \in N}$ sorted in non-increasing order as auxiliary data to the verifier (in addition to the solution (W, X)). More precisely, there exists a verification algorithm $\mathcal{V}(W, X, \mathcal{K})$ that decides stability in time linear in the size of the election (which matches the runtime of verifying that an output of phragmms satisfies PJR [11]). This is because $\mathcal{V}(W, X, \mathcal{K})$ does not have to sort the capacities; instead, it checks that the capacities have been computed correctly and verifies that $\mathcal{K}$ is non-increasing, all in time $O(nm)$.

Theorem 3. *There is a verification process* $\mathcal{V}(W, X, \mathcal{K})$ *that, given an election* $E = (N, P, (A(i))_{i \in N}, b, cost)$ *with* $|N| = n$, $|P| = m$, *a solution in* $\mathcal{WX}(E)$, *and capacities* $\mathcal{K} = (\kappa_i(X))_{i \in N}$ *sorted in non-increasing order decides the stability of* $(W, X, \mathcal{K})$ *and runs in time* $O(mn)$.

To conclude this section, we mention a subtlety in the definition of stability. In our definition, we deem an outcome (W, X) unstable if there is a project $p \in W$ such that the number of voters paying for p could be increased. In the full version of the paper, we consider a weaker notion of stability, which only considers projects not currently included in the outcome. We show that this alternative notion of stability does not in general imply EJR; however, it *does* imply EJR in the special case of multiwinner voting (i.e., unit-cost projects).

4 An Adaptive Method of Equal Shares

Our notion of stability suggests a natural adaptive algorithm, which, given an equal-shares solution (X, W), performs greedy update steps until it reaches a stable outcome.

Greedy Update Step. If (W, X) is unstable, it admits an *update step* $(W, X) \xrightarrow{p} (W', X')$, defined as follows. We identify a project p and a size-t set of voters $V = \{i \in N : p \in A(i), \kappa_i(X) \geq \frac{cost(p)}{t}\}$ that witness the instability of (W, X). Voters in V all approve p; if each of them contributes $\frac{cost(p)}{|V|}$ then p will be fully paid for. However, if $\kappa_i(X) = z_i - \epsilon$, increasing i's load by $\frac{cost(p)}{|V|}$ may result in their load exceeding $\frac{b}{n}$. To overcome this, we determine a set of projects W^- to be removed from W.

To this end, we initialize $W^- = \varnothing$, and iterate through voters in V. For each voter $i \in V$, if $\frac{cost(p)}{|V|} \leq \frac{b}{n} - X_i$, we do nothing: this voter can afford to pay for p from her remaining budget. Similarly, if $p \in W$ and $x_{i,p} > 0$, we do nothing: after the update, this voter's contribution towards p will be smaller than $x_{i,p}$. If $\frac{cost(p)}{|V|} > \frac{b}{n} - X_i$ and $x_{i,p} = 0$, we let q be a project with the highest per-voter price among the ones that i contributes to, i.e., and $\pi(q, X) = \max_{q' \in P} x_{i,q'}$, and set $W^- \leftarrow W^- \cup \{q\}$ (of course, it may be the case that q is already in W^-). After we process all voters in V, we set $W' = (W \backslash W^-) \cup \{p\}$, and define X' by setting

$$x'_{i,p} = \frac{cost(p)}{|V|} \qquad\qquad \text{for all } i \in V,$$

$$x'_{i,p} = 0 \qquad\qquad \text{for all } i \in N \backslash V,$$

$$x'_{i,q} = 0 \qquad\qquad \text{for all } i \in N \text{ and } q \in W^-,$$

$$x'_{i,q} = x_{i,q} \qquad\qquad \text{for all } i \in N, q \in P \backslash (W^- \cup \{p\}).$$

Note that, by construction, if a solution is stable, it does not admit an update step. Moreover, we say that an update step $(W, X) \xrightarrow{p} (W', X')$ is *greedy* if for every other update step $(W, X) \xrightarrow{q} (\overline{W}, \overline{X})$ we have $\pi(p, X') \leq \pi(q, \overline{X})$. That is, a greedy update step selects a project (by adding it or increasing its number of contributors) so as to minimize the per-voter price. The following proposition summarizes the key properties of an update step.

Proposition 1. *Suppose* $(W, X) \in \mathcal{WX}^=(E)$, *and* (W', X') *is obtained from* (W, X) *by an update step. Then* $(W', X') \in \mathcal{WX}^=(E)$ *and* $(W', X') \rhd (W, X)$.

Our notion of an update step now suggests the following procedure.

Start with an arbitrary pair $(W, X) \in \mathcal{WX}^=(E)$, and execute a sequence of greedy update steps from (W, X) until a stable solution (W^*, X^*) is reached. A pseudocode description of this iterative algorithm, which we call the Adaptive Method of Equal Shares (AMES), is given in Algorithm 1. We say that we run AMES *from scratch* if the starting solution (W, X) satisfies $W = \varnothing$, $x_{i,p} = 0$ for all $i \in N, p \in P$.

In more detail, in each iteration our algorithm loops over all projects, and checks if there is a project for which the number of supporters can be increased (lines 5–14); we use the convention that $\max \varnothing = -\infty$. It also keeps track of the best such project (p^*), as measured by the price per voter (lines 9–11). If some such project has been identified, the algorithm iterates through all voters whose contribution towards p is about to increase in a way that exceeds their remaining budget; for each such voter, it identifies one project that this voter is currently paying for, and adds it to the set of projects to be removed from W (lines 16–19). It then updates W and X accordingly (lines 20–23).

We will use the instance E from Example 1 to illustrate the different kinds of update steps AMES can perform.

Suppose the priority order over the projects is $p_1 > p_2 > p_3 > p_4 > p_5$. Consider the stable outcome (W, X) given by $W = \{p_1, p_2, p_3, p_4\}$, $x_{1,p} = x_{2,p} = 3$ for $p \in \{p_1, p_2, p_3\}$, $x_{1,p} = x_{2,p} = 0$ for $p \in \{p_4, p_5\}$, $x_{3,p_4} = 7$ and $x_{3,p} = 0$ for $p \neq p_4$. When initialized on E and (W, X), AMES performs the following steps:

(1) $(W, X) \xrightarrow{p_1} (W^1, X^1)$ where

$$W^1 = W, x^1_{i,p_1} = 2 \text{ and } x^1_{i,p} = x_{i,p} \text{ for } p \neq p_1, i \in N.$$

(2) $(W^1, X^1) \xrightarrow{p_2} (W^2, X^2)$. where

$$W^2 = W^1, x^2_{i,p_2} = 2 \text{ and } x^2_{i,p} = x^1_{i,p} \text{ for } p \neq p_2, i \in N.$$

ALGORITHM 1: AMES (Adaptive Method of Equal Shares)

Input: $E = (N, P, (A(i))_{i \in N}, b, cost), (W, X) \in \mathcal{W}\mathcal{X}^=(E)$
Output: $(W^*, X^*) \in \mathcal{W}\mathcal{X}^*(E)$

1 **repeat**
2     $(\kappa_i)_{i \in N} \leftarrow$ capacities(W, X) (as per (4))
3     $flag \leftarrow$ **false**
4     $\pi^* = +\infty$
5   **for** $p \in P$ **do**
6         $t_p \leftarrow \max\{t \in \mathbb{N} : t > |N_p(X)|, |\{i \in N : p \in A(i), \kappa_i \geq cost(p)/t\}| \geq t\}$
7   **if** $t_p \neq -\infty$ **then**
8             $flag \leftarrow$ **true**
9   **if** $cost(p)/t_p < \pi^*$ **then**
10                 $\pi^* \leftarrow cost(p)/t_p$
11                 $p^* \leftarrow p$
12   **end**
13   **end**
14   **end**
15   **if** $flag =$ **true then**
16         $S \leftarrow \{i \in N : p^* \in A(i), x_{i,p^*} = 0 \text{ and } \pi^* > \frac{b}{n} - X_i\}$
17   **for** $i \in S$ **do**
18   Let $p^-(i)$ be some project from $\arg\max_{p \in W}\{x_{i,p}\}$
19   **end**
20         $W^- \leftarrow \{p^-(i) : i \in S\}$
21         $W \leftarrow (W \cup \{p^*\}) \backslash W^-$
22         $x_{i,p^*} \leftarrow \pi^*$ for all $i \in N$ such that $p \in A(i)$ and $\kappa_i \geq \pi^*$
23         $x_{i,p} \leftarrow 0$ for all $p \in W^-, i \in N$
24   **end**
25 **until** $flag =$ **false**;
26 **return** (W, X)

---

(3) $(W^2, X^2) \xrightarrow[p_4]{p_3} (W^3, X^3)$ where

$$W^3 = W^2 \backslash \{p_4\}, x^3_{i,p_3} = 2 \text{ for } i \in [3], x^3_{3,p_4} = 0,$$
$$\text{and } x^3_{i,p} = x^2_{i,p} \text{ for all other } (i,p) \in N \times P$$

(4) $(W^3, X^3) \xrightarrow{p_4} (W^4, X^4)$ where

$$W^4 = W^3 \cup \{p_4\}, x^4_{i,p_4} = \frac{7}{3} \text{ and } x^4_{i,p} = x^3_{i,p} \text{ for } p \neq p_4, i \in N.$$

(5) $(W^4, X^4) \xrightarrow{p_5} (W^5, X^5)$ where

$$W^5 = W^4 \cup \{p_5\}, x^5_{i,p_5} = \frac{10}{3} \text{ and } x^5_{i,p} = x^4_{i,p} \text{ for } p \neq p_5, i \in N.$$

Update steps (1) and (2) illustrate increasing the number of supporters of a project that is already included in the solution. Update step (3) additionally illustrates the removal of a project (p_4), which enables the algorithm to increase the number of supporters of p_3. Finally, update steps (4) and (5) illustrate simple addition of new projects. Note that after step (2) the transition $(W^2, X^2) \xrightarrow[p_3]{p_4}$ (W', X') would be another valid update step, where

$$W' = W^2 \setminus \{p_3\}, x'_{i,p_4} = \frac{7}{3} \text{ for } i \in N, x'_{j,p_3} = 0 \text{ for } j \in [2], \text{ and}$$

$$x'_{i,p} = x^2_{i,p} \text{ for all other } (i,p) \in N \times P,$$

but it is not greedy, since the update in step (3) is strictly better according to $\rhd$.

The next lemma shows that the per-voter prices of projects are non-decreasing in the order in which they were added.

Lemma 2. *Consider two consecutive greedy update steps* $(W, X) \xrightarrow{p'} (W', X')$ *and* $(W', X') \xrightarrow{p''} (W'', X'')$. *Suppose that* $\pi(p', X') < \pi(p', X)$ *and* $\pi(p'', X'') < \pi(p'', X')$. *Then* $\pi(p', X') \leq \pi(p'', X'')$.

Lemma 2 is the key to showing that, by performing greedy update steps, AMES converges quickly.

Proposition 2. *Suppose that AMES is executed on input* (W, X) *and outputs a solution* (W^*, X^*). *Then it executes at most* $|\{p \in P : \pi(p, X^*) < \pi(p, X)\}|$ *greedy update steps.*

The proof of Proposition 2 shows that in each update step we either add a project from $W^* \setminus W$ or else lower the per-voter price of an existing project in W. A consequence of this result is that we can compute a stable solution in polynomial time, by starting from an arbitrary equal-shares solution and performing greedy update steps.

Theorem 4. *Given an election* $E = (N, P, (A(i))_{i \in N}, b, cost)$ *and an outcome* $(W, X) \in \mathcal{WX}^=(E)$, *Algorithm 1 outputs* $(W^*, X^*) \in \mathcal{WX}^*(E)$ *by performing at most* $|\{p \in P : \pi(p, X^*) < \pi(p, X)\}| = O(m)$ *update steps; each update step can be completed in time* $O(n \log n + nm)$.

4.1   On Using AMES Adaptively

An advantage of our notion of stability (Definition 1) is that it is easily check-able, and hence one can quickly verify that an outcome of AMES provides EJR. In particular, this means that the execution of AMES can be outsourced to a non-trusted, but computationally powerful third party. On the other hand, it is arguably easier to explain that a solution was obtained by running AMES from scratch with a virtual budget b', as opposed to being obtained by computing the solution for the true budget b and then gradually increasing the budget to b' and adapting the solution as necessary.

It turns out that a modification of AMES guarantees that, independently of which equal shares-outcome (W, X) AMES is initialized with, it will output the same solution (W^*, X^*) in either case. That is, we can execute AMES adaptively, starting with budget b and increasing the budget until an exhaustive solution is found, yet explain the result as the output of AMES on the final budget b'.

Consider two stable solutions (W, X) and (W', X') for the election $E = (N, P, (A(i))_{i \in N}, b, cost)$, and suppose that $W \neq W'$ or $X \neq X'$. Let $\pi(X')$ and $\pi(X)$ be the corresponding price-per-voter vectors. Let j be the first index such that $\pi_j(X) = \pi(p, X) \neq \pi(q, X') = \pi_j(X')$ or $\pi_j(X) = \pi(p, X) = \pi(q, X') = \pi_j(X')$ but $p \neq q$. We will now show that the former case cannot occur.

Proposition 3. $\pi(p, X) = \pi(q, X')$ *and so in particular* $p \neq q$.

This shows that the only reason for inconsistency between the outcomes (W, X) and (W', X') can come from ties. Specifically, the following scenario may lead to inconsistency: Suppose we have two voters 1 and 2, projects p_1, p_2, p_3 where v_1 approves p_1 and p_2 and v_2 approves p_2 and p_3. The tie-breaking rule for projects is $p_1 < p_2 < p_3$ and their costs are $cost(p_1) = 1$, $cost(p_2) = 4$, $cost(p_3) = 2$ with a total budget of $b = 5$. AMES selects outcome $\{p_1, p_2\}$ which are respectively fully paid by their unique supporters. If we increase the budget to 6 and then initialize AMES on $\{p_1, p_2\}$ with the corresponding load distribution, then p_2 would not be selected because the capacity of v_2 is strictly less than 2. The outcome remains $\{p_1, p_2\}$. If, however, we run AMES from scratch with budget $b = 6$, the selected outcome is $\{p_1, p_2\}$ with p_1 fully paid by voter 1 and p_2 paid equally by both voters.

A simple resolution to this inconsistency is to allow greedy update steps that add a project and (if necessary) remove projects with higher per-voter price *or* same per-voter price and lower tie-breaking priority. Specifically, we introduce a more refined update step, where the capacities of voters depend on the project under consideration. Given a tie-breaking order $>$ on P, for each voter $i \in N$, $p \in P$, and solution (W, X), we define i's *project-dependent capacity*

$$\kappa_{i,p}(X) = \max\{\kappa_i, \max\{x_{i,p'} : p > p', p' \in W\}\}.$$

We will say that (W, X) is *lexicographically stable* if there is no project $p \in P$ and a positive integer $t > N_p(X)$ such that $|\{i \in N : p \in A_i, \kappa_{i,p}(X) \geq \frac{cost(p)}{t}\}| \geq t$. The *tie-consistent AMES* proceeds in the same way as AMES, but uses lexicographic stability instead of stability.

Project-dependent capacities may raise the concern that capacities now cannot be computed and sorted for all projects in one sweep; that instead the $O(n \log n)$ sorting cost is incurred for each project, leading to an increase in run-time to $O(m^2 n \log n)$. In the proof of the following theorem we provide an implementation of tie-consistent AMES that has run-time $O(mn \log n + m^2 n)$.

Theorem 5. *Let* (W', X') *be the outcome of AMES with budget* $b' < b$. *Let* (W'', X'') *be the outcome of tie-consistent AMES initialized on* (W', X') *with budget* b. *Let* (W, X) *be the outcome of AMES with budget* b. *Then* $W = W''$ *and* $X = X''$. *Furthermore, each update step of tie-consistent AMES can be completed in time* $O(n \log n + nm)$.

4.2 Skipping Budgets

Another potential for gain in efficiency of MES comes directly from our notion of stability. Fix an election $E = (N, P, (A(i))_{i \in N}, b, cost)$, and write $E(b')$ to denote the election obtained from E by changing the budget to b'; thus, $E = E(b)$. Consider a stable solution (W, X) for $E(b)$. It may be the case that (W, X) remains stable for $E(b')$, where $b' > b$. Now, consider a budget b'' with $b < b'' < b'$. The following monotonicity property is easy to verify.

Proposition 4. *If (W, X) is unstable for $E(b'')$, then it is also unstable for $E(b')$.*

It follows that if (W, X) is stable for $E(b')$, then we do not have to check intermediate budgets b'' with $b < b'' < b'$. It is then natural to ask (1) what is the minimum value $b' > b$ such that (W, X) is unstable for $E(b')$, and (2) can we compute this value efficiently? It turns out that for AMES b' is easy to compute; we remark that, in contrast, for MES this is not known to be the case (this is because the solutions output by MES are not necessarily equal-share solutions, which makes MES rather fragile with respect to budget modifications).

Theorem 6. *The minimum $b' > b$ such that (W, X) is unstable for $E(b')$ can be computed in time $O(mn^2 \log n)$.*

5 Experimental Evaluation

We evaluate the potential gain in efficiency of AMES for the completion problem on real-world participatory budgeting data from three Polish cities. We consider PB data from Pabulib [23], an open participatory budgeting library. We select several data sets with a large[1] number of proposed projects and mean ballot size of more than one from three Polish cities: Warsaw, Wroclaw and Lodz. Warsaw holds district-based elections, so we consider three districts of Warsaw in our experiments, while the data from Wroclaw and Lodz comes from city-wide elections. The following table summarizes the properties of these data sets (Fig. 1).

City/District	year	vote count	projects	budget	mean vote length
Wroclaw/city-wide	2018	53,801	39	4,000,000	1.87643
Lodz/city-wide	2020	51,472	151	5,715,627	3.82408
Warszawa/Ursynow	2023	6260	54	6,067,849	11.586
Warszawa/Bielany	2023	4,956	98	5,258,802	11.3999
Warszawa/Praga-Polodnie	2023	8,922	81	7,180,288	11.5092

Fig. 1. Participatory budgeting data from Poland.

---

[1] That is, in the range of 40–150, as opposed to 5–10, as in the PB elections in, e.g., Zabrze.

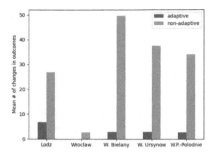 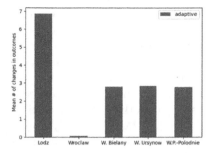

Fig. 2. We repeatedly top up each voter's budget up by 1 and evaluate the average change in the outcome over 50 runs.

We compare the average difference in outcomes when the budget is increased (i.e., each voter gets more money, as is done in the top-up solution for the completion problem for MES). More formally, for each $i \in [50]$, let (W^i, X^i) denote the output of AMES with budget of $\frac{b}{n} + i$ per voter. We compare solutions at step $i - 1$ and step i by calculating the change in outcome as measured by

$$|\{p \in P : \pi(p, X^i) \neq \pi(p, X^{i-1})\}|$$

(that is, how many projects have lower per-voter price plus how many were removed), and we average our results over these 50 runs to obtain a final figure for each data set for the adaptive method. If we run AMES from scratch for each (voter) budget $\frac{b}{n} + i$, then the number of projects added will be exactly $|W^i|$. Therefore, we take the average number of selected projects over the 50 runs as the baseline comparison when using AMES non-adaptively. The results of these experiments are shown in Fig. 2. Consider Warszawa Bielany; in every run we add 50 projects on average, while the average difference between consecutive outcomes (and hence the amount of work required) is just below 3. Similarly, in Lodz, every budget-increasing iteration on average adds over 25 projects, while the average differences amount to less than 7. These results confirm that consecutive outcomes are very similar on average, and hence our budget-adaptive approach offers significant savings over repeatedly executing MES from scratch repeatedly, as measured by the number of iterations that modify the outcome.

6   Conclusion

We introduced a new voting rule for PB with approval utilities. This rule, which we call Adaptive Method of Equal Shares (AMES), can leverage similarities between an equal-shares initialization (W, X) for budget b' and output solution (W^*, X^*) for budget $b'' \geq b'$ when run from scratch, allowing it to find (W^*, X^*) in fewer iterations. This feature is relevant for the completion problem of MES-style methods, as it alleviates the need to run the method repeatedly from scratch for different budgets. The solutions output by AMES satisfy the proportionality axiom EJR, and this can be verified efficiently in time $O(n \log n + mn)$.

Anecdotal evidence suggests that in practice there is a dislike of voting rules that are difficult to explain. One may worry that AMES suffers from this drawback to a greater extent than its non-adaptive cousin MES. However, we emphasize that it suffices to explain the simpler rule, which runs from scratch with a fixed budget $b' \geq b$: the adaptive version is only used for finding b' efficiently, by leveraging the similarity in consecutive outcomes. Thus, AMES is essentially as transparent as the currently used completion method for MES: while the choice of the virtual budget b' may appear "magical", it can be justified by showing that a further virtual budget increase would result in a proposal that is not feasible with respect to b.

To evaluate the potential of our budget-adaptive approach, we analyzed real-world PB data, and showed that solutions of AMES obtained by repeatedly topping up budgets differ very little on average. More broadly, AMES can also be useful when the input election is changed in other ways, e.g., the costs of some projects are updated, or a few voters change their preferences; in such cases, too, checking for updates may be faster than re-computing solutions from scratch.

Our experiments were conducted in Python. To analyze the gain in efficiency by using AMES budget-adaptively in practice, it would be useful to implement fully optimized versions of both adaptive and non-adaptive completion methods in C. This will allow us to compare the constant overhead of an adaptive method with its gain in efficiency from leveraging similarities. Improving upon the runtime of $O(mn^2 \log n)$ to skip budgets (Theorem 6) by removing the quadratic dependency on n is an important direction for future work.

References

1. Aziz, H., Brill, M., Conitzer, V., Elkind, E., Freeman, R., Walsh, T.: Justified representation in approval-based committee voting. Soc. Choice Welfare **48**(2), 461–485 (2017)
2. Aziz, H., Elkind, E., Huang, S., Lackner, M., Sánchez-Fernández, L., Skowron, P.: On the complexity of extended and proportional justified representation. In: AAAI 2018 (2018)
3. Aziz, H., Gaspers, S., Gudmundsson, J., Mackenzie, S., Mattei, N., Walsh, T.: Computational aspects of multi-winner approval voting. In: AAMAS 2015, pp. 107–115 (2015)
4. Aziz, H., Lee, B.E., Talmon, N.: Proportionally representative participatory budgeting: axioms and algorithms. In: AAMAS 2018, pp. 23–31 (2018)
5. Aziz, H., Shah, N.: Participatory budgeting: models and approaches. In: Rudas, T., Péli, G. (eds.) Pathways Between Social Science and Computational Social Science. CSS, pp. 215–236. Springer, Cham (2021). https://doi.org/10.1007/978-3-030-54936-7_10
6. Beck, R., Müller-Bloch, C., King, J.L.: Governance in the blockchain economy: a framework and research agenda. J. Assoc. Inf. Syst. **19**(10), 1 (2018)
7. Brill, M., Peters, J.: Robust and verifiable proportionality axioms for multiwinner voting. In: ACM EC 2023 (2023)
8. Brünjes, L., Kiayias, A., Koutsoupias, E., Stouka, A.-P.: Reward sharing schemes for stake pools. In: EuroS & P'20, pp. 256–275 (2020)

9. Burdges, J., et al. Overview of polkadot and its design considerations. arXiv preprint arXiv:2005.13456 (2020)
10. Cabannes, Y.: Participatory budgeting: a significant contribution to participatory democracy. Environ. Urban. **16**(1), 27–46 (2004)
11. Cevallos, A., Stewart, A.: A verifiably secure and proportional committee election rule. In: AFT 2021, pp. 29–42 (2021)
12. Garg, P.: EOS voting structure encourages centralization (2019). https://cryptoslate.com/eos-voting-structure-encourages-centralization/. Accessed 14-May 2023
13. Grigg, I.: EOS–an introduction. White paper (2017). https://whitepaperdatabase.com/eos-whitepaper
14. Hagerty, A.: History of DPOS governance (2020). https://gist.github.com/cc32d9/db582e349fc8e44aca9e339f23d7f7e8. Accessed 14 May 2023
15. Janson, S.: Phragmén's and Thiele's election methods. arXiv preprint arXiv:1611.08826 (2016)
16. Lackner, M., Skowron, P.: Approval-based committee voting. In: Multi-Winner Voting with Approval Preferences. SpringerBriefs in Intelligent Systems, pp. 1–7. Springer, Cham (2023). https://doi.org/10.1007/978-3-031-09016-5_1
17. Peters, D., Pierczyński, G., Shah, N., Skowron, P.: Market-based explanations of collective decisions. In: AAAI 2021, pp. 5656–5663 (2021)
18. Peters, D., Pierczyński, G., Skowron, P.: Proportional participatory budgeting with additive utilities. In: NeurIPS 2021, pp. 12726–12737 (2021)
19. Peters, D.,Skowron, P.: Proportionality and the limits of welfarism. In: ACM EC 2020, pp. 793–794 (2020)
20. Peters, D., Skowron, P.: Completion of the method of equal shares (2023). https://equalshares.net. Accessed 14 May 2023
21. Phragmén, E.: Sur une méthode nouvelle pour réaliser, dans les élections, la représentation proportionelle des partis. Öfversigt af Kongliga Vetenskaps-Akademiens Förhandlingar **51**(3), 133–137 (1894)
22. Sims, A.: Blockchain and decentralised autonomous organisations (DAOs): the evolution of companies? (2019)
23. Stolicki, D., Szufa, S., Talmon, N.: Pabulib: a participatory budgeting library. arXiv preprint arXiv:2012.06539 (2020)
24. Wampler, B., McNulty, S., Touchton, M.: Participatory Budgeting in Global Perspective. Oxford University Press, Oxford (2021)

Routing MEV in Constant Function Market Makers

Kshitij Kulkarni[1(✉)] ⓘ, Theo Diamandis[2] ⓘ, and Tarun Chitra[3] ⓘ

[1] University of California, Berkeley, Berkeley, CA 94720, USA
ksk@eecs.berkeley.edu
[2] Massachusetts Institute of Technology, Cambridge, MA 02139, USA
tdiamand@mit.edu
[3] Gauntlet, New York, NY 10013, USA
tarun@gauntlet.network

Abstract. Miner Extractable Value (MEV) refers to excess value captured by miners (or validators) from users in a cryptocurrency network. This excess value often comes from reordering users' transactions to maximize fees or from inserting new transactions that front-run users' transactions. One of the most common types of MEV involves a 'sandwich attack' against a user trading on a constant function market maker (CFMM), which is a popular class of decentralized exchange. We analyze game theoretic properties of MEV in CFMMs that we call *routing MEV*, where an aggregate user trade is routed over a network and potentially sandwich attacked. We present examples where the existence of routing MEV both degrades and, counterintuitively, *improves* the quality of routing. We construct an analogue of the price of anarchy for this setting and demonstrate that if the impact of a sandwich attack is localized in a suitable sense, then the price of anarchy is constant (The code for all the numerical experiments in this paper can be found at this link: https://github.com/tjdiamandis/mev-cfmm).

1 Introduction

Public blockchains, including Bitcoin and Ethereum, allow any user to submit a transaction that modifies the shared state of the network. Miners (or validators in proof-of-stake networks)[1] aggregate these transactions into blocks which they propose to the network. Blockchains have rules governing block validity, but the majority do not enforce constraints on transaction ordering *within* a block. As a result, individual miners can propose blocks with a transaction ordering that nets them highest profit, possibly by inserting additional transactions into the block. For example, a miner may observe a user's submitted decentralized exchange (DEX) trade and insert their own trade ahead of and after the user's trade to profit from the user's worse execution price, in a strategy known as a *sandwich attack*. Any type of excess profit that a miner can extract by adjusting the execution of users' transactions is known as Miner Extractable Value (MEV).

---

[1] We use 'miner' in this paper for consistency with the existing literature, *e.g.* [8].

© The Author(s), under exclusive license to Springer Nature Switzerland AG 2024
J. Garg et al. (Eds.): WINE 2023, LNCS 14413, pp. 456–473, 2024.
https://doi.org/10.1007/978-3-031-48974-7_26

There are three principal agents involved in MEV: miners, network users, and MEV searchers. Miners contribute resources to a network in order to win the chance to earn fees by validating transactions. Network users are ordinary users who submit financial transactions to miners to be validated and added to the blockchain. Finally, MEV searchers (or simply, 'searchers') are agents who find profitable opportunities from reordering, inserting, or omitting transactions.

In this paper, we formalize a game theoretic view of MEV as it appears in decentralized exchanges that are implemented as constant function market makers (CFMMs), reviewed in Sect. 2. Specifically, we consider the problem of routing an aggregate trade composed of multiple users' contribution over a network of CFMMs. Our interest is in understanding how the presence of MEV searchers on such a network can impact the quality of routing that users experience via a particular kind of extractable value called a sandwich attack.

Prior Work on MEV. Since MEV was first defined in 2019 [8], miners and searchers have extracted over $650 million [10]. Moreover, the observed types of MEV strategies have grown rapidly [3,6,7,15,16,22]. Thus, it is important to rigorously understand the space of possible MEV strategies and quantify their profitability. In [7,12], the authors quantify sandwich attack profitability for constant product market makers, but they do not provide price of anarchy or worst-case bounds for generic constant function market makers.

MEV in CFMMs and This Paper. Constant function market makers (CFMMs) are decentralized exchanges that have seen widespread use in blockchains [1, 11,21]. Sandwich attacks are by far the most popular type of MEV, with over $500m extracted from users via sandwich attacks [10]. Prior work has focused on analyzing sandwich attacks only in one type of CFMM, the constant product market maker Uniswap [12,23]. As our first contribution, in Sect. 2, we generalize this analysis to *any* CFMM as defined in [1]. To do so, we first define a sandwich attack in terms of the forward exchange function of a CFMM. This definition allows us to utilize the notion of CFMM curvature [2] to explicitly bound the profitability of a sandwich attack, given a particular user trade.

Main Results. Having defined sandwich attacks for CFMMs formally, we aim to answer the following question regarding MEV in CFMMs:

In the case of a network of CFMMs trading multiple assets, how much does the presence of sandwich attackers on the network affect the routing of trades?

We answer this question in Sect. 3 by adapting conventional price of anarchy (PoA) results [17,18,20] to CFMMs. We consider the routing of a single aggregate trade across a network of CFMMs. We define *selfish routing*, in which trades on each path in the network try to get the maximum pro-rata share of the output from that path. This results in the notion of an equilibrium splitting of a trade. We compare selfish routing to optimal routing [4], which seeks to maximize the net output from the network, and analyze the gap between them known as the price of anarchy, in the presence of sandwich attacks. The presence of sandwich attacks shifts both optimal and selfish routing. Our main result shows that the price of anarchy is bounded by a constant for any sized sandwich attack, which we establish using the (λ, μ)-smoothness results of [17]. In addition, we construct a CFMM network that, perhaps counterintuitively, avoids Braess paradox-like [18] behavior after a sandwich attacker is introduced, as this attacker makes the Braess edge more expensive. This example suggests that sandwich attackers can sometimes *improve* the quality of selfish routing.

2   Sandwich Attacks

In sandwich attacks [16], an adversary places orders before and after a user's order in a decentralized exchange to force the user to have a worse execution price. When placing an order, users specify both trade side and a limit price (in the form of a slippage limit). We first introduce constant function market makers.

2.1   Constant Function Market Makers

A *constant function market maker* (CFMM) is a contract that holds some amount of *reserves* $R, R' \geq 0$ of two assets and has a *trading function* $\psi :$ $\mathbf{R}^2 \times \mathbf{R}^2 \to \mathbf{R}$. Users can then submit a *trade* (Δ, Δ') denoting the amount they wish to tender (if negative) or receive (if positive) from the market. The contract then accepts the trade if $\psi(R, R', \Delta, \Delta') = \psi(R, R', 0, 0)$, and pays out (Δ, Δ') to the user.

Curvature. We briefly summarize the main definitions and results of [2] here. Suppose that the trading function ψ is differentiable (as most trading functions in practice are), then the *forward exchange rate* for a trade of size Δ is $g(\Delta) = \frac{\partial_3 \psi(R, R', \Delta, \Delta')}{\partial_4 \psi(R, R', \Delta, \Delta')}$. Here ∂_i denotes the partial derivative with respect to the ith argument, and Δ' is specified by the implicit condition $\psi(R, R', \Delta, \Delta') = \psi(R, R', 0, 0)$; *i.e.*, the trade (Δ, Δ') is assumed to be valid. Additionally, the reserves R, R' are assumed to be fixed. The function g represents the marginal forward exchange rate of a positive-sized trade. We say that a CFMM is α-*stable* if it satisfies $g(0) - g(-\Delta) \leq \alpha\Delta$, for all $\Delta \in [0, M]$ for some $M > 0$. This condition provides a linear upper bound on the maximum price impact that a trade bounded by M can have. Similarly, we say that a CFMM is β-*liquid* if it satisfies $g(0) - g(-\Delta) \geq \beta\Delta$, for all $\Delta \in [0, K]$ for some $K > 0$. One important property of g is that it can be used to compute Δ' [2, §2.1]:

$$\Delta' = \int_0^{-\Delta} g(t)dt.$$

Simple methods for computing α and β in common CFMMs are presented in [2, §1.1] and [1, §4]. We define $G(\Delta) = \Delta'$ to be the *forward exchange function*, which is the amount of output token received for an input of size Δ. Whenever we reference the function $G(\Delta)$ for a given CFMM, we always make clear the reserves associated with that CFMM, or explicitly note that we are abusing notation. We note that $G(\Delta)$ was shown to be concave and increasing in [1]. A forward exchange function $G(\Delta)$ is (ν, κ)-*smooth* if there exists $M > 0$ such that for all $\Delta \in [0, M]$ there exist constants $\nu, \kappa > 0$ such that

$$\kappa\Delta \le G(\Delta) - G(0) \le \nu\Delta. \tag{1}$$

Usually $G(0) = 0$, so these inequalities correspond to a set of bilipschitz bounds on G. We can define an analogous notion of smoothness for the reverse exchange function G^{-1} (and bounds for $\Delta < 0$). It may also be seen that $G(\Delta)$ always also has local quadratic upper and lower bounds, provided that it is smooth.

Slippage Limits. Due to the ambiguity in ordering of trades, when a user submits a trade to a CFMM, they submit two parameters: a trade size $\Delta \in \mathbf{R}$ and a *slippage limit* $\eta \in [0, 1]$. The slippage is interpreted as the minimum output amount that the user is willing to accept as a fraction of $G(\Delta)$. That is, the trade is accepted if the amount in output token the user receives is larger than or equal to $(1 - \eta)G(\Delta)$.

Sandwich Attacks. We generalize prior work analyzing sandwich attacks to CFMMs with two-sided bounds on $g(\Delta)$ and $G(\Delta)$. Recall that a user submits a trade to a CFMM of the form $T = (\Delta, \eta) \in \mathbf{R} \times [0, 1]$, where η is the slippage limit. If a user submits an order that is not tight, then there exists a Δ^{sand} such that $G(\Delta + \Delta^{\mathrm{sand}}) - G(\Delta^{\mathrm{sand}}) > (1 - \eta)G(\Delta)$. The user's order can be made to be tight by finding a Δ^{sand} that satisfies:

$$G(\Delta + \Delta^{\mathrm{sand}}) - G(\Delta^{\mathrm{sand}}) = (1 - \eta)G(\Delta). \tag{2}$$

One can use the equation $G(\Delta + \Delta^{\mathrm{sand}}) - G(\Delta^{\mathrm{sand}}) = (1 - \eta)G(\Delta)$ to numerically solve for the optimal Δ^{sand} by finding the roots of $G(\Delta + x) - G(x) - (1 - \eta)G(\Delta) = 0$. Note that we have abused notation by not explicitly denoting the reserves at the various stages of the sandwich attack in the function $G(\cdot)$. After Δ^{sand} and T are executed, the sandwich attacker sends a trade of $\Delta^{\mathrm{sand}'}$ to recover their initial investment of Δ^{sand} and make a profit. To derive the backward trade $\Delta^{\mathrm{sand}'}$, recall that after sending the initial trade of Δ^{sand}, the sandwich attacker holds $G(\Delta^{\mathrm{sand}})$ of the output token. The attacker thus sells back the amount of output token they hold, $G(\Delta^{\mathrm{sand}})$, which defines $\Delta^{\mathrm{sand}'}$ as follows, in units of input token:

$$\Delta^{\mathrm{sand}'} = \Delta^{\mathrm{sand}} + \Delta - G^{-1}(G(\Delta + \Delta^{\mathrm{sand}}) - G(\Delta^{\mathrm{sand}})), \tag{3}$$

where $G^{-1}(\cdot)$ is the reverse exchange function, *i.e.*, the inverse of G [1]. Therefore, we can define a *sandwich attack* as a triplet of transactions:

$$\Delta^{\mathrm{sand}}(\Delta, \eta), (\Delta, \eta), \Delta^{\mathrm{sand}'}(\Delta, \eta).$$

We emphasize that both $\Delta^{\mathrm{sand}}(\Delta, \eta)$ and $\Delta^{\mathrm{sand}'}(\Delta, \eta)$ are in units of input token, are functions of Δ and η, and solve the Eqs. (2) and (3). If the sandwich attack is executed, the sandwich attacker can make a profit and loss (or PNL) of:

$$\mathrm{PNL}(\Delta, \eta) = \Delta^{\mathrm{sand}'}(\Delta, \eta) - \Delta^{\mathrm{sand}}(\Delta, \eta) \tag{4}$$

$$= \Delta - G^{-1}(G(\Delta + \Delta^{\mathrm{sand}}) - G(\Delta^{\mathrm{sand}})), \tag{5}$$

measured in input token, where $\Delta^{\mathrm{sand}}(\Delta, \eta)$ refers to the solution of the implicit Eq. (2) and $\Delta^{\mathrm{sand}'}(\Delta, \eta)$ refers to the quantity defined in (3). Note that when $\eta = 0$, $\mathrm{PNL}(\Delta, \eta) = 0$ for all Δ, as desired. Frequently, we will abuse notation by dropping the dependence of the sandwich attack on the user trade Δ and the slippage limit η, and just denote the sandwich attack by Δ^{sand} and $\Delta^{\mathrm{sand}'}$.

Bounds on Δ^{sand}. Using the smoothness constants, we can furnish bounds on Δ^{sand}, $\Delta^{\mathrm{sand}'}$, and $\mathrm{PNL}(\Delta, \eta)$ that allow us to quantify the impact of a sandwich attack on a user's trade (proofs of the below claims are in Appendix B of [13]).

Claim. If $\eta \geq 1 - \frac{\kappa}{\nu}$ then we have $\Delta^{\mathrm{sand}}(\eta, \Delta) = O(\eta)\Delta$.

Claim. Suppose that the forward exchange rate $g(\Delta)$ is β-liquid in addition to G being (ν, κ)-smooth. Then there exists $\zeta = 1 + \Theta(\sqrt{1 + \eta})$ such that $\Delta^{\mathrm{sand}} \geq \left(\frac{\nu}{\beta} - \Delta\right)\zeta$.

The first bound demonstrates that the size of a sandwich attack is linear in the slippage limit, provided that the slippage limit is sufficiently larger than a curvature ratio. The second claim (a lower bound on Δ^{sand}), requires that the forward exchange rate $g(\Delta) = G'(\Delta)$ grows sufficiently fast. One sees why we need this extra assumption by analyzing a constant sum market maker which has $\beta = 0$ [1]. In a constant sum market maker, there is no sandwich profit, as there is no price impact when one executes the sequence of trades $(\Delta^{\mathrm{sand}}, \Delta, \Delta^{\mathrm{sand}'})$. Therefore, in order for us to lower bound the sandwich attack size (and profit, which is linear in Δ^{sand} as per 4), we need some non-zero price impact.

Bounds on $\Delta^{\mathrm{sand}'}$. Now, we bound the round trip trade made by the sandwich attacker, $\Delta^{\mathrm{sand}'}$, which satisfies Eq. (3) (note that $\Delta^{\mathrm{sand}'}$ is in units of input token). We prove the following two claims in Appendix C of [13].

Claim. Suppose that $\eta \geq \frac{\nu - \kappa}{\nu}$. Then $\Delta^{\mathrm{sand}'} = O(\eta)\Delta$.

Claim. Suppose that $g(\Delta)$ is β-liquid. Then there exists $\gamma = 1 + \Theta(\sqrt{1 + \eta})$ such that $\Delta^{\mathrm{sand}'} \geq \frac{\nu\gamma}{\beta} - \Delta\left(\gamma + \frac{\eta\nu}{\kappa}\right)$.

The first claim demonstrates that under mild conditions on the slippage limit, we can control the roundtrip profit in terms of linear factors of the slippage limit. Note that in the next bound, we require that $g(\cdot)$ is β-liquid for the same reason as above: there needs to be some excess price impact that the sandwicher can cause for the sandwich to be profitable.

Upper Bound on Sandwich Attacker Profit. Recall that the profit of an attack is defined by $\mathsf{PNL}(\Delta, \eta) = \Delta^{\mathrm{sand}}(\Delta, \eta) - \Delta^{\mathrm{sand}'}(\Delta, \eta)$. To upper bound price impact, we first need to lower bound $\Delta^{\mathrm{sand}'}$. Using the above claims gives us a bound on the profit (4):

$$\mathsf{PNL}(\Delta, \eta) \leq C \max(\eta, \sqrt{1 + \eta})\Delta - \frac{\nu\gamma}{\beta}.$$

This bound implies that, given all of the liquidity and slippage conditions of the claims are met, the sandwich attacker's profit is linear in η. Using the precise constants for the above bounds, one can also compute a 'hurdle rate' in terms of γ which describes minimal conditions for a sandwich attacker to be profitable (proof in Appendix C of [13]):

Claim. If $\left(\eta\left((1 + \frac{\nu}{\kappa}) - (2 - \frac{\kappa}{\nu}) + \gamma\right)\right) \Delta \geq \frac{\nu\gamma}{\beta}$ then $\mathsf{PNL}(\Delta, \eta) \geq 0$.

This simple result can be used by both wallet designers (who are optimizing η for users) and protocol designers (who can control ν and κ) as a way to minimize expected sandwich profit.

3 Routing MEV

Having defined sandwich attacks, we can now introduce the notion of routing MEV on networks of CFMMs. Intuitively, routing MEV is the excess value that a sandwich attacker can extract from a user trades over a network of CFMMs when the trade is routed selfishly versus when it is routed optimally. Given this network of CFMMs, some amount of an input token, and a desired output token, we call a sequence of trades that converts all of the input token to some amount of the output token a *route* through the network. We consider a setting where multiple users wish to trade token A for token B, and trades have been aggregated into a single trade to be routed across a network of CFMMs (this is commonly done by protocols such as CoWSwap [14]). Such an aggregation of user trades is also called an *order flow*. We assume that the output of each path is distributed pro-rata to all users who trade through that path.

To measure the impact of sandwich attacks on this aggregate trade, we define *selfish routing* as the scenario in which users selfishly route their component of the trade to optimize their own output. We will show that selfish routing leads to congestion, *i.e.*, worse prices for users that choose to take routes that other users are also taking. In the case of selfish routing, an equilibrium is an allocation of order flow to routes such that the average price of the output token among all used paths is equal (we will make this precise shortly.) We will compare selfish routing to *optimal routing*, where a 'central planner' can allocate the order flow across paths to maximize the amount of output [4].

We define the *price of anarchy* as the ratio of the output under optimal routing to the output under selfish routing, possibly in the presence of sandwich attackers on the CFMM network. The main result of this section is to prove that the price of anarchy is constant and bounded by constants related to both the slippage limits defined by the user and the liquidity of the CFMM network. Before introducing the formal definitions of the above quantities, we provide two illustrative examples that show cases in which sandwich attacks worsen optimal routing, but, counterintuitively, improve selfish routing on CFMM networks.

3.1   The CFMM Pigou Example

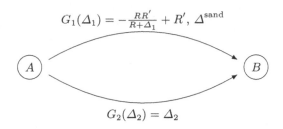

Fig. 1. The CFMM Pigou Network, for $R = 1, R' = 2$, and $\Delta = 1$.

Sandwich Attackers Impact Routing. Here we provide an explicit example of how optimal and selfish routing change when there is a sandwich attacker on the Pigou network shown in Fig. 1. Users desire to trade between tokens A and B and have two CFMMs on which to trade. Suppose that CFMM 1 is a constant product CFMM (*e.g.*, Uniswap) with reserves (R, R'), and CFMM 2 is a constant sum CFMM which always quotes the same forward exchange rate (until its reserves are depleted). We denote the forward exchange functions of these CFMMs by G_1 and G_2 respectively. Uniswap has a forward exchange function of the form $G_1(\Delta) = -\frac{RR'}{R+\Delta} + R'$, and the forward exchange rate is $g_1(\Delta) = \frac{RR'}{(R+\Delta)^2}$. The forward exchange function for the constant sum CFMM is $G_2(\Delta) = c\Delta$, where c is the exchange rate. The forward exchange rate is simply $g_2(\Delta) = c$. The two *paths* through the network, over which we can exchange A for B, are given by the two CFMMs.

Optimal Routing. The optimal routing problem can be written directly as maximizing the amount of token B received, subject to splitting an input amount Δ of token A between the two CFMMs:

$$
\begin{aligned}
\text{maximize} \quad & G_1(\Delta_1) + G_2(\Delta_2) \\
\text{subject to} \quad & \Delta = \Delta_1 + \Delta_2 \\
& \Delta_1, \Delta_2 \geq 0.
\end{aligned}
$$

Forward exchange functions are concave, so this problem is a convex optimization problem. Furthermore, after the optimal trade is made, the marginal forward exchange rates across the two routes will be equal, *i.e.*, $g_1(\Delta_1^*) = g_2(\Delta_2^*)$ where

(Δ_1^*, Δ_2^*) is a solution to the optimization problem. In other words, the optimizer cannot redirect any small amount of flow to another path with a better marginal price. This fact follows directly from the optimality conditions (see Appendix D.1 of [13]).

Selfish Routing. In selfish routing, we view the net trade of size Δ as composed of infinitely many infinitesimal users that act independently. We note that atomic routing [19] may be a more appropriate model but leave exploration of this model in the CFMM context to future work. We assume that each path's output is distributed pro-rata to the users trading over that path, which motivates our equilibrium condition: the average price on each route should be equal. On any path in the Pigou network, the average price is given by $\frac{1}{\Delta_i} \int_0^{\Delta_i} g_i(t)dt = \frac{1}{\Delta_i} G_i(\Delta_i)$. This leads to the equilibrium equation

$$\frac{1}{\Delta_1} G_1(\Delta_1) = \frac{1}{\Delta_2} G_2(\Delta_2),$$

when $\Delta_1, \Delta_2 > 0$, subject to the feasibility condition $\Delta_1 + \Delta_2 = \Delta$. If one of these paths clearly dominates the other in terms of average price for all flow allocation up to Δ, then all users will choose that path.

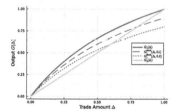

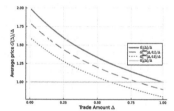

(a) Forward exchange function for each route.

(b) Average price for each route.

Fig. 2. Forward exchange function and average price for the Pigou network example.

Sandwiching. Now, we introduce a sandwich attacker on path 1, the constant product CFMM. We denote the forward exchange function with sandwiching by $G_1^{\mathrm{sand}}(\Delta, \eta)$, which is equal to

$$G_1^{\mathrm{sand}}(\Delta, \eta) = -\frac{(R + \Delta^{\mathrm{sand}})(R' - G_1(\Delta^{\mathrm{sand}})}{R + \Delta^{\mathrm{sand}} + \Delta} + R' - G_1(\Delta^{\mathrm{sand}}),$$

where the optimal sandwich trade is

$$\Delta^{\mathrm{sand}} = (1/2)\left(-(\Delta + 2R) + \sqrt{(\Delta + 2R)^2 + 4(R^2 + R\Delta)\left(\frac{\eta}{1-\eta}\right)}\right).$$

We can compute the optimal and selfish routes in the presence of sandwiching by simply replacing G_1 with G_1^{sand} in the optimal routing problem and in the equilibrium conditions respectively. We provide a derivation of Δ^{sand} in the case of Uniswap in Appendix A of [13].

Numerical Example. We consider an instance of this Pigou network with reserves $(R, R') = (1, 2)$ in the constant product CFMM and $c = 1$ for the constant sum CFMM. Consider one unit of token A traded to token B, with Δ_1 traded through CFMM 1 and $1 - \Delta_1$ through CFMM 2. Without the presence of the sandwich attacker, the optimal route is the x such that

$$g_1'(\Delta_1^\star) = g_2'(1 - \Delta_1^\star) \implies \frac{2}{(1 + \Delta_1^\star)^2} = 1 \implies \Delta_1^\star = \sqrt{2} - 1,$$

which gives a total output of $G_1(\Delta_1^\star) + G_2(1 - \Delta_1^\star) = 4 - 2\sqrt{2} \approx 1.17$. The equilibrium, on the other hand, is for all users to use CFMM 1 (see Fig. 2b), which has a total output of 1. With sandwiching, the top path becomes less desirable. We plot the forward exchange functions and the average price for a number of η's in Fig. 2 as a function of the trade size. (The average price here is also a forward exchange rate, where higher is better.) It is clear that the equilibrium will move towards a more balanced split as the slippage tolerance increases, since order flow will move away from CFMM 1 (see Fig. 3b). In Fig. 3 we show that increasing the slippage tolerance hurts optimal routing until it has the same output as selfish routing, at which point all of the order flow goes over CFMM 2. It follows that the price of anarchy for this network is highest without the sandwich attacker and decreases to 1 once neither the optimal nor the selfish routes use CFMM 1 (see Fig. 3a). We also note that the profitability of the sandwich attacker increases and then decreases due to the competing effects of increasing price slippage but decreasing order flow over CFMM 1 (see Fig. 3c).

3.2   The CFMM Braess Example

Sandwich Attackers Can Improve Selfish Routing. We now construct an example representing an 'inverse Braess paradox' in CFMMs in which the presence of sandwich attackers improves the quality of selfish routing by reducing the price of anarchy. This example demonstrates that while sandwich attacks necessarily worsen the execution of individual trades on a given CFMM, they can improve the *social welfare*, measured as the combined output that users receive over all paths under selfish routing. For the CFMM network in Fig. 4, we introduce the variables $\Delta_i \in \mathbf{R}$, $i = 1, \ldots, 5$ to denote the input into CFMM i. The output of CFMM i is then $G_i(\Delta_i)$.

No Middle CFMM. First we consider the simple case where $G_5(\Delta)$ does not exist; all users must use either the top or the bottom route. In this case, the top and bottom routes are equivalent, and the output of both is strictly concave. As a result, users routing selfishly will split their order flow equally between the two routes. It is easy to see that this equilibrium is, in fact, optimal as well.

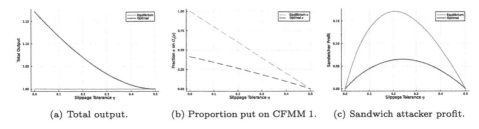

(a) Total output.   (b) Proportion put on CFMM 1.   (c) Sandwich attacker profit.

Fig. 3. Optimal and selfish routing in terms of total output, route taken, and sandwich attacker profit in the Pigou network as the slippage tolerance η varies.

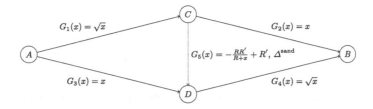

Fig. 4. The CFMM Braess Network.

Optimal Routing. Next we consider optimal routing in the case with the middle CFMM. We wish to maximize the output at node B, *i.e.*, $G_4(\Delta_4) + G_2(\Delta_2)$, subject to the flow conservation constraints implied by this network. We can then find an optimal trade split by solving the following optimization problem:

$$
\begin{aligned}
\text{maximize } & G_4(\Delta_4) + G_2(\Delta_2) \\
\text{subject to } & 1 = \Delta_1 + \Delta_3 \\
& G_1(\Delta_1) = \Delta_2 + \Delta_5 \\
& G_3(\Delta_3) + G_5(\Delta_5) = \Delta_4 \\
& \Delta_i \geq 0 \text{ for } i = 1, \dots, 5.
\end{aligned}
\tag{OPT}
$$

In fact, since all the CFMM functions G_i are concave and increasing, the convex relaxation of this problem, found by relaxing the equality constraints into inequalities, is tight. Furthermore, at optimality, the marginal forward exchange rate across each route will be equal. See [4,9] for details and further discussion of optimal routing.

Selfish Routing. Recall that the equilibrium condition is that the average price across all used paths is equal, and that the trades are feasible. Let α_1, α_2, and α_3 be the order flow on the top ($G_1 \to G_2$), bottom ($G_3 \to G_4$), and middle ($G_1 \to G_5 \to G_4$) paths respectively. Again, we assume that, after a trade has been made, the output is distributed pro-rata to all users of a particular path. The average price condition can be written as

$$
\frac{1}{\alpha_1} \cdot G_2 \left(\frac{\alpha_1}{\alpha_1 + \alpha_3} \Lambda_1 \right) = \frac{1}{\alpha_2} \cdot \frac{\Lambda_3}{\Lambda_3 + \Lambda_5} \cdot G_4(\Lambda_5 + \Lambda_3) = \frac{1}{\alpha_3} \frac{\Lambda_5}{\Lambda_3 + \Lambda_5} \cdot G_4(\Lambda_5 + \Lambda_3),
$$

where Λ_i is the output of CFMM i, for example $\Lambda_1 = G_1(\alpha_1 + \alpha_3)$. (Here we assume all paths are used. A more complete definition is given in the sequel.) The feasibility conditions are the same as those for optimal routing found in OPT. The solution to this system is an equilibrium; no infinitesimal flow can get a better output for its share of the flow (*i.e.*, a better average price) by deviating.

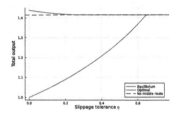
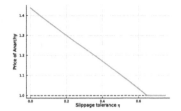

(a) Output for optimal and equilibrium routes.

(b) Price of anarchy for the Braess network.

Fig. 5. Output and price of anarchy in the Braess network example.

Numerical Example. We consider an instance of this Braess network with middle CFMM reserves $(R, R') = (1, 2)$ and one unit of token A traded to token B. Without the middle CFMM, the optimal and equilibrium routes both split the order flow equally between the top and bottom routes, resulting in an output of $2 \cdot \sqrt{0.5} = 1.414$. When we add the middle CFMM, the path through this CFMM has the best average price, even when all the flow is allocated to it. As a result, we observe congestion in selfish routing. This effect is clearly illustrated in Fig. 5a. We see that, despite the addition of the new market, the equilibrium output decreases without sandwiching ($\eta = 0$). As the slippage tolerance increases and users get a worse execution price, they move away from using the middle CFMM. As a result, the optimal routing and selfish routing output converge to the optimal output without the middle CFMM, and the price of anarchy decreases to 1.0, shown in Fig. 5b. Figure 6a illustrates the decrease of the fraction of order flow allocated to this path. In this way, the presence of the sandwich attacker relieves congestion on the network. The opposing effects of increased slippage but less order flow on CFMM 5 again cause the sandwich attacker profit to increase and then decrease with slippage tolerance (see Fig. 6b).

Discussion. This toy example illustrates an 'inverse Braess's paradox'. Like in Braess's paradox, the addition of the additional CFMM decreases the total output at equilibrium of selfish routing (of course, it increases the optimal output). All of the selfish order flow wants to use the middle link, which causes congestion. However, the addition of the sandwich attacker causes this order flow to begin to migrate away from the middle CFMM as users' slippage tolerance increases. The sandwich attacker can be thought of as a 'decentralized traffic controller', where users' best responses to the existence of a sandwich attacker lead to better output across the network.

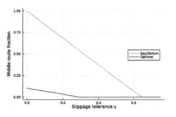

(a) Fraction on the middle link in the Braess network.

(b) Sandwich attacker profit for the Braess network.

Fig. 6. Characteristics of the middle link route in the Braess network example.

3.3   Price of Anarchy

A general formulation for the impact of sandwich attacks on networks of CFMMs follows from the intuition built by the previous two examples. Suppose that we have a graph $G = (V, E)$ where each vertex $A \in V$ denotes a token and each edge $e \in E$ represents a CFMM for trading between pairs of tokens in the graph. Suppose that users want to trade between tokens A and B. Denote the set of paths from A to B as $\mathcal{P}$. Associated to each $e \in E$ is a price function $g_e(\cdot)$ and a corresponding forward exchange function $G_e(\cdot)$ (these are functions of the reserves on edge e, but we suppress this dependence). We first define the general setup of *optimal routing*, in which a central planner is able to maximize the net output from a network of CFMMs for a user trading between a pair of tokens, and *selfish routing*, in which infinitesimal users greedily optimize for their pro-rata share of the output along a path. The outcome of this process is an equilibrium. We then define *routing MEV* as any excess value that can be extracted from adjusting how transactions are executed on this graph (by sandwich attackers). Informally, we prove:

The price of anarchy of selfish routing through a network of CFMMs is bounded by a constant that depends on the slippage η and constants $\kappa, \nu, \alpha, \beta$.

Trade Splittings and Path Outputs. To analyze the price of anarchy, we define the space of trades on a network as $\mathcal{T} = \mathbf{R}_+ \times [0,1]^{|\mathcal{P}|} \times V \times V$. Each trade $T \in \mathcal{T}$ where $T = (\Delta, \eta, A, B)$ specifies an amount, Δ to be traded from vertex A to vertex B along with a slippage limit η_p on every path $p \in \mathcal{P}$. We will abuse notation and utilize $(\Delta_{AB}, \eta_{AB}) \in T$ to refer to $(\Delta, \eta, A, B) \in T$. For a trade (Δ_{AB}, η_{AB}), we define any $\alpha \in \{x \in \mathbf{R}^{|\mathcal{P}|} : \sum_i x_i = \Delta_{AB}\} = \hat{S}_{|\mathcal{P}|}$ to be a *splitting vector* that indicates what fraction of the Δ_{AB} units of token A are routed onto each path. For every path p, we denote the *path forward exchange function* $G_p : \hat{S}_{|\mathcal{P}|} \to \mathbf{R}$. This function gives the output of an amount of the trade $\alpha_p \Delta_{AB}$ placed on path $p \in \mathcal{P}$. We use this function as opposed to the edge forward exchange functions $G_e(\cdot)$ because the conditions for optimal and selfish routing are tractably written in this notation. We note that for any trade splitting α, the path output function $G_p(\cdot)$ may depend on components of α

belonging to paths other than path p, as there may be edges that intersect on multiple paths. First, we define both optimal and selfish routing in the absence of sandwich attacks on the network.

Optimal Routing. We define optimal routing over the trade split α. (Note that this formulation differs from the original formulation of [4]). In terms of this trade split α and the path output function G_p, the optimal routing problem is

$$\text{maximize} \quad \sum_{p \in \mathcal{P}} G_p(\alpha)$$

$$\text{subject to} \quad \sum_{p \in \mathcal{P}} \alpha_p = \Delta_{AB}$$

$$\alpha_p \geq 0.$$

We denote any trade splitting that is a solution to the optimal routing problem by $\alpha^\star$. A modified version of the objective $\sum_{p \in \mathcal{P}} G_p(\alpha)$ (after accounting for the presence of sandwich attacks on the network) will be used when defining the price of anarchy.

Selfish Routing. In order to generalize the notion of selfish routing in Sect. 3.1, we define an equilibrium in terms of the average price on each path. Once again, the interpretation of our equilibrium notion is that a small unit of flow should not want to deviate from the path it has chosen because it has greedily optimized for the share of the output it will receive. A splitting vector $\bar{\alpha} \in \hat{S}_{|\mathcal{P}|}$ is said to be an *equilibrium splitting*[2] if for all $p, p' \in \mathcal{P}$ with $\bar{\alpha}_p > 0$ we have

$$\frac{G_p(\bar{\alpha})}{\bar{\alpha}_p} \geq \frac{G_{p'}(\bar{\alpha})}{\bar{\alpha}_{p'}}.$$

This definition says that over *all* paths on which there is a nonzero trade in equilibrium, the average price must be equal. This equilibrium condition says that an infinitesimal trade (or unit of 'flow') on a path from A to B greedily optimizes for its pro-rata share of the output coming from that path. This situation stands in stark contrast to optimal routing.

Resolving Ambiguities in Splitting. We note that the selfish routing equilibrium over paths described above leaves trades splitting at intermediate nodes in the network ambiguous, so we provide a procedure for resolving this ambiguity; note that this procedure picks one splitting, but that the subsequent results still hold for any equilibrium splitting. Consider a trade Λ, made up of two previous CFMM outputs, that then must be split among two CFMMs, as in Fig. 7. Suppose that trades Δ_1 and Δ_2, corresponding to each path, are incident on G_1, and so this CFMM outputs $\Lambda = G_1(\Delta_1 + \Delta_2)$. We prescribe that this trade is inputted into CFMMs G_2 and G_3 by splitting it according to the pro-rata percentage of the input into G_1 it represents. That is, the input into the top path is $\frac{\Lambda \Delta_1}{\Delta_1 + \Delta_2}$ and the input into the bottom path is $\frac{\Lambda \Delta_2}{\Delta_1 + \Delta_2}$.

---

[2] The existence of such an equilibrium can be established using standard results in nonatomic routing games [18].

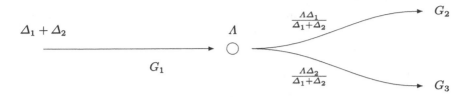

Fig. 7. Resolving ambiguous splits in CFMM networks.

Sandwich Attacks on Graphs. We now move to defining optimal and self-ish routing in the presence of sandwich attackers by introducing the modified path output functions $G_p^{\text{sand}}(\cdot)$ to account for the value captured by sandwich attackers on the CFMM networks. Recall that users provide a slippage limit $\eta_{AB} \in [0,1]^{|\mathcal{P}|}$ over all the paths. However, we need to construct slippage limits for each edge in a path. We solve for the implied slippage limits on each edge in a path to bound sandwich attack size and price impact. Specifically, for a path $p = (e_1, \ldots, e_{|p|}) \in \mathcal{P}$ along with a slippage $\eta_{AB,p}$, we know that for any trade splitting $\alpha \in \hat{S}_{|\mathcal{P}|}$, the minimum acceptable amount from path p is $(1 - \eta_{AB,p})G_p(\alpha)$. First, we denote for a path p, $\Delta_{e_{|p|-k}}$ to be the net trade that enters edge $|p| - k$ on that path, given a trade Δ_{AB} and splitting α. Then, we can define the edge slippage limits for path p, given by the vector $(\eta_{p,e_1}, \ldots, \eta_{p,|p|})$, where $\eta_{p,|p|} = \eta_{AB,p}$ recursively as:

$$G_{e_{|p|-1}}(\Delta_{e_{|p|-1}} + \Delta_{e_{|p|-1}}^{\text{sand}}) - G_{e_{|p|-1}}(\Delta_{e_{|p|-1}}) = (1 - \eta_{AB,p})G_{e_{|p|}}(\Delta_{|p|})$$

$$G_{e_{|p|-k-1}}(\Delta_{e_{|p|-k-1}} + \Delta_{e_{|p|-k-1}}^{\text{sand}}) - G_{e_{|p|-k-1}}(\Delta_{e_{|p|-k-1}}) = (1 - \eta_{p,k})G_{e_{|p|-k}}(\Delta_{|p|-k}).$$
$$(6)$$

The above set of equations for $k = 1, \ldots, |p| - 1$ provide recursions that can be solved from the terminal edge back to edge e_1 to find the corresponding slippage limits and sandwich amounts on every path $p \in \mathcal{P}$. Once these slippage limits have been pinned down on every edge, there is a well-defined notion of a *path* sandwich attack $\Delta_p^{\text{sand}}(\alpha, \eta_{AB})$. Therefore, we can define the modified path output functions $G_p^{\text{sand}}(\alpha, \eta_{AB})$ as the amount of output that the user receives over a path after the sandwich attacker's excess output over that path has been removed:

$$G_p^{\text{sand}}(\alpha, \eta_{AB}) := G_p(\alpha_p + \Delta_p^{\text{sand}}(\alpha, \eta_{AB})) - G_p(\Delta_p^{\text{sand}}(\alpha, \eta_{AB})).$$

We note that the recursion for solving for η_{p,e_i} given in (6) is not guaranteed to return unique slippage limits as it is the difference of convex functions, which means there could be multiple sequences $\eta_{p,e_1}, \ldots, \eta_{p,|p|}$ that could lead to the same optimal output. Using curvature bounds, however, we can bound this recursion and, therefore, provide bounds on Δ_p^{sand}.

Bounds on $\Delta^{\text{sand}}{}_p$. In order to derive bounds on the price of anarchy, we must bound the size of a sandwich attack along a path $\Delta^{\text{sand}}{}_p(\alpha, \eta_{AB})$ by the trade size on that path and by the terminal slippages. The following result says that the size of the sandwich attack along a path p is bounded on both sides by functions that depend on the path length, and the trade size α_p. These functions will be used to derive the upper bound on price of anarchy in Theorem 1:

Proposition 1. *There exist functions* $f(\kappa, \nu, \eta_{AB})$ *and* $g(\kappa, \nu, \eta_{AB})$ *such that*

$$\Delta_p^{sand}(\alpha, \eta_{AB}) \leq f(\kappa, \nu, \eta_{AB})^{|p|} \alpha_p, \tag{7}$$

and

$$\Delta_p^{sand}(\alpha, \eta_{AB}) \geq g(\kappa, \nu, \eta_{AB})^{|p|} \alpha_p, \tag{8}$$

for all paths $p \in \mathcal{P}$.

Using the bounds in Proposition 1 and by that fact that the function G_p has local quadratic bounds (that is, there exist κ', ν' such that $\kappa' \alpha_p^2 \leq G_p(\alpha) \leq \nu' \alpha_p^2$) we immediately get the following bounds on $G_p^{\text{sand}}(\alpha, \eta_{AB})$:

$$G_p^{\text{sand}}(\alpha, \eta_{AB}) \leq \left(\nu' + \nu' f(\kappa, \nu, \eta_{AB})^{|p|} - \kappa' g(\kappa, \nu, \eta_{AB})^{|p|} \right) \alpha_p^2 \tag{9}$$

$$G_p^{\text{sand}}(\alpha, \eta_{AB}) \geq \left(\kappa' + \kappa' g(\kappa, \nu, \eta_{AB})^{|p|} - \nu' f(\kappa, \nu, \eta_{AB})^{|p|} \right) \alpha_p^2. \tag{10}$$

The definition of the modified path output function and the bounds provided on $\Delta_p^{\text{sand}}(\alpha, \eta_{AB})$ will be important in defining the excess price impact realized by the user and the sandwich profit realized by attacker. Proposition 1 (and associated desiderata about analyzing the aforementioned recursions) is described in Appendix D.2 of [13].

Social Welfare and the Price of Anarchy. To define a measure of social welfare, we compare the net output under selfish routing to the net output under optimal routing in the presence of sandwich attacks (of course, such a quantity may also be defined in the absence of such attacks, which leads to a conventional notion of price of anarchy). This quantity indicates how much sandwich attackers degrade the quality of selfish routing relative to optimal routing, which we call routing MEV. Our welfare function takes the sum of the path output functions incorporating sandwiches, $W^{\text{sand}}(\alpha) = \sum_{p \in \mathcal{P}} G_p^{\text{sand}}(\alpha)$, which allows us to formally define the *price of anarchy* for any optimal splitting α^* and any equilibrium splitting $\bar{\alpha}$:

$$\mathsf{PoA}(\Delta_{AB}, \eta_{AB}) = \frac{W^{\text{sand}}(\alpha^\star)}{W^{\text{sand}}(\bar{\alpha})}$$

Our main theorem bounds the price of anarchy as a function of the curvature and slippage parameters of the network. Notably, given sufficient liquidity conditions on the CFMM network, the price of anarchy is upper bounded by a *constant*. We show this result by connecting the output functions of the CFMMs to (λ, μ)-smoothness arguments from [17].

Theorem 1. *Suppose that $f(\kappa, \nu, \eta_{AB}), g(\kappa, \nu, \eta_{AB}) \in O(1 + (\nu\kappa)^{O(1)})^{1/\mathrm{diam}(G)}$.*
Then there exists a function $C(\kappa, \alpha, \beta, \nu, \eta)$ that is constant in the size of the
network graph G such that

$$\mathsf{PoA}(\Delta_{AB}, \eta_{AB}) \leq C(\kappa, \alpha, \beta, \nu, \eta). \tag{11}$$

The proof of this theorem (which is based on (λ, μ)-smoothness results of [17],
and is provided in Appendix D.4 of [13]) relies on the following lemma (the proof
of this lemma can be found in Appendix D.3):

Lemma 1. *For every $\alpha, \alpha' \in \hat{S}_{|\mathcal{P}|}$, we have:*

$$\frac{G_p^{sand}(\alpha)}{\alpha_p}\alpha'_p \geq \lambda G_p^{sand}(\alpha) + \mu G_p^{sand}(\alpha'),$$

for $\lambda = \mu = \dfrac{\left(\nu' + \nu' f(\kappa, \nu, \eta_{AB})^{|P|} - \kappa g(\kappa, \nu, \eta_{AB})^{|P|}\right)}{\left(\kappa' + \kappa' g(\kappa, \nu, \eta_{AB})^{|P|} - \nu' f(\kappa, \nu, \eta_{AB})^{|P|}\right)}.$

One can view the condition of bounds on f and g informally as saying that
provided there is enough liquidity on each edge in the graph (measured by the
μ, κ dependence in f, g), then the price of anarchy from sandwiching is constant.
We note that our result can likely be sharpened and that the constants are not
tight.

4   Conclusion and Future Work

In this paper, we provide a formal description of generic sandwich attacks in arbitrary CFMMs. Using this description, we explicitly compute bounds on sandwich
attack profitability that depend on curvature and liquidity. These bounds allowed
us to analyze a prominent form of CFMM MEV: routing MEV. Paradoxically, we
find that for routing MEV, sandwich attacks can, in certain cases, increase
social welfare for users when user trades are selfishly routed across a network of
CFMMs. We generalize this example to larger networks of CFMMs and showed
that the price of anarchy for routing MEV is constant given sufficient liquidity
on the network graph using (λ, μ)-smoothness results from [17].

Prior works on MEV [5,7,12,15,22] have focused on illuminating either specific profitable attacks or methodologies for the numerical or empirical estimation of MEV. Our work provides a formal algorithmic game theory framework
to reason about the impact of MEV on users. Such results not only provide
asymptotic, theoretical insight into the nature of MEV, but also suggest some
mitigating strategies for protocol developers to properly set slippage limits. We
leave the atomic version of the selfish routing model, and models in which trades
are sequentially executed over the network as future work. We further note that
our results may be used to provide robust bounds (valid for mixed or coarse
correlated equilibria), which we also leave for future work.

References

1. Angeris, G., Agrawal, A., Evans, A., Chitra, T., Boyd, S.: Constant function market makers: multi-asset trades via convex optimization. In: Tran, D.A., Thai, M.T., Krishnamachari, B. (eds.) Handbook on Blockchain. SOIA, vol. 194, pp. 415–444. Springer, Cham (2022). https://doi.org/10.1007/978-3-031-07535-3_13
2. Angeris, G., Chitra, T., Evans, A.: When does the tail wag the dog? Curvature and market making. Cryptoeconomic Syst. **2**(1) (2022)
3. Angeris, G., Evans, A., Chitra, T.: A note on bundle profit maximization (2021)
4. Angeris, G., Evans, A., Chitra, T., Boyd, S.: Optimal routing for constant function market makers. In: Proceedings of the 23rd ACM Conference on Economics and Computation, EC 2022, pp. 115–128. Association for Computing Machinery, New York (2022). https://doi.org/10.1145/3490486.3538336. ISBN 9781450391504
5. Babel, K., Daian, P., Kelkar, M., Juels, A.: Clockwork finance: automated analysis of economic security in smart contracts. arXiv preprint arXiv:2109.04347 (2021)
6. Bartoletti, M., Chiang, J.H., Lluch-Lafuente, A.: A theory of automated market makers in DeFi. In: Damiani, F., Dardha, O. (eds.) COORDINATION 2021. LNCS, vol. 12717, pp. 168–187. Springer, Cham (2021). https://doi.org/10.1007/978-3-030-78142-2_11
7. Bartoletti, M., Chiang, J.H., Lluch-Lafuente, A.: Maximizing extractable value from automated market makers. arXiv preprint arXiv:2106.01870 (2021)
8. Daian, P., et al.: Flash boys 2.0: frontrunning, transaction reordering, and consensus instability in decentralized exchanges. arXiv:1904.05234 [cs] (2019)
9. Diamandis, T., Resnick, M., Chitra, T., Angeris, G.: An efficient algorithm for optimal routing through constant function market makers. arXiv preprint arXiv:2302.04938 (2023)
10. Flashbots Team. MEV explore (2022)
11. Goyal, M., Ramseyer, G., Goel, A., Mazières, D.: Finding the right curve: optimal design of constant function market makers. In: Proceedings of the 24th ACM Conference on Economics and Computation, pp. 783–812 (2023)
12. Heimbach, L., Wattenhofer, R.: Eliminating sandwich attacks with the help of game theory. arXiv preprint arXiv:2202.03762 (2022)
13. Kulkarni, K., Diamandis, T., Chitra, T.: Towards a theory of maximal extractable value I: constant function market makers. arXiv preprint arXiv:2207.11835 (2022)
14. CoW Protocol. Cow protocol overview (2022). https://docs.cow.fi/
15. Qin, K., Zhou, L., Gervais, A.: Quantifying blockchain extractable value: how dark is the forest? arXiv preprint arXiv:2101.05511 (2021)
16. Qin, K., Zhou, L., Livshits, B., Gervais, A.: Attacking the DeFi ecosystem with flash loans for fun and profit. arXiv preprint arXiv:2003.03810 (2020)
17. Roughgarden, T.: Intrinsic robustness of the price of anarchy. J. ACM (JACM) **62**(5), 1–42 (2015)
18. Roughgarden, T.: Selfish Routing and the Price of Anarchy. MIT Press, Cambridge (2005)
19. Roughgarden, T.: Selfish routing with atomic players. In: Proceedings of the 16th Symposium on Discrete Algorithms (SODA), pp. 1184–1185. Citeseer (2004)
20. Roughgarden, T., Syrgkanis, V., Tardos, E.: The price of anarchy in auctions. J. Artif. Intell. Res. **59**, 59–101 (2017)
21. Schlegel, J.C., Kwaśnicki, M., Mamageishvili, A.: Axioms for constant function market makers. Available at SSRN (2022)

22. Zhou, L., Qin, K., Gervais, A.: A2MM: mitigating frontrunning, transaction reordering and consensus instability in decentralized exchanges. arXiv preprint arXiv:2106.07371 (2021)
23. Züst, P., Nadahalli, T., Wattenhofer, Y.W.R.: Analyzing and preventing sandwich attacks in Ethereum. ETH Zürich (2021)

Auction Design for Value Maximizers with Budget and Return-on-Spend Constraints

Pinyan Lu[1,2(✉)], Chenyang Xu[3(✉)], and Ruilong Zhang[4(✉)]

[1] Shanghai University of Finance and Economics, Shanghai, China
lu.pinyan@mail.shufe.edu.cn
[2] Laboratory of Interdisciplinary Research of Computation and Economics (SUFE), Ministry of Education, Shanghai, China
[3] Shanghai Key Laboratory of Trustworthy Computing, East China Normal University, Shanghai, China
cyxu@sei.ecnu.edu.cn
[4] Department of Computer Science and Engineering, University at Buffalo, Amherst, USA
ruilongz@buffalo.edu

Abstract. The paper designs revenue-maximizing auction mechanisms for agents who aim to maximize their total obtained values rather than the classical quasi-linear utilities. Several models have been proposed to capture the behaviors of such agents in the literature. In the paper, we consider the model where agents are subject to budget and return-on-spend constraints. The budget constraint of an agent limits the maximum payment she can afford, while the return-on-spend constraint means that the ratio of the total obtained value (return) to the total payment (spend) cannot be lower than the targeted bar set by the agent. The problem was first coined by [5]. In their work, only Bayesian mechanisms were considered. We initiate the study of the problem in the worst-case model and compare the revenue of our mechanisms to an offline optimal solution, the most ambitious benchmark. The paper distinguishes two main auction settings based on the accessibility of agents' information: *fully private* and *partially private*. In the fully private setting, an agent's valuation, budget, and target bar are all private. We show that if agents are unit-demand, constant approximation mechanisms can be obtained; while for additive agents, there exists a mechanism that achieves a constant approximation ratio under a large market assumption. The partially private setting is the setting considered in the previous work [5] where only the agents' target bars are private. We show that in this setting, the approximation ratio of the single-item auction can be further improved, and a $\Omega(1/\sqrt{n})$-approximation mechanism can be derived for additive agents.

Keywords: Auction Design · Value Maximizers · Return-on-spend Constraints

All authors (ordered alphabetically) have equal contributions. The full version is available at https://arxiv.org/abs/2307.04302.

J. Garg et al. (Eds.): WINE 2023, LNCS 14413, pp. 474–491, 2024.
https://doi.org/10.1007/978-3-031-48974-7_27

1   Introduction

In an auction with n agents and m items, the auctioneer decides the allocation $\mathbf{x} = \{x_{ij}\}_{i \in [n], j \in [m]}$ of the items and the agents' payments $\mathbf{p} = \{p_i\}_{i \in [n]}$. The agent i's obtained value is usually denoted by a valuation function v_i of the allocation; while the agent's utility depends on both the obtained value and the payment made to the auctioneer. Combining the valuation and payment to get the final utility function is a tricky modeling problem.

In the classic auction theory and the vast majority of literature from the algorithmic game theory community, one uses the quasi-linear utility function $u_i = v_i - p_i$, i.e., the utility is simply the obtained value subtracting the payment. This natural definition admits many elegant mathematical properties and thus has been widely investigated in the literature (e.g. [24–26]). However, as argued in some economical literature [3,29], this utility function may fail to capture the agents' behaviors and thus cannot fit reality well in some circumstances. In these circumstances, one usually uses a generic function $u = f(v, p)$ (with monotonicity and possibly convexity properties) to model the utility function. Such treatment is surely general enough, but usually not explicitly enough to get a clear conclusion. In particular, designing non-trivial truthful mechanisms for agents with a generic and inexplicit utility function is difficult.

Is there some other explicit utility function (beyond the quasi-linear one) that appears in some real applications? One simple and well-studied model is agents with budget constraints (e.g. [6,12,16,27]). In this setting, besides the valuation function, an agent i also has a budget constraint B_i for the maximum payment he can make. In the formal language, the utility is

$$u_i := \begin{cases} v_i - p_i & \text{if } p_i \le B_i, \\ -\infty & \text{otherwise.} \end{cases}$$

In mechanism design, the valuation function $v_i(\cdot)$ is considered as the private information for agent i. Thus the auctioneer needs to design a truthful mechanism to incentivize the agents to report their true information. For these models beyond the simplest quasi-linear utility, other parameters might be involved in the agents' utility functions besides the valuation function, such as budget B in the above example. For the mechanism design problem faced by the auctioneer, one can naturally ask the question of whether these additional parameters are public information or private. Both cases can be studied, and usually, the private information setting is more realistic and, at the same time, much more challenging. This is the case for the budget constraint agents. Both public budget and private budget models are studied in the literature (e.g. [8,11,17,18,21]).

Value Maximizer. The above budget constraint agent is only slightly beyond the quasi-linear model since it is still a quasi-linear function as long as the payment is within the budget. However, it is not uncommon that their objective is to maximize the valuation alone rather than the difference between valuation and payment for budget-constrained agents. This is because, in many scenarios, the objective/KPI for the department/agent/person who really decides the bidding strategy is indeed the final value obtained. On the other hand, they cannot collect the remaining unspent money by themselves anyway, and as a result, they do not care about the payment that much as long as it is within the budget given to them. For example, in a company or government's procurement process, the agent may be only concerned with whether the procurement budget can generate the maximum possible value. We notice that with the development of modern auto-bidding systems, value maximization is becoming the prevalent behavior model for the bidders [1,2,4,15]. This motivates the study of value maximizer agents, another interesting explicit non-quasi-linear utility model. In many such applications, there is another return-on-spend (RoS) constraint τ_i for each agent i which represents the targeted minimum ratio between the obtained value and the payment and is referred to as the *target ratio* in the following. Formally, the utility function is

$$u_i := \begin{cases} v_i & \text{if } p_i \leq B_i \text{ and } p_i \tau_i \leq v_i, \\ -\infty & \text{otherwise.} \end{cases} \tag{1}$$

As one can see, the value maximizer's utility function is still a function of v and p but with two additional parameters B and τ, which result from two constraints. Note that the above utility function is identical to that of [5]. Their paper focused on one particular setting where both value and budget are public information, with RoS parameter τ being the only single-dimensional private information. Considering τ as the only private information helps design better auctions, but it may fail to capture wider applications. On top of capturing more practical applications, we consider the setting where all these pieces of information are private, which we call the fully private setting. This makes designing an efficient auction for the problem challenging. With the focus on the fully private setting, we also consider some partially private settings, for which we can design better mechanisms.

There are other definitions of value maximizer in the literature, most of which can be viewed as a special case of the above model [10,23]. For example, there might be no budget constraint ($B = \infty$) or no RoS constraint. Another example is to combine $\frac{v_i}{\tau_i}$ as a single value (function). A mechanism for the fully private setting in our model is automatically a mechanism with the same guarantee in all these other models. That is another reason why the fully private setting is the most general one.

Revenue Maximization and Benchmarks. This paper considers the revenue maximization objective for the auctioneer when designing truthful[1] mechanisms for value maximizers. For the revenue maximization objective, there are usually two benchmarks, called "*first-best*" and "*second-best*". The first-best benchmark refers to the optimal objective we can get if we know all the information. In

---

[1] A mechanism is *truthful* if for any agent i, reporting the true private information always maximizes the utility regardless of other agents' reported profiles, and the utility of any truthtelling agent is non-negative.

our setting, it is $\max_{\mathbf{x}} \sum_i \min \left\{ B_i, \frac{v_i(\mathbf{x})}{\tau_i} \right\}$. For the traditional quasi-linear utility function, the first-best benchmark is simply the maximum social welfare one can generate $\max_{\mathbf{x}} \sum_i v_i(\mathbf{x})$. It is proved that such a benchmark is not achievable or even not approximated by a constant ratio in the traditional setting. Thus the research there is mainly focused on the second-best benchmark. The second-best benchmark refers to the setting where the auctioneer additionally knows the distribution of each agent's private information and designs a mechanism to get the maximum expected revenue with respect to the known distribution. The benchmark in [5] is also this second-best benchmark and they provide optimal mechanism when the number of agents is at most two.

It is clear that the first-best benchmark is more ambitious and more robust since it is prior free. They focus on the second-best in the traditional setting because the first-best is not even approximable. In our new value maximizer agents setting, we believe it is more important to investigate if we can achieve the first-best approximately. Thus, we focus on the first-best benchmark in this paper. This is significantly different from that of [5].

1.1   Our Results

Problem Formulation. The formal description of the auction model considered in the paper follows. One auctioneer wants to distribute m heterogeneous items among n agents. Each agent $i \in [n]$ has a value v_{ij} per unit for each item $j \in [m]$ and a budget B_i, representing the maximum amount of money agent i can pay. The agent also has a RoS constraint τ_i, representing the minimum ratio of the received value (return) to the total payment (spend) that she can tolerate. As mentioned above, several settings of the type (public or private) of $(\mathbf{B}, \mathbf{v}, \boldsymbol{\tau})$ are considered in the paper. Agents are value maximizers subject to their budget constraints and RoS constraints (see Eq. (1) for the formula). The auctioneer aims to design a truthful mechanism that maximizes the total payment.

We investigate our model in a few important auction environments. We studied both indivisible and divisible items, both the single-item and the multiple-item auctions. When there are multiple items, we consider the two most important valuations: unit demand and additive. Unit demand models are the setting where the items are exclusive to each other. Additive models are the setting where an agent can get multiple items and their values add up. We leave the more generic valuation function, such as submodular or sub-additive, to future study. See Table 1 for an overview.

In the fully private information setting, we obtain constant approximation truthful mechanisms for both the single-item auction and the multiple-item auction among unit demand agents. This is quite surprising given the fact that such a constant approximation to the first-best benchmark is proved to be impossible for the classic quasi-linear utility agents, even in the single-item setting. The intuitive reason is that the agent is less sensitive to the payment in the value maximizer setting than in the quasi-linear utility setting, and thus, the auctioneer has a chance to extract more revenue from them. But this does not

Table 1. An overview of our results. We use divisibility "Y" or "N" to represent whether items are divisible or not, and use notation $\Leftarrow$ to express the result can be directly implied by the left one. The notion † in the table means that the approximation ratio is obtained under an assumption, while (∗) implies that the results are deferred to the full version of this paper.

	Divisibility	Fully Private	Partially Private
Single Item	Y	$\frac{1}{52}$ (Theorem 2)	$\frac{1}{2} - \delta$ (∗)
	N	OPT (Theorem 1)	$\Leftarrow$
Multiple Items with Unit Demand Agents	Y	$\Omega(1)$ (Theorem 6)	$\Leftarrow$
	N	$\frac{1}{2}$ (Theorem 3)	$\Leftarrow$
Multiple Items with Additive Agents	Y	$\Omega(1)^{\dagger}$ (∗)	$\Omega(\frac{1}{\sqrt{n}})$ (∗)
	N	Open	Open

imply that designing a good truthful mechanism is easy. Quite the opposite, we need to bring in some new design and analysis ideas since the truthfulness here significantly differs from the traditional one as agents' utility functions are different. For the additive valuation, we provide constant approximation only under an additional large market assumption. This is obtained by observing an interesting and surprising relationship between our model and the model of "liquid welfare for budget-constrained agents".

We also consider the partially private information setting. For the public budget (but private value and target ratio), we obtained an improved constant approximation truthful mechanism for the single-item environment. The improved mechanism for the single-item setting has a much better approximation since we cleverly use the public budget information in the mechanism. For the additive valuation without the large market assumption, we also investigate it in the private target ratio (but public budget and valuation) setting, which is the setting used in [5]. we obtained an $\Omega(\frac{1}{\sqrt{n}})$ approximation truthful mechanism. In the additive setting, an agent may get multiple items, and thus, the payment she saved from one item can be used for other items, which is an impossible case in the unit demand setting. Due to this reason, agents may become somewhat more sensitive to payment, which leads to an $\Omega(\frac{1}{\sqrt{n}})$ approximation.

1.2 Related Works

The most relevant work is [5], in which they also aim to design a revenue-maximizing Bayesian mechanism for value maximizers with a generic valuation and utility function under budget and RoS constraints. As mentioned above, they focus on the setting where each agent's only private information is the target ratio, which is referred to as the partially private setting in our paper. They show that under the second-best benchmark, an optimal mechanism can be obtained for the two-agent case.

Another closely related line of work is "liquid welfare for budget constraint agents" [9,18,21,22]. We observe an interesting and surprising relationship between these two models since the liquid welfare benchmark is almost identical to the first-best benchmark in our setting. Therefore, some algorithmic ideas there can be adapted here. However, there are two significant differences: (i) the objective for the auctioneer is (liquid) welfare rather than revenue. This difference mainly affects the approximation; (ii) the bidders are quasilinear utility (within the budget constraint) rather than value maximizers. This difference mainly affects truthfulness. Observing this relation and difference, some auction design ideas from their literature inspire part of our methods. Furthermore, building deeper connections or ideal black-box reductions between these two models would be an interesting future direction.

The model of budget feasible mechanism [7,13,19,20,28] also models the agent as a value maximizer rather than a quasi-linear utility maximizer as long as the payment is within the budget. The difference is that the value maximizer agent is the auctioneer rather than the bidders.

1.3 Paper Organization

In this version, we focus on unit demand agents and the fully private setting, where all the budgets, valuations, and target ratios are private. We first consider the single-item auction in Sect. 2 and then extend the algorithmic ideas to the multiple items auction for unit demand agents in Sect. 3. Both of the two environments can be constant-approximated. For all other results present in Table 1, due to space limits, we defer them to the full version of the paper.

2 Warm-Up: Single Item Auction

Let us warm up by considering the environment where the auctioneer has only one item to sell. Our first observation is that if the item is indivisible, we can achieve a truthful optimal solution by directly assigning the item to agent k with the maximum $\min\left\{B_k, \frac{v_k}{\tau_k}\right\}$ and charging her that value. Basically, the first price auction with respect to $\min\left\{B_i, \frac{v_i}{\tau_i}\right\}$. The optimality is obvious. For truthfulness, since $\min\left\{B_i, \frac{v_i}{\tau_i}\right\}$ is the maximum willingness-to-pay of each agent i, if someone other than k misreports the profile and gets assigned the item, one of the two constraints must be violated. On the other hand, misreporting a lower profile can only lead to a lower possibility of winning but without any benefit.

Theorem 1. *There exists a truthful optimal mechanism for the single indivisible item auction.*

The above theorem gives some intuition for the divisible item environment. If the indivisible optimum is at least a constant fraction c of the divisible optimum, selling the item indivisibly can give a constant approximation. We refer to this

---

Algorithm 1. Single Item Auction

Input: The reported budgets $\{B_i\}_{i \in [n]}$, the reported value profile $\{v_i\}_{i \in [n]}$, and the reported target ratios $\{\tau_i\}_{i \in [n]}$.

Output: An allocation and payments.

1: Initially, set allocation $x_i \leftarrow 0 \ \forall i \in [n]$ and payment $p_i \leftarrow 0 \ \forall i \in [n]$.

2: With probability of $\frac{9}{13}$, **begin**   ▷ Indivisibly Selling Procedure

3:     Find the agent k with the maximum min $\left\{ B_k, \frac{v_k}{\tau_k} \right\}$, and break the ties in a fixed manner.

4:     Set $x_k \leftarrow 1, p_k \leftarrow \min \left\{ B_k, \frac{v_k}{\tau_k} \right\}$.

5: **end**

6: With probability of $\frac{4}{13}$, **begin**   ▷ Random Sampling Procedure

7:     Randomly divide all the agents with equal probability into set S and R.

8:     Compute the offline optimal solution $\left(\mathbf{z}^S = \{z_i^S\}, \mathbf{p}(\mathbf{z}^S) = \{p_i(\mathbf{z}^S)\} \right)$ of selling the item to the agent subset S.

9:     Set the item's reserve price $r \leftarrow \frac{1}{4} \sum_{i \in S} p_i(\mathbf{z}^S)$.

10:     Let agents in R come in an arbitrarily fixed order. When each agent i comes, use α to denote the remaining fraction of the item, and set $x_i \leftarrow \min \left\{ \frac{B_i}{r}, \alpha \right\}, p_i \leftarrow x_i \cdot r$ if $r \leq \frac{v_i}{\tau_i}$.

11: **end**

12: **return** Allocation $\{x_i\}_{i \in [n]}$ and payments $\{p_i\}_{i \in [n]}$.

---

idea as *indivisibly selling* in the following. In contrast, for the case that the indivisible optimum is smaller than a constant fraction c of the divisible optimum (denoted by OPT in the following), we have min $\left\{ B_i, \frac{v_i}{\tau_i} \right\} \leq c \cdot \text{OPT}$ for any agent i. This property implies that the *random sampling* technique can be applied here. More specifically, we randomly divide the agents into two groups, gather information from one group, and then use the information to guide the item's selling price for the agents in the other group. Since in an optimal solution, each agent does not contribute much to the objective, a constant approximation can be proved by some concentration inequalities based on the above two strategies, we give our mechanism in Algorithm 1.

Theorem 2. *Algorithm 1 is feasible, truthful, and achieves an expected approximation ratio of $\frac{1}{52}$.*

Proof. The feasibility is obvious. Firstly, since $x_i \leq \alpha$ when each agent i comes, $\sum_{i \in [n]} x_i \leq 1$. Secondly, due to $x_i \leq \frac{B_i}{r}$ for each agent i, $p_i = x_i \cdot r \leq B_i$. Thirdly, for each agent i, we have $x_i v_i \geq p_i \tau_i$ because an agent buys some fractions of the item and gets charged only if $r \leq \frac{v_i}{\tau_i}$.

Then we show that regardless of which procedure is executed, the mechanism is truthful. The truthfulness of the first procedure is proved by Theorem 1 directly. For the second procedure, we show that agents in neither S nor R have the incentive to lie. For an agent in S, she will not be assigned anything, and therefore, misreporting her information cannot improve her utility; while for the agents in R, they are also truthtelling because their reported information determines neither the arrival order nor the reserve price, and misreporting a higher $\frac{v_i}{\tau_i}$ (resp. a larger B_i) to buy more fractions of the item must violate the RoS (resp. budget) constraint of agent i.

Finally, we analyze the approximation ratio. Let $(\mathbf{z}^*, \mathbf{p}^*)$ be an optimal solution. Use OPT and ALG to denote the optimal payment and our total payment, respectively. Without loss of generality, we can assume that $p_i^* = x_i^* \cdot \frac{v_i}{\tau_i} \leq B_i$ for any i. Clearly, if there exists an agent l with $p_l^* \geq \frac{1}{36}$OPT, we can easily bound the expected total payment by the first procedure:

$$\mathbb{E}(\text{ALG}) \geq \frac{9}{13} \cdot \min\left\{B_k, \frac{v_k}{\tau_k}\right\} \geq \frac{9}{13} \cdot \min\left\{B_l, \frac{v_l}{\tau_l}\right\} \geq \frac{1}{52}\text{OPT}.$$

Otherwise, we have $p_i^* < \frac{1}{36}$OPT $\forall i \in [n]$. Then according to the concentration lemma proved in [14, Lemma 2], we can establish the relationship between $\sum_{i \in S} p_i^*$ and OPT in the second procedure:

$$\Pr\left[\frac{1}{3}\text{OPT} \leq \sum_{i \in S} p_i^* \leq \frac{2}{3}\text{OPT}\right] \geq \frac{3}{4}. \tag{2}$$

Namely, with probability of at least $3/4$, both $\sum_{i \in S} p_i^*$ and $\sum_{i \in R} p_i^*$ are in $[\frac{1}{3}\text{OPT}, \frac{2}{3}\text{OPT}]$.

Let us focus on the second procedure and consider a subset S such that $\sum_{i \in S} p_i^* \in [\frac{1}{3}\text{OPT}, \frac{2}{3}\text{OPT}]$. We distinguish two cases based on the final remaining fraction of the item. If the item is sold out, our payment is at least $\frac{1}{4}\sum_{i \in S} p_i(\mathbf{z}^S)$. Since $(\mathbf{x}^S, \mathbf{p}(\mathbf{z}^S))$ is the optimal solution of distributing the item among the agents in S, we have

$$\text{ALG} \geq \frac{1}{4}\sum_{i \in S} p_i(\mathbf{z}^S) \geq \frac{1}{4}\sum_{i \in S} p_i^* \geq \frac{1}{12}\text{OPT}.$$

If the procedure does not sell out the item, for any agent $i \in R$ who does not exhaust the budget, $\frac{v_i}{\tau_i} < r = \frac{1}{4}\sum_{i \in S} p_i(\mathbf{z}^S)$. Using $T \subseteq R$ to denote such agents, we have

$$\frac{1}{3}\text{OPT} \leq \sum_{i \in R} p_i^* \leq \sum_{i \in R \setminus T} B_i + \sum_{i \in T} p_i^* \leq \text{ALG} + \sum_{i \in T} \frac{v_i}{\tau_i} x_i^*$$

$$\leq \text{ALG} + \frac{1}{4}\sum_{i \in S} p_i(\mathbf{z}^S) \sum_{i \in T} x_i^* \leq \text{ALG} + \frac{1}{4}\sum_{i \in S} p_i(\mathbf{z}^S)$$

$$\leq \text{ALG} + \frac{1}{4}\text{OPT}.$$

We have $\text{ALG} \geq \frac{1}{12}$OPT from the above inequality.

Thus, in either case, ALG is at least $\frac{1}{12}$OPT under such a subset S. Then according to Eq. (2), we can complete the proof:

$$\mathbb{E}(\text{ALG}) \geq \frac{4}{13} \cdot \frac{3}{4} \cdot \frac{1}{12}\text{OPT} = \frac{1}{52}\text{OPT}.$$

3   Multiple Items Auction for Unit Demand Agents

This section considers the environment where the auctioneer sells multiple items
to unit-demand agents, a set of agents who each desires to buy at most one item.
We extend the results in the last section and show that a constant approximation
can still be obtained. Similar to the study of the single-item auction, Sect. 3.1
starts from the indivisible goods environment and shows a $\frac{1}{2}$-approximation.

For the divisible goods environment, our mechanism is also a random combination of the "indivisibly selling" procedure and the "random sampling" procedure. However, the mechanism and its analysis are much more complicated
than that for single item environment and this section is also the most technical
part of this paper. We describe the indivisibly selling procedure in Algorithm 2.
For the random sampling procedure, the multiple-item setting needs a variant
of greedy matching (Algorithm 3) to compute the reserved prices of each item
and Sect. 3.2 has a discussion about this algorithm. Finally, Sect. 3.3 analyzes
the combined mechanism (Algorithm 5). In order to analyze the approximation
ratio of Algorithm 5, we introduce Algorithm 4, a non-truthful mechanism and
purely in analysis, to bridge Algorithm 5 and Algorithm 3.

3.1   Indivisibly Selling

We first prove the claimed truthful constant approximation in the scenario of
selling indivisible items and then give two corollaries to show the performance
of applying the indivisibly selling idea to distributing divisible items.

Consider the indivisible goods setting. For each agent-item pair (i, j), define
its weight w_{ij} to be the maximum money that we can charge agent i if assigning
item j to her, i.e., $w_{ij} = \min\{B_i, \frac{v_{ij}}{\tau_i}\}$. Since items are indivisible and each
agent only wants to buy at most one item, a feasible solution is essentially a
matching between the agent set and the item set, and the goal is to find a
maximum weighted matching. However, the algorithm to output the maximum
weighted matching is not truthful. We observe that a natural greedy matching
algorithm can return a constant approximation while retaining the truthfulness.
The mechanism is described in Algorithm 2 and the detailed proof is provided
in the full version.

Theorem 3. *Algorithm 2 is feasible, truthful and achieves an approximation
ratio of* $1/2$ *when items are indivisible.*

Consider a feasible solution $\mathbf{z} = \{z_{ij}\}_{i\in[n],j\in[m]}$ (not necessarily truthful) for
multiple **divisible** items auction among unit-demand agents. We assume that
each unit-demand agent i has at most one variable $z_{ij} > 0$, Define $\mathcal{W}_j(\mathbf{z}) :=
\sum_{i:z_{ij}>0} p_i$ to be the total payments related to item j. We observe the following
two corollaries.

Corollary 1. *If solution* $\mathbf{z}$ *is* α-*approximation and for any item* j, $\max_{i\in[n]} \min
\{\frac{v_{ij}}{\tau_i} z_{ij}, B_i\} \geq \beta \cdot \mathcal{W}_j(\mathbf{z})$, *then running Algorithm 2 directly obtains an approxi-
mation ratio of* $\frac{\alpha\beta}{2}$.

Algorithm 2. Indivisibly Selling

Input: The reported budgets $\{B_i\}_{i\in[n]}$, the reported value profile $\{v_{ij}\}_{i\in[n],j\in[m]}$ and the reported target ratios $\{\tau_i\}_{i\in[n]}$.
Output: An allocation and payments.
1: Initially, set allocation $x_{ij} \leftarrow 0 \ \forall i \in [n], j \in [m]$ and payment $p_i \leftarrow 0 \ \forall i \in [n]$.
2: Sort all the agent-item pairs $\{(i,j)\}_{i\in[n],j\in[m]}$ in the decreasing lexicographical order of $\left(\min\{B_i, \frac{v_{ij}}{\tau_i}\}, v_{ij}\right)$ and break the ties in a fixed manner.
3: **for** each agent-item pair (i,j) in the order **do**
4:     If both agent i and item j have not been matched, match them: $x_{ij} \leftarrow 1, p_i \leftarrow \min\{B_i, \frac{v_{ij}}{\tau_i}\}$.
5: **end for**
6: **return** Allocation $\{x_{ij}\}_{i\in[n],j\in[m]}$ and payments $\{p_i\}_{i\in[n]}$.

Corollary 2. *For a constant $\beta \in [0,1]$, define item subset $H(\mathbf{z},\beta) \subseteq [m]$ to be the set of items with $\max_{i\in[n]} \min\{\frac{v_{ij}}{\tau_i} z_{ij}, B_i\} \geq \beta \cdot \mathcal{W}_j(\mathbf{z})$. Running Algorithm 2 directly obtains a total payment at least $\frac{\beta}{2} \sum_{j\in H(\mathbf{z},\beta)} \mathcal{W}_j(\mathbf{z})$ for any $\beta \in [0,1]$.*

3.2 Foundations of Random Sampling

The subsection explores generalizing the random sampling procedure in Algorithm 1 to multiple items auction. We first randomly sample half of the agents and investigate how much revenue can be earned per unit of each item if the auctioneer only sells the items to these sampled agents. Recall that the mechanism does not actually distribute any item to the sampled agents. Then, the auctioneer sets the reserve price of each item based on the investigated revenues and sells them to all the remaining agents. More specifically, let these agents arrive in a fixed order. When an agent arrives, she is allowed to buy any remaining fraction of any item as long as she can afford the reserve price.

It is easy to observe that the mechanism is still truthful according to the same argument in the proof of Theorem 2: for a sampled agent, she will not be assigned anything, and therefore, she does not have any incentive to lie; while for the agents that do not get sampled, they are also truthtelling because neither the arrival order nor the reserve prices are determined by their reported profiles and a fake profile that can improve the agent's obtained value must violate at least one constraint.

The key condition that random sampling can achieve a constant approximation ratio is that the revenue earned by each item among the sampled agents is (w.h.p.) close to its contribution to the objective in an optimal solution or a constant approximation solution; otherwise, there is no reason that the reserve prices are set based on the investigated revenues. Unfortunately, unlike the single-item environment, we cannot guarantee that an optimal solution of the multiple-item auction satisfies this condition. Thus, to obtain such a nice structural property, we present an algorithm based on greedy matching and item supply clipping in Algorithm 3. Note that this algorithm is untruthful, and we only use it to

---

Algorithm 3. Greedy Matching and Item Supply Clipping

Input: The budgets $\{B_i\}_{i\in[n]}$, the value profile $\{v_{ij}\}_{i\in[n],j\in[m]}$ and the target ratios $\{\tau_i\}_{i\in[n]}$.

Output: An allocation and payments.

1: Initially, set allocation $x_{ij} \leftarrow 0$ $\forall i \in [n], j \in [m]$ and payment $p_i \leftarrow 0$ $\forall i \in [n]$.
2: For each agent-item pair $(i,j) \in [n] \times [m]$, define its weight $w_{ij} := \frac{v_{ij}}{\tau_i}$. Sort all the pairs in the decreasing order of their weights and break the ties in a fixed manner.
3: **for** each agent-item pair (i,j) in the order **do**
4:     If agent i has not bought any item and the remaining fraction R_j of item j is more than $1/2$, $x_{ij} \leftarrow \min\{R_j, \frac{B_i}{w_{ij}}\}, p_i \leftarrow w_{ij}x_{ij}$.
5: **end for**
6: **return** Allocation $\{x_{ij}\}_{i\in[n],j\in[m]}$ and payments $\{p_i\}_{i\in[n]}$.

---

simulate the auction among the sampled agents. We first prove that it obtains a constant approximation, and then show several nice structural properties of the algorithm.

Theorem 4. *The approximation ratio of Algorithm 3 is $1/6$.*

Proof. Use $\left(\mathbf{x}^* = \{x_{ij}^*\}, \mathbf{p}^* = \{p_i^*\}\right)$ and $\left(\mathbf{x} = \{x_{ij}\}_{i\in[n],j\in[m]}, \mathbf{p} = \{p_i\}\right)$ to represent the allocations and the payments in an optimal solution and Algorithm 3's solution respectively. We w.l.o.g. assume that $p_i^* = \sum_{j\in[m]} x_{ij}^* w_{ij} \leq B_i$ for any $i \in [n]$. For each item $j \in [m]$, define A_j to be the set of agents who buy some fractions of item j in the optimal solution, i.e., $A_j := \{i \in [n] \mid x_{ij}^* > 0\}$, and then based on $\mathbf{x}$, we partition A_j into three groups:

$$A_j^{(1)} = \{i \in [n] \mid x_{ij} > 0\},$$
$$A_j^{(2)} = \{i \in [n] \mid x_{ij} = 0 \text{ due to } R_j \leq 1/2\},$$
$$A_j^{(3)} = \{i \in [n] \mid x_{ij} = 0 \text{ due to agent } i \text{ has bought another item}\}.$$

Note that if some agent does not buy the item j in $\mathbf{x}$ due to both of the two reasons, we add the agent into an arbitrary one of $A_j^{(2)}$ and $A_j^{(3)}$.

Use OPT and ALG to denote the objective values of the optimal solution and our solution, respectively. Based on the partition mentioned above, we split the optimal objective into three parts:

$$\text{OPT} = \sum_{i\in[n],j\in[m]} x_{ij}^* w_{ij} = \sum_{j\in[m]} \sum_{i\in A_j^{(1)}} x_{ij}^* w_{ij} + \sum_{j\in[m]} \sum_{i\in A_j^{(2)}} x_{ij}^* w_{ij} + \sum_{j\in[m]} \sum_{i\in A_j^{(3)}} x_{ij}^* w_{ij}.$$

In the following, we analyze the three parts one by one and show that each part is at most twice ALG, which implies that ALG is $1/6$ approximation.

Due to the definition of $A_j^{(1)}$, for each (i,j) pair in the first part, Algorithm 3 assigns some fractions of item j to agent i, and therefore, $x_{ij} \geq \min\{\frac{1}{2}, \frac{B_i}{w_{ij}}\}$. Since $x_{ij}^* \leq 1$ and we assume w.l.o.g. that $x_{ij}^* \leq \frac{B_i}{w_{ij}}$, we have

$$\sum_{j\in[m]}\sum_{i\in A_j^{(1)}} x_{ij}^* w_{ij} \le \sum_{j\in[m]}\sum_{i\in A_j^{(1)}} \min\{w_{ij}, B_i\} \le \sum_{j\in[m]}\sum_{i\in A_j^{(1)}} 2w_{ij} \min\{\frac{1}{2}, \frac{B_i}{w_{ij}}\}$$

$$\le \sum_{j\in[m]}\sum_{i\in A_j^{(1)}} 2x_{ij}w_{ij} \le 2\text{ALG}.$$

$$(3)$$

For each item j with non-empty $A_j^{(2)}$, Algorithm 3 must sell at least half of the item, and then due to the greedy property of the algorithm, we have

$$\sum_{i\in A_j^{(2)}} x_{ij}^* w_{ij} \le 2\mathcal{W}_j(\mathbf{x}),$$

recalling that $\mathcal{W}_j(\mathbf{x}) = \sum_{i:x_{ij}>0} p_i$. Thus,

$$\sum_{j\in[m]}\sum_{i\in A_j^{(2)}} x_{ij}^* w_{ij} \le \sum_{j\in[m]} 2\mathcal{W}_j(\mathbf{x}) \le 2\text{ALG}. \tag{4}$$

Finally, for each item j and agent $i \in A_j^{(3)}$, suppose that agent i buys some fractions of item j' in solution $\mathbf{x}$. Due to the greedy property, $w_{ij} \le w_{ij'}$. Hence,

$$x_{ij}^* w_{ij} \le \min\{B_i, w_{ij'}\} \le 2\min\{\frac{B_i}{w_{ij;}}, \frac{1}{2}\}w_{ij'} \le 2x_{ij'}w_{ij'} = 2p_i.$$

Summing over these (i,j) pairs,

$$\sum_{j\in[m]}\sum_{i\in A_j^{(3)}} x_{ij}^* w_{ij} = \sum_{i\in[n]}\sum_{j:i\in A_j^{(3)}} x_{ij}^* w_{ij} \le \sum_{i\in[n]} 2p_i = 2\text{ALG}. \tag{5}$$

Combining Eq. (3), Eq. (4) and Eq. (5) completes the proof.

Remark 1. Note that the item supply clipping parameter $1/2$ in Algorithm 3 can be replaced by any other constant in $(0,1)$. By setting this parameter to be $\frac{\sqrt{2}}{1+\sqrt{2}}$, the algorithm can get an approximation ratio of $3 + 2\sqrt{2}$.

For an agent subset $S \subseteq [n]$, use $(\mathbf{x}^S, \mathbf{p}^S)$ to denote the allocation and the payments if using Algorithm 3 to distribute all the items to agents in S. We claim the following two properties and a corollary. The proof is omitted due to space limits.

Lemma 1. *For any agent subset $S \subseteq [n]$, we have*

- *agent payment monotonicity: $p_i^S \ge \frac{1}{2}p_i, \forall i \in S$.*
- *selling revenue monotonicity: $\mathcal{W}_j(\mathbf{x}^S) \le 2\mathcal{W}_j(\mathbf{x}), \forall j \in [m]$.*

Corollary 3. *Randomly dividing all the agents with equal probability into set S and R, we have*

$$\mathbb{E}\left(\sum_{j\in[m]} \mathcal{W}_j(\mathbf{x}^S)\right) = \mathbb{E}\left(\sum_{i\in S} p_i^S\right) \ge \frac{1}{2}\mathbb{E}\left(\sum_{i\in S} p_i\right) = \frac{1}{4}\sum_{i\in[m]} p_i \ge \frac{1}{4}\sum_{j\in[m]} \mathcal{W}_j(\mathbf{x}).$$

3.3   Final Mechanism

Algorithm 4. (*Auxiliary*) Multiple Items Auction for Unit Demand Agents

---

Input: The reported budgets $\{B_i\}_{i\in[n]}$, the reported value profile $\{v_{ij}\}_{i\in[n],j\in[m]}$ and the reported target ratios $\{\tau_i\}_{i\in[n]}$.

Output: An allocation and payments.

1: Initially, set allocation $x_{ij} \leftarrow 0 \ \forall i \in [n], j \in [m]$ and payment $p_i \leftarrow 0 \ \forall i \in [n]$.
2: With probability of $45/47$, run Algorithm 2.   ▷ Indivisibly Selling Procedure
3: With probability of $2/47$, **begin**   ▷ Random Sampling Procedure
4:     Run Algorithm 3 and use $(\mathbf{z} = \{z_{ij}\}, \mathbf{p}(\mathbf{z}) = \{p_i(\mathbf{z})\})$ to denote the output.
5:     Randomly divide all the agents with equal probability into set S and R.
6:     Run Algorithm 3 on the sampled set S and use $(\mathbf{z}^S = \{z_{ij}^S\}, \mathbf{p}(\mathbf{z}^S) = \{p_i(\mathbf{z}^S)\})$
       to denote the output.
7:     For each item j, set the reserve price $r_j \leftarrow \frac{1}{12}\mathcal{W}_j(\mathbf{z}^S)$.
8:     Let agents in R come in an arbitrarily fixed order, and when each agent i comes,
       find the unique item $k(i)$ that she buys in solution $\mathbf{z}$ and use $R_{k(i)}$ to denote the
       remaining fraction of the item. Set $x_{i,k(i)} \leftarrow \min\{R_{k(i)}, \frac{B_i}{r_{k(i)}}\}, p_i \leftarrow r_{k(i)}x_{i,k(i)}$ if
       $r_{k(i)} \leq w_{i,k(i)}$.
9: **end**
10: **return** Allocation $\{x_{ij}\}_{i\in[n],j\in[m]}$ and payments $\{p_i\}_{i\in[n]}$.

This subsection states the final mechanism, which is a random combination of the indivisibly selling idea and the random sampling idea. To streamline the analysis, we first introduce an *auxiliary mechanism*, which is constant-approximate but not truthful, and then show it can be altered to a truthful mechanism by losing only a constant factor on the approximation ratio.

Theorem 5. *Algorithm 4 obtains a constant approximation ratio.*

Recollect that $H(\mathbf{z}, \beta) := \{j \in [m] \mid \max_{i\in[n]} \min\{\frac{v_{ij}}{\tau_i}z_{ij}, B_i\} \geq \beta \cdot \mathcal{W}_j(\mathbf{z})\}$ defined in Corollary 2. To prove Theorem 5, we partition all the items into two sets: $H(\mathbf{z}, \frac{1}{144})$ and $\bar{H}(\mathbf{z}, \frac{1}{144}) = [m] \setminus H(\mathbf{z}, \frac{1}{144})$. Corollary 2 directly implies that the first procedure (Algorithm 2) guarantees our objective value is at least a constant fraction of $\sum_{j \in H(\mathbf{z}, \frac{1}{144})} \mathcal{W}_j(\mathbf{z})$.

Lemma 2. *The revenue obtained by the first procedure in Algorithm 4 is at least* $\frac{1}{288} \sum_{j \in H(\mathbf{z}, \frac{1}{144})} \mathcal{W}_j(\mathbf{z})$.

For the second procedure, we need to show that $\sum_{j \in \bar{H}(\mathbf{z}, \frac{1}{144})} \mathcal{W}_j(\mathbf{z})$ can be bounded by the total payment obtained by this procedure. The technical lemma is stated below and its detailed proof is deferred to the full version of this paper.

Lemma 3. *The expected revenue obtained by the second procedure in Algorithm 4 is at least*

$$\frac{1}{192} \sum_{j \in \bar{H}(\mathbf{z}, \frac{1}{144})} \mathcal{W}_j(\mathbf{z}) - \frac{7}{96} \sum_{j \in H(\mathbf{z}, \frac{1}{144})} \mathcal{W}_j(\mathbf{z}).$$

Proof (Proof of Theorem 5). Combing Lemma 2, Lemma 3 and the probabilities set in Algorithm 4,

$$\mathbb{E}(\text{ALG}) \geq \frac{45}{47} \cdot \frac{1}{288} \sum_{j \in H(\mathbf{z}, \frac{1}{144})} \mathcal{W}_j(\mathbf{z})$$

$$+ \frac{2}{47} \cdot \left(\frac{1}{192} \sum_{j \in \bar{H}(\mathbf{z}, \frac{1}{144})} \mathcal{W}_j(\mathbf{z}) - \frac{7}{96} \sum_{j \in H(\mathbf{z}, \frac{1}{144})} \mathcal{W}_j(\mathbf{z}) \right)$$

$$= \frac{1}{4512} \sum_{j \in [m]} \mathcal{W}_j(\mathbf{z}) \geq \frac{1}{27072} \text{OPT},$$

where the last inequality used Theorem 4.

Algorithm 5. Multiple Items Auction for Unit Demand Agents

Input: The reported budgets $\{B_i\}_{i \in [n]}$, the reported value profile $\{v_{ij}\}_{i \in [n], j \in [m]}$ and the reported target ratios $\{\tau_i\}_{i \in [n]}$.

Output: An allocation and payments.

1: Initially, set allocation $x_{ij} \leftarrow 0 \ \forall i \in [n], j \in [m]$ and payment $p_i \leftarrow 0 \ \forall i \in [n]$.
2: With probability of 45/53, run Algorithm 2.   ▷ Indivisibly Selling Procedure
3: With probability of 8/53, **begin**   ▷ Random Sampling Procedure
4:    Randomly divide all the agents with equal probability into set S and R.
5:    Run Algorithm 3 on the sampled agent set S and use $\left(\mathbf{z}^S = \{z_{ij}^S\}, \mathbf{p}(\mathbf{z}^S) = \{p_i(\mathbf{z}^S)\} \right)$ to denote the output.
6:    For each item j, set the reserve price $r_j \leftarrow \frac{1}{12} \mathcal{W}_j(\mathbf{z}^S)$.
7:    Let agents in R come in an arbitrarily fixed order. When each agent i comes, use R_j to denote the remaining fraction of each item j, and let her pick the most profitable item

$$k(i) := \arg \max_{j : r_j \leq \frac{v_{ij}}{\tau_i}} v_{ij} \cdot \min\{\frac{B_i}{r_j}, R_j\}.$$

8:    Set $x_{i,k(i)} \leftarrow \min\{\frac{B_i}{r_{k(i)}}, R_{k(i)}\}, p_i \leftarrow \min\{w_{i,k(i)} x_{i,k(i)}, B_i\}$ if item $k(i)$ exists.
9: **end**
10: **return** Allocation $\{x_{ij}\}_{i \in [n], j \in [m]}$ and payments $\{p_i\}_{i \in [n]}$.

Finally, we present our final mechanism in Algorithm 5. The only difference from Algorithm 4 is that in the last step of the second procedure, we let the agent choose any item she wants as long as she can afford the reserve price, and then charge her the maximum willingness-to-pay.

Theorem 6. *Algorithm 5 is feasible, truthful, and constant-approximate.*

Proof. According to Theorem 3, the first procedure is feasible and truthful. For the second procedure, the mechanism is truthful since any agent is charged her maximum willingness-to-pay. Then according to the same argument in the proof of Theorem 2, we can prove the truthfulness.

The following focuses on analyzing the approximation ratio. To this end, we couple the randomness in Algorithm 5 and Algorithm 4. The two algorithms are almost identical to each other except for one line and their randomness can be coupled perfectly. If by the coupling of randomness, both algorithms execute the first procedure, they are exactly identical and thus Lemma 2 also apples to Algorithm 5. Now, by randomness, they both execute the second procedure. In the second procedure, we can further couple the randomness so that they randomly sample the same set S. Conditional on all these (they both execute the second procedure and sample the same set S), we prove that the revenue of Algorithm 5 is at least $\frac{1}{4}$ of that of Algorithm 4.

Let $(\mathbf{x}, \mathbf{p})$ and $(\mathbf{x}', \mathbf{p}')$ be the two solutions respectively of Algorithm 5 and Algorithm 4 under the above conditions. Let ALG and ALG$'$ be their revenues respectively. Use A'_j to denote the agents who buy some fractions of item j in solution $\mathbf{x}'$. According to Algorithm 4,

$$\text{ALG}' = \sum_{j \in [m]} \sum_{i \in A'_j} x'_{ij} r_j.$$

For an item j, if the corresponding revenue in Algorithm 5's solution $\mathcal{W}_j(\mathbf{x}) \geq \frac{1}{2} r_j$, we have $\sum_{i \in A'_j} x'_{ij} r_j \leq 2\mathcal{W}_j(\mathbf{x})$, and then summing over all such items,

$$\sum_{j:\mathcal{W}_j(\mathbf{x}) \geq \frac{1}{2} r_j} \sum_{i \in A'_j} x'_{ij} r_j \leq \sum_{j:\mathcal{W}_j(\mathbf{x}) \geq \frac{1}{2} r_j} 2\mathcal{W}_j(\mathbf{x}) \leq 2\text{ALG}. \tag{6}$$

For each item j with $\mathcal{W}_j(\mathbf{x}) < \frac{1}{2} r_j$, we distinguish three cases for agents in A'_j based on $(\mathbf{x}, \mathbf{p})$: (1) $p_i = B_i$, (2) $p_i < B_i$ and $x_{ij} > 0$, and (3) $p_i < B_i$ and $x_{ij} = 0$.

For case (1), clearly,

$$x'_{ij} r_j \leq B_i = p_i. \tag{7}$$

For case (2), since $\mathcal{W}_j(\mathbf{x}) < \frac{1}{2} r_j$, the remaining fraction of item j is at least $1/2$ when Algorithm 5 let agent i buy, and therefore, $x_{ij} \geq \min\{\frac{1}{2}, \frac{B_i}{r_j}\}$. According to $p_i < B_i$, we have $p_i = w_{ij} x_{ij} \geq r_j \min\{\frac{1}{2}, \frac{B_i}{r_j}\}$. Then, due to $x'_{ij} \leq \min\{1, \frac{B_i}{r_j}\}$,

$$x'_{ij} r_j \leq 2p_i. \tag{8}$$

For case (3), suppose that agent i buys item k in solution $\mathbf{x}$. Since the remaining fraction of item j is at least $1/2$ and the agent always pick the most profitable part in Algorithm 5, we have

$$\min\left\{\frac{1}{2}, \frac{B_i}{r_j}\right\} \cdot v_{ij} \leq x_{ik} v_{ik} \text{ and } \min\left\{\frac{1}{2}, \frac{B_i}{r_j}\right\} \cdot w_{ij} \leq x_{ik} w_{ik}.$$

Again, due to $p_i < B_i$, $r_j \le w_{ij}$ and $x'_{ij} \le \min\{1, \frac{B_i}{r_j}\}$, we have

$$\frac{1}{2}x'_{ij}r_j \le \min\left\{\frac{1}{2}, \frac{B_i}{r_j}\right\} \cdot w_{ij} \le x_{ik}w_{ik} = p_i. \tag{9}$$

Due to Eq. (7), Eq. (8) and Eq. (9), for an item with $\mathcal{W}_j(\mathbf{x}) < \frac{1}{2}r_j$, in either case, we always have $x'_{ij}r_j \le 2p_i$. Thus, summing over all such items and the corresponding agents,

$$\sum_{j:\mathcal{W}_j(\mathbf{x}) < \frac{1}{2}r_j} \sum_{i \in A'_j} x'_{ij}r_j \le \sum_{j:\mathcal{W}_j(\mathbf{x}) < \frac{1}{2}r_j} \sum_{i \in A'_j} 2p_i \le 2\text{ALG}. \tag{10}$$

Combining Eq. (6) and Eq. (10) proves ALG$' \le 4$ALG. Combining this with Lemma 3, we know that The expected revenue obtained by the second procedure in Algorithm 5 is at least

$$\frac{1}{768} \sum_{j \in \bar{H}(\mathbf{z}, \frac{1}{144})} \mathcal{W}_j(\mathbf{z}) - \frac{7}{384} \sum_{j \in H(\mathbf{z}, \frac{1}{144})} \mathcal{W}_j(\mathbf{z}).$$

Further combining with Lemma 2, which we argued also applies to Algorithm 5, we have the expected revenue obtained by Algorithm 5 is at least

$$\frac{45}{53} \cdot \frac{1}{288} \sum_{j \in H(\mathbf{z}, \frac{1}{144})} \mathcal{W}_j(\mathbf{z}) + \frac{8}{53} \cdot \left(\frac{1}{768} \sum_{j \in \bar{H}(\mathbf{z}, \frac{1}{144})} \mathcal{W}_j(\mathbf{z}) - \frac{7}{384} \sum_{j \in H(\mathbf{z}, \frac{1}{144})} \mathcal{W}_j(\mathbf{z})\right)$$

$$= \frac{1}{5088} \sum_{j \in [m]} \mathcal{W}_j(\mathbf{z}) \ge \frac{1}{30528}\text{OPT}.$$

Remark 2. In the proof of Theorem 6, we have not tried to optimize the constants in our analysis in the interests of expositional simplicity. The parameters (e.g. $45/47$ and $1/144$) in our algorithm and analysis can be easily replaced by some other constants in $(0, 1)$ to obtain another constant approximation ratio.

4   Conclusion and Open Problems

We investigate the emerging value maximizer in recent literature but also significantly depart from their modeling. We believe that the model and benchmark proposed in this paper are, on the one hand, more realistic and, on the other hand, friendlier to the AGT community. We get a few non-trivial positive results which indicate that this model and benchmark is indeed tractable. There are also many more open questions left. For additive valuation, it is open if we can get a constant approximation. It is interesting to design a mechanism with a better approximation for the setting of the single item and unit demand since our current ratio is fairly large. We also want to point out that no lower bound is obtained in this model, and thus any non-trivial lower bound is interesting. We get a much better approximation ratio for the single-item environment when valuation and budget are public than in the fully private setting. However, this is not a separation since we have no lower bound. Any separation result for different information models in terms of public and private is interesting.

Acknowledgements. Chenyang Xu and Pinyan Lu were supported in part by Science and Technology Innovation 2030 - "The Next Generation of Artificial Intelligence" Major Project No. 2018AAA0100900. Additionally, Chenyang Xu received support from the National Natural Science Foundation of China (Grant No. 62302166), and the Dean's Fund of Shanghai Key Laboratory of Trustworthy Computing, East China Normal University. Ruilong Zhang was supported by NSF grant CCF-1844890.

References

1. Aggarwal, G., Badanidiyuru, A., Mehta, A.: Autobidding with constraints. In: Caragiannis, I., Mirrokni, V., Nikolova, E. (eds.) WINE 2019. LNCS, vol. 11920, pp. 17–30. Springer, Cham (2019). https://doi.org/10.1007/978-3-030-35389-6_2
2. Babaioff, M., Cole, R., Hartline, J.D., Immorlica, N., Lucier, B.: Non-quasi-linear agents in quasi-linear mechanisms (extended abstract). In: ITCS. LIPIcs, vol. 185, pp. 84:1–84:1. Schloss Dagstuhl - Leibniz-Zentrum für Informatik (2021)
3. Baisa, B.: Auction design without quasilinear preferences. Theor. Econ. **12**(1), 53–78 (2017)
4. Balseiro, S.R., Deng, Y., Mao, J., Mirrokni, V.S., Zuo, S.: Robust auction design in the auto-bidding world. In: NeurIPS, pp. 17777–17788 (2021)
5. Balseiro, S.R., Deng, Y., Mao, J., Mirrokni, V.S., Zuo, S.: Optimal mechanisms for value maximizers with budget constraints via target clipping. In: EC, p. 475. ACM (2022)
6. Balseiro, S.R., Kim, A., Mahdian, M., Mirrokni, V.S.: Budget-constrained incentive compatibility for stationary mechanisms. In: EC, pp. 607–608. ACM (2020)
7. Bei, X., Chen, N., Gravin, N., Lu, P.: Budget feasible mechanism design: from prior-free to Bayesian. In: STOC, pp. 449–458. ACM (2012)
8. Borgs, C., Chayes, J.T., Immorlica, N., Mahdian, M., Saberi, A.: Multi-unit auctions with budget-constrained bidders. In: EC, pp. 44–51. ACM (2005)
9. Caragiannis, I., Voudouris, A.A.: The efficiency of resource allocation mechanisms for budget-constrained users. Math. Oper. Res. **46**(2), 503–523 (2021)
10. Cavallo, R., Krishnamurthy, P., Sviridenko, M., Wilkens, C.A.: Sponsored search auctions with rich ads. In: WWW, pp. 43–51. ACM (2017)
11. Chawla, S., Malec, D.L., Malekian, A.: Bayesian mechanism design for budget-constrained agents. In: EC, pp. 253–262. ACM (2011)
12. Che, Y., Gale, I.L.: The optimal mechanism for selling to a budget-constrained buyer. J. Econ. Theory **92**(2), 198–233 (2000)
13. Chen, N., Gravin, N., Lu, P.: On the approximability of budget feasible mechanisms. In: SODA, pp. 685–699. SIAM (2011)
14. Chen, N., Gravin, N., Lu, P.: Truthful generalized assignments via stable matching. Math. Oper. Res. **39**(3), 722–736 (2014)
15. Deng, Y., Mao, J., Mirrokni, V.S., Zuo, S.: Towards efficient auctions in an auto-bidding world. In: WWW, pp. 3965–3973. ACM/IW3C2 (2021)

16. Devanur, N.R., Weinberg, S.M.: The optimal mechanism for selling to a budget constrained buyer: The general case. In: EC, pp. 39–40. ACM (2017)
17. Dobzinski, S., Lavi, R., Nisan, N.: Multi-unit auctions with budget limits. In: FOCS, pp. 260–269. IEEE Computer Society (2008)
18. Dobzinski, S., Leme, R.P.: Efficiency guarantees in auctions with budgets. In: Esparza, J., Fraigniaud, P., Husfeldt, T., Koutsoupias, E. (eds.) ICALP 2014. LNCS, vol. 8572, pp. 392–404. Springer, Heidelberg (2014). https://doi.org/10.1007/978-3-662-43948-7_33
19. Jalaly, P., Tardos, É.: Simple and efficient budget feasible mechanisms for monotone submodular valuations. ACM Trans. Econ. Comput. 9(1), 4:1–4:20 (2021)
20. Leonardi, S., Monaco, G., Sankowski, P., Zhang, Q.: Budget feasible mechanisms on matroids. Algorithmica 83(5), 1222–1237 (2021)
21. Lu, P., Xiao, T.: Improved efficiency guarantees in auctions with budgets. In: EC, pp. 397–413. ACM (2015)
22. Lu, P., Xiao, T.: Liquid welfare maximization in auctions with multiple items. In: Bilò, V., Flammini, M. (eds.) SAGT 2017. LNCS, vol. 10504, pp. 41–52. Springer, Cham (2017). https://doi.org/10.1007/978-3-319-66700-3_4
23. Lv, H., et al.: Utility maximizer or value maximizer: mechanism design for mixed bidders in online advertising. In: AAAI 2023 (2023)
24. Myerson, R.B.: Optimal auction design. Math. Oper. Res. 6(1), 58–73 (1981)
25. Nisan, N., Ronen, A.: Algorithmic mechanism design. Games Econ. Behav. 35(1–2), 166–196 (2001)
26. Nisan, N., Roughgarden, T., Tardos, É., Vazirani, V.V. (eds.): Algorithmic Game Theory. Cambridge University Press, Cambridge (2007)
27. Pai, M.M., Vohra, R.: Optimal auctions with financially constrained buyers. J. Econ. Theory 150, 383–425 (2014)
28. Singer, Y.: Budget feasible mechanisms. In: FOCS, pp. 765–774. IEEE Computer Society (2010)
29. Zhou, Y., Serizawa, S.: Multi-object auction design beyond quasi-linearity: leading examples. Technical report, ISER Discussion Paper (2021)

Auction Design for Bidders with Ex Post ROI Constraints

Hongtao Lv[1,3], Xiaohui Bei[2], Zhenzhe Zheng[3(✉)], and Fan Wu[3]

[1] School of Software and Joint SDU-NTU Centre for Artificial Intelligence Research (C-FAIR), Shandong University, Jinan, China
`lht@sdu.edu.cn`
[2] School of Physical and Mathematical Sciences, Nanyang Technological University, Singapore, Singapore
`xhbei@ntu.edu.sg`
[3] Department of Computer Science and Engineering, Shanghai Jiao Tong University, Shanghai, China
`{zhengzhenzhe,wu-fan}@sjtu.edu.cn`

Abstract. Motivated by practical constraints in online advertising, we investigate single-parameter auction design for bidders with constraints on their Return On Investment (ROI) – a targeted minimum ratio between the obtained value and the payment. We focus on *ex post* ROI constraints, which require the ROI condition to be satisfied for every realized value profile. With ROI-constrained bidders, we first provide a full characterization of the allocation and payment rules of dominant-strategy incentive compatible (DSIC) auctions. In particular, we show that given any monotone allocation rule, the corresponding DSIC payment should be the Myerson payment with a *rebate* for each bidder to meet their ROI constraints. Furthermore, we also determine the optimal auction structure when the item is sold to a single bidder under a mild regularity condition. This structure entails a randomized allocation scheme and a first-price payment rule, which differs from the deterministic Myerson auction and previous works on ex ante ROI constraints.

Keywords: Return on investment (ROI) · Mechanism design · Myerson auction

1 Introduction

Online advertising auctions are a vital source of revenue for many IT companies, generating billions of dollars of revenue annually. In recent years, with tens of millions of ad auctions being conducted in real-time each day, this large-scale and complex market has prompted modern online advertising platforms to develop auto-bidding services, which allow the advertisers to set up high-level marketing goals for their ad campaigns and then bid on behalf of the advertisers.

In these auto-bidding scenarios, advertisers' financial constraints such as budget and return on investment (ROI) constraints have become critical in auction

J. Garg et al. (Eds.): WINE 2023, LNCS 14413, pp. 492–508, 2024.
https://doi.org/10.1007/978-3-031-48974-7_28

design. While auctions for budget-constrained bidders have been extensively studied in the literature [8,19,22], research on auction design for bidders with ROI constraints is still in its nascent stage. The ROI constraints of advertisers require that the payment cannot be more than a certain fraction of the obtained advertising value. In other words, there is a targeted minimum ratio between the obtained value and the payment for an ROI-constrained bidder. Unlike budget constraints which set a hard limit on payment, ROI constraints establish a payment limit that is linearly related to the allocated value. Previous studies [2,16] have demonstrated that ROI constraints align better with real-world empirical evidence than budget constraints, and it is the aim of this paper to explore how to design auctions with good incentive and revenue guarantees for ROI-constrained bidders.

The existing literature on auction design for ROI-constrained bidders primarily focuses on *ex ante* ROI constraints, which requires an *expected* ROI with respect to the prior value distributions of bidders [6,16]. This approach is suitable for advertisers who participate in a large number of auctions daily and are only concerned with their average spend per unit of value. However, in reality, most ad campaigns experience the "long-tail phenomenon" [9], which means they only receive dozens of or fewer clicks per day. Under these conditions, an auction with ex ante ROI guarantees may have a non-negligible probability of violating the ROI constraints of these ad campaigns over a day. Due to these reasons, in this work, we focus on the *ex post* or *hard* ROI constraints, which ensures that the auction respects the ROI constraints of bidders for any realized value profile. This is a stronger requirement compared to ex ante ROI constraints and addresses the limitation of current auction design methods.

1.1 Our Results

In this work, we examine the design of truthful and optimal auction design for ex post ROI-constrained bidders. We inherit the setting from the classic single-parameter mechanism design and consider the values of bidders as private information and the targeted ROIs as public information. In this single-parameter environment, the ROI constraints can be integrated into the objective function (see Sect. 2 for details), resulting in a transformed utility model: $u_i = M_i v_i x_i - p_i$, where u_i represents the utility of bidder i, v_i is the value, x_i is the allocation quantity, and p_i is the payment. Here $M_i > 1$ is the targeted ROI ratio, which differentiates this model from the classical quasilinear utility model.

We first study the characterizations of truthful auctions with ROI-constrained bidders. Compared to Myerson's characterization of truthful auctions in the single-parameter environment, we show that the monotonicity requirement of the allocation rule remains true for ex post ROI-constrained bidders, but the unique payment rule in [24] should be modified by subtracting a max term, which can be interpreted as a "rebate" equal to the largest "violation" of the Myerson payment to the ROI constraint for all lower valuations. This is a full characterization that completely describes all truthful auctions with ROI-constrained bidders. This result can be proved using similar techniques from

Myerson's analysis. It can also be derived from the following alternative interpretation of the payment rule: note that the ROI-constrained bidder assigns a weight $M_i > 1$ to her obtained value $v_i x_i$ from the allocation, but not to her payment. To not violate the individual rationality (IR) condition, instead of applying a naive approach that charges the bidder M_i times the Myerson payment, we must iteratively apply the Myerson payment increment (multiplied by M_i) in small intervals and truncate the payment at the obtained value whenever necessary.

Next, we turn our focus to the optimal (*i.e.* revenue-maximizing) auction design. The additional max term in our payment rule poses a significant challenge to the optimal auction design, since it is unclear how this term can be incorporated into a modified virtual valuation function as seen in previous literature. Instead, we concentrate on the case of selling a single item to a single bidder. Our main result suggests that under a mild regularity assumption known as *decreasing marginal revenue* (DMR)[1], the optimal auction for selling to a single ex post ROI-constrained bidder employs a randomized scheme. More specifically, the allocation rule $x(\cdot)$ starts with a *first-price* interval, where the payment always matches the obtained value, until it reaches the highest allocation and $x(\cdot)$ becomes constant thereafter. This finding is in contrast to the classic Myerson auction [24] and previous results for bidders with ex ante ROI constraints, where the optimal auctions are always deterministic. It implies that similar to much literature on optimal mechanism design for multi-parameter settings, a slight generalization, such as the inclusion of the M_i term, in the single-parameter setting can lead to randomized optimal auctions.

1.2   Related Work

There are two main threads of studies of auctions with ROI-constrained bidders. The first thread investigates how the bidding strategies of the bidders are affected by the ROI constraints in classic VCG or generalized second price (GSP) auctions [1,3,7,15,18,26,27]. The second thread, which our paper follows, focuses on the *design* of auctions with ROI-constrained bidders, which is of practical interest to many online advertising platforms [5,6,10,15,16,20,21,23,28]. In this line of study, the most related work to ours is [16], which showed empirically that a fraction of the buyers in online advertising are indeed ROI-constrained. They also took the first step towards revenue-maximizing auction design for bidders with *ex ante* ROI constraints. Note that ex ante ROI constraints only require the ROI conditions to be met in expectation and are strictly weaker than ex post ROI constraints. Another recent work [6] considered the scenario where either the value or the ROI constraint is private information of the bidder. They used a similar utility function as ours, but still focus on the concept of ex ante ROI.

---

[1] DMR requires the marginal revenue, $vf(v) + F(v) - 1$ to be non-decreasing in the *value space*. This is different from the usual definition of regularity which requires the same monotonicity but in the *quantile space*. Please see the related work section for a more detailed discussion of their differences and more related works.

Unlike these works, we concentrate on *ex post* ROI constraints, which provide a hard ROI guarantee for bidders in every possible value realization. In [4,12], the authors considered ex post ROI constraints in multiple stages, and assumed that each bidder maintains a fixed bid multiplier among stages, which leads to completely different problems from ours.

One particular line of research focuses on requirements of truthfulness for ex post ROI constraints. Cavallo *et al.* studied the same utility function as ours in [10, Appendix A], and investigated the corresponding payment rules. The main difference is that, they limited their focus on *deterministic* mechanisms for bidders with *identical* ROI constraints, while we consider a more general single-parameter setting in the randomized mechanism domain. Li *et al.* [20] proposed a condition on truthfulness of the ROI information, based on which they provided a mechanism framework using tools from control theory. They took the ROI constraints as private information, instead of the value, which leads to a substantially different problem from ours.

The DMR assumption used in our optimal auction characterization has been widely discussed in the literature. It means that the function $\psi(v) \triangleq vf(v) + F(v) - 1$ is non-decreasing, or equivalently $v \cdot (1 - F(v))$, which is the expected revenue of selling the item at price p, is concave, and this is where the name of this condition comes from. Intuitively, many commonly used distribution functions satisfy this assumption, *e.g.*, uniform distributions, and any distribution of finite support and monotone non-decreasing density. The DMR condition was first proposed in [11] for bidders with budget constraints. In [14], the authors found that the DMR condition is more natural in their setting than the traditionally used notion of *regularity* [24], since DMR precisely removes the requirement of ironing in the *value* space, instead of in the *quantile* space as in [24]. In [13], DMR was discussed comprehensively, and the authors showed that the optimal mechanism is deterministic under the DMR condition in a multi-unit setting with private demands. We refer the reader to their work for concrete examples and more discussion.

2   Preliminaries

We consider a general single-parameter auction environment, which consists of a seller and n bidders $N = \{1, 2, ..., n\}$. Each bidder i has a private valuation t_i per unit of the good. We represent x_i as the quantity of the allocated good to bidder i and p_i as the payment of bidder i. Without loss of generality, we assume the maximum possible allocation is $x_i^{\max} = 1$ and the good is indivisible, that is, x_i denotes the probability of bidder i receiving the good. Besides the allocated value, each bidder also has a return on investment (ROI) constraint M_i, as public information[2], which specifies the minimum targeted ratio between her obtained value and the payment. We assume $1 < M_i < +\infty$ in this work. We note that the ROI constraint is considered in an ex post measure, *i.e.*, it

---

[2] This setting is practical and prevalent in practice, *e.g.*, in online advertising, the targeted ROI typically remains the same over a certain period.

requires that $\frac{t_i x_i}{p_i} \geq M_i$ strictly holds in the outcome of every auction instance. Note that the same model is also adopted in [10, appendix A].

With the above definitions, the utility of bidder i is given by

$$u_i = \begin{cases} t_i x_i - p_i & \text{if } \frac{t_i x_i}{p_i} \geq M_i \\ -\infty & \text{otherwise.} \end{cases} \tag{1}$$

It is worth noting that this is the standard quasilinear utility model with the addition of the ROI constraint. We can further define

$$v_i = \frac{t_i}{M_i},$$

which could be interpreted as the maximum willingness-to-pay of the bidder i per unit of the good. Then, we can rewrite the utility function as

$$u_i = \begin{cases} M_i v_i x_i - p_i & \text{if } v_i x_i \geq p_i \\ -\infty & \text{otherwise.} \end{cases} \tag{2}$$

One can observe that, as M_i is a public constant, v_i and t_i are completely interchangeable. To avoid confusion, we use the term *value* to represent v_i, and *initial value* to represent t_i in the following discussion. Each value v_i is independently drawn from a probability distribution $F_i : [0, v_{\max}] \rightarrow [0,1]$, with a continuous probability density function f_i. While the distributions F_i's are common knowledge, the exact value v_i is known only to the bidder i. We denote $\mathbf{v}$ as the value profile of all bidders, and $\mathbf{v}_{-i}$ as that of all bidders except bidder i.

In an auction, each bidder reports her value as b_i, which is not necessarily equal to v_i. We define $\mathbf{b}$ and $\mathbf{b}_{-i}$ similarly as the notations of $\mathbf{v}$ and $\mathbf{v}_{-i}$. Based on the reported bids, an auction mechanism consists of an allocation rule $x_i(b_i, \mathbf{b}_{-i})$, mapping the bid profile to the allocated quantity to each bidder i, and a payment rule $p_i(b_i, \mathbf{b}_{-i})$, mapping the bid profile to the payment for each bidder. When clear from context, we will omit $\mathbf{b}_{-i}$ in the mappings. We also use $u_i(b_i, v_i)$ to represent the utility of bidder i who has value v_i and bid b_i. In the following discussion, we assume that the allocation rule $x_i(\cdot)$ is always right-differentiable, and there are finite non-differentiable points. When $x_i(\cdot)$ is non-continuous at v, let $x_i(v) = \lim_{z \rightarrow v+} x_i(z)$.

We are interested in auctions that are *dominant-strategy incentive compatible* (DSIC) and *individually rational* (IR).

Definition 1 (Dominant-Strategy Incentive Compatibility, DSIC). *A mechanism is dominant-strategy incentive compatible if and only if*

$$u_i(v_i, v_i) \geq u_i(b_i, v_i), \quad \forall b_i, \mathbf{b}_{-i}, i \in \mathbf{N}.$$

Definition 2 (Individual Rationality, IR). *A mechanism is individually rational if and only if*

$$u_i(v_i, v_i) \geq 0, \quad \forall v_i, \mathbf{b}_{-i}, i \in \mathbf{N}.$$

For ease of notation, we use *truthfulness* to represent the properties of both DSIC and IR in the following sections. In addition, for truthful auctions, we do not distinguish v_i and b_i hereinafter.

The revenue of a truthful auction is defined as

$$\mathsf{rev} = \mathbb{E}_{\mathbf{v}} \left[\sum_{i \in N} p_i(v_i) \right].$$

The aim of this work is to characterize both truthful and revenue-maximizing (optimal) auctions with ex post ROI-constrained bidders.

3   Characterize the Structure of DSIC Auctions

In this section, we present characterizations of the DSIC auctions with ex post ROI constraints. These results generalize the classical Myerson's Lemma [24] for the traditional utility model (*i.e.*, $M_i = 1$), which states that in the single-parameter environment, a mechanism is DSIC if and only if its allocation rule is monotone and the payment scheme follows a unique rule.

Lemma 1 (Myerson's Lemma [24]). *For traditional bidders with $M_i = 1$, a single-parameter mechanism is DSIC if and only if:*

- *[Monotone Allocation Rule] the allocation rule is monotonically non-decreasing, i.e., $x_i(v) \leq x_i(v')$ for all $v < v'$ and bidder i;*
- *[Unique Payment Rule] for each monotonically non-decreasing allocation rule $x_i(\cdot)$, and $p_i(0) = 0$, the payments are given by*

$$p_i(v) = v x_i(v) - \int_0^v x_i(z)\, \mathrm{d}z. \tag{3}$$

Clearly, these results cannot be directly applied to the ROI-constrained bidders, because the payment derived from Myerson's Lemma may violate the ROI constraints. The main result in this section is a complete characterization of the DSIC mechanisms with ROI-constrained bidders. We will see that the monotonicity condition for the allocation remains the same, but the payment rule needs to be modified appropriately to accommodate the ROI constraints.

Theorem 2 (Characterization). *For ex post ROI-constrained bidders, a single-parameter mechanism is DSIC if and only if:*

- *[Monotone Allocation Rule] the allocation rule is monotonically non-decreasing, i.e., $x_i(v) \leq x_i(v')$ for all $v < v'$ and bidder i;*
- *[Unique Payment Rule] for each monotonically non-decreasing allocation rule $x_i(\cdot)$, and $p_i(0) = 0$, the payments are given by*

$$p_i(v) = M_i \widetilde{p}_i(v) - \max_{0 \leq z \leq v} \{ M_i \widetilde{p}_i(z) - z x_i(z) \}, \tag{4}$$

where $\widetilde{p}_i$ is the Myerson payment given in (3).

We can interpret this characterization from two perspectives: First, from the perspective of the initial value t_i of bidder i, it suggests that compared to the classic Myerson auction, an ROI-constrained bidder with value v will need to pay the initial Myerson payment $M_i\widetilde{p}_i(v)$ (recall that $t_i = M_iv_i$), minus a "rebate" which equals to the largest "violation" of the Myerson payment to the ROI requirement when the bidder's valuation is no more than v.

Second, from the perspective of the value v_i, since the ROI-constrained bidder assigns a weight $M_i > 1$ to her obtained value from the allocation, but not to her payment, a naive application that charges the bidder M_i times of the Myerson payment may violate the IR constraint. Therefore, we need to iteratively apply the Myerson payment increment (multiplied by M_i) in small intervals and truncate the payment at the value whenever necessary. These two perspectives are mathematically equivalent, and we adopt the second perspective in the following for exposition convenience.

Next, before proving this theorem, some observations are immediate from this characterization. We omit the proof due to the space limitation.

Proposition 1.

1. *We always have $p_i(v) \leq vx_i(v)$ and $p_i(v) \leq M_i\widetilde{p}_i(v)$ for any bidder i and value v in DSIC mechanisms.*
2. *The payment function $p_i(\cdot)$ is monotonically non-decreasing for any bidder i in DSIC mechanisms.*

Now we proceed to prove Theorem 2. The proof consists of showing the following claims in sequence. It is not difficult to see that these three claims together imply Theorem 2.

1. For ex post ROI-constrainted bidders, if a mechanism is DSIC, then the allocation rule must be monotonically non-decreasing.
2. Any monotonically non-decreasing allocation rule $x_i(\cdot)$ with the payment rule given in (4) produces a DSIC mechanism.
3. Given any monotone allocation rule $x(\cdot)$, the payment rule $p(\cdot)$ such that (x, p) is DSIC, if exists, must be unique.

The analyses of steps (1) and (3) are very similar to the proof of the original Myerson's Lemma, and we omit them due to space limits. Next, we prove step (2). When clear from context, we will drop the subscript i in $x_i(\cdot), p_i(\cdot), u_i(\cdot)$ and M_i as shorthand in the following proofs.

Proof of Step (2). Consider a bidder i with private valuation v and fix the other bids b_{-i}. We examine the utilities of bidder i when she bids her true valuation and when she bids some different value $v' \neq v$. Consider two cases.

- When $v' < v$, we have $\max_{0 \leq z \leq v}\{M\widetilde{p}(z) - zx(z)\} \geq \max_{0 \leq z \leq v'}\{M\widetilde{p}(z) - zx(z)\}$, which implies

$$p(v) - p(v') \leq M(\widetilde{p}(v) - \widetilde{p}(v')).$$

This inequality effectively removes the max term in the payment formula (4) and reduces the problem to that with the Myerson payment. This allows us to apply the standard argument for the Myerson auction to show the DSIC property of our mechanism. We show the analysis below for completeness. We can compute the utility difference of bidder i when she bids v and v', and get

$$
\begin{aligned}
u(v, v) - u(v', v) &= (Mvx(v) - p(v)) - (Mvx(v') - p(v')) \\
&\geq M(vx(v) - \widetilde{p}(v)) - M(vx(v') - \widetilde{p}(v')) \\
&= M \int_0^v x(z)\,\mathrm{d}z - M\left(vx(v') - v'x(v') + \int_0^{v'} x(z)\,\mathrm{d}z \right) \\
&= M\left(\int_{v'}^v x(z)\,\mathrm{d}z - (v - v')x(v') \right) \\
&\geq M\left(\int_{v'}^v x(v')\,\mathrm{d}z - (v - v')x(v') \right) = 0,
\end{aligned}
$$

where the second equality is by plugging in the Myerson payment formula (3). This means bidder i has no incentive to misreport her valuation v as v' in this case.

– When $v' > v$, we examine $\max_{0 \leq z \leq v}\{M\widetilde{p}(z) - zx(z)\}$ and $\max_{0 \leq z \leq v'}\{M\widetilde{p}(z) - zx(z)\}$. There are two possibilities:
  - If these two terms are equal, then we can apply the same argument as in the previous case (and also as in the Mycrson auction analysis) to prove the DSIC property. We omit the details here.
  - If $\max_{0 \leq z \leq v}\{M\widetilde{p}(z) - zx(z)\} < \max_{0 \leq z \leq v'}\{M\widetilde{p}(z) - zx(z)\}$, this means $\arg\max_{0 \leq z \leq v'}\{M\widetilde{p}(z) - zx(z)\} = v^* > v$. Then at valuation v', we should have

$$
\begin{aligned}
p(v') &= M\widetilde{p}(v') - (M\widetilde{p}(v^*) - v^*x(v^*)) \\
&= M\left(v'x(v') - v^*x(v^*) - \int_{v^*}^{v'} x(z)\,\mathrm{d}z \right) + v^*x(v^*)
\end{aligned}
$$

$$
\begin{aligned}
\text{(replace } M \text{ by 1)} \geq v'x(v') - \int_{v^*}^{v'} x(z)\,\mathrm{d}z \\
\geq v'x(v') - \int_{v^*}^{v'} x(v')\,\mathrm{d}z \\
= v'x(v') - (v' - v^*)x(v') \\
=^* x(v') > vx(v').
\end{aligned}
$$

That is to say, when reporting v', the payment of bidder i will be greater than the value she obtains (which is $vx(v')$), therefore violating the IR condition. So bidder i has no incentive to misreport as v' in this case.

$\square$

4   Optimal Auction Design for a Single Bidder

Having obtained the precise characterization of the allocation rule and payment function in the setting with ROI constraints, we now turn to the revenue maximization auction design. Recall that in the Myerson auction [24] and previous works in the ex ante ROI constraints setting [6,16], the revenue maximization problem is reduced to the problem of maximizing (modified) virtual welfare. Unfortunately, with ex post ROI constraints, the payment function characterization (4) involves an additional max term compared to the Myerson payment, and it is unclear how to incorporate this term into a modified virtual valuation formulation. We present the following simple example with a single bidder to demonstrate that, unlike the Myerson auction, the allocation that maximizes the virtual welfare may no longer be optimal with ROI-constrained bidders.

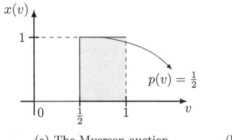
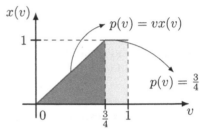

| (a) The Myerson auction | (b) The optimal auction for ROI-constrained bidders |

Fig. 1. The Myerson auction and the optimal auction for one bidder with uniform value distribution over $[0, 1]$ and $M = 2$.

Example 1. *Consider selling a single item to a single bidder with ROI constraint $M = 2$ and valuation for the item v following a uniform distribution $U[0, 1]$. If we disregard the ROI constraint (i.e., let $M = 1$), the virtual valuation of this bidder is $\phi(v) = 2v - 1$, and the optimal Myerson auction, as shown in Fig. 1a, sells the item at price $p = \frac{1}{2}$ with the expected revenue of $\frac{1}{2} \cdot \frac{1}{2} = \frac{1}{4}$.*

However, with the ROI constraint $M = 2$ in presence, this allocation rule $(x(v) = 0$ when $v < 1/2$ and $x(v) = 1$ otherwise) is no longer optimal. As shown in Fig. 1b, the optimal auction, which will be proved in Theorem 5 later in this section, is a randomized auction with the allocation rule given by

$$x(v) = \begin{cases} \frac{4}{3}v & \text{if } v \le \frac{3}{4} \\ 1 & \text{if } v > \frac{3}{4}. \end{cases}$$

This allocation rule would generate an expected revenue of $\frac{3}{8}$, which is higher than $\frac{1}{4}$.

This example already highlights an important feature of the optimal auction with ROI constraints: the allocation and payment may be randomized, even in the simple setting with a single bidder and uniform value distribution. It also suggests that it is difficult to follow the Myerson auction regime and reduce the revenue maximization problem to a welfare maximization problem with some modified virtual valuation. It seems a very challenging problem to obtain a characterization for the optimal auction in this setting. Instead, in this section we focus on the special case when the item is sold to a single bidder. As we will show in the following analysis, this is already a nontrivial and interesting problem to design an optimal auction for a single bidder.

First, we show that with a single ROI-constrained bidder, the max term in the payment formula (4) would be always 0, reducing the payment rule (4) to the standard Myerson payment (multiplied by a factor of M). We omit the proof here for space reasons.

Lemma 3. *In the optimal auction with a single ROI-constrained bidder, let (x, p) be the revenue-maximizing auction, then we have $M\widetilde{p}(v) \leq v \cdot x(v)$ for any value $v \in [0, v_{\max}]$. In other words, $\max_{0 \leq z \leq v}\{M\widetilde{p}(z) - zx(z)\} = 0$ (which is achieved at $z = 0$) for all $v \in [0, v_{\max}]$, and the payment rule reduces to $p(v) = M\widetilde{p}(v)$.*

With Lemma 3 at hand, it seems with a single bidder, we are back to the classic Myerson regime, where the revenue maximization problem can be converted to a welfare maximization problem with respect to the virtual valuation. That is, recall from the Myerson's theorem [24], we have

$$\mathsf{rev} = M \int_0^{v_{\max}} \phi(v)x(v)f(v)\,\mathrm{d}v,$$

where $\phi(v) = \left(v - \frac{1-F(v)}{f(v)}\right)$ is the *virtual valuation*. However, we still have the additional constraint that $M\widetilde{p}(v) \leq v \cdot x(v)$ for every v. This restricts our allocation space and turns the problem into a constrained welfare maximization problem.

In the following, we provide some further characterizations on the structure of the optimal auction with a single ROI-constrained bidder. The proof is omitted because of the space limitation.

Lemma 4. *With a single ROI-constrained bidder, there always exists a revenue-maximizing auction such that for any valuation v, at least one of the following statements holds:*

- *the derivative of allocation rule at the valuation v exists, and $x'(v) = 0$;*
- *the payment follows the first-price rule, i.e., $p(v) = vx(v)$.*

Lemma 4 allows us to focus on auctions with a very specific structure: as long as the allocation is not constant, it always follows the first-price payment rule. In particular, combining with Lemma 3, it implies whenever $x'(v)$ exists and $x'(v) > 0$, we always have $p = vx(v) = M\widetilde{p}(v)$.

Next, we want to obtain a further characterization of the optimal auction with a single bidder. However, to do this would require us to make a mild assumption on the value distribution of the bidder, which is known as the *Decreasing Marginal Revenue (DMR)* condition by [13].

Definition 3 (Decreasing Marginal Revenue, DMR). *The value distribution of a bidder satisfies the condition of decreasing marginal revenue if and only if the function*

$$\psi(v) \triangleq \phi(v)f(v) = vf(v) + F(v) - 1$$

is monotonically non-decreasing.

Note that $\psi(v)$ being non-decreasing is equivalent to the fact that $v \cdot (1 - F(v))$, which is the expected revenue of selling the item at price p, being concave, and this is where the name of this condition comes from. Intuitively, many commonly used distribution functions satisfy this assumption, *e.g.*, uniform distributions, and any distribution of finite support and monotone non-decreasing density. The DMR condition is closely related to the regularity condition but they are incompatible[3]. We refer the reader to [13] for concrete examples and more discussion.

With the assumption of DMR, the optimal auction exhibits an even simpler structure than what is described in Lemma 4, namely that there exist only two intervals in the optimal auction: interval of $(0, D)$ with $x'(v) > 0$ and interval of $(D, v_{\max})$ with $x'(v) = 0$, where D is a threshold valuation between them. This leads to our main theorem in this section, which characterizes the optimal allocation rule and payment rule for a single ROI-constrained bidder.

Theorem 5. *The optimal auction for a single ex post ROI-constrained bidder with a DMR value distribution over $[0, v_{\max}]$ is as follows:*

- *when $v < D$, the allocation is given by*

$$x(v) = \left(\frac{v}{D}\right)^{\frac{1}{M-1}},$$

   and the payment follows the first-price rule, i.e., $p(v) = vx(v)$;
- *when $v \geq D$, the allocation rule is $x(v) = 1$, and the payment is given by $p(v) = D$.*

Here D is a threshold valuation given as follows:

- *if $\int_0^{v_{\max}} \psi(v)v^{\frac{1}{M-1}}\, dv > 0$, then $D = D^*$ such that $\int_0^{D^*} \psi(v)v^{\frac{1}{M-1}}\, dv = 0$;*
- *if $\int_0^{v_{\max}} \psi(v)v^{\frac{1}{M-1}}\, dv \leq 0$, then $D = v_{\max}$.*

---

[3] The regularity condition is equivalent to the expected revenue being concave in the *quantile* space, while the DMR condition means the expected revenue is concave in the *value* space.

This theorem provides an important insight that the optimal auction in the ROI-constrained setting is a randomized mechanism. Note that Myerson's optimal auction in the single-parameter setting is deterministic, but a decent body of works has shown that many generalizations to multi-parameter settings will lead to randomized optimal auctions [17,25]. Theorem 5 indicates that, even in the single-parameter environment, a slight generalization with an ROI constraint to the bidder will also lead to a randomized optimal auction.

We prove the theorem via the following steps. First, we derive the allocation of an optimal auction in an interval $(0, v^*)$ when $x'(v)$ is always positive in that interval. Then, we show in Lemma 6 that there exist only two intervals in the optimal auction: interval of $(0, D)$ with $x'(v) > 0$ and interval of (D, v_{max}) with $x'(v) = 0$. Finally, we will compute the optimal threshold valuation D between these two intervals.

Proposition 2. *If for some $v^* \in (0, v_{max}]$, we have $x'(v) > 0$ for all $v \in (0, v^*)$ in an optimal auction, then $x(\cdot)$ is continuous at v^*, and the allocation rule $x(v)$ for all $v \in [0, v^*]$ is given as:*

$$x(v) = \left(\frac{v}{v^*}\right)^{\frac{1}{M-1}} x(v^*).$$

Proof. We first assume $x(\cdot)$ is continuous at v^*, and we will prove later that, if it is discontinuous, we can improve the revenue without violating the DSIC property. By Lemma 3 and Lemma 4, we get that for all valuations $v \in [0, v^*]$, $M\widetilde{p}(v) - vx(v) = 0$ always holds, that is,

$$M\left(vx(v) - \int_0^v x(z)\,dz\right) - vx(v) = 0.$$

After transposition and derivation, this translates to

$$x(v) - (M-1)vx'(v) = 0.$$

By solving this differential equation with the value of $x(v^*)$ at valuation v^*, we can get

$$x(v) = \left(\frac{v}{v^*}\right)^{\frac{1}{M-1}} x(v^*), \quad \forall v \in [0, v^*]. \tag{5}$$

Next, if $x(\cdot)$ is discontinuous at v^*, we need to replace $x(v^*)$ in (5) with $x(v^{*-})$, i.e., the left limit of $x(\cdot)$ at v^* (recall that we denote $x(v^*)$ as the right limit when it is discontinuous). Since $x(v^{*-}) < x(v^*)$, we can observe that directly using (5) as the allocation rule will increase the revenue without violating the DSIC property, which also makes $x(\cdot)$ continuous at v^*. This concludes the proof. $\square$

Lemma 6. *For a single ROI-constrained bidder with DMR value distribution, if there exist intervals with $x'(v) = 0$ in an optimal auction, then there is exactly one such interval, and it appears in the highest value region.*

Proof. Assume by contradiction that there exist multiple intervals with $x'(v) = 0$ in an optimal auction. Pick $(\underline{v}, \bar{v})$ to be the first such interval. That is, we have $x'(v) > 0$ for all $v \in (0, \underline{v})$, $x'(v) = 0$ for all $v \in (\underline{v}, \bar{v})$, and $\bar{v} < v_{\max}$.

First, we must have $\underline{v} > 0$. That is, the allocation rule cannot start with a flat interval. To see why this is true, assume otherwise that $\underline{v} = 0$. We focus on a point $v' = \bar{v} + \delta$ in the next interval for a sufficiently small $\delta > 0$ such that $v' < M\bar{v}$. Then there are two cases: (1) $x'(v') = 0$ and (2) $x'(v')$ is strictly positive. In the first case, we will have $x(\bar{v}) > 0$, and the Myerson price at $\bar{v}$ will be $\widetilde{p}(\bar{v}) = \bar{v} \cdot x(\bar{v}) < M\widetilde{p}(\bar{v})$, which directly contradicts Lemma 3. In the second case, we look at the payment $p(v')$ at point v'. Note that since $x'(v) > 0$, by Lemma 3 and Lemma 4, we should have $M\widetilde{p}(v') = v'x(v')$. But this cannot happen because

$$M\widetilde{p}(v') = M\left(v'x(v') - \int_0^{v'} x(z)\, \mathrm{d}z\right) = M\left(v'x(v') - \int_{\bar{v}}^{v'} x(z)\, \mathrm{d}z\right)$$
$$> M\left(v'x(v') - (v' - \bar{v})x(v')\right) = M\bar{v}x(v') > v'x(v').$$

Knowing $\underline{v} > 0$, by Proposition 2, we know $x(\cdot)$ is continuous at $\underline{v}$, and the allocation in $[0, \underline{v}]$ is given as:

$$x(v) = \left(\frac{v}{\underline{v}}\right)^{\frac{1}{M-1}} x(\underline{v}), \quad \forall v \in [0, \underline{v}].$$

Next, combined with $p(\bar{v}) = \bar{v}x(\bar{v})$, we have $M\widetilde{p}(\bar{v}) - \bar{v}x(\bar{v}) = M\widetilde{p}(\underline{v}) - \underline{v}x(\underline{v})$, that is,

$$x(\bar{v}) = \frac{M\bar{v} - \underline{v}}{(M-1)\bar{v}} \cdot x(\underline{v}). \tag{6}$$

In order to argue that allocation rule $x(\cdot)$ is not revenue-maximizing, we construct a new allocation rule as

$$\bar{x}(v) = \begin{cases} x(v) & \text{if } v \geq \bar{v} \\ \left(\frac{v}{\bar{v}}\right)^{\frac{1}{M-1}} x(\bar{v}) & \text{if } v \in [0, \bar{v}). \end{cases} \tag{7}$$

That is, we replace the first price interval $[0, \underline{v}]$ and the flat interval $[\underline{v}, \bar{v}]$ in $x(\cdot)$ with a single first-price interval $[0, \bar{v}]$ in $\bar{x}(\cdot)$. Comparing the two allocation rules $x(\cdot)$ and $\bar{x}(\cdot)$, we first note from Lemma 3 and Lemma 4 that $p(\bar{v}) = M\widetilde{p}(\bar{v}) = \bar{v}x(\bar{v}) = \bar{v}\bar{x}(\bar{v})$, which indicates that

$$\int_0^{\bar{v}} x(z)\, \mathrm{d}z = \int_0^{\bar{v}} \bar{x}(z)\, \mathrm{d}z, \tag{8}$$

because both sides equal to $\bar{v}x(\bar{v})(M-1)/M$. Next, we have for $v \in [0, \underline{v}]$,

$$\bar{x}(v) - x(v) = \left(\frac{v}{\bar{v}}\right)^{\frac{1}{M-1}} x(\bar{v}) - \left(\frac{v}{\underline{v}}\right)^{\frac{1}{M-1}} x(\underline{v})$$
$$= v^{\frac{1}{M-1}} \cdot x(\underline{v}) \cdot \left(\left(\frac{1}{\underline{v}}\right)^{\frac{1}{M-1}} - \left(\frac{1}{\bar{v}}\right)^{\frac{1}{M-1}} \frac{M\bar{v} - \underline{v}}{(M-1)\bar{v}}\right).$$

Since v only appears in the first term, $\bar{x}(v) - x(v)$ must be constantly positive or negative in $(0, \underline{v}]$, determined by the last term. Combined with (8) and $x'(v) = 0, \forall v \in (\underline{v}, \bar{v})$, we know it is constantly negative, i.e., $\bar{x}(v) < x(v)$ for all $v \in (0, \underline{v}]$. Therefore, there exists a threshold $v^* \in (\underline{v}, \bar{v})$ such that $x(v) > \bar{x}(v)$ for all $v \in [0, v^*)$ and $x(v) \leq \bar{x}(v)$ for all $v \in [v^*, \bar{v})$. Combining this with Eq. (8), we have

$$\int_0^{v^*} (x(z) - \bar{x}(z))\, dz = \int_{v^*}^{\bar{v}} (\bar{x}(z) - x(z))\, dz > 0. \tag{9}$$

Next, for any allocation rule $x(\cdot)$, we denote

$$\mathsf{rev}_{[0,\bar{v}]}^{x(\cdot)} = M \int_0^{\bar{v}} \psi(z) x(z)\, dz,$$

and we can compare the revenue generated from $x(\cdot)$ and $\bar{x}(\cdot)$ in the interval $[0, \bar{v}]$:

$$\mathsf{rev}_{[0,\bar{v}]}^{x(\cdot)} - \mathsf{rev}_{[0,\bar{v}]}^{\bar{x}(\cdot)} = M \left(\int_0^{v^*} \psi(z)(x(z) - \bar{x}(z))\, dz - \int_{v^*}^{\bar{v}} \psi(z)(\bar{x}(z) - x(z))\, dz \right).$$

Finally, we can see that this difference is always negative due to Eq. (9) and the fact that $\psi(\cdot)$ is non-decreasing.[4] Figure 2 demonstrates the idea of this argument.

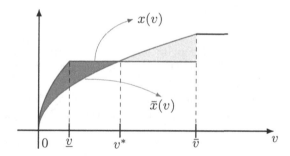

Fig. 2. An illustration for the proof of Lemma 6 with $M = 3$. The blue line denotes an allocation rule $x(\cdot)$ with $x'(v) = 0$ in $(\underline{v}, \bar{v})$ where $\bar{v} < v_{\mathsf{max}}$, and $x(v)$ for $v \in [0, \underline{v})$ is computed by Proposition 2. The red line denotes our constructed allocation rule $\bar{x}(\cdot)$ as given in (7). The point v^* denotes the intersection of $x(\cdot)$ and $\bar{x}(\cdot)$. By Eq. (9) we have the areas of the two shadowed regions are the same. Furthermore, as $\psi(\cdot)$ is non-decreasing, we can conclude that $\bar{x}(\cdot)$ leads to a higher revenue than $x(\cdot)$. (Color figure online)

Therefore, we conclude that $\bar{x}(\cdot)$ generates a higher revenue than $x(\cdot)$, contradicting the fact that (x, p) is an optimal auction. This completes the proof. □

---

[4] We omit an ill-defined special case where $\psi(v) = 0, \forall v \in [0, \bar{v}]$, since in such case $\psi(v)$ will be infinity when $v \to 0$.

We now proceed to complete the proof of Theorem 5.

Proof of Theorem 5. By Lemma 6, we know $x'(v) > 0$ for all valuations $v \in (0, D)$ and $x'(v) = 0$ for all valuations $v \in (D, v_{\mathsf{max}})$. First, we have $x(D) = 1$, since otherwise, the revenue could be improved by setting $x(v) = 1$ for all $v \in [D, v_{\mathsf{max}}]$. Next, we find the optimal threshold D^* that maximizes the overall revenue. By Proposition 2, the allocation rule for valuations $v \in [0, D]$ is

$$x(v) = \left(\frac{v}{D}\right)^{\frac{1}{M-1}} \cdot x(D) = \left(\frac{v}{D}\right)^{\frac{1}{M-1}}, \forall v \in [0, D].$$

And the overall revenue is

$$\mathsf{rev} = M \left(\int_0^D \psi(v) \left(\frac{v}{D}\right)^{\frac{1}{M-1}} \mathrm{d}v + \int_D^{v_{\mathsf{max}}} \psi(v) \, \mathrm{d}v \right).$$

In order to maximize this revenue, we compute its derivative

$$\frac{\mathrm{drev}}{\mathrm{d}D} = -\frac{M}{M-1} \cdot D^{-\frac{M}{M-1}} \cdot \int_0^D \psi(v) v^{\frac{1}{M-1}} \, \mathrm{d}v. \tag{10}$$

Looking at this derivative, we see that the term $-\frac{1}{M-1} \cdot D^{-\frac{M}{M-1}}$ is always negative and $v^{\frac{1}{M-1}}$ is always positive. Furthermore, we have $\psi(0) = -1 < 0$ and $\psi(v_{\mathsf{max}}) > 0$. By the DMR condition, $\psi(\cdot)$ is non-decreasing. Therefore, we have the following two cases:

- If $\int_0^{v_{\mathsf{max}}} \psi(v) v^{\frac{1}{M-1}} \, \mathrm{d}v > 0$, then by letting $\int_0^{D^*} \psi(v) v^{\frac{1}{M-1}} \, \mathrm{d}v = 0$, the revenue will increase with D until D^* and then decrease. Therefore, the optimal solution is achieved at $D = D^*$.
- If $\int_0^{v_{\mathsf{max}}} \psi(v) v^{\frac{1}{M-1}} \, \mathrm{d}v \leq 0$, then the revenue will always increase with D until v_{max}. Therefore, the optimal solution is achieved at $D = v_{\mathsf{max}}$.

This completes the proof.   □

5   Conclusion

In this paper we discuss optimal auction design for bidders who have ex post ROI constraints. We provide characterizations for DSIC auctions and optimal auctions with a single bidder in this setting. We show that the optimal auction may entail a randomized allocation scheme even in the simple single-bidder setting.

There are several important open questions left in this model. The first and foremost one is to characterize the optimal auction with a single item and multiple ROI-constrained bidders. As we have discussed in the paper, this would require us to go beyond the virtual welfare maximization regime in the Myerson auction setting. We believe such characterization could shed light on the mechanism design for bidders with non-quasilinear utility functions and provide useful insights for practical applications such as online advertising. Another direction is to study ex post ROI constraints when the target ratio M_i is also private information of bidder i. This brings the problem to the domain of multidimensional mechanism design, which is often challenging in the mechanism design literature. Here one possible approach is to identify conditions under which some "simple" and deterministic auctions are optimal or close to optimal.

Acknowledgement. We thank Hu Fu, Yiqing Wang, Yidan Xing, Xiangyu Liu and anonymous reviewers for their insightful and helpful suggestions. This work was supported in part by National Key R&D Program of China No. 2021YFF0900800, in part by China NSF grant No. 62220106004, 62322206, 62132018, 62272307, 61972254, 62025204, 62302267, in part by the Natural Science Foundation of Shandong (No. ZR202211150156, ZR2021LZH006), in part by Alibaba Group through Alibaba Innovative Research Program, and in part by Tencent Rhino Bird Key Research Project. The opinions, findings, conclusions, and recommendations expressed in this paper are those of the authors and do not necessarily reflect the views of the funding agencies or the government.

References

1. Aggarwal, G., Badanidiyuru, A., Mehta, A.: Autobidding with constraints. In: Caragiannis, I., Mirrokni, V., Nikolova, E. (eds.) WINE 2019. LNCS, vol. 11920, pp. 17–30. Springer, Cham (2019). https://doi.org/10.1007/978-3-030-35389-6_2
2. Auerbach, J., Galenson, J., Sundararajan, M.: An empirical analysis of return on investment maximization in sponsored search auctions. In: Proceedings of the 2nd International Workshop on Data Mining and Audience Intelligence for Advertising, pp. 1–9 (2008)
3. Babaioff, M., Cole, R., Hartline, J., Immorlica, N., Lucier, B.: Non-quasi-linear agents in quasi-linear mechanisms. Leibniz Int. Proc. Inform. **185** (2021)
4. Balseiro, S., Deng, Y., Mao, J., Mirrokni, V., Zuo, S.: Robust auction design in the auto-bidding world. In: Advances in Neural Information Processing Systems, vol. 34, pp. 17777–17788 (2021)
5. Balseiro, S., Golrezaei, N., Mirrokni, V., Yazdanbod, S.: A black-box reduction in mechanism design with private cost of capital. Available at SSRN 3341782 (2019)
6. Balseiro, S.R., Deng, Y., Mao, J., Mirrokni, V.S., Zuo, S.: The landscape of auto-bidding auctions: Value versus utility maximization. In: Proceedings of the 22nd ACM Conference on Economics and Computation, pp. 132–133 (2021)
7. Borgs, C., Chayes, J., Immorlica, N., Jain, K., Etesami, O., Mahdian, M.: Dynamics of bid optimization in online advertisement auctions. In: Proceedings of the 16th International Conference on World Wide Web, pp. 531–540 (2007)
8. Borgs, C., Chayes, J., Immorlica, N., Mahdian, M., Saberi, A.: Multi-unit auctions with budget-constrained bidders. In: Proceedings of the 6th ACM Conference on Electronic Commerce, pp. 44–51 (2005)
9. Brynjolfsson, E., Hu, Y., Simester, D.: Goodbye pareto principle, hello long tail: the effect of search costs on the concentration of product sales. Manage. Sci. **57**(8), 1373–1386 (2011)
10. Cavallo, R., Krishnamurthy, P., Sviridenko, M., Wilkens, C.A.: Sponsored search auctions with rich ads. In: Proceedings of the 26th International Conference on World Wide Web, pp. 43–51 (2017)

11. Che, Y.K., Gale, I.: Standard auctions with financially constrained bidders. Rev. Econ. Stud. **65**(1), 1–21 (1998)
12. Deng, Y., Mao, J., Mirrokni, V., Zuo, S.: Towards efficient auctions in an auto-bidding world. In: Proceedings of the Web Conference 2021, pp. 3965–3973 (2021)
13. Devanur, N.R., Haghpanah, N., Psomas, C.A.: Optimal multi-unit mechanisms with private demands. In: Proceedings of the 2017 ACM Conference on Economics and Computation, pp. 41–42 (2017)
14. Fiat, A., Goldner, K., Karlin, A.R., Koutsoupias, E.: The fedex problem. In: Proceedings of the 2016 ACM Conference on Economics and Computation, pp. 21–22 (2016)
15. Golrezaei, N., Jaillet, P., Liang, J.C.N., Mirrokni, V.: Bidding and pricing in budget and ROI constrained markets. arXiv preprint arXiv:2107.07725 (2021)
16. Golrezaei, N., Lobel, I., Paes Leme, R.: Auction design for ROI-constrained buyers. In: Proceedings of the Web Conference 2021, pp. 3941–3952 (2021)
17. Hart, S., Reny, P.J.: Maximal revenue with multiple goods: nonmonotonicity and other observations. Theor. Econ. **10**(3), 893–922 (2015)
18. Heymann, B.: Cost per action constrained auctions. In: Proceedings of the 14th Workshop on the Economics of Networks, Systems and Computation, pp. 1–8 (2019)
19. Laffont, J.J., Robert, J.: Optimal auction with financially constrained buyers. Econ. Lett. **52**(2), 181–186 (1996)
20. Li, B., et al.: Incentive mechanism design for ROI-constrained auto-bidding. arXiv preprint arXiv:2012.02652 (2020)
21. Li, J., Tang, P.: Auto-bidding equilibrium in ROI-constrained online advertising markets. arXiv preprint arXiv:2210.06107 (2022)
22. Malakhov, A., Vohra, R.V.: Optimal auctions for asymmetrically budget constrained bidders. Rev. Econ. Design **12**(4), 245–257 (2008)
23. Mehta, A.: Auction design in an auto-bidding setting: Randomization improves efficiency beyond VCG. In: Proceedings of the ACM Web Conference 2022, pp. 173–181 (2022)
24. Myerson, R.B.: Optimal auction design. Math. Oper. Res. **6**(1), 58–73 (1981)
25. Pavlov, G.: Optimal mechanism for selling two goods. BE J. Theor. Econ. **11**(1) (2011)
26. Szymanski, B.K., Lee, J.S.: Impact of ROI on bidding and revenue in sponsored search advertisement auctions. In: Second Workshop on Sponsored Search Auctions, pp. 1–8 (2006)
27. Tillberg, E., Marbach, P., Mazumdar, R.: Optimal bidding strategies for online ad auctions with overlapping targeting criteria, vol. 4, pp. 1–55. ACM, New York (2020)
28. Wilkens, C.A., Cavallo, R., Niazadeh, R.: GSP: the Cinderella of mechanism design. In: Proceedings of the 26th International Conference on World Wide Web, pp. 25–32 (2017)

Nash Stability in Fractional Hedonic Games with Bounded Size Coalitions

Gianpiero Monaco[ID] and Luca Moscardelli[(✉)][ID]

University of Chieti-Pescara, Viale Pindaro 42, 65127 Pescara, Italy
{gianpiero.monaco,luca.moscardelli}@unich.it

Abstract. We consider fractional hedonic games, a natural and succinct sub-class of hedonic games able to model many real-world settings in which agents have to organize themselves in groups called coalitions. An outcome of the game, also called coalition structure, is a partition of all agents into coalitions. Previous work assumed that coalitions can be of any size. However, in many real-world situations, the size of the coalitions is bounded: vehicles, offices, classrooms and project teams are some examples of possible coalitions of bounded size. In this paper, we initiate the study k-fractional hedonic games ($k-FHG$), in which all coalitions have size at most k, by considering Nash stable coalition structures, i.e., outcomes in which no agent can improve her utility by unilaterally changing her own coalition; in particular, we study existence of, convergence to, complexity and efficiency of Nash stable outcomes, and we also provide results about the complexity and approximation of optimal outcomes. We perform a thorough-going analysis of $k-FHG$ for $k = 2, 3, 4$. We remark that, on the one hand, considering these values of k is interesting in itself as many real world scenarios (as some of the aforementioned ones) deal with coalitions of small size; on the other hand, studying $k-FHG$ for small values of k both represents a necessary step for understanding the cases with higher values of k and already constitutes a challenging task.

Keywords: Coalition Formation Games · Fractional Hedonic Games · Nash stability · Price of Anarchy

1 Introduction

Coalition formation is a widely investigated issue in computer science and in particular in artificial intelligence and multi-agent research. Hedonic games, introduced in [17], likely constitute the most important game-theoretic approach to the study of coalition formation problems. They consist of a finite set of agents together with a list of agents' preferences, with the preferences of an agent only depending on the members of the coalition she belongs to. An outcome of such games, also called coalition structure, is a partition of all agents into pairwise disjoint coalitions whose union is equal to the set of agents.

Fractional hedonic games (FHG), introduced in [3] (see also [1]), embody a natural and succinct subclass of hedonic games. In FHG, each agent has a value for any other agent, and the utility that an agent gets for a coalition C is the sum of values she assigns

J. Garg et al. (Eds.): WINE 2023, LNCS 14413, pp. 509–526, 2024.
https://doi.org/10.1007/978-3-031-48974-7_29

to the members of C divided by the size of C. These games model many real-world setting where the aim is forming groups while maximizing the fraction of like-minded members (or people with the same interests). Examples are politicians organizing themselves in parties, countries organizing themselves in international groups, employees forming unions, faculty members being assigned desks in a department with multi-person offices, employees being assigned to project teams, travellers being assigned to vehicles, students being assigned to classrooms or dormitories, etc. Moreover, even the specific setting of *simple symmetric* FHG where agents' values are symmetric and can only take the values 0 and 1, suitably model a basic economic scenario referred to in [1] as Bakers and Millers.

Although most of the previous work on FHG (see Sect. 1.1) assume that coalitions can be of any size, in many real-world situations (e.g., the ones in which coalitions are vehicles, offices, classrooms, project teams, etc.) the size of the coalitions is bounded. Therefore, in this paper we initiate the study of stable coalition structures in k-fractional hedonic games ($k-$FHG) where coalitions have size at most k (where k is a given positive integer) and where agents' values are non-negative. We focus on Nash stable coalition structures, where no agent can improve her utility by unilaterally changing her own coalition. We notice that, while in FHG where agents have non-negative values the grand coalition structure, in which all agents belong to the same coalition, is trivially Nash stable [7], in $k-$FHG the grand coalition structure is in general not feasible, and therefore understanding whether Nash stable coalition structures are guaranteed to exist or not is a more challenging question. Our aim is to characterize the existence, efficiency and complexity of Nash stable outcomes in $k-$FHG. We also study the convergence of dynamics where agents perform improving deviations (in a coalition structure, an agent possesses an improving deviation if she can strictly improve her utility by unilaterally changing her own coalition). It is worth remarking that studying the convergence of dynamics is not only interesting in itself, given that dynamics can be exploited in order both to show the existence of Nash stable coalition structures and to transform any coalition structure in a stable one with, hopefully, no loss or a small loss in terms of efficiency.

1.1   Related Work

Hedonic games were introduced in [17] and then further developed in their present form of versatile model of coalition formation in [6,9,14], where different solution concepts of stability such as Nash stability, core and individual stability are addressed. There are many subsequent papers studying several subclasses of hedonic games under many different perspectives. We suggest [5] for a great survey on the topic.

FHG with no restriction on the size of the coalitions (i.e., a coalition can be composed by any number of agents) were introduced in [3] (see also [1]), in which they were studied under the perspective of core stability: an outcome is core stable if there is no blocking coalition, i.e., no set of agents all improving their utility by forming a new coalition together. In particular, the authors show that the core can be empty even for simple symmetric agents' values and that however, it is not empty for some very specific sub-classes. The authors also consider the complexity of deciding the non-emptiness of the core. Nash stable outcomes were investigated in [7] where the authors provide many results about their existence, efficiency and computability. Specifically, they show that, while in presence of negative agents' values Nash stable outcomes are not guaranteed to exist, with non-negative ones the grand coalition is always Nash stable. Moreover, they show that the dynamics is not convergent even for simple symmetric agents' values. They further study the price of anarchy and stability for several different graph topologies, and provide complexity results on the computation of Nash stable coalition structures. Improved results about the price of stability can be found in [22].

Several computational results about core and individual stability in FHG were presented in [10], while the convergence of simple dynamics based on individual stability were studied in [11]. Local core stability, in which there is a structural constraint on the blocking coalition, has been addressed in [13]. Further relaxed forms of core stability were studied in [19]. FHG were also considered under different perspectives in [4,12,18,20,21].

Hedonic games with restriction on the size of the coalitions were considered in a few papers. In [8], the authors study additively separable hedonic games (ASHG) with fixed size coalitions where the number of coalitions is fixed and each coalition has its own fixed size. In this setting, an outcome is swap stable if no pair of agents can exchange coalitions and improve their utilities. The authors study existence, complexity and efficiency of swap stable outcomes, and that of complexity of a social optimum. In [26], the authors consider ASHG with bounded coalitions by focusing on strategyproof mechanisms. In [21], the authors consider a setting where agents form coalitions that can have bounded size; however, they do not study stable outcomes, but focus on online algorithms with the goal of partitioning agents into coalitions so as to maximize the resulting social welfare. In [24], the authors consider the problem of partitioning a set of agents into a fixed number of coalitions of almost equal size (i.e., having sizes differing by at most one). They study core stable outcomes and envy-freeness and provide (approximated) results for both stability concepts. Finally, in [23] the authors study ASHG with bounded coalitions; given that, in this setting, it is easy to show the existence of Nash stable outcomes by applying techniques already exploited the classical non-bounded case, the authors focus on the existence and computation of the core, as well as on computing coalition structure with high social welfare.

Coalition formation with bounded size coalitions were also studied in [2,16,25] under different perspectives.

1.2 Our Contribution

To the best of our knowledge, this paper initiates the study of Nash stable coalition structures in FHG with bounded size coalitions. We perform a thoroughgoing analysis of $k-$FHG for $k = 2, 3, 4$. We remark that, on the one hand, considering these values of k is interesting in itself as many real world scenarios (e.g., the ones in which coalitions are vehicles, offices, project teams, etc.) deal with coalitions of small size; on the other hand, studying $k-$FHG for $k = 2, 3, 4$ both represents a necessary step for understanding the cases with higher values of k and, somewhat astonishingly, already constitutes a challenging task. As a further consideration, it is worth noticing that, as shown in the conclusions (Sect. 5), there is no inclusion between the sets of Nash stable outcomes for $k-$FHG and $k'-$FHG, with $k \neq k'$. This fact gives evidence of the difficulty of providing general results holding for any value of k. For this reason, it is worth studying $k-$FHG for specific values of k.

We notice that considering coalitions with bounded size makes our model different from the one of classical FHG, that is, many results holding for FHG do not also hold for $k-$FHG and vice-versa. For instance, on the one hand, we show that Nash stable outcomes may not exist in $3-$FHG and $4-$FHG when agents' values are symmetric and non-negative, while their existence is guaranteed for FHG. On the other hand, we show that dynamics in $3-$FHG and $4-$FHG are convergent for simple symmetric agents' values, while they are not in classical FHG.

In Sect. 2.2 we start by showing several preliminary results. Specifically, we show that even when considering simple symmetric games, both problems of computing an optimal coalition structure (we consider the utilitarian social welfare which is defined as the sum of the agents' utilities) and a Nash stable coalition structure with highest social welfare are $\mathcal{NP}$-hard for any $k \geq 3$ (Theorem 1). When considering non symmetric games, Nash stable outcomes are not guaranteed to exist for any $k \geq 2$ even for the case when agents' values are 0 or 1 (Proposition 1). Given this last result, in the rest of the paper we focus on symmetric $k-$FHG. For $k = 2$, we show that any symmetric $2-$FHG is convergent and that any simple symmetric $2-$FHG is polynomially convergent (Theorem 2), that the price of anarchy is unbounded (Proposition 2), and that a Nash stable coalition structure also being an optimal solution can be computed in polynomial time, thus implying that the price of stability is 1 (Proposition 3). Finally, we show that the price of anarchy of simple symmetric $2-$FHG is exactly 2 (Theorem 3 and Proposition 4).

In Sect. 3 we study $3-$FHG. Specifically, we show that when considering symmetric games, Nash stable outcomes are not guaranteed to exist (Proposition 5). Therefore, in the rest of the section, we focus on simple symmetric agents' values. Any simple symmetric $3-$FHG converges in a polynomial number of improving deviations (Theorem 4). Moreover, the price of stability is 1 (Theorem 5) and the price of anarchy is exactly 2 (Theorem 6 and Proposition 6). Finally, it is possible to compute in polynomial time a Nash stable coalition structure $\mathcal{C}$ approximating the optimal outcome by a factor of $\frac{4}{3}$ (Theorem 7).

In Sect. 4 we deal with $4-$FHG. Specifically, we show that, when considering symmetric games, Nash stable outcomes are not guaranteed to exist (Proposition 7). Therefore, in the rest of the section, we focus on simple symmetric agents' values. We show that any simple symmetric $4-$FHG has the polynomially bounded improvement path property (Theorem 8). Moreover, we prove that the price of anarchy is exactly 3 (Theorem 9 and Proposition 8). We also show that it is possible to compute in polynomial time a coalition structure approximating the optimal social welfare by a factor of $\frac{3}{2}$ (Theorem 12). Moreover, we prove that any coalition structure can be transformed in polynomial time in a Nash stable one at the price of losing a factor of $\frac{10}{9}$ on the social welfare (Theorem 10). This is our main technical result which is also exploited to prove that the price of stability is exactly $\frac{10}{9}$ (Theorem 11 and Proposition 9), and that it is possible to compute in polynomial time a Nash stable coalition structure approximating the optimal social welfare by a factor of $\frac{5}{3}$ (Theorem 12).

2   Model and Preliminaries

In this section we introduce fractional hedonic games with bounded size coalitions, in which, given an integer parameter $k \geq 2$, agents can form coalitions of size at most k. For the sake of brevity, we will omit in the following the specification *with bounded size coalitions*.

2.1   Definitions

For any $n \in \mathbb{N}$, we denote by $[n]$ the set $\{1, 2, \ldots, n\}$.

A *k-Fractional Hedonic Game* (k$-$FHG) $\mathcal{G} = (N, (v_i)_{i \in N}, k)$ is a game in which each agent $i \in N$, where $N = [n]$, has a valuation $v_i : N \rightarrow \mathbb{R}^{\geq 0}$, mapping every agent to a real non-negative value, and $k \geq 2$ is an integer parameter representing the maximum possible size of a coalition. We assume that the number of agents is $n \geq 2$. We denote with $v_i^{max}(\mathcal{G}) = \max_{j \in N} v_i(j)$ the maximum valuation of agent i for any other agent $j \in N$ in the game $\mathcal{G}$. We assume that $v_i(i) = 0$ for every $i \in N$. If it holds that $v_i(j) = v_j(i)$ for every $i, j \in N$, i.e., valuations are symmetric, we say that the game is a *Symmetric k-Fractional Hedonic Game* (k$-$SFHG). Finally, if valuations are symmetric and it holds that $v_i(j) \in \{0, 1\}$ for every $i, j \in N$, we say that the game is a *Simple Symmetric k-Fractional Hedonic Game* (k$-$SSFHG).

Graph Representation. A k$-$FHG has a very intuitive graph representation. In fact, it can be expressed by a weighted directed graph $G = (N, E, w)$, where nodes in N represent the agents, and arcs are associated to non-zero valuations. Namely, for any $i, j \in N$, if $v_i(j) > 0$, an arc (i, j) of weight $w((i, j)) = v_i(j)$ belongs to E. Analogously, a k$-$SFHG can be expressed by a weighted undirected graph and a k$-$SSFHG can be expressed by an unweighted undirected graph $G = (N, E)$ in which, for any $i, j \in N$, edge $\{i, j\}$ belongs to E if and only if $v_i(j) = 1$. Given a subset of agents $C \subseteq N$, we denote with $G(C) = (N(C), E(C), w)$ the subgraph of G induced by agents in C. As usual, we denote by K_n a clique of n nodes.

Coalitions and Utilities. A *k-coalition*, for any fixed $k \geq 2$, is a non-empty subset C of N such that $|C| \leq k$, i.e., it is a subset of at most k agents. When clear from the context, we will refer to a k-coalition simply by calling it a coalition. A k-coalition composed by a single agent (i.e., of size 1) is called a *singleton*. In hedonic games, every agent i has a preference order $\succeq_i$ over all coalitions containing herself. Given that the number of possible preference orders is exponential in the number of agents, several interesting classes of games with a succinct representation of these orderings have been considered, with the additively separable hedonic games being perhaps the most famous: In additively separable hedonic games, every agent i has valuations for any other agent and she prefers coalition C to coalition C' (i.e., $|C| \succ_i C'$) whenever the sum of valuations of i for all agents in coalition C is grater than that for agents in C'. Fractional hedonic games represent a variant of additively separable hedonic games in which agents also care about the size of the coalitions: roughly speaking, between two coalitions in which an agents has the same sum of valuations for all other agents, she prefers the one with smallest size. More precisely, the utility of an agent i in coalition C is given by the ratio between the sum of valuations of i for all other agents in C and the size of C; it follows that agent i prefers coalition C to coalition C' whenever her utility in C is greater than the one in C'. Formally, given a coalition C and any agent $i \in C$, let $\delta_C(i) = \sum_{j \in C} v_i(j)$ be the sum of valuations of agent i for every agent belonging to coalition C. The *utility* or *payoff* $\mu_i(C)$ of agent i in coalition C such that $C \ni i$ is equal to $\delta_C(i)$ divided by the total number of agents in the coalition, that is $\mu_i(C) = \frac{\delta_C(i)}{|C|}$. Notice that, for any $k-\mathrm{FHG}$ $\mathcal{G}$, it holds that $\mu_i(C) \leq \frac{|C|-1}{|C|} v_i^{max}(\mathcal{G}) \leq \frac{k-1}{k} v_i^{max}(\mathcal{G})$ for any agent $i \in N$ and coalition C where $|C| \leq k$. Assuming that $C \succ_i C'$ whenever $\mu_i(C) > \mu_i(C')$, the property stating that an agent i, between two coalitions C and C' in which she has the same sum of valuations for all other agents, prefers the one with smallest size, always holds if $\delta_C(i) = \delta_{C'}(i) > 0$. In the spirit of fractional hedonic games, in order to guarantee this property also in the case $\delta_C(i) = \delta_{C'}(i) = 0$, we assume that, in this case, i prefers $|C|$ to $|C'|$ whenever $|C| < |C'|$. Therefore, in a k-fractional hedonic game, the preference order $\succeq_i$ of agent i is induced by her valuations for all other agents as follows: given two coalitions C and C' containing agent i, $C \succ_i C'$ if and only if either (i) $\mu_i(C) > \mu_i(C')$ or (ii) $\mu_i(C) = \mu_i(C') = 0$ and $|C| < |C'|$. An outcome of the game is a *k-coalition structure* $\mathcal{C} = \{C_1, \ldots, C_h\}$. $\mathcal{C}$ is a partition of the agents into h k-coalitions, that is, $\bigcup_{t \in [h]} C_t = N$ and $C_t \cap C_p = \emptyset$ $\forall t, p \in [h]$, with $t \neq p$. When clear from the context, we will refer to a k-coalition structure simply by calling it a coalition structure. We denote by $\mathcal{C}(i)$ the coalition agent i belongs to in coalition structure $\mathcal{C}$. The utility $\mu_i(\mathcal{C}(i))$ of an agent i in coalition structure $\mathcal{C}$ is also denoted by $\mu_i(\mathcal{C})$.

Nash Stability. Given a k-coalition structure $\mathcal{C}$, an agent $i \in N$ and a coalition $C \in \mathcal{C} \cup \{\emptyset\}$ such that $|C| \leq k - 1$, coalition $C \in \mathcal{C}$ is an *improving deviation* for agent i in $\mathcal{C}$ if $C \cup \{i\} \succ_i \mathcal{C}(i)$. Denote with $N_{\mathrm{ID}}(\mathcal{C})$ the set of agents possessing an improving deviation in $\mathcal{C}$. We say that agent i is *stable* in $\mathcal{C}$ if $i \notin N_{\mathrm{ID}}(\mathcal{C})$ and that a coalition structure $\mathcal{C}$ is *Nash stable* if $N_{\mathrm{ID}}(\mathcal{C}) = \emptyset$, that is, if all agents are stable in $\mathcal{C}$. A *dynamics* is a sequence of coalition structures $\langle \mathcal{C}^0, \mathcal{C}^1, \ldots \rangle$ such that, for any $j \geq 0$, $\mathcal{C}^{j+1}$ is obtained by $\mathcal{C}^j$ by an improving deviation of agent a_j selecting coalition $C_j \in \mathcal{C}^j$. We denote with $[\mathcal{C}_{-i}, C] = \mathcal{C} \setminus \{\mathcal{C}(i), C\} \cup \{\mathcal{C}(i) \setminus \{i\}, C \cup \{i\}\}$ the coalition structure obtained from $\mathcal{C}$ when agent i changes her strategy/coalition from $\mathcal{C}(i)$ to C. Therefore, for any $j \geq 0$, $\mathcal{C}^{j+1} = [\mathcal{C}^j_{-a_j}, C_j]$. A game has the *finite improvement path property* if it does not admit a dynamics of infinite length. In this case, we also say that the game is *convergent*. Moreover, if a game always converges after a polynomial number of improving deviations, we say that it is *polynomially convergent* or, equivalently, that it has the *polynomially bounded improvement path property*. We denote with $\mathrm{NS}(\mathcal{G})$ the set of Nash stable coalition structures of $\mathcal{G}$. Clearly, a game possessing the finite improvement path property always admits a Nash stable coalition structure.

Social Welfare. The *social welfare* of a coalition structure $\mathcal{C} = \{C_1, \ldots, C_h\}$ is given by the sum of the agents' utilities, i.e., $SW(\mathcal{C}) = \sum_{i \in N} \mu_i(\mathcal{C})$. By extending the previous definition, given a coalition C, we denote by $SW(C)$ the sum of utilities of the agents belonging to C. Notice that $SW(\mathcal{C}) = \sum_{C \in \mathcal{C}} SW(C) = \sum_{C \in \mathcal{C}} \sum_{i \in C} \mu_i(C)$.

Efficiency. Given a game $\mathcal{G}$, let $\mathcal{C}^*(\mathcal{G})$ be the outcome maximizing the social welfare, and let $\mathrm{NS}(\mathcal{G})$ be the set of Nash stable coalition structures. The *price of anarchy* (respectively, the *price of stability*) of $\mathcal{G}$ is defined as the ratio between the social welfare of the optimal outcome $C^*(\mathcal{G})$ and the one of a *worst* (respectively, *best*) Nash stable outcome. Formally, $\text{POA}(\mathcal{G}) = \max_{C \in \text{NS}(\mathcal{G})} \frac{SW(C^*(\mathcal{G}))}{SW(C)}$ and $\text{POS}(\mathcal{G}) = \min_{C \in \text{NS}(\mathcal{G})} \frac{SW(C^*(\mathcal{G}))}{SW(C)}$.

2.2 Preliminary Results

We first focus on the complexity of computing an optimal solution as well as a Nash stable coalition structure with the highest possible social welfare. We show that, even when considering simple symmetric games, both problems are $\mathcal{NP}$-hard for any $k \geq 3$. On the positive side, optimal solutions for $k = 2$ and approximate ones for $k > 2$ will be provided in the following of this section (see Proposition 3) and in the next sections (see Theorems 7 and 12), respectively.

Theorem 1. *Fixed any $k \geq 3$ and given any* $k-\text{SSFHG}$ *$\mathcal{G}$, the problem of computing a coalition structure maximizing the social welfare is $\mathcal{NP}$-hard. The same holds for the problem of computing a Nash stable coalition structure having the highest possible social welfare.*

We now show that, when considering non symmetric $k-\text{FHG}$, Nash stable outcomes are not guaranteed to exist even for the case when agents' values are 0 or 1. For this reason, in the rest of the paper we will focus on symmetric fractional hedonic games.

Proposition 1. *For any $k \geq 2$, there exists a $k-\text{FHG}$ $\mathcal{G}$ where agents' values can only take the values 0 and 1 such that $\text{NS}(\mathcal{G}) = \emptyset$.*

Proof. Consider the $k-\text{FHG}$ $\mathcal{G}$ whose graph representation is $G = (N, E)$, where G is a directed cycle of $k + 1$ nodes, that is, $|N| = \{1, 2, \ldots, k + 1\}$ and there is an arc $(i, i+1) \in E$, for any $i = 1, \ldots, k$, and also the arc $(k+1, 1) \in E$. Notice that, since k does not divide $|N|$, in any coalition structure $\mathcal{C}$ of $\mathcal{G}$ there must exist a coalition $C \in \mathcal{C}$ such that $|C| \leq k - 1$. Let $j \in C$ be a node with no incoming arc in $G(C)$. Notice that such node j has to exist in C. If $j = 1$, coalition C represents an improving deviation for agent $k + 1$ (notice that agent $k + 1$ has utility zero in C). If $j > 1$, coalition C represents an improving deviation for agent $j - 1$ (notice that agent $j - 1$ has utility zero in C), and this completes the proof. $\qquad\square$

In the following of this section, we provide some preliminary results for the case of $k = 2$. The following theorem shows that, when $k = 2$, any symmetric fractional hedonic game is convergent and that any simple symmetric fractional hedonic game is polynomially convergent. It is a direct consequence of the analogous results by [9] holding for additively separable hedonic games. In fact, $2-\text{SFHG}$ are equivalent to symmetric additively separable hedonic games in which every valuation is divided by 2.

Theorem 2 ([9]). *Any $2-\text{SFHG}$ has the finite improvement path property. Moreover, any $2-\text{SSFHG}$ also possesses the polynomially bounded improvement path property.*

The following proposition shows that for $2-$SFHG the price of anarchy is unbounded.

Proposition 2. *Given any value $M > 0$, there exists a $2-$SFHG $\mathcal{G}$, such that $\text{PoA}(\mathcal{G}) \geq M$.*

We now show that a Nash stable coalition structure also being an optimal solution can be computed in polynomial time, thus implying that the price of stability is 1.

Proposition 3. *Given any $2-$SFHG $\mathcal{G}$, a Nash stable coalition structure C such that $SW(C) = SW(C^*(\mathcal{G}))$ can be computed in polynomial time. Moreover, $\text{PoS}(\mathcal{G}) = 1$.*

In the following, we show that for $2-$SSFHG the price of anarchy is exactly 2. We start by proving an upper bound, and then we also provide a matching lower bound.

Theorem 3. *Given any $2-$SSFHG $\mathcal{G}$, $\text{PoA}(\mathcal{G}) \leq 2$.*

Proof. Consider the graph representation $G = (N, E)$ of $\mathcal{G}$. We start by showing that any Nash stable coalition structure C of $\mathcal{G}$ induces a maximal matching on G. First notice that in any Nash stable outcome there is no 2-coalition composed by two agents i and j such that $v_i(j) = v_j(i) = 0$ (or such that $\{i, j\} \notin E$). Thus, in any Nash stable outcome for $\mathcal{G}$, we have that any coalition contains either an agent in a singleton or two agents i and j such that $\{i, j\} \in E$. Assume, by contradiction, that C does not induce a maximal matching. It means that there exist two agents i and j in singletons such that $\{i, j\} \in E$, thus implying that both agents i and j possess an improving deviation: a contradiction to the fact that C is Nash stable. The claim follows by the well known fact that any maximal unweighted matching is a 2-approximation of a maximum unweighted matching and that a maximum unweighted matching is optimum (see Proposition 3). $\square$

Proposition 4. *There exists a $2-$SSFHG $\mathcal{G}$ such that $\text{PoA}(\mathcal{G}) \geq 2$.*

3   Results for 3-Coalitions

In this section we study $3-$SFHG. We first show that, when considering symmetric games with general non-negative valuations, Nash stable outcomes are not guaranteed to exist. For this reason, in the rest of this section we will focus on $3-$SSFHG.

Proposition 5. *There exists a $3-$SFHG $\mathcal{G}$ such that $\text{NS}(\mathcal{G}) = \emptyset$.*

We notice that, given a $k-$FHG $\mathcal{G}$ having graph representation G and any coalition structure C for $\mathcal{G}$, the set of possible improving deviations for $\mathcal{G}$ starting from C is a subset of the set of possible improving deviations for $\mathcal{G}'$ starting from the same coalition structure C, where $\mathcal{G}'$ is any $k'-$FHG, with $k' > k$, having the same graph representation G (since $k' > k$, it holds that coalition structure C is a feasible coalition structure also for $\mathcal{G}'$). Therefore, Theorem 8 also shows the convergence for $3-$SSFHGs and the following theorem directly follows.

Theorem 4. *Any $3-$SSFHG $\mathcal{G}$ has the polynomially bounded improvement path property.*

In the following, we focus on the efficiency of Nash stable outcomes: we show that for $3-$SSFHG the price of stability is 1 and price of anarchy is exactly 2 (we provide both an upper bound and a matching lower bound).

Theorem 5. *Given any* $3-$SSFHG $\mathcal{G}$ *and any dynamics* $\langle \mathcal{C}^0, \ldots, \mathcal{C}^h \rangle$ *for* $\mathcal{G}$, *it holds that* $SW(\mathcal{C}^h) \geq SW(\mathcal{C}^0)$. *As a consequence,* $\text{POS}(\mathcal{G}) = 1$.

Proof. Consider any $3-$SSFHG $\mathcal{G}$ and any dynamics $\langle \mathcal{C}^0, \ldots, \mathcal{C}^h \rangle$ for $\mathcal{G}$. We now show that no improving deviation can decrease the social welfare, thus proving the first part of the claim. Notice that also the second part of the claim is an immediate consequence of this fact, given that any dynamics starting from an optimal solution (i) leads to a stable outcome by Theorem 8 and (ii) is such that the social welfare of the reached stable outcome is not smaller than the one of the initial outcome, i.e., the optimal social welfare.

Consider agent i performing an improving deviation from coalition $C \cup \{i\}$ to coalition $C' \cup \{i\}$. Clearly, it holds that $\mu_i(C' \cup \{i\}) \geq \mu_i(C \cup \{i\})$. Let C be such that $G(C)$ has n_C agents m_C edges, and C' be such that $G(C')$ has $n_{C'}$ agents $m_{C'}$ edges. Given that, after the considered improving deviation of agent i, the only coalitions modifying their contribution to the social welfare are C and C', in order to show that the social welfare does not decrease, it is sufficient to prove that $SW(C) + SW(C' \cup \{i\}) \geq SW(C \cup \{i\}) + SW(C')$. If $C = \emptyset$, or more generally if $\mu_i(C \cup \{i\}) = 0$, it trivially holds that the improving deviation cannot decrease the social welfare, because adding an edge to a coalition of size at most 2 cannot lower the social welfare: we can assume $n_C > 0$ and $\mu_i(C \cup \{i\}) > 0$. Moreover, if $C' = \emptyset$ it follows that $\mu_i(C \cup \{i\}) = 0$ and therefore we can also assume that $n_{C'} > 0$. We have, assuming $n_C > 0$ and $n_{C'} > 0$,

$$SW(C) + SW(C' \cup \{i\}) \geq SW(C \cup \{i\}) + SW(C')$$
$$\frac{2m_C}{n_C} + \frac{2(m_{C'} + \delta_{C'}(i))}{n_{C'} + 1} \geq \frac{2(m_C + \delta_C(i))}{n_C + 1} + \frac{2m_{C'}}{n_{C'}}$$
$$\frac{m_C}{n_C(n_C + 1)} + \frac{\delta_{C'}(i)}{n_{C'} + 1} \geq \frac{\delta_C(i)}{n_C + 1} + \frac{m_{C'}}{n_{C'}(n_{C'} + 1)}.$$

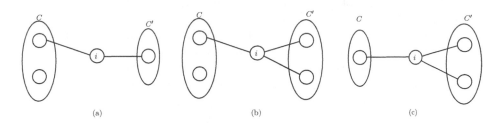

Fig. 1. Possible improving deviations of agent i.

By the last obtained inequality, in order to show that no improving deviation can decrease the social welfare, having fixed n_C, $n_{C'}$, $\delta_C(i)$ and $\delta_{C'}(i)$, it is sufficient to consider the greatest possible value for $m_{C'}$ (i.e., $m_{C'} = \frac{n_{C'}(n_{C'}-1)}{2}$) and the smallest possible value for m_C (i.e., $m_C = 0$). We distinguish among the following disjoint cases.

- $n_C = 2$, $\delta_C(i) = 1$, $n_{C'} = 1$ and $\delta_{C'}(i) = 1$. In this case (see Fig. 1(a)), agent i improves her utility from $\frac{1}{3}$ to $\frac{1}{2}$ and the greatest possible value for $m_{C'}$ is 0. It holds that $0 + 1 = SW(C) + SW(C' \cup \{i\}) \geq SW(C \cup \{i\}) + SW(C') = \frac{2}{3} + 0$.
- $n_C = 2$, $\delta_C(i) = 1$, $n_{C'} = 2$ and $\delta_{C'}(i) = 2$. In this case (see Fig. 1(b)), agent i improves her utility from $\frac{1}{3}$ to $\frac{2}{3}$ and the greatest possible value for $m_{C'}$ is 1. It holds that $0 + 2 = SW(C) + SW(C' \cup \{i\}) \geq SW(C \cup \{i\}) + SW(C') = \frac{2}{3} + 1$.
- $n_C = 1$, $\delta_C(i) = 1$, $n_{C'} = 2$ and $\delta_{C'}(i) = 2$. In this case (see Fig. 1(c)), agent i improves her utility from $\frac{1}{2}$ to $\frac{2}{3}$ and the greatest possible value for $m_{C'}$ is 1. It holds that $0 + 2 = SW(C) + SW(C' \cup \{i\}) \geq SW(C \cup \{i\}) + SW(C') = 1 + 1$.

$\square$

Theorem 6. *Given any* $3-$SSFHG $\mathcal{G}$, $\mathrm{PoA}(\mathcal{G}) \leq 2$.

Proof. Consider any $3-$SSFHG $\mathcal{G}$ and an optimal outcome $\mathcal{C}^* = (C_1^*, \ldots, C_h^*)$ for $\mathcal{G}$ such that any agent with zero utility is in a singleton. Let $\mathcal{C}$ be any Nash stable outcome for $\mathcal{G}$. Recall that, in any Nash stable outcome, every node with zero utility is in a singleton. In the following, we show that, for any $j \in [h]$, $\sum_{i \in C_j^*} \mu_i(\mathcal{C}^*) \leq 2 \sum_{i \in C_j^*} \mu_i(\mathcal{C})$, thus implying the claim.

For any $j \in [h]$, we consider coalition C_j^* and we divide the proof in the following disjoint cases, depending on the graph coalition C_j^* is isomorphic to.

- C_j^* is isomorphic to K_3.
  If every agent in C_j^* has utility greater than zero (i.e., at least $\frac{1}{3}$) in $\mathcal{C}$, than it holds that $\sum_{i \in C_j^*} \mu_i(\mathcal{C}^*) = 2 \leq 2 \sum_{i \in C_j^*} \mu_i(\mathcal{C})$.
  Otherwise, i.e., if there exists agent $i_1 \in C_j^*$ such that $\mu_{i_1}(\mathcal{C}) = 0$, we now show that the other two agents $i_2, i_3 \in C_j^*$ are such that $\mu_{i_2}(\mathcal{C}) = \mu_{i_3}(\mathcal{C}) = \frac{2}{3}$, thus implying that $2 = \sum_{i \in C_j^*} \mu_i(\mathcal{C}^*) \leq 2 \sum_{i \in C_j^*} \mu_i(\mathcal{C}) = 2 \cdot \frac{4}{3}$. In fact, agent i_2 cannot be such that $\mu_{i_2}(\mathcal{C}) \leq \frac{1}{3}$, because in this case she would have an improving deviation by joining the singleton of i_1: a contradiction to the stability of $\mathcal{C}$. Moreover, agent i_2 cannot be such that $\mu_{i_2}(\mathcal{C}) = \frac{1}{2}$, because in this case $|\mathcal{C}(i_2)| = 2$ and agent i_1 would have an improving deviation by joining the coalition of i_2: again a contradiction to the stability of $\mathcal{C}$. The same holds, symmetrically, for agent i_3.
- C_j^* is isomorphic to a path of 3 nodes containing agents i_1, i_2 and i_3.
  In this case, if at least two agents in C_j^* have utility greater than zero (i.e., at least $\frac{1}{3}$) in $\mathcal{C}$, than it holds that $\sum_{i \in C_j^*} \mu_i(\mathcal{C}^*) = \frac{4}{3} \leq 2 \sum_{i \in C_j^*} \mu_i(\mathcal{C})$.
  Otherwise, i.e., if there exist two agents $i_1, i_2 \in C_j^*$ such that $\mu_{i_1}(\mathcal{C}) = \mu_{i_2}(\mathcal{C}) = 0$, we now show that the other agent $i_3 \in C_j^*$ is such that $\mu_{i_3}(\mathcal{C}) = \frac{2}{3}$, thus implying that $\frac{4}{3} = \sum_{i \in C_j^*} \mu_i(\mathcal{C}^*) \leq 2 \sum_{i \in C_j^*} \mu_i(\mathcal{C}) = 2 \cdot \frac{2}{3}$. In fact, let $\bar{i}$ be an agent in C_j^* being adjacent to i_3, i.e., either $\bar{i} = i_1$ or $\bar{i} = i_2$. Agent i_3 cannot be such that $\mu_{i_3}(\mathcal{C}) \leq \frac{1}{3}$, because in this case she would have an improving deviation by joining the singleton of $\bar{i}$: a contradiction to the stability of $\mathcal{C}$. Moreover, agent i_3 cannot

be such that $\mu_{i_3}(\mathcal{C}) = \frac{1}{2}$, because in this case $|\mathcal{C}(i_3)| = 2$ and agent $\bar{i}$ would have an improving deviation by joining the coalition of i_3: again a contradiction to the stability of $\mathcal{C}$.

- C_j^* is isomorphic to K_2 and contains nodes i_1 and i_2 with $\delta_{C_j^*}(i_1) = \delta_{C_j^*}(i_2) = 1$. If every agent in C_j^* has utility greater than zero (i.e., at least $\frac{1}{3}$) in $\mathcal{C}$, than it holds that $\sum_{i \in C_j^*} \mu_i(\mathcal{C}^*) = 1 \leq 2 \cdot \frac{2}{3} \leq 2 \sum_{i \in C_j^*} \mu_i(\mathcal{C})$.

  Otherwise, i.e., if there exists agent $i_1 \in C_j^*$ such that $\mu_{i_1}(\mathcal{C}) = 0$, we now show that the other agent $i_2 \in C_j^*$ is such that $\mu_{i_2}(\mathcal{C}) = \frac{2}{3}$, thus implying that $1 = \sum_{i \in C_j^*} \mu_i(\mathcal{C}^*) \leq 2 \sum_{i \in C_j^*} \mu_i(\mathcal{C}) = 2 \cdot \frac{2}{3}$. In fact, agent i_2 cannot be such that $\mu_{i_2}(\mathcal{C}) \leq \frac{1}{3}$, because in this case she would have an improving deviation by joining the singleton of i_1: a contradiction to the stability of $\mathcal{C}$. Moreover, agent i_2 cannot be such that $\mu_{i_2}(\mathcal{C}) = \frac{1}{2}$, because in this case $|\mathcal{C}(i_2)| = 2$ and agent i_1 would have an improving deviation by joining the coalition of i_2: again a contradiction to the stability of $\mathcal{C}$.

- C_j^* is a singleton. In this case, it trivially holds that $0 = \sum_{i \in C_j^*} \mu_i(\mathcal{C}^*) \leq 2 \sum_{i \in C_j^*} \mu_i(\mathcal{C})$.

$\square$

Proposition 6. *There exists a* $3-$SSFHG $\mathcal{G}$ *such that* POA$(\mathcal{G}) \geq 2$.

We now study the complexity of computing solutions with high social welfare. On the one hand, by Theorem 1 we know that the problem of computing an optimal coalition structure is $\mathcal{NP}$-hard as well as the one of computing the best Nash stable outcome; on the other hand, we provide in the next theorem a $\frac{4}{3}$-polynomial time approximation algorithm for both problems.

Theorem 7. *Given any* $3-$SSFHG $\mathcal{G}$, *it is possible to compute in polynomial time a Nash stable coalition structure* $\mathcal{C}$ *such that* $SW(\mathcal{C}) \geq \frac{3}{4}SW(\mathcal{C}^*)$.

Proof. Consider any $3-$SSFHG $\mathcal{G}$ with its corresponding graph representation $G = (N, E)$ and an optimal outcome $\mathcal{C}^* = (C_1^*, \ldots, C_h^*)$ for $\mathcal{G}$ such that any agent with zero utility is in a singleton.

In [15], a polynomial time algorithm for packing edges and triangles in a graph in order to cover the maximum number of nodes is provided. More formally, the algorithm, given an unweighted undirected graph in input, produces in outputs a set of disjoint cliques of size 2 and 3 that cover the maximum possible number of nodes. Let $\mathcal{C}'$ be the coalition structure corresponding to the output of the aforementioned algorithm applied to graph G, in which a singleton is added for every node of the graph not belonging to any matching or triangle.

Let k_3, k_2 and p_3 be the number of coalitions in $\mathcal{C}^*$ being isomorphic to K_3, K_2 and to a path with 3 nodes, respectively. In the following, we show that there exists a solution for the problem of packing edges and triangles in graph G covering $x = 3k_3 + 2k_2 + 2p_3$ nodes. Therefore, in $\mathcal{C}'$ at least x agents are in coalition isomorphic to K_2 or K_3; since each agent in such coalitions has utility at least $\frac{1}{2}$, it holds that $SW(\mathcal{C}') \geq \frac{x}{2} = \frac{3}{2}k_3 + k_2 + p_3$. On the other hand, it holds that $SW(\mathcal{C}^*) = 2k_3 + k_2 + \frac{4}{3}p_3$. We thus obtain

$$SW(\mathcal{C}^*) = 2k_3 + k_2 + \frac{4}{3}p_3 \leq \frac{4}{3}\left(\frac{3}{2}k_3 + k_2 + p_3\right) \leq \frac{4}{3}SW(\mathcal{C}').$$

The claim then follows by Theorems 8 and 5 guaranteeing that, starting from outcome $\mathcal{C}'$, any dynamics converges in polynomial time to a Nash stable outcome $\mathcal{C}$ such that $SW(\mathcal{C}) \geq SW(\mathcal{C}')$.

It remains to show that there exists a solution for the problem of packing edges and triangles in graph G covering $x = 3k_3 + 2k_2 + 2p_3$ nodes, i.e., covering three nodes for each coalition of $\mathcal{C}^*$ isomorphic to K_3 and two nodes for each coalition of $\mathcal{C}^*$ isomorphic to K_2 or to a path with 3 nodes. For any $j \in [h]$, consider coalition C_j^*:

- If C_j^* is isomorphic to K_3, it is possible to cover all three nodes in C_j^* by a triangle.
- If C_j^* is isomorphic to K_2, it is possible to cover both nodes in C_j^* by an edge.
- If C_j^* is isomorphic to a path of 3 nodes, it is possible to cover two nodes (the central one and any of the remaining nodes) by an edge.

□

4   Results for 4-Coalitions

In this section we study $4-$SFHG. We first show that, when considering symmetric games with general non-negative valuations, Nash stable outcomes are not guaranteed to exist. For this reason, in the rest of this section we will focus on $4-$SSFHG, for which we have already proven in Theorem 8 that they always converge after a polynomial number of improving deviations.

Proposition 7. *There exists a* $4-$SFHG $\mathcal{G}$ *such that* $NS(\mathcal{G}) = \emptyset$.

We now show that simple symmetric games are polynomially convergent.

Theorem 8. *Any* $4-$SSFHG $\mathcal{G}$ *has the polynomially bounded improvement path property.*

Proof. Consider any $4-$SSFHG $\mathcal{G}$ and any dynamics $\langle \mathcal{C}^0, \mathcal{C}^1, \ldots \rangle$ for $\mathcal{G}$. Let $v_j = (\alpha_j, \beta_j^1, \beta_j^2, \beta_j^3)$ with $\alpha_j = \sum_{C \in \mathcal{C}^j} |E(C)|$ being the total number of edges in all coalitions of $\mathcal{C}^j$ and, for any $h = 1, 2, 3$, $\beta_j^h = |\{C|C \in \mathcal{C}^j \wedge |C| \geq h\}|$ being the number of coalitions of size at least h in $\mathcal{C}^j$ (notice that $\beta_j^1 = |\mathcal{C}^j|$). We prove the convergence of the dynamics by showing that, for any $j \geq 0$, it holds that $v_{j+1} >^{\text{lex}} v_j$, i.e., v_{j+1} is lexicographically bigger than v_j[1]. Given that, for $k = 4$, the number of possible vectors can be easily verified to be polynomially bounded (in particular, it is $O(n^5)$), the claim immediately follows.

Consider any $j \geq 0$, with a_j being the agent performing an improving deviation $C_j \in \mathcal{C}^j$ leading from $\mathcal{C}^j$ to $\mathcal{C}^{j+1}$. First of all, notice that, since we are considering coalitions of size at most 4, the sum of valuations of a_j for all agents in coalition C_j that she joins cannot be smaller than the one for all agents in coalition $\mathcal{C}^j(a_j)$ that she leaves, i.e., $\delta_{C_j \cup \{a_j\}}(a_j) \geq \delta_{\mathcal{C}^j(a_j)}(a_j)$. We divide the proof in two disjoint cases.

- If $\delta_{C_j \cup \{a_j\}}(a_j) > \delta_{\mathcal{C}^j(a_j)}(a_j)$, it directly follows that $\alpha_{j+1} > \alpha_j$, thus implying that $v_{j+1} >^{\text{lex}} v_j$.

---

[1] A vector v is lexicographically bigger than vector v' if the first non-zero coordinate of $v - v'$ is positive.

– Otherwise, i.e., if $\delta_{C_j \cup \{a_j\}}(a_j) = \delta_{C^j(a_j)}(a_j)$, we have that a_j leaves coalition $C^j(a_j)$ of size x for joining a coalition C_j of size $y < x - 1$, so that $|C_j \cup \{a_j\}| = y + 1 < x$ (notice that, when $\delta_{C^j(a_j)}(a_j) = 0$, a_j could also form a singleton, i.e., it may be $C_j = \emptyset$). In this case, let $h' = \min\{x - 1, y + 1\}$ (notice that $h' \leq 3$). We have that $\alpha_{j+1} = \alpha_j$, $\beta_{j+1}^h = \beta_j^h$ for all $h < h'$ and $\beta_{j+1}^{h'} > \beta_j^{h'}$, thus implying that $v_{j+1} >^{\text{lex}} v_j$.

$\square$

In the following we show that for $4-$SSFHG the price of anarchy is exactly 3. We start by proving an upper bound, and then we also provide a matching lower bound.

Theorem 9. *Given any* $4-$SSFHG $\mathcal{G}$, $\text{PoA}(\mathcal{G}) \leq 3$.

Proof. Consider any $4-$SSFHG $\mathcal{G}$ with its corresponding graph representation $G = (N, E)$ and an optimal outcome $\mathcal{C}^* = (C_1^*, \ldots, C_h^*)$ for $\mathcal{G}$ such that any agent with zero utility is in a singleton. Let $\mathcal{C}$ be any Nash stable outcome for $\mathcal{G}$. Recall that, in any Nash stable outcome, every node with zero utility is in a singleton. In the following, we show that, for any $j \in [h]$, $\sum_{i \in C_j^*} \mu_i(\mathcal{C}^*) \leq 3 \sum_{i \in C_j^*} \mu_i(\mathcal{C})$, thus implying the claim.

The following *basic property* will be exploited several times in the remainder of the proof: for every edge $\{i_1, i_2\} \in E$, if agent i_1 (resp. i_2) is such that $\mu_{\mathcal{C}}(i_1) = 0$ (resp. $\mu_{\mathcal{C}}(i_2) = 0$), then it holds that $\mu_{\mathcal{C}}(i_2) \geq \frac{1}{2}$ (resp. $\mu_{\mathcal{C}}(i_1) \geq \frac{1}{2}$): in fact, if $\mu_{\mathcal{C}}(i_2) < \frac{1}{2}$ (resp. $\mu_{\mathcal{C}}(i_1) < \frac{1}{2}$), agent i_2 (resp. agent i_1) could perform an improving deviation by joining the singleton of agent i_1 (resp. agent i_2), thus contradicting the stability of $\mathcal{C}$.

For any $j \in [h]$, we consider coalition C_j^* and we divide the proof in the following disjoint cases, depending on the size of coalition C_j^*.

– $|C_j^*| = 4$.
  If every agent in C_j^* has utility greater than zero (i.e., at least $\frac{1}{4}$) in $\mathcal{C}$, than it holds that $\sum_{i \in C_j^*} \mu_i(\mathcal{C}^*) \leq 3 \leq 3 \sum_{i \in C_j^*} \mu_i(\mathcal{C})$.
  Otherwise, i.e., if there exists agent $i_1 \in C_j^*$ such that $\mu_{i_1}(\mathcal{C}) = 0$, we distinguish between two subcases, depending on the number $\delta_{i_1}(C_j^*)$ of neighbours of i_1 in C_j^*:
  - if $\delta_{i_1}(C_j^*) \geq 2$, by the basic property it holds that $\sum_{i \in C_j^*} \mu_i(\mathcal{C}) \geq 2 \cdot \frac{1}{2} = 1$, thus implying that $\sum_{i \in C_j^*} \mu_i(\mathcal{C}^*) \leq 3 \leq 3 \sum_{i \in C_j^*} \mu_i(\mathcal{C})$;
  - otherwise, since i_1 cannot be isolated in C_j^*, it has to hold that $\delta_{i_1}(C_j^*) = 1$ and, therefore, the number of edges in C_j^* is at most 4, which implies that $SW(C_j^*) \leq 2$. Let i_2 be the unique adjacent of i_1 in C_j^* (by the basic property, $\mu_{i_2}(\mathcal{C}) \geq \frac{1}{2}$), and i_3 and i_4 the other two nodes belonging to C_j^*. If both i_3 and i_4 have utility greater then zero (i.e., at least $\frac{1}{4}$) in $\mathcal{C}$, then we are done because $\sum_{i \in C_j^*} \mu_i(\mathcal{C}) \geq \frac{1}{2} + 2 \cdot \frac{1}{4} = 1$ and therefore $\sum_{i \in C_j^*} \mu_i(\mathcal{C}^*) \leq 2 \leq 3 \sum_{i \in C_j^*} \mu_i(\mathcal{C})$. Otherwise, let i_3 such that $\mu_{i_3}(\mathcal{C}) = 0$: If $\{i_3, i_4\} \in E$, by the basic property it holds that $\mu_{i_4}(\mathcal{C}) \geq \frac{1}{2}$ and therefore $\sum_{i \in C_j^*} \mu_i(\mathcal{C}) \geq \frac{1}{2} + \frac{1}{2} = 1$, thus implying $\sum_{i \in C_j^*} \mu_i(\mathcal{C}^*) \leq 2 \leq 3 \sum_{i \in C_j^*} \mu_i(\mathcal{C})$; otherwise, the number of edges in C_j^* is at most 3, which implies that $SW(C_j^*) \leq \frac{3}{2}$ and $\sum_{i \in C_j^*} \mu_i(\mathcal{C}^*) \leq \frac{3}{2} \leq 3 \sum_{i \in C_j^*} \mu_i(\mathcal{C})$.

- $|C_j^*| = 3$.

  If every agent in C_j^* has utility greater than zero (i.e., at least $\frac{1}{4}$) in $\mathcal{C}$, than it holds that $\sum_{i \in C_j^*} \mu_i(\mathcal{C}^*) \leq 2 \leq 3 \sum_{i \in C_j^*} \mu_i(\mathcal{C})$.

  Otherwise, i.e., if there exists agent $i_1 \in C_j^*$ such that $\mu_{i_1}(\mathcal{C}) = 0$, we distinguish between two subcases, depending on the number $\delta_{i_1}(C_j^*)$ of neighbours of i_1 in C_j^*:

  - if $\delta_{i_1}(C_j^*) = 2$, by the basic property it holds that $\sum_{i \in C_j^*} \mu_i(\mathcal{C}) \geq 2 \cdot \frac{1}{2} = 1$, thus implying that $\sum_{i \in C_j^*} \mu_i(\mathcal{C}^*) \leq 2 \leq 3 \sum_{i \in C_j^*} \mu_i(\mathcal{C})$;

  - otherwise, since i_1 cannot be isolated in C_j^*, it has to hold that $\delta_{i_1}(C_j^*) = 1$ and, therefore, the number of edges in C_j^* is at most 2, which implies that $SW(C_j^*) \leq \frac{4}{3}$. Therefore, by the basic property it holds that $\sum_{i \in C_j^*} \mu_i(\mathcal{C}) \geq \frac{1}{2}$, thus implying that $\sum_{i \in C_j^*} \mu_i(\mathcal{C}^*) \leq \frac{4}{3} \leq 3 \sum_{i \in C_j^*} \mu_i(\mathcal{C})$.

- $|C_j^*| = 2$.

  If both agents in C_j^* have utility greater than zero (i.e., at least $\frac{1}{4}$) in $\mathcal{C}$, than it holds that $\sum_{i \in C_j^*} \mu_i(\mathcal{C}^*) = 1 \leq 3 \sum_{i \in C_j^*} \mu_i(\mathcal{C})$.

  Otherwise, i.e., if there exists agent $i_1 \in C_j^*$ such that $\mu_{i_1}(\mathcal{C}) = 0$, by the basic property the other agent $i_2 \in C_j^*$ is such that $\mu_{i_2}(\mathcal{C}) \geq \frac{1}{2}$, again implying that $\sum_{i \in C_j^*} \mu_i(\mathcal{C}^*) = 1 \leq 3 \sum_{i \in C_j^*} \mu_i(\mathcal{C})$.

- $|C_j^*| = 1$. In this case, it trivially holds that $0 = \sum_{i \in C_j^*} \mu_i(\mathcal{C}^*) \leq 3 \sum_{i \in C_j^*} \mu_i(\mathcal{C})$.

  $\square$

Proposition 8. *There exists a* $4-$SSFHG $\mathcal{G}$ *such that* $\mathrm{POA}(\mathcal{G}) \geq 3$.

Our main technical result is represented by the following theorem, stating that any coalition structure can be transformed in polynomial time in a stable one at the price of losing a factor of $\frac{10}{9}$ on the social welfare.

Theorem 10. *Given any* $4-$SSFHG $\mathcal{G}$ *and any coalition structure* $\mathcal{C}$ *for* $\mathcal{G}$, *it is possible to compute in polynomial time a stable coalition structure* $\mathcal{C}'$ *such that* $SW(\mathcal{C}') \geq \frac{9}{10} SW(\mathcal{C})$.

The consequences of Theorem 10 are twofold: it allows us both to provide a (tight) upper bound to the price of stability, and to devise an algorithm computing a Nash stable coalition structure $\frac{5}{3}$-approximating the optimal solution.

We start by showing that for $4-$SSFHGs the price of stability is exactly $\frac{10}{9}$. Next theorem, providing an upper bound to the price of stability, is an immediate consequence of Theorem 10 exploited for transforming in a Nash stable outcome an optimal coalition structure $\mathcal{C}^*$ for the considered $4-$SSFHG $\mathcal{G}$. We also give in Proposition 9 a matching lower bound.

Theorem 11. *Given any* $4-$SSFHG $\mathcal{G}$, $\mathrm{PoS}(\mathcal{G}) \leq \frac{10}{9}$.

Proposition 9. *There exists a* $4-$SSFHG $\mathcal{G}$ *such that* $\mathrm{PoS}(\mathcal{G}) \geq \frac{10}{9}$.

Proof. Consider the $4-$SSFHG $\mathcal{G}$ with 6 agents depicted in Fig. 2(a). On the one hand, there exists outcome $\mathcal{C}'$ (see Fig. 2(b)) with two coalitions having three agents each, for

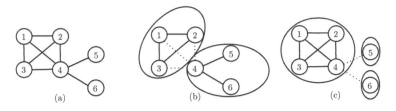

Fig. 2. Game $\mathcal{G}$ proving the lower bound on the price of stability. (a) The graph instance. (b) Outcome $\mathcal{C}'$. (c) Outcome $\mathcal{C}$.

which $SW(\mathcal{C}') = \frac{4}{3} + 2 = \frac{10}{3}$. It follows that $SW(\mathcal{C}^*) \geq \frac{10}{3}$. Notice that outcome $\mathcal{C}'$ is not stable, as agent 4 can improve her utility from $\frac{2}{3}$ to $\frac{3}{4}$ by joining the other coalition.

On the other hand, consider outcome $\mathcal{C}$ (see Fig. 2(c)) with only one coalition of size 4 and two singletons: it holds that $SW(\mathcal{C}) = 3$.

It can be easily verified that $\mathcal{C}$ is Nash stable: the agents in the coalition of size 4 have the maximum possible utility of $\frac{3}{4}$, while all the other agents, even if are experiencing a utility equal to zero, have no improving deviation available.

We now show that there exists no stable outcome $\bar{\mathcal{C}}$ having social welfare greater than 3. Assume, by way of contradiction, that there exists a stable outcome $\bar{\mathcal{C}}$ having social welfare greater than 3. We consider the following disjoint cases.

- $\bar{\mathcal{C}}$ is such that $\mu_5(\bar{\mathcal{C}}) = \mu_6(\bar{\mathcal{C}}) = 0$.
  In this case, since $\bar{\mathcal{C}}$ is stable, agent 4 has to belong to a coalition of size 4 (with agents 1, 2 and 3), otherwise agents 5 and 6 would have an improving deviation (see again Fig. 2(c)). Therefore, we have that $SW(\bar{\mathcal{C}}) = 3$: a contradiction.
- $\bar{\mathcal{C}}$ is such that $\mu_5(\bar{\mathcal{C}}) > 0$ and $\mu_6(\bar{\mathcal{C}}) = 0$.
  In this case, since $\bar{\mathcal{C}}$ is stable, agent 4 has to belong to a coalition of size 4 (with agent 5 and two agents among 1, 2 and 3), otherwise agent 6 would have an improving deviation (see Fig. 3(a)). Therefore, we have that $SW(\bar{\mathcal{C}}) = 2$: a contradiction.
- $\bar{\mathcal{C}}$ is such that $\mu_5(\bar{\mathcal{C}}) = 0$ and $\mu_6(\bar{\mathcal{C}}) > 0$.
  This case the symmetric of the previous one.
- $\bar{\mathcal{C}}$ is such that $\mu_5(\bar{\mathcal{C}}) > 0$ and $\mu_6(\bar{\mathcal{C}}) > 0$, i.e., agents 5 and 6 are in the same coalition C of agent 4.
  In this case, since $\bar{\mathcal{C}}$ is stable, all agents not belonging to coalition C have to be together in the same coalition, and (without considering the symmetric configura-

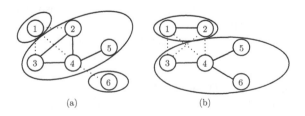

Fig. 3. Candidate stable outcomes $\bar{\mathcal{C}}$ having social welfare greater than 3.

tions) two possible outcomes are possible (see Fig. 2(b) and Fig. 3(b)).
The former outcome is C' that we have shown not to be stable: a contradiction.
The latter outcome is not stable as well, given that agent 3 has an improving devia-
tion toward the coalition of agents 1 and 2: again a contradiction.

The claim follows because $\frac{SW(C^*)}{SW(C)} \geq \frac{10}{9}$. ☐

We finally show that it is possible to compute a coalition structure approximating
the optimal social welfare by a factor $\frac{3}{2}$ and a stable one approximating the optimal
social welfare by a factor $\frac{5}{3}$ (recall that by Theorem 1 these problems are $\mathcal{NP}$-hard).

Theorem 12. *Given any* $4-$SSFHG $\mathcal{G}$, *it is possible to compute in polynomial time
a coalition structure* C *such that* $SW(C^*) \leq \frac{3}{2}SW(C)$ *and a Nash stable coalition
structure* C' *such that* $SW(C^*) \leq \frac{5}{3}SW(C')$.

5 Concluding Remarks and Open Problems

Several worth investigating research directions arise from this work. Concerning the
existence of Nash stable outcomes, we have shown that they are not guaranteed to exist
for $k-$FHG, for any $k \geq 2$, even for the case when agents' values are 0 or 1. Therefore,
we have considered $k-$SFHG and shown that the existence of Nash stable outcomes
is guaranteed for $k = 2$ (in particular, we have shown that dynamics for $2-$SFHG is
convergent) but not for $k = 3, 4$. Finally, we have shown that dynamics for $3-$SSFHG
and $4-$SSFHG games are convergent in a polynomial number of improving deviations.

The first important left open question is studying the existence of Nash stable outcomes of $k-$SFHG when $k \geq 5$. To this respect, as discussed in the introduction, we
now give evidence of the difficulty of providing general results holding for any value
of k. To this aim, we show that there is no inclusion between Nash stable outcomes
of $k-$SFHG and $k'-$SFHG games, with $k \neq k'$: Given any $n \geq 2$, consider the
family of games $\mathcal{G}_2 \in 2-$SSFHG$, \ldots, \mathcal{G}_n \in$ n$-$SSFHG all having as graph representation G an unweighted undirected clique of n nodes. It is easy to see that, for any
$k \in \{2, \ldots, n\}$, any Nash stable outcome of $\mathcal{G}_k$ must be a coalition structure with $\lfloor \frac{n}{k} \rfloor$
coalitions containing exactly k agents each, and possibly one last coalition containing
the remaining $n - k\lfloor \frac{n}{k} \rfloor$ agents. Therefore, we get that the set of Nash stable outcomes
of the NS$(\mathcal{G}_k) \cap$ NS$(\mathcal{G}_{k'}) = \emptyset$ for any $2 \leq k < k' \leq n$. Moreover, the following example shows that it may even happen that, for games having the same graph representation,
Nash stable outcomes exist for some values of k and do not exist for other values of k:
Consider the $3-$SFHG $\mathcal{G}$ of Proposition 5 that admits no Nash stable outcome. It is
easy to see that the $4-$SFHG $\mathcal{G}'$ having the same graph representation of $\mathcal{G}$ admits a
unique Nash stable outcome which is the grand coalition. Furthermore, given the results
of Theorem 2, we have that the set of Nash stable outcomes of the $2-$SFHG $\mathcal{G}''$ having
the same graph representation of $\mathcal{G}$ and $\mathcal{G}'$ is non-empty, but it does not contain the
grand coalition (being not feasible in the $\mathcal{G}''$).

Another interesting future work is to study the convergence of dynamics in
$k-$SFHG with $k \geq 5$. Unfortunately, even in the simple case, the proof of Theorem 8
holding for $4-$SSFHG cannot be extended to bigger values of k, because the exploited

property that in any improving deviation of agent i from coalition C to coalition C', the sum of valuations of i for all agents in coalition C' cannot be smaller than the one for all agents in coalition C no longer holds for $k \geq 5$. Moreover, it is worth noticing that in [7], the authors show a cycle of the dynamics of simple symmetric fractional hedonic games where all the involved coalitions have size at most 33. This clearly implies that dynamics of $k-\text{SSFHG}$ (and thus of $k-\text{SFHG}$) games is not convergent, for any $k \geq 33$ (since the same dynamics is a cycle also for $k-\text{SSFHG}$, for any $k \geq 33$).

Concerning the complexity of computing Nash stable outcomes, we have shown that, for $2-\text{SFHG}$, it is possible to compute in polynomial time a Nash stable coalition structure with the highest social welfare, while, for any $k \geq 3$, the problem is $\mathcal{NP}$-Hard even for $k-\text{SSFHG}$. We have thus proposed polynomial time approximation algorithms for $3-\text{SSFHG}$ and $4-\text{SSFHG}$. It would be important to study which is the best approximation of a Nash stable outcome that can be computed in polynomial time in case the existence of Nash stable outcomes is guaranteed, as well as studying the complexity of checking whether a given game admits a Nash stable outcome in case the existence of Nash stable outcomes is not guaranteed.

More generally, we believe that it is worth to investigate bounded size coalitions also for other subclasses of hedonic games (in this paper we have only considered fractional hedonic games), as well as for other stability notions (core stability, individual stability, etc.).

References

1. Aziz, H., Brandl, F., Brandt, F., Harrenstein, P., Olsen, M., Peters, D.: Fractional hedonic games. ACM Trans. Econ. Comput. **7**(2), 6:1–6:29 (2019)
2. Aziz, H., Brandt, F., Harrenstein, P.: Pareto optimality in coalition formation. Games Econ. Behav. **82**, 562–581 (2013)
3. Aziz, H., Brandt, F., Harrenstein, P.: Fractional hedonic games. In: Proceedings of the 13th International Conference on Autonomous Agents and Multiagent Systems (AAMAS), pp. 5–12 (2014)
4. Aziz, H., Gaspers, S., Gudmundsson, J., Mestre, J., Täubig, H.: Welfare maximization in fractional hedonic games. In: Proceedings of the Twenty-Fourth International Joint Conference on Artificial Intelligence (IJCAI), pp. 461–467 (2015)
5. Aziz, H., Savani, R.: Hedonic games. In: Handbook of Computational Social Choice, pp. 356–376. Cambridge University Press (2016)
6. Banerjee, S., Konishi, H., Sönmez, T.: Core in a simple coalition formation game. Soc. Choice Welfare **18**(1), 135–153 (2001). https://doi.org/10.1007/s003550000067
7. Bilò, V., Fanelli, A., Flammini, M., Monaco, G., Moscardelli, L.: Nash stable outcomes in fractional hedonic games: existence, efficiency and computation. J. Artif. Intell. Res. **62**, 315–371 (2018)
8. Bilò, V., Monaco, G., Moscardelli, L.: Hedonic games with fixed-size coalitions. In: Thirty-Sixth AAAI Conference on Artificial Intelligence, AAAI, pp. 9287–9295 (2022)
9. Bogomolnaia, A., Jackson, M.O.: The stability of hedonic coalition structures. Games Econom. Behav. **38**(2), 201–230 (2002). https://doi.org/10.1006/game.2001.0877
10. Brandl, F., Brandt, F., Strobel, M.: Fractional hedonic games: individual and group stability. In: Proceedings of the 2015 International Conference on Autonomous Agents and Multiagent Systems (AAMAS), pp. 1219–1227 (2015)

11. Brandt, F., Bullinger, M., Wilczynski, A.: Reaching individually stable coalition structures in hedonic games. In: Proceedings of the 35th AAAI Conference on Artificial Intelligence (AAAI), pp. 5211–5218 (2021)
12. Bullinger, M.: Pareto-optimality in cardinal hedonic games. In: Proceedings of the 19th International Conference on Autonomous Agents and Multiagent Systems (AAMAS), pp. 213–221 (2020)
13. Carosi, R., Monaco, G., Moscardelli, L.: Local core stability in simple symmetric fractional hedonic games. In: Proceedings of the 18th International Conference on Autonomous Agents and MultiAgent Systems (AAMAS), pp. 574–582 (2019)
14. Cechlárová, K., Romero-Medina, A.: Stability in coalition formation games. Int. J. Game Theory 29(4), 487–494 (2001). https://doi.org/10.1007/s001820000053
15. Cornuéjols, G., Hartvigsen, D., Pulleyblank, W.: Packing subgraphs in a graph. Oper. Res. Lett. 1(4), 139–143 (1982). https://doi.org/10.1016/0167-6377(82)90016-5
16. Cseh, Á., Fleiner, T., Harján, P.: Pareto optimal coalitions of fixed size. J. Mech. Inst. Des. 4(1), 87–108 (2019)
17. Dreze, J.H., Greenberg, J.: Hedonic coalitions: optimality and stability. Econometrica 48(4), 987–1003 (1980)
18. Elkind, E., Fanelli, A., Flammini, M.: Price of pareto optimality in hedonic games. Artif. Intell. 288, 103357 (2020). https://doi.org/10.1016/j.artint.2020.103357
19. Fanelli, A., Monaco, G., Moscardelli, L.: Relaxed core stability in fractional hedonic games. In: Proceedings of the Thirtieth International Joint Conference on Artificial Intelligence (IJCAI), pp. 182–188 (2021)
20. Flammini, M., Kodric, B., Monaco, G., Zhang, Q.: Strategyproof mechanisms for additively separable and fractional hedonic games. J. Artif. Intell. Res. 70, 1253–1279 (2021). https://doi.org/10.1613/jair.1.12107
21. Flammini, M., Monaco, G., Moscardelli, L., Shalom, M., Zaks, S.: On the online coalition structure generation problem. J. Artif. Intell. Res. 72, 1215–1250 (2021)
22. Kaklamanis, C., Kanellopoulos, P., Papaioannou, K., Patouchas, D.: On the price of stability of some simple graph-based hedonic games. Theor. Comput. Sci. 855, 1–15 (2020)
23. Levinger, C., Azaria, A., Hazon, N.: Social aware coalition formation with bounded coalition size. In: Proceedings of the 2023 International Conference on Autonomous Agents and Multiagent Systems, AAMAS, pp. 2667–2669. ACM (2023)
24. Li, L., Micha, E., Nikolov, A., Shah, N.: Partitioning friends fairly. In: Proceedings of the AAAI Conference on Artificial Intelligence, pp. 5747–5754 (2023)
25. Ng, C., Hirschberg, D.S.: Three-dimensional stable matching problems. SIAM J. Discret. Math. 4(2), 245–252 (1991)
26. Wright, M., Vorobeychik, Y.: Mechanism design for team formation. In: Proceedings of the Twenty-Ninth AAAI Conference on Artificial Intelligence, pp. 1050–1056. AAAI Press (2015)

Improved Competitive Ratio for Edge-Weighted Online Stochastic Matching

Guoliang Qiu[1]($\boxtimes$)(iD), Yilong Feng[2](iD), Shengwei Zhou[2](iD), and Xiaowei Wu[2]($\boxtimes$)(iD)

[1] Shanghai Jiao Tong University, Shanghai, China
guoliang.qiu@sjtu.edu.cn
[2] IOTSC, University of Macau, Taipa, China
{mc15517,yc17423,xiaoweiwu}@um.edu.mo

Abstract. We consider the edge-weighted online stochastic matching problem, in which an edge-weighted bipartite graph $G = (I \cup J, E)$ with offline vertices J and online vertex types I is given. The online vertices have types sampled from I with probability proportional to the arrival rates of online vertex types. The online algorithm must make immediate and irrevocable matching decisions with the objective of maximizing the total weight of the matching. For the problem with general arrival rates, Feldman et al. (FOCS 2009) proposed the **Suggested Matching** algorithm and showed that it achieves a competitive ratio of $1 - 1/e \approx 0.632$. The ratio has recently been improved to 0.645 by Yan (2022), who proposed the **Multistage Suggested Matching** (MSM) algorithm. In this paper, we propose the **Evolving Suggested Matching** (ESM) algorithm and show that it achieves a competitive ratio of 0.650.

Keywords: Online Algorithms · Stochastic Matching · Poisson Arrival

1 Introduction

Motivated by its real-world applications, the online bipartite matching problem has received extensive attention since the work of Karp, Vazirani, and Vazirani [20] in 1990. The problem is defined on a bipartite graph where one side of the vertices are given to the algorithm in advance (aka the *offline* vertices), and the other side of (unknown) vertices arrive online one by one. Upon the arrival of an online vertex, its incident edges are revealed and the online algorithm must make immediate and irrevocable matching decisions with the objective of maximizing the size of the matching. The performance of the online algorithm is measured by the *competitive ratio*, which is the worst ratio between the size of matching computed by the online algorithm and that of the maximum matching,

The project is funded by the Science and Technology Development Fund (FDCT), Macau SAR (file no. 0014/2022/AFJ, 0085/2022/A, 0143/2020/A3 and SKL-IOTSC-2021-2023).

J. Garg et al. (Eds.): WINE 2023, LNCS 14413, pp. 527–544, 2024.
https://doi.org/10.1007/978-3-031-48974-7_30

over all online instances. As shown by Karp et al. [20], the celebrated Ranking algorithm achieves a competitive ratio of $1 - 1/e \approx 0.632$ and this is the best possible for the problem. However, the assumption that the algorithm has no prior information regarding the online vertices and the adversary decides the arrival order of these vertices, is believed to be too restrictive and in fact unrealistic. Therefore, several other arrival models with weaker adversaries, including the random arrival model [16,18,19,21], the degree-bounded model [2,7,23] and the stochastic model [11,14,15,17], have been proposed.

The stochastic model is proposed by Feldman et al. [11], in which the arrivals of online vertices follow a known distribution. Specifically, in the stochastic setting there is a bipartite graph $G = (I \cup J, E)$ that is known by the algorithm, where J contains the offline vertices and I contains the online vertex *types*, where each vertex type $i \in I$ is associated with an arrival rate λ_i. There are $\Lambda = \sum_{i \in I} \lambda_i$ online vertices to be arrived. Each online vertex has a type sampled from I independently, where type $i \in I$ is sampled with probability λ_i / Λ. The online vertex with type i has its set of neighbors defined by the neighbors of i in the graph $G = (I \cup J, E)$. The competitive ratio for the online algorithm is then measured by the worst ratio of $\frac{\mathrm{E[ALG]}}{\mathrm{E[OPT]}}$ over all instances, where ALG denotes the size of matching produced by the algorithm and OPT denotes that of the maximum matching. For the online stochastic matching problem, Feldman et al. [11] proposed the Suggested Matching algorithm that achieves a competitive ratio of $1 - 1/e$ and the Two Suggested Matching that is 0.67-competitive for instances with integral arrival rates. These algorithms are based on the framework that makes matching decisions in accordance to some offline optimal solution x pre-computed on the instance $(G, \{\lambda_i\}_{i \in I})$. Especially, the Two Suggested Matching algorithm employs a novel application of the idea called *the power of two choices* by specifying two offline neighbors and matching one of them if any of these two neighbors is unmatched. The approximation ratio was later improved by a sequence of works [3,6,17,22], resulting in the state-of-the-art competitive ratio 0.7299 by [6] under the integral arrival rate assumption. Without this assumption, the first competitive ratio beating $1 - 1/e$ was obtained by Manshadi et al. [22], who provided a 0.702-competitive algorithm for the problem. The competitive ratio was then improved to 0.706, 0.711, and 0.716 by Jaillet and Lu [17], Huang and Shu [14] and Huang et al. [15], respectively. Notably, Huang and Shu [14] established the asymptotic equivalence between the original stochastic arrival model and the Poisson arrival model in which online vertex types arrive independently following Poisson processes.

The weighted versions of the online stochastic matching problem have also received a considerable amount of attention. In the edge-weighted (resp. vertex-weighted) version of the problem, each edge (resp. offline vertex) is associated with a non-negative weight and the objective is to compute a matching with maximum total edge (resp. offline vertex) weight. For the vertex-weighted version, Jaillet and Lu [17] and Brubach et al. [6] achieved a competitive ratio of 0.725 and 0.7299, respectively, under the integral arrival rate assumption. Without this assumption, Huang and Shu [14] and Tang et al. [24] achieved a competitive ratio of 0.7009 and 0.704 respectively, while the state-of-the-art ratio 0.716 was achieved by Huang et al. [15]. For the edge-weighted version, under the integral arrival rate assumption, competitive ratios 0.667 and 0.705 were proved by Hae-uper et al. [13] and Brubach et al. [6], respectively. Without the assumption, the $(1 - 1/e)$-competitive Suggested Matching algorithm by Feldman et al [11] remained the state-of-the-art until recently Yan [25] proposed an algorithm called Multistage Suggested Matching (MSM) algorithm and showed that it achieves a competitive ratio of 0.645. Regarding hardness results, Huang et al. [15] proved that no algorithm can be 0.703-competitive for the edge-weighted online stochas-tic matching problem. The hardness result separates the edge-weighted version of the problem from the unweighted and vertex-weighted versions, for both of which competitive ratios strictly larger than 0.703 have already been proved. It also separates the problem without the integral arrival rate assumption from that with the assumption, which indicates the difficulty of the problem.

1.1 Our Contribution

We consider the edge-weighted online stochastic matching problem without the integral arrival rate assumption, and propose the Evolving Suggested Matching (ESM) algorithm that improves the state-of-the-art competitive ratio to 0.650.

Theorem 1. *The Evolving Suggested Matching algorithm is 0.650-competitive for edge-weighted online stochastic matching with a sufficiently large number of arrivals.*

Remark 1. The result follows from the asymptotic equivalence between the stochastic arrival model and the Poisson arrival model established by Huang and Shu [14]. For simplicity, we analyze the performance of our ESM algorithm under the Possion arrival model (the formal definition can be found in Sect. 2) instead.

Our work follows the common framework that formulates the matching in the graph $G = (I \cup J, E)$ into a Linear Program (LP) and uses the corresponding pre-computed optimal solution $\boldsymbol{x}$ to guide the design of our online algorithm. It can be shown that if we can design an algorithm that matches each edge $(i, j) \in E$ with probability at least $\alpha \cdot x_{ij}$, then the algorithm is α-competitive. The LP is customized for different problems, and we use the LP proposed by Jaillet and Lu [17] in this paper. By a reduction from [25], it suffices to consider some kernel instances in which all online vertex types have degree at most two. We call an online vertex type with one neighbor a *first-class* type and an online vertex type with two neighbors a *second-class* type. The Multistage Suggested Matching (MSM) algorithm proposed by Yan [25] is a hybrid of the Suggested Matching and the Two-Choice algorithms. In the MSM algorithm, the second-class online vertices follow different strategies at three different stages of the algorithm. Inspired by the MSM algorithm, we propose the Evolving Suggested Matching (ESM) algorithm by introducing an activation function that allows us to have theoretically infinitely many different "stages" in the algorithm.

Technical Contribution 1: Evolving Strategies by Activation Function.
With the activation function, our algorithm is able to evolve the matching strategies in time horizon. As discussed in Sect. 4, the activation function controls how "aggressive" the second-class online vertex types propose to their neighbors and is general enough to capture many of the existing algorithms for the online stochastic matching problem, including the Suggested Matching algorithm, the Two-Choice algorithm and that of Yan [25]. Furthermore, the design of a better online algorithm can thus be reduced to the design of the activation function in a tractable way, which is key to refining the state-of-the-art competitive ratio.

Technical Contribution 2: Fine-grained Correlation Analysis. One particular difficulty in competitive analysis by introducing the activation function is that most of the matching events are now intricately correlated (as opposed to the Suggested Matching algorithm [11]), and in order to fully utilize the power of the activation function (as opposed to the analysis for the MSM algorithm [25]), we need to carefully bound the matching probability of a vertex or an edge conditioned on the matching status of other vertices or edges. Essentially, all analysis for the two proposals algorithm needs to take into account the failure of the first proposal affected by the pairwise correlations. Existing works [14,25] get around this issue by specific relaxation of the correlated events and thus incur some intrinsic loss.

To improve the competitive analysis, it is inevitable to conduct a fine-grained correlation analysis of the matching probability of a vertex or an edge. As shown in our analysis (in Sect. 4), we conjecture that the offline vertices being unmatched up to some time t are positively correlated. However, it is difficult to capture the correlation between the matching events of offline vertices as there are many places where the random decisions are dependent. Instead, we observe that by abstracting the independent arrivals of the "extended online vertices types", we can prove a pseudo-positive correlated inequality available for our analysis. We believe this observation is technically interesting and may lead to further inspiration in future works.

1.2   Other Related Works

For the online bipartite matching problem under the adversarial arrival model, the optimal competitive ratio $1 - 1/e$ was proved in [1,4,8,12] using different analysis techniques (even in the vertex-weighted version). For the problem under random arrivals, it is assumed that the adversary decides the underlying graph but the online vertices arrive following a uniformly-at-random chosen order. For the unweighted version of this model, the competitive ratios 0.653 and 0.696 have been proved by Karande et al. [19] and Mahdian and Yan [21] respectively. For the vertex-weighted version of the problem, the competitive ratios 0.653 and 0.662 were proved by Huang et al. [16] and Jin and Williamson [18], respectively. It is well-known that there exists no competitive algorithm for the edge-weighted online bipartite matching problem [10]. Consequently, the edge-weighted version of the problem has attracted attention when additional assumptions are considered. One significant variant is often referred to as the edge-weighted online bipartite matching problem with free disposal, where each offline vertex can be matched repeatedly, but only the heaviest edge matched to it contributes to the objective. For the edge-weighted version of the problem with free disposal under the adversarial arrival order, Feldman et al. [10] proposed a $1 - 1/e$ competitive algorithm under a large market assumption. For general cases, Fahrbach et al. [9] first demonstrated a competitive ratio of 0.5086, which surpasses the long-standing barrier of 0.5 achieved by the greedy algorithm. The current best competitive ratio for this problem is 0.5368 achieved by Blanc and Charikar [5]. As for the stochastic setting, Huang et al. [15] further improved the competitive ratio from $1 - 1/e$ [11] to 0.706.

2   Preliminaries

We consider the edge-weighted online stochastic matching problem. An instance of the problem consists of a bipartite graph $G = (I \cup J, E)$, a weight function w, and the arrival rates $\{\lambda_i\}_{i \in I}$ of the online vertex types. In the bipartite graph $G = (I \cup J, E)$, I denotes the set of online vertex types and J denotes the set of offline vertices. The set $E \subseteq I \times J$ contains the edges between I and J, where each edge $(i, j) \in E$ has a non-negative weight w_{ij}. In the stochastic model, each online vertex type $i \in I$ has an arrival rate λ_i and $\Lambda = \sum_{i \in I} \lambda_i$. Online vertices arrive one by one and each of them draws its type i with probability $\frac{\lambda_i}{\Lambda}$ independently. Any online algorithm must make an immediate and irrevocable matching decision upon the arrival of each online vertex, with the goal of maximizing the total weight of the matching, subject to the constraint that each offline vertex can be matched at most once. Throughout, we use OPT to denote the weight of the maximum weighted matching of the realized instance; and ALG to denote the weight of the matching produced by the online algorithm. Note that both ALG and OPT are random variables, where the randomness of OPT comes from the random realization of the instance while that of ALG comes from both the realization and the random decisions by the algorithm. The competitive ratio of the algorithm is measured by the infimum of $\frac{E[\text{ALG}]}{E[\text{OPT}]}$ over all problem instances.

Poisson Arrival Model. Instead of fixing the number of online vertices, in the Poisson arrival model, the online vertex of each type i arrives independently following a Poisson process with time horizon $[0, 1]$ and arrival rate λ_i. The independence property allows a more convenient competitive analysis. In this paper, we consider the problem under the Poisson arrival model. Specifically, we show that

Theorem 2. *The Evolving Suggested Matching algorithm is 0.650-competitive for edge-weighted online stochastic matching under the Poisson arrival model.*

Together with the asymptotic equivalence analysis in [14], Theorem 2 implies Theorem 1.

Jaillet-Lu LP. We use the following linear program LP_{JL} proposed by Jaillet and Lu [17] to bound the expected offline optimal value for instance $(G, w, \{\lambda_i\}_{i \in I})$:

$$\text{maximize} \quad \sum_{(i,j) \in E} w_{ij} \cdot x_{ij}$$

$$\text{subject to} \quad \sum_{j:(i,j) \in E} x_{ij} \leq \lambda_i, \qquad\qquad \forall i \in I;$$

$$\sum_{i:(i,j) \in E} x_{ij} \leq 1, \qquad\qquad \forall j \in J;$$

$$\sum_{i:(i,j) \in E} \max\{2x_{ij} - \lambda_i, 0\} \leq 1 - \ln 2, \qquad \forall j \in J;$$

$$x_{ij} \geq 0, \qquad\qquad \forall (i,j) \in E.$$

We use x_i and x_j to denote $\sum_{j:(i,j) \in E} x_{ij}$ and $\sum_{i:(i,j) \in E} x_{ij}$, respectively. Note that for any feasible solution $\boldsymbol{x}$ we have $x_i \leq \lambda_i$ for all $i \in I$ and $x_j \leq 1$ for all $j \in J$. We remark that although the third set of constraints is not linear, they can be transformed into an LP problem by applying the standard technique of introducing auxiliary variables.

Lemma 1 (Analysis Framework). *If an online algorithm matches each edge $(i,j) \in E$ with probability at least $\alpha \cdot x_{ij}$ for any instance $(G, w, \{\lambda_i\}_{i \in I})$, where $\boldsymbol{x}$ is the optimal solution to the above LP, then the algorithm is α-competitive under the Poisson arrival model.*

Proof. Since the algorithm matches each edge $(i,j) \in E$ with probability at least $\alpha \cdot x_{ij}$, the expected total weight of the matching produced by the algorithm is

$$\mathbf{E}[\text{ALG}] = \sum_{(i,j) \in E} w_{ij} \cdot \mathbf{Pr}[(i,j) \text{ is matched by the algorithm}]$$

$$\geq \sum_{(i,j) \in E} w_{ij} \cdot \alpha \cdot x_{ij} = \alpha \cdot P^* \geq \alpha \cdot \mathbf{E}[\text{OPT}],$$

where P^* denotes the objective of the optimal solution $\boldsymbol{x}$. The last inequality follows because by defining x'_{ij} to be the probability that edge (i,j) is included in the maximum weighted matching of the realized instance, we can obtain a feasible solution[1] with objective $\mathbf{E}[\text{OPT}]$. Here, the feasibility of the third set of constraints can be briefly explained as follows: Under the Poisson arrival model, as shown in [15], it holds that for any offline vertex j and any subset of online vertex types S that are adjacent to j, the probability of j getting matched among S is bounded by $1 - e^{-\sum_{i \in S} \lambda_i}$. Together with the converse of Jensen's inequality, the third set of constraints holds. $\square$

---

[1] The proof showing that $\boldsymbol{x}'$ satisfies all constraints (especially the third set of constraints) can be found in the appendix of [14].

Following the above lemma, in the rest of the paper, we focus on lower bounding the minimum of

$$\frac{1}{x_{ij}} \cdot \mathbf{Pr}[(i,j) \text{ is matched by the algorithm}]$$

over all edges $(i,j) \in E$.

Suggested Matching. We first give a brief review of the edge-weighted version of the Suggested Matching algorithm, proposed by Feldman et al. [11]. The algorithm starts from an optimal solution to a linear program[2], in which x_{ij} is the corresponding variable for edge $(i,j) \in E$. The Suggested Matching algorithm proceeds as follows: when an online vertex of type i arrives, it chooses an offline neighbor j with probability x_{ij}/λ_i and propose to j. Note that by the feasibility of solution $\boldsymbol{x}$, the probability distribution is well-defined. If j is not matched, then the algorithm includes (i,j) into the matching. It can be shown that each edge (i,j) will be matched with probability at least $(1 - 1/e) \cdot x_{ij}$ when the algorithm terminates, which implies a competitive ratio of $1 - 1/e$.

Two-Choice. The idea of *power of two choices* was first introduced by Feldman et al. [11], in the algorithm Two Suggested Matching. Generally speaking, the idea is to allow each online vertex to choose two (instead of one) neighbors, and match one of them if any of these two neighbors is unmatched. Formally, the Two-Choice algorithm is described as follows: upon the arrival of each online vertex of type $i \in I$, the algorithm chooses two different offline neighbors j_1, j_2 following a distribution defined by some optimal solution to an LP. If j_1 is unmatched, then the algorithm matches i to j_1; otherwise, the algorithm matches i to j_2 if j_2 is unmatched. We call j_1 and j_2 the *first-choice* and *second-choice* of i, respectively. As introduced, the Two-Choice algorithm has achieved great success in the research field of online stochastic matching problems.

3   Multistage Suggested Matching

As a warm-up, we first briefly review the recent progress on the edge-weighted online stochastic matching problem by Yan [25].

Kernel Instances. We call an instance consisting of graph $G = (I \cup J, E)$, arrival rates $\{\lambda_i\}_{i \in I}$, and a fractional matching $\boldsymbol{x}$ of $\mathrm{LP_{JL}}$ a *kernel* instance if (1) there are only two classes of online vertex types: one with a single offline neighbor j such that $x_{ij} = \lambda_i$, and the other with two offline neighbors j_1, j_2 such that $x_{ij_1} = x_{ij_2} = \frac{1}{2}\lambda_i$; (2) for any offline vertex $j \in J$, we have $x_j = 1$.

Yan [25] showed that if there exists an online algorithm on the kernel instances such that for any edge $(i,j) \in E$, the probability that (i,j) gets

---

[2] The LP used in [11] is similar to the Jaillet-Lu LP defined above but without the third set of constraints.

matched by the algorithm is at least $\alpha \cdot x_{ij}$, then we can transform this algorithm to an α-competitive algorithm for general problem instances. Specifically, for any general problem instances with an optimal fractional matching $\boldsymbol{x}$ of $\mathrm{LP_{JL}}$, we can first assume that $x_i = \lambda_i$ for any online vertex type and $x_j = 1$ for any offline vertex. Otherwise, a satisfying instance can be constructed by introducing some dummy online vertex types and offline vertices with specific fractional matching on the involved dummy edges such that the objective and the feasibility of the fractional matching are preserved. To reduce the number of online vertex types, Yan [25] proposed a split scheme to split each online vertex type into sub-types and pair up the fractional matching on their edges systematically for its feasibility. The split scheme can be simulated by the downsampling of the online vertex types.

In the rest of this paper, we only consider the kernel instances. We call an edge (i, j) a *first-class* edge if $x_{ij} = \lambda_i$, or a *second-class* edge if $x_{ij} = \frac{\lambda_i}{2}$. If the online vertex type has an incident first-class (resp. second-class) edge, we call it a first-class (resp. second-class) online vertex type. For each offline vertex j, we use $N_1(j)$ and $N_2(j)$ to denote the set of first-class and second-class neighbors of j, respectively. Let $y_j := \sum_{i \in N_1(j)} x_{ij}$ be the sum of variables corresponding to first-class edges incident to j. Note that for the kernel instance, y_j also denotes the total arrival rate of first-class neighbors of j. By the feasibility of solution $\boldsymbol{x}$ we have:

$$y_j \leq 1 - \ln 2, \qquad \forall j \in J.$$

Multistage Suggested Matching Algorithm. While the Two-Choice algorithm provides good competitive ratios for the unweighted and vertex-weighted online stochastic matching problems [6,11,14], it cannot be straightforwardly applied to the edge-weighted version of the problem. Specifically, traditional analysis for the unweighted or vertex-weighted versions of the problem focuses on lower bounding the probability of each offline vertex being matched by the algorithm. However, for the edge-weighted version we need to lower bound the probability of each edge being matched (see Lemma 1), and it is not difficult to construct examples of kernel instances in which some (first-class) edge $(i, j) \in E$ is matched with probability strictly less than $(1 - 1/e) \cdot x_{ij}$ in the Two-Choice algorithm. In contrast, the Multistage Suggested Matching (MSM) algorithm proposed by Yan [25] is a hybrid of the Suggested Matching and the Two-Choice algorithms. Specifically, in the MSM algorithm, the first-class vertices always follow the Suggested Matching algorithm while the second-class vertices follow different strategies at different stages of the algorithm. In the first stage of the algorithm, all second-class vertices are discarded; in the second stage the second-class vertices follow the Suggested Matching algorithm; and in the last stage they follow the Two-Choice algorithm. Intuitively speaking, the second-class vertices are getting more and more aggressive in terms of trying to match their neighbors, as time goes by. The design of the first stage is crucial because without this stage the matching probability of a first-class edge $(i, j) \in E$ with $x_{ij} = 1$ will be at most $1 - 1/e$. The design of the third stage is also important because without it the performance of the algorithm will not be better than the Suggested Matching algorithm, e.g., on the second-class edges. By carefully leveraging the portions of the three stages, they show that their algorithm is at least 0.645-competitive.

4   Evolving Suggested Matching Algorithm

In this section, we propose the Evolving Suggested Matching (ESM) algorithm that generalizes several existing algorithms for the online stochastic matching problem, including that of Yan [25]. The algorithm is equipped with a non-decreasing *activation* function $f : [0, 1] \rightarrow [0, 2]$. We first present the algorithm in its general form and then provide a lower bound on the competitive ratio in terms of the function f. By carefully fixing the activation function (in the next section), we show that the competitive ratio is at least 0.650.

4.1   The Algorithm

Our algorithm is inspired by the MSM algorithm from [25], in which second-class edges follow different matching strategies at different stages of the algorithm. The high-level idea of our algorithm is to introduce a non-decreasing *activation* function $f : [0, 1] \rightarrow [0, 2]$ to make this transition happen smoothly. As in [25], we only consider the kernel instances in which each online vertex type is either first-class (having only one neighbor) or second-class (having exactly two neighbors). In the ESM algorithm, when an online vertex of type i arrives at time $t \in [0, 1]$,

- if i is a first-class type, then it proposes to its unique neighbor j. That is, if j is unmatched then we include the edge (i, j) into the matching; otherwise i is discarded.
- if i a second-class type, then it chooses a neighbor j_1 uniformly at random as its *first-choice*, and let the other neighbor j_2 be its *second-choice*. Then with probability $\min\{f(t), 1\}$, i proposes to j_1. If the proposal is made and j_1 is unmatched then the edge (i, j_1) is included in the matching and this round ends; if the proposal is made but j_1 is already matched, then i proposes to j_2 with probability $\max\{f(t) - 1, 0\}$.

The detailed description of the algorithm can be found in Algorithm 1.

We remark that the activation function f controls how "aggressive" the second-class online vertex types propose to their neighbors. For example, when $f(t) = 0$, the second-class online vertex arriving at time t will be discarded immediately without making any matching proposal; if $f(t) = 1$ then it will only propose to its first-choice; if $f(t) = 2$ then it will first propose to its first-choice and if the proposal is unsuccessful, then it will propose to its second-choice. Therefore, with different choices of the activation function, the ESM is general enough to capture many of the existing algorithms for the online stochastic matching problem, including the Suggested Matching algorithm (with $f(t) = 1$ for all $t \in [0, 1]$); the Two-Choice algorithm (with $f(t) = 2$ for all $t \in [0, 1]$) and that of Yan [25] (with $f(t) = 0$ when $t \leq 0.05$; $f(t) = 1$ when $t \in (0.05, 0.75)$ and $f(t) = 2$ when $t \geq 0.75$). In the following, we derive a lower bound on the

---

Algorithm 1: Evolving Suggested Matching algorithm

Input: A kernel instance with graph $G = (I \cup J, E)$, arrival rates $\{\lambda_i\}_{i \in I}$, the optimal solution x to $\mathrm{LP_{JL}}$, and an activation function f.

Output: A matching $\mathcal{M}$.

1 Initialize $\mathcal{M} = \emptyset$ to be an empty matching;

2 **for** *each online vertex of type i arriving at time $t \in [0,1]$* **do**

3   **if** *i is a first-class online vertex type* **then**

        // Propose to its unique first-class neighbor j

4   **if** *j is unmatched* **then**

5              $\mathcal{M} \leftarrow \mathcal{M} \cup \{(i,j)\}$;

6   **else**

7   choose a neighbor j_1 uniformly at random and let j_2 be the other neighbor;

8          $r_1, r_2 \sim \mathtt{Unif}[0,1]$;

9   **if** $r_1 \leq f(t)$ **then**

            // Propose to j_1

10   **if** *j_1 is unmatched* **then**

11                  $\mathcal{M} \leftarrow \mathcal{M} \cup \{(i,j_1)\}$;

12   **else if** $r_2 \leq f(t) - 1$ **then**

                // Propose to j_2

13   **if** *j_2 is unmatched* **then**

14                      $\mathcal{M} \leftarrow \mathcal{M} \cup \{(i,j_2)\}$;

15 **return** $\mathcal{M}$.

---

competitive ratio of the algorithm in terms of the activation function f. A specific choice of f will be decided in the next section to optimize the lower bound on the ratio.

4.2 Extended Online Vertex Types

For convenience of analysis, in the following, we make use of the properties of the Poisson process and present an equivalent description of the ESM algorithm. Specifically, upon the arrival of a second-class online vertex of type i at time t, suppose that j_1 is chosen as the first-choice and j_2 is the second-choice.

- If $r_1 > f(t)$ then we call the online vertex of type $i(\perp, \perp)$, indicating that it will not propose to any of its two choices;
- If $r_1 \leq f(t)$ and $r_2 \geq f(t) - 1$ then we call the online vertex of type $i(j_1, \perp)$, indicating that it will only propose to its first-choice;
- If $r_1 \leq f(t)$ and $r_2 < f(t) - 1$ then we call the online vertex of type $i(j_1, j_2)$, indicating that it will propose to its both choices unless it gets matched.

Recall that type i arrives following a Poisson process with rate λ_i. Hence the aforementioned types arrive following Poisson processes with time-dependent arrival rates (depending on the activation function f). Since the first-choice is chosen uniformly at random between the two neighbors, and r_1, r_2 are uniformly distributed in $[0, 1]$, we can characterize the arrival rates of each specific extended vertex type as follows.

Proposition 1. *For any second-class online vertex type i with neighbors $\{j_1, j_2\}$, at time $t \in [0, 1]$*

– *the arrival rate of type $i(j_1, \perp)$ and $i(j_2, \perp)$ are both $\frac{\lambda_i}{2} \cdot \min\{f(t), 1\} \cdot \min\{2 - f(t), 1\}$;*
– *the arrival rate of type $i(j_1, j_2)$ and $i(j_2, j_1)$ are both $\frac{\lambda_i}{2} \cdot \min\{f(t), 1\} \cdot \max\{f(t) - 1, 0\}$.*

Moreover, the Poisson processes describing the arrivals of online vertex of type $i(j, j')$, for all $i \in I$ and $j, j' \in J \cup \{\perp\}$ are independent.

In the following, we use $i(j, j')$ to describe an extended type, where i is a second-class online vertex and each of j, j' is either $\perp$ or a neighbor of i. Under this independent Poisson process modeling on the arrivals of the extended types, we give an equivalent description of the ESM algorithm in Algorithm 2.

Algorithm 2: ESM algorithm with extend online vertex types

Input: A kernel instance with extended vertex types.
Output: A matching $\mathcal{M}$.
1 Initialize $\mathcal{M} = \emptyset$ to be an empty matching;
2 **for** *each online vertex of type i* **do**
3 | **if** *i is a first-class online vertex type* **then**
4 | | **if** *its neighbor j is unmatched* **then**
5 | | | $\mathcal{M} \leftarrow \mathcal{M} \cup \{(i, j)\}$;
6 | **else**
7 | | Suppose the type is $i(j, j')$;
8 | | **if** $j \neq \perp$ **then**
9 | | | **if** *j is unmatched* **then**
10 | | | | $\mathcal{M} \leftarrow \mathcal{M} \cup \{(i, j)\}$;
11 | | | **else if** $j' \neq \perp$ **then**
12 | | | | **if** *j' is unmatched* **then**
13 | | | | | $\mathcal{M} \leftarrow \mathcal{M} \cup \{(i, j')\}$;

14 **return** $\mathcal{M}$.

In the remaining analysis, we say that an online vertex is of type $i(j, *)$ if its extended type is $i(j, j')$ for some $j' \in J \cup \{\perp\}$. Likewise we define $i(*, *)$, $*(j, *)$, $*(*, j)$, etc. These notations only consider the second class arrivals. Note that for all second-class online vertex type $i \in I$ and any of its neighbor j, the arrival rate of type $i(j, *)$ at time t is $\frac{\lambda_i}{2} \cdot \min\{f(t), 1\}$ and the arrival rate of type $i(*, *)$ at any time is always λ_i. Similarly, for any offline vertex j, the arrival rate of type $*(j, *)$ and $*(*, j)$ are both $1 - y_j$ which corresponds to the second class arrival among its neighbors.

4.3   Matching Probability of Edges

By Lemma 1, to derive a lower bound on the competitive ratio of the algorithm, it suffices to lower bound the probability that an arbitrarily fixed edge $(i, j) \in E$ is matched by the ESM algorithm. Let $M_{ij} \in \{0, 1\}$ be the indicator of whether (i, j) is matched by the algorithm, and $F(t) = \int_0^t f(x) \, dx$. Let $U_j(t) \in \{0, 1\}$ be the indicator of whether the offline vertex j is unmatched at time t, i.e., $U_j(t) = 1$ if and only if j is unmatched at time t. In the following, we provide a lower bound on $\mathbf{Pr}[M_{ij} = 1]$.

First-Class Edge. We first consider the case when (i, j) is a first-class edge. By the design of the algorithm, the edge will be included in the matching by the algorithm if an online vertex of type i arrives, and j is unmatched, because i will always propose to j upon its arrival. Since the arrival rate of type i is λ_i, we immediately have the following.

Lemma 2. *For any first-class edge* $(i, j) \in E$, *we have*

$$\mathbf{Pr}[M_{ij} = 1] = \int_0^1 \lambda_i \cdot \mathbf{Pr}[U_j(t) = 1] \, dt.$$

Second-Class Edge. Now suppose that $(i, j) \in E$ is a second-class edge, and let j' be the other neighbor of i. There are two events that will cause edge $(i, j) \in E$ being matched by the algorithm:

- an online vertex of type $i(j, *)$ arrives, and j is unmatched;
- an online vertex of type $i(j', j)$ arrives, j' is already matched and j is unmatched.

Let $t^* := \sup\{t \in [0, 1] : f(t) \leq 1\}$. We observe that

- at time $t \leq t^*$, the arrival rate of $i(j, *)$ is $\frac{\lambda_i}{2} \cdot f(t)$; after time t^*, the arrival rate is $\frac{\lambda_i}{2}$;
- before time t^*, the arrival rate of $i(j', j)$ is 0; at time $t > t^*$, the arrival rate is $\frac{\lambda_i}{2} \cdot (f(t) - 1)$.

Therefore, we have the following characterization on $\mathbf{Pr}[M_{ij} = 1]$.

$$\mathbf{Pr}[M_{ij} = 1] = \int_0^{t^*} \frac{\lambda_i}{2} \cdot f(t) \cdot \mathbf{Pr}[U_j(t) = 1] \, dt + \int_{t^*}^1 \frac{\lambda_i}{2} \cdot \mathbf{Pr}[U_j(t) = 1] \, dt$$

$$+ \int_{t^*}^1 \frac{\lambda_i}{2} \cdot (f(t) - 1) \cdot \mathbf{Pr}[U_j(t) = 1, U_{j'}(t) = 0] \, dt$$

$$= \int_0^1 \frac{\lambda_i}{2} \cdot f(t) \cdot \mathbf{Pr}[U_j(t) = 1] \, dt - \int_{t^*}^1 \frac{\lambda_i}{2} \cdot (f(t) - 1) \cdot \mathbf{Pr}[U_j(t) = 1, U_{j'}(t) = 1] \, dt.$$

By upper bounding $\mathbf{Pr}[U_j(t) = 1, U_{j'}(t) = 1]$, we derive the following.

Lemma 3. *For any second-class edge $(i, j) \in E$, we have*

$$\mathbf{Pr}[M_{ij} = 1] \geq \int_0^1 \frac{\lambda_i}{2} \cdot f(t) \cdot \mathbf{Pr}[U_j(t) = 1]\, dt$$

$$- \int_{t^*}^1 \frac{\lambda_i}{2} \cdot (f(t) - 1) \cdot e^{-y_j \cdot t^* - (2-y_j) \cdot F(t^*) - 2(t-t^*)}\, dt.$$

Proof. To prove the lemma, it suffices to argue that for any $t > t^*$, we have

$$\mathbf{Pr}[U_j(t) = 1, U_{j'}(t) = 1] \leq e^{-y_j \cdot t^* - (2-y_j) \cdot F(t^*) - 2(t-t^*)}.$$

Observe that if any online vertex of type $i \in N_1(j)$ or $*(j, *)$ arrives before time t, then j will be matched before time t. The same holds for the offline vertex j'. The arrival rate of type $N_1(j) \cup *(j, *)$ is

- at time $x \leq t^*$: $\sum_{i \in N_1(j)} \lambda_i + \sum_{i \in N_2(j)} \frac{\lambda_i}{2} \cdot f(x) = y_j + (1 - y_j) \cdot f(x)$.
- after time t^*: $\sum_{i \in N_1(j)} \lambda_i + \sum_{i \in N_2(j)} \frac{\lambda_i}{2} = y_j + (1 - y_j) = 1$.

Similarly the arrival rate of type $N_1(j') \cup *(j', *)$ is $y_{j'} + (1 - y_{j'}) \cdot f(x)$ at time $x \leq t^*$ and 1 after time t^*. Since $\mathbf{Pr}[U_j(t) = 1, U_{j'}(t) = 1]$ is at most the probability that no online vertex of type $N_1(j) \cup *(j, *)$ or $N_1(j') \cup *(j', *)$ arrives before time t, and the arrivals of types $N_1(j) \cup *(j, *)$ or $N_1(j') \cup *(j', *)$ are independent, we have

$$\mathbf{Pr}[U_j(t) = 1, U_{j'}(t) = 1]$$

$$\leq e^{-\int_0^{t^*} (y_j + (1-y_j) \cdot f(x))\, dx - \int_{t^*}^t 1\, dx} \cdot e^{-\int_0^{t^*} (y_{j'} + (1-y_{j'}) \cdot f(x))\, dx - \int_{t^*}^t 1\, dx}$$

$$= e^{-(y_j + y_{j'}) \cdot t^* - (2 - y_j - y_{j'}) \cdot F(t^*) - 2(t-t^*)}$$

$$\leq e^{-y_j \cdot t^* - (2-y_j) \cdot F(t^*) - 2(t-t^*)},$$

where the last inequality holds since $F(t^*) \leq t^*$ and $y_{j'} \geq 0$.   □

Given Lemma 2 and 3, to provide a lower bound on $\mathbf{Pr}[M_{ij} = 1]$, it remains to lower bound $\mathbf{Pr}[U_j(t) = 1]$ (in terms of the activation function f).

4.4   Lower Bounding $\mathbf{Pr}[U_j(t) = 1]$

In this section, we fix an arbitrary offline vertex j and provide a lower bound on $\mathbf{Pr}[U_j(t) = 1]$. To begin with (and as a warm-up), we first establish a loose lower bound as follows.

Lemma 4. *For any offline vertex $j \in J$ and any time $t \in [0, 1]$, we have*

$$\mathbf{Pr}[U_j(t) = 1] \geq e^{-y_j t - (1-y_j) F(t)}.$$

Moreover, the above equation holds with equality when $t \leq t^*$.

Proof. As before, we characterize the events that will cause j being matched. Observe that if j is matched before time t, then at least one of the following must happen before time t:

- a first-class online vertex type $i \in N_1(j)$ arrives;
- an online vertex of type $*(j, *)$ arrives;
- an online vertex of type $*(*, j)$ arrives.

Note that the third event will only happen after time t^*. Moreover, it will contribute to the matching of vertex j only if the first-choice of the online vertex is matched upon its arrival. However, for the purpose of lower bounding $\mathbf{Pr}[U_j(t) = 1]$, we only look at the arrivals without caring whether a proposal to j is made. Since the combined arrival rate of types $*(j, *)$ and $*(*, j)$ at time $x \in [0, 1]$ is $(1 - y_j) \cdot f(x)$, and the total arrival rate of vertices in $N_1(j)$ is y_j, the lemma follows immediately.   □

It is apparent that the above lower bound on $\mathbf{Pr}[U_j(t) = 1]$ can be improved when $t > t^*$ because, in the above analysis, we use the event that "an online vertex of type $*(*, j)$ arrives" to substitute that "an online vertex of type $*(*, j)$ arrives and its proposal to its first-choice fails". The advantage of this substitution is that now the events that may contribute to j being matched are independent and it is convenient in lower bounding the probability that none of them happens. On the other hand, it is reasonable to believe that via lower bounding the probability that a vertex of type $*(*, j)$ fails in matching its first-choice, a better lower bound on $\mathbf{Pr}[U_j(t) = 1]$ can be derived. However, this requires a much more careful characterization of these events, because some of them are not independent. Specifically, suppose that j is not matched at time x, and an online vertex of type $i(j_1, j)$ arrives. The online vertex will contribute to j being matched only if j_1 is matched. Therefore the contributions of types $i(j_1, j)$ and $i'(j_1, j)$ are not independent random events because they both depend on the matching status of j_1. Moreover, the contributions of types $i(j_1, j)$ and $i'(j_2, j)$ may also be dependent because whether j_1 and j_2 are matched might not be independent, e.g., they might have common neighbors.

Therefore, in order to provide a better lower bound on $\mathbf{Pr}[U_j(t) = 1]$, it is inevitable to take into account the dependence on the random events. To enable the analysis, we introduce the following useful notations.

Definition 1 (Competitor). *We call an offline vertex j' a competitor of j if $N_2(j) \cap N_2(j') \neq \emptyset$. We use $\mathcal{C}(j) = \{j_1, j_2, \ldots, j_K\}$ to denote the set of competitors of j. For each $j_k \in \mathcal{C}(j)$, we use $c_k = \sum_{i \in N_2(j) \cap N_2(j_k)} \frac{\lambda_i}{2}$ to denote the total arrival rate of types $\{i(j_k, *)\}_{i \in N_2(j) \cap N_2(j_k)}$.*

Note that by definition we have $\sum_{k=1}^{K} c_k = 1 - y_j$. In the following, we show the following improved version of Lemma 4 when $t > t^*$.

Lemma 5. *For any offline vertex $j \in J$ and any time $t \in [t^*, 1]$, we have*

$$\mathbf{Pr}[U_j(t) = 1] \geq e^{-y_j \cdot t - (1 - y_j)\left(e^{-F(1)} F(t^*) + (1 - e^{-F(1)}) F(t) + e^{-F(1)}(t - t^*)\right)}.$$

when $F(1) \geq 1$.

Proof. By Lemma 4, we have $\mathbf{Pr}[U_j(t^*) = 1] = e^{-y_j \cdot t^* - (1-y_j)F(t^*)}$. Hence the statement is true when $t = t^*$. Observe that

$$\mathbf{Pr}[U_j(t) = 1] = \mathbf{Pr}[U_j(t) = 1 \mid U_j(t^*) = 1] \cdot \mathbf{Pr}[U_j(t^*) = 1].$$

To prove the lemma, it suffices to show that for all $t > t^*$, we have

$$\mathbf{Pr}[U_j(t) = 1 \mid U_j(t^*) = 1] \geq e^{-y_j(t-t^*) - (1-y_j)\left((1-e^{-F(1)})(F(t)-F(t^*))+e^{-F(1)}(t-t^*)\right)}. \tag{1}$$

For ease of notation, we use $h(t)$ to denote the LHS of the above. Note that $h : [t^*, 1] \to [0, 1]$ is a decreasing function with $h(t^*) = 1$. Fix any $t > t^*$. Conditioned on j being unmatched at time t^*, j will be matched at or before time t if any of the following events happen during the time interval $[t^*, t]$:

- a first-class online vertex $i \in N_1(j)$ arrives;
- an online vertex of type $*(j, *)$ arrives;
- an online vertex of type $*(*, j)$ arrives, and it fails matching its first-choice.

Since the total arrival rate of the first two events is 1 and the arrival rate of $*(j_k, j)$ at time $x > t^*$ is $c_k \cdot (f(x) - 1)$, we have the following

$$\mathbf{Pr}[U_j(t) = 0 \mid U_j(t^*) = 1] = \int_{t^*}^{t} \Bigg(\mathbf{Pr}[U_j(x) = 1 \mid U_j(t^*) = 1]$$

$$+ (f(x) - 1) \cdot \sum_{k=1}^{K} c_k \cdot \mathbf{Pr}[U_{j_k}(x) = 0, U_j(x) = 1 \mid U_j(t^*) = 1] \Bigg) dx.$$

Observe that $\mathbf{Pr}[U_j(t) = 0 \mid U_j(t^*) = 1] = 1 - h(t)$. The above equation implies that

$$1 - h(t) = \int_{t^*}^{t} \left(1 + (f(x) - 1) \cdot \sum_{k=1}^{K} c_k \cdot \mathbf{Pr}[U_{j_k}(x) = 0 \mid U_j(x) = 1] \right) \cdot h(x) \, dx.$$

Solving the above using the standard differential equation, we have

$$h(t) = e^{- \int_{t^*}^{t} g(x) \, dx}, \tag{2}$$

where $g(x) = 1 + (f(x) - 1) \cdot \sum_{k=1}^{K} c_k \cdot \mathbf{Pr}[U_{j_k}(x) = 0 \mid U_j(x) = 1]$. To provide an upper bound on $g(x)$, we first establish the following useful claim.   □

Claim. For all $x \geq t^*$ and $j' \in J \setminus \{j\}$, we have $\mathbf{Pr}[U_{j'}(x) = 0 \mid U_j(x) = 1] \leq 1 - e^{-F(1)}$ when $F(1) > 1$.

Proof. We prove the equivalent statement that $\mathbf{Pr}[U_{j'}(x) = 1 \mid U_j(x) = 1] \geq e^{-F(1)}$. As before, we first list the events that may cause j' being matched:

- a first-class online vertex $i \in N_1(j')$ arrives;
- an online vertex of type $*(j', *)$ arrives;
- an online vertex of type $*(*, j')$ arrives.

We call the above types the *key* types and use $A(x)$ to denote the event that none of the key types arrive before time x. Note that if $A(x)$ happens then j' is guaranteed to be unmatched at time x. Hence we have

$$\mathbf{Pr}[U_{j'}(x) = 1 \mid U_j(x) = 1] = \frac{\mathbf{Pr}[U_{j'}(x) = 1, U_j(x) = 1]}{\mathbf{Pr}[U_j(x) = 1]}$$

$$\geq \frac{\mathbf{Pr}[A(x), U_j(x) = 1]}{\mathbf{Pr}[U_j(x) = 1]} = \frac{\mathbf{Pr}[A(x)] \cdot \mathbf{Pr}[U_j(x) = 1 \mid A(x)]}{\mathbf{Pr}[U_j(x) = 1]}.$$

Since the key types related to event $A(x)$ arrive independently, we have

$$\mathbf{Pr}[A(x)] = e^{-y'_j \cdot x - (1 - y'_j) \cdot F(x)} \geq e^{-y'_j - (1 - y'_j) \cdot F(1)} \geq e^{-F(1)},$$

where the last equality holds from the assumption that $F(1) \geq 1$.

Given the above, it remains to show that

$$\mathbf{Pr}[U_j(x) = 1 \mid A(x)] \geq \mathbf{Pr}[U_j(x) = 1]. \tag{3}$$

The statement can be proved by the coupling argument. Given the same set of randomness, let $\mathcal{M}_1$ and $\mathcal{M}_2$ be the matchings obtained by the ESM algorithm and the one neglecting all arrivals of key types before time x respectively. We next show that $j \notin \mathcal{M}_2$ if $j \notin \mathcal{M}_1$ for any specified randomness, which implies Eq. (3) immediately. Note that $j \notin \mathcal{M}_1$ means

1. no online vertex of type $i \in N_1(j)$, $*(j, *)$ and $*(*, j)$ arrive, or
2. some online vertex of type $i(j'', j)$ arrive but j'' is not matched.

In the first circumstance, it is obvious to have $j \notin \mathcal{M}_2$; Otherwise, it suffices to show that $j'' \notin \mathcal{M}_2$ when $j'' \notin \mathcal{M}_1$ before the time x' that the online vertex of type $i(j'', j)$ arrive. Again, we can repeat a similar argument till the beginning where all vertices are unmatched. Hence the inequality holds and our claim follows immediately. $\square$

Given the above claim, we have

$$g(x) \leq 1 + (f(x) - 1) \cdot \sum_{k=1}^{K} c_k \cdot \left(1 - e^{-F(1)}\right)$$

$$= y_j + (1 - y_j) \cdot \left(\left(1 - e^{-F(1)}\right) \cdot f(x) + e^{-F(1)}\right).$$

Plugging the above upper bound on $g(x)$ into Eq. (2), we obtain Eq. (1), and thus finish the proof of the lemma. $\square$

4.5 Putting Things Together

Plugging the lower bounds on $\mathbf{Pr}[U_j(t) = 1]$ we have proved in Lemma 4 and 5 into Lemma 2 and 3, we obtain the following when the activation function satisfying $F(1) \geq 1$. For any first-class edge $(i,j) \in E$,

$$\frac{\mathbf{Pr}[M_{ij} = 1]}{\lambda_i} \geq \int_0^{t^*} e^{-y_j \cdot t - (1-y_j)F(t)} \, dt$$
$$+ \int_{t^*}^1 e^{-y_j \cdot t - (1-y_j)\left(e^{-F(1)}F(t^*) + (1-e^{-F(1)})F(t) + e^{-F(1)}(t-t^*)\right)} \, dt.$$

$$(4)$$

For any second-class edge $(i,j) \in E$,

$$\frac{\mathbf{Pr}[M_{ij} = 1]}{\lambda_i/2} \geq \int_0^{t^*} f(t) \cdot e^{-y_j \cdot t - (1-y_j)F(t)} \, dt$$
$$+ \int_{t^*}^1 f(t) \cdot e^{-y_j \cdot t - (1-y_j)\left(e^{-F(1)}F(t^*) + (1-e^{-F(1)})F(t) + e^{-F(1)}(t-t^*)\right)} \, dt$$
$$- \int_{t^*}^1 (f(t) - 1) \cdot e^{-y_j \cdot t^* - (2-y_j) \cdot F(t^*) - 2 \cdot (t-t^*)} \, dt.$$

$$(5)$$

By Lemma 1, to show that the ESM algorithm is α-competitive, it remains to design an appropriate activation function f such that for all $y_j \leq 1 - \ln 2$, the RHS of Eqs. (4) and (5) are both at least α and $F(1) \geq 1$. The details of the construction of the activation function are deferred to the full version. Specifically, we have

Lemma 6. *There exists a non-decreasing activation function f with $F(1) \geq 1$ such that*

– *for any first-class edge $(i,j) \in E$, $\frac{\mathbf{Pr}[M_{ij}=1]}{\lambda_i} \geq 0.650$;*
– *for any second-class edge $(i,j) \in E$, $\frac{\mathbf{Pr}[M_{ij}=1]}{\lambda_i/2} \geq 0.650$.*

Together with Lemma 1 and the lower bounds we derived, we prove Theorem 2.

References

1. Aggarwal, G., Goel, G., Karande, C., Mehta, A.: Online vertex-weighted bipartite matching and single-bid budgeted allocations. In: SODA, pp. 1253–1264. SIAM (2011)
2. Albers, S., Schubert, S.: Tight bounds for online matching in bounded-degree graphs with vertex capacities. In: ESA. LIPIcs, vol. 244, pp. 4:1–4:16. Schloss Dagstuhl - Leibniz-Zentrum für Informatik (2022)
3. Bahmani, B., Kapralov, M.: Improved bounds for online stochastic matching. In: de Berg, M., Meyer, U. (eds.) ESA 2010. LNCS, vol. 6346, pp. 170–181. Springer, Heidelberg (2010). https://doi.org/10.1007/978-3-642-15775-2_15

4. Birnbaum, B.E., Mathieu, C.: On-line bipartite matching made simple. SIGACT News **39**(1), 80–87 (2008)
5. Blanc, G., Charikar, M.: Multiway online correlated selection. In: FOCS, pp. 1277–1284. IEEE (2021)
6. Brubach, B., Sankararaman, K.A., Srinivasan, A., Xu, P.: Online stochastic matching: new algorithms and bounds. Algorithmica **82**(10), 2737–2783 (2020)
7. Cohen, I.R., Wajc, D.: Randomized online matching in regular graphs. In: SODA, pp. 960–979. SIAM (2018)
8. Devanur, N.R., Jain, K., Kleinberg, R.D.: Randomized primal-dual analysis of RANKING for online bipartite matching. In: SODA, pp. 101–107. SIAM (2013)
9. Fahrbach, M., Huang, Z., Tao, R., Zadimoghaddam, M.: Edge-weighted online bipartite matching. J. ACM **69**(6), 45:1–45:35 (2022)
10. Feldman, J., Korula, N., Mirrokni, V., Muthukrishnan, S., Pál, M.: Online ad assignment with free disposal. In: Leonardi, S. (ed.) WINE 2009. LNCS, vol. 5929, pp. 374–385. Springer, Heidelberg (2009). https://doi.org/10.1007/978-3-642-10841-9_34
11. Feldman, J., Mehta, A., Mirrokni, V.S., Muthukrishnan, S.: Online stochastic matching: beating 1–1/e. In: FOCS, pp. 117–126. IEEE Computer Society (2009)
12. Goel, G., Mehta, A.: Online budgeted matching in random input models with applications to adwords. In: SODA, pp. 982–991. SIAM (2008)
13. Haeupler, B., Mirrokni, V.S., Zadimoghaddam, M.: Online stochastic weighted matching: improved approximation algorithms. In: Chen, N., Elkind, E., Koutsoupias, E. (eds.) WINE 2011. LNCS, vol. 7090, pp. 170–181. Springer, Heidelberg (2011). https://doi.org/10.1007/978-3-642-25510-6_15
14. Huang, Z., Shu, X.: Online stochastic matching, Poisson arrivals, and the natural linear program. In: STOC, pp. 682–693. ACM (2021)
15. Huang, Z., Shu, X., Yan, S.: The power of multiple choices in online stochastic matching. In: STOC, pp. 91–103. ACM (2022)
16. Huang, Z., Tang, Z.G., Wu, X., Zhang, Y.: Online vertex-weighted bipartite matching: beating 1–1/e with random arrivals. ACM Trans. Algorithms **15**(3), 38:1–38:15 (2019)
17. Jaillet, P., Lu, X.: Online stochastic matching: new algorithms with better bounds. Math. Oper. Res. **39**(3), 624–646 (2014)
18. Jin, B., Williamson, D.P.: Improved analysis of RANKING for online vertex-weighted bipartite matching in the random order model. In: Feldman, M., Fu, H., Talgam-Cohen, I. (eds.) WINE 2021. LNCS, vol. 13112, pp. 207–225. Springer, Cham (2022). https://doi.org/10.1007/978-3-030-94676-0_12
19. Karande, C., Mehta, A., Tripathi, P.: Online bipartite matching with unknown distributions. In: STOC, pp. 587–596. ACM (2011)
20. Karp, R.M., Vazirani, U.V., Vazirani, V.V.: An optimal algorithm for on-line bipartite matching. In: STOC, pp. 352–358. ACM (1990)
21. Mahdian, M., Yan, Q.: Online bipartite matching with random arrivals: an approach based on strongly factor-revealing LPs. In: STOC, pp. 597–606. ACM (2011)
22. Manshadi, V.H., Gharan, S.O., Saberi, A.: Online stochastic matching: online actions based on offline statistics. Math. Oper. Res. **37**(4), 559–573 (2012)
23. Naor, J.S., Wajc, D.: Near-optimum online ad allocation for targeted advertising. ACM Trans. Econ. Comput. **6**(3–4), 16:1–16:20 (2018)
24. Tang, Z.G., Wu, J., Wu, H.: (Fractional) online stochastic matching via fine-grained offline statistics. In: STOC, pp. 77–90. ACM (2022)
25. Yan, S.: Edge-weighted online stochastic matching: beating 1–1/e. CoRR abs/2210.12543 (2022). https://doi.org/10.48550/arXiv.2210.12543

Separation in Distributionally Robust Monopolist Problem

Hao Qiu[1] , Zhen Wang[2,3](✉) , and Simai He[1]

[1] School of Information Management and Engineering,
Shanghai University of Finance and Economics, Yangpu District,
Shanghai 200433, People's Republic of China
qiu.hao@163.sufe.edu.cn, simaihe@mail.shufe.edu.cn
[2] School of Data Science, The Chinese University of Hong Kong,
Shenzhen (CUHK-Shenzhen), Longgang District, Shenzhen 518172,
Guangdong, People's Republic of China
wangzhen@cuhk.edu.cn
[3] University of Science and Technology of China, Hefei 230026,
Anhui, People's Republic of China

Abstract. We consider a monopoly pricing problem, where a seller has multiple items to sell to a single buyer, only knowing the distribution of the buyer's value profile. The seller's goal is to maximize her expected revenue. In general, this is a difficult problem to solve, even if the distribution is well specified. In this paper, we solve a subclass of this problem when the distribution is assumed to belong to the class of distributions defined by given marginal partial information. Under this model, we show that the optimal strategy for the seller is a randomized posted price mechanism under which the items are sold separately, and the result continues to hold even when the buyer has a budget feasibility constraint. Consequently, under some specific ambiguity sets which include moment-based and Wasserstein ambiguity sets, we provide analytical solutions for these single-item problems. Based on the additive separation property, we show the general additive separation problem is a special case of resource allocation problems that can be solved by known polynomial-time algorithms.

Keywords: mechanism design · monopolist problem · distributionally robust optimization

1 Introduction

The monopolist problem has long been a central topic in economic theory, where a seller aims to sell multiple indivisible items to a single buyer. The buyer assigns each item a private value, indicating their willingness to pay. The seller has a prior belief about the distribution of the buyer's value profile and wants to maximize the expected revenue. In the classic setting, the seller is assumed to have full knowledge of the buyer's valuation distribution. That is, the buyer's value profile is governed by a known distribution. If there is only one good, it is well known that the optimal mechanism is a posted price mechanism [27,31].

ⓒ The Author(s), under exclusive license to Springer Nature Switzerland AG 2024
J. Garg et al. (Eds.): WINE 2023, LNCS 14413, pp. 545–562, 2024.
https://doi.org/10.1007/978-3-031-48974-7_31

However, when extending this to a multi-item framework, the optimal mechanism is very difficult to characterize and compute. For general distributions, it has been shown that probabilistic bundling might be necessary and even that the seller might have to offer a menu of infinitely many such bundles [11, 25]. Moreover, the revenue of the optimal auction may be non-monotone [18] when the buyer's values in the prior distribution are moved upwards (in the stochastic dominance sense). These phenomena not only appear for correlation in the multi-item problems but also when the values for the two items are independently distributed. Given these challenges in the simple monopoly problems, it is harder for us to find the optimal mechanism for more complex problems.

Contemporary challenges arise when sellers often face difficulties in accurately obtaining the true distribution of buyers' valuation. For instance, if there is a limited amount of historical sales data for a product, the valuation distribution estimated by the historical data may deviate significantly from the true distribution. Moreover, estimating the valuation distribution becomes even more challenging for a new product since historical data is lacking. These practical considerations restrict sellers to accessing only partial information about the valuation distribution. To address the risks arising from distribution uncertainty, we apply the distributionally robust optimization approach. Specifically, the *distributionally robust monopolist problem* makes conservative decisions by considering the worst-case scenario.

Based on the aforementioned discussions, addressing the complexities of multi-item distributional robustness and uncertainty proves to be an intricate endeavor. However, under specific conditions, this challenge can be broken down since the multi-item problem can be additively separated by multiple single-item problems under some specific settings, which is known as the *additive separation property*. Our work is motivated by Carroll [6], who finds the additive separation property under correlation-robust ambiguity set, as well as Gravin and Lu [16], who extend Carroll's result to budget feasible mechanisms. In our study, we consider a more general ambiguity set, referred to as *marginal ambiguity set*, which captures partial information about marginal distributions. The goal of this paper is to find the additive separation property under the marginal ambiguity set with budget feasibility and solve the multi-item distributionally robust monopolist problem based on additive separation property.

1.1   Related Work

In the classic setting of the monopolist problem, the valuation distribution is known. Myerson [27] proves the optimal mechanism for the single-item problem is a posted-price mechanism. For the general multi-item monopolist problem, finding an optimal mechanism is complex, even in the two-item case. We refer readers to literature reviews in [6, 16, 25] for the challenges that arose in previous research.

Our work follows a trend in the economic literature on robust mechanism design. Most of these works only study the single-item and single-bidder case. Bergemann and Schlag [3] consider the ambiguity under the Prohorov metric and show that the optimum is a posted price mechanism. Du [13] studies robust mechanisms to sell a common-value good with multiple players, and provides a revenue guarantee for this auction. His results lead to an optimal mechanism under the ambiguity set based on mean-preserving spread. Carrasco et al. [5] study the ambiguity set with the first n moments. They characterize the optimal mechanism for the case of the first two moments and derive some characteristics of the optimal mechanism for the general case. Chen et al. [10] evaluated an ambiguity set with asymmetric information and proposed the closed-form expressions of both the optimal robust mechanism and deterministic mechanism. Recently, Chen et al. [9] establish the equivalence between the robust monopolist problem and minimax pricing problem when the ambiguity set is convex. They also provide a unified geometric approach to find the optimal mechanism analytically under several important ambiguity sets. There are also some studies that consider the discrete value distribution. Pınar and Kızılkale [29] characterize the optimal mechanism under the first two moments ambiguity set. Li et al. [24] give an explicit characterization of the optimal mechanism under ambiguity set based on the Wasserstein metric.

Besides the single-item and single-bidder settings, the robust monopolist problem has been explored in multi-item settings. Carroll [6] proposes a new framework where the seller has full knowledge about the marginal distribution of private value for each item. This framework does not contain any information about correlation across different items, and thus is further referred to as the correlation-robust framework. The main result of Carroll [6] states that the correlation-robust monopolist problem is additively separable. This implies the correlation-robust monopolist problem can be additively reduced to multiple single-item classic monopolist problems. Gravin and Lu [16] extend Carroll's framework by adding budget feasibility conditions and find that the property of additive separation still holds with budget feasibility. Except for the correlation-robust framework, Che and Zhong [7] consider the ambiguity with a variety of moment information on subsets of items, and find the partial bundling property on the robust optimality. Koçyiğit et al. [22] study a robust monopoly pricing problem with a minimax regret objective under a rectangular ambiguity set, and show the optimal mechanism is selling items separately.

Another related work is the revenue maximization problem of a single-item auction, in which a seller sells one item to multiple bidders. Some works [2, 19, 34] state the performance of some commonly used mechanisms under the correlation-robust framework. Bei et al. [2] show that the sequential posted price mechanism achieves a constant (4.78) approximation to the optimal correlation-robust mechanism. He et al. [19] establish a lower bound of the performance of the VCG mechanism by using the property of order statistics. Zhang [34] shows that if the seller knows only the marginal distribution of each bidder but not the joint distribution, then the second auction mechanism maximizes worst-case revenue over all dominant strategy incentive compatible mechanisms in the case of two bidders. Besides, Suzdaltsev [33] studies the case when private values are independent and identically distributed, and the seller knows only the mean and an upper bound on either values or variance of value distribution. He shows that the optimal solution to the worst-case expected revenue maximization problem is choosing a reserve price in a second-price auction. For the multi-item multi-buyer robust revenue maximization problem, Bandi and Bertsimas [1] provide an efficient algorithm given the support information. He et al. [20] consider robust revenue-maximizing auctions under various classes of ambiguity sets such as moment-based ambiguity set and correlation-robust ambiguity set, based on an elementary property related to order statistics.

Our work is also related to distributionally robust optimization (DRO). We refer the readers to Rahimian and Mehrotra [30] for a comprehensive review of DRO. There are mainly two types of ambiguity sets studied in DRO: moment-based ambiguity set and discrepancy-based ambiguity set. Other important literature on moment-based ambiguity sets includes [8,12,17,23]. The most popular discrepancy-based ambiguity set is based on the Wasserstein metric. Gao and Kleywegt [15] give the strong duality condition under the Wasserstein ambiguity set. Other research on Wasserstein ambiguity sets includes [4,14,26].

1.2   Our Contribution

- **Additive separation property under general marginal ambiguity set.** We consider a general marginal ambiguity set for the multi-item distributionally robust monopolist problem. Under this ambiguity set, we prove the additive separation property in the context of budget-feasible mechanisms.
- **Analytical solutions for the single-item problem.** We provide analytical solutions for the single-item distributionally robust monopolist problem with budget feasibility under some specific ambiguity sets, including moment-based ambiguity sets and Wasserstein ambiguity sets.
- **Algorithms for the general additive separation problem with budget feasibility.** We show that the objective function of the multi-item problem is separable concave, which can be seen as a special case of resource allocation problems. We can use the Lagrangian multiplier algorithm to solve the multi-item problem by known polynomial-time complexity.

2   Problem Formulation and Preliminaries

We consider the problem of designing a mechanism for selling n heterogeneous items to a single buyer. The buyer assigns each item $i \in [n]$ a private value v_i, which reflects the maximum she is willing to pay. For each item $i \in [n]$, the seller perceives that v_i is drawn from a distribution F_i with support $V_i \subseteq \mathbb{R}_+$. We denote her value profile by $\boldsymbol{v} = (v_1, \ldots, v_n)$, and denote the corresponding joint distribution by F with support $V = \times_{i=1}^n V_i$. The mechanism of the seller is defined by a pair (x, p) where $x : V \rightarrow [0, 1]^n$ is the allocation rule and $p : V \rightarrow \mathbb{R}_+$ is the payment rule. Suppose the buyer has a private value $\boldsymbol{v}$ and submits a bid $\boldsymbol{w}$ to the seller. The seller receives the submitted bid $\boldsymbol{w}$, and outputs an allocation $x(\boldsymbol{w}) = (x_1(\boldsymbol{w}), \ldots, x_n(\boldsymbol{w}))$ and a payment $p(\boldsymbol{w})$. In other words, the seller allocates item i to the buyer with probability $x_i(\boldsymbol{w})$, and the total payment of the buyer is $p(\boldsymbol{w})$.

In this paper, we assume the buyer is risk-neutral and rational, and the buyer's expected utility coincides with the expected profit $\sum_{i=1}^{n} v_i x_i(\boldsymbol{v}) - p(\boldsymbol{v})$ and is additively separable with respect to the items. We focus on direct mechanisms of the form

$$
\mathcal{M} = \left\{ \begin{array}{l} x : V \to [0,1]^n, \\ p : V \to \mathbb{R}_+ \end{array} \middle| \begin{array}{ll} \sum_{i=1}^{n} v_i x_i(\boldsymbol{v}) - p(\boldsymbol{v}) \geq \sum_{i=1}^{n} v_i x_i(\boldsymbol{w}) - p(\boldsymbol{w}) & \forall \boldsymbol{v}, \boldsymbol{w} \in V \\ \sum_{i=1}^{n} v_i x_i(\boldsymbol{v}) - p(\boldsymbol{v}) \geq 0 & \forall \boldsymbol{v} \in V \end{array} \right\}.
$$

The first constraint in $\mathcal{M}$ is known as incentive compatibility (IC), which requires the buyer to submit her true valuation to the seller because the truth-telling bid $\boldsymbol{v}$ results in maximal utility. The second constraint in $\mathcal{M}$ is known as individual rationality (IR), which requires that the utility of the buyer is always nonnegative without regard to her valuation. A mechanism (x, p) is truthful if it satisfies IC and IR constraints.

We denote a random valuation vector by $\tilde{\boldsymbol{v}} = (\tilde{v}_1, \ldots, \tilde{v}_n)$, which follows the joint distribution F, and the marginal distribution F_i. In our settings, we assume the seller does not know the exact joint distribution of $\tilde{\boldsymbol{v}}$. He only knows partial information about the valuation distribution, characterized by an ambiguity set $\mathcal{F}$. In addition, we assume the seller only knows partial distributional information about each item $i \in [n]$, characterized by a convex ambiguity set $\mathcal{F}_i$.

$$
\mathcal{F}_i = \left\{ F_i \in \mathbb{M}(V_i) : F_i \in \mathcal{P}_i \right\},
$$

where $\mathbb{M}(V_i)$ represents the set of all the probability measures with support V_i. $\mathcal{P}_i \subseteq \mathbb{M}(V_i)$ is a convex set which denotes the known distributional information. We assume that the seller knows only the ambiguity set of each item, $\mathcal{F}_1, \ldots, \mathcal{F}_n$, but not how these items are correlated with each other. Formally, let $\mathcal{F}$ be the set of all distributions F that have marginal distribution $F_i \in \mathcal{F}_i$ for each item i, i.e.,

$$
\mathcal{F} = \left\{ F \in \mathbb{M}(V) : F_i \in \mathcal{F}_i, i \in [n] \right\}, \tag{1}
$$

which we referred to as the *marginal ambiguity model*. The marginal ambiguity model is general and encompasses numerous commonly used ambiguity sets studied in the existing literature. Here are a few examples.

(1) **Marginal Moment Model.** The most popular ambiguity set studied in DRO is based on moment information. Our ambiguity set covers the case when the first K moments of each item are known, as defined by

$$
\mathcal{F} = \left\{ F \in \mathbb{M}(V) : \mathbb{E}_{\tilde{v} \sim F} \left[\tilde{v}_i^k \right] = m_{ik}, k \in [K], i \in [n] \right\},
$$

where m_{ik} denotes the kth moment of item i.

(2) **Correlation-robust ambiguity set.** The correlation-robust ambiguity set studied by Carroll [6] and Gravin and Lu [16] is a special case of our marginal ambiguity set (1) by defining

$$
\mathcal{F} = \left\{ F \in \mathbb{M}(V) : \mathbb{E}_{\tilde{v} \sim F} \left[\mathbb{I}\{\tilde{v}_i \leq v\} \right] = F_i(v), v \in V_i, i \in [n] \right\},
$$

where F_i denotes the known marginal distribution of item i.

(3) **Wasserstein ambiguity set.** Li et al. [24] studies monopolist problem under the Wasserstein ambiguity set. The multi-item version of the marginal Wasserstein ambiguity set can be formulated by

$$\mathcal{F} = \{F \in \mathbb{M}(V) : d_\rho(F, G) \le r_i, i = 1, \dots, n\},$$

where G is a reference distribution and $d_\rho(F, G)$ is the Wasserstein distance with $\rho \in [1, \infty]$.

The seller's goal is to design a truthful mechanism (x^*, p^*) that maximizes the worst-case expected revenue over all possible joint distributions within marginal ambiguity set $\mathcal{F}$. Formally, we solve the following max-min problem.

$$\phi = \max_{(x,p) \in \mathcal{M}} \min_{F \in \mathcal{F}} \mathbb{E}_{\tilde{v} \sim F} [p(\tilde{v})]. \tag{2}$$

which is known as the *distributionally robust monopolist problem.*

We are also interested in the budget feasibility property of the mechanism. The budget feasible mechanism is first proposed by Singer [32]. In this mechanism, the buyer's payment is bounded by a total budget $B \in [0, +\infty]$. The buyer is not able to pay for the items when payment $p(v)$ exceeds her budget B. We assume the total budget B is known to the seller. Thus, the seller must apply a truthful mechanism such that

$$p(v) \le B \qquad \forall v \in V. \tag{3}$$

We denote $\varphi(B)$ by the maximum worst-case revenue of the mechanism considering the budget feasibility. Formally,

$$\varphi(B) = \max_{(x,p) \in \mathcal{M}} \min_{F \in \mathcal{F}} \mathbb{E}_{\tilde{v} \sim F} [p(\tilde{v})]$$
$$\text{s.t.} \quad p(v) \le B \qquad \forall v \in V. \tag{4}$$

Note that problem (2) is a special case of problem (4) because $\varphi(B) = \phi$ when $B = +\infty$. In the rest of this paper, we discuss the distributionally robust monopolist problem with budget feasibility (4).

3   Additive Separation

In this section, we show the additive separation property of multi-item distributionally robust monopolist problem (4), that is, the multi-item auction (4) can be additively separated by multiple single-item auctions. For each item $i \in [n]$, we define the single-item distributionally robust monopolist problem with budget feasibility as follows.

$$\varphi_i(B_i) = \max_{(x_i,p_i) \in \mathcal{M}_i} \min_{F_i \in \mathcal{F}_i} \mathbb{E}_{\tilde{v} \sim F_i} [p_i(\tilde{v})]$$
$$\text{s.t.} \quad p_i(v) \le B_i \qquad \forall v \in V_i \tag{5}$$

where the single-item truthful mechanism set is defined by

$$\mathcal{M}_i = \left\{ \begin{array}{ll} x_i : V_i \to [0,1], & vx_i(v) - p_i(v) \geq vx_i(w) - p_i(w) \qquad \forall v, w \in V_i \\ p_i : V_i \to \mathbb{R}_+ & vx_i(v) - p_i(v) \geq 0 \qquad\qquad\qquad\quad \forall v \in V_i \end{array} \right\}.$$

Here, $B_i \in [0, B]$ represents the buyer's budget assigned to item i. The total budget should not exceed B, i.e.,

$$\sum_{i=1}^{n} B_i \leq B.$$

We define the *additive separation problem* by maximizing the total revenue $\sum_{i=1}^{n} \varphi_i(B_i)$ over all feasible $(B_1, \ldots, B_n)$ as follows.

$$\tilde{\varphi}(B) \quad = \quad \max_{B_i \geq 0} \quad \sum_{i=1}^{n} \varphi_i(B_i)$$

$$\text{s.t.} \quad \sum_{i=1}^{n} B_i \leq B. \tag{6}$$

Denote $(B_1^*, \ldots, B_n^*)$ as the optimal solution of problem (6). For each item $i \in [n]$, let (x_i^*, p_i^*) be the optimal mechanism of single-item problem (5) given the budget B_i^*, and F_i^* denote the corresponding worst-case distribution. Our additive separation result is concluded in Theorem 1.

Theorem 1. *The multi-item distributionally robust monopolist problem* (4) *is equivalent to the additive separation problem* (6), *i.e.,* $\varphi(B) = \tilde{\varphi}(B)$. *Moreover,*

$$x^*(\boldsymbol{v}) = (x_1^*(v_1), \ldots, x_n^*(v_n)) \qquad\qquad \forall \boldsymbol{v} \in V,$$

$$p^*(\boldsymbol{v}) = \sum_{i=1}^{n} p_i^*(v_i) \qquad\qquad\qquad\quad \forall \boldsymbol{v} \in V.$$

is an optimal mechanism of multi-item problem (4).

Proof. We first prove that $(x^*(\boldsymbol{v}), p^*(\boldsymbol{v}))$ is a feasible mechanism for multi-item problem (4). Since $p^*(\boldsymbol{v}) = \sum_{i=1}^{n} p_i^*(v_i)$, we have

$$\sum_{i=1}^{n} v_i x_i^*(\boldsymbol{v}) - p^*(\boldsymbol{v}) - \sum_{i=1}^{n} v_i x_i^*(\boldsymbol{w}) + p^*(\boldsymbol{w})$$

$$= \sum_{i=1}^{n} [v_i x_i^*(v_i) - p_i^*(v_i) - v_i x_i^*(w_i) + p_i^*(w_i)] \geq 0,$$

which implies $(x^*(\boldsymbol{v}), p^*(\boldsymbol{v}))$ satisfies the IC constraint. Besides, the IR constraint holds by

$$\sum_{i=1}^{n} v_i x_i^*(\boldsymbol{v}) - p^*(\boldsymbol{v}) = \sum_{i=1}^{n} v_i x_i^*(v_i) - \sum_{i=1}^{n} p_i^*(v_i) = \sum_{i=1}^{n} [v_i x_i^*(v_i) - p_i^*(v_i)] \geq 0.$$

The budget feasibility constraint in problem (4) holds by $p^*(\boldsymbol{v}) = \sum_{i=1}^{n} p_i^*(v_i) \leq \sum_{i=1}^{n} B_i^* \leq B$. Thus, $(x^*(\boldsymbol{v}), p^*(\boldsymbol{v}))$ is a feasible mechanism for multi-item problem (4).

Next, we shall prove that $\varphi(B) \geq \check{\varphi}(B)$. Denote F^B and (x^B, p^B) as the worst-case joint distribution and the optimal mechanism for multi-item problem (4), respectively. Then,

$$\varphi(B) = \mathbb{E}_{\tilde{\boldsymbol{v}} \sim F^B} \left[p^B(\tilde{\boldsymbol{v}}) \right] \geq \mathbb{E}_{\tilde{\boldsymbol{v}} \sim F^B} \left[p^*(\tilde{\boldsymbol{v}}) \right] = \sum_{i=1}^{n} \mathbb{E}_{v \sim F^B} \left[p_i^*(\tilde{v}_i) \right]. \tag{7}$$

The inequality holds due to the feasibility of $(x^*(\boldsymbol{v}), p^*(\boldsymbol{v}))$. On the other hand, F_i^* is the worst-case distribution for single-item problem (5), it leads that

$$\varphi_i(B_i^*) = \mathbb{E}_{\tilde{v} \sim F_i^*} \left[p_i^*(\tilde{v}) \right] \leq \mathbb{E}_{v \sim F^B} \left[p_i^*(\tilde{v}_i) \right]. \tag{8}$$

Combining (7) and (8), it follows that

$$\varphi(B) \geq \sum_{i=1}^{n} \mathbb{E}_{\boldsymbol{v} \sim F^B} \left[p_i^*(\tilde{v}_i) \right] \geq \sum_{i=1}^{n} \varphi_i(B_i^*) = \check{\varphi}(B).$$

Next, we show that $\varphi(B) \leq \check{\varphi}(B)$. Let $\hat{\mathcal{F}}$ be the set of all possible distributions with the marginal distributions $F_1^*, \ldots, F_n^*$, that is,

$$\hat{\mathcal{F}} = \left\{ \hat{F} \in \mathbb{M}(V) : \hat{F}_i = F_i^*(v), i = 1, \ldots, n \right\}.$$

Define

$$\hat{\varphi}(B) \quad = \quad \max_{(x,p) \in \mathcal{M}} \quad \min_{F \in \hat{\mathcal{F}}} \quad \mathbb{E}_{\tilde{\boldsymbol{v}} \sim F} \left[p(\tilde{\boldsymbol{v}}) \right]$$
$$\text{s.t.} \quad p(\boldsymbol{v}) \leq B \quad \forall \boldsymbol{v} \in V. \tag{9}$$

Let $(\hat{x}, \hat{p})$ and $\hat{F}$ be the optimal mechanism and the worst-case joint distribution for problem (9), respectively. Then, from the optimality of F^B in multi-item problem (4) and the fact $\hat{F} \in \hat{\mathcal{F}} \subseteq \mathcal{F}$,

$$\varphi(B) = \mathbb{E}_{\tilde{\boldsymbol{v}} \sim F^B} \left[p^B(\tilde{\boldsymbol{v}}) \right] \leq \mathbb{E}_{\tilde{\boldsymbol{v}} \sim \hat{F}} \left[p^B(\tilde{\boldsymbol{v}}) \right].$$

On the other hand, from the optimality of $(\hat{x}, \hat{p})$ for problem (9), we have

$$\hat{\varphi}(B) = \mathbb{E}_{\tilde{\boldsymbol{v}} \sim \hat{F}} \left[\hat{p}(\tilde{\boldsymbol{v}}) \right] \geq \mathbb{E}_{\tilde{\boldsymbol{v}} \sim \hat{F}} \left[p^B(\tilde{\boldsymbol{v}}) \right].$$

Therefore, we have $\varphi(B) \leq \mathbb{E}_{\tilde{\boldsymbol{v}} \sim \hat{F}} \left[p^B(\tilde{\boldsymbol{v}}) \right] \leq \hat{\varphi}(B)$. From Theorem 4.1 in [16], we know $\hat{\varphi}(B) = \check{\varphi}(B)$. Hence, $\varphi(B) \leq \check{\varphi}(B)$.

Therefore, we get $\varphi(B) = \check{\varphi}(B)$. From the previous discussion, we know that

$$x^*(\boldsymbol{v}) = (x_1^*(v_1), \ldots, x_n^*(v_n)), \qquad p^*(\boldsymbol{v}) = \sum_{i=1}^{n} p_i^*(v_i)$$

is an optimal mechanism for multi-item problem (4), and $\hat{F}$ is the corresponding joint worst-case distribution. $\square$

Remark 1. The correlation-robust ambiguity set studied by Carroll [6] and Gravin and Lu [16] is a special case of our ambiguity set (1). Thus, Theorem 1 is an extension to Theorem 4.1 in Gravin and Lu [16]. Our result is also an extension to Theorem 2.1 in Carroll [6] because $\varphi(B) = \phi$ when $B = +\infty$.

From the additive separation property of multi-item monopolist problem (2), if one can derive the solution of each single-item problem $\varphi_i(B_i)$, then it suffices to solve the additive separation problem (6). In Sect. 4, we consider some classic ambiguity sets and provide analytical solutions to the single-item problems. In Sect. 5, we show how to solve the additive separation problem (6).

4   Single-Item Monopolist Problem

This section gives analytical solutions to the single-item problem (5) under two moment-based ambiguity sets and the Wasserstein ambiguity set. Since the discussion is identical for each item i, we omit the subscript related to product i for brevity in our demonstration. We consider the single-item distributionally robust monopolist problem with a budget constraint.

$$\varphi(B) = \max_{(x,p)\in\mathcal{M}} \min_{F\in\mathcal{F}} \mathbb{E}_{\tilde{v}\sim F}\left[p(\tilde{v})\right]$$
$$\text{s.t.} \quad p(v) \leq B \quad \forall v \in V, \tag{10}$$

We first show the equivalence between the robust mechanism design problem and a minimax pricing problem with a budget penalty in Lemma 1.

Lemma 1. *When $\mathcal{F}$ is convex, we have*

$$\varphi(B) = \min_{F\in\mathcal{F},y\geq 0} \max_{v\in V} v(1 - F(v)) + (B - v)y. \tag{11}$$

Lemma 1 can be derived from von Neumann's minimax theorem. We can interpret the variable y as the *level of penalty*. Note that, when $V = [0, \bar{v}]$ and $B \geq \bar{v}$, we have $v \leq \bar{v} \leq B$. Thus, $y = 0$ is the optimal solution for problem (11). In other words, when the budget is sufficiently large, it imposes no restrictions on the optimal mechanism, implying that the budget constraint does not impose any penalties on the robust revenue. At this point, Lemma 1 reverts to the classic result (Lemma 1 in [9]). When the budget constraint takes effect, there will be an additional linear penalty term $(B - v)y$.

Next, we further analyze problem (11). Define the minimax pricing problem with a fixed level of penalty y by

$$R(y) = \min_{F\in\mathcal{F}} \max_{v\in V} v(1 - F(v)) + (B - v)y. \tag{12}$$

Note that the objective function $v(1 - F(v)) + (B - v)y$ is linear in (F, y). Taking the maximum over v yields a convex function. Thus, the value $\max_{v\in V} v(1 - F(v)) + (B - v)y$ is convex in (F, y). By partial minimization, we further know that $R(y)$ is convex. Based on the above discussion, we introduce the following lemma which plays a key role in solving the robust monopolist problem (10).

Lemma 2. *When $\mathcal{F}$ is a compact convex ambiguity set, we have the optimal value $R(y)$ of problem (12) is convex and*

$$\varphi(B) = \min_{y \in [0,1]} R(y).$$

If we can obtain the closed-form solution of the minimax pricing problem (12), combining the convexity of $R(y)$, we can calculate the optimal solution for problem (10) by the bisection method. In the remainder of this section, we shall present the closed-form expressions of $R(y)$ under three classic ambiguity sets. It helps us to obtain the optimal mechanism and the worst-case distributions for the single-item problem (5) under three classic ambiguity sets.

4.1   Moment Based Ambiguity Set

The most popular ambiguity set in distributionally robust optimization is based on moment information. There are a number of results for the robust single-item monopolist problem under the moment-based ambiguity set. We start from a simple moment-based ambiguity set, in which the seller only knows the mean and support of the buyer's valuation. Next, we provide an extension with additional variance information.

Mean and Support Ambiguity. Define the mean and support ambiguity set by

$$\mathcal{F} = \{F \in \mathbb{M}(V) : \mathbb{E}_{\tilde{v} \sim F}[\tilde{v}] = \mu\}, \tag{13}$$

where the mean $\mu \in (0, \bar{v})$ and the support $V = [0, \bar{v}]$ are known.

Proposition 1. *Suppose $\mathcal{F}$ is defined by the mean and support ambiguity set (13). Let $u(y)$ denote the unique solution in $(0, \mu - \bar{v}y)$ to the equation*

$$\ln\left(\frac{\bar{v}(1-y)}{u(y)}\right) + 1 + \frac{\bar{v}y - \mu}{u(y)} = 0.$$

Define $y^* = \arg\min_{y \in [0, \mu/\bar{v}]} u(y) + By$ and $u^* = u(y^*)$. The optimal revenue is $\varphi(B) = R(y^*) = u^* + By^*$.

The unique worst-case distribution of the single-item problem (10) is

$$F^*(v) = \begin{cases} 0 & 0 \le v < \frac{u^*}{1-y^*}, \\ 1 - \frac{u^*}{v} - y^* & \frac{u^*}{1-y^*} \le v < \bar{v}, \\ 1 & v = \bar{v}. \end{cases}$$

The optimal mechanism is

$$(x^*(v), p^*(v)) = \begin{cases} (0,0) & 0 \le v < \frac{u^*}{1-y^*}, \\ \left(\beta^* \ln\left(\frac{v(1-y^*)}{u^*}\right), \beta^*\left(v - \frac{u^*}{1-y^*}\right)\right) & \frac{u^*}{1-y^*} \le v \le \bar{v}, \end{cases}$$

with $\beta^* = u^*/(\mu - \bar{v}y^* - u^*)$.

Example 1. We solve robust single-item monopolist problem (10) under the mean and support ambiguity set $\mathcal{F}$ with $\bar{v} = 5$. We show the optimal revenue $\varphi(B)$ in Fig. 1 for different values of μ. The worst-case distribution $F^*(v)$ and the optimal mechanism (x^*, p^*) are given in Fig. 2 when the mean is $\mu = 3$.

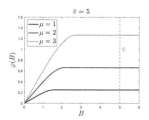

Fig. 1. Optimal revenue under mean and support ambiguity set.

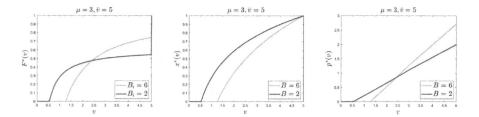

Fig. 2. Optimal solution under the mean and support ambiguity set. Left: worst-case distribution. Middle: optimal allocation. Right: optimal payment.

Mean and Variance Ambiguity. Suppose the seller only knows the mean and variance of the buyer's valuation, the ambiguity set is specified by

$$\mathcal{F} = \left\{ F \in \mathbb{M}(V) : \mathbb{E}_{\tilde{v} \sim F}[\tilde{v}] = \mu, \ \mathbb{E}_{\tilde{v} \sim F}[\tilde{v}^2] = \mu^2 + \sigma^2 \right\} \tag{14}$$

where the mean $\mu \in (0, +\infty)$, the standard deviation $\sigma \in (0, +\infty)$ and the support $V = [0, +\infty)$ are known.

Proposition 2. *Suppose $\mathcal{F}$ is defined by the mean and variance ambiguity set* (14). *Let $(u(y), t(y))$ denote the unique solution in* $\left(0, \frac{(1-y)(\mu^2+\sigma^2) - \sigma\sqrt{(1-y)(\mu^2+\sigma^2)}}{\mu} \right) \times$ $(0, +\infty)$ *to the equations*

$$\ln\left(\frac{(1-y)t(y)}{u(y)} \right) + 1 + \frac{yt(y) - \mu}{u(y)} = 0,$$

$$2t(y) - \frac{u(y)}{1-y} + \frac{y(t(y))^2 - \mu^2 - \sigma^2}{u(y)} = 0.$$

Define $y^* = \arg\min_{y \in [0, \frac{\mu^2}{\mu^2+\sigma^2}]} u(y) + By$ and $(u^*, t^*) = (u(y^*), t(y^*))$. The optimal revenue is $\varphi(B) = R(y^*) = u^* + By^*$.

The unique worst-case distribution of the single-item problem (10) is

$$
F^*(v) = \begin{cases}
0 & 0 \le v < \frac{u^*}{1-y^*}, \\
1 - \frac{u^*}{v} - y^* & \frac{u^*}{1-y^*} \le v < t^*, \\
1 & v \ge t^*.
\end{cases}
$$

The optimal mechanism $(x^*(v), p^*(v))$ is

$$
\begin{cases}
(0,0) & 0 \le v < \frac{u^*}{1-y^*}, \\
\left(\beta_1^* \ln\left(\frac{v(1-y^*)}{u^*}\right) + 2\beta_2^*\left(v - \frac{u^*}{1-y^*}\right), \beta_1^* v + \beta_2^* v^2 + \alpha^*\right) & \frac{u^*}{1-y^*} \le v < t^*, \\
(1, \beta_1^* t^* + \beta_2^* t^{*2} + \alpha^*) & v \ge t^*,
\end{cases}
$$

where $\beta_2^* = 1/2\left(t^* - \frac{u^*}{1-y^*} - t^* \ln\left(\frac{t^*(1-y^*)}{u^*}\right)\right)$, $\beta_1^* = -2t^*\beta_2^*$ and $\alpha^* = -\beta_1^* \frac{u^*}{1-y^*} - \beta_2^*\left(\frac{u^*}{1-y^*}\right)^2$.

Example 2. We solve problem (10) under the mean and variance ambiguity set (14). The optimal revenue $\varphi(B)$ is plotted in Fig. 3 for fixed mean $\mu = 2$ and different variance values. We show the worst-case distribution $F^*(v)$ and optimal mechanism (x^*, p^*) in Fig. 4 when $\mu = 2$ and $\sigma = 1$.

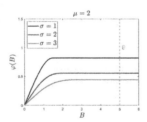

Fig. 3. Optimal revenue under mean and variance ambiguity set.

4.2   Wasserstein Ambiguity

Recently, the Wasserstein ambiguity set is becoming popular in distributionally robust optimization. Suppose the reference distribution is G, we consider all distributions F such that the Wasserstein distance between F and G is no more than a fixed $r > 0$. Formally, for any $\rho \in [1, \infty]$, the type-ρ Wasserstein distance between two distributions F and G is defined by

$$
d_\rho(F, G) = \left(\int_0^1 |\bar{F}^{-1}(q) - \bar{G}^{-1}(q)|^\rho \, dq\right)^{1/\rho}.
$$

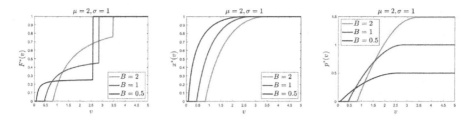

Fig. 4. Optimal solution under mean and variance ambiguity. Left: worst-case distribution. Middle: optimal allocation. Right: optimal payment.

where $\bar{F}$ and $\bar{G}$ are the complementary cumulative distributions function of F and G, respectively. For any given reference distribution G, the type-ρ Wasserstein ambiguity set is defined as the set of all distributions, whose type-ρ Wasserstein distance to G is bounded by a Wasserstein ball radius $r > 0$:

$$\mathcal{F}(r,\rho) = \{F \in \mathbb{M}(V) : \mathbb{E}_{\tilde{v} \sim F}[1] = 1, \ d_\rho(F,G) \le r\}. \tag{15}$$

We first give the analytical form of worst-case distribution and optimal revenue under $\mathcal{F}(r,\rho)$ in Proposition 3 for any $\rho \in [1, +\infty]$ and $r \in (0, +\infty)$, given the support $V = [0, +\infty)$. Furthermore, following Proposition 3, we calculate the optimal mechanism in Proposition 4 under the type-1 Wasserstein ambiguity set.

Proposition 3. *Suppose the ambiguity set $\mathcal{F}(r,\rho)$ is given by (15) with $\rho \in [1, +\infty]$. Let $u(y)$ denote the unique solution in $[0, +\infty)$ to the equation*

$$\left(\int_y^1 \left(\left(\bar{G}^{-1}(q) - \frac{u(y)}{q-y} \right)_+ \right)^\rho dq \right)^{1/\rho} = r.$$

Define $y^* = \arg\min_{y \in [0, h^{-1}(r)]} u(y)$ and $u^* = u(y^*)$, where

$$h(y) = \left(\int_y^1 \left(G^{-1}(q) \right)^\rho dq \right)^{1/\rho}.$$

The optimal revenue is $\varphi(B) = R(y^*) = u^* + By^*$.
The worst-case distribution of the single-item problem (10) is

$$F^*(v) = 1 - \min\left\{ \bar{G}(v), \frac{u^*}{v} + y^* \right\},$$

which is unique when ρ is finite.

Proposition 4 (Chen et al. [9], Proposition 6). *Suppose the ambiguity set is $\mathcal{F}(r,1)$. Assume there are finitely many disjoint intervals $[u_j, w_j] \subseteq [0, B_i]$, $j \in [J]$ on which $\bar{G}(v) > u^*/v + y^*$, where (u^*, y^*) is defined in Proposition 3.*

In addition, $u_0 = w_0 = 0$ and $u_{J+1} = \bar{v}$. The optimal mechanism $(x^*(v), p^*(v))$ of problem (10) is of the form

$$
\begin{cases}
\alpha^* \left(\ln \left(\frac{v}{u_j} \right) + \sum_{k=1}^{j-1} \ln \left(\frac{w_k}{u_k} \right), (v - u_j) + \sum_{k=1}^{j-1} (w_k - u_k) \right), & v \in [u_j, w_j), j \in [J] \\
\alpha^* \left(\sum_{k=1}^{j} \ln \left(\frac{w_k}{u_k} \right), \sum_{k=1}^{j} (w_k - u_k) \right), & v \in [w_j, u_{j+1}), j \in [J]
\end{cases}
$$

where $\alpha^* = 1/\sum_{j \in [J]} \ln(w_j/u_j)$ and $(x^*(\bar{v}), p^*(\bar{v})) = (1, \alpha^* \sum_{k \in [J]} (w_k - u_k))$.

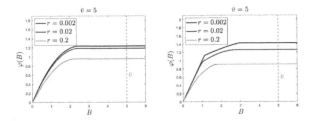

Fig. 5. Left: Optimal revenue when G is uniformly distributed on $[0, 5]$. Right: Optimal revenue when G is an empirical distribution.

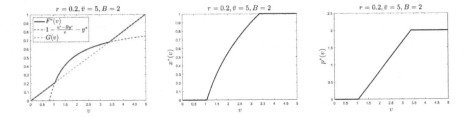

Fig. 6. Optimal solution when G is uniformly distributed on $[0, 5]$. Left: worst-case distribution. Middle: optimal allocation. Right: optimal payment.

Example 3. (1) Suppose G is uniform on support $[0, 5]$. We show the optimality results under the type-1 Wasserstein ambiguity set when $B = 2$. The optimal revenue $\varphi(B)$ is given in the left of Fig. 5. We show the worst-case distribution $F^*(v)$ and optimal mechanism (x^*, p^*) in Fig. 6.

(2) Consider a more complex case in which G is the empirical distribution of four samples $\{1.2, 1.5, 3, 4\}$ supported on $[0, 5]$. We give the optimal revenue $\varphi(B)$ in the right of Fig. 5. The worst-case distribution $F^*(v)$ and optimal mechanism (x^*, p^*) is given in Fig. 7. One can find that the optimal revenue $\varphi(B)$ is not concave over B.

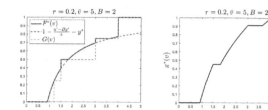
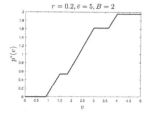

Fig. 7. Optimal solution when G is an empirical distribution. Left: worst-case distribution. Middle: optimal allocation. Right: optimal payment.

5 Algorithm for Multi-Item Monopolist Problem with Budget

We have shown how to calculate the single-item monopolist problem (5) under some specific ambiguity sets in Sect. 4. In this section, based on additive separation property, we provide algorithms for multi-item problem (4) by solving the equivalent formulation (6).

Gravin and Lu [16] show that the optimal revenue function $\varphi_i(B_i)$ of the single-item monopolist problem is monotone increasing, piecewise linear, and concave when the marginal distribution is known. Thus, they provide an efficient greedy algorithm to solve the multi-item problem under the correlation-robust ambiguity set. However, for the single-item distributionally monopolist problem, as examples in Sect. 4 show, the optimal revenue function $\varphi_i(B_i)$ is not guaranteed to be piecewise linear. Hence, we need to design new algorithms to solve the additive separated problem (6). We first show that the optimal revenue function $\varphi_i(B_i)$ is concave over B_i for each item i.

Lemma 3. *When $\mathcal{F}$ is convex, $\varphi_i(B_i)$ is monotone increasing and concave in $B_i \in [0, +\infty)$.*

Lemma 3 implies the formulation (6) is a linear-constrained convex optimization problem. Besides, we can substitute the inequality constraint $\sum_{i=1}^{n} B_i \leq B$ by an equality constraint in problem (6) from Lemma 4. That is, we solve the problem

$$\max_{B_i \geq 0, \forall i} \sum_{i=1}^{n} \varphi_i(B_i) \qquad \text{s.t.} \qquad \sum_{i=1}^{n} B_i = B. \qquad (16)$$

Lemma 4. *For any $B \in [0, +\infty)$, there must exist an optimal solution $(B_1^*, \ldots, B_n^*)$ of problem (4) such that*

$$(B_1^*, \ldots, B_n^*) \in \mathcal{B} = \left\{ (B_1, \ldots, B_n) \in \mathbb{R}_+^n \ \middle| \ \sum_{i=1}^{n} B_i = B \right\}.$$

Note that the objective function of Problem (16) is separable concave, which can be seen as a special case of classic *resource allocation problems* that is widely studied in the literature. It seeks to find an optimal allocation of a fixed amount of budget, so as to maximize the expected revenue incurred by the allocation. Abundant algorithms have been proposed, such as Lagrangian multiplier algorithms and pegging algorithms. We refer readers to Patriksson [28] for the literature review of the resource allocation problem on applications and algorithms. Among all algorithms, the polynomial algorithms proposed by Hochbaum [21] satisfy our settings. For the case when the budget B_i's are integers, the running time of their algorithm is $O(n \log(n) \log(B/n))$; for the continuous case, it takes $O(n \log(n) \log(B/\epsilon))$ steps to get an ϵ-accurate solution.

6   Conclusion

To summarize, we study the multi-item distributionally robust monopolist problems with budget feasibility. We construct a marginal ambiguity model when the valuation distributions are assumed to belong to the class of distributions defined by given marginal partial information. Under this model, we show the additive separation property holds even when the buyer has a budget feasibility constraint. Moreover, we give analytical solutions to the single-item problem under some specific ambiguity sets. Finally, based on the additive separation property and the analytical solutions of single-item problems, we solve the multi-item problem based on polynomial-time algorithms for the resource allocation problem.

There are various interesting directions for future research. First, since the additive separable property is helpful in solving the multi-item monopolist problem, it would be interesting to investigate other forms of the objective function that satisfy the additive separable property. Second, the ambiguity set considered in this paper does not include any correlation information across different items. It would be significant to study the impact of correlation because the correlation across items is very common in real-world products.

Acknowledgements. Zhen Wang received support from the National Science Foundation of China (NSFC) Grant 72301235, the Guangdong Key Lab of Mathematical Foundations for Artificial Intelligence and the Shenzhen Science and Technology Program under Grant ZDSYS20220606100601002. Simai He received support from the Major Program of National Natural Science Foundation of China (NSFC) Grant (72192830,72192832) and Grant 71825003. The authors thank the senior editor, the associate editor, and the three reviewers for constructive comments on the previous drafts of this paper.

References

1. Bandi, C., Bertsimas, D.: Optimal design for multi-item auctions: a robust optimization approach. Math. Oper. Res. **39**(4), 1012–1038 (2014)
2. Bei, X., Gravin, N., Lu, P., Tang, Z.G.: Correlation-robust analysis of single item auction. In: Proceedings of the Thirtieth Annual ACM-SIAM Symposium on Discrete Algorithms, pp. 193–208. SIAM (2019)

3. Bergemann, D., Schlag, K.: Robust monopoly pricing. J. Econ. Theory **146**(6), 2527–2543 (2011)
4. Blanchet, J., Murthy, K.: Quantifying distributional model risk via optimal transport. Math. Oper. Res. **44**(2), 565–600 (2019)
5. Carrasco, V., Luz, V.F., Kos, N., Messner, M., Monteiro, P., Moreira, H.: Optimal selling mechanisms under moment conditions. J. Econ. Theory **177**, 245–279 (2018)
6. Carroll, G.: Robustness and separation in multidimensional screening. Econometrica **85**(2), 453–488 (2017)
7. Che, Y.K., Zhong, W.: Robustly-optimal mechanism for selling multiple goods. In: Proceedings of the 22nd ACM Conference on Economics and Computation, pp. 314–315 (2021)
8. Chen, X., He, S., Jiang, B., Ryan, C.T., Zhang, T.: The discrete moment problem with nonconvex shape constraints. Oper. Res. **69**(1), 279–296 (2021)
9. Chen, Z., Hu, Z., Wang, R.: Screening with limited information: a dual perspective and a geometric approach. Available at SSRN 3940212 (2021)
10. Chen, Z., He, S., Wang, Z., Zheng, M.: Robust monopoly pricing: deterministic and randomized mechanisms under asymmetric information. Available at SSRN 4454460 (2023)
11. Daskalakis, C., Deckelbaum, A., Tzamos, C.: The complexity of optimal mechanism design. In: Proceedings of the Twenty-Fifth Annual ACM-SIAM Symposium on Discrete Algorithms, pp. 1302–1318. SIAM (2014)
12. Delage, E., Ye, Y.: Distributionally robust optimization under moment uncertainty with application to data-driven problems. Oper. Res. **58**(3), 595–612 (2010)
13. Du, S.: Robust mechanisms under common valuation. Econometrica **86**(5), 1569–1588 (2018)
14. Gao, R., Chen, X., Kleywegt, A.J.: Wasserstein distributionally robust optimization and variation regularization. Oper. Res. (2022)
15. Gao, R., Kleywegt, A.: Distributionally robust stochastic optimization with Wasserstein distance. Math. Oper. Res. (2022)
16. Gravin, N., Lu, P.: Separation in correlation-robust monopolist problem with budget. In: Proceedings of the Twenty-Ninth Annual ACM-SIAM Symposium on Discrete Algorithms, pp. 2069–2080. SIAM (2018)
17. Guo, J., He, S., Jiang, B., Wang, Z.: A unified framework for generalized moment problems: a novel primal-dual approach. arXiv preprint arXiv:2201.01445 (2022)
18. Hart, S., Reny, P.J.: Maximal revenue with multiple goods: nonmonotonicity and other observations. Theor. Econ. **10**(3), 893–922 (2015)
19. He, W., Li, J.: Correlation-robust auction design. J. Econ. Theory **200**, 105403 (2022)
20. He, W., Li, J., Zhong, W.: Order statistics of large samples: theory and an application to robust auction design. Technical report, Mimeo (2022)
21. Hochbaum, D.S.: Lower and upper bounds for the allocation problem and other nonlinear optimization problems. Math. Oper. Res. **19**(2), 390–409 (1994)
22. Koçyiğit, Ç., Rujeerapaiboon, N., Kuhn, D.: Robust multidimensional pricing: separation without regret. Math. Program. 1–34 (2022)
23. Lasserre, J.B.: Moments, Positive Polynomials and Their Applications, vol. 1. World Scientific (2009)
24. Li, Y., Lu, P., Ye, H.: Revenue maximization with imprecise distribution. In: Proceedings of the 18th International Conference on Autonomous Agents and Multi-Agent Systems, pp. 1582–1590 (2019)
25. Manelli, A.M., Vincent, D.R.: Multidimensional mechanism design: revenue maximization and the multiple-good monopoly. J. Econ. Theory **137**(1), 153–185 (2007)

26. Mohajerin Esfahani, P., Kuhn, D.: Data-driven distributionally robust optimization using the Wasserstein metric: performance guarantees and tractable reformulations. Math. Program. **171**(1–2), 115–166 (2018)
27. Myerson, R.B.: Optimal auction design. Math. Oper. Res. **6**(1), 58–73 (1981)
28. Patriksson, M.: A survey on the continuous nonlinear resource allocation problem. Eur. J. Oper. Res. **185**(1), 1–46 (2008)
29. Pınar, M.Ç., Kızılkale, C.: Robust screening under ambiguity. Math. Program. **163**, 273–299 (2017)
30. Rahimian, H., Mehrotra, S.: Distributionally robust optimization: a review. arXiv preprint arXiv:1908.05659 (2019)
31. Riley, J., Zeckhauser, R.: Optimal selling strategies: when to haggle, when to hold firm. Q. J. Econ. **98**(2), 267–289 (1983)
32. Singer, Y.: Budget feasible mechanisms. In: 2010 IEEE 51st Annual Symposium on Foundations of Computer Science, pp. 765–774. IEEE (2010)
33. Suzdaltsev, A.: Distributionally robust pricing in independent private value auctions. J. Econ. Theory **206**, 105555 (2022)
34. Zhang, W.: Correlation-robust optimal auctions. arXiv preprint arXiv:2105.04697 (2021)

Target-Oriented Regret Minimization for Satisficing Monopolists

Napat Rujeerapaiboon[1]([✉]) [iD], Yize Wei[2] [iD], and Yilin Xue[1] [iD]

[1] Department of Industrial Systems Engineering and Management,
National University of Singapore, Singapore, Singapore
napat.rujeerapaiboon@nus.edu.sg, yilin.xue@u.nus.edu
[2] Department of Computer Science, National University of Singapore, Singapore,
Singapore
yize.wei@u.nus.edu

Abstract. We study a robust monopoly pricing problem where a seller aspires to sell an item to a buyer. We assume that the seller, unaware of the buyer's willingness to pay, ambitiously optimizes over a space of all individually rational and incentive compatible mechanisms with a regret-type objective criterion. We particularly adopt a robust satisficing approach, which has been touted as a promising alternative of robust optimization, and aim at minimizing the excess regret above the predetermined target level. We interpret our pricing problem both probabilistically and distributionally robustly, and we analytically show that the optimal mechanism involves the seller offering a menu of lotteries that charges a buyer-dependent participation fee and allocates the item with a buyer-dependent probability. Then, we consider two additional variants of the problem where the seller restricts her attention to a class of only deterministic posted price mechanisms and where the seller is relieved from specifying the target regret in advance. Finally, we determine a randomized posted price mechanism that is readily implementable and equivalent to the optimal mechanism, compute its statistics, and quantify the strength of the entailed randomization. Besides, we compare the proposed mechanism with a robust benchmark and numerically find that the former is predominantly superior to the latter in terms of the expected regret and the expected revenue especially when the coefficient of variation of the buyer's value is under a hundred percent.

Keywords: Mechanism design · Monopoly pricing · Regret minimization · Robust optimization · Satisficing

1 Introduction

We study a variant of the monopoly pricing problem where the seller (*'she'*) offers an item to gain maximum benefit. The buyer (*'he'*) attaches a private value ν

This research was supported by the Ministry of Education, Singapore, under its Academic Research Fund Tier 2 grant MOE-T2EP20222-0003.

to the item offered, and to him this value is a personal monetary equivalent of the item. It indicates the maximum amount of money that he is willing to pay. Consequently, if this value was known to the seller, then she could have simply offered the item at a fixed price ν and earned maximally.

However, a prudent buyer would always keep his value strictly private before the seller making her first move in order to not lose his bargaining power. It is therefore impossible for the seller to earn exactly an amount of ν from the sale transaction. The seller designs and broadcasts a mechanism that sets out the instructions on how the buyer can communicate with her. A mechanism consists of three parts: the set of messages for the buyer to choose from, the allocation rule, and the payment rule. Once the buyer selects a message that to him is the most favourable, he will pay an amount indicated by the payment rule to the seller and be allotted the item with a certain probability that is specified by the allocation rule. Modern studies of mechanism design largely simplify due to the influential Revelation Principle [18] which allows the seller to restrict her attention to a subclass of direct mechanisms, where the message set is simply chosen as the set of all possible values of the buyer's private value ν, without any loss of generality and revenue. In the remainder of the paper, we will henceforth use the terms mechanism and direct mechanism interchangeably.

As ν is unknown to the seller, it may be perceived as a random variable that crisply follows a known probability distribution. In this case, [18] and [23] showed that the seller can attain a maximum expected revenue by posting a deterministic price for the item. Although without a doubt this is the most popular variant of the mechanisms currently seen in practice, especially at retail stores, several studies and observations have pointed out the advantages of the seller offering the same item to different buyers at different price points.

The assumption that the seller knows the distribution of the buyer's private value ν itself is not innocuous, and a question concerning the robustness of the so-called optimal mechanism has come under the spotlight [3]. A seller with limited information about ν may consider stepping away from maximizing the expected revenue and instead leverage robust optimization (see, e.g., [2,6]) with an aim to maximize the worst-case revenue. While this approach is methodologically sound, it may recommend a trivial and unrealistically conservative mechanism where the seller keeps the item to herself provided that the uncertainty set for ν contains zero.

To address having limited information and also to avoid being superfluously conservative, [24] introduced a minimax regret as a decision criterion; see also [19, 21] for further justifications. Specifically in a sale transaction, the seller's regret is defined as the difference between the hypothetical revenue that the seller could have earned if she exactly knew ν and the actual revenue which is specified by the payment rule. Recently, it was discovered in [13] that the mechanism which attains the smallest worst-case regret is non-trivial, and it involves a piece-wise logarithmic allocation rule and a piece-wise linear payment rule even if ν can take a value of zero. The minimax regret criterion has also been adopted in several other papers, for example, [4] and [5], with the latter aptly noting that "a deterministic pricing policy exposes the seller to substantial regret, and that the seller can decrease her exposure by offering many prices," which underscores the importance of a randomized pricing strategy when a regret-type objective is used by the seller. Slightly differently, [8] and [27] adopted a worst-case relative regret criterion, also known as a competitive ratio, and again the benefit of randomization remains.

The principle of worst-case regret minimization resonates well with that of robust optimization. A duality technique that is frequently leveraged in the robust optimization literature is indispensable for [13] to analytically derive the optimal mechanism that minimizes the seller's worst-case regret. We refer our readers to [26] for a broader discussion on the link between mechanism design and linear programming as well as the corresponding duality theory.

For the comprehensiveness of our literature review, we remark that a risk-neutral, ambiguity-averse seller may instead use distributionally robust optimization (e.g., [11,17,28]) to maximize her worst-case expected revenue when facing the inadequacy of robust optimization in mechanism design. To achieve this, the seller needs to construct an ambiguity set comprising all probability distributions of ν that are consistent with her prior information. Listing a few important examples, [5] and [14] adopted a neighbourhood of a reference distribution of ν with respect to the Prohorov and the Wasserstein metric, respectively, whereas [20] and [7] considered an ambiguity set that is characterized by the support, the mean, and/or higher order moments. Recently, [9] provided a unified framework for computing the robustly optimal mechanisms under different ambiguity sets of distributions.

Recently, [15] provided a follow-up on an earlier work on globalized robust optimization due to [1] and introduced an alternative to robust optimization known as 'robust satisficing.' When the objective function representing cost (or regret) is uncertain, robust optimization computes a solution that minimizes the worst-case cost. On the contrary, robust satisficing determines a solution that, in the nominal scenario, has a cost under the predetermined target and, in all other scenarios, a cost that only proportionately deviates from the same target. Robust satisficing has been numerically shown to produce statistically better decisions than various state-of-the-art benchmarks (see, e.g., [12,25]), and it also satisfies several axioms from behavioral decision making. Abiding by this new principle, the decision maker needs to specify the 'nominal scenario' and to supply the 'target value,' which represents the acceptability level of the nominal objective. In this paper, we revisit the monopoly pricing problem with a regret objective that was studied by [13], but we formulate the problem using the robust satisficing ideology instead. Computationally, depending on the convexity and linearity properties of the problem (or the lack thereof), robust satisficing solutions could be exactly determined by Fenchel duality [1] or approximately attained by using either a primal decision rule [15] or a dual linear decision rule [22]. To our knowledge, we are the first to *analytically* solve a robust satisficing problem of this level of complexity, that is, the infinite dimensionality of the mechanism design problem and the interaction between the two agents (*i.e.*, the buyer and the seller), without any approximation.

We summarize the main contributions of the paper as follows.

1. We use robust satisficing to formulate a mechanism design problem for a monopolist who has an item to liquidate and a regret minimization objective. The problem takes as input parameters the support of the buyer's value for the item and its nominal value as well as the target regret level that should not be exceeded (that is, the threshold that the seller is willing to tolerate) under the nominal scenario. After mathematically formulating the problem, we derive a probabilistic regret bound based on its solution to provide further justification for this work.
2. We characterize the condition on the problem's input that is equivalent to the problem's feasibility. Whenever it is feasible, we *analytically* propose a candidate solution of the mechanism design problem, which itself is an infinite linear program. We establish that the proposed mechanisms are *optimal* with respect to the problem's given input by developing tight lower bounds of the problem. Besides, we also argue that each of these lower bounds has an intimate relationship with the problem's dual, see our online Appendix B, and we relate a subset of our recommended mechanisms to those from the existing literature on distributionally robust mechanism design.
3. We study two additional variants of the mechanism design problem that are similarly analytically solvable. The first extension assumes that the seller restricts her attention to a class of deterministic posted price mechanisms only, whereas the second relaxes the requirement of the target regret needing specifying.
4. To increase the acceptance and the relevance of the derived optimal mechanism, we interpret it as a randomized posted price mechanism, compute its statistics and compare it with the optimal deterministic posted price mechanism. Besides, we compare our mechanisms with several benchmarks including the worst-case regret minimizing mechanism; see [13], where we numerically show the dominance of our mechanism in terms of the seller's expected regret and expected revenue, especially when the coefficient of variation of the buyer's value falls below 100%. Based on our experiment results, we also provide a guideline on how our pricing model should be calibrated.

The rest of the paper is structured as follows. Section 2 discusses how regret could be leveraged as a decision criterion in both robust optimization and robust satisficing settings. Section 3 derives the optimal mechanism for different ranges of the input parameters. Section 4 considers a restricted problem where only deterministic posted price mechanisms are considered, and Sect. 5 studies another variation of the problem where the seller is relieved from choosing the target regret. Finally, Sect. 6 reports the statistics and the performance of the proposed mechanisms in relation to the benchmarks, and Sect. 7 concludes the paper. All proofs can be found in Appendix A, which is available online.

Notation: For a logical expression $\mathcal{E}$, we define $\mathbb{1}(\mathcal{E}) = 1$ if $\mathcal{E}$ is true; $= 0$ otherwise, and for any real number x, we denote by x^+ its positive part, that is, $x^+ = \max\{x, 0\}$. We adopt the convention for division by zero that $a/0 = \infty$ if

$a > 0$; $= 0$ otherwise. All logarithms have a natural base of e, and it is assumed that $\log(0) = -\infty$. Besides, the set of all (bounded) Borel-measurable functions from $\mathcal{D}$ to $\mathcal{R}$ is denoted by $\mathcal{L}(\mathcal{D}, \mathcal{R})$, and finally the cone of all non-negative Borel measures supported on $\mathcal{A}$ is denoted by $\mathcal{M}_+(\mathcal{A})$.

2 Regret Satisficing in Robust Mechanism Design

We consider a prototypical monopoly pricing problem where the seller aims to sell a single product to a buyer. She perceives the value ν that the buyer privately assigns to the item as a stochastic-free uncertain variable which could take any value in the interval $[0, \overline{\nu}]$. The vanishing lower bound is justified by the observation that no matter what the product offered is, it can be inconsequential to some people, and the upper bound $\overline{\nu}$ could perhaps be estimated from the price of a similar yet superior substitute that is currently available in the market; see [10]. If nothing else is known, a direct mechanism that attains the minimum worst-case regret can be found by solving

$$
\begin{aligned}
&\text{minimize} && \sup_{\nu \in [0, \overline{\nu}]} \nu - m(\nu) \\
&\text{subject to} && q \in \mathcal{L}([0, \overline{\nu}], [0, 1]),\ m \in \mathcal{L}([0, \overline{\nu}], \mathbb{R}) \\
& && q(\nu)\nu - m(\nu) \geq 0 && \forall \nu \in [0, \overline{\nu}] \\
& && q(\nu)\nu - m(\nu) \geq q(\omega)\nu - m(\omega) && \forall \nu, \omega \in [0, \overline{\nu}],
\end{aligned}
\tag{1}
$$

where q and m denote the allocation and the payment rule, respectively. In particular, $q(\nu)$ represents the probability that the buyer with value ν will obtain the item after he makes a payment of amount $m(\nu)$ to the seller. Besides, Problem (1) assumes that the buyer is risk-neutral and that his expected utility coincides with $q(\nu)\nu - m(\nu)$. The two inequalities are known as the *'individual rationality'* and the *'incentive compatibility'* constraints, respectively. A mechanism is said to be individually rational if it ensures that the buyer's expected utility is always non-negative, and it is said to be incentive compatible if the buyer can maximize his expected utility by truthfully reporting his true value ν, which is unknown to the seller when the mechanism is designed. Both of these constraints must hold for a buyer of any value. Finally, the objective function of Problem (1) contains $\nu - m(\nu)$ which we refer to as a *'regret'* of the seller since it characterizes the difference between the hypothetical revenue that the seller could have earned if she precisely knew ν and the actual revenue that is generated by the mechanism (q, m).

As an allocation of the item is probabilistic, the seller announcing a mechanism is similar to the seller offering a lottery: requesting an upfront payment from the buyer without making a definite promise to deliver the winning prize. Though, a mechanism can be as simple as offering a product at a certain price, say $p \in \mathbb{R}$. In this case, we can write down the mechanism as

$$
q(\nu) = \mathbb{1}(\nu \geq p) \quad \text{and} \quad m(\nu) = p\mathbb{1}(\nu \geq p),
$$

and we shall refer to it as a *'deterministic posted price mechanism.'* One can readily verify that such a mechanism is always incentive compatible and individually rational. Note that due to the linearity of the constraints involved, even if the price p is chosen at random, $\tilde{p} \sim \mathbb{P}$, a *'randomized posted price mechanism'* with

$$q(\nu) = \mathbb{E}_{\mathbb{P}}\left(\mathbb{1}(\nu \geq \tilde{p})\right) \quad \text{and} \quad m(\nu) = \mathbb{E}_{\mathbb{P}}\left(\tilde{p}\mathbb{1}(\nu \geq \tilde{p})\right)$$

is too incentive compatible and individually rational. Amongst others, [7] argued that every incentive compatible and individually rational mechanism with a right-continuous allocation rule can be interpreted as a randomized posted price mechanism.

Essentially, [4] solved the above worst-case regret minimization problem, and subsequently [13] provided an extension to this problem in which the seller has multiple items to sell simultaneously. Similarly, both papers showed that the optimal mechanism consists of a piece-wise logarithm allocation rule and a piece-wise linear payment rule.

In this paper, we are going to take a similar yet fundamentally different approach to the seller's worst-case regret minimization problem. Instead of using robust optimization, we adopt a recently proposed *'robust satisficing'* approach and consider the following mechanism design problem.

$$
\begin{aligned}
\text{minimize} \quad & k \\
\text{subject to} \quad & q \in \mathcal{L}([0,\overline{\nu}],[0,1]), \ m \in \mathcal{L}([0,\overline{\nu}],\mathbb{R}), \ k \in \mathbb{R}_{+} \\
& q(\nu)\nu - m(\nu) \geq 0 & \forall \nu \in [0,\overline{\nu}] \quad (\mathcal{P}) \\
& q(\nu)\nu - m(\nu) \geq q(\omega)\nu - m(\omega) & \forall \nu, \omega \in [0,\overline{\nu}] \\
& \nu - m(\nu) \leq \tau + k|\nu - \hat{\nu}| & \forall \nu \in [0,\overline{\nu}].
\end{aligned}
$$

In this formulation, we have additional parameters $\hat{\nu} \in (0,\overline{\nu})$ and $\tau \in \mathbb{R}$. They represent the nominal value of ν and the target regret (see [15]). The target τ represents an admissible upper bound of the seller's regret under the nominal scenario $\nu = \hat{\nu}$. The additional decision variable k characterizes the maximum level of constraint violation of all other scenarios $\nu \neq \hat{\nu}$ in relation to their deviation from the nominal counterpart. Problem $(\mathcal{P})$ seeks the smallest upper bound on the regret $\nu - m(\nu)$ of the form $\tau + k|\nu - \hat{\nu}|$, parameterized by k. This upper bound takes a minimum value when $\nu = \hat{\nu}$ and is increasing when the buyer's value ν further deviates from the nominal value $\hat{\nu}$ in either direction. The inputs $\hat{\nu}$ and $\overline{\nu}$ are to be obtained or statistically estimated from the available data, whereas the target τ is to be specified by the seller to reflect her risk tolerance level. Throughout, we will refer to the last constraint of Problem $(\mathcal{P})$ as the *'satisficing'* constraint. By construction, the optimal objective value of Problem $(\mathcal{P})$ could be infinity if τ is too small, and it decreases as τ increases. As an upper bound on the nominal regret, *i.e.*, $\hat{\nu} - m(\hat{\nu})$, that the seller is willing to tolerate, τ is a trade-off parameter that the seller could explore. She may leverage a smaller τ if she believes that $\hat{\nu}$ is an adequate representative of the unknown ν. Conversely, if ν is considerably uncertain, she may employ a larger

τ to get a smaller optimal k and strengthen the regret bound elsewhere as ν departs from $\hat{\nu}$.

The satisficing constraint allows us to draw the following probabilistic regret bound, which serves as an additional motivation for us to study Problem $(\mathcal{P})$.

Theorem 1. *If $\nu \sim \mathbb{Q}$ and $\mathbb{Q}\left(\nu \in [0, \overline{\nu}]\right) = 1$, then*

$$\mathbb{Q}\left(\nu - m(\nu) < \tau + \delta\right) \geq 1 - \frac{k}{\delta}\, \mathbb{E}_{\mathbb{Q}}\left[|\nu - \hat{\nu}|\right] \tag{2}$$

for any $\delta > 0$ and any (q, m, k) that is feasible in Problem $(\mathcal{P})$.

The regret bound in Theorem 1 conforms with the convention of robust satisficing, which yearns for the smallest k so that the probabilistic guarantee is strongest. The theorem also suggests that, if possible, the seller may want to choose $\hat{\nu}$ to be the median of ν. On the other hand, if $\hat{\nu}$ is chosen as the expected value of ν, then the right-hand side of (2) is completely characterized by the mean absolute deviation of ν. Other choices of $\hat{\nu}$ will be presented in the later part of the paper. The fact that this probabilistic guarantee depends explicitly on the input $\hat{\nu}$ marks a clear distinction between robust satisficing and robust optimization, with the latter focusing on only the worst-case ν [16].

Before solving Problem $(\mathcal{P})$, we state its feasibility condition as follows.

Theorem 2. *Problem $(\mathcal{P})$ is feasible if and only if $\tau > 0$.*

Due to Theorem 2, we henceforth always assume that $\tau > 0$ to avoid trivialities.

3   Optimal Mechanisms

This section contains our main results which are the analytical derivations of the optimal solutions of Problem $(\mathcal{P})$ for different values of the seller's target $\tau > 0$. We will consider in total four cases depending on the relationship between τ and the other inputs to the problem: $\hat{\nu} \in (0, \overline{\nu})$ and $\overline{\nu} > 0$.

I: $\tau \geq \dfrac{\overline{\nu}}{e}$ II: $\tau < \dfrac{\overline{\nu}}{e}$ and $\tau > \hat{\nu}$

III: $\tau < \dfrac{\overline{\nu}}{e}$, $\tau > \left(2e^{-1/2} - 1\right)\hat{\nu}$ and $\tau \leq \hat{\nu}$ IV: $\tau \leq \left(2e^{-1/2} - 1\right)\hat{\nu}$

3.1   Analysis of Case I

Proposition 1. *If $\tau \geq \frac{\overline{\nu}}{e}$, then $(q^{\star}, m^{\star}, 0)$ with*

$$q^{\star}(\nu) = \left(1 + \log\left(\frac{\nu}{\overline{\nu}}\right)\right)^{+} \quad and \quad m^{\star}(\nu) = \left(\nu - \frac{\overline{\nu}}{e}\right)^{+} \tag{3}$$

is optimal in Problem $(\mathcal{P})$.

Note that this optimal mechanism depends on neither τ, as long as it is sufficiently large, nor $\hat{\nu}$. This observation indicates that the regret incurred by the buyer of any value $\nu \in [0, \overline{\nu}]$ is no more than $\frac{\overline{\nu}}{e}$, which is consistent with and encapsulates the analysis by [13].

3.2   Analysis of Case II

Proposition 2. *If* $\hat{v} < \tau < \frac{\overline{v}}{e}$, *then* $(q^\star, m^\star, k^\star)$, *where*

$$q^\star(\nu) = \left(1 + (1 - k^\star)\log\left(\frac{\nu}{\overline{v}}\right)\right)^+ \quad and \quad m^\star(\nu) = (1 - k^\star)\left(\nu - e^{1/(k^\star - 1)}\overline{v}\right)^+ \tag{4}$$

and $k^\star \in (0, 1)$ *is a solution of* $k\hat{v} + (1 - k)e^{1/(k-1)}\overline{v} = \tau$, *is feasible in Problem* $(\mathcal{P})$.

To establish the optimality of mechanism in (4), our strategy is to construct a tight lower bound of Problem $(\mathcal{P})$. This lower bound is expressed as the optimal objective value of the following maximization problem

$$
\begin{aligned}
\text{maximize} \quad & \int_0^{\overline{v}} (\nu - \tau)\,\beta(\nu)\,\mathrm{d}\nu - \tau\overline{v}\beta(\overline{v}) \\
\text{subject to} \quad & \beta \in \mathcal{L}([0, \overline{v}], \mathbb{R}_+) \\
& \int_0^{\overline{v}} |\nu - \hat{v}|\beta(\nu)\,\mathrm{d}\nu + (\overline{v} - \hat{v})\overline{v}\beta(\overline{v}) = 1 \\
& \nu\beta(\nu) \leq \overline{v}\beta(\overline{v}) + \int_\nu^{\overline{v}} \beta(x)\,\mathrm{d}x \quad \forall\nu \in [0, \overline{v}].
\end{aligned}
\tag{$\mathcal{D}$}
$$

Note that Problem $(\mathcal{D})$ implicitly imposes that any feasible β must be integrable; however, β is not necessarily continuous or monotonic.

Proposition 3. *Problem* $(\mathcal{P})$ *is lower bounded by Problem* $(\mathcal{D})$.

We next establish the tightness of the proposed lower bound, which will in turn imply the optimality of the mechanism defined in (4).

Proposition 4. *If* $\hat{v} < \tau < \frac{\overline{v}}{e}$, *then* $\beta^\star$ *which is defined through*

$$\beta^\star(\nu) = \begin{cases} \frac{c}{\nu^2} & \text{if } \nu \in \left[e^{1/(k^\star - 1)}\overline{v}, \overline{v}\right], \\ 0 & \text{if } \nu \in \left[0, e^{1/(k^\star - 1)}\overline{v}\right), \end{cases}$$

where $k^\star \in (0, 1)$ *is defined as in Proposition 2 and*

$$c = \left[\frac{2 - k^\star}{1 - k^\star} - e^{1/(1 - k^\star)}\frac{\hat{v}}{\overline{v}}\right]^{-1},$$

is feasible in Problem $(\mathcal{D})$ *and it attains the objective value of* $k^\star$.

3.3   Analysis of Case III

Proposition 5. *If* $\tau < \frac{\overline{v}}{e}$ *and* $\left(2e^{-1/2} - 1\right)\hat{v} < \tau \leq \hat{v}$, *then* $(q^\star, m^\star, k^\star)$, *where*

$$q^\star(\nu) = \begin{cases} 1 + (1 - k^\star)\log\left(\frac{\nu}{\overline{v}}\right) & \text{if } \nu \in (\hat{v}, \overline{v}], \\ 1 + \log\left(\frac{\nu}{\overline{v}}\right) + k^\star \log\left(\frac{\nu\overline{v}}{\hat{v}^2}\right) & \text{if } \nu \in \left[\left(\frac{\hat{v}^{2k^\star}}{e\overline{v}^{k^\star - 1}}\right)^{1/(k^\star + 1)}, \hat{v}\right], \\ 0 & \text{otherwise}, \end{cases} \tag{5a}$$

and

$$m^\star(\nu) = \begin{cases} (1-k^\star)(\nu - \hat{\nu}) + (1+k^\star)\left(\hat{\nu} - \left(\frac{\hat{\nu}^{2k^\star}}{e^{\overline{\nu}k^\star}-1}\right)^{1/(k^\star+1)}\right) & \text{if } \nu \in (\hat{\nu}, \overline{\nu}], \\ (1+k^\star)\left(\nu - \left(\frac{\hat{\nu}^{2k^\star}}{e^{\overline{\nu}k^\star}-1}\right)^{1/(k^\star+1)}\right) & \text{if } \nu \in \left[\left(\frac{\hat{\nu}^{2k^\star}}{e^{\overline{\nu}k^\star}-1}\right)^{1/(k^\star+1)}, \hat{\nu}\right], \\ 0 & \text{otherwise}, \end{cases}$$

(5b)

and $k^\star \in (0,1)$ *is a solution of* $\tau + k\hat{\nu} = (1+k)\left(\frac{\hat{\nu}^{2k}}{e^{\overline{\nu}k}-1}\right)^{1/(k+1)}$, *is feasible in Problem* $(\mathcal{P})$.

Analogously to Case II, Proposition 6 below certifies the optimality of the mechanism from (5).

Proposition 6. *If* $\tau < \frac{\overline{\nu}}{e}$ *and* $\left(2e^{-1/2} - 1\right)\hat{\nu} < \tau \le \hat{\nu}$, *then* $\beta^\star$ *which is defined through*

$$\beta^\star(\nu) = \begin{cases} \frac{c}{\nu^2} & \text{if } \nu \in \left[\left(\frac{\hat{\nu}^{2k^\star}}{e^{\overline{\nu}k^\star}-1}\right)^{1/(k^\star+1)}, \overline{\nu}\right], \\ 0 & \text{if } \nu \in \left[0, \left(\frac{\hat{\nu}^{2k^\star}}{e^{\overline{\nu}k^\star}-1}\right)^{1/(k^\star+1)}\right), \end{cases}$$

where $k^\star \in (0,1)$ *is defined as in Proposition 5 and*

$$c = \left[e^{1/(1+k^\star)}\left(\frac{\hat{\nu}}{\overline{\nu}}\right)^{\frac{1-k^\star}{1+k^\star}} - \frac{2}{1+k^\star}\log\left(\frac{\hat{\nu}}{\overline{\nu}}\right) - \frac{2+k^\star}{1+k^\star}\right]^{-1},$$

is feasible in Problem $(\mathcal{D})$ *and it attains the objective value of* $k^\star$.

It could be readily observed that the derived optimal mechanism of this case is distinctively similar to the mechanism that maximizes the seller's worst-case expected revenue when the probability distribution governing the buyer's value ν is ambiguous and only known to have certain mean and support [7, 20]. Here, we discuss this phenomenon more deeply by showing that the mechanism we propose is indeed distributionally robust in some sense.

Corollary 1. *For any* $0 < \mu < \frac{2\overline{\nu}}{e}$, *then* $(q^\star, m^\star, k^\star)$ *constructed in Proposition 5 when* $\tau = \hat{\nu} \in \left(0, \frac{\overline{\nu}}{e}\right)$ *and* $\hat{\nu}\left(1 + \log\left(\frac{\overline{\nu}}{\hat{\nu}}\right)\right) = \mu$ *satisfies*

$$(q^\star, m^\star) \in \underset{q,m}{\arg\max}\left\{\inf_{\mathbb{Q} \in \mathcal{Q}(\mu)}\{\mathbb{E}_{\mathbb{Q}}\left[m(\nu)\right]\} : \begin{array}{ll} q \in \mathcal{L}([0,\overline{\nu}],[0,1]),\ m \in \mathcal{L}([0,\overline{\nu}],\mathbb{R}) & \\ q(\nu)\nu - m(\nu) \ge 0 & \forall \nu \in [0,\overline{\nu}] \\ q(\nu)\nu - m(\nu) \ge q(\omega)\nu - m(\omega) & \forall \nu, \omega \in [0,\overline{\nu}] \end{array}\right\},$$

where the ambiguity set $\mathcal{Q}(\mu)$ *comprises of all probability distributions of* ν *such that*

$$\mathbb{Q}(\nu \in [0,\overline{\nu}]) = 1 \quad \text{and} \quad \mathbb{E}_{\mathbb{Q}}[\nu] = \mu.$$

To our best knowledge, this observation shows for the first time the relationship between robust satisficing with a regret objective and distributionally robust optimization with a revenue objective. Besides, we numerically compute Corollary 1's choices of τ and $\hat{\nu}$ for different values of μ, from which it is observed that to bring forth the distributional robustness for the expected revenue, the seller should choose τ and $\hat{\nu}$ to be smaller than a half of μ. Together, Theorem 1 and Corollary 1 highlight the impact of our input parameters in the seller's regret and in her revenue, respectively.

3.4   Analysis of Case IV

Proposition 7. *If $\tau \leq \left(2e^{-1/2} - 1\right)\hat{\nu}$, then $(q^\star, m^\star, k^\star)$, where*

$$
q^\star(\nu) = \begin{cases} 1 & \text{if } \nu \in (\hat{\nu}, \overline{\nu}], \\ 1 + (1 + k^\star)\log\left(\frac{\nu}{\hat{\nu}}\right) & \text{if } \nu \in \left[e^{-1/(k^\star+1)}\hat{\nu}, \hat{\nu}\right], \\ 0 & \text{otherwise,} \end{cases} \tag{6a}
$$

and

$$
m^\star(\nu) = \begin{cases} (1 + k^\star)\left(1 - e^{-1/(k^\star+1)}\right)\hat{\nu} & \text{if } \nu \in (\hat{\nu}, \overline{\nu}], \\ (1 + k^\star)\left(\nu - e^{-1/(k^\star+1)}\hat{\nu}\right) & \text{if } \nu \in \left[e^{-1/(k^\star+1)}\hat{\nu}, \hat{\nu}\right], \\ 0 & \text{otherwise,} \end{cases} \tag{6b}
$$

and $k^\star \in [1, \infty)$ is a solution of $\tau + k\hat{\nu} = (1 + k)e^{-1/(k+1)}\hat{\nu}$, is feasible in Problem $(\mathcal{P})$.

To establish the optimality of the mechanism in (6), we need a different lower bound which is

$$
\begin{aligned}
\text{maximize} \quad & \int_0^{\hat{\nu}} (\nu - \tau)\,\beta(\nu)\,\mathrm{d}\nu - \tau\hat{\nu}\beta(\hat{\nu}) \\
\text{subject to} \quad & \beta \in \mathcal{L}([0, \hat{\nu}], \mathbb{R}_+) \\
& \int_0^{\hat{\nu}} (\hat{\nu} - \nu)\beta(\nu)\,\mathrm{d}\nu = 1 \\
& \nu\beta(\nu) \leq \hat{\nu}\beta(\hat{\nu}) + \int_\nu^{\hat{\nu}} \beta(x)\,\mathrm{d}x \quad \forall \nu \in [0, \hat{\nu}].
\end{aligned} \tag{$\mathcal{D}'$}
$$

Note that Problem $(\mathcal{D}')$ implicitly imposes that any feasible β must be integrable.

Proposition 8. *Problem $(\mathcal{P})$ is lower bounded by Problem $(\mathcal{D}')$.*

We are now ready to argue that the mechanism suggested in Proposition 7 is indeed optimal.

Proposition 9. *If* $\tau \leq \left(2e^{-1/2} - 1\right)\hat{\nu}$, *then* $\beta^{\star}$ *which is defined through*

$$\beta^{\star}(\nu) = \begin{cases} \frac{c}{\nu^2} & \text{if } \nu \in \left[e^{-1/(k^{\star}+1)}\hat{\nu}, \hat{\nu}\right], \\ 0 & \text{if } \nu \in \left[0, e^{-1/(k^{\star}+1)}\hat{\nu}\right), \end{cases}$$

where $k^{\star} \in [1, \infty)$ *is defined as in Proposition 7 and*

$$c = \left[e^{1/(k^{\star}+1)} - \frac{2+k^{\star}}{1+k^{\star}}\right]^{-1},$$

is feasible in Problem $(\mathcal{D}')$ *and it attains the objective value of* $k^{\star}$.

We summarize the analyses of the four cases in the theorem below.

Theorem 3. *Problem* $(\mathcal{P})$ *is solved by the mechanism*

$$\begin{cases} \text{from Proposition 1} & \text{if } \tau \geq \frac{\overline{\nu}}{e}, \\ \text{from Proposition 2} & \text{if } \tau < \frac{\overline{\nu}}{e} \text{ and } \tau > \hat{\nu}, \\ \text{from Proposition 5} & \text{if } \tau < \frac{\overline{\nu}}{e}, \ \tau > \left(2e^{-1/2} - 1\right)\hat{\nu} \text{ and } \tau \leq \hat{\nu}, \\ \text{from Proposition 7} & \text{if } \tau \leq \left(2e^{-1/2} - 1\right)\hat{\nu}. \end{cases}$$

All in all, we show that the optimal solution of Problem $(\mathcal{P})$ can be determined by using a simple line search to find a value of $k^{\star}$ from a certain interval which solves a given characteristic equation. Although for Case IV the upper bound on $k^{\star}$ is not explicitly given, we can still without any loss impose that $k^{\star} \leq \max\left\{\frac{\overline{\nu}}{\tau} - 2, 1\right\}$ which is a valid inequality known from the proof of Theorem 2 in Appendix Λ, which is available online.

4   Optimal Deterministic Posted Price Mechanisms

Our aim for this section is to derive an optimal deterministic posted price mechanism. We denote by $p \in [0, \overline{\nu}]$ the posted price which is to be optimized, and we restrict the allocation and payment rule to $q(\nu) = \mathbb{1}(\nu \geq p)$ and $m(\nu) = p\mathbb{1}(\nu \geq p)$. Under this restriction, Problem $(\mathcal{P})$ reduces to

$$\begin{aligned} & \text{minimize} & & k \\ & \text{subject to} & & p \in [0, \overline{\nu}], \ k \in \mathbb{R}_+ \\ & & & \nu - p\mathbb{1}(\nu \geq p) \leq \tau + k|\nu - \hat{\nu}| \quad \forall \nu \in [0, \overline{\nu}]. \end{aligned} \tag{7}$$

Unlike Problem $(\mathcal{P})$, Problem (7) involves a finite number of decision variables (*i.e.*, p and k). It however still contains an infinite number of constraints, each of which is neither convex nor concave in the posted price p.

Theorem 4. *Problem* (7) *is solved by*

$$p^{\star} = \begin{cases} \tau & \text{if } \tau \geq \frac{\overline{\nu}}{2}, \\ \text{a root of } \frac{p-\tau}{p-\hat{\nu}} = \frac{\overline{\nu}-p-\tau}{\overline{\nu}-\hat{\nu}} \text{ from } (\tau, \overline{\nu}-\tau) & \text{if } \tau < \frac{\overline{\nu}}{2} \text{ and } \tau > \hat{\nu}, \\ \hat{\nu} & \text{if } \tau < \frac{\overline{\nu}}{2} \text{ and } \tau = \hat{\nu}, \\ \max\left\{\hat{\nu} - \tau, \text{ a root of } \frac{p-\tau}{\hat{\nu}-p} = \frac{\overline{\nu}-p-\tau}{\overline{\nu}-\hat{\nu}} \text{ from } (\tau, \hat{\nu})\right\} & \text{if } \tau < \frac{\overline{\nu}}{2} \text{ and } \tau < \hat{\nu}. \end{cases}$$

Based on Theorem 4 and our earlier results, we can now establish the following observations.

1. The mechanism that is optimal in Problem $(\mathcal{P})$ is not necessarily unique. Indeed, when $\tau \geq \frac{\overline{\nu}}{2}$, the optimal objective value of Problem $(\mathcal{P})$ vanishes and it could be solved by a mechanism with either a randomized or a deterministic allocation; see Theorems 3 and 4.
2. The restriction to a class of deterministic posted price mechanisms does not impact the feasibility of Problem $(\mathcal{P})$. That is Problem $(\mathcal{P})$ is feasible if and only if Problem (7) is. Particularly, both problems are feasible if and only if $\tau > 0$.
3. Even though Problems $(\mathcal{P})$ and (7) share the same necessary and sufficient condition for feasibility, the latter can result in a mechanism that is arbitrarily worse than the former. To see this, we may consider a target $\tau \in [\frac{\overline{\nu}}{e}, \frac{\overline{\nu}}{2})$. The optimal objective value of Problem $(\mathcal{P})$ is zero, whereas that of the restricted problem (7) is strictly positive. Suppose otherwise for the sake of contradiction that k can be zero in Problem (7). Its satisficing constraint evaluated at $\nu \uparrow p$ and $\nu = \overline{\nu}$ would then imply that both p and $\overline{\nu} - p$ are smaller than or equal to τ. Therefore, $\tau \geq \frac{\overline{\nu}}{2}$ and this observation contradicts with the admissible range of τ currently considered.

5   Optimal Target-Free Mechanisms

We will now consider another variant of Problem $(\mathcal{P})$ where the seller is relieved from choosing τ:

$$
\begin{aligned}
\text{minimize} \quad & k \\
\text{subject to} \quad & q \in \mathcal{L}([0,\overline{\nu}],[0,1]), \ m \in \mathcal{L}([0,\overline{\nu}],\mathbb{R}), \ k \in \mathbb{R}_+ \\
& q(\nu)\nu - m(\nu) \geq 0 && \forall \nu \in [0,\overline{\nu}] \\
& q(\nu)\nu - m(\nu) \geq q(\omega)\nu - m(\omega) && \forall \nu, \omega \in [0,\overline{\nu}] \\
& \nu - m(\nu) \leq \hat{\nu} - m(\hat{\nu}) + k|\nu - \hat{\nu}| && \forall \nu \in [0,\overline{\nu}].
\end{aligned}
$$
(8)

In other words, Problem (8) is obtained by replacing the target regret τ which is an explicit input parameter of Problem $(\mathcal{P})$ by $\hat{\nu} - m(\hat{\nu})$, which represents the regret of the mechanism (q, m) under the nominal scenario $\nu = \hat{\nu}$. Similar to our solution approach in Sect. 3, we perform a case-by-case analysis depending on the value of $\hat{\nu} \in (0, \overline{\nu})$.

Proposition 10. *If $\hat{\nu} \geq \frac{\overline{\nu}}{e}$, then Problem (8) is solved by the mechanism defined in (3).*

Proposition 11. *If $\hat{\nu} < \frac{\overline{\nu}}{e}$, then Problem (8) is solved by the mechanism defined in (4) where $k^\star \in (0,1)$ satisfies $e^{1/(k^\star - 1)}\overline{\nu} = \hat{\nu}$.*

In fact, there are infinitely many mechanisms that solve Problem (8). Indeed, for any $(q^\star, m^\star, k^\star)$ that is optimal, we can construct a family of optimal mechanisms of the form $(q^\star, m^\star - \delta, k^\star)$, which are parametrized by $\delta \geq 0$. However, when $\delta > 0$, these mechanisms are Pareto inefficient as to the seller they would earn a strictly smaller expected revenue (regardless of the probability distribution that governs ν), and hence they are of less interest.

Last but not least, we present a distributionally robust interpretation for the mechanism provided in Proposition 11.

Corollary 2. *For any* $0 < \mu < \frac{2\overline{v}}{e}$, *then* $(q^\star, m^\star, k^\star)$ *constructed in Proposition 11 when* $\hat{\nu} \in \left(0, \frac{\overline{v}}{e}\right)$ *and* $\hat{\nu}\left(1 + \log\left(\frac{\overline{v}}{\hat{\nu}}\right)\right) = \mu$ *satisfies*

$$(q^\star, m^\star) \in \arg\max_{q,m} \left\{ \inf_{\mathbb{Q} \in \mathcal{Q}(\mu)} \{\mathbb{E}_{\mathbb{Q}}[m(\nu)]\} : \begin{array}{ll} q \in \mathcal{L}([0,\overline{v}], [0,1]), \ m \in \mathcal{L}([0,\overline{v}], \mathbb{R}) \\ q(\nu)\nu - m(\nu) \geq 0 & \forall \nu \in [0,\overline{v}] \\ q(\nu)\nu - m(\nu) \geq q(\omega)\nu - m(\omega) & \forall \nu, \omega \in [0,\overline{v}] \end{array} \right\},$$

where the ambiguity set $\mathcal{Q}(\mu)$ *is defined as in Corollary 1.*

6   Statistics and Performance of the Optimal Mechanisms

In all cases, the optimal allocation rule of Problem $(\mathcal{P})$ is continuous and non-decreasing as well as satisfies $q^*(0) = 0$ and $q^*(\overline{v}) = 1$. Hence, q^* admits an interpretation as a cumulative distribution function that is supported on $[0, \overline{v}]$, and there exists a random variable $\tilde{p} \sim \mathbb{P}$ such that

$$q^*(\nu) = \mathbb{P}(\tilde{p} \leq \nu) = \mathbb{E}_{\mathbb{P}}(\mathbb{1}(\tilde{p} \leq \nu)) \quad \forall \nu \in [0, \overline{v}].$$

Moreover, the optimal mechanism satisfies $m^*(0) = 0$. From Lemma 1 in our Appendix A, we further have

$$m^*(\nu) = q^*(\nu)\nu - \int_0^\nu q^*(x)\,\mathrm{d}x = \int_0^\nu x \,\mathrm{d}q^*(x) = \int_0^{\overline{v}} x\mathbb{1}(x \leq \nu)\,\mathrm{d}q^*(x)$$
$$= \mathbb{E}_{\mathbb{P}}(\tilde{p}\mathbb{1}(\tilde{p} \leq \nu)) \quad \forall \nu \in [0, \overline{v}].$$

We can thus interpret the optimal mechanism (q^*, m^*) as a *'randomized posted price mechanism'* where the price $\tilde{p}$ is drawn from the probability distribution $\mathbb{P}$. Using randomized posted prices offers a distinct advantage to the seller in the sense that it increases the willingness of the buyer to engage in the sale transaction as the buyer has to pay only when he is actually given the item. In contrast, directly implementing the mechanism (q^*, m^*) entails offering a lottery that requires the buyer with a value ν to make a payment of amount $m^*(\nu)$ regardless of whether or not he will obtain the item, which to him can only happen favourably with a probability of $q^*(\nu)$.

We will now carry out a sensitivity analysis to see how the mean price $\mathbb{E}_{\mathbb{P}}(\tilde{p})$ changes with the values of τ and $\hat{\nu}$, and simultaneously we will compare it with the optimal deterministic posted price p^* derived in Sect. 4. Throughout the experiment, we assume that $\bar{\nu} = 1\$ = 100¢$. From Fig. 1, it is seen that when τ is small, $\mathbb{E}_{\mathbb{P}}(\tilde{p})$ and $p^\star$ coincide. As the target τ gets increasingly larger (*i.e.*, as the seller gets more uncertainty-conscious), the variance of the optimal random price becomes more sizeable, which justifies the implementation of the more convoluted mechanism we propose. When τ reaches a certain threshold, the variance of $\tilde{p}$ may drop but stay significant nonetheless. Overall, we find that the optimal random price is, on average, at least as large as the optimal deterministic price; hence, the randomized strategy does not only incur a smaller regret but it also has a potential to extract higher revenue from the buyer. Last but not least, we compute and compare the optimal objective value of Problem $(\mathcal{P})$, denoted by $k^\star$, and that of Problem (7), denoted by $k^\star_{\mathrm{det}}$. These optimal objective values characterize the sensitivity of the regret upper bound with respect to the departure of the buyer's value ν from $\hat{\nu}$. Regardless of $\hat{\nu} \in \{20¢, 40¢, 60¢, 80¢\}$, it can be observed from Fig. 2 (top) that both $k^\star$ and $k^\star_{\mathrm{det}}$ change abruptly when $\tau \le 21.3\% \times \hat{\nu}$. Our recommendation for the seller is therefore to choose the target τ above this level, and the larger τ is suitable for a seller who prefers to have a stronger safeguard against an uprising regret from any scenario $\nu \ne \hat{\nu}$. To further appreciate the benefit of the price dispersion, Fig. 2 (bottom left) shows that the randomized pricing strategy can lead to a substantial reduction of the sensitivity level (from $k^\star_{\mathrm{det}}$ to $k^\star$) of at least 12%, and oftentimes the difference between the two strategies is considerably more pronounced. Figure 2 (bottom right) exemplary visualizes the regret bound $\tau + k|\nu - \hat{\nu}|$, $k \in \{k^\star, k^\star_{\mathrm{det}}\}$, of the two pricing strategies when $\tau = 20¢$ and $\hat{\nu} = 30¢$.

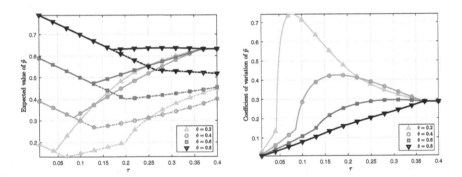

Fig. 1. The mean (left) and the coefficient of variation (right) of the optimal random price $\tilde{p}$ under different combinations of τ and $\hat{\nu}$. The accompanied dashed curves (left) show the optimal deterministic prices $p^\star$.

Next, we make a statistical comparison between the worst-case regret minimization [13] and the proposed robust regret satisficing frameworks. Retaining the names of the optimization techniques adopted, we shall refer to their recommended mechanisms as '*robust*' and '*satisficing*' solutions, respectively. Specifically for this experiment, we work with a target-free model proposed in Sect. 5, we normalize $\bar{\nu}$ to 1\$, and we now model the buyer's value as a Beta

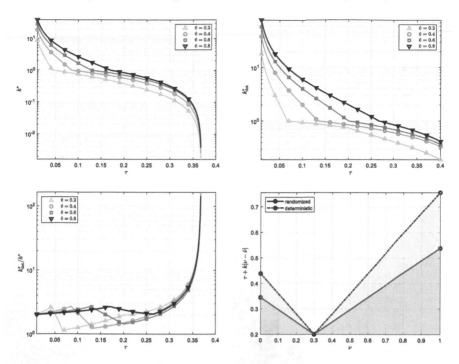

Fig. 2. The comparison between $k^\star$ and $k_{\mathrm{det}}^\star$ (in a logarithmic scale) under different combinations of τ and $\hat{\nu}$ as well as the comparison between the corresponding regret upper bounds.

random variable, $\nu \sim \mathbb{Q}$, with some positive shape parameters s_1 and s_2, and we set $\hat{\nu}$ as $\mu = \mathbb{E}_{\mathbb{Q}}[\nu]$. We consider various combinations of the shape parameters that are consistent with different means from $\{5¢, 10¢, \ldots, 35¢\}$ (as when $\hat{\nu} = \mu > \frac{1}{e} \approx 0.368$, the robust and the satisficing solutions coincide) and coefficients of variation (CV) from $\{0.1, 0.2, \ldots, 2.5\}$. It should be noted that there may be no shape parameters that are compatible with a certain combination of the mean and the CV of ν from their stipulated ranges. In such cases, the corresponding entries in Table 1 are painted black, and there are 137 remaining cells in total. Table 1 (left) reports the relative improvement (in %) in terms of the expected regret of the satisficing solution from Proposition 11 over the robust solution from [13], whereas Table 1 (right) similarly reports the relative improvement (in %) in terms of the expected revenue. We observe that, in an overwhelming majority of instances, our target-free robust satisficing mechanism makes a significant improvement. Though, as one would expect, if the variance of ν becomes exceptionally large, the seller should carefully and cautiously hedge against this uncertainty, and the standing of the robust solution from [13] remains. In light of this, Table 1 markedly defines the region where the satisficing solution is superior to the robust solution and vice versa.

Table 1. Relative improvement (in %) of the satisficing mechanism over its robust counterpart in terms of the expected regret (left) and the expected revenue (right) under different Beta distributions of the buyer's value. The entries marked with a star symbol indicate that the improvement exceeds 100%.

CV \ μ	5¢	10¢	15¢	20¢	25¢	30¢	35¢
0.1	1.3	1.7	2.1	2.5	2.9	3.3	1.9
0.2	2.7	3.5	4.2	5.0	5.7	5.3	2.0
0.3	4.0	5.2	6.3	7.4	7.7	6.0	1.9
0.4	5.3	6.9	8.4	9.4	8.7	6.1	1.8
0.5	6.6	8.6	10.3	10.7	9.2	5.9	1.6
0.6	7.8	10.2	11.8	11.5	9.1	5.5	1.4
0.7	9.0	11.8	12.9	11.7	8.7	4.9	1.1
0.8	10.2	13.1	13.6	11.5	7.9	4.1	0.8
0.9	11.4	14.2	13.8	10.9	6.9	3.1	0.5
1.0	12.5	15.0	13.6	9.9	5.6	1.8	0.0
1.1	13.5	15.5	13.0	8.5	3.8	0.3	-0.7
1.2	14.5	15.6	12.1	6.8	1.7	-1.7	-1.5
1.3	15.3	15.4	10.7	4.6	-1.0	-4.2	-2.5
1.4	16.0	14.9	8.9	1.8	-4.4	-7.6	
1.5	16.5	14.0	6.7	-1.5	-8.7	-11.3	
1.6	16.8	12.8	4.0	-5.7	-14.2		
1.7	17.0	11.3	0.8	-10.8	-20.3		
1.8	17.0	9.4	-3.1	-17.3			
1.9	16.7	7.2	-7.7	-25.3			
2.0	16.3	4.6	-13.3				
2.1	15.7	1.5	-20.2				
2.2	14.9	-2.1	-28.6				
2.3	13.9	-6.3	-38.2				
2.4	12.7	-11.1					
2.5	11.3	-16.7					

CV \ μ	5¢	10¢	15¢	20¢	25¢	30¢	35¢
0.1	★	★	★	★	★	★	92.6
0.2	★	★	★	★	★	★	32.1
0.3	★	★	★	★	★	★	17.5
0.4	★	★	★	★	★	71.4	11.0
0.5	★	★	★	★	★	44.2	7.2
0.6	★	★	★	★	88.9	29.0	4.8
0.7	★	★	★	★	58.1	19.2	3.1
0.8	★	★	★	★	38.8	12.4	1.8
0.9	★	★	★	73.0	25.6	7.4	0.8
1.0	★	★	★	49.4	16.1	3.5	0.0
1.1	★	★	★	32.8	8.8	0.4	-0.7
1.2	★	★	73.3	20.5	3.1	-2.1	-1.3
1.3	★	★	50.2	11.1	-1.5	-4.2	-1.7
1.4	★	★	33.2	3.6	-5.3	-6.0	
1.5	★	★	20.2	-2.5	-8.5	-7.4	
1.6	★	79.9	9.9	-7.5	-11.3		
1.7	★	56.4	1.7	-11.8	-13.3		
1.8	★	38.4	-5.2	-15.4			
1.9	★	24.2	-10.9	-18.5			
2.0	★	12.8	-15.7				
2.1	★	3.5	-19.9				
2.2	★	-4.3	-23.6				
2.3	★	-10.8	-26.5				
2.4	86.1	-16.5					
2.5	64.5	-21.3					

Finally, we recall that, although natural, we do not necessarily have to set $\hat{\nu}$ as μ. For instance, Corollary 2 suggests a value for $\hat{\nu}$, which will be henceforth denoted by $\nu^\dagger \in (0, \mu]$, that enables the seller to earn more on average under the worst-case distribution from the mean-support ambiguity set [7]. We now consider 11 possibilities of $\hat{\nu}$, namely

$$\hat{\nu} \in \left\{ \nu^\dagger + \frac{i}{10}(\mu - \nu^\dagger) : i \in \{0, \ldots, 10\} \right\}.$$

For each of the 137 distributions $\mathbb{Q}$ which we experiment with in Table 1, we determine which value of $\hat{\nu}$ (or equivalently, which $i \in \{0, \ldots, 10\}$) earns the highest expected revenue. From Fig. 3, we conclude that the majority of the optimal i's are either zero or ten highlighting the benefit of the distributionally robust mechanism studied in [7] and our previous choice of the robust satisficing mechanism, respectively, and the seller is recommended to increase the value of $\hat{\nu}$ when CV or μ becomes larger. We also compute the expected revenue of the mechanism that corresponds to the optimal $i \in \{0, \ldots, 10\}$ and compare it with the maximum expected revenue obtained without any restriction besides incentive compatibility and individual rationality, and we find that in more than 93.4% of the test distributions the difference between the two is no more than 10¢.

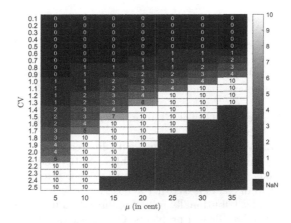

Fig. 3. The value of $i \in \{0, \ldots, 10\}$ corresponding to the maximally-earning value of $\hat{\nu}$ under different Beta distributions of the buyer's value.

7 Conclusions

Inspired by the recent development of the satisficing decision-making paradigm, we propose an innovative way of incorporating regret into the robust monopoly pricing problem. We demonstrate that the resulting optimization problem, which is an infinite linear program, and several of its variants can be solved analytically. We then alternatively express each of our optimal mechanisms as a randomized posted price mechanism since the latter is more widely accepted and readily implementable in practice. We show in our numerical studies both the benefit of randomizing prices and of adopting the satisficing over the traditional robust and distributionally robust approaches. As a by-product, we also give a managerial guideline on how to calibrate the risk-aversion parameter of our problem.

Acknowledgements. We would like to thank the reviewers, Daniel Kuhn and Bahar Taşkesen for their comments on an earlier version of the manuscript. The full paper is available at https://ssrn.com/abstract=4169471.

References

1. Ben-Tal, A., Brekelmans, R., den Hertog, D., Vial, J.: Globalized robust optimization for nonlinear uncertain inequalities. INFORMS J. Comput. **29**(2), 350–366 (2017)

2. Ben-Tal, A., El Ghaoui, L., Nemirovski, A.: Robust Optimization. Princeton University Press, Princeton (2009)
3. Bergemann, D., Morris, S.: Robust mechanism design. Econometrica **73**(6), 1771–1813 (2005)
4. Bergemann, D., Schlag, K.: Pricing without priors. J. Eur. Econ. Assoc. **6**(2–3), 560–569 (2008)
5. Bergemann, D., Schlag, K.: Robust monopoly pricing. J. Econ. Theory **146**(6), 2527–2543 (2011)
6. Bertsimas, D., Brown, D., Caramanis, C.: Theory and applications of robust optimization. SIAM Rev. **53**(3), 464–501 (2011)
7. Carrasco, V., Farinha Luz, V., Kos, N., Messner, M., Monteiro, P., Moreira, H.: Optimal selling mechanisms under moment conditions. J. Econ. Theory **177**, 245–279 (2018)
8. Chen, H., Hu, M., Perakis, G.: Distribution-free pricing. Manuf. Serv. Oper. Manag. **24**(4), 1887–2386 (2022)
9. Chen, Z., Hu, Z., Wang, R.: Screening with limited information: The minimax theorem and a geometric approach. Available at SSRN#3940212 (2021)
10. Cohen, M., Perakis, G., Pindyck, R.: A simple rule for pricing with limited knowledge of demand. Manage. Sci. **67**(3), 1608–1621 (2021)
11. Delage, E., Ye, Y.: Distributionally robust optimization under moment uncertainty with application to data-driven problems. Oper. Res. **58**(3), 595–612 (2010)
12. Hu, B., Jin, Q., Long, Z.: Robust assortment revenue optimization and satisficing. Available at SSRN#4045001 (2022)
13. Koçyiğit, Ç., Rujeerapaiboon, N., Kuhn, D.: Robust multidimensional pricing: separation without regret. Math. Program. **196**, 841–874 (2022)
14. Li, Y., Lu, P., Ye, H.: Revenue maximization with imprecise distribution. In: Proceedings of the 18th International Conference on Autonomous Agents and Multi-Agent Systems, pp. 1582–1590 (2019)
15. Long, Z., Sim, M., Zhou, M.: Robust satisficing. Oper. Res. **71**(1), 61–82 (2022)
16. Mitzenmacher, M., Vassilvitskii, S.: Algorithms with predictions. Commun. ACM **65**(7), 33–35 (2022)
17. Mohajerin Esfahani, P., Kuhn, D.: Data-driven distributionally robust optimization using the wasserstein metric: performance guarantees and tractable reformulations. Math. Program. **171**, 115–166 (2018)
18. Myerson, R.: Optimal auction design. Math. Oper. Res. **6**(1), 58–73 (1981)
19. Perakis, G., Roels, G.: Regret in the newsvendor model with partial information. Oper. Res. **56**(1), 188–203 (2008)
20. Pınar, M., Kızılkale, C.: Robust screening under ambiguity. Math. Program. **163**(1–2), 273–299 (2017)
21. Poursoltani, M., Delage, E.: Adjustable robust optimization reformulations of two-stage worst-case regret minimization problems. Oper. Res. **70**(5), 2906–2930 (2022)
22. Ramachandra, A., Rujeerapaiboon, N., Sim, M.: Robust conic satisficing. Available at SSRN#3842446 (2022)
23. Riley, J., Zeckhauser, R.: Optimal selling strategies: when to haggle, when to hold firm. Q. J. Econ. **98**(2), 267–289 (1983)
24. Savage, L.: The theory of statistical decision. J. Am. Stat. Assoc. **46**(253), 55–67 (1951)
25. Sim, M., Tang, Q., Zhou, M., Zhu, T.: The analytics of robust satisficing: predict, optimize, satisfice, then fortify. Available at SSRN#3829562 (2021)
26. Vohra, R.: Mechanism Design: A Linear Programming Approach. Cambridge University Press, Cambridge (2011)

27. Wang, S., Liu, S., Zhang, J.: Minimax regret mechanism design with moments information. Available at SSRN#3707021 (2020)
28. Wiesemann, W., Kuhn, D., Sim, M.: Distributionally robust convex optimization. Oper. Res. **62**(6), 1358–1376 (2014)

One Quarter Each (on Average) Ensures Proportionality

Xiaowei Wu[ID], Cong Zhang[(✉)][ID], and Shengwei Zhou[ID]

IOTSC, University of Macau, Macau, China
{xiaoweiwu,yc27429,yc17423}@um.edu.mo

Abstract. We consider the problem of fair allocation of m indivisible items to a group of n agents with subsidy (money). Our work mainly focuses on the allocation of chores but most of our results extend to the allocation of goods as well. We consider the case when agents have (general) additive cost functions. Assuming that the maximum cost of an item to an agent can be compensated by one dollar, we show that a total of $n/4$ dollars of subsidy suffices to ensure a proportional allocation. Moreover, we show that $n/4$ is tight in the sense that there exists an instance with n agents for which every proportional allocation requires a total subsidy of at least $n/4$. We also consider the weighted case and show that a total subsidy of $(n-1)/2$ suffices to ensure a weighted proportional allocation.

Keywords: Fair allocation · Proportionality · Subsidy

1 Introduction

We consider the problem of fairly allocating a set of m indivisible items M to a group of n heterogeneous agents N, where an allocation $\mathbf{X}$ is a partition of the items M into n disjoint bundles, each of which is allocated to a unique agent. In this paper, we consider both the allocation of goods and chores. When the items are goods, we use $v_i : 2^M \to \mathbb{R}^+ \cup \{0\}$ to denote the *valuation function* of agent i, and we say that agent i has *value* $v_i(S)$ for the bundle of items $S \subseteq M$. When items are chores, we use $c_i : 2^M \to \mathbb{R}^+ \cup \{0\}$ to denote the *cost function* of agent i, and say that agent i has *cost* $c_i(S)$ for bundle S. We assume that the valuation/cost functions are additive. Among the different notions to measure the fairness of an allocation, envy-freeness [21,28,38] and proportionality [45] are arguably the most well-studied two. An allocation $\mathbf{X} = (X_1, \ldots, X_n)$ if called *envy-free* (EF) if every agent weakly prefers her own bundle than any other bundle, i.e., $v_i(X_i) \geq v_i(X_j)$ for all $i, j \in N$ (for goods) or $c_i(X_i) \leq c_i(X_j)$ for all $i, j \in N$ (for chores). The allocation is called *proportional* (PROP) if every agent receives a bundle at least as good as her proportional share, i.e., $v_i(X_i) \geq v_i(M)/n$ for all $i \in N$ (for goods) or $c_i(X_i) \leq c_i(M)/n$ for all $i \in N$ (for chores). Clearly, every EF allocation is PROP. Unfortunately, when items are indivisible, EF/PROP allocations are

The authors are ordered alphabetically. The project is funded by the Science and Technology Development Fund (FDCT), Macau SAR (file no. 0014/2022/AFJ, 0085/2022/A, 0143/2020/A3 and SKL-IOTSC-2021-2023).

J. Garg et al. (Eds.): WINE 2023, LNCS 14413, pp. 582–599, 2024.
https://doi.org/10.1007/978-3-031-48974-7_33

not guaranteed to exist, e.g., consider allocating a single item to two agents. Existing works have taken two different paths to circumvent this non-existence result, one by considering relaxations of the fairness notions, and the other by introducing money (a divisible good) to eliminate the inevitable unfairness.

Relaxations. Various relaxations of EF have been proposed in the past decades, among which EF1 and EFX are the most popular ones. The concept of *envy-freeness up to one item* (EF1) is proposed by Budish [19], and requires that the envy between any two agents can be eliminated by removing a single item. The notion of *envy-freeness up to any item* (EFX) is proposed by Caragiannis et al. [21] and is defined in a similar manner, but requires that the envy can be eliminated by removing any item from the envied agent (for goods) or the envious agent (for chores). EF1 allocations are guaranteed to exist and can be efficiently computed for goods [38], chores, and even mixture of goods and chores [6, 15]. However, unlike EF1, EFX allocations are known to exist only for some special cases, e.g., see [3, 22, 44] for goods and [8, 36, 46, 47] for chores. Whether EFX allocations always exist remains the most interesting open problem in this research area. Similar to EF1 and EFX, we can relax PROP to *proportionality up to one item* (PROP1) and *proportionality up to any item* (PROPX). Like EF1, PROP1 allocations always exist and can be efficiently computed for goods [11, 25], chores [17] and mixture of goods and chores [9]. For the allocation of goods, Aziz et al. [9] show that PROPX allocations may not exist. In contrast, PROPX allocations for chores always exist and can be efficiently computed [36, 42]. For a more comprehensive review of the existing works, please refer to the recent surveys [2, 7].

Fair Allocation with Charity. One recent approach is that of relaxing the requirement to allocate all available goods. Clearly, if done without any constraints, this makes the problem trivial: simply leaving all goods unallocated results in an envy-free allocation. However, the objective here is to only leave "a few" goods unallocated (e.g., donate them to charity instead), or remove some goods without affecting the maximum possible Nash social welfare by "too much". On this front, Caragiannis et al. [20] showed that it is possible to compute an EFX allocation of a subset of the goods, the Nash welfare of which is at least half of the maximum Nash welfare on the original instance. Feldman et al. [27] further generalized the result to α-EFX and $\frac{1}{\alpha+1}$-fraction of maximum Nash welfare. Chaudhury et al. [24] presented an algorithm that computes a partial EFX allocation, but the number of unallocated goods is at most $n - 1$, and no agent prefers the set of these goods to her own bundle. The latter result was recently improved by Berger et al. [14] who showed that the unallocated goods can be decreased to $n - 2$ in general, and to just one for the case of four agents. Chaudhury et al. [23] showed that a $(1 - \epsilon)$-EFX allocation with $O((n/\epsilon)^{\frac{4}{5}})$ unallocated goods and high Nash welfare can be computed in polynomial time for every constant $\epsilon \in (0, 0.5]$. The result was later improved to $O((n/\epsilon)^{\frac{2}{3}})$ unallocated goods by Akrami et al. [1].

Fair Allocation with Money. Since unfairness is inevitable, a natural idea is to compensate some agents with a subsidy to eliminate envy or achieve proportionality. Specifically, suppose that each agent $i \in N$ receives a subsidy $s_i \geq 0$, we say that the resulting allocation (with subsidy) is EF if for all $i, j \in N$, $v_i(X_i) + s_i \geq v_i(X_j) + s_j$ (for goods);

or $c_i(X_i) - s_i \leq c_i(X_j) - s_j$ (for chores). We are interested in computing an allocation with a small amount of total subsidy to achieve envy-freeness, assuming that each item has value/cost at most one to each agent. For the allocation of goods, Halpern and Shah [32] show that a total subsidy of $m(n-1)$ dollars suffices. The result was then improved to $n-1$ dollars by Brustle et al. [18], who also showed that it suffices to subsidize each agent at most one dollar and that the allocation without subsidy is EF1. Note that $n-1$ dollars are necessary to guarantee envy-freeness: consider allocating a single good with value 1 to n identical agents. Since every EF allocation is PROP, the result of Brustle et al. also holds for achieving proportionality. However, it remains unknown whether $n-1$ dollars are necessary to ensure proportionality. As far as we know, the problem of computing EF or PROP allocations with subsidy has not been considered for the allocation of chores.

1.1 Our Results

Our work aims at filling in the gaps for the fair allocation problem with subsidy, for both goods and chores. Ideally, we would like to compute allocations that are EF or PROP with a small amount of subsidy, and also satisfy some relaxations of EF and PROP without the subsidy. We mainly focus on the allocation of chores, but most of our results extend to the allocation of goods. We use $\mathbf{X} = (X_1, \ldots, X_n)$ to denote an allocation, $\mathbf{s} = (s_1, \ldots, s_n) \in [0,1]^n$ to denote the subsidies and $\|\mathbf{s}\|_1 = \sum_{i \in N} s_i$ to denote the total subsidy. We first show that the result of Brustle et al. [18] (for computing EF allocation for goods) can be straightforwardly extended to the allocation of chores: it suffices to compensate each agent a subsidy at most one dollar to achieve an EF allocation with a total subsidy at most $n-1$.

Result 1. For the allocation of chores, we can compute in polynomial time an EF1 allocation $\mathbf{X}$ and subsidies $\mathbf{s}$ such that $(\mathbf{X}, \mathbf{s})$ is EF, where the total subsidy $\|\mathbf{s}\|_1 \leq n-1$. Moreover, there exists an instance for which every EF allocation requires a total subsidy of at least $n-1$.

The main results of this paper concern the fairness notion of proportionality. Note that unlike envy-freeness, proportionality can be achieved given any allocation and sufficient subsidy, e.g., by setting $s_i = \max\{c_i(X_i) - c_i(M)/n, 0\}$ for all $i \in N$. Therefore the interesting question is how much subsidy is sufficient to guarantee the existence of PROP allocations. Since every EF allocation is PROP, this amount is at most $n-1$. However, whether strictly less subsidy is sufficient remains unknown. In this work, we answer this question by showing that a total subsidy of $n/4$ suffices to guarantee the existence of PROP allocation. We propose polynomial-time algorithms for computing such allocations and subsidies. Moreover, we show that the computed allocations satisfy PROPX when agents have identical additive cost functions, and PROP1 when agents have general additive cost functions.

Result 2. For the allocation of chores to a group of n agents with identical additive cost functions, we can compute in polynomial time a PROPX allocation $\mathbf{X}$ and subsidies $\mathbf{s}$ such that $(\mathbf{X}, \mathbf{s})$ is PROP, where the total subsidy $\|\mathbf{s}\|_1 \leq n/4$. Moreover, there exists an instance with n identical agents for which every PROP allocation requires a total subsidy of at least $n/4$. The same results also hold for weighted agents.

Result 3. For the allocation of chores to a group of n agents with general additive cost functions, we can compute in polynomial time a PROP1 allocation $\mathbf{X}$ and subsidies $\mathbf{s}$ such that $(\mathbf{X}, \mathbf{s})$ is PROP, where the total subsidy $\|\mathbf{s}\|_1 \leq n/4$. When agents have arbitrary weights, the total subsidy is at most $(n-1)/2$.

The above set of results (Result 3) extends straightforwardly to the allocation of goods. Result 2 does not extend to the allocation of goods since PROPX allocations are not guaranteed to exist for goods. Our results settle the problem of characterizing the subsidy required for ensuring proportionality, which is overlooked in existing works. The results indicate that for the fair allocation problem with subsidy, proportionality is strictly cheaper to achieve, compared with envy-freeness. Our results for general additive cost functions are achieved by rounding a well-structured fractional allocation to an integral allocation with subsidy, which is novel and might be useful to solve other research problems in the area of fair allocation.

1.2 Other Related Works

A similar setting introduced by Aziz [5] is the fair allocation with monetary transfers, which allows agents to transfer money to each other. Different from our objective of minimizing the total subsidy (money) to achieve fairness, the main result of the paper is to provide a characterization of allocations that are equitable and envy-free with monetary transfers. Beyond additive valuation functions, Brustle et al. [18] show that an envy-free allocation for goods always exists with a subsidy of at most $2(n-1)$ dollars per agent under general monotonic valuation functions. Barman et al. [10] consider the dichotomous valuations, i.e., the marginal value for any good to any agent is either 0 or 1. They show that there exists an allocation that achieves envy-freeness with a per-agent subsidy of at most 1. Goko et al. [30] consider the fair and truthful mechanism with limited subsidy. They show that under general monotone submodular valuations, there exists a truthful allocation mechanism that achieves envy-freeness and utilitarian optimality by subsidizing each agent at most 1 dollar. Another closely related work is the rent division problem that focuses on allocating $m = n$ indivisible goods among n agents and dividing the fixed rent among the agents [4,26,29,35,40]. More general models where envy-freeness is achieved with money have also been considered in [31,41]. Recently, Peters et al. [43] study the robustness of the rent division problem where each agent may misreport her valuation of the rooms.

When considering money as a divisible good, our setting is similar to the fair allocation of mixed divisible and indivisible items. Bei et al. [13] study the problem of fairly allocating a set of resources containing both divisible and indivisible goods. They propose the fairness notion of envy-freeness for mixed goods (EFM), and prove that EFM allocations always exist for agents with additive valuations. Bhaskar et al [15] show that envy-free allocations always exist for mixed resources consisting of doubly-monotonic indivisible items and a divisible chore, which completes the result of Bei et al. [13]. Recently, Li et al. [37] consider the truthfulness of EFM allocation. They show that EFM and truthful allocation mechanisms do not exist in general and design truthful EFM mechanisms for several special cases. For a more detailed review for the existing works on mixed fair allocation, please refer to the recent survey [39].

1.3   Organization of the Paper

Since the result for computing EF allocation with subsidy for chores is a straightforward extension of the result by Brustle et al. [18], we defer the proofs to the full version of this paper. We first provide the notations and definitions (for the allocation of chores) in Sect. 2. We prove Result 2 (the unweighted case) in Sect. 3. The extension to the weighted setting is included in the full version. We prove Result 3 (the unweighted case) in Sect. 4. The extension to the weighted setting is included in the full version. The extensions of the above results to the allocation of goods are also deferred to the full version. We conclude the paper and discuss the open questions in Sect. 5.

2   Preliminary

In the following, we introduce the notations and the fairness notions for the allocation chores. Those for goods will be introduced in the full version. We assume that the agents are unweighted. The weighted setting will be considered in the full version. We consider the problem of allocating m indivisible chores M to n agents N where each agent $i \in N$ has an additive cost function $c_i : 2^M \to \mathbb{R}^+ \cup \{0\}$. A cost function c_i is said to be *additive* if for any bundle $S \subseteq M$ we have $c_i(S) = \sum_{e \in S} c_i(\{e\})$. For convenience, we use $c_i(e)$ to denote $c_i(\{e\})$. We use $\mathbf{c} = (c_1, ..., c_n)$ to denote the cost functions of agents. We assume w.l.o.g. that each item has cost at most one to each agent, i.e. $c_i(e) \leq 1$ for any $i \in N, e \in M$. An allocation is represented by an n-partition $\mathbf{X} = (X_1, \ldots, X_n)$ of the items, where $X_i \cap X_j = \emptyset$ for all $i \neq j$ and $\cup_{i \in N} X_i = M$. In allocation $\mathbf{X}$, agent $i \in N$ receives bundle X_i. For convenience of notation, given any set $X \subseteq M$ and $e \in M$, we use $X + e$ and $X - e$ to denote $X \cup \{e\}$ and $X \backslash \{e\}$, respectively.

Definition 1 (PROP). *An allocation* $\mathbf{X}$ *is called proportional (PROP) if* $c_i(X_i) \leq \frac{c_i(M)}{n}$ *for all* $i \in N$.

We use PROP_i to denote agent i's proportional share, i.e., $\mathsf{PROP}_i = \frac{c_i(M)}{n}$.

Definition 2 (PROP1). *An allocation* $\mathbf{X}$ *is called proportional up to one item (PROP1) if for any* $i \in N$, *either* $X_i = \emptyset$ *or there exists* $e \in X_i$ *such that* $c_i(X_i - e) \leq \mathsf{PROP}_i$.

Definition 3 (PROPX). *An allocation* $\mathbf{X}$ *is called proportional up to any item (PROPX) if for any* $i \in N$, *either* $X_i = \emptyset$ *or for any item* $e \in X_i$, *we have* $c_i(X_i - e) \leq \mathsf{PROP}_i$.

We use $s_i \geq 0$ to denote the subsidy we give to agent $i \in N, \mathbf{s} = (s_1, \ldots, s_n)$ to denote the set of subsidies, and $\|\mathbf{s}\|_1 = \sum_{i \in N} s_i$ to denote the total subsidy.

Definition 4 (PROPS). *An allocation* $\mathbf{X}$ *with subsidies* $\mathbf{s} = (s_1, \ldots, s_n)$ *is called proportional with subsidies (PROPS) if for any* $i \in N$ *we have* $c_i(X_i) - s_i \leq \mathsf{PROP}_i$.

Given any instance, we aim to find PROPS allocation $\mathbf{X}$ with a small amount of total subsidy. Unlike envy-freeness with subsidy, given any allocation $\mathbf{X}$, computing the minimum subsidy to achieve proportionality can be trivially done by setting

$$s_i = \max\{c_i(X_i) - \mathsf{PROP}_i, 0\}, \qquad \forall i \in N.$$

Therefore, in the rest of this paper, we mainly focus on computing the allocation $\mathbf{X}$. The subsidy to each agent will be automatically decided by the above equation.

Remark: Another natural way to define PROPS is to take the subsidy into consideration when defining the proportionality of each agent, i.e., $\mathsf{PROP}_i = (c_i(M) - \|\mathbf{s}\|_1)/n$. However, it can be easily shown that when allocating $n - 1$ unit cost items to n identical agents, as long as the total subsidy is less than $n - 1$, the agent who receives the maximum cost after subsidy does not satisfy proportionality[1], as there exists an agent with cost 0. Since $n - 1$ total subsidy is sufficient to guarantee envy-freeness (which is stronger than proportionality), as our first result shows, it is not interesting to study this definition. In contrast, under our definition, it can be shown that a total subsidy of $n/4$ suffices to ensure proportionality. Since we did not take the subsidy into the definition of PROP_i, our result can be regarded as measuring the amount of "external help" needed to achieve proportionality.

3   Identical Cost Functions

In this section, we focus on the computation of PROPS allocations when agents have identical cost functions, i.e., $c_i(\cdot) = c(\cdot)$ for all agents $i \in N$. We use $\mathsf{PROP} = c(M)/n$ to denote this common proportionality. Before we present our algorithmic results, we first show a lower bound on the total subsidy required to achieve proportionality.

Lemma 1. *Given any $n \geq 2$, there exists an instance with n agents for which every PROPS allocation requires a total subsidy of at least $n/4$ (when n is even); at least $(n^2 - 1)/(4n)$ (when n is odd).*

Proof. Suppose $n \geq 2$ is even. Consider the instance with n agents and $n/2$ items where each item has cost 1 to all agents. For every agent $i \in N$, her proportional share is $\mathsf{PROP} = 1/2$. Consider any allocation $\mathbf{X}$, and suppose that $k \leq n/2$ agents receive at least one item. Then each of these agents i requires a subsidy of $c(X_i) - 1/2$, which implies $\|\mathbf{s}\|_1 = c(M) - k/2 \geq n/4$. In other words, any PROPS allocation requires a total subsidy of at least $n/4$.

Suppose $n \geq 2$ is odd. Consider the instance with n agents and $(n - 1)/2$ items where each item has cost 1 to all agents. For every agent $i \in N$, her proportional share is $\mathsf{PROP} = (n - 1)/(2n)$. Following a similar analysis as above we can show that the total subsidy required by any PROPS allocation is at least

$$c(M) - \frac{n-1}{2} \cdot \frac{n-1}{2n} = \frac{n-1}{2} \cdot \frac{n+1}{2n} = \frac{n^2 - 1}{4n}.$$

Hence, every PROPS allocation requires a total subsidy at least $(n^2 - 1)/(4n)$.   □

---

[1] For the allocation of goods, consider allocating a single unit value item to n identical agents.

Next, we present an algorithm for computing PROPS allocations that require a total subsidy matching the above lower bound. We use the Load Balancing Algorithm to compute an allocation $\mathbf{X}$ that is PROPX (for n identical agents), and show that the total subsidy required to achieve proportionality is at most $n/4$ (when n is even); at most $(n^2 - 1)/(4n)$ (when n is odd). By re-indexing the items, we can assume w.l.o.g. that the items are sorted in decreasing order of costs, i.e. $c(e_1) \geq c(e_2) \geq \cdots \geq c(e_m)$. During the algorithm, we greedily allocate items $e_1, e_2, \ldots, e_m$ one-by-one to the agent with minimum bundle cost, i.e., $\mathrm{argmin}_{i \in N} c(X_i)$. We summarize the steps of the full algorithm in Algorithm 1.

Algorithm 1: Load Balancing Algorithm

    Input: An instance $(M, N, \mathbf{c})$ with identical agents and $c(e_1) \geq c(e_2) \geq \cdots \geq c(e_m)$.

1 For all $i \in N$, let $X_i \leftarrow \emptyset$;

2 **for** $j = 1, 2, \ldots, m$ **do**

3 | Let $i^* \leftarrow \mathrm{argmin}_{i \in N} c(X_i)$;

4 | Update $X_{i^*} \leftarrow X_{i^*} \cup \{e_j\}$;

    Output: An allocation $\mathbf{X} = \{X_1, \ldots, X_n\}$.

Lemma 2. *The Load Balancing Algorithm (Algorithm 1) computes a PROPX alloca-*tion[2] *given any instance with n agents having identical cost functions.*

Proof. Fix any agent $i \in N$ and let $e_{\sigma(i)}$ be the last item agent i receives, it suffices to show that $c(X_i - e_{\sigma(i)}) \leq \mathsf{PROP}$ since items are allocated in the order of descending costs. Assume otherwise, i.e., $c(X_i - e_{\sigma(i)}) > \mathsf{PROP}$. Then at the moment when item $e_{\sigma(i)}$ was allocated, we have $c(X_j) \geq c(X_i - e_{\sigma(i)}) > \mathsf{PROP}$ for all $j \neq i$, which leads to a contradiction that $c(M) = \sum_{i \in N} c(X_i) > n \cdot \mathsf{PROP} = c(M)$. $\square$

Next, we provide upper bounds on the total subsidy required to make $\mathbf{X}$ a PROPS allocation, which exactly match the lower bounds given in Lemma 1.

Theorem 1. *Given any instance with n agents having identical cost functions, there exists a PROPS allocation with total subsidy at most $n/4$ when n is even, and at most $(n^2 - 1)/(4n)$ when n is odd.*

Proof. Given the allocation $\mathbf{X}$ returned by Algorithm 1, we first partition the agents into two disjoint groups N_1 and N_2 depending on whether the allocation is proportional to her as follows.

$$N_1 = \{i \in N : c(X_i) > \mathsf{PROP}\}, \quad N_2 = \{i \in N : c(X_i) \leq \mathsf{PROP}\}.$$

For all $i \in N_1$, we use $e_{\sigma(i)}$ to denote the last item allocated to agent i and define

$$h_i = \begin{cases} \mathsf{PROP} - c(X_i - e_{\sigma(i)}), & \forall i \in N_1 \\ \mathsf{PROP} - c(X_i), & \forall i \in N_2 \end{cases}.$$

---

[2] Actually, the returned allocation also satisfies a stronger fairness notion EFX.

Note that by Lemma 2, we have $h_i \geq 0$ for all $i \in N$. Moreover, since we only need to subsidize agents in N_1, the total subsidy required to achieve proportionality can be expressed as:

$$\|\mathbf{s}\|_1 = \sum_{i \in N_1} (c(X_i) - \text{PROP}) = c(M) - \sum_{i \in N_2} c(X_i) - \frac{|N_1|}{n} \cdot c(M)$$

$$= \frac{|N_2|}{n} \cdot c(M) - \sum_{i \in N_2} c(X_i) = \sum_{i \in N_2} (\text{PROP} - c(X_i)) = \sum_{i \in N_2} h_i.$$

On the other hand, we show that the sequence $(h_1, \ldots, h_n)$ has some useful properties. First, using the same argument as we have shown in the proof of Lemma 2, for all $i \in N_1$ and $j \in N_2$, we have $c(X_i - e_{\sigma(i)}) \leq c(X_j)$ (because otherwise item $e_{\sigma(i)}$ will not be allocated to agent i). Hence we have $h_i \geq h_j$ for all $i \in N_1$ and $j \in N_2$. By renaming the agents, we can assume w.l.o.g. that $h_1 \geq h_2 \geq \cdots \geq h_n$ (see Fig. 1 for an illustrating example).

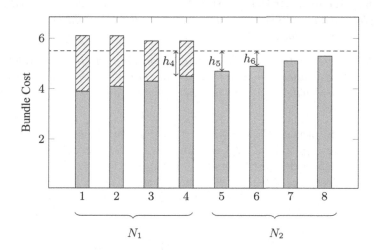

Fig. 1. An illustrating example for the allocation $\mathbf{X}$ with agent groups N_1 and N_2. The dashed areas represent the last items received by agents in N_1. The horizontal dashed line represents the proportional share for all the agents.

Second, by definition we have

$$\sum_{i \in N} h_i = \sum_{i \in N_1} (\text{PROP} - c(X_i) + c(e_{\sigma(i)})) + \sum_{i \in N_2} (\text{PROP} - c(X_i))$$

$$= n \cdot \text{PROP} - \sum_{i \in N} c(X_i) + \sum_{i \in N_1} c(e_{\sigma(i)}) = \sum_{i \in N_1} c(e_{\sigma(i)}) \leq |N_1|,$$

where in the last inequality we use the assumption that each item cost at most one to each agent. Making use of the two properties, we are now ready to prove the theorem: the total subsidy required is

$$\sum_{i \in N_2} h_i \leq \frac{|N_2|}{n} \cdot \sum_{i \in N} h_i \leq \frac{|N_1| \cdot |N_2|}{n} \leq \begin{cases} n/4, & \text{when } n \text{ is even} \\ (n^2 - 1)/(4n), & \text{when } n \text{ is odd} \end{cases},$$

where the first inequality holds since $h_i \geq h_j$ for all $i \in N_1$ and $j \in N_2$. $\square$

4   Agents with General Additive Cost Functions

In this section, we focus on the case when agents have general additive cost functions, i.e., each agent i has cost function c_i, and show that a total subsidy of $n/4$ suffices to ensure proportionality. We first present a reduction showing that if we have an algorithm for computing PROPS allocations for identical ordering instances, we can convert it to an algorithm that works for general instances while preserving the subsidy requirement. Similar reductions are widely used in the computation of approximate MMS allocations [12, 16, 33, 34] and PROPX allocations [36].

Definition 5 (Identical Ordering (IDO) Instances). *An instance is called identical ordering (IDO) if all agents have the same ordinal preference on the items, i.e.,* $c_i(e_1) \geq c_i(e_2) \geq \cdots \geq c_i(e_m)$ *for all* $i \in N$.

Lemma 3. *If there exists a polynomial time algorithm that given any IDO instance computes a PROPS[3] allocation with total subsidy at most* α*, then there exists a polynomial time algorithm that given any instance computes a PROPS allocation with total subsidy at most* α.

Proof. Given any instance $\mathcal{I} = (M, N, \mathbf{c})$, we construct an IDO instance $\mathcal{I}' = (M, N, \mathbf{c}')$ where $\mathbf{c}' = (c_1', \ldots, c_n')$ is defined as follows. Let $\sigma_i(k) \in M$ be the k-th most costly item under cost function c_i. Let $c_i'(e_k) = c_i(\sigma_i(k))$. Thus with cost function $\mathbf{c}'$, the instances $\mathcal{I}'$ is IDO. Note that for all $i \in N$ we have $c_i'(M) = c_i(M)$. Then we run the algorithm for the IDO instance $\mathcal{I}'$ and get a PROPS allocation $\mathbf{X}'$ with subsidy $\mathbf{s}'$ such that $\|\mathbf{s}'\|_1 \leq \alpha$. By definition, for all agent $i \in N$ we have

$$c_i'(X_i') - s_i' \leq \frac{1}{n} \cdot c_i'(M) = \frac{1}{n} \cdot c_i(M).$$

In the following, we use $\mathbf{X}'$ to guide us on computing a PROPS allocation $\mathbf{X}$ with $\mathbf{s}$ for instance $\mathcal{I}$. We show that $c_i(X_i) \leq c_i'(X_i')$ for all $i \in N$, which implies $\|\mathbf{s}\|_1 \leq \|\mathbf{s}'\|_1 \leq \alpha$.

We initialize $X_i = \emptyset$ for all $i \in N$ and let $P = M$ be the set of unallocated items. Sequentially for $j = m, m - 1, \ldots, 1$, we let the agent i who receives item e_j under allocation $\mathbf{X}'$, i.e., $e_j \in X_i'$, pick her favorite unallocated item, i.e., update $X_i \leftarrow X_i + e$ and $P \leftarrow P - e$ for $e = \arg\min_{e' \in P} \{c_i(e')\}$. At the beginning of each round j, we have $|P| = j$. Since e_j is the j-th most costly item under cost function c_i', we must have $c_i(e) \leq c_i'(e_j)$ for the item e agent i picks during round j. Therefore we can establish a one-to-one correspondence between items in X_i and X_i' satisfying the above inequality, which implies $c_i(X_i) \leq c_i'(X_i')$. $\square$

---

[3] The same reduction can also be applied to compute the PROP1 allocation.

With the above reduction, in the following, we only consider IDO instances. Our algorithm has two main steps: we first compute a fractional PROP allocation, in which a small number of items are fractionally allocated; then we find a way to round the fractional allocation to an integral one. Since some agent may have cost exceeding her proportional share after rounding, we offer subsidies to these agents. By carefully deciding the rounding scheme, we show that the total subsidy required is at most $n/4$.

4.1 Computing an Allocation with at Most $n - 1$ Fractional Items

In this section, we use the classic Moving Knife Algorithm to compute a fractional PROP allocation, based on which we compute the PROPS allocation $\mathbf{X}$ with subsidy s.

The Algorithm. For ease of discussion, we interpret the m items as an interval $(0, m]$, where item e_i corresponds to interval $(i - 1, i]$. We interpret every interval as a bundle of items, where some items might be fractional. Specifically, interval $(l, r]$ contains $(\lceil l \rceil - l)$-fraction of item $e_{\lceil l \rceil}$, $(r - \lfloor r \rfloor)$-fraction of item $e_{\lceil r \rceil}$ and integral item e_j for every integer j satisfying $(j - 1, j] \subseteq (l, r]$. The cost of the interval to each agent $i \in N$ is also defined in the natural way:

$$c_i(l, r) = (\lceil l \rceil - l) \cdot c_i(e_{\lceil l \rceil}) + \sum_{j=\lceil l \rceil+1}^{\lfloor r \rfloor} c_i(e_j) + (r - \lfloor r \rfloor) \cdot c_i(e_{\lceil r \rceil}).$$

The algorithm proceeds in rounds, where in each round some agent picks an interval and leaves. We maintain that at the beginning of each round, the remaining set of items forms a continuous interval $(l, m]$. In each round, we imagine that there is a moving knife that moves from the leftmost position l to the right. Each agent shouts if she thinks that the cost of an interval passed by the knife is equal to her proportional share. The last agent[4] who shouts picks the interval passed by the knife and leaves, and the algorithm recurs on the remaining interval. If in some round the knife reaches the end of the interval, any agent who has not shouted picks the whole interval $(l, m]$ and the algorithm terminates. The steps of the full algorithm are summarized in Algorithm 2.

Algorithm 2: The Moving Knife Algorithm

Input: The interval $(0, m]$ corresponding to all items M, agents N, cost function
$\quad$ $\mathbf{c} = (c_1, \ldots, c_n)$.

1 Initialize $X_i^0 \leftarrow \emptyset$ for each $i \in N$, and $l \leftarrow 0$;
2 **while** $l \neq m$ **do**
3 $\quad$ Let $r_i \leftarrow \max\{r \leq m : c_i(l, r) \leq c_i(M)/n\}$ for all $i \in N$;
4 $\quad$ Let $i^* \leftarrow \operatorname{argmax}\{r_i\}$;
5 $\quad$ Update $X_{i^*}^0 \leftarrow (l, r_{i^*}]$;
6 $\quad$ Update $N \leftarrow N \setminus \{i^*\}, l \leftarrow r_{i^*}$;

Output: Fractional allocation $\mathbf{X}^0 = (X_1^0, \ldots, X_n^0)$.

---

[4] In this paper we break tie arbitrarily but consistently, e.g., by agent id.

It has been shown by Aziz et al. [6] that the Moving Knife Algorithm computes PROP allocations for divisible chores. Therefore we have the following lemma immediately. For completeness, we give a short proof. By renaming the agents, we can assume w.l.o.g. that agents are indexed by their picking order, i.e., agent i is the i-th agent who picks and leaves.

Lemma 4. *Algorithm 2 computes fractional PROP allocations in polynomial time.*

Proof. Let $\mathbf{X}^0 = (X_1^0, \ldots, X_n^0)$ be the allocation returned by the algorithm. Consider agent n who receives the last bundle. If X_n^0 is empty then the allocation is clearly PROP to agent n. Suppose X_n^0 is not empty. For any other agent $i \neq n$ we must have $c_n(X_i^0) \geq \frac{1}{n} \cdot c_n(M)$ since at the round when agent i picked a bundle, agent i shouts not earlier than agent n. Hence we have

$$c_n(X_n) = c_n(M) - \sum_{i \neq n} c_n(X_i) \leq c_n(M) - \frac{n-1}{n} \cdot c_n(M) = \frac{1}{n} \cdot c_n(M) = \text{PROP}_n.$$

For any other agent $i \neq n$, the design of the algorithm ensures that she receives at most her proportional share. Thus the algorithm computes a (complete) fractional allocation that is PROP to all agents.   □

Example 1. Consider the following instance $\mathcal{I}^*$ with $n = 4$ agents, $m = 6$ items, and costs shown in Table 1, where $\epsilon > 0$ is arbitrarily small. Note that the instance is IDO and we have $\text{PROP}_1 = 1.5 - \epsilon, \text{PROP}_2 = 1.25 - \epsilon$ and $\text{PROP}_3 = \text{PROP}_4 = 1$.

Table 1. Example instance $\mathcal{I}^*$ with 4 agents and 6 items.

	e_1	e_2	e_3	e_4	e_5	e_6
agent 1	1	1	1	1	1	$1 - 4\epsilon$
agent 2	1	1	1	1	$1 - 4\epsilon$	0
agent 3	1	1	1	1	0	0
agent 4	1	1	1	1	0	0

After running Algorithm 2 on instance $\mathcal{I}^*$, we obtain the following fractional allocation (see Table 2 and Fig. 2), where the number in each cell corresponds to the fraction of item the agent receives.

Table 2. The fractional allocation returned by the algorithm.

	e_1	e_2	e_3	e_4	e_5	e_6
agent 1	1	$0.5 - \epsilon$	0	0	0	0
agent 2	0	$0.5 + \epsilon$	$0.75 - 2\epsilon$	0	0	0
agent 3	0	0	$0.25 + 2\epsilon$	$0.75 - 2\epsilon$	0	0
agent 4	0	0	0	$0.25 + 2\epsilon$	1	1

Since each agent receives a continuous interval in the Moving Knife Algorithm, there are at most $n - 1$ cutting points. Hence in the fractional allocation $\mathbf{X}^0$, there are at most $n - 1$ items that are fractionally allocated. We call these items *fractional* items. Note that the number of fractional items can be strictly less than $n - 1$, e.g., some item may get cut into three or more pieces.

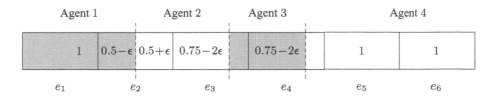

Fig. 2. Illustration the fractional allocation returned for instance $\mathcal{I}^*$.

4.2   Rounding Scheme and Subsidy

In this section, we study different rounding schemes to turn the fractional allocation integral, and upper bound the total subsidy required to achieve proportionality. We call a fractional item *rounded* to some agent i if, in the integral allocation, the item is fully allocated to agent i. In the following, we consider the case when there are exactly $n - 1$ fractional items in $\mathbf{X}^0$ and we leave the case with less than $n - 1$ fractional items to Sect. 4.3. Note that the analysis of these two cases are very similar and the upper bounds we derive follow the same formulation.

Given a fractional allocation $\mathbf{X}^0$ returned by the Moving Knife Algorithm, we use $e_1, \ldots, e_{n-1}$ to denote the $n - 1$ fractional items that are ordered by the time they are allocated. In other words, for all $i \in \{1, \ldots, n - 1\}$, item e_i is shared by agents i and $i + 1$. We denote by x_i the fraction of item e_i agent i holds; consequently $1 - x_i$ is the fraction agent $i+1$ holds. Our goal is to round the fractional allocation $\mathbf{X}^0$ to an integral allocation $\mathbf{X}$ in which each fractional item e_i is rounded to either agent i or $i + 1$. The rounding result can be represented by a vector $\hat{x} = (\hat{x}_1, \ldots, \hat{x}_{n-1})$, where $\hat{x}_i \in \{0, 1\}$ is the indicator of whether item e_i is rounded to agent i. Under the rounding result $\hat{x}$, for each agent $i \in N$, the subsidy s_i required to guarantee proportionality is given by (for convenience we let $x_0 = \hat{x}_0 = x_n = \hat{x}_n = 1$)

$$s_i = \max\{(x_{i-1} - \hat{x}_{i-1}) \cdot c_i(e_{i-1}) + (\hat{x}_i - x_i) \cdot c_i(e_i), 0\}.$$

Next, we introduce two rounding schemes: *up rounding* and *threshold rounding*.

- **Up Rounding:** For each $i \in \{1, \ldots, n - 1\}$, we set $\hat{x}_i = 1$, i.e., round each item e_i to agent i.
- **Threshold Rounding:** For each $i \in \{1, \ldots, n - 1\}$, we set $\hat{x}_i = 1$ if $x_i \geq 0.5$ and $\hat{x}_i = 0$ otherwise. In other words, we greedily round each e_i to the agent who holds a larger fraction of e_i.

Example 2. For the allocation in Table 2, we have $x_1 = 0.5 - \epsilon, x_2 = x_3 = 0.75 - 2\epsilon$.

- Using up rounding, we obtain an allocation with $X_1 = \{e_1, e_2\}, X_2 = \{e_3\}, X_3 = \{e_4\}$ and $X_4 = \{e_5, e_6\}$, which implies $s_1 = 0.5 + \epsilon, s_2 = s_3 = s_4 = 0$.
- Using threshold rounding, we have $X_1 = \{e_1\}, X_2 = \{e_2, e_3\}, X_3 = \{e_4\}$ and $X_4 = \{e_5, e_6\}$, which implies $s_1 = s_3 = s_4 = 0$ and $s_2 = 0.75 + \epsilon$.

In the following, we show that at least one of the above two rounding schemes computes a PROPS allocation with total subsidy at most $n/4$, and prove the following.

Theorem 2. *Given the fractional allocation* $\mathbf{X}^0$ *with* $n-1$ *fractional items returned by Algorithm 2, there exists a rounding scheme that returns an integral PROPS allocation* $\mathbf{X}$ *with total subsidy at most* $n/4$.

Upper Bounding the Total Subsidy Required by Up Rounding. In the following, we derive a formula that upper bounds the total subsidy required by the up rounding in terms of $\{x_1, \ldots, x_{n-1}\}$. We first show the following.

Lemma 5. *Under up rounding, the subsidy* $\mathbf{s} = (s_1, \ldots, s_n)$ *satisfies:*

- $s_1 \leq 1 - x_1$, $s_n = 0$;
- $s_i \leq \max\{x_{i-1} - x_i, 0\}$ *for all* $i \in \{2, \ldots, n-1\}$.

Proof. From the rounding scheme, we directly have $s_1 = (1 - x_1) \cdot c_1(e_1) \leq 1 - x_1$ and $s_n = 0$. Now fix any $i \in \{2, \ldots, n-1\}$. Since item e_{i-1} is rounded to agent $i-1$ and item e_i is rounded to agent i in the integral allocation $\mathbf{X}$, we have

$$s_i = \max\{(x_{i-1} - 1) \cdot c_i(e_{i-1}) + (1 - x_i) \cdot c_i(e_i), 0\}$$
$$\leq \max\{(x_{i-1} - x_i) \cdot c_i(e_{i-1}), 0\} \leq \max\{x_{i-1} - x_i, 0\},$$

where the first inequality follows from $c_i(e_{i-1}) \geq c_i(e_i)$ (since the instance is IDO) and the second inequality follows from $c_i(e_{i-1}) \leq 1$. $\qquad\square$

Lemma 6. *There exists a sequence of indices* $1 \leq j_1 < i_2 < j_2 < \cdots < i_z < j_z \leq n-1$ *such that*

$$\sum_{i=1}^n s_i \leq (1 - x_{j_1}) + (x_{i_2} - x_{j_2}) + \cdots + (x_{i_z} - x_{j_z}).$$

Proof. By Lemma 5, we can upper bound the total subsidy by $\sum_{i=1}^n s_i \leq 1 - x_1 + \sum_{i=2}^{n-1} \max\{x_{i-1} - x_i, 0\}$. For convenience, we introduce $x_0 = 1$. For each $i \in \{1, 2, \ldots, n-1\}$, we use $a_i \in \{0, 1\}$ to indicate whether $x_{i-1} > x_i$. In other words, we have $\max\{x_{i-1} - x_i, 0\} = x_{i-1} - x_i$ if $a_i = 1$ and $\max\{x_{i-1} - x_i, 0\} = 0$ otherwise. Note that $a_1 = 1$ and if we have $a_l = 1$ for all $l \in \{i, i+1, \ldots, j\}$, then we have

$$\sum_{l=i}^j s_l \leq \sum_{l=i}^j (x_{l-1} - x_l) = x_{i-1} - x_j.$$

Therefore, given any $(a_1, \ldots, a_{n-1}) \in \{0, 1\}^{n-1}$, we can break the sequence into several segments of consecutive 1's by removing the 0's. Specifically, let j_1 be the first index such that $a_{j_1+1} = 0$; let i_2 be the first index after j_1 such that $a_{i_2+1} = 1$; let j_2 be the first index after i_2 such that $a_{j_2+1} = 0$, etc. (see Fig. 3 for an example). In other words, the sub-sequences $\{a_1, \ldots, a_{j_1}\}$, $\{a_{i_2+1}, \ldots, a_{j_2}\}, \ldots, \{a_{i_z+1}, \ldots, a_{j_z}\}$ are the maximal segments of consecutive 1's of the sequence $(a_1, \ldots, a_{n-1}) \in \{0, 1\}^{n-1}$.

Therefore we can identify the indices $0 = i_1 < j_1 < i_2 < j_2 < \cdots < i_z < j_z \leq n-1$ and by a telescoping sum for each segment $[i_x, j_x]$, where $x \in \{1, 2, \ldots, z\}$, we obtain the claimed upper bound. $\qquad\square$

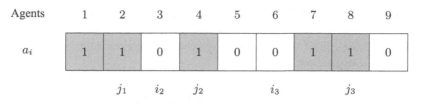

Fig. 3. An example for identifying the indices $j_1, i_2, j_2, \ldots, i_z, j_z$. In the example, we have $n = 10$ and $z = 3$. The total subsidy is upper bounded by $(1 - x_2) + (x_3 - x_4) + (x_6 - x_8)$.

Upper Bounding the Total Subsidy Required By Threshold Rounding. In the following, we derive a formula that upper bounds the total subsidy required by the threshold rounding in terms of $\{x_1, \ldots, x_{n-1}\}$. We use a charging argument that charges money to the fractional items $e_1, \ldots, e_{n-1}$: we charge each fractional item e_i an amount of money $p_i = \min\{x_i, 1 - x_i\}$. We show that the total charge to the fractional items is sufficient to pay for the subsidy.

Lemma 7. *For all $i \in \{1, 2, \ldots, n - 1\}$, let $p_i = \min\{x_i, 1 - x_i\}$. Then under the threshold rounding we have $\|\mathbf{s}\|_1 \le \sum_{i=1}^{n-1} p_i$.*

Proof. Fix any agent $i \in N$. It suffices to show that the money we charge the fractional items that are allocated to i in $\mathbf{X}$ is at least s_i. Suppose item e_{i-1} is rounded to agent i, then we have $x_{i-1} < 0.5$ and thus $p_{i-1} = x_{i-1}$. Note that the inclusion of (the integral) item e_{i-1} to X_i incurs an increase in subsidy by at most $x_{i-1} \cdot c_i(e_{i-1}) \le x_{i-1} = p_{i-1}$. In other words, the money p_{i-1} we charge item e_{i-1} is sufficient to pay for the subsidy incurred. Similarly, if item e_i is rounded to agent i, then the money $p_i = 1 - x_i$ we charge e_i is sufficient to pay for the subsidy $(1 - x_i) \cdot c_i(e_i)$ incurred.   □

Putting the Two Bounds Together. Finally, we put the upper bounds we derived for the two rounding schemes together to prove Theorem 2. By Lemma 6, there exists $1 \le j_1 < i_2 < j_2 < \cdots < i_z < j_z \le n - 1$ such that the total subsidy required by the up rounding is at most

$$(1 - x_{j_1}) + (x_{i_2} - x_{j_2}) + \cdots + (x_{i_z} - x_{j_z}). \tag{1}$$

By Lemma 7, the total subsidy required by the threshold rounding is at most $\sum_{i=1}^{n-1} p_i$, where $p_i = \min\{x_i, 1 - x_i\}$. In order to combine the two upper bounds, we apply the following relaxation on p_i:

- For each $i \in \{i_2, \ldots, i_z\}$, we use $p_i \le 1 - x_i$;
- For each $i \in \{j_1, \ldots, j_z\}$, we use $p_i \le x_i$;
- For every other i, we use that $p_i = \min\{x_i, 1 - x_i\} \le 0.5$.

Applying the above upper bounds, we upper bound the total subsidy required by the threshold rounding by

$$x_{j_1} + (1 - x_{i_2} + x_{j_2}) + \cdots + (1 - x_{i_z} + x_{j_z}) + (n - 2z) \cdot \tfrac{1}{2}. \tag{2}$$

Summing the above two upper bounds (1) and (2), we have that the total subsidy required by the two rounding schemes combined is at most $z + (n - 2z) \cdot \frac{1}{2} = \frac{n}{2}$. Therefore, at least one of the two rounding schemes requires a total subsidy at most $n/4$, which proves Theorem 2.

4.3   When the Number of Fractional Items Is Less Than $n - 1$

Next, we consider the case when the number of fractional items is strictly less than $n - 1$. This can only happen when some agents receive only one item in $\mathbf{X}^0$, and the item is fractional. By applying an analysis similar to Sect. 4.2, we show that the total subsidy required is at most $n/4$. Suppose that some item $e = e_i = \cdots = e_j$ is shared by agents $\{i, \ldots, j + 1\}$, e.g., item e is cut $j - i + 1$ times. We define x_i to be the fraction agent i holds for item e and $x_l - x_{l-1}$ to be the fraction agent l holds for item e, for all $l \in \{i + 1, \ldots, j\}$. We redefine the up rounding and threshold rounding scheme as follows.

- In the up rounding, we set $\hat{x}_i = 1$, and consequently we have $\hat{x}_{i+1} = \cdots = \hat{x}_j = 0$. Furthermore, we have $s_{i+1} = \cdots = s_j = 0$. Therefore, it is easy to verify that Lemma 5 and 6 still hold[5].
- In the threshold rounding we set $\hat{x}_t = 1$ for the agent $t \in \{i, \ldots, j + 1\}$ who holds the largest fraction of item e. We charge item e an amount of money $p = \min\{x_i, 1 - x_j\} + \frac{j-i}{2}$.[6] It suffices to show that p is enough to pay for the subsidy item e incurs under threshold rounding to prove an analogous version of Lemma 7. This is true because item e is cut into $j - i + 2$ fractions $\{x_i, x_{i+1} - x_i, \ldots, x_j - x_{j-1}, 1 - x_j\}$ and agent t holds the largest fraction, which implies that the total fraction not held by agent t (before rounding) is at most $p = \min\{x_i, 1 - x_j\} + \frac{j-i}{2}$.

Therefore we can upper bound the total subsidy required by the two rounding schemes in a similar way as in Lemma 6 and 7. Following the same analysis as in Sect. 4.2, we can prove the following.

Theorem 3. *Given the fractional allocation $\mathbf{X}^0$ with less than $n - 1$ fractional items returned by Algorithm 2, there exists a rounding scheme that returns an integral PROPS allocation $\mathbf{X}$ with total subsidy at most $n/4$.*

4.4   Guaranteeing PROP1 Without Subsidy

Finally, we show that the PROPS allocation $\mathbf{X}$ our algorithm computes is PROP1 to all agents without subsidy. Fix any agent $i \in N$, we show that there exists $e \in X_i$ such that $c_i(X_i - e) \leq \mathsf{PROP}_i$. Recall that $\mathbf{X}$ is obtained by rounding the fractional PROP allocation returned by the Moving Knife Algorithm, using either up rounding or

---

[5] While we define the variables x_i's differently for some agents, since they require 0 subsidy, the corresponding variables do not appear in the final upper bound on the total subsidy.

[6] Note that when $j = i$, the item is cut exactly once and the charging is the same as we defined in Sect. 4.2. Thus we can interpret the charging to item e as paying 0.5 more money for every extra cut.

threshold rounding. If no item or at most one item is rounded to agent i, the allocation PROP1 to her because by removing the item (if any) that is rounded to agent i, the remaining bundle has cost at most $c_i(X_i^0) \leq \mathsf{PROP}_i$. Hence it remains to consider the case when both items e_{i-1} and e_i are rounded to agent i[7]. Note that this only happens in the threshold rounding in which we greedily round each item. Therefore, we have $x_{i-1} < 0.5$ and $x_i \geq 0.5$. After removing the item e_{i-1} we have

$$c_i(X_i) - c_i(e_{i-1}) \leq c_i(X_i) - x_{i-1} \cdot c_i(e_{i-1}) - (1 - x_i) \cdot c_i(e_{i-1})$$
$$\leq c_i(X_i) - x_{i-1} \cdot c_i(e_{i-1}) - (1 - x_i) \cdot c_i(e_i) = c_i(X_i^0) \leq \mathsf{PROP}_i.$$

where the first inequality follows from $x_{i-1} \leq 0.5$ and $1 - x_i \leq 0.5$, the second inequality follows from IDO instances, and the last inequality follows from Lemma 4.

5 Conclusion and Open Questions

In this paper, we provide a precise characterization for the total subsidy required to achieve proportionality (for both the allocation of goods and chores). We show that a total subsidy of $n/4$ suffices to ensure the existence of a PROP allocation, and this is the (nearly) optimal guarantee. As we will show in the full version, the above results extend to the allocation of goods, and partially to the weighted case.

Our work leaves many interesting questions open. The first obvious open question is to complete our work by filling in the gaps in our results. For example, for an odd number of agents having general additive cost functions, our lower bound $(n^2-1)/(4n)$ and upper bound $n/4$ do not match each other. This comes from the choice of rounding schemes: it can be shown that there exists an instance for which both up rounding and threshold rounding require a total subsidy of $n/4$, even when n is odd. Whether we can improve the upper bound by introducing more rounding schemes is an interesting open problem. Another natural open question is to improve the upper bound $(n-1)/2$ we prove for the weighted case (in full version), e.g., to $n/4$. We remark that generalizing our analysis for the unweighted setting (in Sect. 4) to the case when agents have arbitrary weights might be highly non-trivial. One major difficulty comes from the computation of a well-structured fractional PROP allocation. In fact, if we require that the bundle for each agent must form a continuous interval (as Moving Knife Algorithm computes), then such allocation might not exist (see full version for an example). It remains unknown whether other fractional PROP allocation, e.g., the one returned by the Fractional Bid-and-Take we showed in full version, admits rounding schemes that require a total subsidy at most $n/4$. Finally, we believe that it would be interesting to further extend our results to the setting of mixed items (goods and chores), or to study other fairness criteria with subsidy, e.g., Maximin Share (MMS). It would be very interesting to investigate whether a total subsidy strictly less than $n/4$ is sufficient to guarantee the existence of MMS allocations.

---

[7] Note that this can only happen when item e_{i-1} is cut exactly once: if e_{i-1} is cut at least twice then the fraction held by agent $i - 1$ must be at least $1 - x_{i-1}$ (otherwise agent i will shout later), which implies that item e_{i-1} will not be rounded to agent i.

References

1. Akrami, H., Alon, N., Chaudhury, B.R., Garg, J., Mehlhorn, K., Mehta, R.: EFX: a simpler approach and an (almost) optimal guarantee via rainbow cycle number. In: EC, p. 61. ACM (2023)
2. Amanatidis, G., et al.: Fair division of indivisible goods: recent progress and open questions. Artif. Intell. **322**, 103965 (2023)
3. Amanatidis, G., Birmpas, G., Filos-Ratsikas, A., Hollender, A., Voudouris, A.A.: Maximum nash welfare and other stories about EFX. Theor. Comput. Sci. **863**, 69–85 (2021)
4. Aragones, E.: A derivation of the money rawlsian solution. Soc. Choice Welfare **12**, 267–276 (1995)
5. Aziz, H.: Achieving envy-freeness and equitability with monetary transfers. In: AAAI, pp. 5102–5109. AAAI Press (2021)
6. Aziz, H., Caragiannis, I., Igarashi, A., Walsh, T.: Fair allocation of indivisible goods and chores. Auton. Agents Multi Agent Syst. **36**(1), 3 (2022)
7. Aziz, H., Li, B., Moulin, H., Wu, X.: Algorithmic fair allocation of indivisible items: a survey and new questions. SIGecom Exch. **20**(1), 24–40 (2022)
8. Aziz, H., Lindsay, J., Ritossa, A., Suzuki, M.: Fair allocation of two types of chores. In: AAMAS, pp. 143–151. ACM (2023)
9. Aziz, H., Moulin, H., Sandomirskiy, F.: A polynomial-time algorithm for computing a pareto optimal and almost proportional allocation. Oper. Res. Lett. **48**(5), 573–578 (2020)
10. Barman, S., Krishna, A., Narahari, Y., Sadhukhan, S.: Achieving envy-freeness with limited subsidies under dichotomous valuations. In: IJCAI, pp. 60–66. ijcai.org (2022)
11. Barman, S., Krishnamurthy, S.K.: On the proximity of markets with integral equilibria. In: AAAI, pp. 1748–1755 (2019)
12. Barman, S., Krishnamurthy, S.K.: Approximation algorithms for maximin fair division. ACM Trans. Econ. Comput. **8**(1), 5:1–5:28 (2020)
13. Bei, X., Li, Z., Liu, J., Liu, S., Lu, X.: Fair division of mixed divisible and indivisible goods. Artif. Intell. **293**, 103436 (2021)
14. Berger, B., Cohen, A., Feldman, M., Fiat, A.: Almost full EFX exists for four agents. In: AAAI, pp. 4826–4833. AAAI Press (2022)
15. Bhaskar, U., Sricharan, A.R., Vaish, R.: On approximate envy-freeness for indivisible chores and mixed resources. In: APPROX-RANDOM. LIPIcs, vol. 207, pp. 1:1–1:23. Schloss Dagstuhl - Leibniz-Zentrum für Informatik (2021)
16. Bouveret, S., Lemaître, M.: Characterizing conflicts in fair division of indivisible goods using a scale of criteria. Auton. Agents Multi Agent Syst. **30**(2), 259–290 (2016)
17. Brânzei, S., Sandomirskiy, F.: Algorithms for competitive division of chores. CoRR abs/1907.01766 (2019)
18. Brustle, J., Dippel, J., Narayan, V.V., Suzuki, M., Vetta, A.: One dollar each eliminates envy. In: EC, pp. 23–39. ACM (2020)
19. Budish, E.: The combinatorial assignment problem: approximate competitive equilibrium from equal incomes. In: BQGT, p. 74:1. ACM (2010)
20. Caragiannis, I., Fanelli, A.: On approximate pure Nash equilibria in weighted congestion games with polynomial latencies. In: ICALP. LIPIcs, vol. 132, pp. 133:1–133:12. Schloss Dagstuhl - Leibniz-Zentrum für Informatik (2019)
21. Caragiannis, I., Kurokawa, D., Moulin, H., Procaccia, A.D., Shah, N., Wang, J.: The unreasonable fairness of maximum Nash welfare. ACM Trans. Econ. Comput. **7**(3), 12:1–12:32 (2019)
22. Chaudhury, B.R., Garg, J., Mehlhorn, K.: EFX exists for three agents. In: EC, pp. 1–19. ACM (2020)

23. Chaudhury, B.R., Garg, J., Mehlhorn, K., Mehta, R., Misra, P.: Improving EFX guarantees through rainbow cycle number. In: EC, pp. 310–311. ACM (2021)
24. Chaudhury, B.R., Kavitha, T., Mehlhorn, K., Sgouritsa, A.: A little charity guarantees almost envy-freeness. SIAM J. Comput. **50**(4), 1336–1358 (2021)
25. Conitzer, V., Freeman, R., Shah, N.: Fair public decision making. In: EC (2017)
26. Edward, S.F.: Rental harmony: Sperner's lemma in fair division. Am. Math. Mon. **106**(10), 930–942 (1999)
27. Feldman, M., Mauras, S., Ponitka, T.: On optimal tradeoffs between EFX and nash welfare. arXiv preprint arXiv:2302.09633 (2023)
28. Foley, D.: Resource allocation and the public sector. Yale Economic Essays, pp. 45–98 (1967)
29. Gal, Y.K., Mash, M., Procaccia, A.D., Zick, Y.: Which is the fairest (rent division) of them all? J. ACM **64**(6), 39:1–39:22 (2017)
30. Goko, H., et al.: Fair and truthful mechanism with limited subsidy. In: AAMAS. pp. 534–542. International Foundation for Autonomous Agents and Multiagent Systems (IFAAMAS) (2022)
31. Haake, C.J., Raith, M.G., Su, F.E.: Bidding for envy-freeness: a procedural approach to n-player fair-division problems. Soc. Choice Welfare **19**, 723–749 (2002)
32. Halpern, D., Shah, N.: Fair division with subsidy. In: Fotakis, D., Markakis, E. (eds.) SAGT 2019. LNCS, vol. 11801, pp. 374–389. Springer, Cham (2019). https://doi.org/10.1007/978-3-030-30473-7_25
33. Huang, X., Lu, P.: An algorithmic framework for approximating maximin share allocation of chores. In: EC, pp. 630–631. ACM (2021)
34. Huang, X., Segal-Halevi, E.: A reduction from chores allocation to job scheduling. In: EC, p. 908. ACM (2023)
35. Klijn, F.: An algorithm for envy-free allocations in an economy with indivisible objects and money. Soc. Choice Welfare **17**(2), 201–215 (2000)
36. Li, B., Li, Y., Wu, X.: Almost (weighted) proportional allocations for indivisible chores. In: WWW, pp. 122–131. ACM (2022)
37. Li, Z., Liu, S., Lu, X., Tao, B.: Truthful fair mechanisms for allocating mixed divisible and indivisible goods. In: IJCAI, pp. 2808–2816. ijcai.org (2023)
38. Lipton, R.J., Markakis, E., Mossel, E., Saberi, A.: On approximately fair allocations of indivisible goods. In: EC, pp. 125–131. ACM (2004)
39. Liu, S., Lu, X., Suzuki, M., Walsh, T.: Mixed fair division: A survey. CoRR abs/2306.09564 (2023)
40. Maskin, E.S.: On the fair allocation of indivisible goods. In: Arrow and the Foundations of the Theory of Economic Policy, pp. 341–349. Springer (1987). https://doi.org/10.1007/978-1-349-07357-3_12
41. Meertens, M., Potters, J., Reijnierse, H.: Envy-free and pareto efficient allocations in economies with indivisible goods and money. Math. Soc. Sci. **44**(3), 223–233 (2002)
42. Moulin, H.: Fair division in the age of internet. Annu. Rev. Econ. **11**, 407–441 (2018)
43. Peters, D., Procaccia, A.D., Zhu, D.: Robust rent division. In: NeurIPS (2022)
44. Plaut, B., Roughgarden, T.: Almost envy-freeness with general valuations. SIAM J. Discret. Math. **34**(2), 1039–1068 (2020)
45. Steihaus, H.: The problem of fair division. Econometrica **16**, 101–104 (1948)
46. Yin, L., Mehta, R.: On the envy-free allocation of chores. CoRR abs/2211.15836 (2022)
47. Zhou, S., Wu, X.: Approximately EFX allocations for indivisible chores. In: IJCAI, pp. 783–789. ijcai.org (2022)

Two-Sided Capacitated Submodular Maximization in Gig Platforms

Pan Xu[✉]

New Jersey Institute of Technology, Newark, NJ 07102, USA
pxu@njit.edu

Abstract. In this paper, we propose three generic models of capacitated coverage and, more generally, submodular maximization to study task-worker assignment problems that arise in a wide range of gig economy platforms. Our models incorporate the following features: (1) Each task and worker can have an arbitrary matching capacity, which captures the limited number of copies or finite budget for the task and the working capacity of the worker; (2) Each task is associated with a coverage or, more generally, a monotone submodular utility function. Our objective is to design an allocation policy that maximizes the sum of all tasks' utilities, subject to capacity constraints on tasks and workers. We consider two settings: offline, where all tasks and workers are static, and online, where tasks are static while workers arrive dynamically. We present three LP-based rounding algorithms that achieve optimal approximation ratios of $1 - 1/e \sim 0.632$ for offline coverage maximization, competitive ratios of $(19 - 67/e^3)/27 \sim 0.580$ and 0.436 for online coverage and online monotone submodular maximization, respectively.

1 Introduction

The gig economy has flourished due to the popularity of smartphones over the last few decades. Examples range from ride-hailing services such as Uber and Lyft to food delivery platforms like Postmates and Instacart, and freelancing marketplaces such as Upwork. These platforms have leveraged the widespread use of smartphones to efficiently connect service providers and customers, creating a significant number of convenient and flexible job opportunities. At its core, the gig economy involves a system with service requesters (tasks) and service providers (workers), where completing each task contributes revenue to the system. Therefore, a fundamental issue in revenue management is designing a matching policy between tasks and workers that maximizes the total revenue from all completed tasks. A common strategy is assigning a weight to each task, representing the profit obtained from completing it, and formulating the objective as the maximization of a linear function that represents the total revenue

Pan Xu was partially supported by NSF CRII Award IIS-1948157. The author would like to thank the anonymous reviewers from WINE 2023 for their valuable comments. A full version of this paper can be seen here: https://arxiv.org/abs/2309.09098.

over all completed tasks [7,23,37]. However, this paper proposes a different approach by associating each task with a specific monotone submodular function
over the set of workers, thus updating the objective to the maximization of the
sum of utility functions over all tasks. Let's consider the following motivating
examples.

Assigning Diverse Reviewers to Academic Papers. Ahmed et al. [3] have
studied how to assign a set of relevant and diverse experts to review academic
papers/proposals. As mentioned there, experts' diversity plays a key role in
maintaining fairness in the final decision (acceptance or rejection of the paper). A
general approach there is to select a ground set of K features describing reviewers
(*e.g.*, affiliations, research focuses, and demographics). For each reviewer j, we
label it with a binary vector $\chi_j \in \{0,1\}^K$ such that $\chi_{j,k} = 1$ iff reviewer j covers
feature k. For each paper-feature pair (i,k), we associate it with a non-negative
weight $w_{i,k}$ reflecting the degree of importance of feature k to paper i. Under
this configuration, our goal of forming a diverse team of experts for each paper
can be formulated as $\max \sum_i g_i(S_i)$, where S_i is the set of reviewers assigned to
paper i, and $g_i(S_i) = \sum_k w_{i,k} \min \left(1, \sum_{j \in S_i} \chi_{j,k}\right)$ denotes the total weight of all
covered features for paper i.

Multi-skilled Task-Worker Assignments. Consider a special class of crowdsourcing markets featuring that every task and worker is associated with a set
of skills [5,8,15]. A natural goal there is to assign each task to a set of workers
such that the task has as many skills covered as possible. A typical approach is
as follows. We identify a ground set of K skills; each task i and each worker j
are labeled with a binary vector $\chi_i, \chi_j \in \{0,1\}^K$, where $\chi_{i,k} = 1$ and $\chi_{j,k} = 1$
indicate that skill k is requested by task i and possessed by worker j, respectively. Thus, our goal can be formulated as $\max \sum_i g_i(S_i)$, where S_i is the set of
workers assigned to task i, and $g_i(S_i) = \sum_k \chi_{i,k} \cdot \min \left(1, \sum_{j \in S_i} \chi_{j,k}\right)$ denoting
the number of skills requested and covered for task i. More generally, suppose
for each task-skill pair (i,k), we associate it with a non-negative weight $w_{i,k}$
reflecting the importance of skill k to task i. Our generalized objective then can
be reformulated as $\max \sum_i g_i(S_i)$, where $g_i(S_i) = \sum_k w_{i,k} \min \left(1, \sum_{j \in S_i} \chi_{j,k}\right)$
denoting the total weight of all covered skills for task i.

Diversity Maximization Among Online Workers. In many real-world freelancing platforms, it is highly desirable to assign a set of diverse workers to
each task (*e.g.*, labeling an image and soliciting public views for a particular
political topic). This becomes particularly prominent in the context of health-
care when we need to crowdsource a set of highly diversified online volunteers
for medical trials. As reported in [31], biases in health data are common and
can be life-threatening when the training data feeding machine-learning algorithms lack diversity. Ahmed et al. [2] considered an online diverse team formation problem, where the overall goal is to crowdsource a team of diverse online
workers for every (offline) task. They defined a set of K features (or clusters)
reflecting workers' demographics and thus, each work j can be modeled by a
binary vector $\chi_j \in \{0,1\}^K$ such that $\chi_{j,k} = 1$ indicates that work j has feature k (*i.e.*, belongs to cluster k). They associated each task-feature pair (i,k) a

non-negative weight $w_{i,k}$ reflecting the utility of adding one worker with feature k to the team for task i. They proposed a utility function on each task i as $g_i(S_i) = \sum_k \sqrt{w_{i,k} \cdot \sum_{j \in S_i} \chi_{j,k}}$, where S_i is the team of workers assigned to task i. The overall objective in [2] is then formulated as $\sum_i g_i(S_i)$.

Here are some similarities and dissimilarities among the three examples above. First, all objectives can be formulated as the maximization of an unweighted/weighted coverage function or a more general monotone submodular function. Second, each agent in the system is associated with a finite capacity. In the context of the paper-reviewer assignment, each reviewer j has a matching capacity $b_j \in \mathbb{Z}+$ that captures the maximum number of papers reviewer j can handle, while each paper i also has a capacity $b_i \in \mathbb{Z}+$ reflecting the number of reviewers we should allocate for paper i due to the shortage of available reviewers. Similar capacity constraints exist in a wide range of real-world crowdsourcing markets: each task i and worker j practically have a matching capacity b_i and b_j, respectively, where b_i models the limited budget or copies of task i, and b_j reflects the working capacity of worker j. Third, the first application differs from the other two in the arrival setting of agents. Consider the paper-reviewer assignment for a big conference, for example. Generally, all information about tasks (i.e., the papers to review) and workers (i.e., available reviewers) is accessible before any matching decisions, and thus, the setting is called static or offline. In contrast, in most real-world crowdsourcing markets, only part of the agents are static such as tasks, whose information is known in advance, while some agents like workers join the system dynamically [23,37]. Inspired by all the insights above, we propose three generic models as follows. *Throughout this paper, we state our models in the language of matching tasks and workers in a typical crowdsourcing market.*

OFFline Capacitated Coverage Maximization (OFF-CCM). Suppose we have a bipartite graph $G = (I, J, E)$, where I and J denote the sets of tasks and workers, respectively. An edge $e = (i, j)$ indicates the feasibility or interest for worker j to complete task i. We have a ground set $\mathcal{K}$ of K features. Each worker j is captured by a binary vector $\chi_j \in \{0,1\}^K$ such that $\chi_{jk} = 1$ iff it covers feature k. Each task i has a weight vector $\mathbf{w}_i = (w_{ik})$, where $w_{ik} \in [0,1]$ reflects the importance/weight of feature k with respect to i. Each worker j (and task i) has an integer capacity b_j (b_i), which means that it can be matched with at most b_j (b_i) different tasks (workers). Here b_j reflects the working capacity of worker j, while b_i models the number of copies or budget of task i. Consider an allocation $\mathbf{x} = (x_{ij}) \in \{0,1\}^{|E|}$, where $x_{ij} = 1$ with $(i,j) \in E$ means j is assigned to i. For each $i \in I$, let $\mathcal{N}_i = \{j \in J, (i,j) \in E\}$ be the set of neighbors of i; similarly for $\mathcal{N}_j$ with $j \in J$. We say $\mathbf{x}$ is *feasible* or *valid* iff $\sum_{i' \in \mathcal{N}_j} x_{i',j} \le b_j$ and $\sum_{j' \in \mathcal{N}_i} x_{i,j'} \le b_i$ for all $i \in I$ and $j \in J$. We define the utility of task i under $\mathbf{x}$ as $g_i(\mathbf{x}) = \sum_{k \in \mathcal{K}} w_{ik} \cdot \min(1, \sum_{j: \chi_{jk}=1} x_{ij})$, *i.e.*, the total sum of weights of features covered under $\mathbf{x}$, and the total resulting utility under $\mathbf{x}$ as $g(\mathbf{x}) = \sum_{i \in I} g_i(\mathbf{x})$, *i.e.*, the sum of utilities over all tasks. Note that under the offline setting, an input instance can be specified as $\mathcal{I} = (G, \{\chi_j\}, \{\mathbf{w}_i\}, \{b_i, b_j\})$, which is fully accessible. We aim to compute a feasible allocation $\mathbf{x}$ such that $g(\mathbf{x})$ is maximized.

ONline Capacitated Coverage Maximization (ON-CCM). The basic setting here is the same as OFF-CCM. Specifically, we assume all information of $(G, \{\chi_j\}, \{\mathbf{w}_i\}, \{b_i, b_j\})$ is fully known to the algorithm, but that is only part of the input. The graph $G = (I, J, E)$ in our case should be viewed as a compatible graph, where I and J denote the sets of *types* of *offline* tasks and *online* workers, respectively. Tasks are static, while workers arrive dynamically following a *known independent identical distribution* (KIID) as specified as follows. We have a finite time horizon T, and during each round $t \in \{1, 2, \ldots, T\}$ one single worker (of type) $\hat{j}$ will be sampled (called $\hat{j}$ arrives) with replacement such that $\Pr[\hat{j} = j] = r_j/T$ for all $j \in J$ with $\sum_{j \in J} r_j/T = 1$. Here r_j is called the *arrival rate* of j. Note that the arrival distribution $\{r_j/T\}$ is assumed *independent* and *invariant* throughout the T rounds, and it is accessible to the algorithm. The KIID arrival setting is mainly inspired by the fact that we can often learn the arrival distribution from historical logs [29,38]. Upon the arrival of every online worker j, we have to make an *immediate and irrevocable* decision: either reject j or assign it to at most b_j neighbors from $\mathcal{N}_j$ (subject to capacity constraints from tasks as well). Our goal is to design an allocation policy ALG such that $\mathbb{E}[g(\mathbf{X})]$ is maximized, where $\mathbf{X}$ is the allocation (possibly random) output by ALG, and where the expectation is taken over the randomness in the online workers' arrivals and that in ALG.

ONline Capacitated Submodular Maximization (ON-CSM). The setting is almost the same as ON-CCM except that each task $i \in I$ is associated with a general *monotone submodular* utility function g_i over the ground set of $\mathcal{N}_i$ (the set of neighbors of i). WLOG assume $g_i(\emptyset) = 0$ for all $i \in I$.

Apart from crowdsourcing markets, capacitated coverage maximization and general submodular maximization have applications in promoting diversity in other domains. For instance, they are used in crowdsourcing test platforms, where a diverse set of users is needed to test mobile apps [36]. These models also find applications in online recommendations [21,32] and document clustering [1], where diversity is desired.

2   Preliminaries and Main Contributions

Throughout this paper, we assume the total number of online arrivals, denoted by T, is significantly large ($T \gg 1$), and *part of our results are obtained after taking $T \to \infty$*. This assumption is commonly made and practiced in the study of online-matching models with maximization of linear objectives under KIID [9, 19,22,24,30].

Approximation Ratio. For NP-hard combinatorial optimization problems, a powerful approach is *approximation algorithms*, where the goal is to design an efficient algorithm (with polynomial running time) that guarantees a certain level of performance compared to the optimal solution. In the case of a maximization problem like OFF-CCM, we denote an approximation algorithm and its performance as ALG, and an optimal algorithm with no running-time constraint and its performance as OPT. We say that ALG achieves an approximation ratio of at least $\rho \in [0,1]$ if $\mathsf{ALG} \geq \rho \cdot \mathsf{OPT}$ for any input instances.

Competitive Ratio (CR). Competitive ratio is a commonly used metric to evaluate the performance of online algorithms. For a given algorithm ALG and an (online) maximization problem like ON-CCM as studied here, we denote the expected performance of ALG on an instance $\mathcal{I}$ as $\mathsf{ALG}(\mathcal{I})$, where the expectation is taken over the randomness in the arrivals of workers and that in ALG. Similarly, we denote the expected performance of a *clairvoyant optimal* as $\mathsf{OPT}(\mathcal{I})$. We say that ALG achieves a CR of at least $\rho \in [0,1]$ if $\mathsf{ALG}(\mathcal{I}) \geq \rho \cdot \mathsf{OPT}(\mathcal{I})$ for any input instance $\mathcal{I}$. It is important to note that ALG is subject to the real-time decision-making requirement, while a clairvoyant optimal OPT is exempt from that constraint. Thus, CR captures the gap in expected performance between ALG and OPT due to the instant-decision requirement. In the example below, we provide an instance of ON-CCM and demonstrate that the natural algorithm Greedy achieves a CR of zero. By definition, Greedy will assign every arriving worker j to at most its b_j *safe* neighbors (those that still have the capacity to admit workers) following a decreasing order of the marginal contribution to task utility functions resulting from adding j.

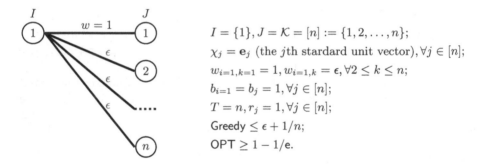

$I = \{1\}, J = \mathcal{K} = [n] := \{1, 2, \ldots, n\};$

$\chi_j = \mathbf{e}_j$ (the jth stardard unit vector), $\forall j \in [n]$;

$w_{i=1,k=1} = 1, w_{i=1,k} = \epsilon, \forall 2 \leq k \leq n$;

$b_{i=1} = b_j = 1, \forall j \in [n]$;

$T = n, r_j = 1, \forall j \in [n]$;

$\mathsf{Greedy} \leq \epsilon + 1/n$;

$\mathsf{OPT} \geq 1 - 1/e.$

Fig. 1. A toy example where Greedy achieves a competitive ratio of zero for ON-CCM.

Example 1. Consider an instance of ON-CCM shown in Fig. 1. We have a star graph $G = (I, J, E)$ with $|I| = 1$, $|J| = |\mathcal{K}| = n$, and $b_i = b_j = 1$ for $i = 1$ and all $j \in [n] \doteq \{1, 2, \ldots, n\}$. Each worker (of type) j covers a single feature $k = j$ for all $j \in [n]$. The weight $w_k \doteq w_{i=1,k} = 1$ if $k = 1$ and $w_k = \epsilon$ if $1 < k \leq n$. Set $T = n$ and $r_j = 1$ for all $j \in [n]$, i.e., during each time $t \in [n]$, a worker j will arrive uniformly at random. In our context, Greedy can be interpreted as assigning whatever arriving worker to the only task in the system.

Observe that at time $t = 1$, with respective probabilities of $1/n$ and $1 - 1/n$, it is the worker (of type) $j = 1$ and $1 < j \leq n$ that will arrive. By the nature of Greedy, we will end up with a total utility of 1 and ϵ with respective probabilities of $1/n$ and $1 - 1/n$. Thus, we claim that $\mathbb{E}[\mathsf{Greedy}] = 1/n \cdot 1 + (1 - 1/n) \cdot \epsilon$. Recall that the clairvoyant optimal OPT has the privilege to optimize its decision after observing the full arrival sequence of all workers. Note that with probability $1 - 1/e$, worker $j = 1$ will arrive at least once. Thus, OPT will end up with a total utility of 1 and ϵ with respective probabilities of $1 - 1/e$ and $1/e$. Therefore, we claim that the expected performance of OPT should be $\mathbb{E}[\text{OPT}] = 1 \cdot (1 - 1/e) + \epsilon \cdot 1/e$. By definition, we conclude that Greedy achieves a CR of zero when $\epsilon \rightarrow 0$ and $n \rightarrow \infty$ (i.e., it has arbitrarily bad performance).

Connections to Existing Models. In the offline setting (OFF-CCM), the closest model, to the best of our knowledge, is Submodular Welfare Maximization (SWM) introduced by Vondrák et al. [35]. The basic setting is as follows (rephrased in our language): We have a complete bipartite graph $G = (I, J)$, where each task i is associated with a non-negative monotone submodular utility function g_i over the ground set of J. Each task i has an unbounded capacity ($b_i = \infty$), and each worker j has a unit capacity ($b_j = 1$). The goal is to compute a *partition* S_i of J such that $\sum_{i \in I} g_i(S_i)$ is maximized. Vondrák et al. [35] demonstrated that SWM can be represented as a special case of maximizing a general monotone submodular function subject to a matroid constraint, and they provided a randomized continuous greedy algorithm that achieves an optimal approximation ratio of $1 - 1/e$. In comparison to SWM, our offline setting (OFF-CCM) considers a special case where each g_i is a weighted coverage function. Furthermore, we generalize SWM in the following three ways: (**F1**) G can be any bipartite graph (not necessarily complete), (**F2**) the capacity b_i of task i can be any positive integer instead of $b_i = \infty$, and (**F3**) the capacity b_j of worker j can be any positive integer instead of $b_j = 1$. Among these three new features, (**F2**) is perhaps the most non-trivial one since it essentially imposes a new partition-matroid constraint on all feasible allocations. *It might be tempting to consider OFF-OPT as a special case of SWM by creating b_j copies for each worker $j \in J$ and introducing an uncapacitated version of the utility function for each task $i \in I$ as $\tilde{g}_i(S) = \max_{S': S' \subseteq S \cap \mathcal{N}_i, |S'| \leq b_i} g_i(S')$ for any $S \subseteq J$. However, this reduction only yields a $(1 - 1/e)^2$-approximate algorithm, which is significantly worse than what we present here, as shown in Theorem 1.*

Regarding the online setting, Kapralov et al. [25] examined the online version of SWM under KIID, whether known or unknown. They demonstrated that Greedy achieves an *optimal* competitive ratio (CR) of $1 - 1/e$. It is worth noting that the introduction of the two new features (**F2**) and (**F3**) to SWM each brings about significant algorithmic challenges to both ON-CCM and ON-CSM scenarios. The example presented in Example 1 suggests that after (**F2**) is introduced to online SWM, Greedy experiences a substantial decrease in performance, shifting from being optimal (with a CR of $1 - 1/e$) to being arbitrarily bad (with a CR of zero). Meanwhile, (**F3**) indicates that a worker can have a non-unit capacity, leading to the involvement of multiple assignments upon her arrival. This further complicates the design and analysis of algorithms.

Glossary of Notations. A glossary of notations used throughout this paper is shown in Table 1.

Table 1. A glossary of notations used throughout this paper.

$[n]$	Set of integers $\{1, 2, \ldots, n\}$ for a generic integer n
$G = (I, J, E)$	Input graph, where I (J) is the set of task (worker) types
$\mathcal{K}$	Ground set of features with $|\mathcal{K}| = K$
$\chi_j \in \{0, 1\}^K$	Characteristic vector of worker j with $\chi_{jk} = 1$ indicating j covers k
$w_{ik} \in [0, 1]$	Weight of feature k with respect to task (of type) i
b_i (b_j)	Matching capacity on task i (worker j)
$\mathcal{N}_i$ ($\mathcal{N}_j$)	Neighbors of task i (worker j) in the input graph with $\mathcal{N}_i \subseteq J$ ($\mathcal{N}_j \subseteq I$)
E_i (E_j)	Set of edges incident to task i (worker j) in the input graph.
T	Time horizon, $i.e.$, the number of online rounds with $T \gg 1$
r_j	Arrival rate of j such that j arrives with probability r_j/T every time.
g_i	Generic monotone submodular utility function for task i with $g_i(\emptyset) = 0$.
e	Natural base with $e \sim 2.718$
e	An edge $e \in E$

Main Contributions. In this paper, we propose three generic models of capacitated coverage and, more generally, submodular maximization, to study task-worker assignment problems that arise in gig economy platforms. For each of the three models, we construct a linear program (LP) that provides a valid upper bound for the corresponding optimal performance. Using these benchmark LPs as a foundation, we develop dependent-rounding-based (DR-based) sampling algorithms. Below are our main theoretical results. *Throughout this paper, all fractional values are estimated with accuracy to the third decimal place.*

Theorem 1. *[Section 3] There is a polynomial-time algorithm that achieves an* optimal *approximation ratio of* $1 - 1/e$ *for OFF-CCM.*

Remarks on Theorem 1. Note that OFF-CCM can be viewed as a special case of maximization of a monotone submodular function subject to two *partition* matroids.[1] Meanwhile, OFF-CCM strictly generalizes the classical Maximum Coverage Problem (MCP), which can be cast as a special case of OFF-CCM with one single task. Feige [18] showed that MCP is NP-hard and cannot be approximated within a factor better than $1 - 1/e$ unless $P = NP$. This suggests the optimality of the approximation ratio of $1 - 1/e$ in Theorem 1. Observe that $1 - 1/e$ is larger than the current best ratio for maximization of a monotone submodular function subject to two *general* matroids, which is $1/2 - \epsilon$ due to the work of [28]. We believe our technique can be of independent interest and perhaps can be generalized to study the case of two or multiple general matroids.

---

[1] The sum of weighted coverage utility functions over all tasks can be viewed as one single monotone submodular function over the ground set of all edges.

Theorem 2. *[Section 4] There is an algorithm that achieves a competitive ratio of at least* $(19 - 67 \cdot e^{-3})/27 \sim 0.580$ *for ON-CCM. In particular, it achieves an optimal competitive ratio (CR) of* $1 - 1/e$ *for ON-CCM when every task has no capacity constraint.*

Theorem 3. *[Full Version] There exists an algorithm that achieves a competitive ratio of at least* 0.436 *for ON-CSM when every task has constant capacity.*

Remarks on Theorems 2 and 3. (1) ON-CCM with feature **(F2)** off (*i.e.*, no capacity constraint on tasks with $b_i = \infty$) still *strictly* generalizes online SWM introduced in [25] due to features **(F1)** and **(F3)**. That being said, our algorithm achieves an optimal CR of $1 - 1/e$ for ON-CCM with **(F2)** off, which matches the best possible CR for online SWM that was achieved by Greedy [25]. (2) Both ON-CCM and ON-CSM strictly generalize the classical online (bipartite) matching under KIID when each task has a unit capacity (*i.e.*, all $b_i = 1$). For online matching under KIID, Manshadi et al. [30] showed upper bounds on the competitive ratio over all possible adaptive and non-adaptive algorithms, which are 0.823 and $1 - 1/e \sim 0.632$, respectively. These two upper bounds apply to the 0.580-competitive algorithm in Theorem 2 and the 0.436-competitive algorithm in Theorem 3.[2] (3) The constant-task-capacity assumption in Theorem 3 is mainly inspired by real-world gig economy platforms, where the capacity of every task is relatively small due to a finite number of copies and/or a limited budget [2,4].

Main Techniques. Our algorithms for the offline and online settings both invoke the technique of dependent rounding (DR) as a subroutine, which was introduced by Gandhi et al. [20]. Recall that $G = (I, J, E)$ is bipartite. Let E_ℓ be the set of edges incident to ℓ for $\ell \in I \cup J$. Suppose each edge $e \in E$ is associated with a fractional value $z_e \in [0, 1]$. DR is such a rounding technique that takes as input a fractional vector $\mathbf{z} = (z_e) \in [0, 1]^{|E|}$ and outputs a random binary vector $\mathbf{Z} = (Z_e) \in \{0, 1\}^{|E|}$ which satisfies the following properties. **(P1) Marginal distribution:** $\mathbb{E}[Z_e] = z_e$ for all $e \in E$; **(P2) Degree preservation:** $\Pr[Z_\ell \in \{\lfloor z_\ell \rfloor, \lceil z_\ell \rceil\}] = 1$ for all $\ell \in I \cup J$, where $Z_\ell \doteq \sum_{e \in E_\ell} Z_e$ and $z_\ell \doteq \sum_{e \in E_\ell} z_e$; **(P3) Negative correlation:** For any $\ell \in I \cup J$, $S \subseteq E_\ell$ and $z \in \{0, 1\}$, $\Pr[\wedge_{e \in S}(Z_e = z)] \leq \prod_{e \in S} \Pr[Z_e = z]$.

As for the algorithm design in the online settings, we have devised two specific linear programs (LPs) to upper bound the performance of the clairvoyant optimal solution (OPT) for ON-CCM and ON-CSM, respectively. Our algorithms follow a straightforward approach: we first solve the LP and extract the marginal distribution among all edges in OPT, and then utilize it to guide the online actions. The main challenge lies in bounding the performance gap between OPT and our algorithms. To tackle this challenge, we leverage the technique of swap rounding as introduced by Chekuri et al. [13] and incorporate ideas from contention resolution schemes [14] to upper bound the gap between the multilinear relaxation

---

[2] Algorithms mentioned in Theorems 2 and 3 are both non-adaptive.

and the concave closure for a monotone submodular function. Additionally, we propose two specific Balls-and-Bins models to facilitate our competitive analysis.

Other Related Work. Our offline version can be viewed as a strict special case of maximization of a general monotone submodular function (a sum of weighted coverage functions here) subject to ℓ-matroids ($\ell = 2$ here). For this problem, Lee et al. [28] gave a *local-search* based algorithm that achieves an approx. ratio of $1/(\ell+\epsilon)$ for any $\ell \geq 2, \epsilon > 0$. Sarpatwar et al. [34] studied a more general setting which has an intersection of k matroids and a single knapsack constraint. Karimi et al. [26] studied a stochastic version of offline weighted-coverage maximization but without capacity constraint.

There is a large body of studies on different variants of submodular maximization problems. Here we list only a few samples that have considered some online arrival settings. Dickerson et al. [16] studied a variant of online matching model under KIID, whose objective is to maximize a single monotone submodular function over the set of all matched edges. They gave a 0.399-competitive algorithm. Esfandiari et al. [17] studied an extension of SWM under adversarial, where each offline agent is associated with two monotone submodular functions. They proposed a parameterized online algorithm, which is shown to achieve good competitive ratios simultaneously for both objectives. Korula et al. [27] considered SWM under the random arrival order: they showed that Greedy can beat $1/2$, which is the best competitive ratio possibly achieved for SWM under adversarial. Rawitz et al. [33] introduced the Online Budgeted Maximum Coverage problem, where we have to select a collection of sets to maximize the coverage, under the constraint that the cost of all sets is at most a given budget. They considered the adversarial arrival setting but under preemption, *i.e.*, irrevocable rejections of previously selected sets are allowed. Anari et al. [6] proposed a robust version of online submodular maximization problem, where during each round of the online phase, we need to select a set and then followed by the adversarial arrivals of k monotone submodular functions. The goal is to maximize the sum of min values over the k functions during the whole online phase. There are several other studies that have considered online maximization of a general submodular function under the adversarial arrival order; see, *e.g.*, [10,12].

3   Offline Capacitated Coverage Maximization

Throughout this section, we use OPT to denote both the optimal algorithm and the corresponding performance. For each edge $e \in E$, let $x_e = 1$ indicate e is matched in OPT. Let $\mathcal{F} \doteq \mathcal{I} \times \mathcal{K}$ be the collection of all task-feature pairs. Recall that for each node $\ell \in I \cup J$, E_ℓ denotes the set of edges incident to ℓ. For each $f = (i, k) \in \mathcal{F}$, let $w_f := w_{i,k}$, and $z_f = 1$ indicate that feature k is covered for task i in OPT; let $E_f = \{e = (i, j) \in E_i : \chi_{jk} = 1\}$, the subset of edges incident to i that can help cover feature k. Consider the below relaxed linear program

(LP).

$$\max \sum_{f \in \mathcal{F}} w_f \cdot z_f \tag{1}$$

$$z_f \leq \min \left(1, \sum_{e \in E_f} x_e\right) \qquad \forall f \in \mathcal{F} \tag{2}$$

$$\sum_{e \in E_i} x_e \leq b_i \qquad \forall i \in I \tag{3}$$

$$\sum_{e \in E_j} x_e \leq b_j \qquad \forall j \in J \tag{4}$$

$$0 \leq x_e, z_f \leq 1 \qquad \forall e \in E, f \in \mathcal{F}. \tag{5}$$

Lemma 1. *The optimal value of LP (1) is a valid upper bound on the optimal performance for OFF-CCM.*

Note that when we require all $x_e, z_f \in \{0, 1\}$, LP (1) is reduced to an IP program whose optimal value is exactly equal to OPT. That's why the relaxed LP (1) offers a valid upper bound for OPT. Our DR-based approximation algorithm is formally stated below.

ALGORITHM 1: A dependent-rounding (DR) based algorithm for OFF-CCM (ALG1).

1: Solve LP (1) to get an optimal solution $\mathbf{x}^* = (x_e^*)$.
   Apply dependent rounding (DR) in [20] to $\mathbf{x}^*$ and let $\mathbf{X}^* = (X_e^*)$ be the random binary vector.
2: Match all edges e with $X_e^* = 1$.

Observe that $\mathbf{X}^* = (X_e^*)$ will be a feasible allocation with probability one. Consider a given $\ell \in I \cup J$, we have $\sum_{e \in E_\ell} X_e^* \leq \lceil \sum_{e \in E_\ell} x_e^* \rceil \leq b_\ell$. The first inequality is due to (**P2**) from DR, while the second follows from the feasibility of $\mathbf{x}^*$ to LP (1).

Proof of Theorem 1. We first show that $\mathbb{E}[\mathsf{ALG1}] \geq (1-1/e)\mathsf{LP}(1)$. By Lemma 1, we have $\mathbb{E}[\mathsf{ALG1}] \geq (1 - 1/e)\mathsf{OPT}$, which establishes the approximation ratio. Consider a given pair $f = (i, k)$, and let $Z_f = 1$ indicate that feature k is covered for task i in ALG1. Observe that $Z_f = 1$ iff at least one edge $e \in E_f$ is matched in ALG1.

$$\mathbb{E}[Z_f] = \Pr\left[\vee_{e \in E_f}(X_e^* = 1)\right] = 1 - \Pr\left[\wedge_{e \in E_f}(X_e^* = 0)\right]$$

$$\geq 1 - \prod_{e \in E_f} \Pr[X_e^* = 0] \quad (\text{due to } (\mathbf{P3}) \text{ of DR})$$

$$= 1 - \prod_{e \in E_f}(1 - x_e^*) \quad (\text{due to } (\mathbf{P1}) \text{ of DR}).$$

Let $\alpha \doteq \sum_{e \in E_f} x_e^*$. Note that $\prod_{e \in E_f}(1 - x_e^*) \leq e^{-\alpha}$. Therefore, $\mathbb{E}[Z_f] \geq 1 - e^{-\alpha}$. Let z_f^* be the optimal value on f in LP (1). The optimality of z_f^* suggests that $z_f^* = \min(1, \alpha)$. Thus, $\frac{\mathbb{E}[Z_f]}{z_f^*} \geq \frac{1 - e^{-\alpha}}{\min(1, \alpha)} \geq 1 - 1/e$, where the second inequality is tight when $\alpha = 1$. Therefore, we claim that $\mathbb{E}[Z_f] \geq (1 - 1/e) \cdot z_f^*$ for all $f \in \mathcal{F}$. By linearity of expectation, we have

$$\mathbb{E}[\text{ALG1}] = \mathbb{E}\Big[\sum_{f \in \mathcal{F}} w_f \cdot Z_f\Big] \geq (1 - 1/e) \cdot \sum_{f \in \mathcal{F}} w_f \cdot z_f^* \geq (1 - 1/e) \cdot \text{OPT}.$$

Thus, we establish our result. The optimality of the ratio follows from the work of Feige [18], which showed that unless $P = NP$, the classical Maximum Coverage Problem (MCP) cannot be approximated within a factor better than $1 - 1/e$, while MCP can be cast as a special case of OFF-CCM. □

4   Online Capacitated Coverage Maximization

4.1   A Benchmark Linear Program (LP) and a DR-Based Algorithm

We present a specific benchmark LP for ON-CCM when every task is associated with a weighted coverage utility function. For each edge $e = (i, j) \in E$, let $x_e \in [0, 1]$ the probability that j is assigned to i in a clairvoyant optimal (OPT). Note that for coverage utility functions, OPT will have no incentive to assign j to i more than once when worker j has multiple online arrivals. Similar to OFF-CCM, let $\mathcal{F} = I \times \mathcal{K}$ be the collection of all task-feature pairs. For each $f = (i, k) \in \mathcal{F}$, let z_f denote the probability that feature k is covered for task i in OPT. Recall that for each $f = (i, k)$, $w_f := w_{i,k}$, and E_f denotes the set of edges incident to i that can help cover feature k. Consider the following LP.

$$\max \sum_{f \in \mathcal{F}} w_f \cdot z_f \tag{6}$$

$$z_f \leq \min\Big(1, \sum_{e \in E_f} x_e\Big), \qquad \forall f \in \mathcal{F} \tag{7}$$

$$\sum_{e \in E_i} x_e \leq b_i, \qquad \forall i \in I \tag{8}$$

$$x_e \leq r_j, \qquad \forall e \in E_j, \forall j \in J \tag{9}$$

$$\sum_{e \in E_j} x_e \leq b_j \cdot r_j, \qquad \forall j \in J \tag{10}$$

$$0 \leq x_e, z_f \leq 1, \qquad \forall e \in E, f \in \mathcal{F}. \tag{11}$$

Note that Constraint on each j in LP (6) differs from that in LP (1). Though the two LPs look similar, they serve essentially different purposes.

Lemma 2. *The optimal value of LP (6) is a valid upper bound on the expected performance of a clairvoyant optimal (OPT) for ON-CCM.*

Proof. By definitions of $\{w_f, z_f\}$, we can verify that Objective (6) encodes the exact expected performance of OPT. It would suffice to justify all constraints in LP (6) for OPT. Constraint (7) follows from the definitions of $\{x_e\}$ and $\{z_f\}$. As for Constraint (8): The left-hand side (LHS) is equal to the expected number of workers assigned to task i; thus, it should be no larger than b_i, which is the matching capacity of task i. Constraint (9): the probability that each given edge $e = (i, j)$ gets assigned (x_e) should be no more than that j arrives at least once, which is $1 - e^{-r_j}$; thus, $x_e \leq 1 - e^{-r_j} \leq r_j$. For Constraint (10): Note that the LHS is equal to the expected number of times worker j gets assigned; thus, it should be no more than $r_j \cdot b_j$, which is the expected number of online arrivals of j multiplied by its capacity. The last Constraint (11) is true since $\{x_e, z_f\}$ are all probability values. □

Let $\{x_e^*, z_f^*\}$ be an optimal solution to the benchmark LP (6). Our DR-based sampling algorithm, denoted by ALG2, is formally stated in Algorithm 2.

ALGORITHM 2: A dependent-rounding (DR) based sampling algorithm for ON-CCM (ALG2).

1: **Offline Phase:**
2: Solve LP (6) to get an optimal solution $\mathbf{x}^* = (x_e^*)$.
3: **Online Phase:**
4: **for** $t = 1, \ldots, T$ **do**
5:     Let an online worker (of type) j arrives at time t. Apply dependent rounding (DR) in [20] to the vector $\{x_e^*/r_j | c \subset E_j\}$, and let $\{X_e^* | e \in E_j\}$ be the random binary vector output.
6:     Match the edge $e = (i, j) \in E_j$ if $X_e^* = 1$, e is not matched before, and i's capacity remains.
7: **end for**

Remarks on ALG2. (i) In Step (5), the vector $\{x_e^*/r_j | e \in E_j\}$ satisfies that each entry $x_e^*/r_j \in [0, 1]$ due to Constraint (9) and the total sum of all entries should be no more than b_j from Constraint (10). (ii) In Step (6), we are guaranteed that the capacity of worker j will never be violated since $\sum_{e \in E_j} X_e^* \leq \lceil \sum_{e \in E_j} x_e^*/r_j \rceil \leq b_j$ thanks to (**P2**) of DR. (iii) Due to potentially multiple arrivals of worker j, edge $e \in E_j$ can be possibly matched before upon the arrival of j at t. We can ignore e in this case since it adds nothing for a second match.

4.2 Proof of the Unconstrained Task-Capacity Case in Theorem 2

Proof. In this case, we can simply ignore the concern that if i has reached the capacity in Step 6 of ALG2. Consider a given pair $f = (i, k) \in \mathcal{F}$, and let $Z_f = 1$ indicate feature k is covered for task i in ALG2. Observe that $Z_f = 1$ iff one edge $e = (ij) \in E_f$ arrives (*i.e.*, j arrives) and e gets rounded ($X_e^* = 1$) at that time.

Consider a given time t and assume $Z_f = 0$ at (the beginning of) t. We see that $Z_f = 1$ at the end of t with probability equal to

$$\sum_{e=(i,j)\in E_f} \Pr\left[j \text{ arrives}\right] \cdot \Pr[X_e^* = 1] = \sum_{e=(ij)\in E_f} (r_j/T) \cdot (x_e^*/r_j)$$

$$= \sum_{e\in E_f} x_e^*/T \doteq \alpha/T, \tag{12}$$

where $\alpha \doteq \sum_{e\in E_f} x_e^*$. Here the first equality on (12) is partially due to (**P1**) of DR. Note that $Z_f = 0$ iff Z_f never gets updated to 1 over all the T rounds. Thus,

$$\Pr[Z_f = 1] = 1 - \Pr[Z_f = 0] = 1 - (1 - \alpha/T)^T \geq 1 - e^{-\alpha}.$$

Let z_f^* be the optimal value in LP (6) and we see that $z_f^* = \min(1, \alpha)$. Therefore,

$$\frac{\mathbb{E}[Z_f]}{z_f^*} \geq \frac{1 - e^{-\alpha}}{\min(1, \alpha)} \geq 1 - 1/e,$$

where the last inequality is tight when $\alpha = 1$. Thus, we claim that $\mathbb{E}[Z_f] \geq (1 - 1/e)z_f^*$ for all $f \in \mathcal{F}$. By linearity of expectation, we have

$$\mathbb{E}[\text{ALG2}] = \mathbb{E}\left[\sum_{f\in\mathcal{F}} w_f Z_f\right] \geq (1 - 1/e) \sum_{f\in\mathcal{F}} w_f z_f^*$$

$$= (1 - 1/e) \cdot \text{LP} - (6) \geq (1 - 1/e) \cdot \text{OPT},$$

where LP-(6) and OPT denote the respective optimal values of the linear program and a clairvoyant optimal, and the last inequality follows from Lemma 2. Thus, we claim that ALG2 achieves a CR at least $1 - 1/e$. The optimality is due to the work [25], which showed that no algorithm can beat $1 - 1/e$ for online SWM under KIID even when all tasks take unweighted coverage functions.   □

4.3   Proof of the General Case in Theorem 2

We consider ON-CCM when each task i is associated with an arbitrary integer capacity b_i.

Theorem 4. *For each $f = (i, k) \in \mathcal{F}$, let $Z_f = 1$ indicate that feature k is covered for task i in* ALG2.

$$\mathbb{E}[Z_f] \geq 0.580 \cdot \min\left(1, \sum_{e\in E_f} x_e^*\right), \quad \forall f \in \mathcal{F}.$$

We defer the proof of Theorem 4 to the next section and first give a full proof of Theorem 2 based on Theorem 4.

Proof. For ease of notation, we use LP-(6) to refer to both the LP itself and the corresponding optimal value. By definition of all tasks' utility functions and the linearity of expectation, we have

$$\mathbb{E}[\mathsf{ALG2}] = \sum_{f \in \mathcal{F}} w_f \cdot \mathbb{E}[Z_f] \geq 0.580 \sum_{f \in \mathcal{F}} w_f \cdot \min\left(1, \sum_{e \in E_f} x_e^*\right)$$

$$\geq 0.580 \sum_{f \in \mathcal{F}} w_f \cdot z_f^* = 0.580 \cdot \mathrm{LP} - (6).$$

Note that the second inequality above is due to Constraint (7) of LP-(6). By Lemma 2, we claim that ALG2 achieves a CR at least 0.580. Thus, we complete the first part of Theorem 2. □

Proof of Theorem 4. Focus on a given task-feature pair $f = (i, k) \in \mathcal{F}$. Observe that feature k will be covered for i iff one edge $e \in E_f$ is matched before task i exhausts its capacity b_i. Note that during each round t, ALG2 will match an edge $e = (i, j) \in E_f$ with probability $\sum_{e=(i,j)\in E_f}(r_j/T) \cdot (x_e^*/r_j) = \sum_{e \in E_f} x_e^*/T \doteq p/T$; meanwhile, ALG2 will match an edge $e \in (E_i - E_f)$ with probability $\sum_{e \in (E_i - E_f)} x_e^*/T \doteq q/T$. By the nature of ALG2, it will keep on sampling edges from E_f and $(E_i - E_f)$ with respective probabilities p/T and q/T during each round until either b_i edges are matched or we reach the last round T. Note that from Constraint (8) of LP-(6), we have $p + q \leq b_i$. Let $Z = 1$ indicate that at least an edge $e \in \mathcal{N}_f$ gets matched by the end. The result in Theorem 4 can be equivalently stated as $\mathbb{E}[Z] \geq 0.580 \cdot \min(1, p)$, where $p := \sum_{e \in E_f} x_e^*$.

Let us treat the task i as a bin with a capacity $b := b_i$ and edges from E_f and $E_i - E_f$ as two types of balls. Then we can restate the question above alternatively as a Balls-and-Bins problem as follows.

A Balls-and-Bins Model (BBM). Suppose we have one single bin and two types of balls, namely type I and type II. We have T rounds and during each round $t \in [T]$, one ball of type I and type II will be sampled with respective probabilities p/T and q/T (with replacement)[3]; and with probability $1 - (p + q)/T$, no ball will be sampled. Here we assume $T \gg b \geq 1$ and $0 \leq p, q \leq b$ and $p + q \leq b$. The bin has a capacity of b in the way that the sampling process will stop either the bin has b balls (copies will be counted) or we reach the last round $t = T$. Let Z indicate that at least one ball of type I will be sampled by the termination of BBM. We aim to prove that $\mathbb{E}[Z] \geq 0.580 \cdot \min(1, p)$. We split the whole proof into the following two lemmas.

Lemma 3. *If $p \geq 1$, we have $\mathbb{E}[Z] \geq 0.580$.*

Lemma 4. *If $p < 1$, we have $\mathbb{E}[Z] \geq 0.580 \cdot p$.*

---

[3] Note that ALG2 may match multiple different edges in E_f, though every single edge will be matched at most once. That's why a ball of type I should be sampled with replacement.

We present a full proof of Lemma 3 here and defer that of Lemma 4 to the full version.

Proof of Lemma 3. Suppose $p \geq 1$ and $q \leq b - p \leq b - 1$. To minimize the probability that a ball of type I gets sampled, we can verify that the adversary will arrange $p = 1$ and $q = b - 1$. Let $A_{I,t} = 1$ and $A_{II,t} = 1$ indicate a ball of type I and a ball of type II get sampled at t, respectively. Let $Z_t = 1$ indicate that a ball of type I gets sampled for the first time at time t. Suppose $\mathrm{Bi}(\cdot, \cdot)$ and $\mathrm{Pois}(\cdot)$ represent a random variable following a binomial and Poisson distributions, respectively.

$$\mathbb{E}[Z] = \sum_{t=1}^{T} \mathbb{E}[Z_t]$$

$$= \sum_{t=1}^{T} \mathbb{E}[A_{I,t}] \Pr[A_{I,t'} = 0, \forall t' < t] \cdot \Pr\left[\sum_{t'<t} A_{II,t} \leq b - 1 | A_{I,t'} = 0, \forall t' < t\right]$$

$$= \sum_{t=1}^{T} \frac{1}{T}\left(1 - \frac{1}{T}\right)^{t-1} \Pr\left[\mathrm{Bi}\left(t - 1, \frac{q}{T-1}\right) \leq b - 1\right].$$

The equality in the last line can be justified as follows. Assume $A_{I,t'} = 0, \forall t' < t$, which means that a ball of type I never gets sampled before t. Conditioning on that, a ball of type II will get sampled during each round $t' < t$ with a probability $(q/T)/(1 - p/T) = q/(T - 1)$. Therefore,

$$\mathbb{E}[Z] \geq \sum_{t=b+1}^{T} \frac{1}{T}\left(1 - \frac{1}{T}\right)^{t-1} \Pr\left[\mathrm{Bi}\left(t - 1, \frac{q}{T-1}\right) \leq b - 1\right]$$

$$= \sum_{t=b}^{T-1} \frac{1}{T}\left(1 - \frac{1}{T}\right)^{t} \sum_{\ell=0}^{b-1} \binom{t}{\ell}\left(\frac{q}{T-1}\right)^{\ell}\left(1 - \frac{q}{T-1}\right)^{t-\ell}$$

$$= \sum_{\ell=0}^{b-1} \frac{q^{\ell}}{\ell!} \sum_{t=b}^{T-1} \frac{1}{T}\left(1 - \frac{1}{T}\right)^{t} \frac{t \cdot (t-1) \cdots (t-\ell+1)}{(T-1)^{\ell}}\left(1 - \frac{q}{T-1}\right)^{t-\ell}$$

$$= \sum_{\ell=0}^{b-1} \frac{q^{\ell}}{\ell!} \int_{0}^{1} \zeta^{\ell} e^{-(q+1)\zeta} d\zeta - O(1/T) = \int_{0}^{1} d\zeta \sum_{\ell=0}^{b-1} \frac{(\zeta q)^{\ell}}{\ell!} e^{-q\zeta} e^{-\zeta} - O(1/T)$$

$$= \int_{0}^{1} d\zeta \cdot e^{-\zeta} \Pr\left[\mathrm{Pois}(\zeta q) \leq b - 1\right] - O(1/T)$$

$$= \int_{0}^{1} d\zeta \cdot e^{-\zeta} \Pr\left[\mathrm{Pois}(\zeta q) \leq q\right] - O(1/T) \doteq H(q) - O(1/T).$$

Recall that $T \gg 1$ and thus, we can ignore the term of $O(1/T)$. Note that $q = b - 1$, which takes integer values only. When q is small, we can use Mathematica to verify that

$$\min_{q \in \{0,1,\dots,100\}} H(q) = H(2) = \frac{1}{27}\left(19 - \frac{67}{e^3}\right) \sim 0.580.$$

Now we try to lower bound $H(q)$ for large q values. Applying the upper tail bound of a Poisson random variable due to the work of [11], we have $\Pr[\text{Pois}(\zeta q) > q] \leq \exp\left(\frac{-q(1-\zeta)^2}{2}\right)$. Therefore,

$$
\begin{aligned}
H(q) &= \int_0^1 d\zeta \cdot e^{-\zeta} \Pr\left[\text{Pois}(\zeta q) \leq q\right] \\
&\geq \int_0^1 d\zeta \cdot e^{-\zeta} \left(1 - \exp\left(\frac{-q(1-\zeta)^2}{2}\right)\right) \doteq H_L(q).
\end{aligned}
$$

We can verify that (i) $H_L(q)$ is an increasing function of q when $q > 0$; and (ii) $H_L(100) \geq 0.582$. Thus, we have $H(q) \geq H_L(q) \geq 0.582$ for all $q > 100$. Finally, we conclude that $\mathbb{E}[Z] \geq H(q) \geq 0.580$ for all non-negative integer q. This establishes the final result.   □

5   Conclusions and Future Work

In this paper, we have proposed three generic models of capacitated submodular maximization inspired by practical gig platforms. Our models feature the association of each task with either a coverage or a general monotone submodular utility function. We have presented specific LP-based rounding algorithms for each of the three models and conducted related approximation-ratio or competitive-ratio analysis. In the following, we discuss a few potential future directions. First, we can explore generalizing the capacity constraint to something more general, such as a matroid. Second, it would be interesting to refine the upper bounds (or establish hardness results) for the benchmark LPs in online settings. Can we narrow or close the gap between the upper and lower bounds in terms of the competitive ratio, similar to what has been achieved in the offline setting?

References

1. Abbassi, Z., Mirrokni, V.S., Thakur, M.: Diversity maximization under matroid constraints. In: Proceedings of the 19th ACM SIGKDD International Conference on Knowledge Discovery and Data Mining, pp. 32–40 (2013)
2. Ahmed, F., Dickerson, J., Fuge, M.: Forming diverse teams from sequentially arriving people. CoRR abs/2002.10697 (2020)
3. Ahmed, F., Dickerson, J.P., Fuge, M.: Diverse weighted bipartite b-matching. arXiv preprint arXiv:1702.07134 (2017)
4. Ahmed, F., Dickerson, J.P., Fuge, M.: Diverse weighted bipartite b-matching. In: Proceedings of the 26th International Joint Conference on Artificial Intelligence, pp. 35–41 (2017)
5. Anagnostopoulos, A., Becchetti, L., Castillo, C., Gionis, A., Leonardi, S.: Online team formation in social networks. In: Proceedings of the 21st International Conference on World Wide Web, pp. 839–848 (2012)
6. Anari, N., Haghtalab, N., Naor, J., Pokutta, S., Singh, M., Torrico, A.: Robust submodular maximization: offline and online algorithms. arXiv preprint arXiv:1710.04740 (2017)

7. Assadi, S., Hsu, J., Jabbari, S.: Online assignment of heterogeneous tasks in crowd-sourcing markets. In: Third AAAI Conference on Human Computation and Crowd-sourcing (2015)
8. Barnabò, G., Fazzone, A., Leonardi, S., Schwiegelshohn, C.: Algorithms for fair team formation in online labour marketplaces. In: Companion Proceedings of The 2019 World Wide Web Conference, pp. 484–490 (2019)
9. Brubach, B., Sankararaman, K.A., Srinivasan, A., Xu, P.: Online stochastic matching: new algorithms and bounds. Algorithmica **82**, 2737–2783 (2020)
10. Buchbinder, N., Feldman, M., Schwartz, R.: Online submodular maximization with preemption. In: SODA (2015)
11. Canonne, C.: A short note on Poisson tail bounds (2020). http://www.cs.columbia.edu/ccanonne/files/misc/2017-poissonconcentration.pdf. Accessed 01 Feb 2020
12. Chan, T., Huang, Z., Jiang, S.H.C., Kang, N., Tang, Z.G.: Online submodular maximization with free disposal: randomization beats 1/4 for partition matroids. In: Proceedings of the Twenty-Eighth Annual ACM-SIAM Symposium on Discrete Algorithms, pp. 1204–1223. Society for Industrial and Applied Mathematics (2017)
13. Chekuri, C., Vondrak, J., Zenklusen, R.: Dependent randomized rounding via exchange properties of combinatorial structures. In: 2010 IEEE 51st Annual Symposium on Foundations of Computer Science, pp. 575–584. IEEE (2010)
14. Chekuri, C., Vondrák, J., Zenklusen, R.: Submodular function maximization via the multilinear relaxation and contention resolution schemes. CoRR abs/1105.4593 (2011). http://arxiv.org/abs/1105.4593
15. Cheng, P., Lian, X., Chen, L., Han, J., Zhao, J.: Task assignment on multi-skill oriented spatial crowdsourcing. IEEE Trans. Knowl. Data Eng. **28**(8), 2201–2215 (2016)
16. Dickerson, J.P., Sankararaman, K.A., Srinivasan, A., Xu, P.: Balancing relevance and diversity in online bipartite matching via submodularity. In: Proceedings of the AAAI Conference on Artificial Intelligence, vol. 33, pp. 1877–1884 (2019)
17. Esfandiari, H., Korula, N., Mirrokni, V.: Bi-objective online matching and sub-modular allocations. In: Advances in Neural Information Processing Systems, pp. 2739–2747 (2016)
18. Feige, U.: A threshold of ln n for approximating set cover. J. ACM **45**, 634–652 (1998)
19. Feldman, J., Mehta, A., Mirrokni, V., Muthukrishnan, S.: Online stochastic matching: beating 1−1/e. In: 50th Annual IEEE Symposium on Foundations of Computer Science, 2009. FOCS'09, pp. 117–126. IEEE (2009)
20. Gandhi, R., Khuller, S., Parthasarathy, S., Srinivasan, A.: Dependent rounding and its applications to approximation algorithms. J. ACM (JACM) **53**(3), 324–360 (2006)
21. Ge, M., Delgado-Battenfeld, C., Jannach, D.: Beyond accuracy: evaluating recommender systems by coverage and serendipity. In: Proceedings of the Fourth ACM Conference on Recommender Systems, pp. 257–260 (2010)
22. Haeupler, B., Mirrokni, V.S., Zadimoghaddam, M.: Online stochastic weighted matching: improved approximation algorithms. In: Chen, N., Elkind, E., Koutsoupias, E. (eds.) WINE 2011. LNCS, vol. 7090, pp. 170–181. Springer, Heidelberg (2011). https://doi.org/10.1007/978-3-642-25510-6_15
23. Ho, C.J., Vaughan, J.W.: Online task assignment in crowdsourcing markets. In: Twenty-Sixth AAAI Conference on Artificial Intelligence (2012)
24. Jaillet, P., Lu, X.: Online stochastic matching: new algorithms with better bounds. Math. Oper. Res. **39**(3), 624–646 (2013)

25. Kapralov, M., Post, I., Vondrák, J.: Online submodular welfare maximization: greedy is optimal. In: SODA (2013)
26. Karimi, M., Lucic, M., Hassani, H., Krause, A.: Stochastic submodular maximization: the case of coverage functions. In: Advances in Neural Information Processing Systems, pp. 6853–6863 (2017)
27. Korula, N., Mirrokni, V., Zadimoghaddam, M.: Online submodular welfare maximization: greedy beats 1/2 in random order. SIAM J. Comput. **47**(3), 1056–1086 (2018)
28. Lee, J., Sviridenko, M., Vondrák, J.: Submodular maximization over multiple matroids via generalized exchange properties. Math. Oper. Res. (MoR) **35**, 95–806 (2010)
29. Li, Y., Fu, K., Wang, Z., Shahabi, C., Ye, J., Liu, Y.: Multi-task representation learning for travel time estimation, pp. 1695–1704. KDD '18 (2018)
30. Manshadi, V.H., Gharan, S.O., Saberi, A.: Online stochastic matching: online actions based on offline statistics. Math. Oper. Res. **37**(4), 559–573 (2012)
31. Novorol, C.: https://ai-med.io/ai-biases-ada-health-diversity-women/ (2018). Accessed 20 Sept 2019
32. Puthiya Parambath, S.A., Usunier, N., Grandvalet, Y.: A coverage-based approach to recommendation diversity on similarity graph. In: Proceedings of the 10th ACM Conference on Recommender Systems, pp. 15–22 (2016)
33. Rawitz, D., Rosén, A.: Online budgeted maximum coverage. In: 24th Annual European Symposium on Algorithms (ESA 2016). Schloss Dagstuhl-Leibniz-Zentrum fuer Informatik (2016)
34. Sarpatwar, K.K., Schieber, B., Shachnai, H.: Interleaved algorithms for constrained submodular function maximization. arXiv preprint arXiv:1705.06319 (2017)
35. Vondrák, J.: Optimal approximation for the submodular welfare problem in the value oracle model. In: Proceedings of the Fortieth Annual ACM Symposium on Theory of Computing, pp. 67–74 (2008)
36. Xie, M., Wang, Q., Cui, Q., Yang, G., Li, M.: CQM: coverage-constrained quality maximization in crowdsourcing test. In: 2017 IEEE/ACM 39th International Conference on Software Engineering Companion (ICSE-C), pp. 192–194. IEEE (2017)
37. Xu, P., Srinivasan, A., Sarpatwar, K.K., Wu, K.L.: Budgeted online assignment in crowdsourcing markets: theory and practice. In: Proceedings of the 16th Conference on Autonomous Agents and MultiAgent Systems, pp. 1763–1765. International Foundation for Autonomous Agents and Multiagent Systems (2017)
38. Yao, H., et al.: Deep multi-view spatial-temporal network for taxi demand prediction, pp. 2588–2595. AAAI '18 (2018)

Price Cycles in Ridesharing Platforms

Chenkai Yu[1][(⊠)], Hongyao Ma[1], and Adam Wierman[2]

[1] Columbia University, New York, NY, USA
cyu26@gsb.columbia.edu
[2] California Institute of Technology, Pasadena, CA, USA

Abstract. In ridesharing platforms such as Uber and Lyft, it is observed that drivers sometimes collaboratively go offline when the price is low, and then return after the price has risen due to the perceived lack of supply. This collective strategy leads to cyclic fluctuations in prices and available drivers, resulting in poor reliability and social welfare. We study a continuous-time, non-atomic model and prove that such online/offline strategies may form a Nash equilibrium among drivers, but lead to a lower total driver payoff if the market is sufficiently dense. Furthermore, we show how to set *price floors* that effectively mitigate the emergence and impact of price cycles.

1 Introduction

Ridesharing platforms have enjoyed great success in recent years. In addition to mobile apps, which enable more efficient matching between riders and drivers, platforms such as Uber and Lyft also employ dynamic "surge" pricing to balance supply and demand in real-time. This guarantees that rider waiting times do not exceed a few minutes [45], achieving a desirable trade-off between (i) the cost of maintaining open driver supply and (ii) the cost of long waiting time for riders and pickup time for drivers [13,52].

In the presence of surge pricing, and with the "real-time flexibility" to decide when and where to drive [17,30], drivers often strategically optimize to increase their earnings instead of accepting all trip dispatches from the platforms. For example, many drivers cherry-pick based on trip lengths or destinations [15,27, 39]. When prices fail to be sufficiently *smooth* in space and time, drivers also decline trips to chase surge prices in neighboring areas or go offline before the end of large events in anticipation of a price hike [38].

Increasingly, the strategic behavior of drivers is not limited to individuals. Collective strategic behavior of groups of drivers supplying the same region has been reported in recent years. A prominent example is drivers all going offline at the same time and returning only after the price has surged due to the perceived lack of supply. As reported by ABC News [23,47]:

> Every night, several times a night, Uber and Lyft drivers at Reagan
> National Airport simultaneously turn off their ride share apps for a minute

J. Garg et al. (Eds.): WINE 2023, LNCS 14413, pp. 618–636, 2024.
https://doi.org/10.1007/978-3-031-48974-7_35

or two to trick the app into thinking there are no drivers available—creating a price surge. When the fare goes high enough, the drivers turn their apps back on and lock into the higher fare.

Such strategies have also been discussed in online forums by drivers from many cities. For example, when discussing the strategy of leaving the app off until it surges 2.5x at the Los Angeles International Airport, a driver posted [50]

"Can someone create a huge sign with a posterboard that says 'stay logged off until 2.5x'. Like people taking turns holding the sign right at the entrance..."

When adopted by drivers supplying the same region, these "online/offline" strategies induce cyclic fluctuations in prices and the number of available drivers, i.e. *price cycles*. Specifically, after all drivers have gone offline, the platform will increase trip prices due to the perceived lack of supply. Once the price has risen high enough, the drivers sign back on at the same time. This abundance of supply leads to a decrease in trip prices, eventually prompting drivers to sign off again.

Such price cycles undercut the platforms' mission of providing reliable transportation to riders [36,48]—when prices are increasing and drivers are not accepting dispatches, even riders with a high willingness to pay may be unable to get access to reliable service. Further, from the drivers' perspective, while this behavior is sometimes described as drivers "collectively gaming the system" in order to "regain their autonomy" (see e.g. [40]), it is not clear whether it actually leads to increased total earnings for drivers in practice.

Despite press attention, the emergence and mitigation of price cycles has not been studied from the perspective of the design and operation of ridesharing platforms. In this paper, we seek insight into questions including:

(i) Is it an equilibrium for all drivers to adopt these online/offline strategies?
(ii) Do drivers collectively benefit from such strategies?
(iii) What are the impacts on riders and the efficiency of the platform?
(iv) How to avoid the emergence and reduce the impact of price cycles?

Addressing these questions is challenging because such collective strategies of drivers are inherently dynamic, even when the market condition is stationary over time. Understanding price cycles and the market equilibria thereby requires characterizing drivers' best-response in non-stationary, dynamic settings.

Contributions. In this work, we answer the four questions highlighted above using a continuous-time model of pricing and matching by a ridesharing platform in a specific region (e.g. a city center).[1] We show that it can be a Nash equilibrium among drivers to collectively withhold supply and accept trips only after prices have risen. For markets that are sufficiently dense, however, we prove that the

---

[1] This focus on a single region models situations where price cycles tend to emerge in ridesharing platforms, e.g. when a majority of riders are leaving restaurants and bars in the city center at the end of the evening.

resulting price cycles *reduce*, instead of improve, the total payoff of drivers. To mitigate the adverse impact on driver payoffs, as well as on social welfare, we show how a platform can introduce "price floors" (i.e. lower bounds on trip prices) and effectively prevent the emergence of stable price cycles.

Related Work. The literature on ridesharing platforms is rapidly growing. Empirical works have studied topics ranging from consumer surplus [12,19] and the labor market of drivers [29,30] to the flexible work arrangements [16,17, 51]. Theoretical modelling works have analyzed the optimal growth of two-sided platforms [24,34], competition between platforms [2,25,32], operations in the presence of autonomous vehicles [33,41], and utilization-based minimum wage regulations [4].

This paper connects to the literature on pricing and matching in ridesharing platforms. [10] and [8] study revenue-optimal pricing when supply and demand are imbalanced in space. [5,22] and [3] focus on matching between riders and drivers and the pooling of shared rides, considering the online arrival of supply and demand in space. Further, [31,44] and [42] design policies that dispatch drivers from areas with relatively abundant supply, while [11] and [43] look at the role of information availability and transparency in platform design. Dynamic pricing [6], state-dependent dispatching [7,14], driver admission control [1] and capacity planning [9] are also studied using queueing-theoretical models.

In regard to dynamic "surge" pricing, [13] and [52] demonstrate the importance of dynamic pricing in maintaining the spatial density of open driver supply, which improves operational efficiency by reducing waiting times for riders/pickup times for drivers. Further, empirical studies have demonstrated the effectiveness of dynamic surge pricing in improving reliability and efficiency [28], increasing driver supply during high-demand times [18], as well as creating incentives for drivers to relocate to higher surge areas [35]. [26] discuss cycles of volatile driver supply, leading to increased average waiting times for riders and pickup costs for drivers. Those cycles are due to dynamic pricing with only two prices (high and low), rather than the strategic behavior of drivers.

Work to this point on strategic behaviors of drivers has focused on individual drivers. [20] discuss learning-by-doing and the gender earnings gap; [38] propose origin-destination based pricing that is welfare optimal and incentive compatible in the presence of spatial imbalance and temporal variation of supply and demand; [27] show that additive instead of multiplicative "surge" pricing is more incentive aligned when prices need to be origin-based only; [46] study pricing in the presence of driver location preferences; [15] demonstrate how to use drivers' waiting times to align incentives and reduce inequity in earnings when some trips are necessarily more lucrative than the others due to operational constraints.

2   Preliminaries

We consider a continuous-time model, focusing on the pricing and matching for trips originating from one specific region, such as a city center. Rider demand

and driver supply are stationary and non-atomic.[2] Specifically, drivers arrive at random locations in the region (e.g. where they completed their previous trip) at a rate of $\lambda_d > 0$, joining existing drivers in the region upon arrival. Riders arrive in the region with rate $\lambda_r > 0$, each looking for a trip out of the region originating from a random location in the region.

At time t, the platform sets a price $p = p(t) \geq 0$ for all riders who request a trip at time t and all drivers who are dispatched at time t. Upon the arrival of a rider, the platform dispatches the rider's trip to the closest available driver in the region if the rider accepts the current price. After a driver has dispatched and accepted a rider trip, it takes some time for the driver to pick up the rider. We denote this *pickup time* as $\eta(n)$, where $n = n(t)$ is the number of available drivers in the region at time t. When there is a higher density of available drivers in the region, a rider is more likely to be matched with a nearby driver, and hence the average pickup time is shorter. We adopt the form

$$\eta(n) = \tau n^{-\alpha}, \tag{1}$$

where $\tau > 0$ can be interpreted as the average pickup time when there is a single driver in the region, and α is typically between $1/3$ and $1/2$ in practice [52].[3]

Riders' value for a trip is a random variable V with cumulative distribution function (CDF) $F(v)$.[4] After being matched with a driver, a rider incurs a *waiting cost* of $c_r \geq 0$ per unit of time while waiting to be picked up. As a result, a rider with realized value $V = v$ has an expected utility $v - p - c_r \eta(n)$ from a trip when there are n available drivers and her trip price is p. Hence, a rider requests a trip if and only if $v \geq p + c_r \eta(n)$. We refer to $p + c_r \eta(n)$ as the *effective cost* of a trip for the riders. With n available drivers, the *total rider demand per unit of time* at price p is $\lambda_r \Pr[V \geq p + c_r \eta(n)] = \lambda_r \bar{F}(p + c_r \eta(n))$, where $\bar{F}(v) \triangleq 1 - F(v)$.

Drivers' opportunity cost is $c_d \geq 0$ per unit of time, modeling the value of their forgone outside option (e.g. driving elsewhere in the city for the same platform) while waiting for a trip dispatch in this region or driving to pick up a rider. We assume that there is no heterogeneity in drivers' continuation earnings after they have completed a trip, and that drivers do not have preference over destinations. A driver's expected *net payoff* after accepting a trip at price p is therefore $p - c_d \eta(n)$, if there are n available drivers in the region at the time of dispatch.[5] At any point in time, an available driver can decide to remain online to accept trip dispatches or temporarily go offline to wait for a future time to

---

[2] Although rider demand may fluctuate substantially during the course of a day, the price cycles typically last for a few hours, (e.g. the evening rush [47]), and our model is focused on this timescale, during which demand is relatively stable.

[3] In theory, if the dispatch is to the closest drivers, then the pickup time follows (1) with $\alpha = 1/2$ when demand and supply are both uniformly distributed in space [9].

[4] A rider's value for a trip represents the rider's willingness to pay minus the costs incurred by a driver while completing the trip, e.g. the costs of time, fuel, wear and tear, etc. Accordingly, the price p in this paper is the payment made by the riders minus a "base payment" that is paid directly to the drivers to cover their costs.

[5] We assume that drivers' decisions on whether to accept trips are based on the *expected* pickup time $\eta(n)$, implicitly assuming that drivers are unable to cherry-

come online again. A strategic driver will act to optimize her utility (i.e. total payoff) from the trip she accepts from the current region, which equals the net payoff from the trip minus the cost she incurred waiting for this trip dispatch.

We consider a setting where the platform has full information about the pickup time, riders' and drivers' arrival rates and waiting costs, the rider value distribution, and the number of online drivers. However, the platform does not price trips based on the number of drivers that are offline waiting to return at a later time. Drivers have the same information about market conditions in the region and this is common knowledge among drivers. Additionally, drivers are also aware of the number of offline drivers, reflecting the fact that the offline-online behavior is typically coordinated among drivers when cycles emerge.

2.1   System Dynamics and Steady States

We now characterize the evolution of the number of available drivers in the region, $n(t) \geq 0$, over time $t \in \mathbb{R}$. Assume for now that drivers are non-strategic, staying online after arrival without leaving the region (we model drivers' strategic behavior in Sect. 3). Given price $p(t)$, the total rider demand is $\lambda_r \bar{F}(p(t) + c_r \eta(n(t)))$ per unit of time, thus the derivative of $n(t)$ can be written as

$$\dot{n}(t) = \lambda_d - \lambda_r \bar{F}(p(t) + c_r \eta(n(t))). \tag{2}$$

The system state is characterized by $(p(t), n(t))$, where $p(t)$ is determined by the platform. The goal of designing $p(t)$ is to drive the system to a steady state that maximizes social welfare, i.e. the total rider utility minus driver cost.

Definition 1 (Socially Optimal Steady State (SOSS)). *A system state (p, n) is said to be steady if the number of drivers is not changing, i.e. $\lambda_d - \lambda_r \bar{F}(p + c_r \eta(n)) = 0$. The socially optimal steady state is the steady state that maximizes social welfare per unit of time:*

$$w \triangleq \lambda_r \bar{F}(p + c_r \eta(n))\big(\mathbb{E}[V|V \geq p + c_r \eta(n)] - (c_r + c_d)\eta(n)\big) - c_d n. \tag{3}$$

We can obtain intuition for (3), by noting that $\lambda_r \bar{F}(p + c_r \eta(n))$ is the rider demand per unit of time;[6] $\mathbb{E}[V|V \geq p + c_r \eta(n)]$ is the expected value of a trip for riders who request a trip at price p when there are n available drivers; $(c_r + c_d)\eta(n)$ is the total cost incurred by the rider and the driver due to the pickup time of a trip; and $c_d n$ is the cost of the available drivers waiting to be dispatched.

A closed-form characterization of the SOSS is provided as follows.

---

pick rider trips based on their distance to the pickup locations. This is in line with practice, as drivers are often penalized for declining trip dispatches [37,49].

[6] We assume that the matching rate is zero if there is no available driver in the region, regardless of riders' cost of time c_r. Technically, this effectively assumes that $c_r \eta(n(t)) = \infty$ when $c_r = 0$ and $n(t) = 0$ (in which case $\eta(n(t)) = \infty$).

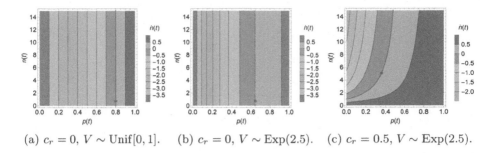

(a) $c_r = 0$, $V \sim \text{Unif}[0, 1]$.   (b) $c_r = 0$, $V \sim \text{Exp}(2.5)$.   (c) $c_r = 0.5$, $V \sim \text{Exp}(2.5)$.

Fig. 1. Contour plot of $\dot{n}(t)$ at different states (p, n) for Example 1. The bold red line is the set of steady states (i.e. $\dot{n}(t) = 0$). The dot on the line indicates the socially optimal steady state (p^*, n^*).

Proposition 1. *The socially optimal steady state* (p^*, n^*) *is given by*

$$n^* = \left(\frac{\alpha \tau \lambda_d (c_r + c_d)}{c_d} \right)^{\frac{1}{\alpha+1}}, \tag{4}$$

$$p^* = \bar{F}^{-1} \left(\frac{\lambda_d}{\lambda_r} \right) - c_r \eta(n^*). \tag{5}$$

See the full version of this paper [53] for the proof of this result. Note that the optimal number of available drivers n^* given by (4) is independent of the distribution of riders' values. Intuitively, n^* achieves an optimal trade-off between $(c_r + c_d)\eta(n)$, the total cost incurred due to the pickup time, and $c_d n$, the cost of the available drivers waiting to be dispatched.

Example 1. Consider a region where drivers arrive at a rate of $\lambda_d = 1$, riders arrive at a rate of $\lambda_r = 5$, and the pickup time is parameterized by $\alpha = 1/2$ and $\tau = 3\sqrt{3}/4$. We show in the figures a variety of settings: drivers' opportunity cost is $c_d = 0.03$ per unit of time, riders' cost of time is either $c_r = 0$ or $c_r = 0.5$, and riders' value distribution is either $\text{Unif}[0, 1]$ or $\text{Exp}(2.5)$.

Figure 1 shows contour lines of $\dot{n}(t)$, the set of steady states, and the socially optimal steady state (SOSS). Each contour line has the same value of $\dot{n}(t)$ and thus the same effective cost for riders $p + c_r \eta(n)$. They are riders' indifference curves. The line of steady states partitions the state space into two regions: The number of drivers decreases over time in the left region and increases in the right region. The solid dot indicates the SOSS (p^*, n^*). Comparing Figs. 1b and 1c, we see that when waiting is more costly for the riders, it is optimal to maintain a higher number of available drivers in order to reduce the pickup time.

2.2   Platform Pricing

Given Proposition 1, we can concretely define the goal of the platform. Specifically, we model a platform that aims to swiftly steer the system to the socially

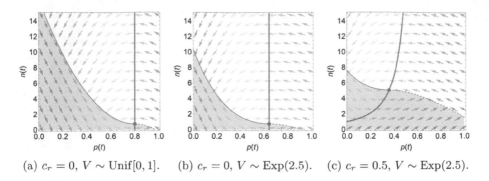

(a) $c_r = 0$, $V \sim \text{Unif}[0, 1]$.   (b) $c_r = 0$, $V \sim \text{Exp}(2.5)$.   (c) $c_r = 0.5$, $V \sim \text{Exp}(2.5)$.

Fig. 2. Pricing policy and state evolution for settings in Example 1.

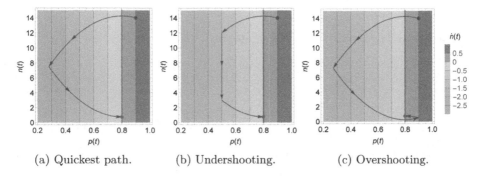

(a) Quickest path.   (b) Undershooting.   (c) Overshooting.

Fig. 3. Three paths from state $(0.9, 14)$ to the SOSS $(0.8, 0.75)$.

optimal steady state without changing the price too rapidly. In practice, platforms limit the rate of price changes because (i) erratic prices lead to poor experiences for riders [21]; (ii) it takes time for drivers to move from one region to another, thereby rendering quick price changes ineffective to incentivize driver movement; and (iii) the underlying market condition typically changes slower than the state, i.e. the number of open cars. In our model, we account for the need for "smooth" prices by limiting the rate of price increase to be at most $\ell_+ > 0$ and the rate of decrease to be at most $\ell_- > 0$, i.e. $-\ell_- \le \dot{p}(t) \le \ell_+$.

Consider an unexpected shock on the supply or the demand side that results in too many available drivers in the region than that under the SOSS. The platform should first decrease the price to encourage greater demand from riders than the arrival of drivers, and then increase the price back to the SOSS price. Similarly, if there are too few drivers available, the platform should first increase the price to reduce the rider demand, building up the pool of available drivers, before decreasing the price back to the SOSS price. Under such a policy, Fig. 2 illustrates how the state $(p(t), n(t))$ evolves over time for settings studied in Example 1 when the size of the price adjustments are limited by $\ell_- = \ell_+ = 0.1$, and Fig. 3a illustrates a trajectory from an initial state to the SOSS.

The state space (except the SOSS) can be divided into two regions corresponding to whether the platform should increase or decrease the prices. We define the boundaries $C_+, C_- \subseteq \mathbb{R}^2$ as parametric curves. Starting from any state $(p, n) \in C_+$, the trajectory eventually converges to the SOSS (p^*, n^*) if the platform continues to increase the price at the highest rate ℓ_+. Similarly, C_- is the set of states leading to (p^*, n^*) if the platform decreases the price at a rate of ℓ_-. Formally, the boundaries $C_+, C_- \subseteq \mathbb{R}^2$ are defined as follows.

– Let $(p_+(t), n_+(t))$ be the solution to the following differential equations:

$$\dot{p}(t) = \ell_+, \quad \dot{n}(t) = \lambda_d - \lambda_r \bar{F}\big(p(t) + c_r \eta(n(t))\big), \quad p(0) = p^*, \quad n(0) = n^*.$$

Define parametric curve $C_+ \triangleq \{(p_+(t), n_+(t)) \in \mathbb{R}^2 \,|\, t \in (-p^*/\ell_+, 0)\}$. Note that the lower bound on t is due to the fact that prices cannot go below zero.
– Let $(p_-(t), n_-(t))$ be the solution to the following differential equations:

$$\dot{p}(t) = -\ell_-, \quad \dot{n}(t) = \lambda_d - \lambda_r \bar{F}\big(p(t) + c_r \eta(n(t))\big), \quad p(0) = p^*, \quad n(0) = n^*.$$

Define parametric curve $C_- \triangleq \{(p_-(t), n_-(t)) \in \mathbb{R}^2 \,|\, t \in (-\infty, 0)\}$.

Given the definitions above, $C_+ \cup C_- \cup \{(p^*, n^*)\}$ divide the state space into two regions. In the upper (or lower) region, the platform first decreases (or increases) the price, and then starts to increase (or decrease) the price again when the system state reaches C_+ (or C_-).

Definition 2. *The platform's pricing policy $\pi : \mathbb{R}^2 \to \mathbb{R}$ is mapping from a state (p, n) to the derivative of the price $\dot{p}$ given by:*

$$\dot{p} = \pi(p, n) = \begin{cases} 0 & \text{if } (p, n) = (p^*, n^*), \\ \ell_+ & \text{if } \exists n' \geq n, (p, n') \in C_+ \text{ or } \exists n' > n, (p, n') \in C_-, \\ -\ell_- & \text{if } \exists n' < n, (p, n') \in C_+ \text{ or } \exists n' \leq n, (p, n') \in C_-. \end{cases} \quad (6)$$

Example 1 (Continued). Suppose that the rates for price changes are limited to $\ell_+ = \ell_- = 0.1$. Figure 2 shows the pricing policy for the three scenarios analyzed in Fig. 1. The vector field indicates the direction of state evolution, and the arrow colors represent the speed with which the number of drivers changes. In the upper region and on the dashed line C_-, the platform reduces the price. In the lower region and on the solid line C_+, the platform raises the price.

When rider cost $c_r = 0$ and rider value $V \sim \text{Unif}[0, 1]$, Fig. 3 provides three example paths from an initial state $(p, n) = (0.9, 14)$ to the SOSS $(p^*, n^*) = (0.8, 0.75)$. Figure 3a illustrates the trajectory under the pricing policy (6), taking $1 + 6\sqrt{3} \approx 11.4$ units of time. In Fig. 3b, prices were held at 0.5 without further decreasing, leading to a slower return to the SOSS, taking 13 units of time. On the other hand, when prices continued to decrease after reaching C_+ in Fig. 3c, there are fewer than n^* drivers when the price rose to p^*, again leading to a slower return to the SOSS, taking $3 + 4\sqrt{7} \approx 13.6$ units of time.

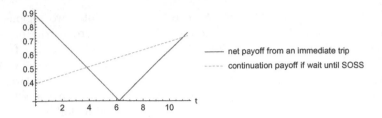

Fig. 4. Driver payoff over time along the path in Fig. 3a.

3 Stable Price Cycles

In this section, we demonstrate how price cycles emerge when drivers collectively adopt *online/offline strategies*, going offline at the same time, and coming back online again only after prices have risen due to the seemingly low supply.

Before formally defining an online/offline strategy, we first illustrate why it can be beneficial for a driver to go offline temporarily and withhold supply. Consider, for example, the path from the initial state $(0.9, 14)$ back to the SOSS as shown in Fig. 3a. Let $t = 0$ be the time where the system state is at $(0.9, 14)$, Fig. 4 illustrates the net payoff $p(t) - c_d\eta(n(t))$ a driver gets from accepting a trip at time t, assuming that the rest of the drivers are non-strategic, stay online and accept dispatches at all times. Intuitively, drivers get a low net payoff from accepting trips during times with low prices.

Instead of staying online and accepting the first dispatch from the platform, consider a driver who stays offline until the system reaches the SOSS. At the SOSS, each driver waits an average of n^*/λ_d units of time for a trip dispatch. The *continuation payoff* of a driver at the SOSS, i.e. the net payoff from the trip minus the cost the driver incurred waiting for the dispatch, is therefore:

$$u^* \triangleq p^* - c_d\eta(n^*) - c_d\frac{n^*}{\lambda_d}. \tag{7}$$

Since it takes $1 + 6\sqrt{3} \approx 11.4$ units of time for the system to return to the SOSS, at time t, the continuation payoff of a driver who stays offline until the SOSS is approximately $u^* - c_d(11.4 - t)$. This is also illustrated in Fig. 4. As we can see, if the rest of the drivers stay online and accept all dispatches from the platform, a driver can benefit from declining trip dispatches during times when prices are low (from $t \approx 4$ to $t \approx 11$, in this example). This illustrates that pricing trips to quickly return to the SOSS can incentivize drivers to strategically go offline and wait for a better price.[7]

3.1 Online/Offline Strategies

In Sect. 2, we define $n(t)$ as the *number of online drivers* on the platform, i.e. those who are in the region and are available for dispatch at time t. Here, we

---

[7] Note that "remaining offline until the SOSS" is not the optimal strategy a driver can adopt in this scenario. See Sect. 4 for a formal discussion on drivers' best response.

additionally denote $n_0(t) \geq 0$ as the *number of offline drivers* in the region at time t, i.e. those waiting to return at a later moment. Let $N(t) \triangleq n(t) + n_0(t)$ be the *total number of drivers* at time t. The platform sets trip prices based on $n(t)$ but not $n_0(t)$. A *driver strategy* σ determines, for any time t, the probability $\sigma(t) \in [0, 1]$ with which the driver is online and accepts dispatches from the platform; thus the driver is offline at time t with probability $1 - \sigma(t)$.

In practice, drivers coordinate on a kind of simple "threshold strategies" [47, 50], going offline when prices drop below some $\underline{p} > 0$ and returning online after the price has risen above some $\bar{p} > \underline{p}$. Such strategies depend on the price level p and whether the price is increasing or decreasing, and are affected by the number of online drivers n indirectly through the platform's pricing policy $\dot{p} = \pi(p, n)$.

Definition 3 (Online/offline threshold strategies). *Under an online/ offline threshold strategy with thresholds $(\bar{p}, \underline{p})$, a driver switches online after the price rises above $\bar{p}$, and goes offline when the price drops below $\underline{p}$. Formally,*

$$\sigma_{\bar{p},\underline{p}}(p, n) = \begin{cases} 1 & \text{if } (p \geq \bar{p}) \text{ or } (p > \underline{p} \text{ and } \pi(p, n) \leq 0), \\ 0 & \text{if } (p \leq \underline{p}) \text{ or } (p < \bar{p} \text{ and } \pi(p, n) > 0), \end{cases} \tag{8}$$

where π is the platform's pricing policy as in (6).

When all drivers coordinate and adopt the same online/offline threshold strategy, at any point in time t, either all drivers are online or all are offline. Such strategies may lead to cyclic fluctuations in prices and the number of available drivers on the platform. Consider, for example, the scenario analyzed in Fig. 3a. When drivers employ an online/offline threshold strategy with $(\underline{p}, \bar{p}) = (0.3, 0.9)$, the state evolves as follows (see Fig. 5 for an illustration).

- Starting from the initial state $(0.9, 14)$, the platform decreases the price in order to accelerate the dispatch of excess driver supply. The number of drivers gradually decreases after a small initial increase. Once the price drops to $\underline{p} = 0.3$, all drivers switch offline. With no available drivers online, the platform starts to increase the price at a rate of ℓ_+.
- At the time the price reaches $\bar{p} = 0.9$, all offline drivers re-emerge and start to accept dispatches again. With a large number of drivers accumulated during the price increase period, the state (p, n) will likely reside above C_-. As a result, the platform will again start to decrease the trip prices at a rate of ℓ_- and potentially repeat the same dynamic if the price drops again to $\underline{p}$.

We describe such cyclic fluctuations of prices using times t_0, t_1 and t_2, where

- t_0 is a time when the price is the lowest (i.e. $p(t_0) = \underline{p}$),
- t_1 is the first time after t_0 s.t.the price reaches the peak (i.e. $p(t_1) = \bar{p}$), and
- t_2 is the first time after t_1 s.t.$p(t_2) = \underline{p}$, which is exactly one period after t_0.

Under the platform's pricing policy (6), we have $t_1 - t_0 = (\bar{p} - \underline{p})/\ell_+$ and $t_2 - t_1 = (\bar{p} - \underline{p})/\ell_-$, thus the *period length* is $T \triangleq t_2 - t_0 = (\bar{p} - \underline{p})(1/\ell_+ + 1/\ell_-)$.

Further, we use $u(t, \sigma, \sigma')$ to denote the *continuation payoff* of a driver from time t onward, when the driver adopts strategy σ while others employ strategy σ'. It is defined as the expected price of the trip the driver eventually accepts, minus the total cost of time the driver incurs:

$$u(t, \sigma, \sigma') \triangleq \mathbb{E}[p(t + \Delta) - c_d(\Delta + \eta(n(t + \Delta)))]. \tag{9}$$

Here, Δ is a random variable representing the amount of time a driver waits before accepting a trip dispatch when the driver adopts strategy σ and when every other driver employs strategy σ'. The first term in the expectation $p(t+\Delta)$ is the trip price at the time the driver accepts the dispatched trip. The second term $c_d(\Delta + \eta(n(t + \Delta)))$ is the total opportunity cost of time incurred by the driver waiting for the dispatch and picking up the rider.

With this notation in hand, we now formally define a stable price cycle.

Definition 4 (Stable price cycle). *Suppose that the rates of price increase and decrease are limited by ℓ_+ and ℓ_-. An online/offline threshold strategy $\sigma_{\bar{p}, \underline{p}}$ forms a stable price cycle with a maximum of $\hat{n} > 0$ drivers if the associated trip price $p(t)$ and the number of online drivers $n(t)$ satisfy $\max_{t' \in [t, t+T]} n(t') = \hat{n}$ for any t, as well as the following conditions.*

(C1) (Market Clearing) The cycle repeats itself, i.e. the net accumulation of drivers over a period is zero. Formally, for any time t,

$$\lambda_d T = \int_t^{t+T} \lambda_r \bar{F}(p(t') + c_r \eta(n(t'))) \, dt'; \tag{10}$$

(C2) (Driver Best Response) The online/offline strategy $\sigma_{\bar{p}, \underline{p}}$ forms a Nash equilibrium among drivers, meaning that it is not useful for a driver to unilaterally deviate to any feasible (and potentially non-threshold) strategy σ:

$$u(t, \sigma_{\bar{p}, \underline{p}}, \sigma_{\bar{p}, \underline{p}}) \geq u(t, \sigma, \sigma_{\bar{p}, \underline{p}}), \quad \forall \sigma, \forall t. \tag{11}$$

We denote a stable price cycle succinctly by $(\ell_+, \ell_-, \underline{p}, \bar{p}, \hat{n})$ for simplicity of notation. As an illustration, we return to the settings analyzed in Example 1.

Example 1 (Continued). Consider the setting where the arrival rates of drivers and riders are $\lambda_d = 1$ and $\lambda_r = 5$, respectively. The pickup time is parameterized by $\alpha = 1/2$ and $\tau = 3\sqrt{3}/4$. The value of riders is uniformly distributed in $[0, 1]$, and the time costs of drivers and riders are $c_d = 0.03$ and $c_r = 0$. Figure 5 illustrates a stable cycle under a pricing policy with $\ell_+ = \ell_- = 0.1$, when all drivers adopt an online/offline threshold strategy $\sigma_{\bar{p}, \underline{p}}$ with $\underline{p} = 0.3$ and $\bar{p} = 0.9$. Figure 5a shows the trajectories induced by the online/offline threshold strategy (solid blue lines) and the trajectories planned by the platform's pricing policy (dotted blue lines). The vertical red dashed lines indicate drivers' collectively going online and offline. The green lines are the indifference curves of the *drivers*. Figure 5b shows the number of offline drivers $n_0(t)$ and online drivers $n(t)$ over time t. Figure 5c shows the trip price $p(t)$, a driver's net payoff $p(t) - c_d \eta(n(t))$,

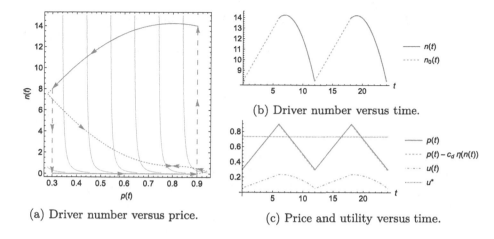

(a) Driver number versus price.

(b) Driver number versus time.

(c) Price and utility versus time.

Fig. 5. The "symmetric" cycle in Example 1.

a driver's continuation payoff during the cycle $u(t)$, and a driver's continuation payoff at SOSS u^*. Drivers go offline at time $t_0 = 0$, then switch online at time $t_1 = 6$, and go offline again at time $t_2 = 12$.

Under strategy $\sigma_{\bar{p},\underline{p}}$, all drivers, including those who arrive during the price increase, choose to remain offline when the trip price surges from $\underline{p} = 0.3$ to $\bar{p} = 0.9$. Drivers opt to go online once the price reaches a level of 0.9. After noting the inflow of drivers, the platform subsequently reduces the price in response to the expanded driver supply. There are more drivers going online at $\bar{p}$ than drivers going offline at $\underline{p}$, because drivers continue to arrive as the price increases.

We formally establish the stability of these price cycles in the next section. Intuitively, drivers do not have an incentive to go online before others do as the price increases, since with a very small $n(t)$, drivers' pickup time to pick up a rider will be very long. Further, if c_d is appropriately high, going offline before other drivers do (as the price decreases) to wait for the next peak price is not better than remaining online and accepting dispatches from the platform.

4   Existence of Stable Price Cycles

In this section, we provide sufficient conditions (that can be easily satisfied and verified) for the existence of stable price cycles induced by online/offline threshold strategies. We prove that while cycles may form an equilibrium among drivers, they are not necessarily beneficial— for markets that are sufficiently dense, drives get a lower average utility than the socially optimal steady state.

In order to prove the existence of stable price cycles, we return to Definition 4, which has two conditions. Condition (C1) on market clearance is easy to verify, while (C2) on drivers' best response is more difficult to approach. Thus, the first step in our analysis is to develop a sufficient condition for (C2).

We start by introducing some notation. When supply and demand are uniformly distributed in space, dispatching the closest driver for each trip request corresponds to dispatching available drivers uniformly at random. Denote

$$\rho(t) \triangleq \frac{\lambda_r \bar{F}(p(t) + c_r \eta(n(t)))}{n(t)}. \tag{12}$$

$$h(t) \triangleq p(t) - c_d \eta(n(t)) - \frac{c_d}{\rho(t)}. \tag{13}$$

$\rho(t)$ is the rate at which a particular driver is dispatched at time t, and $h(t)$ can be interpreted as the continuation payoff of a driver in a stationary environment where the price and number of drivers are fixed at $p(t)$ and $n(t)$. Finally, let

$$u(t) \triangleq \sup_\sigma u(t, \sigma, \sigma_{\bar{p},\underline{p}}) \tag{14}$$

denote a driver's optimal continuation payoff at time t, when all other drivers adopt strategy $\sigma_{\bar{p},\underline{p}}$.

Theorem 1. *Suppose a price cycle $(\ell_+, \ell_-, \underline{p}, \bar{p}, \hat{n})$ is market-clearing, i.e. Condition (C1) holds. Then, Condition (C2) holds, meaning that the online/offline strategy $\sigma_{\bar{p},\underline{p}}$ forms a Nash equilibrium among drivers and the cycle is stable, if the following two conditions are satisfied:*

$$\forall t \in (t_1, t_2), \ \frac{d}{dt}(p(t) - c_d \eta(n(t))) < c_d, \tag{C2.1}$$

$$p(t_0) - c_d \eta(N(t_0)) + c_d(t_1 - t_0) > \max_{t \in (t_1, t_2)} h(t). \tag{C2.2}$$

We provide a detailed discussion and the proof of this theorem in the full version of this paper [53], where we also apply this result and verify that the price cycle in Fig. 5 is a stable cycle.

Next, we turn to the question of whether drivers benefit from stable price cycles. The following result shows that under mild conditions, the total payoff to drivers under any stable cycle is lower than that under the SOSS. In fact, drivers not only incur a higher total waiting cost, but also receive a lower total payoff from trips since many drivers are dispatched after the price dropped below p^*.

Theorem 2. *Assume that the distribution of rider value satisfies*

$$\frac{d^2}{dv^2} \frac{1}{\bar{F}(v)} \geq 0, \tag{15}$$

i.e. $\bar{F}(v)^{-1}$ is convex. The total payoff to drivers in any stable cycle is lower than that in the socially optimal steady state (SOSS), if

$$\frac{\ell_-}{\ell_+} + 1 > \frac{\bar{F}(p^* - c_d \eta(n^*))}{\bar{F}(p^* + c_r \eta(n^*))}. \tag{16}$$

See the full version of this paper [53] for the proof. Condition (15) holds for all the examples in this paper as well as for many common distributions, e.g. uniform, exponential, and Pareto with finite mean. Moreover, the difference between the numerator and denominator of the RHS of (16) is driven by $c_d\eta(n^*)$ and $c_r\eta(n^*)$, the costs incurred by drivers and riders due to the pickup time under the SOSS. Intuitively, for denser markets, with higher n^* and lower $\eta(n^*)$, the RHS of (16) is closer to 1, so (16) is more likely to hold. For a typical market with $\ell_+ = \ell_-$, a violation of (16) implies

$$\frac{\ell_-}{\ell_+} + 1 = 2 < \frac{\bar{F}(p^* - c_d\eta(n^*))}{\bar{F}(p^* + c_r\eta(n^*))},$$

which further implies $\bar{F}(p^* - c_d\eta(n^*)) > 2\bar{F}(p^* + c_r\eta(n^*))$. What this means is that fixing the number of available drivers at n^*, rider demand will *double* when trip price reduces from p^* to $p^* - (c_r + c_d)\eta(n^*)$ (in which case the effective cost of the trip of the riders reduces from $p^* + c_r\eta(n^*)$ to $p^* - c_d\eta(n^*)$). This is not very likely for markets where the pickup time is only a couple of minutes and thus the total pickup cost $(c_r + c_d)\eta(n^*)$ is relatively small.

5   Avoiding Price Cycles

In the previous section, we showed that online/offline threshold strategies can induce stable price cycles but may lead to a lower total utility for drivers compared to the SOSS. This raises the question: how can a platform reduce the chances of price cycles emerging and/or minimize their impact? While there are many suggested solutions to this problem, we'll explain why most of them might not work effectively in real-world scenarios.

A first idea is to impose a *price cap*. Intuitively, drivers may not have much incentive to manipulate if there is little room for the prices to grow. However, this approach will not be effective. First, there exist stable cycles that cannot be eliminated even using a price cap as low as the SOSS price. Moreover, in the presence of stochasticity in practice, limiting the extent to which prices can rise can be very inefficient, since the region may run out of supply, leading to long waiting times for riders / pickup time for drivers [13].

A second approach is to allow *cherry-picking* based on trip details. One reason for a driver to not accept dispatches while the other drivers are offline is that, with few drivers online, the loss of density in space leads to a very long average pickup time. Allowing drivers to decline trip dispatches partially mitigates this issue, since an online driver may choose to only accept rider trips originating from close-by locations. However, incentivizing drivers to remain online in this way is not desirable, since allowing frequent declines leads to excess strategizing on the part of drivers and a loss of reliability for riders.[8]

---

[8] In practice, many platforms discourage cherry-picking by pushing drivers offline for declining multiple trip dispatches [49].

A third approach is to use prices that are more *smooth* in time, i.e. reducing ℓ_+ and ℓ_-. Intuitively, this smoothing reduces the incentive to go offline by increasing the time it takes for prices to rise, thereby the cost drivers incur when waiting for a better price. However, the degree of smoothing required to eliminate cycles with this approach is not practical. As long as $\ell_+ \geq c_d$, waiting offline can be profitable since the rate at which prices may increase is higher than the opportunity cost to drivers. Reducing ℓ_+ to below c_d is not realistic, since when drivers make an average of \$20 per hour, $c_d \approx \$1/3$ per minute. Restricting the price increase to less than \$1/3 per minute substantially limits the platform's ability to respond to changes in market conditions in real time, leading to significant market inefficiency.

Finally, the approach that we propose to reduce the likelihood and impact of price cycles is to introduce a *price floor*. Intuitively, a lower bound on trip prices prevents drivers' payoffs from remaining online from dropping too low, thereby reducing drivers' incentives to go offline and wait for a better price. However, the fundamental question about the efficacy of this approach is: *Can the price floor be set in a way that eliminates all stable price cycles, while still allowing welfare-optimal market clearing? If so, where should the price floor be set?*

Our main result for this section characterizes the market conditions and the set of price floors that ensure no stable price cycles exist.

Theorem 3. *Given any market condition that satisfies*

$$\frac{\ell_-}{\ell_+} > \frac{\lambda_r}{\lambda_d}\bar{F}(p^*) - 1, \tag{17}$$

the platform can ensure no stable price cycles exist and maintain feasibility of the SOSS by setting a price floor p_{floor} such that

$$\bar{F}^{-1}\left(\frac{(\ell_- + \ell_+)\lambda_d}{\ell_+\lambda_r}\right) < p_{\text{floor}} < p^*. \tag{18}$$

See the full version of this paper [53] for the proof. This theorem provides a range of potential price floors that can eliminate stable cycles for markets that satisfy (17).[9] To interpret the market condition in (17), note from (5) that at the SOSS (p^*, n^*), we have $p^* = \bar{F}^{-1}(\lambda_d/\lambda_r) - c_r\eta(n^*)$. Therefore, for markets that are dense, i.e. markets with higher number of available drivers n^* and lower pickup time $\eta(n^*)$, $\frac{\lambda_r}{\lambda_d}\bar{F}(p^*) - 1$ is closer to 0, and (17) is more likely to hold.

In practice, a violation of (17) is not very likely. Consider the typical platforms with $\ell_- = \ell_+$. A violation of (17) implies that

$$\frac{\ell_-}{\ell_+} = 1 < \frac{\lambda_r}{\lambda_d}\bar{F}(p^*) - 1, \tag{19}$$

---

[9] Note that the theorem only rules out stable cycles we study in the paper, which correspond to online/offline strategies as in Definition 3. Technically, it does not eliminate equilibria where drivers coordinate on other more complicated collective strategies. However, the threshold strategies we study are aligned with news reports and driver forum discussions. Moreover, they do not require further communication among drivers once the thresholds have been agreed upon.

which further implies $\bar{F}(p^*) > 2\lambda_d/\lambda_r$. Compared to $\bar{F}(p^* + c_r\eta(n^*)) = \lambda_d/\lambda_r$, we know that in a market that violates (17), eliminating waiting time (i.e. reducing riders' effective cost for a trip from $p^* + c_r\eta(n^*)$ to p^*) will lead to a doubling of rider demand. This is not likely for dense markets where riders wait for only a few minutes for the drivers to pick them up.

For a typical platform with $\ell_- = \ell_+$, observe that the lowest effective price floor $\bar{F}^{-1}\left(\frac{(\ell_-+\ell_+)\lambda_d}{\ell_+\lambda_r}\right) = \bar{F}^{-1}\left(\frac{2\lambda_d}{\lambda_r}\right)$ is the effective cost of trips at which rider demand doubles compared to that under the SOSS. As a result, Theorem 3 provides a simple rule of thumb that a platform can eliminate stable cycles by imposing a lower bound on prices at the point where rider demand doubles. To demonstrate the use of price floors, we return to our running example.

Example 1. (Continued). Consider the two economies studied in Example 1. Condition (17) holds because the RHS of (17) is always zero due to the simplifying assumption that $c_r = 0$. For the economy illustrated in Fig. 5, any lower bound on prices p_{floor} between $\bar{F}^{-1}\left(\frac{(\ell_-+\ell_+)\lambda_d}{\ell_+\lambda_r}\right) = 0.6$ and $p^* = 0.8$ ensures that there can be no stable price cycles.

6   Concluding Remarks

Price cycles resulting from the collective online/offline strategies of drivers are a very visible source of inefficiency in today's ridesharing platforms and undercut the reliability of service for riders. In this paper, we study the emergence of such price cycles and discuss approaches to mitigate their impact. Our results show that the price cycles may form an equilibrium among drivers, but counterintuitively lead to a lower total payoff for drivers in markets that are sufficiently dense. Furthermore, we provide sufficient conditions to guide the design of price floors that effectively prevent the emergence of stable price cycles.

This paper is the first to provide a characterization of the existence and mitigation of price cycles, and as such, many important questions remain to be addressed. Our model focuses on a single region and assumes that drivers' arrival rate and opportunity cost are exogenous. It would be interesting to understand the market dynamics in multiple regions supplied by a fixed pool of drivers. Further, we have assumed that the supply and demand are stationary and known to the platform, in which case welfare-optimal pricing is possible. It is practically relevant to relax these assumptions and analyze a stochastic and potentially non-stationary setting, where the platform needs to make decisions online without complete information on the market conditions.

References

1. Afèche, P., Liu, Z., Maglaras, C.: Ride-hailing networks with strategic drivers: the impact of platform control capabilities on performance. Manuf. Serv. Oper. Manage. **25**(5), 1890–1908 (2023)

2. Ahmadinejad, A., Nazerzadeh, H., Saberi, A., Skochdopole, N., Sweeney, K.: Competition in ride-hailing markets. In: Web and Internet Economics (2019)
3. Aouad, A., Saritaç, Ö.: Dynamic stochastic matching under limited time. In: ACM EC, pp. 789–790 (2020)
4. Asadpour, A., Lobel, I., van Ryzin, G.: Minimum earnings regulation and the stability of marketplaces. Manuf. Serv. Oper. Manage. **25**(1), 254–265 (2023)
5. Ashlagi, I., Burq, M., Dutta, C., Jaillet, P., Saberi, A., Sholley, C.: Maximum weight online matching with deadlines. arXiv preprint arXiv:1808.03526 (2018)
6. Banerjee, S., Johari, R., Riquelme, C.: Pricing in ride-sharing platforms: a queueing-theoretic approach. In: ACM EC (2015)
7. Banerjee, S., Kanoria, Y., Qian, P.: State dependent control of closed queueing networks. In: Abstracts of the 2018 ACM International Conference on Measurement and Modeling of Computer Systems (2018)
8. Besbes, O., Castro, F., Lobel, I.: Surge pricing and its spatial supply response. Manage. Sci. **67**, 1350–1367 (2020)
9. Besbes, O., Castro, F., Lobel, I.: Spatial capacity planning. Oper. Res. **70**(2), 1271–1291 (2022)
10. Bimpikis, K., Candogan, O., Saban, D.: Spatial pricing in ride-sharing networks. Oper. Res. **67**(3), 744–769 (2019)
11. Cai, D., Bose, S., Wierman, A.: On the role of a market maker in networked Cournot competition. Math. Oper. Res. **44**(3), 1122–1144 (2019)
12. Castillo, J.C.: Who benefits from surge pricing? SSRN 3245533 (2020)
13. Castillo, J.C., Knoepfle, D., Weyl, G.: Surge pricing solves the wild goose chase. In: ACM EC, pp. 241–242 (2017)
14. Castro, F., Frazier, P., Ma, H., Nazerzadeh, H., Yan, C.: Matching queues, flexibility and incentives. arXiv preprint arXiv:2006.08863 (2020)
15. Castro, F., Ma, H., Nazerzadeh, H., Yan, C.: Randomized FIFO mechanisms. In: ACM EC, pp. 60 (2022)
16. Chen, K.M., Ding, C., List, J.A., Mogstad, M.: Reservation wages and workers' valuation of job flexibility: Evidence from a natural field experiment. Technical report, National Bureau of Economic Research (2020)
17. Chen, M.K., Rossi, P.E., Chevalier, J.A., Oehlsen, E.: The value of flexible work: evidence from Uber drivers. J. Polit. Econ. **127**(6), 2735–2794 (2019)
18. Chen, M.K., Sheldon, M.: Dynamic pricing in a labor market: surge pricing and flexible work on the Uber platform (2015). https://www.anderson.ucla.edu/faculty_pages/keith.chen/papers/SurgeAndFlexibleWork_WorkingPaper.pdf
19. Cohen, P., Hahn, R., Hall, J., Levitt, S., Metcalfe, R.: Using big data to estimate consumer surplus: the case of Uber. NBER Working Paper No. 22627 (2016)
20. Cook, C., Diamond, R., Hall, J., List, J.A., Oyer, P.: The gender earnings gap in the gig economy: evidence from over a million rideshare drivers. Technical report, National Bureau of Economic Research (2018)
21. Dholakia, U.M.: Everyone hates Uber's surge pricing-here's how to fix it. Harvard Bus. Rev. **21**) (2015)
22. Dickerson, J., Sankararaman, K., Srinivasan, A., Xu, P.: Allocation problems in ride-sharing platforms: online matching with offline reusable resources. In: Proceedings of the AAAI Conference on Artificial Intelligence, vol. 32 (2018)
23. Dustin is Driving: Uber- i can't believe drivers admitted to doing "surge club" (2019). https://youtu.be/SYkLhXMsZ8I. Accessed 31 Jan 2021
24. Fang, Z., Huang, L., Wierman, A.: Prices and subsidies in the sharing economy. In: Performance Evaluation (2019)

25. Fang, Z., Huang, L., Wierman, A.: Loyalty programs in the sharing economy: optimality and competition. In: Performance Evaluation (2020)
26. Freund, D., van Ryzin, G.: Pricing fast and slow: limitations of dynamic pricing mechanisms in ride-hailing (2021). SSRN 3931844
27. Garg, N., Nazerzadeh, H.: Driver surge pricing. Manage. Sci. **68**(5), 3219–3235 (2022)
28. Hall, J., Kendrick, C., Nosko, C.: The effects of Uber's surge pricing: a case study. The University of Chicago Booth School of Business, Technical report (2015)
29. Hall, J.V., Horton, J.J., Knoepfle, D.T.: Labor market equilibration: evidence from Uber. New York University Stern School of Business, Technical report (2017)
30. Hall, J.V., Krueger, A.B.: An analysis of the labor market for Uber's driver-partners in the united states. NBER Working Paper No. 22843 (2016)
31. Kanoria, Y., Qian, P.: Blind dynamic resource allocation in closed networks via mirror backpressure. In: ACM EC (2020)
32. Lian, Z., Martin, S., van Ryzin, G.: Labor cost free-riding in the gig economy (2022). SSRN 3775888
33. Lian, Z., van Ryzin, G.: Capturing the benefits of autonomous vehicles in ride-hailing: the role of dispatch platforms and market structure (2022). SSRN 3716491
34. Lian, Z., Van Ryzin, G.: Optimal growth in two-sided markets. Manage. Sci. **67**, 6862–6879 (2021)
35. Lu, A., Frazier, P.I., Kislev, O.: Surge pricing moves Uber's driver-partners. In: ACM EC, p. 3 (2018)
36. Lyft: pricing when it's busy (2017). https://web.archive.org/web/20170216214743/https://help.lyft.com/hc/en-us/articles/213818898-Prime-Time-for-Passengers. Accessed 16 Feb 2017
37. Lyft: acceptance rate - Lyft help (2021). https://help.lyft.com/hc/en-us/articles/115013077708-Acceptance-rate. Accessed 30 Jan 2021
38. Ma, H., Fang, F., Parkes, D.C.: Spatio-temporal pricing for ridesharing platforms. In: ACM EC, p. 583 (2019)
39. Marshall, A.: Uber changes its rules, and drivers adjust their strategies (2020). https://www.wired.com/story/uber-changes-rules-drivers-adjust-strategies/amp. Accessed 19 Feb 2020
40. Möhlmann, M., Zalmanson, L.: Hands on the wheel: navigating algorithmic management and Uber drivers'. In: 38th ICIS Proceedings (2017)
41. Ostrovsky, M., Schwarz, M.: Carpooling and the economics of self-driving cars. In: ACM EC, pp. 581–582 (2019)
42. Özkan, E., Ward, A.R.: Dynamic matching for real-time ride sharing. Stochast. Syst. **10**(1), 29–70 (2020)
43. Pang, J.Z., Fu, H., Lee, W.I., Wierman, A.: The efficiency of open access in platforms for networked Cournot markets. In: IEEE INFOCOM 2017, pp. 1–9 (2017)
44. Qin, Z., et al.: Ride-hailing order dispatching at DiDi via reinforcement learning. INFORMS J. Appl. Anal. **50**(5), 272–286 (2020)
45. Rayle, L., Shaheen, S., Chan, N., Dai, D., Cervero, R.: App-based, on-demand ride services: Comparing taxi and ride sourcing trips and user characteristics in San Francisco. University of California Transportation Center, Technical report (2014)
46. Rheingans-Yoo, D., Kominers, S.D., Ma, H., Parkes, D.C.: Ridesharing with driver location preferences. In: Proceedings of the 28th International Joint Conference on Artificial Intelligence (2019)
47. Sweeney, S.: Uber, Lyft drivers manipulate fares at Reagan national causing artificial price surges (2019). https://wjla.com/news/local/uber-and-lyft-drivers-fares-at-reagan-national. Accessed 31 Jan 2021

48. Uber: Uber community guidelines (2016). https://www.uber.com/blog/chicago/uberaccess/. Accessed 25 Oct 2020
49. Uber: Auto-offline has changed for the better (2017). https://www.uber.com/en-GB/blog/auto-offline-has-changed-for-the-better/. Accessed 31 Jan 2021
50. UberPeople.net users: Stay logged off at lax, until it surges (2016). https://www.uberpeople.net/threads/stay-logged-off-at-lax-until-it-surges.102047/. Accessed 10 Aug 2021
51. Xu, Z., AMC Vignon, D., Yin, Y., Ye, J.: An empirical study of the labor supply of ride-sourcing drivers. Transp. Lett. **14**, 352–355 (2020)
52. Yan, C., Zhu, H., Korolko, N., Woodard, D.: Dynamic pricing and matching in ride-hailing platforms. Naval Res. Logistics (NRL) **67**(8), 705–724 (2020)
53. Yu, C., Ma, H., Wierman, A.: Price cycles in ridesharing platforms. arXiv preprint arXiv:2202.07086 (2022)

Improved Truthful Rank Approximation for Rank-Maximal Matchings

Jinshan Zhang[1], Zhengyang Liu[2], Xiaotie Deng[3], and Jianwei Yin[1]([✉])

[1] School of Software Technology, Zhejiang University, Ningbo, China
zhangjinshan@zju.edu.cn, zjuyjw@cs.zju.edu.cn
[2] Beijing Institute of Technology, Beijing, China
zhengyang@bit.edu.cn
[3] Peking University, Beijing, China
xiaotie@pku.edu.cn

Abstract. In this work, we study truthful mechanisms for the rank-maximal matching problem, in the view of approximation. Our result reduces the gap from both the upper and lower bound sides. We propose a lexicographically truthful (LT) and nearly Pareto optimal (PO) randomized mechanism with an approximation ratio $\frac{2\sqrt{e}-1}{2\sqrt{e}-2} \approx 1.77$. The previous best result is 2. The crucial and novel ingredients of our algorithm are preservation lemmas, which allow us to utilize techniques from online algorithms to analyze the new ratio. We also provide several hardness results in variant settings to complement our upper bound. In particular, we improve the lower bound of the approximation ratio for our LT and PO mechanism to $18/13 \approx 1.38$. To the best of our knowledge, it is the first time to obtain a lower bound by utilizing the linear programming approach along this research line.

1 Introduction

Graduate college admissions in China proceed sequentially in tiers (Kamada and Kojima 2015). Chinese colleges are categorized into different tiers in decreasing prestige: Key colleges (e.g., National "985" project universities) belong to the first tier and admit students first; ordinary colleges belong to the second tier; and vocational training colleges are included in the third tier. There exists a huge gap between different tiers. Graduates propose their preferences according to these categories. Once assignments in the first tier are finalized, admissions in the second tier start, and so on. Hence, the desirable outcome of this procedure is a rank-maximal matching[1] (Ghosal et al. 2019; Huang and Kavitha 2012; Irving et al. 2006).

---

[1] A rank-maximal matching is a matching where the maximum possible number of applicants are matched to their first choice tie, and subject to that condition, the maximum possible number are matched to their second choice tie, and so on.

J. Garg et al. (Eds.): WINE 2023, LNCS 14413, pp. 637–653, 2024.
https://doi.org/10.1007/978-3-031-48974-7_36

Motivated by the above scenario, we consider the problem of allocating a set of objects to agents, where each agent has a preference order over a subset of objects with ties. In this case, each agent can be allocated at most one object, and each object can be allocated to at most one agent[2]. We consider the rank-maximal matching as an ideal outcome of our problem. Our objective is to approximate such a matching efficiently.

An optimal criterion to quantify the efficiency of the allocations, which is frequently used in the literature, is Pareto optimality (Abdulkadiroğlu and Sönmez 1998; Bogomolnaia and Moulin 2001; Hylland and Zeckhauser 1979), in the sense that no one can improve her outcome without affecting others. Precisely, a matching is *Pareto optimal* if there does not exist another matching such that in the new matching no one gets worse (obtains a less preferred object) and at least one agent gets strictly better (obtains a strictly preferred object). We also adopt such a concept as the criteria to measure the outcome of our mechanism. Note that it is simple to check whether the rank-maximal matching is Pareto optimal.

Based on the decomposition theory by Dulmage and Mendelsohn (1958), Irving et al. (2006) present a deterministic $O(m \min(n+d, d\sqrt{n}))$ time algorithm to compute the rank-maximal matching, where n and m are the numbers of agents and objects respectively, and d is the number of ties used in a rank-maximal matching. However, the agents may not reveal the preference order truthfully to the market maker under their algorithm. An important issue of mechanism design is to aggregate agents' private information and incentivize them to report their orders truthfully. Thus one may sacrifice either rank-maximality or Pareto optimality of the outcome. In this paper, we consider a trade-off between them by lowering the rank-maximality of the outcome while almost retaining the Pareto optimality. We use the rank approximation ratio (Chakrabarty and Swamy 2014) (related to each rank) to quantify the outcome compared to the rank-maximal matching.

We focus on randomized truthful mechanisms. In fact, as shown in Chakrabarty and Swamy (2014), no deterministic truthful mechanism can have a rank approximation ratio better than $\Omega\left(\frac{\log \log n}{\log \log \log n}\right)$. In ordinal settings when there is no money transfer, variants of truthfulness are proposed in the literature such as universal truthfulness (UT) (Krysta et al. 2019), stochastic dominance truthfulness (SDT) (Bogomolnaia and Moulin 2001), lexicographical truthfulness (LT) (Chakrabarty and Swamy 2014) and weak truthfulness (WT) (Bogomolnaia and Moulin 2001). For detailed definitions of these concepts, we refer to Chakrabarty and Swamy (2014). We know that UT $\subsetneq$ SDT $\subsetneq$ LT $\subsetneq$ WT, where UT $\subsetneq$ SDT means that if a mechanism is UT, then it must be SDT but not vice versa. There is a nearly LT (without PO property) mechanism in Chakrabarty and Swamy (2014) that achieves a rank approximation ratio 2. We follow this

---

[2] Our algorithm and analysis can be easily extended to the setting where each object j can be allocated to at most $b(j)$ applicants, e.g., graduate college admission. We can make $b(j)$ copies of object j, add them to each agent's preferences and each copy can be allocated to at most one agent.

line of research and present a fully LT mechanism which is nearly Pareto optimal (PO) in this paper. The rank approximation ratio of our mechanism is 1.77. We leave the rank approximation for UT and SDT mechanism open.

The prominent mechanisms with the above truthfulness fail to achieve a better rank approximation ratio concerning the rank-maximal matching e.g., SDTM mechanism (Krysta et al. 2019) (which is UT), Serial Dictatorship Mechanism (Abdulkadiroğlu and Sönmez 1998) (which is UT and SDT), and Probability Serial Mechanism (Bogomolnaia and Moulin 2001) (which is LT and WT) all have the rank approximation ratio at least $\Omega(\sqrt{n})$ (Chakrabarty and Swamy 2014). Very recently Zhang (2022) shows a tight bound $\Theta(\sqrt{n})$ for Probability Serial Mechanism. To overcome these difficulties, we design a new mechanism in this paper that is both lexicographically truthful and almost Pareto optimal. Meanwhile, the proposed mechanism achieves a better rank approximation ratio.

Our Results. We present an LT mechanism that is Pareto optimal with high probability and achieves a rank approximation ratio at most $\frac{2\sqrt{e}-1}{2\sqrt{e}-2} \approx 1.77$. For the lower bounds, we show that no lexicographically truthful and Pareto optimal mechanism can achieve a rank approximation ratio better than $\frac{18}{13} \approx 1.38$ and no lexicographically truthful mechanism can have the ratio better than $\frac{4}{3} \approx 1.33$. In addition, we show that our mechanism can do no better than $\frac{e}{e-1} \approx 1.58$.

Technical Overview. Here we briefly show the high-level outline of our algorithm and informal descriptions of the ideas behind the proofs as well. The first step of our algorithm is to transfer the problem of approximating the rank matching to that of maximum match by the decomposition theorem by Dulmage and Mendelsohn (1958). Next we utilize the algorithm **TMHA** (Algorithm 2) to tackle the problem. To achieve a better ratio and the property of LT, we also leverage the randomness with two additional routines (Algorithms 3 and 4). To attain the improved approximation ratio, we present a new structural lemma (Lemma 4) to demonstrate how the matching in the optimal rank maximal matching evolves with respect to different agent orders. This lemma enables us to use the charging map (a widely used method in the online matching, e.g., Krysta and Zhang (2016)) to build injective maps from 'bad events' (pairs of the agent order and the unmatched agent) to 'good events' (pairs of the agent order and the matched agent). By charging differently between 'bad events' and 'good events', we establish the ratio of $\frac{2\sqrt{e}-1}{2\sqrt{e}-2} \approx 1.77$.

Another technical contribution lies at linear programming approach to establish an improved lower bound $\frac{18}{13}$ of LT and Pareto optimal mechanisms, which significantly simplify the previous method based on Yao's principle (Krysta et al. 2019). To the best of our knowledge, this is the first time to use linear programming to build such a bound, which could be of independent interest.

Related Work. When not all objects are acceptable to all agents, the Probabilistic Serial mechanism (a LT mechanism (Chakrabarty and Swamy 2014)) gives an approximation of $1 - 1/e$ for max size (Bogomolnaia and Moulin 2015). However, as aforementioned, its rank approximation ratio is at least $\Omega(\sqrt{n})$ (Chakrabarty and Swamy 2014). Schulman and Vazirani (2015) extend PS to the setting where objects have copies and agents can be allocated to multiple copies of different objects and present an LT, Pareto efficient, envy-free, and time efficient mechanism. Since their mechanism is reduced to PS when multiple copies and multiple allocations are not allowed, the rank approximation ratio of their mechanism is also at least $\Omega(\sqrt{n})$.

As aforementioned, the algorithm by Irving (2003) is based on the decomposition theorem due to Dulmage and Mendelsohn (1958) which is also the basis for our algorithm. Following the same result, Aziz and Sun (2021) propose a new combinatorial algorithm to compute the school choice function and add new insights into the combinatorial structure of constrained rank-maximal matchings. Very recently, Aigner-Horev and Segal-Halevi (2022) also leverage this theorem to provide a polynomial-time algorithm for finding an envy-free matching of maximum cardinality.

Variants of rank-maximal matching problems have attracted increasing attention of researchers from computer science or computational social choice (Aziz et al. 2018; Erdem et al. 2020; Galichon et al. 2021; Ghosal et al. 2019; Huang and Kavitha 2012; Kavitha and Shah 2006; Nimbhorkar and Rameshwar 2019; Paluch 2013; Peters 2022). In Ghosal and Paluch (2018), the interesting manipulation strategies for the rank-maximal matching problem are studied. Aziz et al. (2019) employ rank-maximality as the efficiency notions and give a constrained round robin algorithm to achieve the required welfare level. Belahcene et al. (2021) explore fairness and optimality issues related to the rank-maximality and notions of popularity optimality. Hosseini et al. (2021) design efficient algorithms to check if a given matching is necessarily Pareto optimal or necessarily rank-maximal, and to check whether such a matching exists given top-k partial preferences. They also study online algorithms for eliciting partial preferences adaptively, and show bounds on their competitive ratio.

2 Preliminaries

2.1 The Model

We have a set $N = [n] := \{1, 2, \ldots, n\}$ of agents and a set $A = \{a_1, a_2, \ldots, a_m\}$ of objects. Each agent $i \in N$ has a preference order P_i over a subset of objects. We consider the model as a matching problem, in the sense that each agent can be allocated to at most one object and each object can be allocated to at most one agent. Agents' preference orders may have ties, that is, a subset of objects are not distinguishable by some agent. Given $i \in N, j \leq m$, we denote by C_j^i the jth indifference class (tie) in the ith preference order. We denote by r_i the number of non-empty indifference class (tie) for ith preference list, where $0 < r_i \leq m$. Note that C_j^i could be $\emptyset$ when $j > r_i$. Let r denote the highest index of non-empty indifference class of all the agents i.e., $r = \max\{j \mid C_j^i \neq \emptyset\}$.

The object $a \in C_j^i$ is ranked j in agent i's preference order or called a jth ranked object of agent i, denoted by $\text{rank}(i, a) = j$. We write $a \succeq_i b$ to denote that agent i prefers object a to object b. Similarly, $a \succ_i b$ and $a \simeq_i b$ mean that agent i strictly prefers object a to object b and is indifferent between object a and object b, respectively. For convenience, we write $P_i = \left(C_1^i \succ_i C_2^i \succ_i \cdots \succ_i C_m^i\right)$ as $P_i = \left(C_1^i, C_2^i, \ldots, C_m^i\right)$. With a bit abuse of notation, we also use P_i to denote the accepted set of agent i, i.e., $P_i = \bigcup_{j \in [m]} C_j^i$. Let $P = (P_1, P_2, \ldots, P_n)$ denote the joint preference profile of all the agents. We also denote by P_{-i} the joint preference profile of agents except agent i.

Let $G = (V, E)$ be the underlying bipartite graph of the above setting, where $V = N \cup A$ and $E = \{(i, a) \mid a \in P_i\}$. We use $G_j = (V, E_j)$ to denote the subgraph of G with objects ranking at most j involved i.e., $E_j = \{(i, a) \mid \mathrm{rank}(i, a) \leq j\}$. We use $G^j = \left(V, E^j\right)$ to denote the subgraph of G with objects ranking j involved, i.e., $E^j = \{(i, a) \mid \mathrm{rank}(i, a) = j\}$. We denote by $I = (N, A, P)$ an instance of our problem and $\mathcal{I}$ the set of all the instances. Let Π denote the set of all the permutations of agents. Note that for any permutation $\pi \in \Pi$ and agent $i \in [n], \pi(i)$ is the position of agent i in the agent order π.

2.2   Matchings

A (deterministic) *matching* μ in a graph G is standard i.e., a set of edges in G such that no two edges share a common agent or a common object. The *cardinality* of a matching is the number of edges it contains. We denote by $|\mu|$ the cardinality of the matching μ. A *maximum matching* is a matching with the maximum cardinality. A *maximal matching* is the one where no edge can be added to increase its cardinality without violating matching constraints.

Given a matching μ, a *signature* of μ is $(s_1(\mu), s_2(\mu), \ldots, s_m(\mu))$, where s_j denotes the number of agents who are matched to their jth ranked objects in μ. A matching μ has a higher signature than μ' if the signature of μ is lexicographically higher than that of μ', i.e., there exists $j \in [m]$ such that $s_j(\mu) > s_j(\mu')$ and $s_\ell(\mu) = s_\ell(\mu')$, for any $\ell \leq j - 1$. A *rank-maximal matching* is the one that has the lexicographically highest signature. Note that by definition, the signature of any rank-maximal matching is the same.

Given a matching μ, a path is called an *alternating path* w.r.t. μ if the edges along the path alternately in and not in (or not in and in) μ. An *augmenting path* $\mathcal{P}$ w.r.t. μ is an alternating path w.r.t. μ with two ending vertices not matched in μ. Augmenting μ along $\mathcal{P}$ means that we obtain a new matching $\mathcal{P} \setminus \mu$. We use $\mu(i)$ to denote the object matched to agent i in the matching μ and $\mu^{-1}(a)$ to denote the agent matched to object a in the matching μ.

A deterministic matching μ is *Pareto optimal* (PO) if there is no other matching μ' s.t. $\mu'(i) \succeq_i \mu(i)$, for any $i \in [n]$ and $\mu'(i') \succ_{i'} \mu(i')$, for some $i' \in [n]$. Let Π denotes the set of all the permutations of agents. Given an order $\pi \in \Pi$ of agents, we say a matching μ is strictly *lexicographically π-better* than μ' if there exists a $k \in [n]$ such that $\mu\left(\pi^{-1}(i)\right) \simeq_{\pi^{-1}(i)} \mu'\left(\pi^{-1}(i)\right), 1 \leq i \leq k - 1$ and $\mu(\pi^{-1}(k)) \succ_{\pi^{-1}(k)} \mu'(\pi^{-1}(k))$. A matching μ is called a lexicographically π-maximal matching if there is no matching μ' such that μ' is strictly lexicographically π-better than μ. Note that a lexicographically π-maximal matching is Pareto optimal. The symmetric difference of two matchings μ and μ' is defined as $\mu \oplus \mu' = (\mu \setminus \mu') \cup (\mu' \setminus \mu)$.

We call two matchings μ, μ' *equivalent* (denoted by $\mu \simeq \mu'$) if in these two matchings, the matched agents are the same and the matched objects for each agent are in the same indifference class of this agent, i.e., $\mu(i) \simeq_i \mu'(i)$, where we allow $\emptyset \simeq_i \emptyset$, for any $i \in [n]$. Let $\mathrm{CL}(\mu)$ denote the class including all the matchings equivalent to μ.

A *randomized* matching is a distribution over deterministic matchings. Similarly, a randomized matching is Pareto optimal if it is a distribution over deterministic Pareto optimal matchings.

Our goal in this paper is to propose new mechanisms that find optimal matchings. A deterministic mechanism ϕ is a map from any instance I to a matching $\phi(I) = \mu$. A randomized mechanism ϕ is a map from any instance I to a distribution over deterministic matchings. For any instance I, we use μ_{rm}^I to denote a rank-maximal matching of the underlying graph G of I. The rank approximation ratio $\alpha(\phi)$ of a mechanism ϕ is defined as

$$\alpha(\phi) = \max_{I \in \mathcal{I}, k \leq r} \frac{\sum_{j \leq k} s_j \left(\mu_{\mathrm{rm}}^I \right)}{\sum_{j \leq k} \mathbb{E} \left[s_j(\phi(I)) \right]},$$

where the expectation is taken over all the randomness of the mechanism. The competitive ratio $c(\phi)$ of a mechanism ϕ is defined by

$$c(\phi) = \max_{I \in \mathcal{I}} \frac{\left| \mu_{\mathrm{max}}^I \right|}{\mathbb{E}[|\phi(I)|]},$$

where μ_{max}^I denotes a maximum matching on the underlying graph G of I.

Given a mechanism ϕ and the bidding profile $P = (P_i, P_{-i})$, we use $\phi_{ia}(P)$ to denote the probability that agent i is allocated to object a under ϕ and P. Let $\phi_{ij}(P) = \sum_{a \in C_j^i} \phi_{ia}(P)$ denote the probability that the objects in the class C_j^i are allocated to agent i under ϕ and P. Let $P' = (P_i', P_{-i})$. A mechanism ϕ is called *lexicographically truthful* (LT) if for any agent i, preference profile P and P', one of the following holds:

- There exists $k \leq r$ such that $\phi_{ik}(P) \geq \sum_{a \in C_k^i} \phi_{ia}(P')$ and $\phi_{ij}(P) = \sum_{a \in C_j^i} \phi_{ia}(P')$, $j \leq k - 1$;
- $\phi_{ij}(P) = \sum_{a \in C_j^i} \phi_{ia}(P')$, for any $j \leq r$.

3   LT Mechanism **MRAND**

The high level description of our mechanism MRAND is as follows. We first use Dulmage and Mendelsohn decomposition to transform the approximation of rank-maximal matching into that of the maximum matching problem. In particular, we utilize Algorithm 1 given by Irving et al. (2006) to construct a graph G_j's such that each rank-maximal matching up to rank j in G_j corresponding

---

Algorithm 1: Algorithm for a Rank-Maximal Matching

Input: An instance (N, A, P)
Output: A rank-maximal matching μ_r
Initialize $G'_1 = G^1, \mu_1$ any maximum matching in G'_1
for $j = 1, 2, \ldots, r - 1$ **do**

> Partition the vertices in G'_j into three disjoint sets, V^e_j, V^o_j and V^u_j as defined before w.r.t. maximum matching μ_j i.e, V^e_j, V^o_j and V^u_j are the vertices that can be reached from a free vertex by an even, odd (or unreachable) alternating path w.r.t. μ_j.
> Delete all edges incident to a vertex in $V^o_j \cup V^u_j$ from E^ℓ, $\forall \ell > j$. V^o_j and V^u_j are the nodes that are matched by every maximum matching of G'_j. Delete all edges in G'_j connecting two vertices in V^o_j or a vertex in V^o_j with a vertex in V^u_j. These are the edges that are not used by any maximum matching of G'_j. Add the edges in E^{j+1} to G'_j. Call the resulting graph G'_{j+1}.
> Determine a maximum matching μ_{j+1} in G'_{j+1} by augmenting μ_j. (Note that μ_j is still contained in G'_{j+1})

---

to a maximum matching in G'_j (Proposition 1). Note that G'_j is the result of Dulmage and Mendelsohn decomposition, an intermediate graph with desired property carried over. Second, we run TMHA over G'_js to greedily generate the maximal matching in G'_j, see Algorithm 2. Thirdly, in order to obtain better approximation ratio w.r.t. to the maximum matching in G'_j, we adopt the randomization over the order of agents (a widely used technique in online matching), see Algorithm 3. Finally, to achieve lexicographically truthfulness of RAND, we run RAND and a carefully-designed mechanism FIX with proper probabilities, see Algorithm 4. The probability $f\left(\frac{\lambda^j}{|C^i_j|}\right)$ in Algorithm 5 is made up through careful calculations in order to make MRAND LT.

Before giving Algorithm 1, we need some preparations. Let μ be a maximum matching in a bipartite graph G'. Then, the nodes of vertices of G' can be partitioned into three disjoint sets: V^e, V^o, V^u. Vertices in V^e, V^o and V^u are called even, odd, and unreachable respectively. V^e (resp. V^o) are the vertices that can be reached in G' from a free (unmatched in μ) vertex by an even (resp. odd) alternating path (with respect to μ), and V^u are the vertices that can not be reached from a free vertex by any alternating path.

Proposition 1 (Irving *et al.* (2006)). *For every $j \leq r$, every rank-maximal matching in G_j is a maximum matching in G'_j.*

Let $f(y) = \sqrt{y}$ and $\lambda = \min\left\{\frac{1}{16}, m^{-3}\right\}$. Note that $\sum_{j \in [m]} f\left(\frac{\lambda^j}{|C^i_j|}\right) |C^i_j| = \sum_{j \in [r_i]} \sqrt{\lambda^j |C^i_j|} \leq \frac{\sum_{j \in [r_i]} |C^i_j|}{m\sqrt{m}} = \frac{1}{\sqrt{m}} \leq 1$, which means that FIX is well defined. We have the following theorem.

Theorem 1. *The Mechanism MRAND is lexicographically truthful.*

Algorithm 2: Truthful Mechanism for MHA (TMHA)

Input: Agents N; Objects A; Preference list profile P; Order π
Output: Matching μ
Let $G = (N \cup A, E)$, $E \leftarrow \emptyset$, $\mu \leftarrow \emptyset$.
for *each agent $i \in N$ in the order of π* **do**
 Let $\ell \leftarrow 1$
 Step (*): **if** $C_\ell^i \neq \emptyset$ **then**
  $E \leftarrow E \cup \{(i,a) : a \in C_\ell^i\}$; // all new edges are non-matching edges
  Run any algorithm on G and obtain a maximum cardinality matching μ'
  if $|\mu'| = |\mu| + 1$ **then**
   modify μ to μ'; // i must be provisionally allocated some $a \in C_\ell^i$
   and (i,a) is now a matching edge
  else
   $E \leftarrow E \setminus \{(i,a) : a \in C_\ell^i\}$; $\ell \leftarrow \ell + 1$; Go to Step (*)

Return μ; //each matched agent is allocated his matched object

---

Algorithm 3: Randomized Approximate Algorithm (RAND)

Input: An instance (N, A, P)
Output: A rank-maximal matching μ_r
Using Algorithm 1 to construct the graph G_r';
Select an order π of agents uniformly at random from Π;
Initialize $\mu_0 = \emptyset$ and $G_0' = \emptyset$
for $j = 1, 2, \ldots, r - 1$ **do**
 According to the order π, use mechanism TMHA to match agents in G_j'
 (obtained in Algorithm 1) based on μ_{j-1} until no agents can be matched.
 Call the resulting maximal matching (in G_j') μ_j.

The main idea behind the proof of Theorem 1 lies in that MRAND comprises of two mechanisms RAND and FIX. We utilize the truthfulness of TMHA in RAND and delicate decomposition property of function f in FIX to control the probabilities of truthfulness and lies.

Proof. For any agent i, any preference orders P_i, P_i' and P_{-i}. Let k be the first indifference class such that $C_k^i \neq C_k'^i$ and $C_j^i = C_j'^i$, $j \leq k - 1$, where $C_j'^i$ denotes jth indifference class of agent i in P'. Denote RAND by ϕ and FIX by ϕ' and MRAND by ϕ''. Note that $\phi'' = (1 - \epsilon)\phi + \epsilon\phi'$. Recall that $\phi_{ia}(P)$ denotes the probability of allocating object a to agent i under ϕ and $P = (P_i, P_{-i})$. Consider any order $\pi \in \Pi$ of agents and any object $a \in \bigcup_{j < k-1} C_j^i$, by process of RAND. Note that the two resulting graphs $G_j's$, $j \leq k - 1$ under P and P' are the same. By the process of TMHA, we know a is allocated to agent i under P and π if and only if an object $a' \simeq_i a$ is allocated to agent i under P' and π. This gives $\phi_{ij}(P) = \phi_{ij}(P')$, $j \leq k - 1$. Similarly by truthfulness of TMHA, we have $\phi_{ik}(P) \geq \sum_{a \in C_k^i} \phi_{ia}(P')$. Meanwhile, for ϕ', we always have $\phi_{ij}'(P) = \phi_{ij}'(P')$, for any $j \leq k$. To show ϕ'' is LT, it is sufficient to show

---

Algorithm 4: Modified RAND (MRAND)

Input: An instance (N, A, P)
Output: A randomized matching
With probability $1 - \epsilon$, run RAND;
With probability ϵ, run another fixed algorithm FIX.

Algorithm 5: FIX

Input: An instance (N, A, P)
Output: A randomized matching
Select an agent (e.g., agent i) uniformly at random from all the agents and

allocate her one of her jth ranked objects with the equal probability $f\left(\frac{\lambda^j}{|C_j^i|}\right)$ if

$C_j^i \neq \emptyset$.

---

that $\phi_{ik}''(P) > \sum_{a \in C_k^i} \phi_{ia}''(P')$. As we know that $\phi_{ik}(P) \geq \sum_{a \in C_k^i} \phi_{ia}(P')$, by defintion of LT, it is sufficient to show that $\phi_{ik}'(P) > \sum_{a \in C_k^i} \phi_{ia}'(P')$ We observe the following fact of the function f.

Claim. For any $y_1, y_2 \in [m]$, $f\left(\lambda^j(y_1 + y_2)\right) > f\left(\lambda^j y_1\right) + f\left(\lambda^{j+1} y_2\right)$.

Proof. It is sufficient to show that $\sqrt{y_1 + y_2} > \sqrt{y_1} + \sqrt{\lambda y_2}$. This is equivalent to $1 > \lambda + 2\sqrt{\lambda \frac{y_1}{y_2}}$. Since $\lambda = \min\left\{\frac{1}{16}, \frac{1}{m^3}\right\}$ and $\frac{y_1}{y_2} \leq m$, $\lambda + 2\sqrt{\lambda \frac{y_1}{y_2}} \leq \frac{1}{16} + \min\left\{\frac{2}{m}, \frac{\sqrt{m}}{2}\right\} < 1$.

Since $C_k^i \neq C_k^{\prime i}$, we suppose, the objects in C_k^i are distributed in s different indifferent classes in P_i'. Suppose these indifferent classes are $C_{k_1}^{\prime i}, C_{k_2}^{\prime i}, \ldots, C_{k_s}^{\prime i}$, where $k_\ell \geq k$, $\ell \in [s]$ and $k_\ell + 1 \leq k_{\ell+1}$, $\ell \in [s - 1]$. Let n_ℓ denote the number of objects of C_k^i in $C_{k_\ell}^{\prime i}$, $\ell \in [s]$. Then, $\left|C_k^i\right| = \sum_{\ell \in [s]} n_\ell$. By using Claim 3 recursively, we have

$$\sum_{a \in C_k^i} \phi_{ia}'(P') = \sum_{\ell \in [s]} \sum_{a \in C_k^i \cap C_{k_\ell}^{\prime i}} f\left(\frac{\lambda^{k_\ell}}{|C_{k_\ell}^{\prime i}|}\right)$$

$$= \sum_{\ell \in [s]} n_\ell f\left(\frac{\lambda^{k_\ell}}{|C_{k_\ell}^{\prime i}|}\right) \leq \sum_{\ell \in [s]} n_\ell f\left(\frac{\lambda^{k_\ell}}{n_\ell}\right) = \sum_{\ell \in [s]} f\left(\lambda^{k_\ell} n_\ell\right)$$

$$= \left[f\left(\lambda^{k_s} n_s\right) + f\left(\lambda^{k_{s-1}} n_{s-1}\right)\right] + \sum_{\ell \in [s-2]} f\left(\lambda^{k_\ell} n_\ell\right)$$

$$< f\left(\lambda^{k_{s-1}}(n_s + n_{s-1})\right) + \sum_{\ell \in [s-2]} f\left(\lambda^{k_\ell} n_\ell\right) \quad \text{By Claim 3}$$

$$< \cdots < f\left(\lambda^{k_1}\left(\sum_{\ell \in [s]} n_\ell\right)\right) \leq f\left(\lambda^k \left|C_k^i\right|\right) = \phi_{ik}'(P)$$

This shows that MRAND is lexicographically truthful.

4   The Rank Approximation Ratio of **MRAND**

In this section, we are going to analyse the rank approximation ratio of the Mechanism MRAND. Since MRAND runs RAND with probability at least $1 - \epsilon$, the expected matching size of MRAND is at least $1 - \epsilon$ the expected size of the matching generated by RAND. Therefore, an α rank approximation ratio of RAND gives a $(1 + \epsilon)\alpha$ rank approximation ratio of MRAND. This requires us to analyse the rank approximation ratio of RAND. By Proposition 1, we have the following relation between the rank approximation ratio and the competitive ratio.

Lemma 1. *The rank approximation ratio of* RAND *is at most its competitive ratio.*

Proof. From the definition of the rank approximation ratio and Proposition 1, we can see that, for any $j \leq r$, $\sum_{k \leq j} s_k \left(\mu^I \right)$ is the size of the rank-maximal matching in the graph G_j, which is a maximum matching in G'_j. Let $\phi(G_j)$ denote the matching obtained by RAND over G_j under some agent order π. By the process of RAND, $\phi(G_j) = \phi(G'_j)$. Thus, $\sum_{k < j} \mathbb{E}\left[s_k(\phi(I)) \right] = \sum_{k \leq j} \mathbb{E}\left[s_k\left(\phi(G_j) \right) \right] = \sum_{k \leq j} \mathbb{E}\left[s_k\left(\phi(G'_j) \right) \right] = \mathbb{E}\left[|\phi(G'_j)| \right]$. Denote by μ_j a maximum matching in G'_j. Then the rank approximation ratio now becomes

$$\max_{j \leq r} \frac{|\mu_j|}{\mathbb{E}\left[|\phi\left(G'_j \right)| \right]}$$

where the expectation is taken over all the random permutations of the agents. Recall that for any graph G, the competitive ratio of $c(\phi)$ is defined by $c(\phi) = \max_G \frac{|\mu(G)|}{\mathbb{E}[|\phi(G)|]}$, where $\mu(G)$ denotes a maximum matching on G. Therefore, the rank approximation ratio

$$\max_{j \leq r} \frac{|\mu_j|}{\mathbb{E}\left[|\phi\left(G'_j \right)| \right]} = \max_{j \leq r} \frac{|\mu\left(G'_j \right)|}{\mathbb{E}\left[|\phi\left(G'_j \right)| \right]}$$

$$\leq \max_{G} \frac{|\mu(G)|}{\mathbb{E}[|\phi(G)|]} = c(\phi).$$

Although Lemma 1 provides a way to analyze the rank approximation ratio by calculating the competitive ratio, in the worst case, the competitive ratio is 2 in the general graph as shown in the following example.

Example 1. Consider an instance with agents $N = \{1, \ldots, n\}$ and objects $A = \{a_1, \ldots, a_n\}$, where n is an even number. The preference order of agent i is $(a_1, a_2 \ldots, a_i)$, for $i \in \left[\frac{n}{2} \right]$, and for agent $\frac{n}{2} + 1 \leq i \leq n$, the preference order for agent i is $\left(a_{i - \frac{n}{2}}, a_2, a_3, \ldots, a_i \right)$. It is easy to see that the size of the unique maximum matching (by allocating object a_i to agent i, $i \in [n]$) in this instance is n. The size of the unique rank-maximal matching (by allocating object a_1 to agent 1, a_i to agent $i + \frac{n}{2}, 2 \leq i \leq \frac{n}{2}$, and object $a_{\frac{n}{2}+1}$ to agent $\frac{n}{2} + 1$) is $\frac{n}{2} + 1$. The expected size of the matching obtained by $RAND$ is $\frac{n+1}{2}$, which shows that the competitive ratio is at least $\frac{2n}{n+1} = 2 - o(1)$. Note that the rank approximation ratio in this instance is $\frac{n+2}{n+1} = 1 + o(1)$, which is optimal.

Next, we show that the bad example in Example 1 can be avoided. In fact, we can assume that all the agents are matched in a rank-maximal (greedy) matching, which means that the rank-maximal matching is also a maximum matching. We need the following properties (Proposition 2 and Lemma 2). Since RAND can generate a matching for any given order $\pi \in \Pi$, with a bit abuse, we also refer RAND to a deterministic algorithm for any given order $\pi \in \Pi$. For any $\pi \in \Pi$, let μ^π denote the matching obtained by RAND. For any $\pi \in \Pi$, let π_{-i} denote the order (permutation) by removing agent i from the order while keeping the order of other agents unchanged. We use π_i^k to denote the order (permutation) of agents by inserting agent i to the kth position of the order π_{-i}, e.g., $\pi_i^k(i) = k$. We use ν^π to denote the matching obtained by mechanism TMHA under the agent order π.

Proposition 2. *Given any order $\pi \in \Pi$ of agents, the matching generated by* TMHA *is lexicographically π-maximal and thus Pareto optimal, which implies that* RAND *has the same property.*

By Proposition 2, the matching generated by RAND is Pareto optimal. Since MRAND runs RAND with probability with $1 - \epsilon$, the matching generated by MRAND is Pareto optimal with probability $1 - \epsilon$ (nearly PO).

Lemma 2. *There is an injective map f from $\mathrm{CL}(\mu^\pi)$ to $\mathrm{CL}(\mu^{\pi-i})$ such that for any $\mu \in \mathrm{CL}(\mu^\pi)$, the symmetric difference of $\mu \oplus f(\mu)$ is an alternating path starting from agent i assuming that $\mu \oplus f(\mu) \neq \emptyset$.*

Proof. For any order $\pi \in \Pi$ of agents, by the process of RAND, there exists an order $\sigma(\pi)$ of agents, such that the agents are matched to their best objects among the remaining objects sequentially (i.e., agents are matched by TMHA under the order $\sigma(\pi)$). That is $\mu^\pi = \nu^{\sigma(\pi)}$. Note that $\sigma(\pi_{-i}) = \sigma(\pi)_{-i}$. Thus, $\mathrm{CL}(\mu^\pi) = \mathrm{CL}(\nu^{\sigma(\pi)})$ and $\mathrm{CL}(\mu^{\pi-i}) = \mathrm{CL}(\nu^{\sigma(\pi)-i})$. By Lemma 11 in Krysta and Zhang (2016), the lemma holds for $\mathrm{CL}(\nu^{\sigma(\pi)})$ and $\mathrm{CL}(\nu^{\sigma(\pi)-i})$, which completes the proof.

Remark 1. In addition, the indices of the agent order along the alternative path is increasing according to $\sigma(\pi)$ (see Lemma 11 in Krysta and Zhang (2016)), which will be used in the proof of Lemma 4. We emphasize that the map from π to $\sigma(\pi)$ in the above proof is not injective. Hence, the ratio of $\frac{e}{e-1}$ of randomization of TMHA (Krysta and Zhang 2016) can not be applied directly to that of RAND. The following example shows this fact that the matching size generated by RAND is different from that generated by random TMHA. For instance, suppose there are three agents 1,2,3 with preference list $(a \succ_1 b), (a \succ_2 b)$ and $(b \succ_3 c)$. The resulting matching size by running RAND is 2 (since b is matched to 3 and a is either matched to 1 or 2). While the resulting matching by running random THMA is $\frac{7}{3}$ (since in current example, random THMA is the Random Serial Dictatorship Mechanism). We need new insight into the structure of RAND, thus two preserving lemmas are ready to be proposed.

By the process of RAND, it can run for any order π of agents. Thus, for any given order $\pi \in \Pi$, we also say (deterministic) RAND runs over π. By Lemma 2, we can suppose w.l.o.g. that there exists a rank-maximal matching in the graph G such that all the agents are matched (otherwise we can find a rank-maximal matching in graph G and remove the agents who are not matched in this rank-maximal matching and their linking edges. After doing this, the expected matching size of RAND can not be increased and the graph has a matching such that all the agents are matched). Let μ^* denote this matching in G. Let $R_i^* = \text{rank}\,(i, \mu^*(i))$ denote the rank of $\mu^*(i)$ in agent i's preference list, $i \in [n]$. Given an order of agents π, recall that μ^π denotes the matching obtained by RAND under agent order π and ν^π the matching generated by TMHA under the agent order π. We present two preservation lemmas. The first preservation lemma (Lemma 3) roughly states that if an agent (e.g. agent i) is not matched in the tth position of π in μ^π, then her matched object $\mu^*(i)$ in the rank-maximal matching must be matched (preserved) to some agent i_1 in μ^π; furthermore if the position of i_1 in π is more than t, then the matched object $\mu^*(i_1)$ of i_1 in the rank-maximal matching must also be matched (preserved) to some other agent i_2 in μ^π. These new structural lemmas are crucial for establishing the approximation ratio in Theorem 2.

Lemma 3 (Preservation Lemma). *If agent i is not matched in μ^π and $\pi(i) = t(\,for\ some\ t \in [n])$, then the object $\mu^*(i)$ must be matched to some agent i_1 in μ^π and*

- *if $\text{rank}\,(i_1, \mu^*(i)) = R_i^*$, then $\pi\,(i_1) \leq t - 1$;*
- *if $\text{rank}\,(i_1, \mu^*(i)) < R_i^*$, then $\mu^*\,(i_1)$ is also matched to some agent i_2 in μ^π and $R_{i_1}^* = \text{rank}\,(i_1, \mu^*\,(i_1)) \leq \text{rank}\,(i_1, \mu^*(i))$.*

Proof. Let $a = \mu^*(i)$. Since agent i is not matched in μ^π, then a must be matched to some agent i_1 in μ^π by the process of RAND (Otherwise agent i will be matched to object a in RAND).

Case (i). If $\text{rank}\,(i_1, a) = R_i^*$, then object a is matched in the R_i^*th round of RAND. In this round, when we run TMHA mechanism, since i_1 is matched to a and agent i is not in this round, we know that the order of agent i_1 is prior to agent i in π. As $\pi(i) = t$, then $\pi\,(i_1) \leq t - 1$.

Case (ii). Suppose $\text{rank}\,(i_1, a) < R_i^*$. We must have $R_{i_1}^* = \text{rank}\,(i_1, \mu^*\,(i_1)) \leq \text{rank}\,(i_1, a)$. Otherwise suppose $R_{i_1}^* > \text{rank}\,(i_1, a)$, and let $\mu = \mu^* \cup \{(i_1, a)\} \setminus \{(i_1, \mu^*\,(i_1)), (i, a)\}$. Then μ has a higher signature than μ^*, a contradiction to that μ^* is a rank-maximal matching.

(a). Now if $R_{i_1}^* = \text{rank}\,(i_1, \mu^*\,(i_1)) = \text{rank}\,(i_1, a)$, note that $\mu^*\,(i_1) \neq a$, and suppose $\mu^*\,(i_1)$ is not matched in μ^π. By process of RAND, we know that there exists an order $\sigma(\pi)$ of agents such that $\mu^\pi = \nu^{\sigma(\pi)}$. The matching $\nu^{\sigma(\pi)} \cup \{(i_1, \mu^*(i_1)), (i,a))\} \setminus \{(i_1, a)\}$ is strictly lexicographically $\sigma(\pi)$-better than $\nu^{\sigma(\pi)}$ (under the order $\sigma(\pi)$ of agents), a contradiction that $\nu^{\sigma(\pi)}$ is a lexicographically $\sigma(\pi)$-maximal matching (Proposition 2).

(b). If $R^*_{i_1} < \operatorname{rank}(i_1, a)$, and $\mu^*(i_1)$ is unmatched, then the matching $\nu^{\sigma(\pi)} \cup \{(i_1, \mu^*(i_1))\} \setminus \{(i_1, a)\}$ is strictly lexicographically $\sigma(\pi)$-better than $\nu^{\sigma(\pi)} = \mu^\pi$ (under the order $\sigma(\pi)$ of agents), a contradiction to that $\nu^{\sigma(\pi)}$ is a $\sigma(\pi)$-maximal matching. Hence, $\mu^*(i_1)$ must also be matched to some agent i_2 in μ^π.

Based on this preservation lemma, we obtain the Strong Preservation Lemma where the preservation property holds for both μ^π and $\mu^{\pi^k_i}$, given any $k \in [n]$. The proof frequently uses Proposition 2 and a slightly stronger version of Lemma 2.

Lemma 4 (Strong Preservation Lemma).

If agent i is not matched in μ^π and $\pi(i) = t$ (for some $t \in [n]$), then for any $k \in [n]$, the object $\mu^*(i)$ must be matched to an agent i_1 in $\mu^{\pi^k_i}$ and if $\pi^k_i(i_1) \geq t + 1$, then $\operatorname{rank}(i_1, \mu^*(i)) < R^*_i$ and $R^*_{i_1} = \operatorname{rank}(i_1, \mu^*(i_1)) \leq \operatorname{rank}(i_1, \mu^*(i))$ and $\mu^*(i_1)$ is also matched to some agent i_2 in $\mu^{\pi^k_i}$.

By Lemma 4, we will use charging map arguments in Aggarwal et al. (2011) to show that the competitive ratio of RAND is $\frac{2\sqrt{e}-1}{2\sqrt{e}-2} \approx 1.77$. This strong preservation property allows us to build the relation from 'bad' events to 'good' events such that the probability of bad events is upper bounded by that of the good events. We thus have the following theorem.

Theorem 2. *The expected matching size of RAND for agents with ties is at least*

$$T_n = \frac{2n}{2 - \beta(n)} \left(\frac{1}{2n+1} + 1 - \beta(n) \right) > \frac{2\sqrt{e}-2}{2\sqrt{e}-1} \cdot n$$

where $\beta(n) = \left(1 - \frac{1}{2n+1}\right)^{n-1}$.

Theorem 3. *The rank approximation ratio of the Mechanism MRAND is at most $\frac{2\sqrt{e}-1}{2\sqrt{e}-2} \approx 1.77$.*

Proof. For any $n > 0$, we can select $2\epsilon = 1 - \frac{\frac{2\sqrt{e}-2}{2\sqrt{e}-1}n}{T_n} > 0$ by Theorem 2. The reverse of competitive ratio of RAND is at least $\frac{T_n}{n}$. By Lemma 1 and previous arguments that an α rank approximation ratio of RAND gives an $(1+\epsilon)\alpha$ rank approximation ratio of MRAND, the reverse of rank approximation ratio of MRAND is at least $\frac{T_n}{n(1+\epsilon)} > \frac{T_n}{n}(1 - 2\epsilon) = \frac{2\sqrt{e}-2}{2\sqrt{e}-1}$. Thus, the rank approximation ratio of MRAND is at most $\frac{2\sqrt{e}-1}{2\sqrt{e}-2} \approx 1.77$

5   Lower Bounds

A lower bound of RAND is given as follows. This implies the same lower bound for MRAND within a factor of $1 - \epsilon$.

Theorem 4. *The rank approximation ratio of* RAND *is at least* $\frac{e}{e-1} \approx 1.58$.

Proof. The triangle instance (e.g., the agent i 's preference order is $(a_1, a_2, \ldots, a_i)$, $i \in [n]$) has a rank-maximal matching with the cardinality n (i.e., allocate a_i to agent i, $i \in [n]$). RAND runs exactly the same as Random Serial Dictatorship on this triangle instance. Hence, by the result in Krysta et al. (2019), the matching size generated by RAND is $\frac{en}{e-1} - o(n)$, which gives the rank approximation ratio of RAND is at least $\frac{e}{e-1} - o(1)$.

Theorem 5. *No lexicographically truthful mechanism can achieve a competitive ratio less than* $\frac{4}{3}$, *in particular, no one can achieve a rank approximation ratio less than* $\frac{4}{3}$, *either.*

Proof. Consider the instance with two agents: the preference order of agent 1 is (a, b) and that of agent 2 is also (a, b). Let (x_1, x_2) and (y_1, y_2) be the assigned probability vector for agent 1 and agent 2 from a lexicographically truthful mechanism respectively, where x, y denote the allocation probability from item a and b respectively. We have $x_1 + x_2 \leq 1$, $x_1 + y_1 \leq 1$, and $x_2 + y_2 \leq 1$, $y_1 + y_2 \leq 1$. W.l.o.g., suppose $x_1 \leq x_2$, which gives $x_1 \leq \frac{1}{2}$. Now suppose agent 1's true preference list is (a), then total assigned probability to agent 1 is at most $x_1 \leq \frac{1}{2}$ (by lexicographical truthfulness) and the total assigned probability to agent 2 is at most 1 . Thus, the expected matching size is at most $\frac{3}{2}$, while the rank-maximal matching size is 2 (by assigning item a to agent 1 and b to agent 2 . Hence, the competitive ratio is at least $\frac{3/2}{2} = \frac{4}{3}$ and the rank approximation ratio is also at least $\frac{4}{3}$.

In Krysta et al. (2019), they proved that there is no universally truthful and Pareto optimal mechanism that can achieve a competitive ratio less than $\frac{18}{13}$. In the following theorem, we strengthen their results by proving that there is no lexicographically truthful and Pareto optimal mechanism that can achieve a competitive ratio less than $\frac{18}{13}$ (Recall that UT $\subsetneq$ LT). The proof of Theorem 6 was based on the linear programming approach, providing new methods and insights for showing lower bounds beyond Yao's principle (Yao 1977). We believe our proof technique for Theorem 6 can be applied to show a better lower bound for LT and Pareto optimal mechanism, which will be emphasized later. The proof uses a similar idea as done in Theorem 5, however, in a more general way. The main idea is to transform the problem of searching a lower bound into linear programming by restricting the problem into a well-defined subproblem.

Theorem 6. *No lexicographically truthful, Pareto optimal mechanism can achieve a competitive ratio less than* $\frac{18}{13}$, *in particular, no one can achieve a rank approximation ratio less than* $\frac{18}{13}$, *either.*

Remark 2. In the proof of Theorem 6, we consider the case when $n = 3$, it is obvious that this methodology can be used for any n and the size of linear programming will become $O\left(n^{n+2}\right)$ (however, very sparse). It is interesting to know to what extension can this methodology can be generalized (in particular, it is worthy to know about case $n = 4$, which seems can be solved by computer programming). Another way to extend this approach is to theoretically analyze the objective value of such linear programming, although currently, we are not aware of how to do it.

6 Conclusion and Future Work

In this paper, we present a lexicographically truthful and nearly Pareto optimal mechanism that achieves the rank approximation ratio $\frac{2\sqrt{e}-1}{2\sqrt{e}-2} \approx 1.77$ with respect to the rank-maximal matching. We demonstrate the lower bound that no LT and Pareto optimal mechanism can achieve a rank approximation ratio better than $\frac{18}{13} \approx 1.38$. Several variants of lower bounds are also provided.

It is interesting to know whether our approach can be applied to design mechanisms for other related optimal criteria, e.g., maximum cardinality rank-maximal matching (Huang and Kavitha 2012; Irving 2003), generous/fair matching (Huang and Kavitha 2012; Irving 2003), weight-maximal matching (Huang and Kavitha 2012), capacitated rank-maximal matching (Paluch 2013). We are also interested in knowing whether fairness (Belahcene et al. 2021) can be compatible with LT to approximate the rank-maximal matching. Another interesting question as pointed out previously would be whether the lower bound can further be improved by linear programming methodology as done in Sect. 5.

Acknowledgments. This work was supported by the National Key Research and Development Program of China (2022YFF0902005). Jianwei Yin was supported by the National Science Fund for Distinguished Young Scholars (No. 61825205). Jinshan Zhang was supported by the Key Research and Development Jianbing Program of Zhejiang Province (2023C01002), Hangzhou Major Project and Development Program (2022AIZD0140) and Yongjiang Talent Introduction Programm (2022A-236-G). Xiaotie Deng was supported by the National Natural Science Foundation of China (No. 62172012). Zhengyang Liu was supported by the National Natural Science Foundation of China (No. 62002017) and Beijing Institute of Technology Research Fund Program for Young Scholars. We are also grateful for the valuable comments from the anonymous reviewers.

References

Abdulkadiroğlu, A., Sönmez, T.: Random serial dictatorship and the core from random endowments in house allocation problems. Econometrica **66**(3), 689–701 (1998)

Aggarwal, G., Goel, G., Karande, C., Mehta, A.: Online vertex-weighted bipartite matching and single-bid budgeted allocations. In: Proceedings of the 22nd ACM-SIAM Symposium on Discrete Algorithms (SODA), pp. 1253–1264 (2011)

Aigner-Horev, E., Segal-Halevi, E.: Envy-free matchings in bipartite graphs and their applications to fair division. Inf. Sci. **587**, 164–187 (2022)

Aziz, H., Sun, Z.: Multi-rank smart reserves. In: Proceedings of the 22nd ACM Conference on Economics and Computation (EC), pp. 105–124 (2021)

Aziz, H., Luo, P., Rizkallah, C.: Rank maximal equal contribution: a probabilistic social choice function. In: Proceedings of the 32nd AAAI Conference on Artificial Intelligence (AAAI), vol. 32 (2018)

Aziz, H., Huang, X., Mattei, N., Segal-Halevi, E.: The constrained round robin algorithm for fair and efficient allocation (2019). arXiv preprint arXiv:1908.00161

Belahcene, K., Mousseau, V., Wilczynski, A.: Combining fairness and optimality when selecting and allocating projects. In: Proceedings of the 30th International Joint Conference on Artificial Intelligence (IJCAI) (2021)

Bogomolnaia, A., Moulin, H.: A new solution to the random assignment problem. J. Econ. Theory **100**(2), 295–328 (2001)

Bogomolnaia, A., Moulin, H.: Size versus fairness in the assignment problem. Games Econ. Behav. (GEB) **90**, 119–127 (2015)

Chakrabarty, D., Swamy, C.: Welfare maximization and truthfulness in mechanism design with ordinal preferences. In: Proceedings of the 5th Conference on Innovations in Theoretical Computer Science (ITCS), pp. 105–120 (2014)

Dulmage, A.L., Mendelsohn, N.S.: Coverings of bipartite graphs. Can. J. Math. **10**, 517–534 (1958)

Erdem, E., Fidan, M., Manlove, D., Prosser, P.: A general framework for stable roommates problems using answer set programming. Theory Pract. Log. Program. **20**(6), 911–925 (2020)

Galichon, A., Ghelfi, O., Henry, M.: Stable and extremely unequal (2021). arXiv preprint arXiv:2108.06587

Ghosal, P., Paluch, K.E.: Manipulation strategies for the rank-maximal matching problem. In: Wang, L., Zhu, D. (eds.) Computing and Combinatorics. COCOON 2018. LNCS, vol. 10976, pp. 316–327. Springer, Cham (2018). https://doi.org/10.1007/978-3-319-94776-1_27

Ghosal, P., Nasre, M., Nimbhorkar, P.: Rank-maximal matchings-structure and algorithms. Theor. Comput. Sci. (TCS) **767**, 73–82 (2019)

Hosseini, H., Menon, V., Shah, N., Sikdar, S.: Necessarily optimal one-sided matchings. In: Proceedings of the 35th AAAI Conference on Artificial Intelligence (AAAI), vol. 35, pp. 5481–5488 (2021)

Huang, C.-C., Kavitha, T.: Weight-maximal matchings. In: Proceedings of the 2nd International Workshop on Matching Under Preferences (MATCH-UP), vol. 12, pp. 87–98 (2012)

Hylland, A., Zeckhauser, R.: The efficient allocation of individuals to positions. J. Polit. Econ. **87**(2), 293–314 (1979)

Irving, R.W., Kavitha, T., Mehlhorn, K., Michail, D., Paluch, K.E.: Rank-maximal matchings. ACM Trans. Algorithms (TALG) **2**(4), 602–610 (2006)

Irving, R.W.: Greedy matchings, 2003. Technical report Tr-2003-136, University of Glasgow

Kamada, Y., Kojima, F.: Efficient matching under distributional constraints: theory and applications. Am. Econ. Rev. **105**(1), 67–99 (2015)

Kavitha, T., Shah, C.D.: Efficient algorithms for weighted rank-maximal matchings and related problems. In: Asano, T. (ed.) ISAAC 2006. LNCS, vol. 4288, pp. 153–162. Springer, Heidelberg (2006). https://doi.org/10.1007/11940128_17

Krysta, P.J., Zhang, J.: House markets with matroid and knapsack constraints. In: Proceedings of the 43rd International Colloquium on Automata, Languages, and Programming (ICALP) (2016)

Krysta, P., Manlove, D., Rastegari, B., Zhang, J.: Size versus truthfulness in the house allocation problem. Algorithmica **81**(9), 3422–3463 (2019)

Nimbhorkar, P., Rameshwar, V.A.: Dynamic rank-maximal and popular matchings. J. Comb. Optim. **37**(2), 523–545 (2019)

Paluch, K.: Capacitated rank-maximal matchings. In: Proceedings of the 8th International Conference on Algorithms and Complexity (CIAC), pp. 324–335 (2013)

Peters, J.: Online elicitation of necessarily optimal matchings. In: Proceedings of the 36th AAAI Conference on Artificial Intelligence (AAAI), vol. 36, pp. 5164–5172 (2022)

Schulman, L.J., Vazirani, V.V.: Allocation of divisible goods under lexicographic preferences. In: Proceedings of the 35th IARCS Conference on Foundations of Software Technology and Theoretical Computer Science (FSTTCS) (2015)

Yao, A.C.-C.: Probabilistic computations: toward a unified measure of complexity. In: Proceedings of the 18th Symposium on Foundations of Computer Science (SFCS), pp. 222–227 (1977)

Zhang, J.: Tight social welfare approximation of probabilistic serial. Theor. Comput. Sci. **934**, 1–6 (2022)

Reallocation Mechanisms Under Distributional Constraints in the Full Preference Domain

Jinshan Zhang[1], Bo Tang[3], Xiaoye Miao[2]($\boxtimes$), and Jianwei Yin[1]

[1] School of Software Technology, Zhejiang University, Ningbo, China
zhangjinshan@zju.edu.cn, zjuyjw@cs.zju.edu.cn
[2] Center for Data Science, Zhejiang University, Hangzhou, China
miaoxy@zju.edu.cn
[3] Databricks Inc., San Francisco, CA, USA

Abstract. We study the problem of reallocating indivisible goods among a set of agents in one-sided matching market, where the feasible set for each good is subject to an associated distributional *matroid* or *M-convex* constraint. Agents' preferences are allowed to have ties and not all the agents have initial endowments. We present feasible, Pareto optimal, strategy-proof mechanisms for the problems with matroid or *M*-convex constraints. Strategy-proofness is proved based on new structural properties over first choice graphs, which should be of independent interest. These mechanisms strictly generalize the best-known mechanism for non-strict preferences [21] with all desired properties carried over.

Keywords: Mechanism Design · Matroid · Distributional Constraints · Top Trading Cycle · Strategy-proof · Full Preference Domain

1 Introduction

The division of indivisible goods to agents according to their preference via a truthful procedure is a fundamental problem of resource allocation in multi-agent systems at the frontier between Economics and Computer Science. The *House Exchange* problem is a prominent example pioneered by Shapley's work [23], and extended by Abdulkadiroğlu and Sönmez [2] to match individuals and institutions across various markets in practice, such as students and schools, doctors and hospitals, and workers and firms. Nevertheless, many real matching markets are subject to distributional constraints. These constraints frequently involve restrictions on the quantities of agents from one side of the market that can be matched with specific subsets on the other side. Prominent examples of this include "regional cap" policy in Japanese medical residency matching, which regulates the geographical distribution of medical residents across hospitals in the country. Traditional theory cannot be applied to such distributional settings because it has assumed away those constraints.

J. Garg et al. (Eds.): WINE 2023, LNCS 14413, pp. 654–671, 2024.
https://doi.org/10.1007/978-3-031-48974-7_37

To integrate these constraints, we consider the Good Exchange Model (GEM), a general one-sided matching market model that (re)allocates indivisible goods among a set of agents with initial endowments or not, under distributional constraints in the full preference domain. Each good in GEM has certain copies depending on constraints, e.g., the copies can be arbitrary for the matroid constraint. This general model unifies many existing models in one-sided matching markets such as doctoring hiring [13], dormitory exchange [2], campus housing [7], and school admission [26]. In particular, the distributional constraints we consider are *matroid* and *M-convex* constraints, which involve allocating goods among a set of agents in a one-sided matching market where each good's feasible set is subject to an associated *matroid* or *M-convex* constraint. Aforementioned regional cap constraints, for instance, are going to be demonstrated to be reducible to laminar matroid constraints, while this paper presents more practical examples with matroid constraints in subsequent paragraphs. These distributional constraints possess unique characteristics and create significant challenges in designing strategy-proof mechanisms.

Suzuki et al. [26] assume that distributional constraints are represented as an M-convex set and design a strategy-proof and Pareto-optimal mechanism for the school admission problem based on the Top Trading Cycles (TTC) algorithm. This work is a strict extension of Hamada et al. [9], who only consider individual minimum and maximum quotas. While M-convex sets are similar to matroids, they differ in that matroids cannot be expressed as an M-convex set due to their heredity. Reallocation problems subject to distributional M-convex constraints have many practical applications, as noted by Suzuki et al. [26]. However, their work assumes that all preferences are strict, which is overly restrictive in many practical settings. As discussed in [6], there are practical and technical reasons for studying the full preference domain, which is also recognized as the natural preference domain for many allocation problems in economic and AI settings. Therefore, a natural question arises of whether these results for strict preferences extend to the more general case of the full preference domain. In this paper, we answer this question positively by presenting a strategy-proof mechanism for GEM (including the school admission problem) with distributional M-convex constraints in the full preference domain.

1.1 Our Results

For GEM with matroid constraints, we design an individual rational, Pareto optimal, strategy-proof mechanism in full preference domain. In addition, our mechanism runs in polynomial time given the matroid oracle can be called in $O(1)$ time. Based on the mechanism for matroid constraints, we present a feasible, Pareto optimal, strategy-proof mechanism for GEM with distributional M-convex constraints in the full preference domain. In other words, we generalize Top Trading Cycle (TTC) mechanism [23] to GEM under matroid or M-convex constraints in the full preference domain.

1.2   Difficulties and Techniques

A significant technical challenge when working with the complete preference domain under matroid constraints is creating an effective priority rule. This rule will determine the selection of the first-choice graph within the indifference graph. Unfortunately, the natural rule, as discussed by Saban and Sethuraman [21], does not provide a feasible solution for the matroid setting across the full preference domain.

In order to surmount this difficulty, we propose a nuanced, dynamic priority rule, with each agent having their own specific rule. The reasoning behind this priority rule can be broken down into two key components: (1) The individual priority rule is primarily constructed to handle the feasibility of the matroid constraints. (2) The dynamically updated common agent rule, aimed at promoting strategy-proofness, prioritizes unsatisfied and untraded satisfied agents in subsequent exchanges. Additionally, proving strategy-proofness under matroid constraints unveils a novel structure (as shown in Part 1 and 2 of the proof of Theorem 3). This structure goes beyond the path preserving property and could be of independent interest. Furthermore, it may find utility in solving other related problems.

1.3   Related Work

Top Trading Cycle (TTC) is an algorithm for trading indivisible items without using money. It was developed by Gale and published by Scarf and Shapley [23]. There has been a significant amount of research work investigating the generalization of the TTC mechanism, where an object can be allocated to a set of agents. TTC is further generalized to the Hierarchical Exchange mechanism [18] and to the Trading Cycles mechanism [20]. TTC has been applied to a school choice problem [1], as well as assigning teachers to schools [7,27]. Recently, TTC has attracted increasing attention from AI researchers [4,17,24,25]. Several recent papers have explored the generalizations of the TTC mechanism for the full preference domain, e.g. [11,19,21]. Our work for GEM with matroid or M-convex constraints are both strict extensions of the results in [11,19,21].

Our paper is at the intersection of discrete mathematics and economics. This paper is not the first to study distributional matroid constraints in the matching market. Matroid constraints have been extensively investigated in two-sided matching market. Huang [10] considers the academic hiring problem with laminar matroid constraints in two-sided matching settings and presents a polynomial-time algorithm for it. Fleiner and Kamiyama [8] generalize the laminar matroid constraints to be associated with both applicators and institutes and answer an open question posted by Huang. Later the general matroid associated with each institute is considered in [14,15] for different two-sided market objectives. Recently popular matching under this matroid setting is investigated and guaranteed within the polynomial-time [16]. Typical distributional constraints for two-sided matching market are summarized in the survey [5]. It is very interesting to see to what extent can TTC be generalized over these distributional constraints.

The mechanism for M-convex constraint shares almost the same structure as that for the matroid constraint, we thus first present preliminaries and mechanism with proofs for GEM with the matroid constraint and leave the demonstration and discussion of the mechanism for GEM with M-convex constraint at the end.

2   Preliminaries

2.1   Matroids

We use the shorthand notation $X + y := X \cup \{y\}$ and $X - y := X \setminus \{y\}$. A matroid is a pair $M = (U, \mathcal{I})$, where U is a ground set and $\mathcal{I}$ is a collection of subsets of U (called independent sets) satisfying the following three properties: (1)[**non-emptiness**] The empty set is in $\mathcal{I}$; (2)[**heredity**] Every subset of an independent set is also independent; (3)[**exchange**] For any two sets $X, Y \in \mathcal{I}$ such that $|X| < |Y|$, then there exists an element $s \in Y \setminus X$ such that $X + s \in \mathcal{I}$. The rank of a matroid M denoted by r_M is the maximum size of an independent set in the matroid. Similarly, the rank of a subset X denoted by $r_M(X)$ is the maximum size of an independent subset of X.

For any matroid $M = (U, \mathcal{I})$ and two sets $X, Y \subseteq U$, let $\Delta(X, Y)$ denote the (directed) bipartite graph with left vertices $X \setminus Y$ and right vertices $Y \setminus X$ such that there is an edge between a left vertex s and a right vertex t, i.e., the edge (t, s) if $X - s + t \in \mathcal{I}$.[1] We use the following propositions about matroids in proofs.

Proposition 1 (Corollary 39.12a [22]). *Let $M = (U, \mathcal{I})$ be a matroid and let $X, Y \in \mathcal{I}$ with $|X| = |Y|$. Then $\Delta(X, Y)$ contains a perfect matching.*

Proposition 2 (Theorem 39.13 [22]). *Let $M = (U, \mathcal{I})$ be a matroid and let $X \in \mathcal{I}$. Let $Y \subseteq U$ be such that $|X| = |Y|$ and such that $\Delta(X, Y)$ contains a unique perfect matching. Then $Y \in \mathcal{I}$.*

Proposition 3 (Corollary 39.13a [22]). *Let $M = (U, \mathcal{I})$ be a matroid and let $X \in \mathcal{I}$. Let $Y \subseteq U$ be such that $|X| = |Y|$ and $r_M(X \cup Y) = |X|$, and such that $\Delta(X, Y)$ contains a unique perfect matching. Let $s \notin X \cup Y$ with $X + s \in \mathcal{I}$. Then $Y + s \in \mathcal{I}$.*

Throughout this paper, we view the matroids associated with items are given by matroid oracles that can be used to determine a set is independent or not. Both mechanisms designed here can output an allocation by only using polynomial queries to these matroid oracles.

Next, we present some commonly encountered examples of matroid instances and their practical applications in real-life scenarios.

---

[1] In fact, in the following paragraph, the edge may have a direction from t to s, i.e., the directed edge (t, s) when referring to the first choice graph or indifference graphs. However, we will use its undirected version when we refer to Proposition 1 2 3 without causing any confusion.

Definition 1 (Uniform matroid). *Let* $M = (U, \mathcal{I})$ *be a matroid and* $k \leq |U|$ *be a fixed non-negative integer.* M *is called a* $k-$*uniform matroid iff for each subset* $B \subseteq U$, $B \in \mathcal{I}$ *whenever* $|B| \leq k$.

Original Market [23]: The original market in [23] can be viewed as an 1-uniform matroid instance.

Definition 2 (Transversal matroid). *Let* $M = (U, \mathcal{I})$ *be a matroid and* $\mathcal{Q} = \{Q_1, Q_2, \cdots, Q_n\}$ *be a family of subsets of* U. *A set* T *is called a transversal of* $\mathcal{Q}$ *if there exist distinct elements* $q_i \in Q_i$, $i = 1, 2, \cdots, n$ *such that* $T = \{q_1, q_2, \cdots, q_n\}$. *A set* T *is called a partial transversal of* $\mathcal{Q}$ *if* T *is a transversal of some subfamily* $\{Q_{i_1}, Q_{i_2}, \cdots, Q_{i_k}\}$ *of* $\mathcal{Q}$. *Let* $\mathcal{I}$ *be the collection of all partial transversals of* $\mathcal{Q}$, $M = (U, \mathcal{I})$ *is called a transversal matroid (induced by* $\mathcal{Q}$).

Transversals are closely related to matchings in bipartite graphs. This can be seen with the following basic construction of a bipartite graph $G = (V, E)$ associated with a family $\mathcal{Q} = \{Q_1, \cdots, Q_n\}$ of subsets of a set U. Here $V = \{1, \cdots, n\} \cup U$, $E = \{(i, u) | u \in Q_i, i = 1, \cdots, n\}$, assuming that U is disjoint from $\{1, \cdots, n\}$. Then a set T is a transversal of $\mathcal{Q}$ if and only if G has a matching μ of size n such that T is the set of vertices in U covered by μ.

Type-specific Condition [2,7]: In the problem of reallocating dormitory rooms or campus housing, students may have a specific requirement like separate bathrooms, disabled accessibility, etc. So a set of occupants is feasible if and only if there is a matching between the students in the set and the rooms in the dormitory. This corresponds to a traversal matroid with students as left nodes and rooms as right nodes[2].

Definition 3 (Partition matroid). *Let* $M = (U, \mathcal{I})$ *be a transversal matroid induced by* $\mathcal{Q}$. *If* $\mathcal{Q}$ *forms a partition of* U, *then* M *is a partition matroid.*

Capacity Constraints with Type-specific Quotas [3]: Consider the school choice model with capacity constraints and type-specific quotas, i.e. the number of students with the same type should be less than a type specified threshold. Such a constraint corresponds to a partition matroid truncated with a cardinality constraint.

Definition 4 (Laminar matroid). *Let* $M = (U, \mathcal{I})$ *be a matroid and* $\mathcal{Q} = \{Q_1, Q_2, \cdots, Q_n\}$ *be a family of subsets of* U. $\mathcal{Q}$ *is called a laminar family if for any* Q_i, $Q_j \in \mathcal{Q}$, *either* $Q_i \cap Q_j = \emptyset$ *or* $Q_i \subseteq Q_j$ *or* $Q_j \subseteq Q_i$. *There is a function* $c : \mathcal{Q} \to Z^+$ *from* $\mathcal{Q}$ *to positive integral numbers. Suppose* $\mathcal{Q}$ *is a laminar family and* $X \in \mathcal{I}$ *if* $X \cap Q_i \leq c(Q_i)$, $i = 1, 2, \cdots, n$, *then* $M = (U, \mathcal{I})$ *is called a laminar matroid.*

---

[2] One may argue that this problem can be transferred to the classical matching market by asking students to report full preference lists over all rooms instead of dormitories. However, it is not realistic in real-life applications to ask the students to report such lists over numerous rooms. See more discussion in [2,7].

Hierarchy Constraints [8,10,13]: Consider doctor hiring problem [10,13] in which a number of hospitals are hiring doctor members from a pool of applicants with different specialties. A hospital classifies the applicants based on their medical specialties (or any other criterion), each specialty class set consists of doctors belonging to this specialty class. Typically, the specialty class sets form a tree-like structure, see Fig 1 for example. In fact, all the classes in this tree-like structure form a laminar family. That is, let $\mathcal{Q} \subseteq 2^U$ be the set of all the specialty classes, where U denotes the set of all the doctors, for any two specialty class sets $Q_1, Q_2 \in \mathcal{Q}$, it holds that $Q_1 \cap Q_2 = \emptyset$ or $Q_1 \subseteq Q_2$ or $Q_2 \subseteq Q_1$. The hospital will set an upper bound $c(Q)$ on the number hired for each specialty class set $Q \in \mathcal{Q}$. Let X be the set of doctors hired by the hospital, then $|X \cap Q| \leq c(Q)$, $\forall Q \in \mathcal{Q}$. In other words, the collection of the feasible sets (i.e., the possible sets of doctors hired by the hospital) together with the ground set U forms a laminar matroid and each feasible set is an independent set of this matroid.

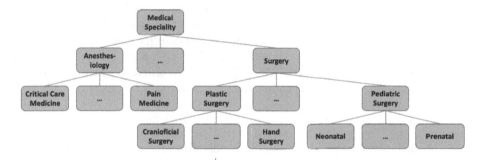

Fig. 1. Medicine Specialties Tree-like Hierarchy.

Region Cap Constraints [12]: Consider the "regional cap" policy implemented in Japanese medical residency matching. This policy mandates an upper limit on the total number of residents assigned per region, effectively imposing a regional cap. Implemented to address the skewed geographical distribution of doctors, favoring urban areas over rural regions, this policy serves as a regulatory measure. The regional caps inherently create a tree-like region hierarchy of constraints that correspond to a laminar matroid. Policies sharing mathematical similarities with the regional cap policy are prevalent in diverse contexts, such as graduate school admissions in China, college admissions in various European countries, residency match processes in the U.K., and teacher assignments in Scotland.

2.2   The Matching Market

We consider a market where a set of n agents N compete for a set of m objects A. Each object $a \in A$ with certain copies[3] can be assigned to a set of agents, that

---

[3] the number of copies is infinite when the constraint is matroid, and finite when the constraint is M-convex.

must be an independent set of a matroid M_a. Each agent $i \in N$ may be initially endowed with an object $\omega(i)$. We assume the initial endowments are feasible, e.g., satisfying the matroid constraints. In addition, each agent is interested in obtaining at most one object and has a preference list, not necessarily strict, over a set of objects including his endowment. For an agent who does not have initial objects, we assign a dummy object to them and add the dummy object to their preference lists with the lowest preference. We write $a \succ_i b$ to denote that agent i strictly prefers object a to b, and write $a \simeq_i b$ to denote that i is indifferent between a and b. We also use $a \succeq_i b$ to denote that agent i either strictly prefers a to b or is indifferent between them, and say that i weakly prefers a to b. Let $P = (P_1, \cdots, P_n)$ denote the joint preference list profile of all agents. We write P_{-i} to denote the joint preference list profile of all agents except agent i. We use C_j^i to denote the jth indifference class of agent i for any $i \in [n]$, $j \in [m]$, e.g., $P_i = (C_1^i \succ_i C_2^i \succ_i \cdots \succ_i C_m^i)$, also denoted by $P_i = (C_1^i, C_2^i, \cdots, C_m^i)$. Given an agent order R and a set of agents $S \subseteq N$, we use R_S to denote the order of R restricted on S, i.e., for any agent $i, j \in S$, $i \succ_{R_S} j$ if and only if $i \succ_R j$. For any disjoint sets S and Q, we use (R_S, R_Q) to denote the agent order over all the agents in $S \cup Q$ by a concatenation of preference orders R_S and R_Q. An instance I is completely defined by the tuple (N, A, M, ω, P) where $M = (M_a)_{a \in A}$ is the matroid vector of all objects. Clearly, the initial endowment ω should be a feasible assignment subject to the matroid constraints M.

2.3   Allocations and Mechanisms

Given an instance $I = (N, A, M, \omega, P)$, an allocation μ is a mapping from N to A. We use $\mu(i)$ to denote the object agent i obtains in μ and $S_a(\mu)$ to denote the set of agents assigned to object a in μ. We say an agent i strictly or weakly prefers an allocation μ to another allocation μ' if he strictly or weakly prefers $\mu(i)$ to $\mu'(i)$. An allocation is said to be *feasible* for the instance I if $S_a(\mu)$ is an independent set of the matroid M_a for all $a \in A$. An allocation μ is *individually rational* if for any agent $i \in N$, $\mu(i) \succeq_i \omega(i)$. An allocation μ is *Pareto-optimal* if there does not exist a feasible allocation μ' such that all agents weakly prefer μ' to μ and at least one of them strictly prefers μ'.

Given (N, A, M, ω, P), a mechanism ϕ is an algorithm that outputs allocations by taking preference profile P as input. A mechanism is Pareto-optimal if it returns a Pareto-optimal allocation. Similarly, we can define a mechanism to be individually rational. A mechanism is *strategy-proof* (or *truthful*) if agents always find it in their best interests to declare their preferences truthfully, no matter what other agents declare, i.e., for every agent i and every possible declared preference list P_i' for i, $\phi_i(P_i, P_{-i}) \succeq_i \phi_i(P_i', P_{-i})$ for any profile P_i, P_{-i}, where ϕ_i denotes the allocation of agent i under the mechanism ϕ.

3   Mechanism Under Matroid Constraints

We say an agent is satisfied if he is currently allocated one of his top objects in his current preference, otherwise, he is unsatisfied. The indifference graph

---

Algorithm 1: Algorithm under Matroid Constraints

Input: An instance $I = (N, A, M, \omega, P)$, a common agent priority order R and
a common object priority order O

Output: An allocation π

1 Initialize $\pi \leftarrow \emptyset$; $\mu \leftarrow \omega$; agent (prior) order R; object (prior) order O;

2 **while** *not all agents in N are assigned (matched) in π* **do**

    /* Construct the indifference graph $\text{IG}(\mu, \pi)$ */

3   Let V be set of all unassigned agents in π;

4   Initialize the indifference graph $\text{IG}(\mu, \pi)$ with vertex set V and empty edge set E;

5   **for** *each agent $i \in V$* **do**

6   Let C_ℓ^i be the most preferred indifference class of i in P_i;

7   **while** $S_a(\pi) + i \notin \mathcal{I}_a$ *for any object $a \in C_\ell^i$* **do**

8   Remove C_ℓ^i from i's preference list P_i; $\ell \leftarrow \ell + 1$;

9   **for** *each object $a \in C_\ell^i$* **do**

10   Add edge (i, j) into E for all $j \in V$ such that $S_a(\mu) + i - j \in \mathcal{I}_a$;

11   **if** *there is a terminal sink in $\text{IG}(\mu, \pi)$* **then**

        /* Implement the terminal sinks in $\text{IG}(\mu, \pi)$ */

12   **for** *each terminal sink X in $\text{IG}(\mu, \pi)$* **do**

13   **for** *each agent $i \in X$* **do**

14   Add $(i, \mu(i))$ into π; Remove agent i from IG;

15   **else**

        /* improvement phase: implementing a trading cycle in $G(\mu, \pi)$ */

16   Construct the first choice graph $G(\mu, \pi)$ from $\text{IG}(\mu, \pi)$ by Algorithm 2;

17         $X \leftarrow$ a (consistently) selected cycle in $G(\mu, \pi)$;

18   **for** *each agent $i \in X$* **do**

19   Let a be the object owned by agent that is pointed to by agent i in X;

20   Replace $(i, \mu(i))$ by (i, a) in μ;

21         $R \leftarrow (R_{\overline{X}}, R_X)$;

IG is defined as follows. There is an edge between (i, j) if and only if there exists an object a which is one of i's favorite objects and $S_a(\mu) + i - j \in \mathcal{I}_a$ (Line 5–10 in Algorithm 1). We say a set of agents S is a terminal sink if all agents in S are satisfied and there is no edge going out from this component in IG. In other words, no agents in a terminal sink would like to trade their objects with agents outside the terminal sink. We say a terminal sink is traded if every agent in this terminal sink is assigned in π, where π is the final assigned matching of Algorithm 1. We trade a terminal sink whenever there exists one in the indifference graph IG (Line 11–14 in Algorithm 1)

The high level description of our mechanism is as follows: The indifference graph IG provides the possibilities for the agent to be satisfied when he trades with any agent he points to in IG. We then need to select a unique agent for each agent to point to in IG from so many possibilities, that is, select the first choice graph FCG in IG. Such a first choice graph should satisfy some combinatorial properties to guarantee feasibility and strategy-proofness meanwhile maintains that (1) each agent points to exactly one agent (2) at least one cycle is formulated (3) at least one unsatisfied agent should be in the cycle. Here, (1), (2) guarantee the cycle trading to happen while (3) guarantees each trading is an improvement of the allocation. Hence, the main obstacle needed to conquer is the selection of FCG in IG.

A simple and natural way to select FCG is to break ties according to a common agent order R [21]. However, we find this method is not even sufficient to give a feasible allocation. Consider four agents $\{1, 2, 3, 4\}$ with three objects $\{a, b, c\}$. Suppose a and c can only be assigned to one agent and b can be allocated to any pair of agents except $\{1, 2\}$. The initial endowment is $\{c, a, b, b\}$ and the preferences are $P_1 = b \succ_1 c \succ_1 a$, $P_2 = a \simeq_2 b \succ_2 c$, $P_3 = b \simeq_3 c \succ_3 a$, $P_4 = a \simeq_4 b \succ_4 c$. Consider the common agent order is $1, 2, 4, 3$ and the object order is a, b, c. Clearly, agent 1 is the only unsatisfied agent and will point to 4. By running the above method, next, 3 points to 1 (since 3 is adjacent to 1 in the indifference graph), next, 2 points to 3 (since 2 adjacent to 3 in the indifference graph) and next, 4 points to 2 (since 4 adjacent to 2 in the indifference graph) in the FCG. It is not hard to see that implementing this trading cycle $(1 \to 4 \to 2 \to 3 \to 1)$ will result in an unfeasible allocation since b is allocated to $\{1, 2\}$ violating the feasibility of M_b.

We now introduce an ordering phase, as one of our main contributions, when constructing FCG to overcome the above obstacle. We first need to define the individual order R_i for each agent i to breaking ties. Given an agent order R, an object order O and an allocation μ, we define an individual order $R_i = R_i(R, O, \mu)$ of agent i over all the agents as follows: $j \succ_{R_i} k$ if (1) agent $\mu(j) \succ_i \mu(k)$, (2) $\mu(j) \simeq_i \mu(k)$ and $\mu(j) \succ_O \mu(k)$ (3) $\mu(j) = \mu(k)$ and $j \succ_R k$ holds.

The selection of the FCG from IG comprises of two ordering phases: ranking process and picking process. The ranking process provides a new order for each agent according to their distance to the previously labeled red agents and their previous order R (Line 3–7 in Algorithm 2), where all the unsatisfied agents are labeled red with highest priority (Line 2 in Algorithm 2). The order of agents after ranking process provide an order under which we select FCG (i.e., each agent according to this order is going to select their pointing agent in the next phase). The next phase is picking process, each agent according to the ranking order will point to its highest priority agent w.r.t. current $R_i(R, O, \mu)$ among previously green labeled agents (Line 14–16 in Algorithm 2), where unsatisfied agents are firstly labeled green and point to highest priority agent w.r.t. current $R_i(R, O, \mu)$ among all the agents (Line 11–13 in Algorithm 2). These two phases guarantee FCG to satisfy previous properties (1)(2)(3). Hence, at least one unsatisfied agent will become satisfied after a cycle trading (Line 16–20 in Algorithm 1). Once an agent is satisfied, he will remain satisfied until being traded in a terminal sink. More importantly, we will show that our selection rule of FCG maintains feasibility and strategy-proofness.

Algorithm 2: Selection of the First Choice Graph $G(\mu, \pi)$

Input: An indifference graph $IG(\mu, \pi)$ and an agent priority R and an object priority O

Output: A first choice graph $G(\mu, \pi)$

1 $V \leftarrow$ the set of agents not matched in π; $S \leftarrow$ the set of satisfied agents in V;

2 Label all the unsatisfied agents (denoted by $U = V \setminus S$) in V red;

/* Reorder the agents to guarantee matroid feasibility   */

3 **while** *there is an agent in V that is not labeled red* **do**

4 $\quad$ Let AL be the set of all the agents adjacent to the set of red labeled vertex in V;

6 $\quad$ Let $i \in$ AL who has the highest agent priority order w.r.t. R;

7 $\quad$ Label agent i red;

8 Reorder R_S to be the order of the above ranking order (the agent order of labelling red) in while;

10 $R \leftarrow (R_U, R_S)$;

11 **for** *each agent $i \in U$* **do**

12 $\quad$ Add $(i, j) \in IG(\mu, \pi)$ into $G(\mu, \pi)$, where j has the highest priority w.r.t. $R_i(R, O, \mu)$;

13 Label all agents in U green;

14 **for** *each agent $i \in S$ in the order of R_S* **do**

15 $\quad$ $j \leftarrow$ the highest priority green labeled agent w.r.t. $R_i(R, O, \mu)$;

16 $\quad$ Add $(i, j) \in IG(\mu, \pi)$ into $G(\mu, \pi)$; Label agent i green;

To maintain strategy-proofness, we also need to alter the order of agents (i.e., R) after implementing every trading cycle. We move all trade agents to the end of R after their trades (Line 21 in Algorithm 1).

We illustrate how Algorithm 1 and 2 work in the previous example. When running Algorithm 1, After the ranking process in Algorithm 2, the common agent order is changed as $(1, 3, 2, 4)$ since the adjacent chain is $(1 \to 4 \to 2 \to 3 \to 1)$. Then the final resulting trading cycle in the picking procedure of Algorithm 2 is $(1 \to 3 \to 1)$ (note that 2 and 4 both point to 3), which gives the desired allocation.

The trading order of cycles might affect the final allocation. So we need to use a consistent way to pick a cycle to trade if there are multiple cycles in FCG. For completeness, we present a consistent way below.

Given the common agent priority order R and object order O, the rule is as follows: the priority of any terminal sink is higher than any trading cycle, that is whenever there is a terminal sink in the indifference graph, before we construct the first choice graph, we trade the highest terminal sinks. The priority of terminal sinks in the trading order is according to the lexicographical order of the agent's initial common agent priority order $R = R_1$, i.e., the terminal sinks

X_1's trading order is higher than terminal sink X_2's trading order if the highest agent priority order among agents in X_1 is higher than that in X_2 (note that the agents in X_1 and X_2 are disjoint).

If there is no terminal sink in the indifference graph, by the definition of the terminal sink and Algorithm 2 of selecting FCG, then there must exist trading cycles containing unsatisfied agents. When we construct the first choice graph, we obtain many trading cycles in the first choice graph. We say a trading cycle X_1's priority order is higher than cycle X_2 (since they are trading cycles, the unsatisfied agents in X_1 and X_2 are both non empty) if the highest agent priority order among the unsatisfied agents in X_1 is higher than that in X_2 according to the initial common agent priority order $R = R_1$. Hence, in Algorithm 1, we only trade the highest priority cycle. As we will show that (see the proof of Lemma 3), the untraded cycle in the first choice graph will remain in the first choice graph until they trade. Next, we prove that Algorithm 1 is feasible, individually rational, and Pareto-optimal. We say an agent trades at time t if he trades his current endowment in the tth iteration of the outer while loop in Algorithm 1. Similarly, we use μ_t and π_t to be the corresponding μ and π at time t. Clearly, $\mu_1 = \omega$ and $\pi_1 = \emptyset$. We also use $\text{IG}(\mu_t, \pi_t)$ to denote the indifference graph computed from μ_t and π_t. Let $G(\mu_t, \pi_t)$ denote the first choice graph selected by Algorithm 2 in $\text{IG}(\mu_t, \pi_t)$. Let R_t denote the agent order before a cycle is traded (i.e., the agent order after the ranking process at time t).

Theorem 1. *Algorithm 1 is feasible.*

Proof. We show this by induction on the time t. For the base case $t = 1$, the feasibility of μ_1 follows from the fact that $\mu_1 = \omega$. For the inductive step from t to $t + 1$, by the trading process of Algorithm 1, since $S_a(\mu_t) = S_a(\mu_{t+1})$, for any object $a \in A$, when a terminal sink is traded at time t, we thus need only to consider the case when a cycle X is traded at time t. To see that Algorithm 1 is feasible, note that when we construct first choice graph in Algorithm 2, the first green labeled agent always has better agent priority than the latter green labeled agent and the agent e.g., agent i points to the highest priority agents among the green labeled agents who own the highest priority object of agent i's favorite indifference class. At time t, for any object a, let $B = S_a(\mu_t) \setminus S_a(\mu_{t+1})$, $D = S_a(\mu_{t+1}) \setminus S_a(\mu_t)$. By the third case of definition of individual agent order $R_i(R_t, O, \mu_t)$ in the selection of FCG in IG in Algorithm 2 at time t, agents in D has the same individual agent order restricted on B, which are all equal to current R_t restricted on B, i.e., for any $i \in D$, $R_i(R_t, O, \mu_t)$ restricted on B is the same as R_t restricted on B.

Case 1: $|B| = |D|$. By Proposition 2, we only need to show $\Delta(B, D)$ contains a unique perfect matching. Let $\lambda = |B \setminus D| = |D \setminus B|$ and $B \setminus D = (x_1, x_2, \ldots, x_\lambda)$ ordered in the priority order R. We sort $D \setminus B = (y_1, y_2, \ldots, y_\lambda)$ such that (y_k, x_k) is an edge in $G(\mu_t, \pi_t)$ for all $k = 1, 2, \ldots, \lambda$. By the process of Algorithm 2, x_k is the highest priority agent in R_t who is connected with y_k in $\Delta(B, D)$. So there is no edge between (y_k, x_ℓ) in $\Delta(B, D)$ for all $k < \ell$ by the selection rule of the edge (y_k, x_k) in Algorithm 2. Thus, $\Delta(B, D)$ contains a unique perfect matching. The independence of D follows from the induction hypothesis that D is independent.

Case 2: $|B| < |D|$. Let i (one unsatisfied agent) be the highest priority unassigned agent in the priority order R_t at time t and j be the agent who points to i in the trading cycle implemented at time t. Recall that in $G(\mu_t, \pi_t)$, all agents except j in the cycle point to agents who own their most preferred objects. Since $|B| < |D|$, agent i does not hold the most preferred object of j. So we must have $|B| = |D| - 1$ (since every assignment in Algorithm 2 is carried over the cycle X, the highest priority agent can only be pointed to by one agent in the cycle X), $j \in D \setminus B$ and $B + j \in \mathcal{I}_a$. By using Proposition 3 on sets B and $D - j$, it suffices to show $\Delta(B, D - j)$ contains a unique perfect matching and $r_{M_a}(B \cup (D - j)) = |B|$. The first condition can be proved by similar arguments in Case 1. The second condition holds because for any agent $k \in (D - j) \setminus B$, $B + k \notin \mathcal{I}_a$, otherwise k will point to i.

Case 3: $|B| > |D|$. Let i be the highest priority unassigned agent in the priority order R_t at time t. Similarly to Case 2, we must have $|B| = |D| + 1$ and $i \in B \setminus D$. By using Proposition 2 on sets B and $D + i$, we can show $D + i \in \mathcal{I}_a$ by similar arguments in Case 1. Then the feasibility of D follows from the heredity of independence in matroids. Combining all cases, we complete the induction and prove the feasibility of Algorithm 1. □

Theorem 2. *Algorithm 1 is individually rational and Pareto-optimal.*

Proof. Individual rationality follows from the process of Algorithm 1 directly since the agent can be satisfied before his endowment is removed from his preference list. Let μ be the finally matching. Suppose μ' is a Pareto improving matching of μ. Let agent i be the first trading agent in Algorithm 1 such that $b = \mu'(i) \succ_i \mu(i) = a$. Suppose agent i first trades at time t, by the process of Algorithm 1, there exists a time $t' < t$ such that $S_b(\pi_{t'}) + i \notin \mathcal{I}_b$. By definition of i, we have $\mu(j) \simeq_j \mu'(j)$ for all j who trades in a terminal sink before agent i. By the definition of terminal sink (none of these agents would like to trade his object), we have $S_b(\pi_{t'}) \subseteq S_b(\mu_{t-1}) \subseteq S_b(\mu')$. Then, $S_b(\mu') = S_b(\mu') + i \notin \mathcal{I}_b$ (since $S_b(\pi_{t'}) + i \notin \mathcal{I}_b$), a contradiction to the feasibility of μ'. Therefore, μ is Pareto-optimal. □

Next, we are going to prove the path preserving proposition (Proposition 4). We divide the proof of this proposition into several simple lemmas (Lemma 1, Lemma 2, Proposition 4).

Lemma 1. *Given an unassigned agent $i \in N$, let a be the most preferred item by i at time t in Algorithm 2. If $i + S_a(\mu_t) \notin \mathcal{I}_a$ and the agent that i points to in $G(\mu_t, \pi_t)$ does not trade at time t, then $i + S_a(\mu_{t+1}) \notin \mathcal{I}_a$.*

Lemma 2. *Given an unassigned agent $i \in N$, let a be the most preferred item by i at time t in Algorithm 2 and j be the agent i points to in $G(\mu_t, \pi_t)$. There is a path from agent i to an unsatisfied agent i' in FCG $G(\mu_t, \pi_t)$ and i' is not traded at time t. For any agent $j' \in S_a(\mu_{t+1}) \cap S_a(\mu_t)$ who is prior than j in the agent priority order R_{t+1} (i.e., $j' \succ_{R_{t+1}} j$), $S_a(\mu_{t+1}) - j' + i \notin \mathcal{I}_a$.*

Proposition 4. *In Algorithm 1, if there is a path* $(i_1, i_2, \cdots, i_k)$ *in the FCG* $G(\mu_t, \pi_t)$ *at some time* t, *where* i_k *is an unsatisfied agent. Then the path will remain in the FCG at any time* $t' \geq t$ *until agent* i_k *trades.*

Proposition 5. *If agent* i *obtains object* a *when he first trades in Algorithm 1 at time* t, *then in any following trading of agent* i *in Algorithm 1, he will obtain an object* $b \simeq_i a$.

Proof. This proposition follows from the process of Algorithm 1.   □

Let P_i be agent i's true preference list and P_i' be his misreporting preference list. μ_t and π_t be the matchings (allocations) by running Algorithm 1 under $P = (P_i, P_{-i})$ at time t. Similarly, μ_t' and π_t' are the matchings by running Algorithm 1 under $P' = (P_i', P_{-i})$. Roughly, the notations without (resp. with) apostrophe denote the notations under P (resp. P'). We use $\mathcal{P}$ and $\mathcal{P}'$ to denote the process of Algorithm 1 under preference profile P and P' respectively. Let T and T' denote the time when agent i first trades in $\mathcal{P}$ and $\mathcal{P}'$ respectively. Let $t = \min\{T, T'\}$. We use R_t and R_t' to denote the temporary common agent priority order at time t in $\mathcal{P}$ and $\mathcal{P}'$ respectively.

Lemma 3. *For each time* $0 < t' < t$, *the same terminal sinks and trading cycles occur in* $\mathcal{P}$ *and* $\mathcal{P}'$. *Furthermore, the indifference graphs* $\mathrm{IG}(\mu_{t'}, \pi_{t'})$ *(resp.* $G(\mu_{t'}, \pi_{t'})$*) and* $\mathrm{IG}(\mu_{t'}', \pi_{t'}')$ *(resp.* $G(\mu_{t'}', \pi_{t'}')$*) will only differ in the outgoing edge of agent* i, $t' \leq t$.

Suppose agent i obtains object $\mu(i) = a \in C_j^i$ in $\mathcal{P}$. Let $P_i'' = (C_1^i, C_2^i, \cdots, C_{j-1}^i, a, C_j^i \setminus a, C_{j+1}^i, \cdots, C_m^i)$. Note that the notations associated with double apostrophe denote notations under preference profile $P'' = (P_i'', P_{-i})$.

Lemma 4. *Suppose for any agent* i *and every preference list* P_i *of* i, *when agent* i *reports* P_i'', *he can still obtain object* a, *then Algorithm 1 is strategy-proof and vice versus.*

By Lemma 4, in order to show that Algorithm 1 is strategy-proof, it is sufficient to show that when agent i bids $P_i' = (C_1^i, C_2^i, \cdots, C_{j-1}^i, a, C_j^i \setminus a, C_{j+1}^i, \cdots, C_m^i)$, where he obtains object a when he bids his true preference list $P_i = (C_1^i, C_2^i, \cdots, C_{j-1}^i, C_j^i, C_{j+1}^i, \cdots, C_m^i)$, he still obtains object a. Hence, in the following, we suppose $P_i' = (C_1^i, C_2^i, \cdots, C_{j-1}^i, a, C_j^i \setminus a, C_{j+1}^i, \cdots, C_m^i)$. Let $D_t'(i)$ denote the set of agents that have a directed path (pointing) to agent i in the first choice graph at time t in $\mathcal{P}'$. Recall that T and T' are the first trading time of agent i in $\mathcal{P}$ and $\mathcal{P}'$ respectively. With Proposition 4, Lemma 3 and 4, we are still far from proving strategy-proofness of Algorithm 1. In fact we need new insights of the structures of FCG, which are Part 1, 2 in the proof of Theorem 3. These novel combinatorial structures may present insights for other related problems, which should be of independent interest.

Theorem 3. *Algorithm 1 is strategy-proof.*

Proof. By Lemma 3 and Proposition 5, we know that agent i will not obtain any object in the set $\bigcup_{s<j} C_s^i$ at any time in $\mathcal{P}'$ since he trades and obtains an object in C_j^i at some time in $\mathcal{P}$. If $T' \leq T$ and agent i does not obtain object a in $\mathcal{P}'$ in μ', we know that there exists a time $t \leq T'$ such that $S_a(\pi_t') + i \notin \mathcal{I}_a$ in $\mathcal{P}'$. By Lemma 3, we know that $S_a(\pi_t) + i \notin \mathcal{I}_a$ in $\mathcal{P}$, which means that agent i will not obtain object a in μ in $\mathcal{P}$. Therefore, if $T' \leq T$, we have $\mu'(i) = a$. Next, we show that it is impossible that $T' > T$. Hence, in the following, we suppose $T' > T$. By Lemma 3, we can suppose Algorithm 1 simultaneously starts at time T in $\mathcal{P}$ and $\mathcal{P}'$ respectively. We will inductively show the following two properties:

Part 1: for any time $T \leq t \leq T'$, consider the first choice graph $G(\mu_t', \pi_t')$. There exists a time $s \geq T$ in $\mathcal{P}$ such that for any $(j, k) \in G(\mu_t', \pi_t')$ and $j \notin D_t'(i)$, then $(j, k) \in G(\mu_s, \pi_s)$ and $\mu_s(j) = \mu_t'(j)$ and $\mu_s(k) = \mu_t'(k)$ and the temporary favorite indifference classes for j and k are the same in both $\mathcal{P}$ and $\mathcal{P}'$. The agents in the trading cycles occurring in $\mathcal{P}$ not in $\mathcal{P}'$ before time s are both in $D_t'(i)$.

Part 2: any terminal sink or trading cycle formulated in the first choice graph $G(\mu_t', \pi_t')$ at any time $T \leq t \leq T'$ in $\mathcal{P}'$ is also in $G(\mu_s, \pi_s)$ in $\mathcal{P}$ at some time $s > T$, their trading orders are the same.

Note that the above Part (1) implies Part (2) by our following proof for Part (1). Since agent i does not trade during time $T \leq t < T'$ in $\mathcal{P}'$, i.e., $D_t'(i)$ will increase according to t in $\mathcal{P}'$. We need only to show that Part (1) holds. Note that the trading order of terminal sinks and trading cycles in $\mathcal{P}'$ during $T \leq t \leq T'$ is the same as they do in $\mathcal{P}$ during some time intervals in $\mathcal{P}$ by our tie breaking rule among cycles and terminal sinks. Before inductively proving Part (1), let's see what Part (1) and (2) imply. Since $\mu(i) = a$, at time T in $\mathcal{P}$, we have $S_a(\pi_T) + i \in \mathcal{I}_a$ (otherwise agent i will not obtain object a in $\mathcal{P}$ in his final trading due matroid feasibility). By Lemma 3, we know that $S_a(\pi_T') + i \in \mathcal{I}_a$. At time T, a is agent i's only favorite object in $\mathcal{P}'$. Therefore, agent i will point to an agent who is currently endowed with object a at time T, i.e., $\mu_T'(i') = a$ or to the highest priority agent k in R_T'. We consider the set $S_a(\mu_t')$, $T \leq t \leq T'$. We show that agent i will always point to an agent endowed with object a or to the highest priority agent k (and obtains a when he trades) in the first choice graph $G(\mu_t', \pi_t')$, $T \leq t \leq T'$. This implies that agent i will obtain object a when he trades at time T', which proves Theorem 3. By Part (2), we always have that $S_a(\pi_t') \subseteq S_a(\pi_s) - i$, $T \leq t \leq T'$, for some $s \geq T$. Hence, at time T', $S_a(\pi_{T'}') + i \in \mathcal{I}_a$, by the construction of indifference graph, we know that agent i will point to at least one agent endowed with object a in $\mathrm{IG}(\mu_{T'}', \pi_{T'}')$ in $\mathcal{P}'$ at time T'. Hence, he will point to an agent endowed with object a or to the highest priority agent in $R_{T'}'$ in the first choice graph $G(\mu_{T'}', \pi_{T'}')$ in $\mathcal{P}'$ at time T'. This means agent i obtains a when he trades in $\mathcal{P}'$ at time T'. As a is the single object in its indifference class in P_i', by Proposition 5, we have $\mu'(i) = a$.

Next, we are ready to prove Properties (1) and (2) inductively. The base case $t = T$ holds by Lemma 3. Suppose for the base $T \leq t < T'$, Part (1) and (2) hold. Now for the case $t + 1 \leq T'$. When a terminal sink or trading cycle X occurs at time t in $\mathcal{P}'$, by Part (2), this terminal sink or trading cycle X also occur at some time s in $\mathcal{P}$. All the tradings occur before time t in $\mathcal{P}'$ have occurred in $\mathcal{P}$ before time s by induction hypothesis.

For any agent j,

Case 1: If there is no directed path from agent j to an agent in X in $G(\mu'_t, \pi'_t)$, then agent j will point to the same agent in $G(\mu'_{t+1}, \pi'_{t+1})$ as he points to in $G(\mu'_t, \pi'_t)$ e.g., $(j, k) \in G(\mu'_t, \pi'_t)$ and $(j, k) \in G(\mu'_{t+1}, \pi'_{t+1})$. Similarly, in $\mathcal{P}$, we have $(j, k) \in G(\mu_s, \pi_s)$ by induction, and by the proof of Lemma 3 $(j, k) \in G(\mu_{s+1}, \pi_{s+1})$. Moreover, $\mu_{s+1}(j) = \mu'_{t+1}(j) = \mu'_t(j)$ and $\mu_{s+1}(k) = \mu'_{t+1}(k) = \mu'_t(k)$.

Case 2: Suppose there is a directed path from agent j to an agent in X and $j \notin D'_{t+1}(i)$. Let $(j, k) \in G(\mu'_{t+1}, \pi'_{t+1})$, note that $k \notin D'_{t+1}(i)$ by definition of $D'_{t+1}(i)$. Suppose k is unsatisfied. By induction hypothesis, $\mu_{s+1}(k) = \mu'_{t+1}(k) = b$ (since $k \notin D'_{t+1}(i)$, k is not traded in the trading cycle involved with agent i). Consider the set $S_b(\mu'_{t+1})$ and $S_b(\mu_{s+1})$, by induction hypothesis, the set $S_b(\mu_{s+1}) - S_b(\mu'_{t+1})$ is obtained by trading cycles or terminal sinks in $\mathcal{P}$ that do not occur in $\mathcal{P}'$. In addition, the set $S_b(\mu_{s+1}) - S_b(\mu'_{t+1})$ is obtained through trading with $S_b(\mu'_{t+1}) - S_b(\mu_{s+1})$ in $\mathcal{P}$. By Proposition 4, $(j, k) \in IG(\mu_{s+1}, \pi_{s+1})$. Note that, for every agent in $S_b(\mu_{s+1}) - S_b(\mu'_{t+1})$ after he trades in a non terminal sink, his common order will be first put into the last common agent priority order (Line 21 of Algorithm 1) and then using the red labeled order to obtain the new common agent order. Hence, if k is an unsatisfied agent in both μ'_{t+1} and μ_{s+1}, then $(j, k) \in G(\mu_{s+1}, \pi_{s+1})$ by the red labeled order (k is the highest priority agent in R_{s+1}) by Proposition 4.

Case 3: Suppose there is a directed path from agent j to an agent in X and $j \notin D'_{t+1}(i)$. Let $(j, k) \in G(\mu'_{t+1}, \pi'_{t+1})$ and suppose k is a satisfied agent. As above, by induction hypothesis, $\mu_{s+1}(k) = \mu'_{t+1}(k) = b$ (We can use induction to suppose that agent k points to the same agent in the first choice graph $G(\mu_{s+1}, \pi_{s+1})$ and $G(\mu'_{t+1}, \pi'_{t+1})$. Agent k is endowed with highest priority object for agent j among all the green labeled agents in both $\mathcal{P}$ at time $t+1$ and in $\mathcal{P}'$ at time $s+1$ due to $S_c(\pi'_{t+1}) \subseteq S_c(\pi_{s+1})$ since $\pi'_{t+1} \subseteq \pi_{s+1}$, for any object c. Therefore, when we search the agent that is selected to be pointed to by agent j among all the green labeled agents in both $IG(\mu_{s+1}, \pi_{s+1})$ and $IG(\mu'_{t+1}, \pi'_{t+1})$ in the selection of FCG in Algorithm 2, agent j points to agent k in $G(\mu'_{t+1}, \pi'_{t+1})$. Then agent j will point to either agent k or another agent in the set $S_b(\mu_{s+1}) - S_b(\mu'_{t+1})$ in $G(\mu_{s+1}, \pi_{s+1})$. Note that $S_b(\mu_{s+1}) - S_b(\mu'_{t+1}) \subseteq D'_{t+1}(i)$ by induction from Part (1). Recall that the set $S_b(\mu_{s+1}) - S_b(\mu'_{t+1})$ is obtained through trading with $S_b(\mu'_{t+1}) - S_b(\mu_{s+1})$. Since agent k is satisfied, by the pointing rule in Algorithm 2, every agent in $S_b(\mu'_{t+1}) - S_b(\mu_{s+1})$ (actually in the set $D'_{t+1}(i)$) has higher agent priority order than agent k in R'_{t+1}. Agent k is the highest priority green labeled agent for agent j in $IG(\mu'_{t+1}, \pi'_{t+1})$. Agent j points to k in $G(\mu'_{t+1}, \pi'_{t+1})$ in $\mathcal{P}'$ by Proposition 4, after $S_b(\mu_{s+1}) - S_b(\mu'_{t+1})$ trades with the set $S_b(\mu'_{t+1}) - S_b(\mu_{s+1})$. We can view the process as follows. First the cycle involved agent k is traded in both $\mathcal{P}$ and $\mathcal{P}'$, then agent j points to agent k in both $\mathcal{P}$ and $\mathcal{P}'$. Now edge (j, k) is in the first choice graphs in both $\mathcal{P}$ and $\mathcal{P}'$. The edge (j, k) will be preserved in the first choice graph in $\mathcal{P}$ by sequentially trading the set $S_b(\mu_{s+1}) - S_b(\mu'_{t+1})$ with $S_b(\mu'_{t+1}) - S_b(\mu_{s+1})$ (in the trading order as they do in $\mathcal{P}$). We only need to consider the subset in $S_b(\mu_{s+1}) - S_b(\mu'_{t+1})$ that trades before agent k trades in $\mathcal{P}$ since the agents in $S_b(\mu_{s+1}) - S_b(\mu'_{t+1})$ trades after agent k trades will have inferior priority than agent k in R_{s+1} (due to Line 21 of Algorithm 1), which means that j will not point to these agents in the first choice graph $G(\mu_{s+1}, \pi_{s+1})$. Next we show that agent j will not point to any agent in $S_b(\mu_{s+1}) - S_b(\mu'_{t+1})$ (that trades before agent k's trading in $\mathcal{P}$) in the first choice graph after their tradings. W.l.o.g. suppose every agent in $S_b(\mu_{s+1}) - S_b(\mu'_{t+1})$ trades before agent k in $\mathcal{P}$. By Proposition 4 and the induction hypothesis, it is not difficult to see that agent j will not point to any agent that trades with agent in $S_b(\mu'_{t+1}) - S_b(\mu_{s+1})$ since these agents' orders in any temporary order after trading are still higher than agent k's order (after agent k's trading, agent k's order will put into the last order) as these cycles occur before agent k trades in $\mathcal{P}$. Hence, agent j will still point to agent k (not the agent in the set $S_b(\mu_{s+1}) - S_b(\mu'_{t+1})$) in the first choice graph $G(\mu_{s+1}, \pi_{s+1})$. The property that temporary favorite indifference classes for j and k are the same in both $\mathcal{P}$ and $\mathcal{P}'$ follows directly from the induction hypothesis. Therefore, Part (1) holds for the case $t + 1 \leq T'$, which completes the induction. □

Finally, we calculate the running time of Algorithm 1. Suppose for any set $S \subseteq N$ and object $a \in A$, checking whether $S \in \mathcal{I}_a$ or not can be done in $O(1)$ time.

Theorem 4. *For any instance $I = (N, A, M, \omega, P)$, Algorithm 1 terminates in time $O(rn^2(\log n + \theta))$ where r is maximum rank among all matroids, i.e. $r = \max_{a \in A} r_{M_a}$ and $\theta = \max_{i \in [n], j \in [m]} |C_j^i|$ is the maximum size of indifference classes.*

Proof. (Sketch) We will maintain a set AL which will store the all vertices currently pointing to a labeled vertex. Using a Fibonacci heap to implement AL, we can add vertices in $O(1)$, obtain the next vertex that must be labeled in $O(1)$ and delete in $O(\log n)$. Terminal sink can be found in $|E| = rn^2$ time, where $|E|$ is the number of edges of the indifference graph IG. The overall running time by simple calculations is $O(rn^2(\log n + \theta))$. □

4 Conclusion

We study the problem of reallocating indivisible goods among a set of strategic agents under matroid and M-convex constraints and provide Pareto-optimal, individually rational and strategy-proof mechanisms for both constraints in the full preference domain. The future works include developing mechanisms in the full preference domain that can work for a class of constraints that is much broader as discussed in [5] for two-sided markets.

Acknowledgments. This work was supported by the National Key Research and Development Program of China (2022YFF0902005). Jianwei Yin was supported by the National Science Fund for Distinguished Young Scholars (No. 61825205). Xiaoye Miao was supported by the National Natural Science Foundation of China under Grant No. 62372404, the Zhejiang Provincial Natural Science Foundation for Distinguished Young Scholars under Grant No. LR21F020005, and the Fundamental Research Funds for the Central Universities under Grant No. 2021FZZX001-25. Jinshan Zhang was supported by the Key Research and Development Jianbing Program of Zhejiang Province (2023C01002), Hangzhou Major Project and Development Program (2022AIZD0140) and Yongjiang Talent Introduction Programm (2022A-236-G). We are also grateful for the valuable comments from the anonymous reviewers.

References

1. Abdulkadiroğlu, A., Che, Y.K., Pathak, P.A., Roth, A.E., Tercieux, O.: Minimizing justified envy in school choice: the design of new orleans' oneapp. Technical report, National Bureau of Economic Research (2017)
2. Abdulkadiroğlu, A., Sönmez, T.: House allocation with existing tenants. J. Econ. Theory **88**(2), 233–260 (1999)
3. Abdulkadiroğlu, A., Sönmez, T.: School choice: a mechanism design approach. Am. Econ. Rev. **93**(3), 729–747 (2003)
4. Athanassoglou, S., Sethuraman, J.: House allocation with fractional endowments. Int. J. Game Theory **40**(3), 481–513 (2011)
5. Aziz, H., Biró, P., Yokoo, M.: Matching market design with constraints. In: Proceedings of the AAAI Conference on Artificial Intelligence (2022)
6. Bogomolnaia, A., Deb, R., Ehlers, L.: Strategy-proof assignment on the full preference domain. J. Econ. Theory **123**(2), 161–186 (2005)
7. Combe, J., Tercieux, O., Terrier, C.: The design of teacher assignment: theory and evidence. Rev. Econ. Stud. **89**(6), 3154–3222 (2022)
8. Fleiner, T., Kamiyama, N.: A matroid approach to stable matchings with lower quotas. Math. Oper. Res. **41**(2), 734–744 (2016)
9. Hamada, N., Hsu, C.L., Kurata, R., Suzuki, T., Ueda, S., Yokoo, M.: Strategy-proof school choice mechanisms with minimum quotas and initial endowments. Artif. Intell. **249**, 47–71 (2017)
10. Huang, C.C.: Classified stable matching. In: Proceedings of the Twenty-First Annual ACM-SIAM Symposium on Discrete Algorithms, pp. 1235–1253. SIAM (2010)
11. Jaramillo, P., Manjunath, V.: The difference indifference makes in strategy-proof allocation of objects. J. Econ. Theory **147**(5), 1913–1946 (2012)
12. Kamada, Y., Kojima, F.: Efficient matching under distributional constraints: theory and applications. Am. Econ. Rev. **105**(1), 67–99 (2015)
13. Kamada, Y., Kojima, F.: Stability and strategy-proofness for matching with constraints: a necessary and sufficient condition. Theor. Econ. **13**(2), 761–793 (2018)
14. Kamiyama, N.: Pareto stable matchings under one-sided matroid constraints. SIAM J. Discret. Math. **33**(3), 1431–1451 (2019)
15. Kamiyama, N.: On stable matchings with pairwise preferences and matroid constraints. In: Proceedings of the 19th International Conference on Autonomous Agents and Multiagent Systems, pp. 584–592 (2020)
16. Kamiyama, N.: Popular matchings with two-sided preference lists and matroid constraints. Theor. Comput. Sci. **809**, 265–276 (2020)

17. Lesca, J., Fujita, E., Sonoda, A., Todo, T., Yokoo, M.: A complexity approach for core-selecting exchange with multiple indivisible goods under lexicographic preferences. In: Proceedings of the AAAI Conference on Artificial Intelligence (2015)
18. Pápai, S.: Strategyproof assignment by hierarchical exchange. Econometrica **68**(6), 1403–1433 (2000)
19. Plaxton, C.G.: A simple family of top trading cycles mechanisms for housing markets with indifferences. In: Proceedings of the 24th International Conference on Game Theory at Stony Brook (2013)
20. Pycia, M., Ünver, M.U.: Incentive compatible allocation and exchange of discrete resources. Theor. Econ. **12**(1), 287–329 (2017)
21. Saban, D., Sethuraman, J.: House allocation with indifferences: a generalization and a unified view. In: Proceedings of the EC '13, pp. 803–820. ACM (2013)
22. Schrijver, A.: Combinatorial Optimization: Polyhedra and Efficiency, vol. 24. Springer, Berlin, Heidelberg (2003)
23. Shapley, L., Scarf, H.: On cores and indivisibility. J. Math. Econ. **1**(1), 23–37 (1974)
24. Sikdar, S., Adali, S., Xia, L.: Mechanism design for multi-type housing markets. In: Proceedings of the AAAI Conference on Artificial Intelligence, vol. 31 (2017)
25. Sun, Z., Hata, H., Todo, T., Yokoo, M.: Exchange of indivisible objects with asymmetry. In: Twenty-Fourth International Joint Conference on Artificial Intelligence (2015)
26. Suzuki, T., Tamura, A., Yokoo, M.: Efficient allocation mechanism with endowments and distributional constraints. In: Proceedings of the 17th International Conference on Autonomous Agents and MultiAgent Systems, pp. 50–58 (2018)
27. Terrier, C.: Matching practices for secondary public school teacher France, MiP country profile 20 (2014)

Abstracts

How Good Are Privacy Guarantees? Platform Architecture and Violation of User Privacy

Daron Acemoglu[1], Alireza Fallah[1,2]($\boxtimes$) (iD), Ali Makhdoumi[3] (iD),
Azarakhsh Malekian[4] (iD), and Asuman Ozdaglar[1] (iD)

[1] Massachusetts Institute of Technology, Cambridge, MA 02139, USA
[2] Simons Laufer Mathematical Sciences Institute (formerly MSRI) & University of
California Berkeley, Berkeley, CA 94720, USA
`afallah@{mit,berkeley}.edu`
[3] Duke University,Durham, NC 27708, USA
[4] University of Toronto, Toronto, ON M5S 3E6, Canada

Abstract. Many platforms deploy data collected from users for a multitude of purposes. While some are beneficial to users, others are costly to their privacy. The presence of these privacy costs means that platforms may need to provide guarantees about how and to what extent user data will be harvested for activities such as targeted ads, individualized pricing, and sales to third parties. In this paper, we build a multi-stage model in which users decide whether to share their data based on privacy guarantees. We first introduce a novel *mask-shuffle* mechanism and prove it is Pareto optimal—meaning that it leaks the least about the users' data for any given leakage about the underlying common parameter. We then show that under any mask-shuffle mechanism, there exists a unique equilibrium in which privacy guarantees balance privacy costs and utility gains from the pooling of user data for purposes such as assessment of health risks or product development. Paradoxically, we show that as users' value of pooled data increases, the equilibrium of the game leads to lower user welfare. This is because platforms take advantage of this change to reduce privacy guarantees so much that user utility declines (whereas it would have increased with a given mechanism). Even more strikingly, we show that platforms have incentives to choose data architectures that systematically differ from those that are optimal from the user's point of view. In particular, we identify a class of pivot mechanisms, linking individual privacy to choices by others, which platforms prefer to implement and which make users significantly worse off. The full paper is available at: https://dx.doi.org/10.2139/ssrn.4333457.

Keywords: Online platforms · Privacy · Game theory · Stackelberg game · Mask-shuffle mechanism

J. Garg et al. (Eds.): WINE 2023, LNCS 14413, p. 675, 2024.
https://doi.org/10.1007/978-3-031-48974-7

Best-of-Both-Worlds Fairness in Committee Voting

Haris Aziz$^{(\boxtimes)}$, Xinhang Lu, Mashbat Suzuki, Jeremy Vollen, and Toby Walsh

UNSW Sydney, Kensington, Australia
{haris.aziz,xinhang.lu,mashbat.suzuki,j.vollen,t.walsh}@unsw.edu.au

Abstract. The best-of-both-worlds paradigm advocates an approach that achieves desirable properties both ex-ante and ex-post. We launch a best-of-both-worlds fairness perspective for the important social choice setting of approval-based committee voting. To this end, we initiate work on ex-ante proportional representation properties in this domain and formalize a hierarchy of properties including Individual Fair Share (IFS), Unanimous Fair Share (UFS), Group Fair Share (GFS), and their stronger variants. We establish their compatibility with well-studied ex-post concepts such as extended justified representation (EJR) and fully justified representation (FJR). Our first main result is a polynomial-time algorithm that simultaneously satisfies ex-post EJR, ex-ante GFS and ex-ante Strong UFS. Subsequently, we strengthen our ex-post guarantee to FJR and present an algorithm that outputs a lottery which is ex-post FJR and ex-ante Strong UFS, but does not run in polynomial time.

Keywords: Committee voting · Approval preference · Randomization and approximation

This work was partially supported by the NSF-CSIRO grant on "Fair Sequential Collective Decision-Making" and the ARC Laureate Project FL200100204 on "Trustworthy AI".

The full version of the paper is available at https://arxiv.org/pdf/2303.03642.pdf.

Fair Division with Subjective Divisibility

Xiaohui Bei[1]([✉]), Shengxin Liu[2], and Xinhang Lu[3]

[1] Nanyang Technological University, Singapore, Singapore
xhbei@ntu.edu.sg
[2] Harbin Institute of Technology, Shenzhen, China
sxliu@hit.edu.cn
[3] UNSW Sydney, Kensington, Australia
xinhang.lu@unsw.edu.au

Abstract. The classic fair division problems assume the resources to be allocated are either divisible or indivisible, or contain a mixture of both, but the agents always have a predetermined and uncontroversial agreement on the (in)divisibility of the resources. In this paper, we propose and study a new model for fair division in which agents have their own *subjective divisibility* over the goods to be allocated. That is, some agents may find a good to be indivisible and get utilities only if they receive the *whole* good, while others may consider the same good to be divisible and thus can extract utilities according to the fraction of the good they receive. We investigate fairness properties that can be achieved when agents have subjective divisibility. First, we consider the *maximin share (MMS) guarantee* and show that the worst-case MMS approximation guarantee is at most 2/3 for $n \geq 2$ agents and this ratio is *tight* in the two- and three-agent cases. This is in contrast to the classic fair division settings involving two or three agents. We also give an algorithm that produces a 1/2-MMS allocation for an arbitrary number of agents. Second, we adapt the notion of *envy-freeness for mixed goods (EFM)* to our model and show that EFM is incompatible with non-wastefulness, a rather weak economic efficiency notion. On the positive side, we prove that an EFM and non-wasteful allocation always exists for two agents if at most one good is discarded.

Keywords: Fair division · Subjective divisibility · MMS · EFM

This work was partially supported by ARC Laureate Project FL200100204 on "Trustworthy AI", by the National Natural Science Foundation of China (No. 62102117), by the Shenzhen Science and Technology Program (No. RCBS20210609103900003), by the Guangdong Basic and Applied Basic Research Foundation (No. 2023A1515011188), and by the CCF-Huawei Populus Grove Fund (No. CCF-HuaweiLK2022005).
The full version of the paper is available at https://arxiv.org/pdf/2310.00976.pdf.

J. Garg et al. (Eds.): WINE 2023, LNCS 14413, p. 677, 2024.
https://doi.org/10.1007/978-3-031-48974-7

The Incentive Guarantees Behind Nash Welfare in Divisible Resources Allocation

Xiaohui Bei[1]($\boxtimes$) (iD), Biaoshuai Tao[2](iD), Jiajun Wu[3], and Mingwei Yang[4](iD)

[1] Nanyang Technological University, Singapore 639798, Singapore
xhbei@ntu.edu.sg
[2] Shanghai Jiao Tong University, Shanghai 200240, China
bstao@sjtu.edu.cn
[3] Huawei Technologies, Shanghai 200125, China
wujiajun17@huawei.com
[4] Stanford University, Stanford, CA 94305, USA
mwyang@stanford.edu

Abstract. We study the problem of allocating divisible resources among n agents, hopefully in a fair and efficient manner. With the presence of strategic agents, additional incentive guarantees are also necessary, and the problem of designing fair and efficient mechanisms becomes much less tractable. While there are flourishing positive results against strategic agents for homogeneous divisible items, very few of them are known to hold in cake cutting.

We show that the Maximum Nash Welfare (MNW) mechanism, which provides desirable fairness and efficiency guarantees and achieves an *incentive ratio* of 2 for homogeneous divisible items, also has an incentive ratio of 2 in cake cutting. Remarkably, this result holds even without the free disposal assumption, which is hard to get rid of in the design of truthful cake cutting mechanisms.

Moreover, we show that, for cake cutting, the Partial Allocation (PA) mechanism proposed by Cole et al. [1], which is truthful and $1/e$-MNW for homogeneous divisible items, has an incentive ratio between $[e^{1/e}, e]$ and when randomization is allowed, can be turned to be truthful in expectation. Given two alternatives for a trade-off between incentive ratio and Nash welfare provided by the MNW and PA mechanisms, we establish an interpolation between them for both cake cutting and homogeneous divisible items.

Finally, we show that any envy-free cake cutting mechanism with the connected pieces constraint has an incentive ratio of $\Theta(n)$.

The full version is available at https://arxiv.org/abs/2308.08903.

Keywords: Fair division · Mechanism design · Incentive ratio

Reference

1. Cole, R., Gkatzelis, V., Goel, G.: Mechanism design for fair division: allocating divisible items without payments. In: EC, pp. 251–268. ACM (2013)

© The Author(s), under exclusive license to Springer Nature Switzerland AG 2024
J. Garg et al. (Eds.): WINE 2023, LNCS 14413, p. 678, 2024.
https://doi.org/10.1007/978-3-031-48974-7

Information Design for Spatial Resource Allocation

Ozan Candogan[1] and Manxi Wu[2]([envelope])

[1] Booth School of Business, University of Chicago, Chicago, USA
ozan.candogan@chicagobooth.edu
[2] Operations Research and Information Engineering, Cornell University, Ithaca, USA
manxiwu@cornell.edu

In many operational settings, the resources that serve the jobs are spatially distributed, and mismatch between the jobs and resources locations causes inefficiencies. Such a mismatch gets exacerbated when resources are self-interested independent contractors, who strategically decide whether or not to provide service and how to relocate from one region to another. Platforms typically address this with two strategies: adopt surge pricing based on location demand or share information of demand distribution. While pricing has been extensively studied, using information to influence self-interested resources has received much less attention. This paper seeks to bridge this gap, focusing on designing optimal information mechanisms for spatial resource allocation.

We consider an undirected network model, where each node represents a location, initially endowed with resources. These resources can reposition between connected nodes, with the edges associated with repositioning costs. The price of the service in each region depends on the number of resources after repositioning and is influenced by a random demand shock (state) that shifts the price curve. The platform collects a commission for facilitating matches between the resources and the jobs, which is modeled as a constant fraction of the generated revenues. The resources are self-interested and they try to maximize their payoffs. Thus, by appropriately revealing information about the state, the platform can influence the repositioning strategies of the resources and improve revenue.

We first show that the repositioning game is a potential game, and the equilibrium repositioning strategies can be computed by solving a convex program. Building on this result, we focus on characterizing the conditions under which the optimal information mechanism is *monotone partitional*. Such mechanism partitions the state space into subintervals, where states in each subinterval are either pooled into a unique signal or fully revealed.

Our main theorem characterizes practically relevant conditions on the market sizes of different locations (characterized in terms of the intercepts of the price curves), under which a simple monotone partitional information mechanism is optimal. This mechanism reveals state realizations below a threshold and above a second (higher) threshold, and pools all states in between and maps them to a

J. Garg et al. (Eds.): WINE 2023, LNCS 14413, pp. 679–680, 2024.
https://doi.org/10.1007/978-3-031-48974-7

unique signal realization. We also find that full information provision is optimal when the commission rate set by the platform is sufficiently high. Finally, we develop an algorithmic approach to compute the optimal information structure and near-optimal monotone information structures in general settings.[1]

---

[1] The full paper can be found at https://arxiv.org/pdf/2307.08040.pdf.

Do Private Transaction Pools Mitigate Frontrunning Risk?

Agostino Capponi[1], Ruizhe Jia[1], and Ye Wang[2($\boxtimes$)]

[1] Columbia University, 2960, Broadway, NY, USA
{ac3827,rj2536}@columbia.edu
[2] University of Macau, Avenida da Universidade, Macao, China
wangye@um.edu.mo

Originally conceived as the foundational technology for digital currencies, the blockchain infrastructure has evolved to enable a wide range of financial services through the use of smart contracts. Nonetheless, this evolution has raised major concerns, with frontrunning attacks being a particularly noticable one. In public blockchain systems, transactions are visible to all network nodes, making it possible for attackers to observe pending transactions and frontrun them. These attacks result in a transfer of wealth from victims to blockchain validators, and the value gain by the latter is also referred to as the maximal extractable value (MEV) [2], diminish allocative efficiency, as block space is allocated to wasteful transactions which do not generate value. Most recently, private transaction submission pools have been proposed as a mechanism to mitigate frontrunning [1]. Transactions submitted through the private pool are not publicly observable, and thus cannot be frontrun. We analyze the welfare impact of private pools on public blockchains, and propose a game-theoretic model to understand the market conditions under which it is incentive compatible for validators to monitor these private pools. Our findings reveal that, in the Nash equilibrium, private pools are only partially adopted by validators, but fully used by attackers. Although they neither completely eliminate frontrunning attacks nor reduce fees, they improve the allocative efficiency of blockspace, and raise aggregate welfare. We show that welfare is maximzed if all validators were to monitor the private pool, as in such situation no block space would be allocated to frontrunning attackers.

References

1. Capponi, A., Jia, R., Wang, Y.: Blockchain private pools and price discovery. In: AEA Papers and Proceedings, vol. 113, pp. 253–256. American Economic Association (2023)
2. Daian, P., et al.: Flash boys 2.0: frontrunning in decentralized exchanges, miner extractable value, and consensus instability. In: 2020 IEEE Symposium on Security and Privacy (SP), pp. 910–927 (2020)

Available at https://eprint.iacr.org/2023/1461. This work was supported by the grant from the FDCT, Macau SAR (File no. 0129/2022/A).

J. Garg et al. (Eds.): WINE 2023, LNCS 14413, p. 681, 2024.
https://doi.org/10.1007/978-3-031-48974-7

Faster Ascending Auctions via Polymatroid Sum

Katharina Eickhoff[1]([✉]) [ID], Britta Peis[1] [ID], Niklas Rieken[1] [ID],
Laura Vargas Koch[2] [ID], and László A. Végh[3] [ID]

[1] School of Business and Economics, RWTH Aachen University, Aachen, Germany
{eickhoff,peis,rieken}@oms.rwth-aachen.de
[2] Department of Mathematics, ETH Zürich, Zürich, Switzerland
laura.vargas@math.ethz.ch
[3] Department of Mathematics, London School of Economics and Political Science,
London, UK
L.Vegh@lse.ac.uk

Keywords: Ascending auction · Walrasian prices · Gross substitutes ·
Polymatroid sum · Push-relabel

We study ascending auctions for computing minimal Walrasian prices (i.e. prices
that admit a *stable* allocation of *all* items) in markets where indivisible items
with multiplicities are sold to buyers with individual, non-decreasing, and gross
substitutes valuations.

Finding the sets on which the ascending auction increases the prices boils
down to a submodular function minimization problem. We provide a fast and
simple algorithm to minimize this submodular function. From a dual viewpoint,
it corresponds to a polymatroid sum problem, and using this, we give a push-
relabel algorithm. This significantly improves on the previously best running
time of Murota, Shioura and Yang [2]. Our algorithm is a special implementation
of the more general push-and-relabel framework by Frank and Miklós [1].

As our second main contribution, we show that for gross substitutes valuations, the component-wise minimal competitive prices are the same as the
minimal Walrasian prices. This enables us also show that the minimal Walrasian prices react naturally to changes in demand and supply, i.e. they can only
increase if supply decreases, or demand increases.

The full paper is available at http://arxiv.org/abs/2310.08454.

K.E. is funded by the Deutsche Forschungsgemeinschaft (DFG, German Research Foundation) – 2236/2 and L.A.V. received funding from the European Research Council
(ERC) under the European Union's Horizon 2020 research and innovation programme
(grant agreement no. ScaleOpt- 757481).

References

1. Frank, A., Miklós, Z.: Simple push-relabel algorithms for matroids and submodular flows. Jpn. J. Ind. Appl. Math. **29**, 419–439 (2012). https://doi.org/10.1007/s13160-012-0076-y
2. Murota, K., Shioura, A., Yang, Z.: Computing a Walrasian equilibrium in iterative auctions with multiple differentiated items. In: Cai, L., Cheng, S.-W., Lam, T.-W. (eds.) ISAAC 2013. LNCS, vol. 8283, pp. 468–478. Springer, Heidelberg (2013). https://doi.org/10.1007/978-3-642-45030-3_44

Dynamic Multinomial Logit Choice Model with Network Externalities: A Diffusive Analysis

Qing Feng[1]([⊠]) [iD] and Zizhuo Wang[2] [iD]

[1] School of Operations Research and Information Engineering, Cornel University,
Ithaca, USA
qf48@cornell.edu
[2] School of Data Science, The Chinese University of Hong Kong, Shenzhen
(CUHK-Shenzhen), China
wangzizhuo@cuhk.edu.cn

Abstract. We study a dynamic multinomial logit choice model with network externalities. In our model, a continuum of customers stay in a market for an infinite horizon. We adopt a random utility model in which the expected utility of each product is determined by its intrinsic value and the market share at the current period. In each period, for a customer, if the highest utility among products is higher than that of the no-purchase option, then this customer will purchase the product with the highest utility and leave. Otherwise, this customer will not purchase and will stay in the market in the next period. This process converges to a final market share. It presents a characterization of the diffusion of products among customers under network externalities (thus we call the model "diffusive model").

We study the properties of the diffusive model and compare it with a previously proposed MNL model with network externalities based on an equilibrium idea (we call it the "equilibrium model") in the literature. We find that under mild conditions the choice probabilities under the diffusive model are more balanced, and the total choice probability of all products is smaller compared to the equilibrium model, among other properties. Then we study the optimal pricing and assortment optimization problem under the diffusive model. For the optimal pricing problem, we propose an approximation scheme of the final market shares and establish efficient algorithms. For the assortment optimization problem, we give a lower bound for the performance of revenue-ordered assortments, propose a new class of assortments (k-proximity assortments), and study their optimality theoretically and numerically. We also verify the estimation power of the proposed model using real data.

Zizhuo Wang's research is partly supported by the National Science Foundation of China (NSFC) Grant 72150002, the Guangdong Key Laboratory of Mathematical Foundations for Artificial Intelligence and the Shenzhen Science and Technology Program under Grant ZDSYS20220606100601002.

J. Garg et al. (Eds.): WINE 2023, LNCS 14413, pp. 684–685, 2024.
https://doi.org/10.1007/978-3-031-48974-7

A full version of the paper is available at https://papers.ssrn.com/ abstract=3939717.

Keywords: Choice model · Network externalities · Pricing · Assortment optimization

PRINCIPRO: Data-Driven Algorithms for Joint Pricing and Inventory Control Under Price Protection

Qing Feng[1]([⊠])(iD) and Ruihao Zhu[2](iD)

[1] Cornell University School of Operations Research and Information Engineering, 206 Rhodes Hall, Hoy Rd, 14853 Ithaca, NY, USA
qf48@cornell.edu
[2] Cornell University SC Johnson College of Business, 14853 Ithaca, NY, USA
ruihao.zhu@cornell.edu

Abstract. We study the impact of price protection guarantee in joint pricing and inventory control under initially unknown and general customer demand. In this problem, the seller periodically makes a pair of pricing and inventory decision in the hope of maximizing total revenue. Meanwhile, the price protection guarantee endows the customers who purchased a product the right to receive a refund from the seller if she lowers the price during the price protection period (defined as a certain time interval after the purchase date). We propose PRINCIPRO, a novel online learning algorithm for joint PRicing and INventory Control under prIce PROtection, to achieve the optimal (up to logarithmic factors) regret for this setting. PRINCIPRO marries phased exploration (on pricing) and classic inventory learning. It explores each price in batches to reduce refund while simultaneously utilizing an inventory learning algorithm to identify the corresponding best order-up-to level. Critically, we exploit the geometric landscape in the inventory learning algorithm (originally intends for maximizing expected revenue rather than estimating its value) to construct tight confidence intervals for each price's maximal expected revenue. This enables PRINCIPRO to efficiently eliminate sub-optimal prices over time with only minimal assumptions on customer demand models. As a complement, we also develop an alternative end-to-end approach that directly translates an inventory learning algorithm's regret bound to confidence intervals, which can be of independent interest. Our work reveals that the optimal regret rate is a piecewise-linear function w.r.t. the length of the price protection period M. That is, when the value of M is moderate, the optimal regret rate grows linear as M; Otherwise, it stays in the respective lower and upper limits. To the best of our knowledge, this is the first work that delineates the role of price protection in the statistical complexity of online learning for joint pricing and inventory control[1].

Keywords: Dynamic pricing · Inventory control · Online learning · Price protection

---

[1] The complete paper is available at https://papers.ssrn.com/abstract=4511384.

J. Garg et al. (Eds.): WINE 2023, LNCS 14413, p. 686, 2024.
https://doi.org/10.1007/978-3-031-48974-7

Substitutes Markets with Budget Constraints: Solving for Competitive *and* Optimal Prices

Simon Finster[1] , Paul W. Goldberg[2] , and Edwin Lock[3]([⊠])

[1] CREST-ENSAE, 5 Avenue Henry Le Chatelier, 91120 Palaiseau, France
`simon.finster@ensae.fr`
[2] Department of Computer Science, University of Oxford, Parks Road, Oxford OX1 3QD, UK
`paul.goldberg@cs.ox.ac.uk`
[3] Nuffield College, University of Oxford, New Road, Oxford OX1 1NF, UK
`edwin.lock@economics.ox.ac.uk`

Abstract. Markets with multiple divisible goods have been studied widely from the perspective of revenue and welfare. In general, it is well known that envy-free, revenue-maximal outcomes can result in lower welfare than competitive equilibrium. In our market, buyers have quasilinear utilities with linear substitutes valuations and budget constraints, and the seller must find prices and an envy-free allocation that maximise revenue or welfare. This mirrors markets such as ad auctions and auctions for the exchange of financial assets.

We prove that the unique competitive equilibrium prices are also envy-free revenue-maximal. This coincidence of maximal revenue and welfare is surprising and breaks down even when buyers have piecewise-linear valuations. We present a novel characterisation of the set of 'feasible' prices (no excess demand), a non-convex set that we show to exhibit the lower semi-lattice structure. We demonstrate that elementwise-minimal prices maximise revenue and welfare. To prove welfare optimality, we adapt an existing algorithm for Fisher markets. Our procedure scales down any non-minimal feasible prices, maintaining feasibility, thus providing an algorithm for finding this unique price vector.

Our market is also called a 'quasi-Fisher' market. In contrast to standard Fisher markets, buyers spend nothing if prices are too high, making revenue maximisation an interesting objective for the seller. The market is also equivalent to an 'arctic product-mix auction' with zero seller costs, an auction developed to exchange financial assets with limited supply in Iceland.

We thank Paul Klemperer, Zaifu Yang, and anonymous reviewers for insightful discussions and comments. During the work on the final version of the paper, Goldberg and Lock were supported by a JP Morgan faculty fellowship. Work on the final version of this project was also supported by the National Science Foundation under Grant No. DMS-1928930 and by the Alfred P. Sloan Foundation under grant G-2021-16778, while Finster was in residence at the Simons Laufer Mathematical Sciences Institute (formerly MSRI) in Berkeley, California, during the Fall 2023 semester.

A full version of this paper can be found at https://arxiv.org/abs/2310.03692.

Keywords: Envy-freeness · Revenue maximisation · Competitive equilibrium · Budget constraints · Fisher market · Product-mix auction

Sequential Recommendation and Pricing Under the Mixed Cascade Model

Pin Gao[1], Yicheng Liu[1,2](✉), Chenhao Wang[1], and Zizhuo Wang[1]

[1] School of Data Science, The Chinese University of Hong Kong, Shenzhen
(CUHK-Shenzhen), China
{gaopin,wangzizhuo}@cuhk.edu.cn,
{yichengliu,chenhaowang}@link.cuhk.edu.cn
[2] Shenzhen Research Institute of Big Data, Shenzhen, China

Abstract. In many recommender systems, consumers usually make purchasing decisions over multiple pages. Complementing previous research, we employ a mixed cascade model to describe consumer decisions. Based on a real-world dataset, we find that the proposed model outperforms other multi-page choice models in terms of predictability. Due to the computational difficulty of estimating the distribution of different preference lists in the proposed model, we investigate the corresponding robust assortment optimization problem when the distribution is unknown and demonstrate the performance guarantees of this robust solution when the distribution is known. The results suggest that a sequential revenue-ordered recommendation can perform very well, demonstrating the limits of personalization and the diminishing marginal contribution of consumer patience to platform revenue. Lastly, we show that the corresponding joint pricing and assortment optimization problem is NP-hard, and develop some heuristics with constant-ratio performance guarantees.

Keywords: Cascade model · Sequential recommendation · Assortment optimization · Approximation

---

Link to the full paper: https://papers.ssrn.com/sol3/papers.cfm?abstract_id=4382163.
Pin Gao's research is partly supported the National Natural Science Foundation of China [Grants 72201234 and 72192805] and the Research Grant Council of Hong Kong (the Collaborative Research Fund) [Grant C6032-21G]. Zizhuo Wang's research is partly supported by the National Science Foundation of China (NSFC) Grant 72150002, The Guangdong Key Lab of Mathematical Foundations for Artificial Intelligence and the Shenzhen Science and Technology Program under Grant ZDSYS20220606100601002.

J. Garg et al. (Eds.): WINE 2023, LNCS 14413, p. 689, 2024.
https://doi.org/10.1007/978-3-031-48974-7

Best-Response Dynamics in Tullock Contests with Convex Costs

Abheek Ghosh[(✉)] [iD]

Department of Computer Science, University of Oxford, Oxford, UK
abheek.ghosh@cs.ox.ac.uk

Abstract. We study the convergence of best-response dynamics in Tullock contests with convex cost functions (these games always have a unique pure-strategy Nash equilibrium). We show that best-response dynamics rapidly converges to the equilibrium for homogeneous agents. For two homogeneous agents, we show convergence to an ϵ-approximate equilibrium in $\Theta(\log \log(1/\epsilon))$ steps. For $n \geq 3$ agents, the dynamics is not unique because at each step $n - 1 \geq 2$ agents can make non-trivial moves. We consider the model proposed by Ghosh and Goldberg (2023), where the agent making the move is randomly selected at each time step. We show convergence to an ϵ-approximate equilibrium in $O(\beta \log(n/(\epsilon\delta)))$ steps with probability $1 - \delta$, where β is a parameter of the agent selection process, e.g., $\beta = n^2 \log(n)$ if agents are selected uniformly at random at each time step. We complement this result with a lower bound of $\Omega(n + \log(1/\epsilon)/\log(n))$, which is applicable for any agent selection process.

One of the novel contributions of the paper is the introduction and analysis of the *discounted-sum* dynamics, which we use in the analysis of the best-response dynamics for $n \geq 3$ agents. This dynamics proceeds as follows: starting from an initial state $\boldsymbol{x}_0 \in \mathbb{R}^n$, at each time t, the dynamics picks a coordinate $i_t \in [n]$ and updates the value at the i_t-th coordinate with the negative discounted sum of the values at the remaining coordinates, i.e., $x_{t+1,i_t} = -\beta_t \sum_{j \neq i_t} x_{t,j}$. The discount factor $\beta_t \in [0, B]$, where $B < 1$, can be picked arbitrarily, possibly adversarially, and can depend upon t, i_t, and $\boldsymbol{x}_t$. We show that this dynamics rapidly converges to 0 using a potential function.

Keywords: Contests · Best-response dynamics · Learning in games

Full version is available at https://arxiv.org/abs/2310.03528.

MNL-Prophet: Sequential Assortment Selection Under Uncertainty

Vineet Goyal[1] , Salal Humair[2] , Orestis Papadigenopoulos[1]([⊠]) ,
and Assaf Zeevi[1]

[1] Columbia University, New York, NY, USA
vgoyal@ieor.columbia.edu, opapadig@columbia.edu, assaf@gsb.columbia.edu
[2] Amazon, Inc., Sammamish, WA, USA
salal@amazon.com

Assortment selection is a core problem in the area of revenue management due to its applications in retail and online advertising. Given a universe of potential products, each associated with a unit revenue, and a discrete choice model capturing the consumers' demand for different alternatives, the objective is to compute an assortment (i.e., subset) to offer to a customer in order to maximize the expected revenue of the seller. The problem has been widely studied under various combinations of demand models and feasibility constraints on the assorted goods. In several modern applications, however, the decision of whether to include or not a product in the assortment often has to be made within a short time interval and without accurate knowledge of all the other alternatives; in such cases, existing offline approaches become inapplicable.

We consider a stochastic variant of assortment selection, where the parameters that determine the revenue and (relative) demand of each product are jointly drawn from some known distribution, independently across products. A decision-maker sequentially observes the products, together with their realized parameters, in an arbitrary and unknown order. Upon observing each product, s/he decides instantaneously and irrevocably whether to include it or not in the constructed assortment. The objective is to maximize the expected total revenue of the resulting assortment, competing against an offline algorithm that foresees all the parameter realizations and computes the optimal assortment.

We provide simple threshold-based online policies for the unconstrained and cardinality-constrained versions of the problem, in the case where the demand is captured by the well-known Multinomial Logit (MNL) choice model. In particular, we prove a worst-case optimal $(1 + \gamma)$-competitive guarantee for the unconstrained setting, where $\gamma \in [0, 1]$ is the expected probability that the consumer purchases (any item) in the optimal assortment; for cardinality constraints, we prove a worst-case optimal 2-competitive guarantee. Interestingly, our unconstrained policy is oblivious to the realized parameters of the products that are related to demand. We extend our results to a natural class of substitutable choice models satisfying certain conditions and design near-optimal policies for the case of knapsack constraints. Finally, we discuss interesting connections to the Prophet Inequality problem, which is already subsumed by our setting. A full version of our work can be found at https://arxiv.org/abs/2308.05207.

Fair Incentives for Repeated Engagement
(Extended Abstract)

Daniel Freund[1] and Chamsi Hssaine[2]([⊠])

[1] MIT Sloan School of Management, Cambridge, MA 08544, USA
dfreund@mit.edu
[2] University of Southern California, Marshall School of Business, Los Angeles, CA
90089, USA
hssaine@marshall.usc.edu

Keywords: Stochastic models · Algorithmic fairness · Platform churn

This work aims to understand the role of fairness in the optimal design of unconditional monetary incentives for repeated engagement. We consider a model wherein agents join a system in each discrete-time period, and receive a possibly random reward to remain in the system in the next period. At the end of the period, agents probabilistically make a decision based on the reward received in the period, or leave once and for all. Agents are partitioned into types defined by (i) a departure function that maps rewards received to the probability of departing, assumed to be non-increasing in the reward paid out, and (ii) the rate at which they join the system. The decision-maker collects revenue associated with the number of agents in the system in each period, and incurs the cost of incentivizing these agents to stay. The goal of the decision-maker is to determine an optimal sequence of reward distributions to maximize her long-run average profit, subject to two fairness requirements: (i) agents must be paid from the same reward distribution in each period, and (ii) different agent types must experience the same reward distribution on average, over a long enough time horizon.

Informal Theorem 1. There exists an asymptotically optimal fair policy that is static, and whose relative loss versus the optimal policy converges to 0 at a rate of $\mathcal{O}(\frac{1}{\sqrt{\theta}})$. Under mild additional conditions on the revenue function, this rate improves to $\mathcal{O}(\frac{1}{\theta})$.

While this static heuristic is conceptually simple, it requires us to solve an a priori intractable high-dimensional nonconvex optimization problem. In our second main technical contribution, we show a structural property of the problem that allows us to efficiently compute its optimal solution.

Full version of paper can be found at https://arxiv.org/pdf/2111.00002.pdf.

J. Garg et al. (Eds.): WINE 2023, LNCS 14413, pp. 693–694, 2024.
https://doi.org/10.1007/978-3-031-48974-7

Informal Theorem 2. For any instance, the static heuristic places positive weight on *at most* two rewards.

This allows for an optimal solution to be found via exhaustive search over pairs of rewards, and then solving a KKT condition consisting of a single equation in one variable for each pair. Thus, our main algorithmic contribution is an efficiently computable and asymptotically optimal fair policy.

Markov Persuasion Processes with Endogenous Agent Beliefs

Krishnamurthy Iyer[1](✉)(iD), Haifeng Xu[2](iD), and You Zu[1](iD)

[1] Industrial and Systems Engineering, University of Minnesota, Minneapolis, USA
{kriyer,zu000002}@umn.edu
[2] Department of Computer Science, University of Chicago, Chicago, USA
haifengxu@uchicago.edu

We consider a dynamic Bayesian persuasion setting where a single long-lived sender persuades a stream of "short-lived" agents (receivers) by sharing information about a payoff-relevant state. The state transitions are Markovian conditional on the receivers' actions, and the sender seeks to maximize the long-run average reward by committing to a (possibly history-dependent) signaling mechanism. Such problems are common in platform markets, where the platform seeks to achieve desirable long-term revenue and welfare outcomes by influencing the actions of users. While most previous studies of Markov persuasion consider exogenous agent beliefs that are independent of the chain, we study a more natural variant with *endogenous* agent beliefs that depend on the chain's realized history. A key challenge to analyzing such settings is to model the agents' partial knowledge about the history information. To address this challenge, we analyze a Markov persuasion process (MPP) under various information models that differ in the amount of information the receivers have about the history of the process. Specifically, we formulate a general partial-information model where each receiver observes the history with an ℓ period lag (for $\ell \geq 0$). Our technical contribution starts with analyzing two benchmark models, i.e., the full-history information model (i.e., $\ell = 0$) and the no-history information model (i.e., $\ell = \infty$). We establish an ordering of the sender's payoff as a function of the informativeness of agent's information model (with no-history as the least informative), and develop efficient algorithms to compute optimal signaling mechanisms for these two benchmarks. For general values of the lag ℓ, we present the technical challenges in finding an optimal signaling mechanism, where even determining the right dependency on the history becomes difficult. Restricting the dependence on the history to a given length, we formulate the sender's problem as a bilinear optimization program. To bypass the resulting computational complexity, we use a robustness framework to design a "simple" *history-independent* signaling mechanism that approximately achieves optimal payoff when ℓ is reasonably large.

The full paper is available at https://arxiv.org/abs/2307.03181.

Acknowledgements. Haifeng Xu is supported in part by an NSF Award CCF-2132506, an Army Research Office Award W911NF-23-1-0030, and an Office of Naval Research Award N00014-23-1-2802.

© The Author(s), under exclusive license to Springer Nature Switzerland AG 2024
J. Garg et al. (Eds.): WINE 2023, LNCS 14413, p. 695, 2024.
https://doi.org/10.1007/978-3-031-48974-7

Stochastic Online Fisher Markets: Static Pricing Limits and Adaptive Enhancements

Devansh Jalota$^{(\boxtimes)}$ and Yinyu Ye

Stanford University, Stanford, CA 94305, USA
{djalota,yyye}@stanford.edu

Abstract. In a Fisher market, agents (users) spend a budget of (artificial) currency to buy goods that maximize their utilities while a central planner sets prices on capacity-constrained goods to clear the market. However, the efficacy of pricing schemes in achieving an equilibrium outcome in Fisher markets typically relies on complete knowledge of users' budgets and utility functions and requires that transactions happen in a static market where all users are present simultaneously.

Motivated by these practical considerations, in this work, we study an online variant of Fisher markets, wherein budget-constrained users with privately known utility and budget parameters, drawn i.i.d. from a distribution $\mathcal{D}$, enter the market sequentially. In this setting, we develop an algorithm that adjusts prices solely based on observations of user consumption, i.e., revealed preference feedback, and achieves a regret and capacity violation of $O(\sqrt{n})$, where n is the number of users and the good capacities scale as $O(n)$. Here, our regret measure is the optimality gap in the objective of the Eisenberg-Gale program between an online algorithm and an offline oracle with complete information on users' budgets and utilities. To establish the efficacy of our approach, we show that any uniform (static) pricing algorithm, including one that sets expected equilibrium prices with complete knowledge of the distribution $\mathcal{D}$, cannot achieve both a regret and capacity violation of less than $\Omega(\sqrt{n})$. While our revealed preference algorithm requires no knowledge of the distribution $\mathcal{D}$, we show that if $\mathcal{D}$ is known, then an adaptive variant of expected equilibrium pricing achieves $O(\log(n))$ regret and constant capacity violation for discrete distributions. Finally, we present numerical experiments to demonstrate the performance of our revealed preference algorithm relative to several benchmarks[1].

Keywords: Fisher market · Online learning · Convex optimization · Regret · Revealed preference

---

[1] (This work was supported by the Stanford Interdisciplinary Graduate Fellowship (SIGF)).

Link to the full paper: https://arxiv.org/pdf/2205.00825.pdf.

The Colonel Blotto Game on Measure Spaces

Siddhartha Visveswara Jayanti[(✉)] [ID]

Google Research, Atlanta, GA 30309, USA
sjayanti@google.com

Abstract. The Colonel Blotto Problem proposed by Borel in 1921 has served as a widely applicable model of budget-constrained simultaneous winner-take-all competitions in the social sciences. Applications include elections, advertising, R&D and more. However, the classic Blotto problem and variants limit the study to competitions over a finite set of discrete battlefields. In this paper, we extend the classical theory to study multiplayer Blotto games over arbitrary measurable battlegrounds, provide an algorithm to efficiently sample equilibria of symmetric "equipartionable" Generalized Blotto games, and characterize the symmetric fair equilibria of the Blotto game over the unit interval.

Keywords: Blotto · Lotto · Measure space · Nash equilibrium · Multiplayer · Characterization · Algorithm · Equipartitionable

Contributions

1. We introduce the novel *Measure Space Blotto* and *Lotto* games which model multiplayer budget-constrained competitions over arbitrary measurable battlegrounds.
2. We extend Boix-Adserá, Edelman, and Jayanti's theory for multiplayer standard Blotto to solve for equilibria in a large class of Measure Space Blotto games. In particular, we compute mixed Nash equilibria for the class of symmetric "equipartitionable" Measure Space Blotto games for any number of players $k \in (1, \infty)$. The class includes Interval Blotto, Blotto on the Sphere, as well as the case of multiplayer Blotto that has been solved for $k > 3$ players. Furthermore, We design an efficient, $O(k)$ time algorithm for each player to sample an equilibrium strategy.
3. We characterize the symmetric fair equilibria of Interval Blotto for an arbitrary number of players. We believe the characterization is the technical highlight of our paper.

A full version of this paper is available at https://arxiv.org/abs/2104.11298

Partially supported by the Department of Defence through the NDSEG fellowship.

Assortment Optimization in the Presence of Focal Effect: Operational Insights and Efficient Algorithms

Bo Jiang[1], Zizhuo Wang[2], Chenyu Xue[1(✉)], and Nanxi Zhang[3]

[1] Research Institute for Interdisciplinary Sciences, Shanghai University of Finance and Economics and Key Laboratory of Interdisciplinary Research of Computation and Economics, Shanghai University of Finance and Economics, Shanghai, China
`jiang.bo@mail.shufe.edu.cn, xcy2721d@gmail.com`
[2] School of Data Science, The Chinese University of Hong Kong, Shenzhen (CUHK-Shenzhen), China
`wangzizhuo@cuhk.edu.cn`
[3] Ivey Business School, University of Western Ontario, London, ON, Canada
`nzhang@ivey.ca`

Abstract. This paper considers the scenario where the assortment provided by the seller can influence customer's evaluation of item utility. A possible consequence is that certain items in an assortment become "star items" to customers, and customers over-evaluate their utilities. We call such a phenomenon the *focal effect*. Kovach and Tserenjigmid (2022) propose a focal Luce model (FLM) to describe customers choice in the presence of the focal effect. FLM can be viewed as a new variant of the classic Luce model (also known as the multinomial logit choice model) equipped with a "focal machine" to describe the focal effect. The merit of FLM lies in its great flexibility to capture customer's different psychologies that lead to different customers choice behaviors. As a result, FLM can serve as a general framework to cover many practical scenarios. Although the assortment optimization problem under the FLM is NP-hard in general, we still identify some structures of the optimal assortment that lead to operational insights of the model. We find that the assortment optimization can be solved in polynomial time by imposing assumptions on the model's "focal machine", and these assumptions are satisfied in many practical scenarios. Furthermore, this polynomial-solvability is reservedeven for the joint assortment and pricing optimization problem.

Link to the full paper: https://papers.ssrn.com/sol3/papers.cfm?abstract_id=4504023. The first author's research is partly supported by the National Natural Science Foundation of China [Grants 11831002, 72150001, and 72171141] and the Program for Innovative Research Team of Shanghai University of Finance and Economics. The second author's research is partly supported by the National Science Foundation of China (NSFC) Grant 72150002, The Guangdong Key Lab of Mathematical Foundations for Artificial Intelligence and the Shenzhen Science and Technology Program under Grant ZDSYS20220606100601002.

J. Garg et al. (Eds.): WINE 2023, LNCS 14413, pp. 698–699, 2024.
https://doi.org/10.1007/978-3-031-48974-7

Therefore, the great flexibility of FLM in capturing the focal effect in many scenarios does not come at a high operational tractability cost.

Keywords: Focal effect · Assortment optimization · Focal luce model

On Hill's Worst-Case Guarantee for Indivisible Bads

Bo Li[1], Hervé Moulin[2], Ankang Sun[1]([✉]), and Yu Zhou[1]

[1] The Hong Kong Polytechnic University, Hong Kong, China
{comp-bo.li,ankang.sun}@polyu.edu.hk, yu-phd.zhou@connect.polyu.hk
[2] University of Glasgow, Glasgow, UK
herve.moulin@glasgow.ac.uk

Abstract. When allocating objects among agents with equal rights, people often evaluate the fairness of an allocation rule by comparing their received utilities to a benchmark *share* – a function only of her own valuation and the number of agents. This share is called a *guarantee* if for any profile of valuations there is an allocation ensuring the share of every agent. When the objects are indivisible goods, Budish [J. Political Econ., 2011] proposed MaxMinShare, i.e., the least utility of a bundle in the best partition of the objects, which is unfortunately not a guarantee. Instead, an earlier pioneering work by Hill [Ann. Probab., 1987] proposed for a share the worst-case MaxMinShare over all valuations with the same largest possible single-object value. Although Hill's share is more conservative than the MaxMinShare, it is an actual guarantee and its computation is elementary, unlike that of the MaxMinShare which involves solving an NP-hard problem. We apply Hill's approach to the allocation of indivisible bads (objects with disutilities or costs), and characterise the tight closed form of the worst-case MinMaxShare for a given value of the worst bad. We argue that Hill's share for allocating bads is effective in the sense of being close to the original MinMaxShare value, and there is much to learn about the guarantee an agent can be offered from the disutility of her worst single bad. Furthermore, we prove that the monotonic cover of Hill's share is the best guarantee that can be achieved in Hill's model for all allocation instances.

Keywords: Fair division · Indivisible bads · Hill's share

The authors are ordered alphabetically. The full version of this paper can be found in https://arxiv.org/abs/2302.00323. This work is funded by NSFC under Grant No. 62102333, HKSAR RGC under Grant No. PolyU 25211321, and GDSTC under Grant No. 2023A1515010592.

J. Garg et al. (Eds.): WINE 2023, LNCS 14413, p. 700, 2024.
https://doi.org/10.1007/978-3-031-48974-7

Prophet Inequality on I.I.D. Distributions: Beating $1 - 1/e$ with a Single Query

Bo Li[1] , Xiaowei Wu[2]([✉]) , and Yutong Wu[3]

[1] The Hong Kong Polytechnic University, Hong Kong, China
comp-bo.li@polyu.edu.hk
[2] University of Macau, Macau, China
xiaoweiwu@um.edu.mo
[3] The University of Texas at Austin, Austin, USA
yutong.wu@utexas.edu

Abstract. In this work, we study the single-choice prophet inequality problem, where a gambler faces a sequence of n online i.i.d. random variables drawn from an unknown distribution. When a variable reveals its value, the gambler needs to decide irrevocably whether or not to accept the value. The goal is to maximize the competitive ratio between the expected gain of the gambler and that of the maximum variable. It is shown by Correa et al. (EC 2019) that when the distribution is unknown or only $o(n)$ uniform samples from the distribution are given, the best an algorithm can do is $1/e$-competitive. In contrast, when the distribution is known (Correa et al., EC 2017) or $\Omega(n)$ uniform samples are given (Rubinstein et al., ITCS 2020), the optimal competitive ratio 0.7451 can be achieved. In this paper, we study a new model in which the algorithm has access to an oracle that answers quantile queries about the distribution and investigate to what extent we can use a small number of queries to achieve good competitive ratios. We first use the answers from the queries to implement the threshold-based algorithms and show that with two thresholds our algorithm achieves a competitive ratio of 0.6786. Motivated by the two-threshold algorithm, we design the observe-and-accept algorithm that requires only a single query. This algorithm sets a threshold in the first phase by making a query and uses the maximum realization from the first phase as the threshold for the second phase. It can be viewed as a natural combination of the single-threshold algorithm and the algorithm for the secretary problem. By properly choosing the quantile to query and the break-point between the two phases, we achieve a competitive ratio of 0.6718, beating the benchmark of $1 - 1/e$.

Keywords: Prophet inequality · IID distributions · Quantile query

The authors are ordered alphabetically. The full version of the paper can be found at https://arxiv.org/abs/2205.05519. Xiaowei Wu is funded by the Science and Technology Development Fund (FDCT), Macau SAR (file no. 0014/2022/AFJ, 0085/2022/A, 0143/2020/A3 and SKL-IOTSC-2021-2023). Bo Li is funded by the National Natural Science Foundation of China (No. 62102333) and Hong Kong SAR Research Grants Council (No. PolyU 25211321).

J. Garg et al. (Eds.): WINE 2023, LNCS 14413, p. 701, 2024.
https://doi.org/10.1007/978-3-031-48974-7

Allocating Emission Permits Efficiently via Uniform Linear Mechanisms

Xingyu Lin[1]($\boxtimes$) and Jiaqi Lu[2]

[1] School of Data Science, The Chinese University of Hong Kong, Shenzhen 518172, China
xingyulin@link.cuhk.edu.cn
[2] School of Data Science, School of Management and Economics, The Chinese University of Hong Kong, Shenzhen 518172, China
lujiaqi@cuhk.edu.cn

Abstract. Cap and trade (CAT) is a popular approach to regulate polluters, and has been widely implemented in practice. One central problem in operating CAT systems is the initial allocation of permits. In this work, we study how to initially allocate emission permits in a CAT system where firms also compete in a production market a la Cournot. We provide efficiency guarantee of simple uniform linear allocation mechanisms in the broad class of component-wise concave mechanisms. It was well accepted in the literature that the equilibrium consumer surplus and social welfare are independent of the initial allocation of emission permits in a deterministic system without trading frictions, and that CAT and Tax systems are equivalent. However, the initial allocation mechanisms previously considered in the literature were largely restricted to constant ones that do not take the firms' current production quantities as input. We show that, by allowing more general allocation mechanisms that are component-wise concave in the firm's production decision, which capture many realistic allocation rules including lump sum allocations (such as grandfathering), output-based allocations (either top-down or bottom-up), etc., the system's equilibrium outcome will no longer be independent of the initial allocations.

In particular, we consider N firms operating under Cournot competition with profit-maximizing production decisions q. The regulator needs to choose a permit allocation mechanism $\Phi_i(q)$ that determines the number of permits to allocate to firm i as a function of the N firms' production decision vector q. We consider any component-wise concave mechanisms and show that when the firms differ only in their abatement abilities, uniform linear permit allocation mechanisms, i.e., $\Phi_i(q) = \alpha q_i$ for some constant α, achieve Pareto efficiency in consumer surplus and total pollution, which also achieve efficiency in total emission reduction. With this result, the regulator's infinite-dimensional policy design question can be reduced to a single-dimensional one, and the original N-firm system can be equivalently represented by a monopoly so that no firm's

J. Lu gratefully acknowledges financial support from Natural Science Foundation of China (NSFC) [Project 72192805].

private information is required. In addition, we show that the corresponding CAT system under uniform linear allocation is equivalent to a generalized carbon Tax regime, which sheds new light on the previously established equivalency between CAT and Taxes. Numerical experiments show that the benefit of uniform linear mechanisms compared to lump sum ones can be large. We also explain when the efficiency of uniform linear allocation mechanisms might fail and give managerial insight into the design of allocation coefficient in practically used output-based allocation methods.

A complete version of the paper can be found at https://ssrn.com/abstract=4536951.

Collective Search in Networks

Niccolò Lomys[(✉)] [iD]

CSEF and Universitá degli Studi di Napoli Federico II, Via Cintia 21, Monte
Sant'Angelo, 80126 Naples, Italy
niccolomys@gmail.com

Abstract. I study the dynamics of collective search in networks.
Bayesian agents act in sequence, observe the choices of their connections, and privately acquire information about the qualities of different
actions via sequential search. If search costs are not bounded away from
zero, maximal learning occurs in sufficiently connected networks where
individual neighborhood realizations weakly distort agents' beliefs about
the realized network. If search costs are bounded away from zero, maximal learning is possible in several stochastic networks, including almost-
complete networks, but generally fails otherwise. When agents observe
random numbers of immediate predecessors, the learning rate, the probability of wrong herds, and long-run efficiency properties are the same
as in the complete network. The density of indirect connections affects
convergence rates. Network transparency has short-run implications for
welfare and efficiency.

The full paper is available here.

Keywords: Networks · Bayesian learning · Search · Speed and
efficiency of social learning

This work was done in part while visiting the Simons Institute for the Theory of
Computing at UC Berkeley (program on Data-Driven Decision Processes). Funding
from the European Research Council (ERC) under the European Union's Horizon
2020 research and innovation program (grant agreements № 714147 and № 714693)
is gratefully acknowledged.

J. Garg et al. (Eds.): WINE 2023, LNCS 14413, p. 704, 2024.
https://doi.org/10.1007/978-3-031-48974-7_7

The Limits of School Choice with Consent

Josué Ortega[1][(✉)] and Gabriel Ziegler[2]

[1] Queen's University Belfast, Belfast, UK
j.ortega@qub.ac.uk
[2] The University of Edinburgh, Edinburgh, UK

Abstract. The deferred acceptance (DA) algorithm is used all around the world to assign pupils to schools due to its stability and strategy-proofness. However, DA can generate an assignment that is Pareto inefficient for students and, in some cases, it may assign all students to their worst or second worst school. The inefficiency of DA has been also documented in real life.

[1] proposed an influential modification of DA, known as efficiency-adjusted deferred acceptance (EADA), which fixes the Pareto inefficiency of DA. A vast literature has now shown that EADA is not only Pareto efficient, but also satisfies a plethora of desirable properties both theoretically and in the lab. For these reasons, EADA is to be implemented in Flanders.

Despite these good properties, we show that EADA can generate allocations that are significantly rank-inefficient and unequal. In particular, we prove that for any number of students, there exists a school choice problem where half of the students are allocated to a school in the bottom half of their preferences according to EADA, even though a different allocation assigns all but one student to their top school, and no student to a school worse than their second choice.

Furthermore, we analyze random, iid one-to-one matching problems. In our main Theorem, we approximate the asymptotic expected average rank in EADA, which is in the order of $\log(n!)/n$. This result has an interesting corollary: the quality of student placement in EADA is asymptotically the same as in DA, which implies that the size of the efficiency gains of EADA over DA vanish when the number of students is large. This result explains the small efficiency gains of EADA versus DA documented empirically [2].

The full paper is available at www.josueortega.com.

Keywords: School choice · Rank-efficiency · Efficiency-adjusted DA

References

1. Kesten, O.: School choice with consent. Q. J. Econ. **125**(3), 1297–1348 (2010)
2. Ortega, J., Klein, T.: The cost of strategy-proofness in school choice. Games Econom. Behav. **141**, 515–528 (2023)

© The Author(s), under exclusive license to Springer Nature Switzerland AG 2024
J. Garg et al. (Eds.): WINE 2023, LNCS 14413, p. 705, 2024.
https://doi.org/10.1007/978-3-031-48974-7

Binary Mechanisms Under Privacy-Preserving Noise

Farzad Pourbabaee[1] and Federico Echenique[2]($\boxtimes$)

[1] Division of the Humanities and Social Sciences, Caltech, 1200 E. California Blvd, Pasadena, CA 91125, USA
far@caltech.edu
[2] Department of Economics, UC Berkeley, 530 Evans Hall, Berkeley, CA 94720, USA
fede@econ.berkeley.edu

Abstract. We study mechanism design for public-good provision under a noisy privacy-preserving transformation of individual agents' reported preferences. The setting is a standard binary model with transfers and quasi-linear utility. Agents report their preferences for the public good, which are randomly "flipped," so that any individual report may be explained away as the outcome of noise. We study the tradeoffs between preserving the public decisions made in the presence of noise (noise sensitivity), pursuing efficiency, and mitigating the effect of noise on revenue.

Keywords: Mechanism design · Boolean analysis · Privacy

Extended Abstract. We study the problem of public-good provision in a setting where an individual's type for the public good is a binary variable (i.e., agent i's type is $x_i \in \{-1, +1\}$). Moreover, individuals have quasilinear utilities over the provision outcome $f \in \{0, 1\}$ and the monetary transfer t_i. We propose adding calibrated noise to the messages sent by the individuals to the planner, thereby protecting their privacy. The noisy transformation is simply a flip from the message $m_i \in \{-1, +1\}$ sent by the agent i to $-m_i$, that happens independently across individuals with the common noise probability of δ. Therefore, the planner instead of receiving the true vector of types $x = (x_1, \ldots, x_n)$ will receive the noisy vector $y = (y_1, \ldots, y_n)$, where n is the number of agents, and y_i is the noisy version of x_i. For a given social choice function, say $f : \{-1, +1\}^n \to \{0, 1\}$, we are interested in the *noise sensitivity*, i.e., $\mathsf{P}_\delta(f(x) \neq f(y))$, that represents how robust is this function to the noise. We then study the conventional economic objectives such as implementability, revenue and surplus. In particular, we show that increasing the noise probability δ, that leads to a better protection of agents' privacy, comes at the cost of lower revenue and social surplus. Therefore, there is a natural tradeoff between privacy protection and the standard economis measures (revenue and surplus). Our main finding is an asymptoticcharacterization

© The Author(s), under exclusive license to Springer Nature Switzerland AG 2024
J. Garg et al. (Eds.): WINE 2023, LNCS 14413, pp. 706–707, 2024.
https://doi.org/10.1007/978-3-031-48974-7

of the allocation rule f that has the minimum noise sensitivity while raising a target revenue level, that turns out to follow a *linear threshold function*. The leading tool in our proof is an Isoperimetric inequality on the Boolean hypercube. The full paper can be found at https://arxiv.org/abs/2301.06967.

Learning Non-parametric Choice Models with Discrete Fourier Analysis

Haoyu Song[1]([⊠]), Hai Nguyen[2], and Thanh Nguyen[1]

[1] Purdue University, West Lafayette, IN, USA
{song522,nguye161}@purdue.edu
[2] ETH Zurich, Zurich, Switzerland
haihoang.nguyen@inf.ethz.ch

Abstract. Choice modeling provides important information for many economic and operational decisions. Recently, non-parametric choice models, such as decision forests and random rankings, receive a great deal of attention due to their universality. It is known that all rational choice functions (RUM) can be expressed as a distribution over rankings, and more impressively, *any* choice function, whether rational or not, can be written as a decision forest. However, a major obstacle is that the parameter space of those non-parametric choices is factorial in the number of products, and as a result practitioners often rely on heuristics to estimate non-parametric choice models. These heuristics lack theoretical guarantee on sample/computational efficiency and can also be far from optimal in practice.

To address this challenge, we propose a solution using discrete Fourier analysis. We demonstrate that any choice model can be approximated with a small number of parameters in the Fourier domain. Then combining the classical Goldreich-Levin algorithm with hashing, we design a provably sample-efficient, active-learning algorithm to estimate those choice models. The algorithm does not require an explicit description of the targeted choice model and needs at most $poly(logn, \frac{1}{\epsilon})$ queries of data to estimate any choice function up to ϵ accuracy. We also complement the theoretical results with computational studies. The experimental results show that on average, Fourier methods obtain a reduction of RMSE (root mean squared error) by more than 20% and MAPE (mean absolute percentage error) by more than 10% when compared against MNL and two common heuristics (random sampling and heuristic column generation) for non-parametric choice estimation. A link to the full-paper can be found at https://web.ics.purdue.edu/~nguye161/WINE2022_Fourier.pdf

Keywords: Non-parametric choice estimation · Discrete fourier analysis · Active learning

J. Garg et al. (Eds.): WINE 2022, LNCS 14413, p. 708, 2024.
https://doi.org/10.1007/978-3-031-48974-7

Threshold Policies with Tight Guarantees for Online Selection with Convex Costs

Xiaoqi Tan[1(✉)], Siyuan Yu[1], Raouf Boutaba[2], and Alberto Leon-Garcia[3]

[1] University of Alberta, Edmonton, AB, Canada
{xiaoqi.tan,syu3}@ualberta.ca
[2] University of Waterloo, Waterloo, ON, Canada
rboutaba@uwaterloo.ca
[3] University of Toronto, Toronto, ON, Canada
alberto.leongarcia@utoronto.ca

Abstract. This paper provides threshold policies with tight guarantees for online selection with convex cost (OSCC). In OSCC, a seller wants to sell some asset to a sequence of buyers with the goal of maximizing her profit. The seller can produce additional units of the asset, but at non-decreasing marginal costs. At each time, a buyer arrives and offers a price. The seller must make an immediate and irrevocable decision in terms of whether to accept the offer and produce/sell one unit of the asset to this buyer. The goal is to develop an online algorithm that selects a subset of buyers to maximize the seller's profit, namely, the total selling revenue minus the total production cost. Our main result is the development of a class of simple threshold policies that are logistically simple and easy to implement, but have provable optimality guarantees among all deterministic algorithms. We also derive a lower bound on competitive ratios of randomized algorithms and prove that the competitive ratio of our threshold policy asymptotically converges to this lower bound when the total production output is sufficiently large. Our results generalize various online search, pricing, and auction problems, and provide a new perspective on the impact of non-decreasing marginal costs on real-world online resource allocation problems.

The full version of this paper is available at: https://arxiv.org/pdf/2310.06166.pdf.

Keywords: Online algorithms · Competitive analysis · Resource allocation

J. Garg et al. (Eds.): WINE 2023, LNCS 14413, p. 709, 2024.
https://doi.org/10.1007/978-3-031-48974-7

Best Cost-Sharing Rule Design for Selfish Bin Packing

Changjun Wang[1]([✉])[iD] and Guochuan Zhang[2][iD]

[1] Academy of Mathematics and Systems Science, Chinese Academy of Sciences,
Beijing 100190, China
wcj@amss.ac.cn

[2] College of Computer Science, Zhejiang University, Hangzhou, China
zgc@zju.edu.cn

Abstract. In selfish bin packing, each item is regarded as a selfish
player, who aims to minimize the cost-share by choosing a bin it can
fit in. To have a least number of bins used, cost-sharing rules play an
important role. The currently best known cost-sharing rule has a *price
of anarchy* (*PoA*) larger than 1.45, while a general lower bound $4/3$ on
PoA applies to any cost-sharing rule under which no items have the
incentive to move unilaterally to an empty bin. In this paper, we propose a novel and simple cost-sharing scheme for selfish bin packing that
satisfies many nice and natural properties. In the scheme, by selecting
a proper threshold parameter, we obtain a cost-sharing rule that always
admits a Nash equilibrium being the optimal packing, which implies the
price of stability (*PoS*) of the game is one. Furthermore, under this new
rule, the game's *PoA* matches the general lower bound of $4/3$, thus completely resolving this game. Besides, the well-known bin packing algorithm *BFD* (Best-Fit Decreasing) is shown to achieve a strong Nash
equilibrium under the cost-sharing rule with another threshold parameter, implying that a stable packing with an asymptotic approximation
ratio of $11/9$ can be produced in polynomial time. As an extension of the
designing framework, we further study a variant of the selfish scheduling
game, and design a best coordination mechanism achieving $PoS = 1$ and
$PoA = 4/3$ as well.

Keywords: Bin packing game · Cost-sharing · Price of anarchy ·
Tight bound

A full version of this work can be found at http://arxiv.org/abs/2204.09202.

Changjun Wang was partially supported by National Natural Science Foundation
of China (No. 11971046). Guochuan Zhang was supported by Science and Technology Innovation 2030 - "New Generation Artificial Intelligence" Major Project
(No. 2018AAA0100902), China.

J. Garg et al. (Eds.): WINE 2023, LNCS 14413, p. 710, 2024.
https://doi.org/10.1007/978-3-031-48974-7

Most Equitable Voting Rules

Lirong Xia$^{(\boxtimes)}$

RPI, Troy, NY 12180, USA
xialirong@gmail.com

Abstract. In social choice theory, *anonymity* (all agents being treated equally) and *neutrality* (all alternatives being treated equally) are widely regarded as *"minimal demands"* and *"uncontroversial"* axioms of equity and fairness. However, the *ANR impossibility*—there is no voting rule that satisfies anonymity, neutrality, and resolvability (always choosing one winner)—holds even in the simple setting of two alternatives and two agents. How to design voting rules that optimally satisfy anonymity, neutrality, and resolvability remains an open question.

We address the optimal design question for a wide range of preferences and decisions that include ranked lists and committees. Our conceptual contribution is a novel and strong notion of *most equitable refinements* that optimally preserves anonymity and neutrality for any irresolute rule that satisfies the two axioms. This is a strong notion of optimality and is not guaranteed to exist by definition, because it requires that a most equitable refinement of $\bar{r}$ achieves the same or higher ANR satisfaction than *every* refinement of $\bar{r}$ at *every* preference profile. Consequently, a most equitable refinement of $\bar{r}$ also has the same or higher probability to satisfy ANR than *every* refinement of $\bar{r}$ under *every* distribution over agents' preferences. We are not aware of a similar notion in the literature and prove that, surprisingly, it always exists.

Our technical contributions are twofold. First, we characterize the conditions for the ANR impossibility to hold under general settings with m alternatives and n voters, where preferences and decisions can be ranked lists or committees—it holds if and only if a *partition condition* is met, i.e., there exists an integer partition $\boldsymbol{m}$ of m that satisfies a *sub-vector constraint*, and a *change-making constraint*, which requires that n can be made up by coins whose denominations depend on $\boldsymbol{m}$. This characterization provides a novel angle that unifies existing ANR impossibility characterizations and resolves the open question for general social choice settings. We also characterize *at-large ANR impossibility*, i.e., the ANR impossibility holds for every sufficiently large n. As a corollary, the ANR impossibility holds for the 2021 New York City Democratic mayoral primary elections. Second, we propose the *most-favorable-permutation (MFP)* tie-breaking to obtain a most equitable refinement and design a polynomial-time algorithm to compute MFP when agents' preferences are full rankings, addressing computational challenges identified in previous work.

© The Author(s), under exclusive license to Springer Nature Switzerland AG 2024
J. Garg et al. (Eds.): WINE 2023, LNCS 14413, pp. 711–712, 2024.
https://doi.org/10.1007/978-3-031-48974-7

Our work builds upon notation and principles of *algebraic voting theory* and the group theoretic framework for analyzing anonymity and neutrality.

The full version of this paper can be found at https://arxiv.org/abs/2205.14838.

Keywords: Social choice · Anonymity · Neutrality · Tie-breaking

Near-Optimal Dynamic Pricing in Large Networks

Ozan Candogan[1] , Yuwei Luo[2], and Mingxi Zhu[3]([✉])

[1] University of Chicago, 5807 S Woodlawn Ave, Chicago, IL 60637, USA
[2] Stanford University, 655 Knight Way, Stanford, CA 94305, USA
[3] Georgia Institute of Technology, 800 W Peachtree St NW, Atlanta, GA 30332, USA
mingxi.zhu@scheller.gatech.edu

Abstract. We consider networks generated by a branching process and assume that each agent/node has either low or high value for a product offered by a seller. We assume that the type of each agent coincides with that of her predecessor in the network with probability $p > 1/2$ (motivated by the homophily commonly documented in social networks). The seller sequentially offers prices to the agents in the network. Offering a low price to an agent guarantees a sale, but does not reveal the type of the agent, which in turn decreases the seller's continuation payoff. The natural dynamic program for obtaining an optimal pricing policy appears to be intractable, and myopically setting prices is highly suboptimal. We provide simple policies that are "in-between" the optimal and myopic policies. These policies partition the network into "active sets" and take optimal actions in each such set of nodes while ignoring the rest of the network. We establish that by choosing small active sets the seller can achieve tractable approaches that guarantee near-optimal revenues. For instance, when each active set is defined in terms of a node and its one-hop neighborhood, this approach guarantees a lower-bound revenue loss, which scales with the reciprocal of the expectation of the degree compared with that achieved by the optimal solution of the dynamic program. Moreover, by considering richer active set structures (e.g., that consist of a node plus its neighborhood of a few hops, or that are constructed adaptively), the payoff loss can be made smaller (albeit at an increased computational cost). Dynamic decision problems are notoriously hard in networked systems, and our paper provides an approach based on taking locally optimal decisions on active sets. This novel approach could be fruitful in other dynamic decision problems related to networks.

Full paper url: https://drive.google.com/file/d/1ymXepBZMCXSRuy9X1Lb30kEpgpp4Xum4/view.

© The Author(s), under exclusive license to Springer Nature Switzerland AG 2024
J. Garg et al. (Eds.): WINE 2023, LNCS 14110, p. 713, 2024.
https://doi.org/10.1007/978-3-031-48974-7

Author Index

© The Editor(s) (if applicable) and The Author(s), under exclusive license
to Springer Nature Switzerland AG 2024
J. Garg et al. (Eds.): WINE 2023, LNCS 14413, pp. 715–717, 2024.
https://doi.org/10.1007/978-3-031-48974-7

Printed in the United States
by Baker & Taylor Publisher Services